# Advances in
# Social and
# Organizational Factors

# Advances in Human Factors and Ergonomics Series

**Series Editors**

**Gavriel Salvendy**
*Professor Emeritus*
*School of Industrial Engineering*
*Purdue University*

*Chair Professor & Head*
*Dept. of Industrial Engineering*
*Tsinghua Univ., P.R. China*

**Waldemar Karwowski**
*Professor & Chair*
*Industrial Engineering and*
*Management Systems*
*University of Central Florida*
*Orlando, Florida, U.S.A.*

## 3rd International Conference on Applied Human Factors and Ergonomics (AHFE) 2010

Advances in Applied Digital Human Modeling
*Vincent G. Duffy*

Advances in Cognitive Ergonomics
*David Kaber and Guy Boy*

Advances in Cross-Cultural Decision Making
*Dylan D. Schmorrow and Denise M. Nicholson*

Advances in Ergonomics Modeling and Usability Evaluation
*Halimahtun Khalid, Alan Hedge, and Tareq Z. Ahram*

Advances in Human Factors and Ergonomics in Healthcare
*Vincent G. Duffy*

Advances in Human Factors, Ergonomics, and Safety in Manufacturing and Service Industries
*Waldemar Karwowski and Gavriel Salvendy*

Advances in Occupational, Social, and Organizational Ergonomics
*Peter Vink and Jussi Kantola*

Advances in Understanding Human Performance: Neuroergonomics, Human Factors Design, and Special Populations
*Tadeusz Marek, Waldemar Karwowski, and Valerie Rice*

## 4th International Conference on Applied Human Factors and Ergonomics (AHFE) 2012

Advances in Affective and Pleasurable Design
*Yong Gu Ji*

Advances in Applied Human Modeling and Simulation
*Vincent G. Duffy*

Advances in Cognitive Engineering and Neuroergonomics
*Kay M. Stanney and Kelly S. Hale*

Advances in Design for Cross-Cultural Activities Part I
*Dylan D. Schmorrow and Denise M. Nicholson*

Advances in Design for Cross-Cultural Activities Part II
*Denise M. Nicholson and Dylan D. Schmorrow*

Advances in Ergonomics in Manufacturing
*Stefan Trzcielinski and Waldemar Karwowski*

Advances in Human Aspects of Aviation
*Steven J. Landry*

Advances in Human Aspects of Healthcare
*Vincent G. Duffy*

Advances in Human Aspects of Road and Rail Transportation
*Neville A. Stanton*

Advances in Human Factors and Ergonomics, 2012-14 Volume Set:
Proceedings of the 4th AHFE Conference 21-25 July 2012
*Gavriel Salvendy and Waldemar Karwowski*

Advances in the Human Side of Service Engineering
*James C. Spohrer and Louis E. Freund*

Advances in Physical Ergonomics and Safety
*Tareq Z. Ahram and Waldemar Karwowski*

Advances in Social and Organizational Factors
*Peter Vink*

Advances in Usability Evaluation Part I
*Marcelo M. Soares and Francisco Rebelo*

Advances in Usability Evaluation Part II
*Francisco Rebelo and Marcelo M. Soares*

# Advances in
# Social and
# Organizational Factors

Edited by
## Peter Vink

**CRC Press**
Taylor & Francis Group
Boca Raton London New York

CRC Press is an imprint of the
Taylor & Francis Group, an **informa** business

CRC Press
Taylor & Francis Group
6000 Broken Sound Parkway NW, Suite 300
Boca Raton, FL 33487-2742

First issued in paperback 2019

© 2012 by Taylor & Francis Group, LLC
CRC Press is an imprint of Taylor & Francis Group, an Informa business

No claim to original U.S. Government works

ISBN-13: 978-1-4398-7019-8 (hbk)
ISBN-13: 978-0-367-38107-3 (pbk)

**Visit the Taylor & Francis Web site at**
**http://www.taylorandfrancis.com**

**and the CRC Press Web site at**
**http://www.crcpress.com**

# Table of Contents

## Section II.  Ergonomics in Industrial Quality

## Section IX.  Changes at the Organizational Level

**Section X. New Ways of Work**

## Section XI.  User Experience, Comfort and Emotion

# Preface

The National Academy of Engineering Committee on Engineering's Grand Challenges has identified in 2008 14 areas awaiting engineering solutions in the 21st century (see http://www.engineeringchallenges.org/cms/8996.aspx). These 14 areas were defined by a panel of technology and engineering luminaries. The goal of the panel was to create a slate of Grand Challenges for scientists and engineers. Among these challenges five are directly related to the content of this book:

1. Restore and improve urban infrastructure
2. Advance health informatics
3. Prevent terror
4. Improve cyberspace
5. Advanced personalized learning

This means that the contributions in this book are of a high value for the future of the world. These challenges have a more technological point of view as they are defined from an engineering perspective. The Lund Declaration prepared the future European Grand Societal Challenges (Chuberre & Lioli, 2010). So, these are more focused on societal issues. They identified six themes calling for solutions to tackle them. Two of these are closely related to the content of this book:

6. Ageing societies: As the life duration of people increases, this raises numerous issues among which economic, social inclusion and accessibility.
7. Security: How to improve the security of European citizens and their goods within but also outside Europe?

For these two challenges (numbered 6 and 7), this book has also useful papers that contribute to the solution of these grant challenges. Thus, ergonomics contributes both to important societal and engineering challenges.
In this book we have interesting papers on urban infrastructures (related to the first grant challenge mentioned above) within the section "perception and design of spaces" on for instance the shaping of urban spaces (paper of Horn). In this section also papers on interiors of stadiums, museums and kitchens are found. Interesting papers on the advance health informatics, the second challenge, can be found in the section "ergonomics industrial quality". The paper on health management systems by Konczal et al. shows that positive effects on the manufacturing process are to be expected. This section also consists of papers on warning systems in cars and voice-based interfaces. The third grant challenge "prevent terror" is closely related to the papers on human factors in counter-terrorism and on interpreting deceptive behaviors which can be found in the section "human factors in terrorism". Improving cyberspace (number 4) is tackled in the section "enterprise ICT and work". Many ideas are described to improve enterprise resource planning systems. For instance, Hankiewicz states that instead of a good interface it is important that a user understands the decision flows and data locations. Additionally, he mentions that the system should not replace human intelligent work, but it should replace the repetitive handlings. The fifth challenge is tackled in many papers. In the

section "learning and training" one paper on careers by Kröll stresses the importance of lifelong learning, while others emphasis the adaptation of courses to the individual needs, which is in the heart of "personalized learning". The "flexible work force and work scheduling" section is also related to advanced learning as Gamber and Zülch et al describe systems to be aware of the work-life balance and educate workers by these systems.

Attention to "Ageing", the sixth challenge, is given in many papers. In the section "adapting for special groups," Sanders shows how to optimize the job design for adult workers. Also, other special groups get attention. Nawakowski focuses on good furniture for children and Piardi et al. show how a ship can be designed taking into account all special groups. Attention to safety and security, the seventh challenge, can be found in many sections. In the section "ship design," the paper of Begovic and Bertorello tackles the safety of a high speed ship. In the section "changes at organizational level" Ritz describes that trust is related to safety in team performance and Novak et al present a model for homeland security.

Themes that play a role in all challenges can be found in "changes at organizational level" and in the section "new ways of work". Akasaka and Okada prescribe knowledge to manage change and Ellegast et al show a measurement system to see whether people have more physical activities, which is related to health informatics. The section on user experience is not related to the grant challenges. However, making people happy and healthy is certainly an issue of importance. Bonenberg uses the term emotional ergonomics and Lin and Lin studied the user experience in relation to fashion design.

This book contains a total of 84 interesting papers by authors from 17 countries. The organizers would like to thank all the authors for their contributions. Each of the chapters were either reviewed by the members of the editorial board or germinated by them. For these our sincere thanks and appreciation goes to the members of the board listed below.

P. Hasle, Denmark
D. Horn, USA
S.-L. Hwang, Taiwan
J. Kantola, Korea/Finland
B. Kleiner, USA
L. Pacholski, Poland

M. Robertson, USA
S. Saito, Japan
M. Smith, USA
H. Vanharanta, Finland
Z. Wisniewski, Poland
R. Yu, China

We sure hope this book contributed to increase in knowledge in the field of social and organizational Ergonomics and that you find the papers in this book interesting and helpful to you and your work.

April 2012

Peter Vink
The Netherlands

Editor

Reference

Nicolas CHUBERRE and Konstantinos LIOLIS (2010). Contribution to Grand Societal Challenges. European Technology Platform, 30th April 2010

# Section I

Perception and Design of Spaces

# CHAPTER 1

# Desirable Features of Contemporary Domestic Kitchen

*Jerzy Charytonowicz, Dzoana Latala*

Wroclaw University of Technology
Wroclaw, Poland
jerzy.charytonowicz@pwr.wroc.pl
dzoanalatala@wp.pl

## ABSTRACT

Contemporary kitchen is an indispensable part of every household. As a result of technical progress, the domestic kitchen has evolved into the most important functional area in the apartment. The diversity of solutions and models of contemporary domestic kitchen gives unlimited possibilities of designing. However, what features should the kitchen have in order to fulfill the expectations of the more and more demanding users? The answer to that question is:
- the friendly kitchen – adapted to individual needs of a diverse group of users; people of special needs, mentally and physically fit and disabled;
- the intelligent kitchen – thanks to devices equipped with information and response systems, which are able to choose and take up previously programmed functions and intelligent reactions in specific cases. Thanks to the so called "intelligent" system, efficient managing of the entire household and controlling the individual devices efficiency is possible;
- the ecological kitchen – equipped with devices saving water, electricity etc., designed with care of the natural environment and the users' "pockets";
- the safe kitchen – designed in such way to limit and eliminate accidents and health hazard by applying "safe in use" devices, equipped with all kind of security systems, with ergonomic, easy to use shapes and control panels, with care for the safety of the youngest users and those with special needs (disabled and advanced in years);
- the open kitchen – most frequently connected with the living room area, creating one shared multifunctional area, or partly connected with the living room by

an opening in the wall. "The openness" of the kitchen makes it possible to introduce interesting, trendy and well designed appliances and accessories on the market, which are willingly exposed by users.

Characteristic features of the above mentioned kitchen types fulfill the ergonomic guidelines of designing the functional area. It is possible to apply a particular type of kitchen as separate solution, however, only grouping together the distinctive features of friendly, intelligent, ecological, safe and open kitchen, makes the formation of the most desirable kitchen possible – the kitchen adapted to users' expectations and complied with the ergonomic designing criteria.

**Keywords**: domestic kitchen, technical progress, ergonomics

# 1 INTRODUCTION

Contemporary kitchen is an indispensable part of every household. Each house has a place for food preparation, even if the household members eat out. Contemporary kitchen has changed during the last few years. It is no longer a tight, closed workspace, where only one person – the housewife – prepares meals. The equipment has changed, both appliances and furniture. It is possible to find solutions, which are completely new or improved compared to those used previously in almost all the fields of kitchen equipment. The omnipresent technical progress took control over that amply equipped, full of diverse hardware functional area in the apartment. One may boldly state, that as a result of technical progress, the domestic kitchen has evolved into the most important functional area in the apartment. A lot of kitchen models, kitchen utensils, furniture and the like, appear on sale. The variety of solutions and models of contemporary domestic kitchen gives unlimited possibilities of designing. Selecting the kitchen equipment individually or buying it as a set, it is possible to create space worth exhibiting, which is simultaneously encouraging the household members to spend free time in it, and causing the house works less onerous, more safe, and thanks to the equipment and arrangement of the kitchen, they nearly become pleasure. Contemporary kitchen seems to encourage all users to work and spend free time there in many ways. The style of the kitchen is an individual matter. Nonetheless, what features should it have in order to fulfill the expectations of the increasingly demanding users? The answer to that question will be the presentation of the characteristic features of the so called friendly, intelligent, ecological, safe and open kitchen.

# 2 THE FRIENDLY KITCHEN

The progressing phenomenon of an ageing society and the increasing number of civilization diseases' victims, determinates the escalating number of disabled people. Contemporary society observes the disabled needs with growing interest, especially the issue of adapting the residential environment, including the kitchen area, to their special needs in order to make self-reliant existence in the society possible for them, as well as to provide them with a certain degree of independence in daily living. Each kitchen user can become disabled as a result of an accident, illness or because of ageing, so it is necessary to adapt the kitchen to the special needs of disabled as much as possible. It has been proven that equipment fitted to disabled needs, is also comfortable and safe in use for a fit person (Skaradzinska, 1991). An optimally designed kitchen takes into consideration the needs of those two groups of users, and conduces the disabled people's integration in the residential environment. Such a kitchen area is called a friendly kitchen – in equal degree to the disabled and fit users. The technical progress of kitchen appliances, especially electric, enabled to apply boosters like the systems adapting the height of hanging countertops, cabinets or sinks, to special needs of disabled users. An example of disabled - friendly cooker is the LiftMatic, its lower element is vertically movable and descends onto the worktop like a lift in order to facilitate taking out the cooked dishes conveniently (Siemens…, 2012). Another example is the Cyber Fridge, which can be opened by voice commands, without the necessity of reaching the door, which is comfortable for fit users and very useful for people with disabled upper limbs (TechKnow…, 2012). The kitchen cupboards and fridges with slide drawers make it easier to reach the inside both for the disabled on a wheelchair and fit users. The cargo cupboards facilitate both side access, just like a cooker with slide baking trays. The Institute of Industrial Design dealt with the designing of furniture sets, compatible to the special needs of the disabled on wheelchairs. To make moving and using the kitchen area by the disabled easier, model sets of kitchen furniture were created there. The spaces under countertops and sinks were adapted to a deep approach of a wheelchair, yet the slide elements became open-work and not limiting visibility (Skaradzinska, 1991). Projects of units in the form of integrated kitchen islands are trying to meet the expectations of the disabled kitchen users. The Japanese solution of the kitchen unit called Kurumono, is the result of a research on disabled people's needs (Kurumono, 2012). The designers created an integrated kitchen island, which reduces the need of movement while performing kitchen works, thanks to the adjustability of the module to individual needs of the user through movable sections. While designing the "friendly" kitchen, there is a need to take into consideration the presence of the youngest users. For them the kitchen should be friendly as well, so that children can learn some kitchen works under the adults' supervision. The kitchen area should be safe for toddlers that move on their own. An example of a "friendly" kitchen equipment is the cooker with a special, heat resistant pane, which excludes a child's accidental burning. It's also an adult friendly solution, and it became almost a standard in each cooker. The friendly kitchen should fulfill the following criteria:

- accessibility – of maneuver, communication and reach areas,
- the space organization – the proper equipment planning,
- the easiness of operation – the intuitive markings, switches, ergonomic shapes facilitating grasp etc.,
- usage of new technologies,
- modularity and multifunctionality.

# 3     THE INTELLIGENT KITCHEN

The technical progress is connected with the computerization development of all fields of life, with regard to domestic kitchen area. Thanks to the development of intelligent technologies, connected with each other and cooperating in order to make the work easier and to raise the quality of household members life, new model of the kitchen, called "the intelligent" kitchen, came into existence. Systems used to connect home appliances to the Internet (e.g. IHA), allow to control and manage individual devices and program them, to remember settings, optimum parameters of microclimate etc. Tactile steering panels (e.g. Icebox Flipscreen, Server Customer 15 ProFace, the Foot) are used to control the refrigerator from a distance, cookers etc., however some devices (e.g. cooker Timo Connect) can be controlled by mobile phone (TechKnow…, 2012; Dubrawski, 2009; Jordan, 2005). Kitchen devices are able to inform the user about breakdown, need for supplies replenishment, like The Cyber Fridge refrigerator, which remembers users preferences and sends appropriate reports. Thanks to the optical wand, products are being located inside the fridge, what makes finding them much easier (TechKnow…, 2012). The function of leaving the message "on the fridge", in the modern form appears in the Screenfridge model, in which the built-in camera allows to record a short movie, and shows it at any given time. The intelligent dishwashers are able to select the proper amount of water for quantity of dishes and the kind of detergent, driers are switching off when the dishes are ready to take out (Jordan, 2005). The kitchen module Kurumono is able to remember combinations of users movements and details about his diet, can remind about the need to replenish the supplies (Kurumono, 2012). The intelligent kitchen faucets are equipped with special sensors allowing the non-touch use, and also with systems adjusting the water parameters to user, e.g. strength of the stream and water temperature (Kitchen and technology, 2009). The small kitchenware also has intelligent trademarks, of what the Oster mixer is an example - with the intelligent system of turnover blades, which eliminates the non shredded particles of foodstuffs. Similarly the food processor called Essence by Philips, is able to choose the optimum speed and the manner of processing the ingredients (TechKnow…, 2012; Kitchen and technology, 2003). The intelligent kitchen hoods can sense the presence of undesirable smells and then switch on in order to eliminate them. The following properties are characteristic of the intelligent kitchen:
-     the comfort of usage,
-     the interlink and operation in the network (the Internet),
-     the facility of usage (intuitiveness),
-     the safeness of usage (alerts, notifications),
-     managing a few devices simultaneously,
-     no need of supervision (automation of processes),
-     the possibility of appliances and processes supervision "from a distance",
-     the quality check - appliances, foodstuffs (in the refrigerator),
-     the selection of optimum processes depending on the circumstances.

The intelligent kitchen was created as an effect of the demand for devices independently controlled, taking decisions instead of the user and such with possibility of control from the distance, in order to save the user's time and energy for other activities in this more and more bustling life. It looks like the intelligent kitchen will appear in every household sooner or later.

## 4    THE ECOLOGICAL KITCHEN

A trend for leading on "the eco type" products among kitchen equipment manufacturers is prevailing. It is encouraging customers to get devices, which "pay back" in the prospect of time - thanks to the energy and water savings plan. It's an environment-friendly action at the same time. The economical dishwashers and kitchen batteries are both aimed at comforting and minimizing costs of consumed water and electricity. Characteristic features of eco-friendly dishwashers (e.g. ActiveWater by Bosh) are: systems of fast heating water, the half – load system, system of drying the dishes with minerals of zeolite, the aquaSensor selecting the program of washing, and even two independent drawers which reduce electricity and water consumption for the half less (Kitchen and technology, 2009). A model called "one-glass-dishwasher" by Electrolux is the most ecological solution, it works in the closed cycle and uses only one glass of water (Eco…, 2012). Eco-friendly kitchen batteries are able to control the flow of water with electronic movement sensors, which activate the water jet (E - go, Kludi), buttons like QuickStop - for determining the amount of water, and S-Pointer Eco which is able to activate the perlator (Kitchen and technology, 2006). The electric kitchen appliances, especially the fridges and the cookers, need less and less electricity (e. g. energetic class "AA", refrigerators with No frost systems). Separate chilling drawers on the slides eliminate the necessity to open the whole fridge; it's a great convenience and energy saving (Kitchen and technology, 2009). Similar solution can be seen in the cookers (two section). Thanks to the Twin Convection technology, the Oven by Samsung is able to bake much faster (60cm…, 2012). During the normal use of the kitchen, as well as in case of a fire, plastics which kitchen accessories are made of, no longer emit harmful substances so as the formaldehyde - thanks to the technological progress (Zielinska, 2012). In the ecological kitchen the segregation of waste is obvious. The most popular solutions are the rubbish bins on slides. The great facility is the supporting system for opening drawers with the knee or the hip by gently touching the fronts of the cupboards (e.g. Servo-Drive by Blum) (Systems…, 2012). Compact home composters solve the problem of organic waste; organic waste is thrown inside the electronically controlled device by Panasonic, in order to use it as fertilizer for plants after their disintegration (New…, 2012). Economical and environmental systems appear in individual cooking appliances. The Whirpool company proposes the kitchen connecting them into the new model - Green Kitchen - equipped with devices using the natural resources entirely. Intelligent systems are able to qualify the cleanness of used water, and filtrate it in order to make it drinkable.

The recuperation and exploitation of the warmth and coolness, which are wasted in traditional kitchen, happen there (Suchodolska, 2008). The French ecological kitchen called Ekokook, makes use of kitchen waste; organic waste is processed into compost, and inorganic waste is being crushed and nailed together into blocks, so that they take less space. Used water from the sink is kept, then filtrated and used for watering plants and for the dishwasher (Ekokook…, 2012). The kitchen, in which the energy and water saving systems are used, in which the recycling products are installed, where the segregation of waste is obvious, and in a similar way the natural environment is cared for, has typical features of the "ecological" kitchen.

## 5    THE SAFE KITCHEN

Contemporary With regard to the potentially dangerous equipment and because of the character of activities being performed there, the kitchen area is the most dangerous functional area in the apartment. Hot temperatures being generated during the cooking are an unquestionable threat. Induction heating plates heat up at a direct contact with the bottom of the dish, which excludes burns by visible fire from burners. Gas cookers have securities of different kinds, such as security system against children and the gas rings self-control system: automatic starting of self-lighting gas in case of its going out, cutting off a gas supply after flooding the gas ring e. g. flameTronic by Siemens (flameTronic…, 2012). So called cold doors in ovens, protecting against burnings, are a standard solution. The neighborhood of water and the electricity in the kitchen can generate dangerous situations. The dishwasher with the aquaStop function cuts off the safety valve by the tap in case of any leakage, and the kitchen is protected against flooding (Kitchen and technology, 2009). Different kinds of sockets with security systems were invented in response to threats of supertensions, electric shocks and fires, e.g. the waterproof, covered and severed when both plug pins are inside the socket, with an extractor, integrated power strips under the kitchen cupboards, extendable table top power strips etc. House smoke and gas detectors allow to monitor the kitchen in case of a fire, gas leakage or carbon dioxide emission. The safe kitchen is one, where all users can easily use the kitchen equipment, without any adverse positions, which can generate accidents - e.g. fridge drawers on slides located at a handy height, dishwashers of the increased height, adjusted to countertops, hoods shaped and hanging in such a way that eliminates bums, e.g. movable hood Miele, or cookers located in a column on a comfortable height (Kitchen and technology, 2006). The safety in the kitchen involves the elimination of harmful substances, to which all kinds of mould belong to. Antibacterial and completely smooth coatings support the safety in the kitchen area. They are applicable, among others, for kitchen worktops (Microban cover built in into the structure of the Silestone top), for kitchen sinks (Signus, Schock of Nanogranit Cristadur ® material with ions of silver), The appliances which can be dismantled and washed in the dishwasher are conducive to cleanness, e.g. Quick Clean kitchen sink overflow (Kitchen and technology, 2009; Camouflage…, 2012).

With regard to presence of water, one of the floor selection criteria is its anti-sliding, easy to clean and antibacterial surface. To provide one's safety of kitchen works it is necessary to assure the proper lighting of worktops, sinks and cooker tops. There are many models of lighting strips mounted under the upper kitchen cupboards. Similarly ventilation hoods and absorbers have an illumination dispatched to cooker tops. The kitchen "safe" for the youngest users has security systems like locks preventing the opening of the cupboards and taking down the hot dishes from the cooker tops. Among kitchenware and equipment it is possible to find models whose control panels are protected against children, and knobs located beyond the reach of the children. Similar locks are applied in cabinets where the trash cans are kept and in waste bins.

## 6    THE OPEN KITCHEN

Along with the technical progress, multiple technical conveniences appeared, supporting the fusion between the kitchen area and the daily area in a household. This "opening" supports the strengthening of contacts between family members, it is also an advantage to join functional areas in an apartment, which is particularly important in small apartments. The open kitchen favors the contemporary trends to cook and prepare meals together with family members and friends. Kitchen appliances became more attractive and is worth showing to the guests. An electric kitchen hood is an example of kitchen equipment which have changed under the influence of the technological progress, and enabled the so-called open kitchen to come into existence. Its invention and continuously improved forms enable to eliminate unpleasant or intense odors, which accompany the thermal treatment (cooking, baking etc.). Contemporary ventilation hoods assume visually interesting forms, can also be equipped with additional functions (e.g. LCD screens). Kitchen batteries, as visible accessories of the kitchen equipment, gained interesting forms and became the decoration of the kitchen. Similarly the refrigerator and the oven, often located higher than up to now, become visible. Kitchen furniture more and more often resembles representative furniture, and refer to furnishing of the living room. Appropriately arranged lighting of the open kitchen allows to conceal or to expose chosen functional areas and thanks to that can affect the arrangement of the domestic daily area. Lighting covers applied in the kitchen are as attractive as in the living room. A so-called kitchen island is a common element of kitchen equipment, and it can determine the border between areas: of the kitchen and the living room in a shared multifunctional room. A kitchen, where the storage area is not closed, and kitchen utensils, appliances and dishes are located "at hand" – cutlery, knives, pans and the like hang on rails or are put on shelves, can also be called an open kitchen. A seeming chaos is a storage system is in fact tidied up system of keeping appliances, dishes and the cutlery, visible and ready to use (Smuga, 2008). The contemporary open kitchen is pro-social, just as the prehistorical bonfire unites the family by the collective preparation and consumption of meals, simultaneously ensuring the family comfort of performing kitchen works.

# 7    CONCLUSIONS

The distinctive features of the types of the kitchen mentioned above fulfill the ergonomic guidelines of designing the functional area - they are designed and produced so that it is possible to adapt them to individual preferences of the user, and its equipment and the way of arrangement are aimed at the comfort and safety of the user. The "friendly", "intelligent" and "ecological" types of kitchen - also possess the features of the "safe" kitchen. However, each of them can be designed as an "open" kitchen. It is possible to apply each of those types of kitchens as separate solutions for the kitchen area, however, only integrating their distinctive features will allow to create the most desired kitchen, adapted for the most required expectations of the user.

# REFERENCES

Skaradzinska, M., 1991. Designer's handbook - disabled problems. Warsaw: The Institute of Industrial Design.

"Siemens Liftmatic", Accessed February 14, 2012, http://www.appliancist.com/builtin_ovens/siemens-liftmatic-oven.html.

"TechKnow: Smart Appliances", Accessed February 18, 2012, http://archive.metropolis.co.jp/tokyo/416/tech.asp.

"Kurumono", Accessed February 14, 2012, http://www.archeworks.org/projects/kurumono/1_home.htm.

Dubrawski, A., 2009. Kitchen and technology: *Technology in the kitchen*. No. 2(37). Bialystok: Medius s.c.

Jordan, A., M.: "Kitchen and Bath show", Accessed June 2, 2005, http://www.sfgate.com/cgibin/article.cgi?f=/c/a/2005/05/25/HOGNJCR44D1.DTL.

"New Electrolux Screen Fridge", Accessed February 15, 2012, http://www.appliancist.com/refrigerators/electrolux-screen-fridge.html.

Kitchen and technology: *Water Economy*. No. 4 (39), 2009. Bialystok: Medius s.c.

Kitchen and technology: *Thinking robot*. No. 1 (12), 2003. Bialystok: Medius s.c.

Kitchen and technology: *Fast, clean and ecologically*. No. 1 (36), 2009. Bialystok: Medius s.c.

"Eco friendly Electrolux dishwasher by Alexey Danilin", Accessed February 15, 2012, http://www.igreenspot.com/eco-friendly-electrolux-dishwasher-by-alexey-danilin/.

Kitchen and technology: *Water under control*. No. 3 (26), 2006. Bialystok: Medius s.c.

Kitchen and technology: *Three times Liebherr*. No. 1 (36), 2009. Bialystok: Medius s.c.

"60cm electric, 65L capacity Twin Convection™ oven", Accessed February 15, 2012, http://www.samsung.com/au/consumer/home-appliances/built-in-cooking-appliances/oven/BT621FSST/XSA.

Zielinska A.: "Whether building materials can harm?", Accessed February 15, 2012, http://ladnydom.pl/budowa/1,106566,2567621.html.

"Systems of waste segregation with Servo-Drive support", Accessed February 17, 2012, http://www.technologia.meblarstwo.pl/artykul/13746-systemy_segregacji_odpadow_ze_wspomaganiem_servo_drive.

"New and Future Kitchen Composters", Accessed February 17, 2012, http://www.insideurbangreen.org/2009/03/kitchen-composters-.html.

Suchodolska, M., 2008. House&Interior: Future kitchen, No. 6 (153/2008). Warsaw: Edipresse Polska.

"Ekokook Kitchen System", Accessed February 18, 2012, http://www.greenmuze.com/green-your/space/2183-the-ekokook-kitchen-system.html.

"flameTronic: electronic gas hob by Siemens", Accessed February 18, 2012, http://www.youtube.com/watch?v=Zew3digHM2g.

Kitchen and technology: *Movable hood*. No. 4 (27), 2006. Bialystok: Medius s.c.

"Camouflage colors", Accessed February 18, 2012, http://www.dobrzemieszkaj.pl/a/3199,Maskujace_kolory,,kuchnie.

Smuga, E., 2008. Kitchen and technology: *Within Reach*. No 4(35). Bialystok: Medius s.c.

# CHAPTER 2

# An Exploration about Interior Ambience Based on User-centered Design Approach

*Cherng-Yee Leung\*    Kuan-Jen Chen\*\**

\* Department of industrial design, Tatung University, Taipei, Taiwan

Email: leung@ttu.edu.tw

\*\*Department of industrial design, Tatung University, Taipei, Taiwan

Email : kjc@ttu.edu.tw

## ABSTRACT

Nowadays, interior ambience is frequently required in the interior design field to give people some affective and pleasure experiences. However, how people feel the interior ambience is hardly studied. Base on the user-centered design approach, before designing, the users' opinions should be understood.

This study aimed to explore the users' ideas about the interior ambience. 8 interior designers and 17 non-professionals from 2 different families and 1 party group of 7 friends were invited to 3 different decorated restaurants. After each dinner, they were asked to describe the restaurant space and ambience in terms of physical cues and their personal feeling. The recorded data were coded and analyzed by qualitative approach. The results showed that regarding physical subjects and space layout, the descriptions between professionals and non-professional were almost the same except the words they used. However, regarding the interior ambience, the difference between them is significant, especially, different ages and different education levels of participants had quite different feelings even for the same space. By categorizing, 3 dimensions of interior ambience were figured out: 1) environments, 2) ages, and 3) ambiences. The implications of the results were discussed. Hopefully, the findings could be a good reference and a good starting point for interior design profession to further understand the issue about interior ambience.

**Keywords:** Interior ambience, Interior design, User-centered Design

# 1    INTRODUCTION

The interior design is a rational creativity, which activities base on  science function and art performance for creating a rational interior environment with spiritual and material life (Wang, 1984). Interior design can be regarded as the physical and non-physical arrangement of interior spaces to meet the users' psychological and physical comfort, improving the quality of life and work efficiency (Li, 1984). With the enhancement of economic prosperity and living standards, modern interior designs not only pay attention to the arrangements of physical objects, but also stresses on aesthetic and environmental design. especially in creating an ambience, such as: Romantic, Luxurious, Japanese Style,…, however, what is ambience? What are its constituent elements? A few of literatures mentioned about the ambiences. An interior space is usually design for some group people to use at the same time, especially a public space. Stations and offices are typical examples of public spaces where multi-users use them. They are usually a big challenge for an interior designer to design spaces to suit all the user's comfort and needs. The possible ways of finding the best designs for every user should involve from the concept of the User-centered Design (eg, Garrett, 2002). Before applying the concept of the User-Centered Design, one must accommodate the users opinion and their point-of-view. Therefore, the purpose of this study is to explore the different views on ambience. Using data collected from a field research composed of each individual user and groups of users, one can strengthen the understanding between interior design and ambience design.

# 2    LITERATURE REVIEW

Promoting the concept of the User-centered Design, interior design focuses on private work areas, storage, and other personal spaces. Conversely, the outcome of the atmospheric interior design should not only focus on private areas, but also on public areas. This study will investigate the feelings on the groups of users who share a place using qualitative analysis (Krishef, 1991). One must not only take feelings and the needs of users under consideration, but also to the elements of indoor ambience (Leung & Chen, 2009).

# 3    METHOD

In this study, using different ambience of three restaurants, feelings of specific members in the dining area is analyzed. Restaurant, A, B, and C (Table 1) are with different ambiences, Restaurant A surveyed 8 family members, Restaurant B surveyed 10 family members, and Restaurant C surveyed 7 companions. Interviews and questionnaires have been used through the survey.

The interviews used the following points: the restaurant in a visual evaluation, the restaurant in a non-visual evaluation, overall feeling of the restaurant's ambience, whether the consumer is satisfied about the restaurant's service, pricing and value, the quality of food the main factor that contributes to a consumer's satisfaction or not, effect of the shape and color and material on the ambience, age contribution to the mood of the ambience, effect on non-visual interior spaces, and using professional description to describe the ambience.

**Table 1 Restaurant, A, B, and C with different ambiences**

| | |
|---|---|
|  | Restaurant A<br>Location: Sizzler saladbar ,Taiwan<br>Respondents: eight family members<br>Investigation of a restaurant ambience A<br>Area: near7000sqft.<br>All-you-can-eat Buffet<br>Middle price<br>Less service from restaurant<br>B1 level but sky light from windows. |
| <br> | Restaurant B<br><br>Investigation of a restaurant ambience B<br>Location: Japanese cuisine restaurant ,Taiwan<br>Respondents: 10 family members<br>Area: near600sqft.<br><br>Japanese cuisine<br><br>Mid-low price<br><br>Full service from restaurant<br><br>1$^{st}$ floor. |
| | Restaurant C<br><br>Investigation of a restaurant ambience C<br>Location: Paradise Restaurant ,Taiwan<br>Respondents: 7 members of friends<br>Area: near9000sqft.<br>Combo cuisine<br>Hi price<br>Less service from restaurant<br>6$^{st}$ floor. |

# 4    RESULTS AND ANALYSIS

Collected data was manipulated by coding, sorting, classification as follows:

After surveying restaurant A, B, C., the result of each member's description of ambience is able to list by their different ages and the users, and can be divided into professional and non- professional members (Table 2-1, Table 3-1, Table 4-1). Also their descriptions are caused by the different situations of the environments which are able to influence users' positive feelings and negative feelings (Table 2-2, Table 3-2, Table 4-2).

Regarding restaurant A, After coding the users' descriptions, the comparison were shows by the level of ages, gender, positive and negative feelings. There're several keywords as follows:

Cheap (A-1), Noisy (A-2), Clean (A-3), (A-4)High Class, Health (A-5), Bright (A-6), Dim (A-7), not High-Level (A-8).

*Table 2-1  List of gender, Level of ages, professional and non- professional*

| age group | Male (age) | Female (age) | Non- professional | professional |
|---|---|---|---|---|
| Child | 10 | - | 1 person | |
| Teen-ager | 16 | - | 1 person | |
| Adult | 38 48 | 42 44 | 3 people | 1 person |
| Elder | 77 | 75 | 2 people | |

Table 2-2 List of gender, Level of ages, professional and non- professional

| Restaurant A Descriptions | Positive feelings | | Negative feelings | |
|---|---|---|---|---|
| | male | female | male | female |
| Child | A-6 | - | A-2 | - |
| Teen-ager | A-2 A-7 | - | A-8 | - |
| Adult | A-1 | A-1 A-4 | A-2 | A-8 |
| Elder | A-4 A-3 A-6 | A-4 A-5 A-3 | A-2 | A-7 A-8 |

Regarding restaurant B, After coding the users' descriptions, the comparison were shown by the level of ages, gender, positive and negative feelings. There're several keywords as follows:

Cheap (B-1), Japanese style fitted (B-2), Delicious (B-3), Dim (B-4),Noisy(B-5), Kindness (B-6), None too harsh colors (B-7), Too close (B-8)。

Table 3-1 List of gender, Level of ages, professional and non-professional

| Age group | male (age) | female (age) | Non-professional | professional |
|---|---|---|---|---|
| Child | 10 | - | 1 person | |
| Teen-ager | 16 | - | 2 people | |
| Adult | 38 48 52 | 42 44 | 4 people | 1 person |
| Elder | 77 | 75 | 2 people | |

Table 3-2 List of gender, Level of ages, professional and non-professional

| Restaurant B Descriptions | Positive feelings | | Negative feelings | |
|---|---|---|---|---|
| | male | female | male | female |
| Child | B-3 | - | B-5 | - |
| Teen-ager | B-3 | - | B-4 | - |
| Adult | B-1 B2 B-3 B-7 | B-1 B-7 B-3 B-6 B-8 | B-5 | B-5 |
| Elder | B-1 B-7 | B-1 B-7 | B-5 B-8 | B-8 |

## 4.3 Restaurant C

### 4.3.1.Environment

Regarding restaurant A, After coding the users' descriptions, the comparison shows by the level of ages, gender, positive and negative feelings. There're several keywords as follows:
Price worthy (C-1), No style present (C-2), Delicious (C-3), Bright Lighting (C-4), Crowded (C-5), Luxurious (C-6), High-Class Restaurant (C-7), Good Ambience (C-8)

Table 4-1  List of gender, Level of ages, professional and non- professional

| Age group | male (age) | female (age) | Non-professional | professional |
|-----------|-----------|--------------|------------------|--------------|
| Adult | 47<br>48<br>49 | 22<br>25<br>43<br>44 | 1 person | 6 people |

Table 4-2  List of gender, Level of ages, professional and non- professional

| Restaurant C Descriptions | Positive feelings | | Negative feelings | |
|---------------------------|--------|--------|--------|--------|
| | male | female | male | female |
| Adult | C-1 | C-1 | C-7 | C-7 |
| | C-3 | C-3 | C-2 | C-5 |
| | | C-4 | | C-8 |
| | | C-6 | | |

# 5    CONCLUSIONS AND DISCUSSIONS

Results of this study shows three factors are needed to affect the individual user or group users' value to space ambient (Descriptions from these users are non-professional, professional descriptions.) include shapes, colors, and materials.The survey results show that the description of the feelings are divided into three Dimensions: (1) Environment (2) Age (3) Ambience. (Figure 1, Table 4).

The three factors are:

**(1) Environment**

From personal food preference and dinning style to each restaurant's service, they are great difference among each user.

Each individual user has subjective point of view to the atmosphere.

In case C, the professionals' descriptions were similar. However, the tastes of each user in the same group are not exactly the same. Their opinions and perceptions are always different.

**(2) Age**

Concerns on the values of ambient space vary between ages. Elders seem to take the prices of different restaurants as standards for their observations.

**(3) Ambience**

The environment ambiences always depend on the users' emotion. Different interior space provides different values. As mentioned above, personal mood and taste about the interior ambient changes the perspective for specific interior spaces.

The field research above clearifies the users' feeling about interior ambience, which relies on user's tastes. This study will analyze the feelings of group users on specific interior ambience to provide details of different perspectives to the interior designers.

Comparing gender, age and environment values, there were different views under the same interior space. E.g. The elder often criticized the noise level and the crowdedness. Females hesitate to point out negative opinions. Nonetheless, the overall ambience descriptions are similar. In addition, some descriptions from non-professionals are very similar to the descriptions of professional designers. This can grant designers greater precision in designing interior spaces.

The conclusions of this study might not be able to take all the user's views in consideration, due to lack of restaurant choices. In future studies, more restaurant interior ambiences should be considered.

Table 5  List of gender, ages, Value of RestaurantA,B,C. Types of Ambience.

| Ambience | | A | | | | B | | | | C | | | |
|---|---|---|---|---|---|---|---|---|---|---|---|---|---|
| | | Good | Average | Below Average | Poor | Good | Average | Below Average | Poor | Good | Average | Below Average | Poor |
| Price worthy | F/M | | | | | | | | | 22/49 | | | |
| cheap | F/M | 42/38 | | | | 42 75/52 77 | | | | | | | |
| neat | F/M | | 75/77 | | | | | | | | | | |
| None style present | F/M | | | | | | | | | | | | /48 |
| Delicious | F/M | | | | | 44/38 16 10 | | | | 43/49 | | | |
| Too close | F/M | | | | | | | 44 77/75 | | | | | |
| Bright Lighting | F/M | | /77 | | | | | | | 44/ | | | |
| Higher Quality | F/M | 75 42/77 | | | 75 44/16 | | | | | | | 43/49 | |
| Nice Ambience | /FM | | | | | | | | | | 22/ | 44/ | |
| Japanese style fitted | F/M | | | | | | /48 | | | | | | |
| Crowded | F/M | | | | | | | | | | | | 44/ |
| Clean & Health | F/M | | 75/ | | | | | | | | | | |
| Noisy | F/M | | | | | | | 10 44/48 77 | | | | | |
| Dim | F/M | | | 75/16 | | | | | /16 | | | | |
| Friendly service | F/M | | 42/ | | | | | | | | | | |
| No too harsh colors | F/M | 44 75/52 77 | | | | | | | | | | | |
| Luxuriou | F/M | | | | | | | | | 25/ | | | |

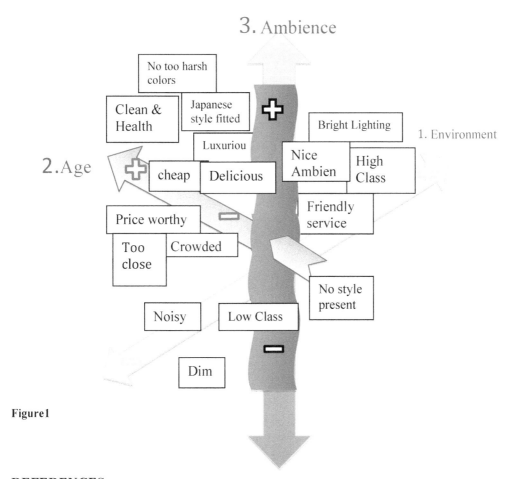

**Figure1**

## REFERENCES

Wang, Jian-Chu. (1986). *Interior Design Theory.* Taiwan I Fung Tang, Taipei, Taiwan.

Philip, Kotler (1973). *Atmopherics as* a *Marketing Tool, Journal of Retailing,*Winter1973/74.

Li, Wan-Wan (1985). Interior Envirement Design, Architecture Sea, Dung Da Publishment

Garrett, J.J. (2002). *The element of User Experience : User-Centered Design for the web.*New Rider, Berkely,CA

Krishef, C.H. (1991). *Fundamental Approaches to Single Subject Design ad Analysis.* Malabar, FL:Keieger Publishing.

Leung, Cherng-Yee and Chen, Kuan-Jen, "A Study on the Descriptions of Interior Spaces by Visually Impairment People," *Proceedings of Special Education and Art Therapy 2008 International Conference*, Taichung University, May 2009, pp. (98/05)

# CHAPTER 3

# Architecture as an Expression of Equity Polices in Manufacturing Plants in Poland

*Magdalena Baborska-Narozny*

Wroclaw University of Technology
Wroclaw, Poland
magdalena.baborska-narozny@pwr.wroc.pl

## 1  ABSTRACT

Equity means freedom from bias of favoritism. This paper indicates how architecture may be used to contribute to equitable attitude a company may voluntarily present towards all its stakeholders. It also shows how architecture may be used to strengthen traditional favoritism of certain groups among the stakeholders. Traditional division between blue and white collars is still the main division that industrial architecture follows or tries to diminish. There is a visible trend in the highly developed countries to eliminate differences in architectural setting designated to office and production staff. Such equitable design is regarded by some clients and architects as highly desirable and beneficial as leading to greater integration of the crew, employees well being and even increase in their productivity. Some examples claimed by the clients and architects as successful in that respect are depicted. Even in short descriptions of their designs, architects always include information on the integrative methods if they have introduced any. Six case study examples from Poland are presented encompassing the whole spectrum of attitudes and architectural solutions. The paper also indicates a need for further research into evaluating the employees' real reactions towards introduction or lack of architectural equity means.

**Keywords**: industrial architecture, architectural equity means

## 2   INTRODUCTION

Equitable means dealing fairly and equally with all concerned. In case of production plants 'all concerned' are the company's stakeholders i.e. the owners, employees, clients, local communities, suppliers. Certain basics of all stakeholders protection are covered by the law through building codes and regulations concerning health and safety. Different treatment of e.g.: employees according to their qualifications, skills and abilities is not unlawful however. The paper focuses on naming and describing lawful architectural gestures that at design stage differentiate or blend various groups of employees, employees and the clients. Traditional division between blue and white collars is still the main division that industrial architecture follows or tries to diminish. There is a visible trend in the highly developed countries to eliminate differences in architectural setting designated to office and production staff. Such equitable design is regarded by some clients and architects as highly desirable and beneficial as leading to greater integration of the crew, employees well being and even increase in their productivity. Some examples claimed by the clients and architects as successful in that respect are depicted in architectural magazines (Schittich, 2003). Even in short descriptions of their designs, architects always include information on the integrative methods if they have introduced any. Among most popular solutions listed are: enabled mutual visual or even oral contact between production and office zone, shared cafeteria, shared entrance, common finishing standard in production and office toilets, unified façade design of different functions. In Poland the majority of industrial premises does not follow that trend. Nevertheless six case study examples have been identified by the author, encompassing the whole spectrum of attitudes and architectural solutions.

Through its landscape or environmental impact industrial activity may 'discriminate' local communities, influencing their potential for future development. The latter problem is particularly socially and economically sensitive and complex in a situation of structural unemployment or one factory being the only major employer for the whole town or neighborhood. The author has identified and described win-win strategies applied by local communities and industrial investors in Switzerland and Germany in her previous writings (Baborska-Narozny, 2008, 2010). Long term loses resulting from neglecting the well-being of local communities are described through a Polish case study analysis in a recent report on challenges of sustainable development in Poland (Struminska-Kutra, 2010). Developing and executing 'equity polices' for the coexistence of local communities with industrial investors would be highly desirable. The scope of this paper does not cover a deeper insight into this problem.

## 3   THE METHODOLOGY

The research is based on the analysis of site layouts and architectural designs of manufacturing plants recently built in Poland, interviews with the architects, clients

and on-site visits. The choice of case-study examples was precedent:
literature study of plants identified in highly developed countries as good pr
The research carried out so far indicates need for further insight into evalua
employees' real reactions towards introduction or the lack of architectura
means.

# 4 CASE STUDIES

Most industrial buildings built in Poland in the last two decades are ov
foreign companies, who bring their business models and expectations to
Architectural representation of those models reveals a whole spectrum of i
relations among the employees groups despite one major common characteri
companies tend to locate assembly plants in Poland and most of pro
employees are low skilled. The R&D activities are performed outside of
Four of six case study examples described belong to this category: Fraba, C
Bosch and MCC UWE. One example i.e.: Forma plant, is a joint venture o
and foreign capital and Bell is a Polish enterprise with its own R&D sect
case studies are arranged from the most equity oriented towards th
hierarchical.

Table 1 Comparison of architectonic equity features applied in the six
analysed factories.

| Architectonic feature\company name | Fraba | Ostervig | Bosch | Forma | Bell |
|---|---|---|---|---|---|
| Shared entrance for clients/office/production | v | x | v | v | x |
| Entrance/-s on the exposed side/-s of the site for all employees | v | v | v | v | x |
| Shared cafeteria/relaxing spaces | v | v | v | x | x |
| Same standard of finishes for facade and interior of office/production/social functions | v | v | v | x | x |
| Visual contact of office/production zones | v | v | x | x | x |
| Oral contact of office/production zones | v | x | x | x | x |

| Executive office expressed through particular architectural features | x | x | x | x | v | x |
|---|---|---|---|---|---|---|
| Offices as an external focal point of a building | x | x | x | v | v | v |

## 4.1   One team – total transparency and information sharing

Fraba Conistics in Slubice, Poland is a manufacturing plant built in 2006 within Kostrzynsko-Slubicka special economic zone as a Polish branch of German based Fraba Group – a future oriented company that focuses on introducing innovations into manufacturing. Research and Development are the core activities of German headquarters located in Cologne, while Polish assembly plant concentrates on delivering products. The group also has facilities in New Jersey, US. The building in Slubice of over 2000 sq. m was designed by German architectural office BeL with Anne Bernhardt and Jorg Leeser as chief architects. The form and functional layout of Fraba in Slubice is extraordinary and has no match in any other factory built in Poland so far. Its whole concept is based on the company's policy towards integration and mutual contact between all employees regardless their qualifications. The maximum number of the crew is limited to 70, as the owners believe that exceeding this quantity leads to anonymity and weakens cooperation among the employees. Currently there are almost 50 employees in Slubice:

1 engineer, 3 production managers and 6 assistant coordinators and 39 unskilled employees doing manual labor within one step production process. One step production means that each worker manually assembles the whole product by him/herself using simple tools. The production does not cause noise or pollution. The proportion of men and women varies as the work can be performed by both genders. There is no typical distinction of office and manufacturing space. Though they are separated with a PVC anti-static curtain it is transparent and does not reach the ceiling height. Thus both visual and oral contact between the two functional zones are maintained. What's more both employee groups use same toilets, kitchen and social areas. Even for the clients who pay visits to the factory there is no separated meeting space. Instead there is a table within the common social space to make arrangements and sign agreements. The plan of the factory is circular with the entrance/delivery area cut into the round external wall with a right angle. The roof remains unchanged thus a canopy over the entrance is created. Within the circular hall three smaller circles are loosely inserted. Those circles contain ladies/gents toilets, showers and technical rooms. From the outside they are covered with lockers, benches and kitchen cupboards and appliances. No dedicated changing rooms are required as the employees only change shoes and put anti-static shirts on top of their home clothes. In result, all employees merge and mutual contact is

enabled. Daylight is generously distributed throughout the white painted top-lit interior. There are only two glazed openings at eye-level: one in the social area and the other by the entrance. Marzena Kaluzna, working in the factory since it's very beginning as a person responsible among other duties for the recruitment and HR declares there is a general appreciation and feeling of pride among the employees, associated with the bold architecture of the factory. The clients and visitors are also always impressed. As one of such visitors reported "From the building structural design, to interior layout, and the production lines, I was not only impressed but overwhelmed. I felt as though I was entering a high-tech NASA operation" (Long, 2011). The building's distinct character is particularly visible when seen in it's context of neighboring rectangular factory sheds typical for any special economic zone across the world.

The non-hierarchical space embodies the philosophy of the Fraba group: total information, no hierarchical levels, nobody reports to nobody. The site is not fenced – another gesture to prove openness and total information policy of Fraba. The link between the clients vision of his company's profile and its spatial form seems direct. The building questions the strong monopoly of typical industrial architectural patterns. No post-occupancy evaluation has been carried out so far. It would thus require further research to establish the link between the low fluctuation rates of employees, high job satisfaction and architectural features of the Fraba building in Slubice.

## 4.2   Integration and mutual respect of two distinct groups of employees

**Ostervig** factory built in the year 1999 in Stanislawow is an assembly plant of cable connections owned by a Danish company Molex. The Ostervig site is located in a currently developing industrial neighborhood along a local road only 15 km away from the Old Town of Polish capital – Warsaw. The factory building of 2270 sq. m is unusual through its modest, subtle elegance. It was designed by Kurylowicz architectural office with Pawel Grodzicki as the project architect. It's external form is nothing of unusual: a simple rectangular block, yet the plan layout and facade proportions and detailing make it stand out of its category. The plan is highly disciplined geometry subordinate to a square figure. The whole building has a square plan that is divided into four squares: three of them are devoted to production and mechanical services and the fourth is further divided into four squares. One of them is an internal courtyard, the three remaining house offices, changing rooms, shared canteen and shared toilets. The entrances are separate for production and office employees. Their architectural setting and exposition on site is similar though. The finishing materials for the facade are the same around the building and include: aluminum cladding, 'profilit' translucent glazing alternated with narrow transparent windows. The same applies to the internal facade of the courtyard. The proportion of glass and metal on the facade varies according to the function it hides. The windows link visually the production with the offices and both zones with the outside world. The architects declare that creation of a pleasant working environment was one of main design objectives. Architectonic equity measures can be found across the design.

## 4.3 Threefold hierarchy of architectural form describing vertical hierarchy of employees.

**Bell** company is a Polish manufacturer of coloured cosmetics. The firm was invented, founded and is owned and managed by Krzysztof Palyska. The design of a 3590 sq. m factory was appointed in the year 2002 to Juvenes office with main architects Slawomir Stankiewicz and Krzysztof Tyszkiewicz. The factory in Jozefow near Warsaw was opened in 2004. The site for the factory is exceptionally serene: it is at the end of a blind alley adjacent at two sides to a pine forest. The architecture clearly shows the vertical hierarchy of the organization. The owner and CEO being one person is at the top, the office and R&D employees are lower in the structure and the production employees are at the base. Lots of architectural features are introduced to represent this hierarchy. The office employees have a separate, glazed entrance at the front of the building right next to the parking. Over the entrance there is a fully glazed loggia adjacent to the owner's office with two cantilevered balconies one extending towards the forest the other overlooking the parking area. The entrance leads to a double height hall. The owner enters the building through the office zone and so do the clients. Production employees have their entrance round the corner, next to dispatch zone, through standard door through security desk into a corridor leading to a staircase and locker rooms. At design stage architects proposed same finishing materials for the exterior of office and production: an extensive use of wood, ceramic tiles and glass. The built factory is covered in wood and tiles only in the office zone, the rest is plaster. Most of the interiors adjacent to external walls, including the production areas are generously day-lit through windows. The unquestionable architectural focal point is the owner's office at '*piano nobile*' - the first floor. The roof over his office is about 1.5 meter higher than over the rest of the building. The ceiling and floor are finished with wood. The executive office extends ca. 2 meters beyond the footprint of the ground floor giving insight into dispatch zone outside. But the most striking feature is the fully glazed wall opening views towards the pine trees. Into its aluminum structure a ca. 4m wide custom made oval wooden frame window is inserted. This impressive space has no math in otherwise rationally organized and finished interiors.

## 4.4 Architecture exemplifying the ambiguity of attitudes among the owners on their intended relations among the employees.

**Forma** is a small join venture enterprise of two owners: Polish and Danish, specialising in production of composite and granite tops. Forma has one factory built in 2009 within Swidnica subzone of Walbrzych Special Economic Zone. It was designed by S+M architectural office with Pawel Spychala as the leading

architect. Both owners agreed on the main decision to build the manufactu
within Atlas Ward building system and to make the offices a focal point
building, custom-made according to architectural commission (Baborska-?
2011). Each owner had his own presumptions however concerning the
among the employees. At different stages of design of Forma plant the infl
each of them prevailed: the Danish owner influenced mostly the formulatic
brief and initial stages. Concept design thus included double height sky l
recreation roof terrace and spacious cafeteria with an exit to recreation area
Only the terrace was given up in following design stages. He also presume(
office and production employees would integrate and have a common
spacious relaxing cafeteria with an exit to a fenced outdoor leisure area. Tl
co-owner who was to run the premises himself had a decisive vote at final
design, construction and handover. In his opinion the offices and pr
employees should not merge and socialize in common leisure spaces. Thi:
of opinion on the functioning of the factory came at the stage when certai
solutions could not have been altered, though some changes were introd
result spacious well lit leisure space with an exit to the outside that was de:
link the offices and production zones is built but it is not to be usec
production employees. They have a small, separate social room for thems(
out of the major leisure space. The functional disposition of the offices
production zones make the visual or oral contact impossible. The common
remained common. The standard of finishes both outside and inside is diff
production and office spaces. The maximum employment planned for tl
space is 20 people and another 20 for the production. Currently there are 5 r
the offices and 10 in production.

## 4.5 Corporate policy of all employee groups' equity integrated into strict guidelines for corporate architecture

The international Bosch Group is represented in Poland by four compan
of them is **Robert Bosch** Sp. z o.o. who owns an almost 20 ha greenfiel
Mirkow near Wroclaw. The first stage of investment there, was designed in
arch. Krzysztof Tetera form Tetera Projektowanie Sp. Z o.o. architectural
was further extended in 2009 by the same architect. Bosch has developed
architectural standards that all its plants across the world follow. The s
cover in detail various issues like modular system for construction and
subdivisions based on a six meter unit, buildings height, company colors, tl
materials and detailing, the use of light, technological layout, shared
entrance and indoor spaces for both office and production employees. A
unified architectonic setting for all the functional parts of the complex i
building is thus characteristic for all Robert Bosch plants including the
Mirkow. The construction technology used for external walls and floor :
different in administration - social part and production hall. These differe
not visible outside. The finishing material and window openings det:

characteristic and unified all around the building. So is the attic height (but in the technical superstructure over part of the hall). Only the two (three after extension) equivalent entrance doors receive different setting and thus become focal points in the 84 m (132 m) long entrance façade. Though windows disposition responds primarily to functional requirements the result is not random as it complies with legible structured guidelines of façade layout.

## 4.6    Visible architectural distinction of two categories of employees: blue and white collars.

**MCC UWE** automotive sub-assemblies factory was opened in November 2008 within special economic zone Invest Park in subzone Olawa in the Lower Silesia. The plant manufacturing ventilation and refrigeration devices in Olawa is owned by MCC (Mobile Climate Control) group founded 36 years ago. MCC has its headquarters and one plant in Sweden and also production plants in Canada, China and Poland. The Polish factory was designed by PM Group Polska – a branch of an international design and engineering company. A further expansion of the 6200 sq. m building was planned as possible on a 2.5 hectare site in industrial neighborhood of Olawa. The factory was designed to employ 100 people: 25 in the offices and 75 in the production (20 women and 80 man). Low cost of labor was one of the decisive factors when choosing Poland as a location for a new factory, as claims Remigiusz Gromiec, area manager of UWE Polska. (Miskiewicz, 2006).

On the opening of the Olawa factory the executive director of MCC Clas Gunneberg gave the chief director of the Uwe factory in Olawa - Jacek Krajcer, a statue by a Swedish sculptor Bertila Valliena entitled "the Ypsylon Ranger" saying: Ypsylon is the twentieth letter of the Greek alphabet and at the same time analogue to number 40. Those symbols are connected with the message this sculpture conveys, that is our goals that our enterprise in Olawa should reach: sales at the level of 40 million zloty together with 20% margin of profit. The employees should also remember, that a ranger is a symbol of both care and control by the concern. At each and every moment they should thus expect a visit of someone who can not only advise but also check the quality and efficiency of the work carried out" (Trybulski, 2008).

The MCC company published on its www a "Human Resources policy" in which it declares: "Our view on Human Resources is founded on our firm commitment to all individuals' equal values regardless of disability, age, gender, sexual preferences, race, color, or ethnic belonging" ("Corporate Responsibility..."). However as it was mentioned in the introduction to this paper it is not unlawful to treat employees differently according to their skills and qualifications. The site layout, architectural form and functional disposition of the MCC plant in Olawa clearly distinguishes and separates two different groups of employees: office and production. The clients are welcome in the office part which is located in the most exposed point of the site on the axis of the only road linking the site with a major public road. When entering the site one sees the generously glazed offices and the steel blank walls of the production hall wrapped around them. At the entrance to the offices there is a paved area with a patch of greenery and flag stands. The parking is directly linked with this entrance area. The majority of employees that work in the production have their own entrance at the back of the building separated from the parking by a loading dock that is circumvented by a pavement. The production hall is top lit. All social areas are separate for the production and office employees, and finished according to different standards. There are no visual links between the two parts of the factory. One fire-door links the two zones. Such solution willfully chosen by the client is strongly supported by Polish fire-safety regulations that tend to divide offices and production into separate fire zones. The resulting working environment follows and strengthens the stereotype gap between the blue and the white collars.

## 5   CONCLUSIONS

Despite a common characteristic of most industrial investments in Poland – assembly plants who employ low skilled workers to perform simple tasks, a whole range of intended relations among a company's stakeholders and particularly among the employees can be observed through the analysis of certain architectural solutions. Architecture has the potential not only to facilitate or impede certain behavior, but also to convey messages and ideas concerning the structure of a company, it's business model and the respect for certain groups affected. The message can be inscribed in the introduction of higher and lower standards for architectural finishes, the location of entrances depending on the position within the organization, etc. Fraba, an organization with all architectural equity features introduced, claims in its values statement: "We are not a democracy, where everybody has a say in every decision, but a high performing team, where everybody has the right and the obligation to do what he is best at" ("Fraba Organization/Values"). Creation of an environment that stimulates and unveils the positive contribution each employee may bring to the company is at the antipodes to the Bentham's Panoptikon environment where the surveillance, inspection and discipline are perceived as the only drivers for employees activity. At the end of 18th century a philosopher Jeremy Bentham believed his invention was applicable "…to any sort of establishment in which persons of any description are to be kept under inspection in particular to penitentiary houses, work-houses, houses of industry, manufactories" (Bentham, 1995). Though strongly hierarchical organizations prevail, implementing equity policies into industry and expressing them through architecture is not within a category of utopia but is already implemented and tested in business practice. It is however an open question still whether the introduction of architectonic equity means implies diminishing negative phenomena like workplace mobbing.

# 6   ACKNOWLEDGMENTS

The author would like to acknowledge her gratitude to all who provided her with insight into design process and its results in the analyzed case-study buildings. In particular to Jorg Leeser, Pawel Spychala, Pawel Grodzicki and Mariusz Kolwzan.

# 7   REFERENCES

Baborska-Narozny, M. 2008. Industry and infrastructure under green roof - different backgrounds, various solutions, *Architectus* 2/2008: 61-68.

Baborska-Narozny, M. 2010. Przemyslowka rozwazna i romantyczna, *Z:A* 5/2010: 12-19.

Baborska-Narożny, M. 2011. Exposed or Disguised? The hierarchy of form and function in case study analysis of recent industrial architecture in Lower Silesia, *Architektus*, 1(29)2011: 47-54.

Bentham, J. 1995. *The Panopticon Writings*. ed. Bozovic, M. London: Verso, Accessed March 1, 2012, http://cartome.org/panopticon2.htm.

"Corporate Responsibility the MCC Way," Accessed March 1, 2012, www.mcc-hvac.com

"Fraba     Organization/Values,"     Accessed     March     1,     2012, www.fraba.com/us/ORGANIZATION/ORGANIZATION_AAA_Organization_base.ht ml

Kusztra, M. 2006. Cosmetics Factory. *Architektura* 01/2006; 42-47.

Long, Ch. 2011. Fraba Inc. - Innovation in products, people and production, *International Door & Operator Industry* 44: 40-44.

Schittich, Ch. ed. 2003. Factory Building near Warsaw. *Detail* 9/2003; 938-939.

Miskiewicz, M. 2006. Kolejny inwestor wybuduje fabryke w Olawie, *Gazeta.pl* Accessed March 1, 2012, http://wiadomosci.gazeta.pl/kraj/1,34309,3548266.html

Struminska-Kutra, M. 2010. Zaklad produkcji plyt wiorowych a srodowisko I społecznosc lokalna [in:] ed. Kronenberg, J. Bergier, T. Wyzwania zrownowazonego rozwoju w Polsce, Fundacja Sendzimira, Krakow 2010: 39-44.

Trybulski, K. 2008. Potop szwedzkich zakladow, *Gazeta powiatowa* 15.12.2008, Accessed March 1, 2012, http://gazeta.olawa.pl/archiwum2/aktualnosci,wiecej,2386.htm.

CHAPTER 4

# Paradise Ambiance in Interactive Art: A Case Study of the National Palace Museum in Taipei

*Hui-Yun Yen\*, Rungtai Lin\*\**

Chinese Culture University\*, National Taiwan University of Art\*\*
Taipei, Taiwan
pccu.yhy@gmail.com

## ABSTRACT

In the modern era, people show the superiority of interactive artworks, because of advanced technology, artists presented his work by technology more general. The visitors play interactive with work and experience meaning of work by the interactive interface. In recent years, interactive art more command that display not just in art gallery that display in daily life space. Thus, this study aims at enhancing the spiritual reflection in the art work through the Paradise Ambiance. The artworks are embodied at interactive space. The design of the "Paradise Ambiance" is considered to be a medium to take the people into a place for taking refuge and a retreat away from the turmoil of the world. This paper aims to enrich the discussion "Paradise Ambiance" of the audience in the interactive art. The authors choose the interactive artworks are *Four Seasons of the National Palace Museum in Taipei* as a case study, and that reveals a specialized interface and open-ended approach to interactive art making. This case study serves as a vehicle for examination of the real world challenges posed by interactive art and its outcomes.

**Keywords**: Paradise Ambiance, Interactive Art, Museum

# 1    INTRODUCTION

The appearance of the audience in the philosophical domains of museology and design is an important concurrence (Macdonald, and Sharon, 2006). With the adoption of advanced technology, the spaces of the contemporary or 'new museum' are wired and mutable, possessing many forms of digital technology including new transdisciplinary cultural practices (Mills, Simon, Stewart, Gavin and Thomas, Sue, 2006). With advances in technology and the works were welcomed with interactive, the scope of interactive art has expanded to include a wide range of methodologies and outcomes. For example, there have been a number of public arts and exhibits that reach a wide public audience. Newer ubiquitous technologies, such as digital museums, have also served as a platform for public interactive art. Museum researchers have documented the typical dwell time at exhibits as approximately 30 seconds, and visitors are unlikely to be able to fully explore the concepts, phenomena, history, or scientific relevance behind each exhibit in a single visit. ( Sherry His, and Holly Fait, 2005 ), but artworks of interactive design can keep audiences staying for long term and help audiences explore the concepts, phenomena, history of Artworks.

Touching content from such interactive art is also a social act, a display in itself; we can more easily interact with friends or strangers. Audiences' reactions and the content are visible. The authors choose the interactive artworks are *Four Seasons of the National Palace Museum in Taipei* as a case study. This paper aims to enrich the discussion "Paradise Ambiance" of the audience in the interactive art. The purposes of this study are twofold. First, a study of "Paradise Ambiance" consisted of a questionnaire that was conducted in conjunction with its features. Second, analyze and list the features of case-interactive act works. Finally, explore the issues "Paradise Ambiance" approaches in two different results. Furthermore findings about the co-experiential aspects of the exhibition audience inspired the visualization of the audience experience enabled by an interactive artifact.

# 2    NEW MEDIA WITHIN THE MUSEUM

Museum spaces nowadays are increasingly augmented with digital technology. While some systems primarily provide context-sensitive, dynamic, and multimodal information (Oppermann and Specht, 1999). Museums have recently developed a strong interest in technology since they are more than ever before in the orbit of leisure industries. They are faced with the challenge of designing appealing exhibitions, handling large volumes of visitors, and conserving precious artwork. They look at technology as a possible partner to help achieve a balance between leisure and learning, as well as to help them be more effective in conveying story and meaning (F. Sparacino, G. Davenport and A. Pentland, 2000). Everyday cyberspaces do exist, generated by telematic communication through networked and mobile digital media, and in the dynamic software worlds of videogames. Whilst they are thoroughly enmeshed in, and accessed from, everyday life, they generate new modes of communication, new games, new opportunities for identity play, and new relationships between the human and technological. The virtual and the actual are intertwined, and each is all the more interesting for it (Martin Lister, Jon Dovey, Seth Giddings, Iain Grant and Kieran Kelly, 2008). Technology can shape the museum contents of a break with traditional patterns. For reasons such as noted above, the museum not only display traditional artworks but also display interactive artworks. From a Human-centred design approach Robertson et al have used extensive field observation of audience behaviour in museums and galleries to develop design tools to be used in the creation of interactive exhibits (Brigid Costello, Lizzie Muller, Shigeki, Amitani and Ernest Edmond, 2005). The most common approach to an exhibit is to walk up to it and try to figure out what to do with it. The exhibits support a combination of "play" and "science" (Margaret Fleck, Marcos Frid, Tim Kindberg, Rakhi Rajani, Eamonn O'Brien-Strain and Mirjana Spasojevic, 2002). Such "play" can lead to "experience", especially the theme of the exhibit about culture or history. The interactive art exhibit is an interesting environment because its physical environment already requires much of the user's personal resources, e.g. eyes, hands, mental attention.

## 2.1 Interactive Art

Interactive art is a new media-based activity with audiences. Presence, mapping, action, learning and vividness of interactive experience are identified by the author for the purpose of establishing basic elements of spatial experience in interactive artworks. This is necessary to emphasize the triadic relationship of the user, the interactive artwork and the artist in establishing a conceptual context involving strategies for interactive public art (Rogala, M, 2010).

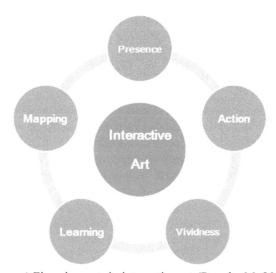

Figure 1 Five elements in interactive art (Rogala, M, 2010)

## 2.2 Co-Creative

Interactive art is crossover art and science. Interactive artworks were created by multimedia and display different results for audiences because they co-creative with audiences in the moment of the exhibition. Through the process, something likes impressions, experience and new meaning has been taken by Co-creative of interactive art and audiences (shown in Figure2).

Figure 2 Co-Creative of Interactive Arts and Audiences

## 3      EXPERIENCE DESIGN OF PARADISE AMBIANCE

## 3.1 Experience Design

Schmitt (1999) proposes five experiences: sense, feel, think, act, and relate. The sense experience includes aesthetics and sensory qualities. The five senses are the basic conditions for experience design. In current parlance, this creation and organization of useful representation in multiple media, in multiple ways, across various user situations, is termed "experience design." (Nathan Shedroff, 2001). The museum also seeks to stage experiences, here the concept of 'experience' suggest a human centred approach that has seen a shift in audiences role from passive recipient to an active participant in the overall scheme of the design process (Anita Kocsis, 2009).

## 3.2 Paradise Ambiance

Cultural interactive experiences can be said to cross the spectrum as can the telling of private stories, both enabling co-creation of content and user-led experiences (Watkins, Jerry and Russo, Angelina, 2005). "Paradise Ambiance" in this paper means a quietude and comfort lifestyle and likes LOHAS lifestyle. In China, Paradise Ambiance also means the ancients' lifestyle. People love for nature and spend much time for Leisure life. The *tianshi* (quietude and comfort) lifestyle merely refers to four activities: tranquilly touring valleys, keeping a peaceful mind via meditation, participating in associations and literary activities, and chatting about Buddhism (Chih-ho Wu, 2002). Nowadays, people sought their carefree and leisure life by means of some activities of quietude and comfort had become the periodic cultural phenomenon of life. In addition, many people in Taiwan like visiting the National Palace Museum in Taipei because they can experience "Paradise Ambiance" through the artworks or the artworks of interactive design.

**(1) The questionnaire survey**

The authors choose nine features (shown in Table 1) of "Paradise Ambiance" to a questionnaire survey, in descending order of the description of "Paradise Ambiance".

- 46 adult female;
- 6 adult male;

The adults ranged in age from about 20. All of the adults were university students. All users were fluent speakers of Chinese with no major disabilities.

Table 1  Features of Paradise Ambiance

| The Features of Paradise Ambiance (世外桃源的屬性) | | |
|---|---|---|
| Feature 1 | Nostalgia to popular views | 思古幽情 |
| Feature 2 | Literary atmosphere | 文人氣息 |
| Feature 3 | Atmosphere of Slow Food | 慢食氛圍 |
| Feature 4 | Totally cut off from the world | 與世隔絕 |
| Feature 5 | Ancient objects of art | 古色古香 |
| Feature 6 | Vigorous and Greenish | 綠意盎然 |
| Feature 7 | Aspire to truth and simplicity | 返璞歸真 |
| Feature 8 | Natural organic | 自然有機 |
| Feature 9 | Secluded and Refined | 清幽雅致 |

**(2) The findings**

Our study yielded the following findings (shown in Figure 3). The features of "Paradise Ambiance" on the first three ranking are about nature, quietude and comfort.

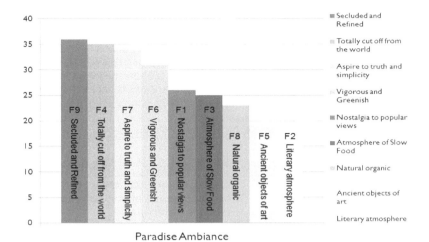

Figure 3 Illustration of the features' ranking of "Paradise Ambiance" .

# 4    FOUR SEASONS OF THE NATIONAL PALACE MUSEUM IN TAIPEI

## (1) The content analysis

The authors choose and make content analysis for the cases of the National Palace Museum in Taipei and they are a series of the same topic-Four Seasons. Four Seasons are four paintings and they are the representative works of China artists. Qing dynasty of Lang Shining (Giuseppe Castiglione)-Sprint , Song dynasty of Feng Dayou-Summer, Yuan Dynasty of Zhao Mengfu-Autumn and Ming Dynasty of Wen Zhengming-Winter (shown in Table 2, 3, 4 and 5) .

Table 2 Four Seasons of the National Palace Museum in Taipei- Spring Birth

| Spring Birth | Spring brings back all the verdure of the Earth; Blossoms burst forth on branches with their songs. |
|---|---|
|  | The images here of flowering crab apple, magnolia, peony, carnation, and peach blossom are representative works of Lang Shining (Giuseppe Castiglione) in fine brushwork and bright colors. The combination of Chinese and Western painting methods forms a unique style to portray the charming poses of birds and realistic finesse of blossoms. Combining voice-activated interaction and animation that brings the moving beauty of spring to life in the paintings, this installation is completed in |

| | using audience participation, with voice activation to control the animation and recreate the appearance of movement. With the birds and blossoms swaying in the breeze, you will become intoxicated by their eternal charm and enchanted by the magical power of Castiglione's spirited naturalism. |
|---|---|

Table 3  Four Seasons of the National Palace Museum in Taipei- Summer Lotus

| Summer Lotus | **The green of lotus leaves stretches to the heavens; Lotus blossoms shine with their red in the sunshine.** |
|---|---|
|  | The lotus pond in full bloom as conveyed by the Song dynasty artist Feng Dayou is filled with lotus leaves and blossoms in radiant bloom swaying gently in the breeze. Some are buds awaiting to open and bring further life to the scene. Ducks swim leisurely about the pond as colorful butterflies dance in the air with swallows in flight for a scene full of life unfolding in a visual feast for the eyes of summer at its height. The construction of a cyber-physical interactive space allows the myriad imagery of breezy summer lotuses filling a pond come to life with your steps activating the emergence of fish swimming and duckweed gathering and dissipating with the ripples on the water's surface. At the same time, the breeze of life follows the cycle of time and day to issue forth boundless life and vitality. New media technology applied to a sense of realism allows you to use digital methods to feel the vibrant life of a summer's day. |

Table 4  Four Seasons of the National Palace Museum in Taipei- Autumn Colors

| Autumn Colors | **A clear stream passes by mountains in green; Clear skies and limpid waters melt in the autumn hues.** |
|---|---|
|  | Zhao Mengfu used his recollections to paint "Autumn Colors on the Qiao and Hua Mountains," taking a romanticized and lyrical approach to sketching scenery as a present for his good friend, Zhou Mi, to relieve Zhou's longing for his ancestral home. The scenery has just turned to autumn with red leaves appearing here and there. Cottages are |

<table>
<tr>
<td></td>
<td>scattered as goats roam around. Hardworking fishermen seem to take advantage of the pleasing weather and make their way home with a full catch. Womenfolk in the cottages come out in preparation for their return, making this scenery even more enticing. This installation uses color and timing to interpret the message of autumn. Using digital body sensor interaction, you can use your hand to wave over the painting and move your hands to initiate close-ups, revealing the original delicate beauty of "Autumn Colors on the Qiao and Hua Mountains." It is like the magic in which the painter picks up and wields the brush to transmit the friendship of old as understood by fellow literati.</td>
</tr>
</table>

**Table 5 Four Seasons of the National Palace Museum in Taipei- Winter Snow**

| Winter Snow | The dawn appears with makeup of bright fleshy snow. |
| --- | --- |
| <br> | Faintly seen in the painting is a small mountain path with travelers winding upwards layer-by-layer. Following the figures in Wen Zhengming's painting, together we pass through layers of deep snow and cross a frozen river to appreciate the scenery of a frozen land in China's north. This installation uses privacy glass and lighting to create a space filled with changing moods in this wintry scene. The variations from thick to thin snow and differences between day and night are suggested by changes in white, yellow, violet, and indigo lighting, creating an atmosphere of changes in life while giving the space a pervading sense of realism. Entering the snow-filled area and experiencing perceptive interactive technology, audiences can place themselves into a completely new awareness of mountain passes in deep snow. |

## (2) The finding

Comparison and analysis of Four Seasons' contents (shown in Table 6), we had some results. The cases of "Spring Birth" and "Summer Lotus" are close shot, "Autumn Colors" and "Winter Snow" are long shot. At long shot, we can see more stuff and feel more emotion; at short shot, we can see the detail stuffs of more interactive. Interact with the audiences and let them stay long time are very important factors for the exhibit. All of the cases of Four Seasons have the feature of "Secluded and Refined", that was also the first ranking of the questionnaire survey of "Paradise Ambiance".

**Table 6 Comparison and analysis of Four Seasons' contents**

| Season (Theme) | Subject Content | Interactive Mode | Original Form | Meaning-Paradise Ambiance Feature (analysis based on visual) |
|---|---|---|---|---|
| Spring Birth | Birds, Flowers, Butterflies | Voice-Activated | Painting | Vigorous and Greenish<br>Natural organic<br>Secluded and Refined |
| Summer Lotus | Lotus, Birds, Butterflies, Carassius auratu, Moon, Water, Teals, Moon | Somatosensory Interactive (Body moving and Wind flowing) | Painting | Vigorous and Greenish<br>Natural organic<br>Secluded and Refined |
| Autumn Colors | Trees, Fence, Houses, Animals, River, Ships, Mountain, People | Hand Recognition | Painting | Nostalgia to popular views<br>Totally cut off from the world<br>Ancient objects of art<br>Aspire to truth and simplicity<br>Secluded and Refined |
| Winter Snow | Mountain, Snow, Ships, People, Horses, Houses | Sensor, Polyvision LCG | Painting | Nostalgia to popular views<br>Totally cut off from the world<br>Ancient objects of art<br>Aspire to truth and simplicity<br>Secluded and Refined |

# 5    CONCLUSIONS

To conclude, the present study is preliminary research on Paradise Ambiance in interactive art, but its relevance to experience design can also be seen. The experience design seemed to construct interactive experiences that benefit from natural interactions, compelling communication. A major finding is that the cases (Four Seasons) of experience design through eyes, hands, mental attention to teach audiences' hearts and experience the ancient life in China. The findings of this research should lead to a theme, feature and mode taxonomy of interactive art which will allow us to study classification of "Paradise Ambiance" of interactive art. It can be reasoned that extended to any theme of Ambiance, and extended to any field of Creative Industries. But it remains unclear what kind of ambiance suiting what kind of interaction. This is the first study about "Paradise Ambiance" in interactive art. This study will extend the research to this field.

# ACKNOWLEDGMENTS

Our deepest respect and gratitude go to The Bright Ideas Design Co and The National Palace Museum in Taipei, for their expert information.

# REFERENCES

Macdonald and Sharon. 2006. A Companion to Museum Studies, Oxford, Blackwell, p.321.

Mills, Simon, Stewart, Gavin and Thomas, Sue. 2006. An End to the New Re-assessing the Claims for New Media Writing(s), Convergence: The International Journal of Research into New Media Technologies, 12(4), pp.371-373, p.372.

Anita Kocsis. 2009. A phenomenological design analysis of audience experience in digitally augmented exhibitions, The 2th IASDR World Conference, pp.123-124.

Sherry His and Holly Fait. 2005. RFID Enhances Visitors' Museum Experience at The Exploratorium, September 2005/Vol. 48, No. 9 Communications of the ACM, p.64.

Nathan Shedroff. 2001. Experience Design 1, 2001 American Institute of Graphic Arts | Virginia Commonwealth University Center for Design Studies, p.304.

Brigid Costello, Lizzie Muller, Shigeki, Amitani and Ernest Edmond. 2005. Understanding the Experience of interactive Art: Iamascope in Beta_space,

The Second Australasian Conference on Interactive Entertainment, University of Technology, Sydney, Australia.

Chih-ho Wu. 2002. Leisurely *Shanshui* Lifestyle of the Ming Dynasty, 漢學研究第20 卷第1 期, pp.101-129.

Oppermann, R., and Specht. 1999. M.: A nomadic Information System for Adaptive Exhition Guidance. Proc. ICHIM'99.

F. Sparacino, G. Davenport and A. Pentland. 2000. Media in performance:

Interactive spaces for dance, theater, circus, and museum exhibits, IBM SYSTEMS JOURNAL, VOL 39, NOS 3&4, 2000, pp.479-510.

Martin Lister, Jon Dovey, Seth Giddings, Iain Grant and Kieran Kelly. 2008. New Media: a critical introduction. This edition published in the Taylor & Francis e-Library, 2008, pp239-242.

Margaret Fleck, Marcos Frid, Tim Kindberg, Rakhi Rajani, Eamonn O'Brien-Strain and Mirjana Spasojevic. 2002. From Informing to Remembering: Deploying a Ubiquitous System in an Interactive Science Museum, Mobile Systems and Services Laboratory HP Laboratories Palo Alto HPL-2002-54, p.3.

Schmitt, Bernd H. 1999. Experiential Marketing: How to Get Customers to Sense, Feel, Think, Act, Relate to Your Company and Brands. New York: The Free Press.

Watkins, Jerry and Russo, Angelina. 2005. Digital Cultural Communication: designing co-creative new media environments . In Candy, Linda, Eds.

*Proceedings Creativity and Cognition 2005*, pages pp. 144-149.

Rogala, M. 2010. The virtual and the vivid: Reframing the issues in interactive

arts. *Technoetic Arts: A Journal of Speculative Research* 8: 3, pp. 299–309, doi: 10.1386/tear.8.3.299_1

"Four Seasons of the National Palace Museum in Taipei" Accessed January 10, 2012, http://www.npm.gov.tw/exh100/npm_digital/en3.html

# CHAPTER 5

# Laboratory Kitchen and "Existenzminimum" Dwellings

*Jadwiga Urbanik*

Wroclaw University of Technology
Wroclaw, Poland
jadwiga.urbanik@pwr.wroc.pl

## ABSTRACT

In the interwar period the most basic and economical type of living space, the so-called Existenzminimum, was regarded as the most pressing topic to be addressed by modern domestic architecture. Each of the Werkbund model housing estates featured versions of the Existenzminimum units intended for sequential manufacturing in the future.

The Existenzminimum flat, which satisfies man's basic social, biological, and technological needs became the principal topic of the 2nd international CIAM congress organized in Frankfurt am Main in 1929. The RFG (State Research Society for Cost Efficiency in Building and Housing) was a government agency established to stimulate and finance research on rational planning and support model housing developments.

Homes by avant-garde architects featured empty, uncluttered rooms, strictly functional objects, lightweight machine-made furniture which would not take up too much space: everything not serving a practical purpose was removed.

The area where the revolutionary changes affecting the average household were made most clearly visible was the rapid evolution of traditional kitchen. The so-called Frankfurt kitchen, designed by Margarete Schütte-Lichotzky in accordance with the science of ergonomics and presented in Frankfurt, was the first in the series of innovative solutions.

In the interwar period small laboratory kitchens were presented and popularized in houses of model experimental Werkbund estates in Germany, Switzerland and Austria.

**Keywords:** Existenzminimum dwelling, Frankfurt kitchen, Werkbund, CIAM, the twenties and thirties

# 1    INTRODUCTION

In the period before World War I, providing affordable housing for the less affluent was not a major concern. Both the availability and quality of housing was inadequate.

The German Werkbund joined in the efforts to solve the housing problem by proposing a programme aimed at providing functional and affordable (small and cheap) housing within a short time. The programme was addressed to all countries suffering from housing shortages in the aftermath of the war. Throughout Europe architects addressed the issue but it was Germany that assumed the lead.

The Weimar Republic, although facing immense economic difficulties, began to tackle the housing problem, making affordable public housing a high priority With its economy ruined by the war and burdened with huge reparations imposed on Germany, the country undertook considerable efforts to develop and present model solutions (Syrkus, 1984). After 1924, foreign capital replenished the German economy under the Dawes Plan and the situation improved, ushering "the golden twenties" (Goldene Zwanziger Jahre) (Rieger, 1976). In many towns and cities the local government became dominated by social democratic circles and supported municipal housing policies, initiating major public housing projects for the broad masses.

# 2    EXISTENZMINIMUM DWELLING

Le Corbusier wrote: "The house is a machine for living in. Baths, sun, hot-water, cold-water, warmth at will, conservation of food, hygiene and beauty in a sense of good proportion.[...] The house: a shelter from the cold and hot weather; from the rain, thieves, and curious passers-by, a container for light and sun. [...] How many rooms? One for cooking, one for eating, one for working, one for bathing, one for sleeping, or a certain number of areas designated for cooking, working and private life. This is the standard of living space" (Rutkowski, 1932).

Aimed at the reduction of construction costs per housing unit, the architects developed standardized layouts to make a more rational distribution of available space. The simplest way to achieve this was to limit the size of an individual flat and thus the concept of "Existenzminimum" whose aim was to reduce the size of each room as much as possible without impairing its basic function, was born. Dutch architects, like Jacobus Johannes Pieter Oud, Mart Stam and others, became leading specialists in this area. They envisioned very small living units located in row-houses featuring the family room and kitchen on the ground floor and sleeping rooms on the first floor. The Kiefhoek housing estate in Rotterdam is a good example of this conception. Each dwelling unit was extremely narrow – approximately 4m.

Research was done on the ergonomics of the living space. The kitchen underwent a particularly drastic metamorphosis. The famous Frankfurt Kitchen was designed by Margarete Schütte-Lichotzky in 1926 (*Neues Bauen, neues Gestalten...*, 1984). The function analysis in the kitchen was carried out what resulted in a design placing appliances and utensils in an optimal way and limiting effort while working in the kitchen.

Equally innovative, the extremely small laboratory kitchens designed by Jacobus Johannes Pieter Oud at Weissenhof estate in Stuttgart, were disciplined studies in ergonomics and functionalism but contained everything necessary in a modern kitchen. His example was followed by others: Ludwig Mies van der Rohe, Walter Gropius, Bruno Taut and Le Corbusier.

Thus, by the 1930s, the revolutionized kitchen caught on throughout Europe. There was also a marked tendency to depart from the isolated kitchen and integrate it into the dwelling's space. Over several decades, the kitchen was transformed from a large room fitted with cupboards, cabinets, tables, chairs into a small, convenient laboratory finally to disappear as an isolated utility room.

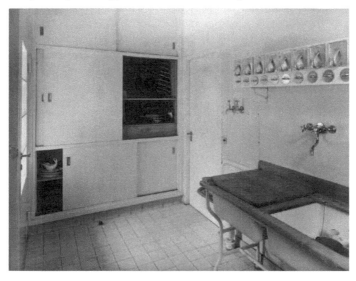

Figure 1 Kitchen, designed by Theo Effenberger, 1929. (From Wroclaw Museum of Architecture, Mat IIIb 533-14.)

In 1926 a national research institution, called Reichsforschungsgesellschaft für Wirtschaftlichkeit im Bau- und Wohnungwesen (RFG; Imperial Research Society for Cost-Efficiency in Building and Housing), was founded to develop optimal housing solutions, organize and finance relevant research and support the development of modern housing estates. (*Trotzdem modern...*, 1994). Members of the RFG's architects who tried to implement its guidelines in their own designs were distinguished.

The RFG worked to define optimal layouts for new developments, standardize the sizes of apartments and outline the most functional floor plans. It strove to implement new building materials and technologies as well as structural solutions that would reduce construction costs. It also aimed to improve hygienic conditions by ensuring adequate light and air through the advantageous alignment of the buildings. Hence, the RFG's interest in linear layout ("Zeilenbauweise").

Eventually the RFG developed guidelines defining the optimal standard size of the apartment in relation to the size of the family: 45 m$^2$, 57 m$^2$, 70 m$^2$ (Nielsen, 1994).

The RFG guidelines specified other dimensions, such as the height of rooms (2.8 m), the size of the kitchen (5m$^2$ in an apartment of 45m$^2$ but no smaller than 6$^2$ in larger apartments), living/dining room (14m$^2$ in the smallest apartment and 16m$^2$ in larger ones). Furthermore, bed-space was specified as: 5 - 7m$^2$ in the smallest apartment (Anzivino, 1989).

Figure 2 Row house designed by Jacobus Johannes Pieter Oud, Weissenhof housing estate, Stuttgart 1927. (From Die Form 1927, p. 271.)

In experimental housing estates, flats of various size and floor plans were tested with regard to their functionality. Aleksander Klein conducted research concerning functional layouts for small flats for the RFG. (Kononowicz, 1995).

The RFG's activity also influenced a number of housing estates in Germany that were not under the organization's direct supervision. Oftentimes they were designed by architects who accepted the Rfg guidelines; some were the organization's members and felt obliged to implement its programme.

Gustav Wolf, an RFG member, defined the criteria for a rational flat: "for every person, a separate bed; for every family, a private bathroom; reduction of the space used solely for housekeeping and sleeping to enlarge the central living area." (Kononowicz, 1995). The living room and the kitchen formed the centre of the living space. This was an entirely new approach to living space design. The living room adjoining the laboratory kitchen or the so-called Wohnküche (living kitchen) became the space where the family could integrate at the dining table: in small flats it was the only room large enough for communal gatherings.

The Deutscher Werkbund and its counterparts in Switzerland, Austria, and Czechoslovakia held a number of exhibitions to present contemporary developments in domestic architecture and construction. The experimental estates were to test the functional premises of the New Architecture and present new designs as well as construction methods for apartment blocks and modest houses that could be built inexpensively on a large scale. The aim was to reduce construction costs while optimizing the effect.

8 model experimental estates were built in Europe: Weissenhof in Stuttgart (1927), Nový Dům in Brno (1928), WuWA in Wroclaw (Breslau, 1929), Dammerstock in Karlsruhe (1929), Eglisee in Basel (1930), Neubühl in Zürich (1931), Lainz in Vienna (1932), Baba in Prague (1932).

It was emphasized that the size and number of rooms should correspond with the residents' basic needs: shelter, eating, sleeping, personal hygiene, washing, and cooking.

In the Weissehof estate in Stuttgart Mies van der Rohe elaborated on the idea of unified living space that could be divided up by means of lightweight mobile walls, screens, and shelving units: the rooms were interconnected, with no doors between them. His was the most convincing proposal of the conceptions of open layout seen to date. The house by Mies van der Rohe attracted enormous interest but it was the complex of row-houses by Jacobus Johannes Pieter Oud and Mart Stam that emerged as the most fitting answer to the exhibition's professed objective. The dwellings for working class families were based on a very simple and economical layout, reflecting the Dutch home's tradition and the principle of functionalism: the ground floor was occupied by a unified living space and a kitchen while the first floor accommodated three sleeping rooms.

In Wroclaw Hans Scharoun presented a new type of hotel-like building for childless couples and single people.

The building's right and left wings together comprised of 48 small duplex (split-level) apartments while the central section accommodates a restaurant, a lobby, and the recreational space complete with a roof garden. It was written that "the apartments have been calculated for two people so precisely that the arrival of an unexpected guest would render the living quarters impossible to use" (Norwerth, 1929).

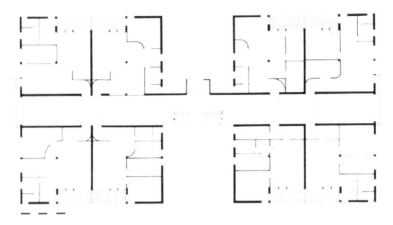

Figure 3 High-rise tenement block of flats, typical layout of the floor, designed by Adolf Rading. (Drawn by J.Urbanik)

The building's skeleton frame construction allowed for considerable flexibility in developing the plan so that on each floor there were eight flats of the same size (57 m$^2$) but with different layouts. Implementing the "communal living" idea, the architect provided a number of communal rooms to be used by all inhabitants. The building was intended as a prototype for large housing estates composed of high-risers (Szymański, 1992). It seems possible that both Scharoun and Rading were inspired by American solutions, the so-called "boarding houses" or "apartment houses", or the Soviet "communal houses" (Rutkowski, 1932).

Figure 4 Tenement row housing of the WUWA estate, designed by Emil Lange, Ludwig Moshamer, Heinrich Lauterbach, Moritz Hadda, Paul Häsler, Theo Effenberger. (From Wroclaw Museum of Architecture, Mat IIIb 533-6.)

In Basel Scherrer and Meyer and the studio of Artaria & Schmidt designed the 'back-to-back' row-houses presenting a very consistent approach to the economical use of space despite certain errors in location and layout.

In almost all model Werkbund estates Existenzminimum dwellings were presented in multi family or row houses.

Interior design and furnishings provided another focus point. The new living space required a new and innovative approach. The compact and functional modern apartment of the "Existenzminimum" type, promoted by architects and public authorities, could not accommodate the traditional heavy and cumbersome furniture, which just did not fit (literally and metaphorically) into the restrained modern interior. "There is no room here for furniture pieces from before World War I" – wrote Gustav Wolf (Wolf, 1929). The pastel walls stripped of any ornamentation provided an ideal backdrop for simple furniture pieces in bright vivid colors.

Figure 5 Living-cum-dining room connected to the kitchen via an opening in the wall to facilitate the serving of meals, designed by Heinrich Lauterbach, WUWA estate, 1931. (From Wroclaw Museum of Architecture, Mat IIIb 1037-1.)

They favored functional built-in or fitted furniture: closets, bookshelves and kitchen cabinets, which saved space and did not clutter the room. In a small apartment, a window pierced in the wall between the kitchen and the living room, through which meals could be served, was as important as an ergonomically arranged bathroom. Heinrich Lauterbach emphasized that the closer the fit between the apartment's layout and the inhabitant's needs, the easier the job of the furniture designer. The problem was that the architects designed for an anonymous person, projecting his supposed needs and preferences.

The old-fashioned *horror vacui* that afflicted traditional interiors was eliminated. The architects believed that the young generation, accustomed to sports and physical activity needed an uncluttered apartment. Furniture was to offer comfort without occupying much space (physically as well as visually). The creed was to use well-designed, mass-produced furnishings which were affordable and offered considerable flexibility in interior arrangement. The owner could furnish the apartment to suit both his needs and his pocket, thus giving the space an individual character. Most model interiors featured light chairs of tubular steel or plywood that offered hygiene, affordability, and an inherent resiliency that provided comfort without the need for excessive expenditures.

Figure 6 Kitchen with an opening in the wall to facilitate the serving of meals, designed by Moritz Hadda, WUWA estate, 1929. (From Rzeczy piękne 1929, p. 73.)

Trying to adjust architectural design to human needs, the architects also tried to adjust man to modern architecture. The method of analyzing the function of particular spaces/areas of the apartment disregarded the inhabitants individual characteristics. The modern functional interior was suited to the rational, well-organized resident. Anything not regarded as "modern" and everything that did not fit into the simple, hygienic interior was eliminated. The architects envisioned the ideal user of modern architecture.

Some architects active in the 1920s were willing to consider the needs of the prospective residents.

Ernst May's activity in Schlesiche Heimstätte – housing cooperative in Wroclaw and Silesia in the years 1919-25, was the first stage of applying new parameters concerning housing construction. It was here where he suggested "Egsistenzminimum" ("Borsig" type) for the first time. He suggested laying out a flat where the kitchen consisting of a dwelling part and a cooking niche, being the prototype of a later Frankfurt kitchen, would be the central place of it. Margarete Schutte-Lichotzky cooperated with May in Silesia where she promoted a small kitchen laboratory in "Schlesisches Heim".

Ernst May's activity in Frankfurt-am-Main was the model approach to community housing on a grand scale. While designing flats for working class families, he would construct life-sized models and presented them, fully furnished, to the target group of prospective residents to hear their opinions on new floor plans and design solutions in order to understand their preferences. He regarded this stage as absolutely essential to the design process and had great respect for the opinion of people who would live in his buildings for many years to come. He declared that modern architects, with their mostly middle-class background, had no understanding of the needs and preferences of the working class (Syrkus, 1984).

The successive meetings of CIAM (Congrès Internationaux d'Architecture Moderne) provided an important forum for Europe's avant-garde architects to present and discuss their ideas. Particularly important in promoting innovative perspectives on domestic architecture and living space design was the 2nd CIAM congress organized in Frankfurt-am-Main in 1929 and accompanied by the *Die Wohnung für das Existenzminimum* exhibition featuring plans of small flats from new housing estates established in various European countries.

The venue must have been purposefully selected because the city served as a testing ground for new urban planning, domestic architecture, and technological innovations. Ernst May insisted that rational planning must have scientific foundations and biological, psychological and sociological research was needed to establish the minimal living space depending on the family's size.

In Frankfurt-am-Mein, Ernst May, a pioneer of the community housing idea in Europe, initiated the construction of many housing estates of the new type. They were part of the larger development project in the valley of the Nidda River. The rigorous implementation of the new master plan for the city resulted in the construction of a complex of modern housing estates in Frankfurt: Praunheim, Römerstadt, Westhausen, Bornheimer Hang, Riederwald, Hellerhof, Niederrad (Bruchfeldstraße) and Heimat (Riedhof-West).

## 3    CONCLUSIONS

The solution of small dwelling was the main task for architects in interwar period.

Adressing the 2nd CIAM congress, Ernst May declared: "Our activity will focus on providing each individual with his due 'housing share' in the best possible way. Many years will pass before, as a result of international co-operation between civilized nations, solutions are found to ameliorate the housing problem that would guarantee each individual, housing standards defined as an acceptable minimum" (Syrkus, 1976).

Small flats are still being looked for on the housing market and the so-called kitchen laboratory – Frankfurt kitchen is still popular. After W.W. II in the East European countries it was a standard solution in mass housing estates.

## REFERENCES

Anzivino, G. 1989. Il Dammerstock di Karlsruhe. Microstoria di una "Siedlung", *QUASAR - Quaderni di Storia dell'Architettura e Restauro*: 53-74.

Berning, M., M. Braum, E. Lütke-Daldrup. 1990. *Berliner Wohnquartiere - ein Führer durch 40 Siedlungen*. Berlin.

*Ciam Dokumente 1928–1939*, ed. M. Steinmann, Basel, Stuttgart 1979.

Cramer, J., Gutschow, N. 1984. *Bauausstellungen. Eine Architekturgeschichte des 20.Jahrhunderts*. Stuttgart.

Kononowicz, W. 1995. Ewolucja osiedla mieszkaniowego we Wrocławiu okresu Republiki Weimarskiej - Księże Małe. In: *Architektura Wrocławia*, Vol.2. *Urbanistyka*, ed. J. Rozpędowski, Wrocław.

*Les cites de Ernst May - Guide d'architecture des cites nouvelles de Francfort (1926-1930)*, Frankfurt am Main 1988.

*Neues Bauen, neues Gestalten. Das neue Frankfurt, die neue stadt - eine Zeitschrift zwischen 1926 und 1933*, ed. H. Hirdina, Dresden 1984.

Nielsen, Ch. 1994. *Die Versuchsiedlung der Werkbundausstellung „Wohnung und Werkraum", Breslau 1929*. Bonn.

Norwerth, E. 1929. Wystawa mieszkaniowa we Wrocławiu, *Architektura i Budownictwo*: 319-336.

Rieger, H. J. 1976. *Die farbige Stadt. Beiträge zur Geschichte der farbigen Architektur in Deutschland und der Schweiz 1910-1939*, Zürich.

Rutkowski, Sz. 1932. *Osiedla ludzkie*. Warszawa, Kraków.

Syrkus, H. 1976. *Ku idei osiedla społecznego 1925-1975*. Warszawa.

Syrkus, H. 1984. *Społeczne cele urbanizacji*, Warszawa.

Szymanski, B. 1992. *Der Architekt Adolf Rading (1888-1957) - Arbeiten in Deutschland bis 1933*. München.

*Trotzdem modern. Die wichtigsten Texte zur Architektur in Detschland 1919-1933*, ed. K. Hartmann, Wiesbaden 1994.

Urbanik, J. 1993. Toward functionalist dwelling, successes and failures – model Werkbund Estates. Proceedings of the Second International DOCOMOMO Coference 1992, Dessau 1993: 36–41.

Urbanik, J. 2009. *WUWA 1929-1980, Wrocławska Wystawa Werkbundu*. Wrocław.

*Werkbund Ausstellung "Die Wohnung" 1927*. 1927, Exhibition Catalogue. Stuttgart.

Wolf, G. 1929. Sonderausgabe „Wohnung und Werkraum", *Breslauer Illustrierte Zeitung*.

# CHAPTER 6

# Intercultural Differences in the Formation of Space of the Courtroom

*Grazyna Hryncewicz-Lamber, Marek Lamber*
Wroclaw University of Technology
Wroclaw, Poland
grazyna.hryncewicz-lamber@pwr.wroc.pl

## ABSTRACT

The courtroom is a space of law enforcement and authority as well as the arena of precisely drawn circulation routes and lines of sight. Although the authority of law is always represented by the judge, whose bench rises above the courtroom floor, the drama of the trial involves diverse spatial practices in different countries and types of courts. The two systems of legislation – European and Anglo – American have formed distinctive spatial solutions that reflect the social order, with characteristic positions of the judge and/or the jury as well as the barristers, the prosecutors, the plaintiff, the suspect and the public within the spectacle of adjudication. There are most interesting differences of the seating arrangement, lines of sight in the courtroom and overall spatial organization of the movement in the adjoining area, that seem to spring from the historic development of the above mentioned systems of law.

**Keywords**: courtroom design, architecture, historic courts, lines of sight

## 1. Introduction

While entering a modern courthouse one is overwhelmed by the efficiency of spatial organization. The arrangement of justice facilities, from the entrance to one's prescribed seat in the courtroom, is designed under the governing principles of security, categorization and filtering of several classes of users (Graham, 2003, Mulcahy, 2011). It seems that the courthouse and its core – the adjudication room are the epitome of pure usefulness, safety and spatial organization based on rationality. If it were so, the courtrooms would have had similar solutions of the use of space worldwide. Visible differences occur though, due to different court procedures adopted in various law systems: they derive from the historic patterns and cultural background of the lawgivers and architects. The significance of the spatial order of the courtroom is discussed by the constitutional rights activists. Nowadays some lawyers raise the question of the proper use of the courtroom space in connection with ideas as fundamental as the representation of law and the due process. Some elements of the furnishings, like the dock, the witness stand are scrutinized to eradicate the notions of inequality and spatial discomfort of people.

## 2. A brief overview of the history of the law courts

Since its beginnings the law was considered either the responsibility of the sovereign or the community. It was performed for a long time in the open: the imagery of law is abundant with examples of rulers, such as Charlemagne, king Salomon or biblical prophet Deborah in their seats of power under symbolic trees. Ancient Greeks as well as barbaric German and Slavic people adhered to the tradition of gatherings organized to solve legal problems by voting or acclamation, performed in sacred open spaces. The need for a designated place for law grew alongside the idea of the city: the oldest medieval courtrooms were either the jury chambers in town halls or independent structures, raised above ground to combine commerce and law (Berlin's Gerichtslaube, courthouses in Maidstone, England). City – state invented the jury: in the medieval times – a permanent body of senior citizens, administering justice to their fellow-burghers, independent from the feudal rulers. Justice was brought to the rest of predominantly rural society by itinerant judges, using suitable rooms in an inn (hence: Inns of Court), or a castle, that provided dignified space of a hall. The oldest buildings devoted entirely to the law survive in France: the Palais de Justice in Rouen of the 15th century and the Palais du Parlement in Rennes (S. de Brosse 16th century), one of them recently refurbished still serves as a courthouse.

The transactions of law required at least a central table, around which the jury, and later the professional barristers, gathered and a dais for the judge. A very characteristic 17th century images picture the court as an area around a large central table, surrounded by a temporary enclosure, formed by benches and barred from the public; with a high seat for the judge (the frontispiece of Pracktycke der Nederlansche Rechten[…] by B. v. Zutphen 1655, reproduced in: Resnick, Curtis, 2011). Those elements: the dais, the enclosure, that later became the well of court and the bar survived till the end of 20th century. The central table and the seating arrangement became subject to numerous changes, since the 18th century brought the Enlightment's ideas of professionalization and systematization of court procedures, as well as the idea of the return to the source of the Roman Law. The architecture evoking Rome's serene monumentality (courts in Aix-en-Provence by Ledoux), characterized by clear functionality (Robert Smirke's country halls), in the course of the 19th century developed into a new typology of the palace of justice (Brussels, Joseph Poelaert).The need for a more sophisticated architecture indicated the expansion of the law as a social practice and a profession (Graham, 2003, Mc Namara, 2004). The 19th century was the era of the elaboration of very complex judicial buildings, that exemplify the ideas of segregation, functional zoning, security, and express the legal power. The 19th century was the time of the post-revolutionary reform of European justice. Until then many countries (several German Duchies, Poland, Russia) had feudal systems of law, often without permanent seats of justice. It changed after the French pattern at the beginning of 19th century.

Figure 1  Interior of the Assize Court in the Palais de Justice in Paris (From: Handbuch der Architektur, IV Teil., 7Halbband: Gebaude fur Verwaltung, Rechtspflege und Gesetzgebung, 1887, Darmstadt, Verlag  Arnold Bergstrasser, p. 186)

**Common Law and the Napoleonic Code as driving forces behind the diversity of the courtroom designs.**

There are two major law systems: the adversarial – British and American with no written code, characterized by oral procedure, and inquisitorial – European, based on the Code Napoleon (Fischer Taylor, 1993).

The systems differ in the organization of the trial: the adversarial system is connected with active role of the barristers and the litigants, who argue in the presence of an inactive judge; while in the inquisitorial – the judge's role is active. The adversarial system, until well into the 19th century did not employ prosecution or a state attorney, both used the jury. These main dissimilarities were the key factor of spatial differentiation between courtroom designs, as specific demarcation of space was  followed by different location of the furnishings. It is the 19th century when the court furnishings became permanent, and very elaborate. The British courtroom amplified the enclosure of separate classes of users: the judiciary, the barristers, solicitors, the suspect within an elevated dock, and the public in a gallery, while a more open French courtroom barreau (well) became a stage with the involved parties visible to the public in a theater-like space (Fischer Taylor, ).

Figure 2  Criminal Court in the Assizes Building in Durham, seen from the gallery for public (From: Handbuch der Architektur, IV Teil., 7 Halbband: Gebaude fur Verwaltung, Rechtspflege und Gesetzgebung, 1887, Darmstadt, Verlag  Arnold Bergstrasser, p. 187)

### The 20th and 21st centuries

The spatial solutions introduced then lasted with minor changes until the World War II. The notion of democracy of the court proceedings and the equality of rights of the litigants brought about changes, that started to permeate courtroom design in mid- 20th century. Since then, the meaning of the courtroom changed: an unbiased place of the transactions of law, secure and effective, conveying the "principles of transparency, accessibility and civic engagement" (Greene, 2006) is sought for. As a German author puts it: a courtroom may turn into a "bulwark of intimidation", in seeking of safety measures (Gerhart Laage quoted in Klemmer, Wassermann, Wessel, 1993), hence the governing policy should be communication – friendly and functionally transparent place of justice. While the courtroom itself simplifies visually, its ancillary areas still tend to attain complexity and substantial dimensions, additional changes being introduced thanks to extensive use of computerized equipment, such as multimedia court recording and remote appearances technology (Lederer, 2006).

The withdrawal from the 19th century typologies of vertical and heavily partitioned courtrooms was followed by a strong tendency to make the same viewing and access conditions possible for all participants of the adjudication and the public. Inequality of the sightlines and visibility axes of the room, seemingly a purely cultural invention, is still present in the design of furnishings of the courtroom. At the beginning of the $21^{st}$ century the separation of user groups – symbolic and physical – inherent in the courtroom design is seen by some as a threat for the fair trial (Mulcahy, 2011), demarcation of space between the court and the defendant is often challenged as being a tool of exclusion that may affect the due process.

## 3. The criminal court procedures and their participants

The court procedures can be roughly divided into three major types: civil, criminal and administrative. The most elaborate and complex spatial solutions are usually inherent to the practicing of criminal law. The criminal process involves the presence of at least three classes of the people involved: the judiciary, the suspect in holding, and the public. Separate routes within the building are ascribed to these three categories. In the contemporary courtroom design special provisions are made to accommodate the waiting witnesses outside the courtroom, victims and children, whose safety and well-being is provided for.

**The jury**

Lay juries are part of legal procedures in different systems of law and symbolize the idea of the citizens participation in the transactions of law. The jury of layman represents an old tradition of the common law in English-speaking countries; it was a part of the history of the city – state adjudication, and found its place in the French Revolutionary courtroom. French, German and Polish law systems involve the representation of the citizens in the mixed juries, formed with the participation of professional judges and lay people.

The role of the jury in the procedure places them in between of the practitioners of law – judges and barristers, and the public (Graham, 2003). This is manifested by their physical presence at the stage of the court proceedings: in standard American or British criminal courtroom they sit in a designated jury box, within the well of court, and are granted a separate deliberation room. In the historic British courtrooms the jury box was an arrangement of steeply ascending seats, allowing for good sight and hearing of the procedure. The jury box lined one of the longer walls of the almost square room, opposite to the seating for witnesses and at a right angle to the elevated judge's bench. The pre-reform French and German spatial organization was similar in this aspect, less elaborate as to the height of the jury box, but with distinctive wooden partitions. In the French and German systems of 19th and the beginning of 20th century the jury box faced the defense box and the suspect dock, situated behind it (Fischer Taylor, 1993, Handbuch der Architektur,

1887). Nowadays in the French assize court the jurors and judges share the slightly heightened space of the bench and common deliberation room; in both Germany and Poland, if the lay jurors appear at court, they are seated likewise. The role of jurors in Europe equates them with the judges in the trial procedures, so their place in the courtroom and adjoining space is not underlined by any distinction.

The number of jury members is such that, if seating is provided in a row, there is no non-verbal communication between them. The American and British juries do not maintain visual contact during the courtroom proceedings, as they usually sit in ascending rows, some French courtrooms also provide seating in two rows for large juries. Very often though, the French juries sit in front of a slightly curved table for better communication and eye contact.

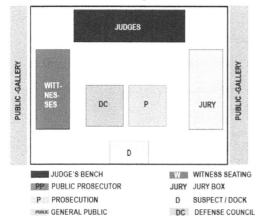

Figure 3  A comparative sketch of two courtroom types of the 19th Century, left: adversarial (after Old Bailey), Right: Inquisitorial (After Paris Palace De Justice)

### The Judge

As it was mentioned above, the judge's role in court is dependent on the system of justice. In the adversarial trial the judge presides over the courtroom, interprets justice, represents the power of state and instigates the order. Therefore the judge in the American courtroom is seated at an elevated place, behind a table with a screen in front to assure privacy and security. The witness is placed at the judges side, so there is a good verbal contact, but it is the jury who have to see the witness's face, not the judge. Instead, the judge faces the litigants and their solicitors in the same manner as at the civil trial. The American courtroom has two basic variations as to the location of the bench: it is placed either in a corner of the room or centrally, against the end wall. A slight difference occurs in the historic British courtrooms still in use, where the  bench is located opposite an elevated suspect's dock, lining the far end of the court's well (between the public and the well) so the judge can maintain eye contact with the accused.

In the inquisitorial system the judges investigate the case, ask questions, therefore the witness stand faces the bench, which in assize courts is shared by the judges and jurors. In historic French courts the line of sight of the judges was very long, as the well, unlike the British and American, was not filled by the advocates' seating, there were only the tables for corpora delicti and movable lecterns in it. The judges could see the witnesses' seating at the bar, and behind them, the public. In courts with larger numbers of judges their bench attained curvilinear shape for better visual contact. Some historic German examples show uniform horseshoe judges' tables, shared by them with the prosecutor and court clerks.

The judges seating is the most conservative part of the courtroom: the American courtroom standards continue to require, that: "while seated, the judge's eye level should be higher than of a standing person of average height" (Michigan Court Facility Standards, 2000) .

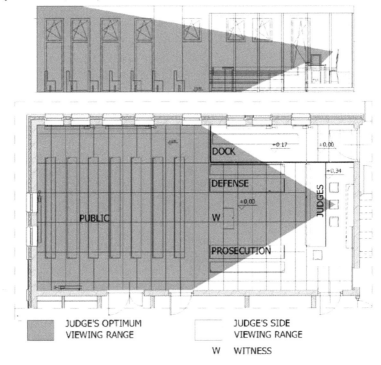

Figure 3 Viewing range of the judge in one of the criminal courtrooms in Zabkowice Slaskie Court, designed by G. Hryncewicz –Lamber, L. Rubik, P. Zarzycki et al., occupancy permit - 2012.

### The prosecutor

In the inquisitive procedure the place of the prosecutor was located on the dais, but was not quite equal to the judges. The prosecutor's table was located at the corner of the room , askew, this position enabling a good, though not face-to-face view of the defense and prisoners. The French judiciary includes public prosecutors, hence this distinction of this public servant, who beside a heightened seat was furnished with a substantial lectern. The German courtroom often introduced a small difference of height: the seat for the Richter was one step higher than the Stadtanwalt.

The idea of adversarial system did not take into account a position of public prosecutor: there were the magistrates or unpaid investigating judges and local sheriffs, who were granted a place at the bench, but their role in the trial was marginal. The case was instigated by the victim's advocate, who took seat in the well or stood at the bar. After the introduction of public prosecution in the British system, the prosecutor simply took seat of the victim's advocate, facing the judge.

### The advocates

The practice of law was always regarded as an exclusive trade, therefore the position of the barristers was high in the society, esp. those in the civil law. The system of law accommodated their professional pride by positioning them in the well of the court, side by side with their equal opponents in the case. Behind them sat the solicitors, clients being cut off by the bar in the British courtroom. The same seating scheme was adopted for the criminal trial; characteristic to the 19th century adversarial system was the fact, that the defendant was not supposed to speak at court (Graham 2003).

The French courtroom in the 19th century adopted a different spatial solution: here at the criminal trial, the advocates sat in front of their clients, in full view of the jury, at the right angle to the judge. The French introduced to the criminal court a possibility of the victim's representation, the "partie civile", who in the 19th century courts sat alongside the defence (Fischer Taylor, 1993). Both spatial arrangements were supposed to be geometrically symmetrical, but the exclusion of the suspect is a characteristic feature of the British courtroom.

Nowadays French, German and Polish systems accommodate the defense and prosecution, accompanied by the representatives of the victim in a uniform manner. The parties sit at the opposite sides of the court well, while the witness is placed in between them, facing the judge. In this manner two major axes are of sight are established: the immediate eye contact of the judges and the witness is obtained, while the prosecution and defense face each other. The British and American advocates continue to be seated side by side, while in Britain the prisoner's dock is situated at end of the room, not allowing for visual or verbal contact between the defendant and the advocate (Mulcahy 2007).

### The suspect in holding

The dock is one of the most criticized elements of courtroom furnishings. Its presence in the courtroom symbolizes the bias against the suspect in holding (Mulcahy, 2011), which is why the use of physical barrier between the suspect and the defense counsel was refuted by the American and German jurists (Klemmer, Wassermann, Wessel, 1995). It is still present in Polish and British places of law, its location though is different in the two above mentioned countries. In Great Britain many of the new courts continue the tradition of placing the dock at far end of the courtroom, while in Poland the dock is located behind the seats of defense counsel.

### The witness

The witness's stand is located to provide best sightline for the judge or jury. In American and British courts it is placed below the judge's bench, American

60

standard recommend to place the witness at the same side of the room as the jury, while the British practice is often the opposite (Mucahy 2007). The European courts adopted a uniform practice of witness placement. The solution of placing the witness between the litigants is sometimes questioned as giving the witness a sense of seclusion, awkwardness, even threatening (Asian Legal Resource Centre, 2006).

## 4. Conclusion

All parties involved in the adjudication process meet in the courtroom in state of stress, hence the need for supervised restrictive space. Psychological experiments confirm, that human behavior changes in different environment: a calming effect of traditional solemn, dignified, and very high courtroom is therefore expected. Apart of the notion of the democratic procedure, which requires equality of the sight, hearing and access conditions within the courtroom for all the involved, there is also a problem of the natural lighting. Researches in environmental psychology put emphasis on two positive aspects of the presence of windows in a room: natural light and the eye contact with the outside world, which is referred to as a key aspect of emotional comfort (Bell et al., 2003). Despite the fact that law enforcement is not supposed to evoke lay people's spatial discomfort (Resnick, Curtis, 2011), security zoning of the places of law often prevails over the human factors. Day lighting gives way to functional priorities and circulation demands, even though it is acknowledged, that: "carefully designed natural lighting is desirable within the courtroom (…) because there may be positive psychological effects on litigants and other participants (…)".(Gruzen, Daskalis, Krasnow, 2006). Architects seek methods to mediate between daylight, proper sightlines, and security in the courtroom design.

A very interesting spatial solution for a jury court (Schwurgericht) was introduced in the extension of the Landgericht Hannover in Germany in the 1992. As it is important for the senior judge and jurors to see each other, the architects Storch and Ehlers decided to introduce conference – type spatial organization to the courtroom, supported by the German Ministry of Justice. The seating for the judiciary and litigants is flat, without any dais, the parties (the prosecution, the jury and judges, the defense, and the witness) sit at similar curvilinear tables that form three quarters of a roundtable, and there is no visual barrier between them and the public, as the public seating is ascending like an auditorium (Klemmer, Wassermann, Wessel, 1995). The room is naturally lit from one side. This arrangement of space seems to be an optimum for the hearing, sight, and spatial comfort of the participants of adjudication process. It departs, though, from the notion of a dignified and symbolic place of adjudication, towards a semi-theatrical didactic space.

The courtroom aforementioned did not allow for the input of multimedia techniques in the courtroom design. In the experimental projects for the criminal courtroom of the future, planned for multimedia and communication, the seating for the litigants as well as other parties, is placed at such angles as facilitate seeing, and forms a conference – type space, as in the McGlothlin Courtroom at William and

Mary Law School. The location of the litigants tables allows for side view of the opponents, with better viewing conditions for the security and remote counsel. The space is lit by a skylight. Since it is possible for witnesses to appear at court from different places, such as their natural environment, or a special room in a place of detention, it is less stressful and intimidating to testify (Lederer, 2006).

The two attitudes towards courtroom design aiming at improved communication within the courtroom seem to account for a better, more humane spatial order of the place of law, in two different ways: one is to imitate a conference room with natural light with as little distinctions of rank and security segregation as possible, the other – is to deploy to the newest communication technology to eradicate the need for the secured courtroom in future.

## REFERENCES

Asian Legal Resource Centre: What really happens to witnesses in Thailand? Some cases, (2006) http://www.article2.org/mainfile.php/0503/233/ ( 2012-02-29);

Bell, A., Greene, T.C., Fisher, J.D., Baum, A.: Psychologia srodowiskowa (2003), Gdansk: Gdanskie Wydawnictwo Psychologiczne,

Graham Claire: Ordering Law: the Architectural and Social History of the English Law Court to 1914, 2003, Aldershot, England, Ashgate,

Gruzen, J., Daskalis, C., Krasnow,P.: "The Geometry of a Courthouse Design", 2006 in: Flanders, S. (ed.): Celebrating the Courthouse. A Guide for Architects, Their Clients, and the Public, New York, London, W.W. Norton&Co

Klemmer K., Wassermann R., Wessel T. M.: Deutsche Gerichtsgebaude. Von den Dorflinde ueber den Justitzpalast zum Haus des Rechts, 1993, Munich, Verlag C.H. Beck,

Landauer, T. v., ESchmitt, E., Wagner, H.: Gerichtshäuser, Straf- und Besserungs-Anstalten, in: Gebäude für Verwaltung, Rechtspflege und Gesetzgebung; Militärbauten. Darmstadt : Verlag von Arnold Bergsträsser, 1887 (Handbuch der Architektur, tl. 4, Entwerfen, Anlage und Einrichtung der Gebäude, Halbband 7)

Lederer, F. L.: "The Courtroom in the Age of Technology", 2006 in: Flanders, S. (ed.): Celebrating the Courthouse. A Guide for Architects, Their Clients, and the Public, New York, London, W.W. Norton&Co

Michigan Court Facility Standards Project, 2000, online: http://courts.michigan.gov/scao/resources/standards/fc_section3.pdf (2012-02-29)

Mulcahy L.: Legal Architecture: Justice, Due Process and the Place of Law, 2010 London, Routledge,;

Mulcahy L.: (2007) 'Architects of Justice: the politics of court house design', Social and Legal Studies, Vol. 16(3) SAGE Publications, online: http://www2.warwick.ac.uk/fac/soc/pais/research/gcrp/resources/architects_of_justice_t he_politics_of_courtroom_design.pdf (2012-02-29)

Mc Namara M. J.: From Tavern to Courthouse. Architecture and Ritual in American Law 1658 – 1860, , 2004, Baltimore-London, The Johns Hopkins University Press;

Resnick J., Curtis, D.: Representing Justice. Invention, Controversy, and Rights in City-States and Democratic Courtrooms, 2011, New Haven and London, Yale University Press

Taylor, Katherine Fischer: In the theater of criminal justice : the Palais de Justice in Second Empire Paris, 1993, Princeton, Princeton University Press,.

CHAPTER 7

# Using Mental Models and Ergonomics to Analyze Chinese Opera Performing Skills

*Wang, Tai-Jui\*, Lin, Rung-Tai\*\**

Chinese Culture University\*, National Taiwan University of Arts\*\*
Taipei, Taiwan
tyraywang@gmail.com

## ABSTRACT

The Peking opera, Taiwanese opera, Hakka opera, and other varieties of traditional Chinese opera, demonstrate performances in movement of a stylized program shared by homogeneous contents of interoperability. Performers of Chinese opera must have their learning fostered through the processes of realization, memorization, internalization, transformation, and representation for every performance. Theoretically, there are two main types of symbolized performance styles. Chinese opera experts implied that "the performing styles conveyed by the performer are tangible, but the performance methods are intangible". Therefore, how a performer uses intangible experience to manipulate tangible knowledge and skills is the priority of fundamental training processes in the Chinese opera curriculum. Ergonomics is used as an example in one special tangible-stylized program of Chinese opera performing skills for the female warrior (Wudan role) who is adept at rapidly handling weapons in battle such as deflecting tasseled spears with her foot. Thus, integrating ergonomics and mental model methodologies might produce a significant breakthrough application of pedagogic issues and props design in traditional Chinese opera training, and provide a means to analyze creative concepts for those intangible-stylized programs in further.

**Keywords**: Chinese Opera, Ergonomics, Mental Model, Virtualization

# 1    INTRODUCTION

The history of Chinese Opera is not clear, but performances combining song and dance were popular as far back as the Warring States period (403-222 BC). A few centuries later during the Han Dynasty (206 BC - AD 219), storytelling blended with song and dance to transform into an elementary form of musical drama. Chinese Opera demonstrates a unique art form in front of a live audience, refining and combining the best of several different opera styles from more than 300 different regional opera styles in China. Although special visual effects, such as backgrounds, props, and lighting, convey the setting and feel of a scene, they are not intended to create an actual sense of reality. Rather than attempting to approximate reality, Chinese Opera has evolved into a stylized, abstract, and specific coherent art form (see fig. 1), much like ballet or modern dance. Using minimal props and effects, the Chinese Opera applies an extensive display of symbols to cue the audience to every action.

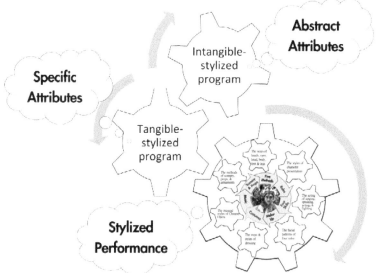

Figure 1: A Stylized, Abstract, and Specific Coherence Art Form of Chinese Opera Performance

As shown in Figure 1, there are two main types of symbolized performing styles. Chinese opera experts implied that the performing styles conveyed from the performer are tangible, but the performance methods are intangible. The movements of a tangible-stylized program are controlled by the formation of an intangible-stylized program. Using the concept of computer programming as a metaphor, the movements of a tangible-stylized program are similar to computer programming languages. On the other hand, the formation of an intangible-stylized program is similar to compiling a series of computer languages as software for the final application.

The goal of this research is to achieve training effectiveness as great as in olden times using modern methods of Mental Models and Ergonomic Analysis. This study focuses on how performers use their mental sets and human factors to manipulate their tangible-stylized skills and on the knowledge of skills, which are the priority of the fundamental training processes in the Chinese opera curriculum. Such research is still in its infancy, but it may have a contribution to make towards unraveling the mystery of developing information and communication technologies (ICTs) and educational technologies (ETs) that might have the potential to provide Chinese opera a productive context in the digital learning of the 21st century.

Thus, the practical purpose of the proposed methodologies is demonstrated through "Mental Models" and "Ergonomics", which might analyze for creative concepts for those intangible-stylized programs in further to approaching the purpose of this discussion in the steps of future research. While research on these questions is still at a beginning stage, hoping it is hoped that this study will have broad implications in a number of performing arts areas.

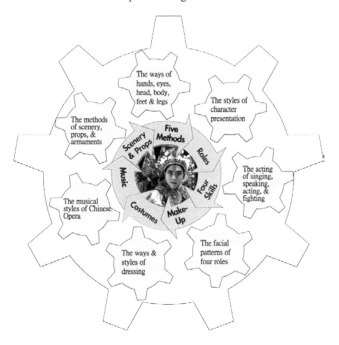

Figure 2: Seven Main Factors of Chinese Opera Performance

## 2    LITERATURE REVIEW

As shown in Figure 2, there are seven main factors - "Five Methods", "Roles", "Four Skills", "Make-Up", "Costumes", "Music", "Scenery & Props" – which make Chinese Opera into a stylized performance. However, the theoretical approach of this research focuses on only three of these seven main factors; "Five Methods",

"Costumes", and "Scenery & Props". These factors concern the manner of hands, eyes, head, body, feet & legs, costumes, and props that present the movements of the Chinese opera performer's body capabilities.

Nowadays, as to the so-called special techniques and great stunts in Chinese opera performance, the instructor should personally demonstrate step-by-step. They are arrived at through long-term practice but to force the students into correct form by cruel physical punishments is decidedly counter-productive and impermissible. Moreover in recent years, young students have come to resent the hard work involved in the five types of disciplines in Chinese opera training curriculum. Lacking perseverance, their human-body-oriented skills are lack-luster, loose and desultory, and complex routines become more and more scarce. When these sequences are not perfectly timed and tightly knit in execution, they fail to win appreciation from the audience and the standard of Chinese opera steadily declines. So, how to achieve the effectiveness as great as in the old time?

Therefore, the aim of this paper is directed towards three types of human-body-oriented disciplines, which can be described as "basic exercises" (Ji-ben-gong, 基本功), "mat-work acrobatics" (Tan-zi-gong, 毯子功), and "hand-prop martial routines" (Ba-zi-gong, 把子功); and two types of performing-props-oriented disciplines, which can be described as "costume exercises" (Fu-shi-gong, 服飾功), and "props exercises" (Dao-ju-gong, 道具功). They are all important foundations in the training for traditional Chinese Opera curriculum (see fig. 2-1).

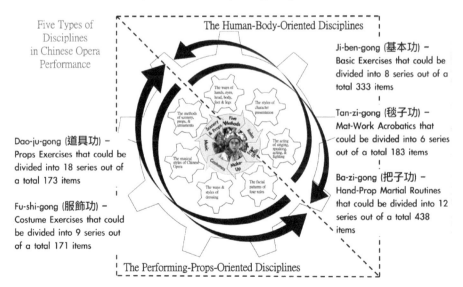

Figure 2-1: Five types of disciplines in Chinese opera performance

## 2.1 The Human-body-oriented disciplines

Performers must first master all the physical training before they can embark on the performance of the stylized programs. This is because every single movement in Chinese Opera performance, whether major, like martial sequences or tumbling acrobatics, or minor, like the lifting of a hand or the flicking of a foot, all have their prescribed and very specific styles and programs of execution and the performer cannot perform properly without the strictest and minutest training.

### 2.1.1 Ji-ben-gong (基本功) - Basic Exercises

According to Yu, Han-don (2006), there are about 8 series out of a total 333 items in the orientation of "Ji-ben-gong". It is the basic physical training required for the performance of Chinese Opera. For such performance utilizes the entire performer's body, where gesturing and acrobatics is each exactly defined by tradition both in method and in symbolic choreography. So, in expressive movements or gestures, any single part of the body must be rigorously and minutely trained in order to attain the perfect stances and routines.

### 2.1.2 Tan-zi-gong (毯子功) - Mat-Work Acrobatics

According to Yu, Han-don (2006), there are about 6 series out of a total 183 items in the orientation of "Tan-tzu-kung". For the students' protection, Mat-work acrobatics which include somersaults, rolls, falls, leaps and vaults are practiced entirely on mats or a thick carpet - hence the name "tan-zi" (literally carpet) - much as gymnastic training on mats in the West. The varieties of such movements and their basic forms are all rooted in fundamental movements. And yet the actors often develop myriad variations as a result of their particular physical endowments and hard work.

### 2.1.3 Ba-zi-gong (把子功) - Hand-Prop Martial Routines

According to Yu, Han-don (2006), there are about 12 series out of a total 438 items in the orientation of "Ba-zi-gong". It is the discipline of performing martial routines with hand-props, weapons of various types or of fighting barehanded. The combat routines of "Ba-zi-gong" are extremely numerous and complex. They undergo myriad variations with the different weapons, as in sword (dao[刀]) versus sword, spear (qiang[槍]) versus spear, double sword, single sword, long spear, double spear, club (gun[棍]), hoop (quan[圈]), etc. Each weapon has its own particular routine and when two persons are in combat, there are various sets of fighting routines.

## 2.2    The Performing-props-oriented Disciplines

The performers must master all the performing-props-oriented disciplines training simultaneously. This is because every performer in Chinese Opera performances has to wearing many and very specific stylized costumes, and carrying certain props for the classical character. So, without the strictest and minutest training, actors cannot perform properly with their props.

### 2.2.1   Fu-shi-gong (服飾功) - Costume Exercises

According to Yu, Han-don (2006), there are about 9 series out of a total 171 items in the orientation of "Fu-shi-gong". Costumes generally refer to what a performer wears on stage. They can be traced back to the middle 14th century and change gradually and continually. In general, a performer's costume primarily designates his or her role on the stage no matter when or where the action takes place. The costume has to distinguish a character's sex and status at first glance. In terms of symbolism, Chinese opera costumes may well be regarded as having the main function of marking off people from all walks of life, be it noble or humble, civil or military, as well as in or out of office.

### 2.2.2   Dao-ju-gong (道具功) - Props Exercises

According to Yu, Han-don (2006), there are about 18 series out of a total 173 items in the orientation of "Dao-ju-gong". Armaments are usually made of wood, rattan or bamboo except some few swords, knives and clubs. They are wrapped with cloth stripes or painted with golden paint on the blade or top of the arms. When the actors begin waving the arms, they look like real arms and make the performance vivid. Another kind of prop, the horsewhip stands for a horse. The audience admires the posture of the cast riding on it like on a real horse. The audience is concerned about is what happened to the characters. It is not important to the audience whether the devices are real or not.

68

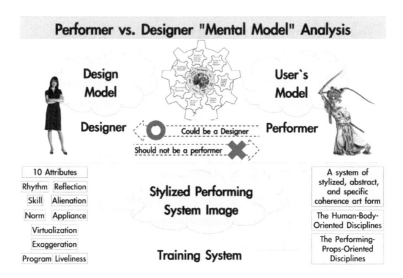

Figure 2-3: Performer vs. Designer "Mental Model" Analysis (redesigned from Norman, 1988)

## 2.3 "Mental Model"

According to Norman's "Mental Model" (Norman, 1988), there are ten attributes to let the designer understand how the performance can be stylized and systemized in Chinese opera. These are the findings of "Rhythm", "Reflection", "Skill", "Alienation", "Norm", "Appliance", "Virtualization", "Exaggeration", "Program", and "Liveliness". So, designers do not have to be Chinese opera performers to design the props for the performance. On the other hand, as the performer has already mastered the training system, a performer really could be a designer of both the props and the movements of Chinese opera performance as shown in figure 2-3, "Performer vs. Designer Mental Model Analysis".

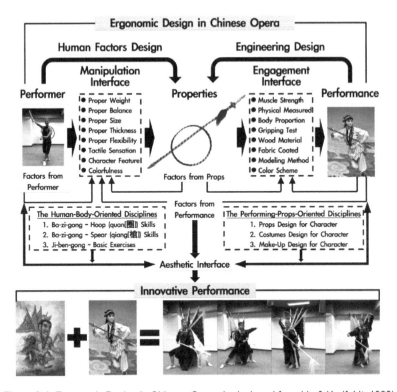

Figure 2-4: Ergonomic Design in Chinese Opera (redesigned from Lin & Kreifeldt, 1999)

## 2.4 "Ergonomics"

According to Lin and Kreifeldt (1999), "ergonomics becomes even more relevant when it is necessary for a user to fulfill operational requirements in more complex systems". So, this study is intended to use an ergonomic analysis model for analyzing the correlation of user, tool and task in Chinese opera performance. Figure 2-4 shows the possible new ways of dressing and performing,

No such ideas have ever existed in the field of Chinese opera training and performance. So, because I am a well-trained performer and experienced instructor at a Chinese opera school, there is no opposition to the ideas coming from not only the training system but also from the performing stylized system image. But for the designers, they only can get to know Chinese opera through the 10 attributes by circumstances or through long-term participant experiences coming from the system image.

## 3 A BLENDED RESEARCH FRAMEWORK

This study is designed to explore the identification, classification, and application of Chinese opera learning, and designing ways of both the props and the movements of Chinese opera performance. So, for the sake of providing a visual picture on the distinguish factors among sociology, physiology, and psychology, consider the graphic representation in the figure 3. It proposes the following different methods for a future research framework.

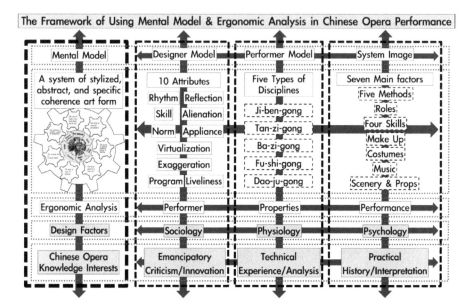

Figure 3: The Framework of Using Mental Model & Ergonomic Analysis in Chinese Opera

"The Framework of Using Mental Model & Ergonomic analysis in Chinese Opera Performance" is a pre-survey to identify the factors that support or impede the acceptance of the "Designer Model"& "Performer Model" as designing tools at all Chinese opera performances. As can be seen, the Framework is designed for representing the above theoretical positions with the flow chart shown as the figure of the construction methods in framework. It highlights three different points of views among three approaches of knowledge areas in sociology, physiology, and psychology, which are worth making about how to understand traditional Chinese opera performing skills, the methods of mental model, and the ergonomic analysis methodologies into a blended system. Thus, the framework is a correlation processes among learning, performing, and designing, which are a system of Chinese opera profession.

## 4    FINDING AND DISCUSSION

A two-phase field study was designed to explore the observation of the formal performance and informal exercises about the wudan role in a combat with other performers. This survey was conducted to identify the factors that how the wudan role control or manipulate such high-risk performances. So, in terms of the relationships among performers' body-oriented skills and performing-props-oriented effectiveness, the findings were what was expected from my many years experiences that the female warrior character deflecting tasseled spears with her foot in a battle, is not so simple as just putting one in hard training as people imagine.

To summarize the observations, the findings indicate that the performer's height, the length and weight of props, the distance in between performers, the jumping methods of the wudan role, the inclined projectile motion of throwing tasseled spears, and the probability of miss-throwing have a definite effect on the performance being successful or not. (See Fig. 4) This seems to indicate the qualitative results that the factors reflect a highly positive attitude towards the techniques of body-oriented and props-oriented skills. However, in the absence of statistically signify results, no definite conclusion can be drawn.

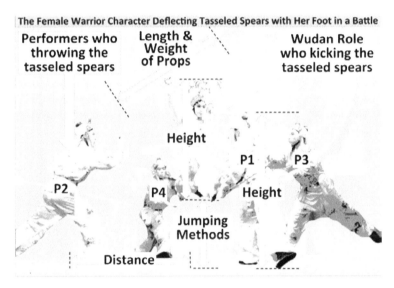

Figure 4: The Female Warrior Character Deflecting Tasseled Spears with Her Foot in a Battle

# 5    CONCLUSION

Given the exploratory nature of this study, any teaching implications based on these prelim findings should be treated with caution. Several pedagogical implications can be drawn from this study. We might reinforce training in the ways of virtual routines for the ways of reality, such as "3D Motion Capture" of ergonomic analysis for the human-body-oriented performing skills, and "Digital Interactive Arts Installation" as the teaching aids of interactive training activities, are all need to be finding out in detailed methods.

Thus, integrating ergonomics and mental model methodologies might produce a significant breakthrough application of pedagogic issues and props design in traditional Chinese opera training, and provide a means to analyze creative concepts for those intangible-stylized programs in further.

## REFERENCES

Chu, Wen-tsien, 2004. *Chinese Opera Subject Outline*. Beijing: Culture and Arts Publishing House. pp. 497□503.

Jiao, Ju-yin, 1988. *Jiao Juyin Set - Volume IV*. Beijing: Culture and Arts Publishing House.

Wan, Fon-chu, 2005. *The Training Methods of Chinese Opera Acting*. Beijing: Chinese Drama Publishing House. pp. 1□42.

Yu, Han-don, 2006. *Dictionary of the Stagecraft of Traditional Chinese Drama*. Beijing: Chinese Drama Publishing House.

Norman, Donald A., 1988. *The Design of Everyday Things.* New York: Currency. pp.190.

Lin, Rungtai & Kreifeldt, John G., 1999. Ergonomics in wearable computer design. *International Journal of Industrial Ergonomics* 27 (2001) 259-269.

Wang, T. J. & Yang, S. C. "Chinese Opera – Peking Opera Website." Accessed Feb. 7, 2012, http://jingju.koo.org.tw/

Wang, Tai-Jui, 2010. The feasibility of deepening educational research: 3D motion capture data analysis and digital archiving technique applied to traditional Chinese opera performance movements. *2010 TELDAP International Conference Proceedings*. 168-177, Academia Sinica, Taipei, Taiwan.

CHAPTER 8

# Perceptual Mechanisms of Transparency Recognition as Measures of Increased Human Spatial Orientation

*Marcin Brzezicki*

Wroclaw University of Technology
Wroclaw, Poland
marcin.brzezicki@pwr.wroc.pl

## ABSTRACT

Accidentally walking into a transparent pane usually results from improper highlighting or failure to comply with safety requirements. The main cause of these unfortunate events is the impairment of clear pane perception in certain lighting circumstances. Basic methods of transparency recognition, which have been determined using medical and psychological tests, are analyzed in this paper in the context of spatial orientation and obstacle identification. Different perceptual mechanisms are discussed for static and dynamic conditions, including both transmitted and reflected light. Those used in stasis include the recognition of transparency on a figure-ground principle based on differences perceived in the luminance values, the presence (or absence) of reflection on the surface of a transparent pane and distortion of the course of the light ray. Other mechanisms are used in dynamic perception, as additional physical properties of transparent surfaces arise with the introduction of motion; these include the motion parallax of transmitted and mirrored objects and the dynamic change in reflectance due to changes in the observation angle. In everyday life, the key factor in transparency recognition seems to be the ratio of transmitted to reflected light, as a backlit pane tends to decrease in visibility. Another important issue is the question of the observer's field of vision.

**Keywords**: transparency perception, architectural façade, user safety

# 1    INTRODUCTION

The ability to perceive light-transmitting surfaces is supposed to have evolved based on human contact with transparency that occurs in nature, for instance, in the form of fog or smoke. This natural transparency is generally not associated with the presence of a physical barrier; therefore, no evolutionarily conditioned mechanism for danger alert has developed. Water surfaces, which are light-transmitting mediums, may be exceptions; however, these are dangerous only in cases of excessive speed. Natural transparency, which all humans have evolved to perceive, is heterogeneous due to the random dispersion of particles. This important feature also greatly influences the perception of contemporary uniform transparent surfaces. As this problem has already been widely explored, it will be omitted here (Singh and Anderson, 2002).

The launch of higher-quality light-transmitting materials during the 19th and 20th centuries brought new challenges for the human perceptual system. New methods of manufacturing eased the production of large-scale, thin, uniformly transparent and faultless panes of glass, acrylic, polycarbonate or micro-thin foils. In a built-up environment, those materials are present in numerous transparent surfaces in the form of doors, openings, partitions, shop windows, wind screens and even acoustic barriers. The quality of material and degree of "transparency" is so high that a perceptual mistake could present a considerable threat to human safety. Currently, accidentally walking into a glass pane rarely results in serious injury because of the common application of safety standards; blunt trauma is still possible, though, as a result of collision with the hard surface. Safety glass, if broken, breaks into small, dull pieces, although substantial force is required.

# 2    PERCEPTION OF TRANSPARENCY

Perception of transparency is a complex process that is subject to two mechanisms: optical processes performed by the eye (stage one) when perceiving the appropriate light signals, and neural processes performed in visual cortex (stage two) during the interpretation of visual signals. In the process, the human perceptual mechanism constructs "two distinct surfaces", the transparent object and the one located behind it, on the basis of information that would otherwise be used to construct only one plane (Singh and Anderson, 2002a). This remarkable ability is called "perceptual decomposition". In this process, certain combinations of visual signals are decoded into the two surfaces, one of which is recognized as transparent. The mechanisms of transparency perception are currently being researched by many vision scientists and psychologists. This discipline is under constant development.

# 3    "IMAGE-BASED" AND "SCENE-BASED" MECHANISMS

In general, the mechanisms of transparency perception can be divided into two groups, based on the based on the visual information processes that are required.

"Image-based" mechanisms are mainly used in static observation, as they rely on the luminance differences in the image reaching the retina, or the "out of context principle" (explained below). "Scene-based" mechanisms take into account the 3-dimensional arrangement of objects, thus requiring the observer's motion to gather information from various points of view to verify if transparency is present. This distinction was originally made through a study of motion reported by Stoner & Albright (Stoner and Albright, 1998); it is also demonstrated here.

# 4    STATIC PERCEPTION MECHANISM

Mechanisms of static transparency perception are based on the assessment of the pane's appearance within its context. It is supposed that this group relies on the presence of a cue, addressed as the X-junction. The X-junction occurs where four adjacent regions with characteristic luminance levels are located next to each other (see Fig. 1). It is supposed that "X-junctions are the single most important monocular cue for transparency" but there is also a possibility that this mechanism is guided by the process of perceiving "shadows and opaque occlusions" (Keil and Wilson, 2001). As a result, it seems that luminance is a basic parameter used by the human neural system to recognize transparency; oculomotor cues (such as accommodation) are void in the case of transparency perception, where the lens can not focus on the transparent pane. Given the above, the following perceptual mechanisms are in effect:

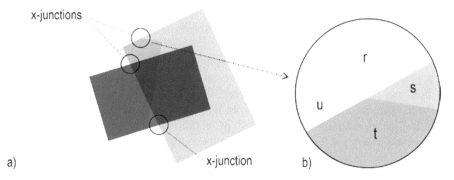

Figure 1  The "X-junction cue": a) illustrated in the example of three overlaid transparent surfaces, b) with enlarged sampled adjacent regions of different luminance: r, s, t and u.

## 4.1    Absorbance

An elementary mechanism in the recognition of transparency is based on the absorbance of the transparent material. An object that differs in luminance from its background is recognized separately, based on the figure-ground or depth-from-occlusion principle (Kiyohisa, Itoô and Sunagaô, 2009). Transparent panes differ from their backgrounds in luminance because every transparent surface absorbs a portion (whether small or large) of the light that it transmits. As a result, the luminance of the background behind the transparent pane is attenuated, and this change in luminance is a main indicator of light transmission (see Fig. 2). The transparent pane is properly recognized when the human perception mechanism locates the pane between the observer and background, based on differences in surface luminance (X-junction might be relevant here). Recently published results of research conducted by Singh and Anderson of MIT revealed that the "visual system uses Michelson contrast as a critical image variable to initiate precepts of transparency and to assign transmittance to transparent surfaces" (Singh and Anderson, 2001). The reverse mechanism also exists and could be used to produce the illusion of transparency, based only on properly chosen color regions.

Figure 2 Absorbance-based mechanisms of transparency perception. The figure shows the visual field in two versions: a) partially covered with transparent surface, and b) totally overlaid with transparent surface where no transparency could be perceived (Water Pumping Station Schönhauserstraße, by Kaspar Kraemer Architekten BDA, 2008).

The absorbance-based mechanism can be applied only when the reference area is present, meaning that apart from the obscured image, the observer must see a clear background. When the entire visual field is covered with a transparent pane, the human visual system is unable to compare the levels of luminance and properly judge the existence and orientation of a transparent surface. Rudolf Arnheim observed that "…if the shape of physically transparent surface coincides with the shape of the ground, no transparency is seen" (Arnheim, 2004). This example may seem to be rare, but it is common in a built-in environment. Every architectural facade or glazing observed from the inside of a building is an example of an equal area of transparent surface and background. The lack of a reference surface makes the whole pane unnoticeable by an observer.

## 4.2   Reflection presence

Every transparent material with a surface smooth enough to let the light through without deflection simultaneously reflects light, due to its physical properties at the micro level (the imperfections of the material's surface are smaller than the wavelength of light). Commonly used transparent panes are usually so smooth that they both transmit and reflect light; thus, the image perceived by the observer is composed of both images transmitted and reflected (virtual image) but in different proportions.

Specular reflection and the resulting virtual image seem to be important in the perception of transparency. Reflected objects are often located outside of their natural contexts (the "out of context" cue). This superposition of images is believed to be a strong cue for human perceptual mechanisms, especially when the elicited results are peculiar (e.g., a tree in a swimming pool, or a burning candle inside the glass). Contrary to expectation, lack of reflection increases the notion of transparency but simultaneously decreases one's ability to perceive the transparent surface (Fig. 3). Whereas this first factor might be desirable in adding so-called 'architectural transparency' to the building, the second might be dangerous because it impedes the human ability to locate surrounding barriers.

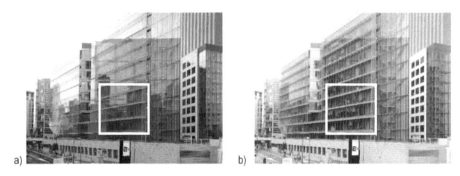

Figure 3 Figure shows the comparison of two photographs of Le Triangle de l'Arche building in Paris (by Valode & Pistre Architectes, 2002): a) unmodified scene, b) the same scene shot using a polarized light filter. The absence of reflection visibly decreases the capability to perceive transparent barrier – observe the region marked with the square.

## 4.3    Distortion detection

Image distortion is recognized through a comparison of a viewed object with the prototype stored in human memory, called the mental representation. Thin, flat transparent panes distort the course of the light ray, but the distortion is relatively small and can hardly be used to recognize transparency. Curved panes, due to their thinness, usually only distort images reflected on the pane, while the observation of the transmitted images remain unaffected. Transmission distortions are visible in static perception when thicker or uneven transparent surfaces are used (e.g., ornamental glass or glass bricks). Distortion reveals the presence of a light-transmitting surface, as distortions seem to be important determinants of transparency recognition. It could be speculated that neural mechanisms assume the presence of a surface every time a distorted image is received (see Fig. 4). Discrepancies between transmitted and reflected images might be used as additional cues to recognize transparency (Homa et al, 1973).

Figure 4 Photographs illustrate various possible light distortions that clearly indicate the presence of light-transmitting surfaces: a) glass bricks, b) ornamental glass, c) glass channel sections.

## 5    DYNAMIC PERCEPTION MECHANISMS

The dynamic perception, in general, greatly improves human spatial orientation, due to the change in viewpoint and perspective in "scene-based" perception. A sole observer's displacement can resolve many questionable and ambiguous signals received by the eye. The inability to judge the location of a transparent pane based solely on oculomotor signals has resulted in the development of other mechanisms that detect and properly spatially locate transparent surfaces. The most important are briefly described below:

## 5.1    Motion parallax, or "pictorial shift"

In dynamic perception, the transmitted image changes according to the position of the observer; it apparently moves behind the transparent pane. The ratio of this motion is proportional to the distance from the observer: objects located closer move more significantly, while those located farther move less. The sole common parallax phenomenon is in no way conclusive in case of transparency perception. If an observer looks through a rectangular hole in the wall of a building, the result is the same as if he looked through an opening that is glazed (except that the image luminance would be attenuated, although this could be noticed by comparison alone). Only when transmitted and reflected images are superposed on the surface of the plane and move in certain relation to each other can transparency be recognized with a higher degree of certainty (Fig. 5). Overlapping images could also be observed in a static condition, but the difference in shift, as well as the additional obstruction of objects (the observer is always located between some object and their virtual image), could be considered one of the determinants facilitating and amplifying the notion of transparency.

## 5.2    Dynamic parameter change

Apart from the mutual displacement of images, values of the transparent pane's reflectance and transmittance would differ as well. Dielectric materials (most transparent materials are dielectric) tend to change those parameters significantly; transmission drops gradually, while the reflection rises when a viewing angle

ranging from 40 degrees to normal – a line perpendicular to the surface of the pane
– is exceeded (Wiggington, 2002).

Figure 5  Motion parallax with virtual displacement of object reflected vs. transmitted: a schematic drawing that illustrated apparent replacement of the house and tree images.

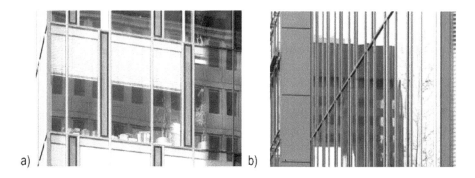

Figure 6  Dynamic parameter change, illustrated by the same building's facade observed from different points of view (this is the same facade; please observe the mullion and transom pattern). Tetragon Office Building in Frankfurt (by Bundschuh/Baumhauer GmbH, 2003).

This variability might be a crucial indicator for our neural perception system. A gradual transition of the pane from the reflection to transmission state might be a good example here. When the observation of a pane begins at an oblique angle, the pane appears to be reflective (from this point of view, the observer cannot tell the difference between a mirror and a transparent surface, as both look alike) (Fig. 6b). As the angle of viewing decreases, reflection is reduced and the transparent quality of the pane progressively appears (Fig. 6a). In this case, transparency cannot be properly recognized in static perception; the change of viewpoint is necessary to resolve whether the pane is a flat mirror or a transparent surface. Without the motion of the observer, proper identification would not be possible.

Perceptual mechanisms working in stasis (described in part 4) might be enforced by dynamic perception. In case of the absorbance-based mechanism, static perception might provide inconclusive information for the observer's brain. Dynamic perception is a conclusive cue in this case, but only when observer's motion, followed by the change in point of view, will result in the reference area coming into the field of view and contribute to the correct recognition of its transparency. The results of other research show that there still are perceptual mechanisms governing transparency that we do not fully understand.

# 6     RISK-INCREASING FACTORS

Perception of transparency constitutes a challenge for the human perceptual system, mainly due to the lack of clearly defined signs that determinate the spatial orientation of light-transmitting barriers. Perceptual mechanisms described above would work in proper lighting conditions and with all of the reflectance and luminance cues. In everyday practice, the critical parameters of transparency recognition might be restricted to the ratio of transmitted to reflected light and the presence of a reference area. In many cases, unfortunately, the mechanisms of transparency perception are mutually ineffective, regardless of the type of observation (static or dynamic).

The potential hazard is increased when special anti-reflective coatings are used in shopping windows in department stores, banks and restaurants. The lack of reflection on pane surfaces makes windows hard to perceive, potentially becoming unnoticed barriers. The reason for this application is obvious and understandable. Reflection-free transparent panes are aimed at improving clients' visual contact with products and lowering the consumption of energy in the building by decreasing the need for artificial lighting ("Amiran…", 2012). However, reducing reflection makes the backlit panes visually disappear more easily, especially in daylight, when an increased amount of illumination coming from outside alludes the users, customers and clients.

Another important but underestimated factor is the quality and cleanness of transparent panes. Panes devoid of any markings or dirt spots that are very well maintained are hard to identify when backlit. The author of this paper **does not advocate low hygienic standards**, but the fact is that poorly maintained transparent surfaces are easier to perceive.

# 7     EXISTING AND SUGGESTED SAFETY MEASURES

Architectural transparency and user safety seem to be contradictory. While designers and their clients want to achieve the highest possible degree of light transmittance, an awareness that the invisible barrier should be visually perceivable is increasing. In optical terms, this aim could be fulfilled only by local suppression of transparency. There are advanced methods of solving this problem with animals in mind (e.g., applying a UV-reflective coating that makes the transparent surface visible to birds while maintaining transparency to the human eye, called ornilux glazing). In the case of people, the warning may need to be communicated in the same "language" in which perception occurs, using the visible light spectrum.

The most common safety measure is to apply markings (also called manifestations – see Fig. 7a) at eye level ("Glazing – safety in relation to impact…", 1994). Those markings are obligatory on glazings in some countries, while in some other countries obligatory only in cases of openings (doors and windows). Markings are meant to interrupt the clear view of the world on the other side of the pane by "contrasting visually with the background observed through the glass in all lighting conditions", thus alerting the user of the pane's presence. This method has become standard in many countries (UK), but is still not entirely accepted from an aesthetic point of view. It destroys the impression of transparency by dividing large transparent panes into segments, thus interrupting the view.

In contrast, this suppression of transparency, once accepted by designers and building owners, might be creatively utilized by gifted architects; new architectural and optical qualities have arisen, including "veiling", "delaying" and "slowing down the light" (Riley, 1995). Reduction of the area of the transparent envelope might also be beneficial for purposes of energy conservation (lower solar gain).

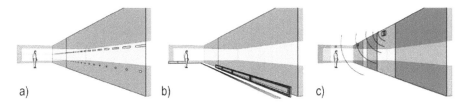

Figure 7  Possible safety measures: a) manifestations applied on the glass panes, b) physical obstacles blocking possible passage, though not the view and c) microwave sensors for approach-warning systems.

Other possible safety measures include the proper placement of physical obstacles of different sizes and locations to alert observers to the transparent pane's presence. The most protruding could take the form of curb-type railings or plates. This type of railing blocks possible passage, but due to the relatively low height does not block the view. The application of these kinds of railings should be undertaken with care to not create additional hazards after also solving issues of disabled citizens' access to the building. An alternative method uses horizontally positioned grooved paving tiles along the transparent facade. The tiles produce a clearly recognizable sound, not only when touched with a visually impaired individual's cane (which is what such tiles are intended for) but also when stepped upon, due to changes in pattern and friction (Fig. 7b).

In extreme cases (there is no record of this type of application thus far), the type of microwave sensor commonly used for automatic doors could be used as an integral element of a warning system utilizing bright flashing lights to alert people. The wide range of technologies available that allow for rapid change of a pane's properties might be utilized as well (e.g., LCD glazing, Priva-Light, glass-encased LEDs), potentially with very good results. A possible system employing sensors to measure the position and speed of an approaching person in close proximity of the barrier would adjust the glass, for instance from a transparent to an opaque state, in case of risky behavior (such as excessive gait speed). This instantly appearing barrier would be undoubtedly perceived by the observer, followed by a change in direction (Fig. 7c).

Rough guidelines for the avoidance of transparent pane collisions:
- Analyze lighting conditions and predict potential hazards (e.g., large-scale backlit panes)
- Take every lighting condition into account, including daytime and nighttime lighting
- Carefully select the type of glass used, avoiding the use of reflection-free glass on pathways and in trafficked areas such as doors and shopping windows
- Use proper manifestations when required by building regulations
- Creatively use facade patterns to improve both the esthetic of the facade and recognition of the barrier.

## ACKNOWLEDGMENTS

The author of this paper thanks prof. E. Trocka-Leszczynska, the Dean of Faculty of Architecture, Wroclaw University of Technology for financial support and prof. J. Charytonowicz for encouragement to submit the paper to AHFE 2012.

## REFERENCES

"AMIRAN® – the glass you can't see", Accessed February 2, 2012, http://www.us.schott.com/mexicana/spanish/download/amiran_us.pdf

Arnheim, R. 2004. Art and visual perception: a psychology of the creative eye, Berkeley CA.: University of California Press.

"Glazing – safety in relation to impact, opening and cleaning". 1994. British Standard BS 6262: Part 4: 2005 and Section N1 of Approved Document N.

Homa, D., J. Cross, D. Cornell, D. Goldman, S. Schwartz. 1973. Prototype abstraction and classification of new instances as a function of number of instances defining the prototype. *Journal of Experimental Psychology* 101, 116–122.

Keil, F.C., R.A. Wilson. 2001. The MIT Encyclopedia of the Cognitive Sciences. Cambridge, MA.: MIT Press.

Kiyohisa, Y., H. Itoô, S. Sunagaô. "Effect of orientation on depth-order and lightness perception in perceptual transparency." Paper presented at 32nd European Conference on Visual Perception, Regensburg, 2009.

Riley, T. 1995. Light construction. New York, N.Y.: The Museum of Modern Art.

Singh, M., L. B. Anderson. 2002. Toward a Perceptual Theory of Transparency. *Psychological Review* 109: 492–519.

Singh, M., L. B. Anderson. 2002a. Perceptual assignment of opacity to translucent surfaces: The role of image blur. *Perception* 31: 531–552.

Stoner, R., Th.D. Albright. 1998. Luminance Contrast Affects Motion Coherency in Plaid Patterns by Acting as a Depth-from-occlusion Cue, *Vision Research* 38: 387–401.

Wiggington, M. 2002. Glass In Architecture, London: Phaidon Press.

# CHAPTER 9

# The Issue of the Range of Vision in Design of Grandstands at the Contemporary Stadiums

*Zdzislaw Pelczarski*

Facultuy of Architecture
Bialystok University of Technology
Grunwaldzka 11/17
15-893 Bialystok, Poland

## ABSTRACT

The main attention in modern design has been focused on spectacular roofing systems and external facilities. At the same time, parameters of visibility inside stadiums can hardly be said to have been going hand in hand, especially considering the visibility standards established about half century ago.

The principal interest of presented study lies in the issues related to the determinants of visual perception, and among them especially, to the range of human sight and its influence for shape and size of stadium interior. The quality of sight depends, first of all, on distance between observer and the furthest laid object of observation field.

**Keywords**: stadium design, stadium architecture, range of view. spectator zone

## 1    INTRODUCTION

After nearly two thousand years since the times of great ancient structures for public spectacles such as the Coliseum, rapid development of their modern counterparts can be observed. This phenomenon is a result of gradual development of their architectural form that has been evolving for more or less one hundred years. The size and shape of the field of game, which at the same time is the field of observation, is established solely as a derivative of the game rules, with no regard to the factors conditioning visual comfort of the gathered crowds. These game rules, having been invented over one hundred years ago, did not account for the entertaining character of the field action, or for the need of co-existence between the field and huge spectator stands.

Alterations in shape and size of the arena finished in the thirties of the 20th century, and all subsequent evolution of the form of the stadium related only to spectator stands and other construction elements. As a result, the conflict between the rectangular shape of the field and the oval or circular shape of the stands still lingers. No less important implication of this situation is the prevailing drastically low visibility standard, which approaches the human eye perception limits, but nevertheless is tolerated by norms out of necessity.

The issues presented in this paper are part of the, conducted by the author, studies on optimizing the spectator's zones at modern stadiums. They are based on many years of practical experience arising from the performance of his duties of the general designer of the reconstruction of the *Silesian Stadium* in Chorzow.

## 2 THE RANGE OF HUMAN SIGHT

For the human eye critical angle of seeing of two points as separate objects equals $0°1'$ (Lapaczewska, 1986). European Standard EN 13200-1 specifies two values of maximum range of sight for soccer stadiums: 190m, as a maximum value and 150m, as recommended one (EN 13200-1, 2003). In case of the first, angular height of the retinal image of soccer ball is $\psi=0°4'$, what means only four time more than critical minimal angle of view. In case of the second one its value rises to $0°5'$.

## 2.1 Theoretical Basis

The spatial seeing we owe to the phenomenon of refraction which have been used by the nature to construct the lens in the optical system of the eye, as well as to the anatomical structure of the eye, consisting of a constant distance between the lens and retina. This last fact makes that as a result of simple laws of trigonometry, retinal image size of the actual observed object is a function of the distance of the object from the eye (Figure 1-A). The observed object farther from the eye, the angle between the rays of view of the extreme points of that object (the viewing angle) is smaller, and thus, it is also smaller retinal image of the object. This simple mechanism allows the recognition of spatial depth and governs the rules of optical perspective (Dyba, 1979).

Figure 1-B illustrates the relationship between the angular height of a player in the retinal image and the distance between the viewer and observed object. Retinal image size can be expressed in the angular value. The measure is then called. viewing angle, the angle between, intersecting at the geometric center of the

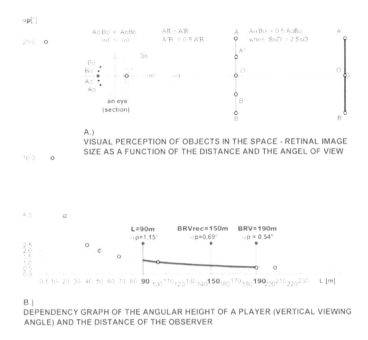

A.)
VISUAL PERCEPTION OF OBJECTS IN THE SPACE - RETINAL IMAGE
SIZE AS A FUNCTION OF THE DISTANCE AND THE ANGEL OF VIEW

B.)
DEPENDENCY GRAPH OF THE ANGULAR HEIGHT OF A PLAYER (VERTICAL VIEWING
ANGLE) AND THE DISTANCE OF THE OBSERVER

Figure 1  Retinal image size of the observed object - geometric determinants; (author)

*Legend: ao,( ao') - Vertical viewing angle of the object AB, (A'B '); **Ss** - Geometric center of the eye lens; **Ao** –Retinal image of point A; **αp** –Angular height of a player; **L** -Distance of the observer; **L = 90m**- Closest possible distance between the observer and the furthest player on the pitch, αp = 0.69°; **BRV=190m** - Boundary of the field of maximum range of view of soccer pitch, angular height of a player, αp = 0,54°; **BRVrec=150m** - Boundary of the field of recomended range of view, αp = 0,69°;*

lens, rays of view of the extreme points of the object. Viewing angles decrease with distance of the object from the eye of the observer.

Dynamics of changes in the size of the vertical viewing angle of a player figure depending on the distance from the observer. In zone of small distances (5-35m) occurs a sharp drop in angular height of the object from 20° to 3° (with average decline rate 0,57°/1m). In zone 35-100m the reduction of height is 3° to 1° (with average decline rate 0,03°/1m). Zone 100-190m is characterized by the decrease of hight of 1 ° to 0.5 ° (average rate of decline 0.006°/1m). Expressing these viewing angles as the multiplicity of the minimum angle of view of the ball (ψ40or 0,067°), obtained values are respectively from 16,5ψ to 7,5ψ.

As shown in Figure 1-B, the minimal angular heights of players on the pitch for all the viewers, regardless of the place occupied, are in the range of very low values. This is due directly to the fact that the dimensions of the playing field, concerning its details, considerably exceeds the capabilities of the human visual perception.

Figure 2  Spatial relationships between the pitch and the spectators zone in the interior of a typical contemporary soccer stadium; (photo: author)

## 2.2    The Readability of Informational Signs and Symbols

One of the important problems associated with a range of visibility in the interior of a modern stadium is a system of the identification of players with numbers placed on their sports clothing (Figure 2, Figure 3). Readability of the numbers, considering the distance between an observer and the playing field is a fundamental parameter which allows the identification of individual players, and watching the competition.

Rules of the International Federation of Football Associations (FIFA) regulate the size of the identification numbers placed on players uniforms. From the year 2005 dimensions of basic number have been increased. Currently its height should be in the range 25-35cm (FIFA, 2005). Readability of informational graphic characters requires that the angular size of their outer contour achieves a minimum of 0°5' (Neufert, 2005; Nixdorf, 2008).

From a distance specified by the limites of  maximum range of vision (190m), the minimum character height should be 28cm. Limit values for viewing angles are reliable only with proper lighting and transparency of the air. When these conditions are not met the increase of the character size of 1.5 to 2 times is required. In the practice of ergonomics to provide readability of letters and numbers applies the optimal viewing angle $\psi$=08', which is more than three times larger than the minimum ($\psi$=05') (Ziobro, 1989). This parameter c orresponds to the algorithm H=L/200, where H is the height of the letters and L is the distance of the observer (Rosner, 1985).

Accepted by FIFA values only to a limited extent meet the minimum anatomical conditions of visual perception. The practical experience shows that they are not sufficient to ensure full comfort of observation and the readability of informational signs and symbols.

Figure 3  Illustration of the key elements in the field of observation, which are the ball and the identification number of a player; (photo: author archives)

Graphic character readability issues are also important because of the visual communication systems, in the form of electronic information screens, and various stationary boards.

# 3 THE RANGE OF VIEW AND ITS IMPACT ON THE DESIGN OF STADIUMS

As mentioned earlier, an acceptable standard of visibility in the design of the audience for most team sports, both called sports of a large fields or a small fields, regardless of the shape and size of the used ball is determined by the minimal ball view angle ($\psi$), which is equal to $0°4'$.(Brzuchowski, 1982). Consequently, these regulations limit the distance of the farthest rows of the audience. In other words, this parameter is determined by calculating a distance of which the ball, or other movable prop (such as hockey puck) can be seen in the angular size equal to the view angle $\psi$. Standard EN 13200-1 gives two values of the maximum range of view for football: the absolute limit value equal to 190m and150m equal to the recommended value. The minimum viewing angle of the ball from a distance 150m is $\psi=0°5'$(the same as minimum angle of the readability of graphic signs). Figure 4 shows a diagram of the method for calculating the maximum range of view for the soccer stadia (John and Sheard,1997) It is also the illustration of relationship between viewers area set by the maximum and by the recommended range of view. As seen the audience area contained in the circle of recommended range of view ($Rrec=150m$) is only 35% of the entire viewers zone. So it is the reason for which

88

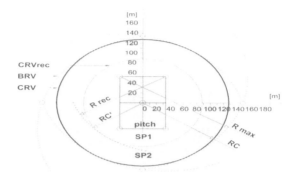

Figure 4 Field of maximum range of view of the soccer pitch (according to the European Standard: EU-13200-1); R max = 190m, recomended: R rec = 150m; (author)

**Legend: SP-1** *Spectator zone designated by the optimal, recommended range of view Rrek= 150m; covers only 35% of the audience;* **Sp-2** *Spectator zone comprised between the recommended range of view Rrec=150m and boundary of the field of maximum range of view limited by Rmax=190m; covers up to 65% of the audience;* **BRV** *–as in Figure 1;* **CRV** *- Circle of the maximum range of view (radius RC);* **CRVrec** *- Circle of the recomended range of view (RC').*

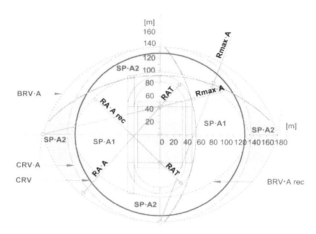

Figure 5 Field of maximum range of view of the athletics arena (according to the European Standard: EU-13200-1); RmaxA = 230m, recomended: RrecA=190m; (author)

**Legend:** **SP·A-1** *-Spectator zone designated by the optimal, recommended range of view for athletics Rrec·A = 190m; covers only 30% of the audience;* **SP·A-2** *-Spectator zone comprised between the recommended range of view Rrec·A=190m and boundary of the field of maximum range of view, limited by Rmax·A=230m; covers up to 70% of the audience;* **RA·A** *- Radius of the arc of the maximum range of view for athletics (BRV·A); RA·A = Rmax·A – RAT;* **RAT-** *Radius of* *the outer arc of the athletic track;* **BRV·A** *-Boundary of the field of maximum range of view for athletics;* **CRV·A** *- Circle of the maximum range of view for athletics;* **CRV**-*as in Figure 4.*

the recommended range of vision (150m) is not widely used, causing a dramatic reduction in capacity of the audience zone.

A separate issue are the requirements for the range of view for the athletic stadiums. The standards set in this case the maximum range of view equal to 230m and 190m as the recommended. The author has developed for this type of stadia own way of setting limits of the maximum range of view (Figure 5). Is worth noting the shape of the visibility field, which is strongly elongated in the direction of the transverse axis of the arena. Its furthest point, lying on this axis, is at a distance of almost 180m from the geometric center of the arena, ie. 40m more than the corresponding point on the longitudinal axis (Pelczarski, 2009). The results of these considerations are surprising, when compared to the shape of the audiences realized at the biggest of this type of the audience eg. the Olympic Stadium in Sydney, where a very complex auditorium was located on the longitudinal axis. Another interesting conclusion is that when applying the recommended range of view (RmaxA=190m) vanish completely the spectators zones on the longitudinal axis.

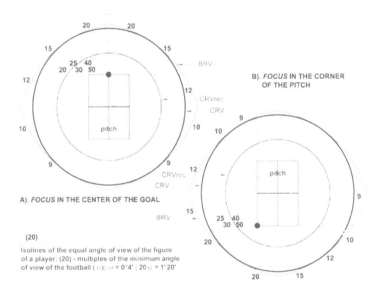

Figure 6  Isolines of an equal angle of view of the figure of player (AVP) on the football pitch. AFP are expressed as multiples of the minimum angel of view of the ball (ψ = 0°4'); (author)

*Legend: All symbols as in Figure 4.*

Points of the eye, which maintain the same distance from the observed object, create isolines of equal angel of view, which is a circle with a radius equal to this distance. This regularity allows you to create grid of isolines of uniform viewing angles in the form of concentric circles (Figure 6). Such graphs, made for the characteristic points of the arena, allow to determine the visibility conditions of these points from different areas of the viewers zone. They are a kind of visibility maps which are very useful design tool. They show, among others extreme differences in quality of viewing a goalkeeper figure from the opposite fields located behind the closer and farther goal. In the first case the figure of a goalkeeper seen from the farthest row has 20 times the size of the minimum angle ψ, while considering the corresponding row on the opposite side of the pitch, this value decreases up to about 9. These differences are even greater when it comes to a corner point of the pitch.

With this method you can also state that the field of spectator zone with very good visibility, set by isoline of the minimum angle of view approximately equal to 17ψ equal to 1,1°covers only about 30% of its whole surface.

The consequence of high-capacity spectator zone is the height of elevation of the extreme points of the eye (H=40 to 50m). This causes that the real viewing lines of the furthest points of the observation field are characterized by significant slope, Figure 7. To determine the required range of view (for football Lmax = 190m), an important is the actual length of the line, running in the space fronm the farthest point of the eye (PE), elevated to a height HPE, to the farthest corner of the pitch.

Figure 7 The maximum range of vision of football (Lmax = 190m) determined by the elevation hight of points of the eye in the top row and minimal angle of view of the ball (ψ = 0°4'). *Stadion Slaski* in Chorzow, Poland 2005; (photo: P. Oles, WOSiR Achives)

Figure 8 shows, developed by the author, method of determining maximum position of the eye in the farthest row, including the height of its elevation (Pelczarski,

2009). For extreme values of elevation of the point of eye (HOkn=45m) radius of the circle of the viewing range Rmax" is reduced by 5,4m, relative to the value charged to the level of the arena. This corresponds to the surface of the auditorium where you can fit about 8000 seats. Should be noted that up to the height of about 25m the differences are small, but growing rapidly beyond this value.

Reducing the radius Rmax" affects the substantial decrease of the surface of the spectators zone located in the field of good vision, which in turn affects the capacity expressed by number of places. In conclusion it should be emphasized that, in the design practice is rarely used the parameter of radius R" using the simply radius at the level of pitch and thus when calculating range of view the slope of the line of sight is not taken into account .

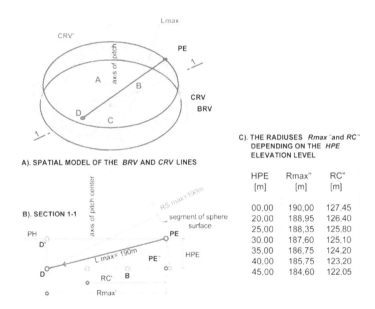

C). THE RADIUSES *Rmax* "and *RC*" DEPENDING ON THE *HPE* ELEVATION LEVEL

| HPE [m] | Rmax" [m] | RC" [m] |
|---------|-----------|---------|
| 00,00   | 190,00    | 127,45  |
| 20,00   | 188,95    | 126,40  |
| 25,00   | 188,35    | 125,80  |
| 30,00   | 187,60    | 125,10  |
| 35,00   | 186,75    | 124,20  |
| 40,00   | 185,75    | 123,20  |
| 45,00   | 184,60    | 122.05  |

A). SPATIAL MODEL OF THE *BRV* AND *CRV* LINES

B). SECTION 1-1

Figure 8  Method for determining the positions of the points of eye (PE) in the highest row of the spectator zone; (author)

**Legend: SR -** *Radius of a spher, the surface of which is the locus of points equidistant from the farthest corner of the pitch; The maximum range of view of football (Lmax = 190m) is determined by the four segments of the surface of such a sphere;* **PH** *-The horizontal plane through the ophthalmic line of the highest row, which is a set of points of eye (PE), located at an altitude of HPE;* **HPE** *-elevation height of points of the eye in the top row;* **Rmax"** *-The radius of an arc of range of view on the particular plane HPE;* **RC"** *-The radius of a circle inscribed into arcs of the range of view for a particular plane HPE;* **PE** *–point of eye; The other symbols as in Figure 1.*

# 4 CONCLUSIONS

Considering the partial conclusions presented in this paper, there is a good reason to formulate one of the most important propositions concerning future development of the interior of the modern stadium. In brief, the interior should be shaped by actual relationship between the audience and arena. The relationship must prioritize adherence to visibility standards, which, in turn, must be adjusted to anatomical features of human eye perception, consequently causing the shape and size of the arena to be an outcome of these standards. This will trigger the need to revise the rules now known team games and sports of a large field. Can also lead to the creation of entirely new types of sports.

# REFERENCES

Brzuchowski, J., 1982. Widownie, *Urzadzenia sportowe – planowanie, projektowanie, budowa, uzytkowanie*, wyd. IV, edited by Romuald Wirszyllo, Arkady, Warszawa: 221.

Dyba, K., 1979. *Perspektywa linijna*, Wyd. Politechniki Wroclawskiej, Wroclaw: 3-9.

EN 13200-1 (2003), The European Standard which has the status of a Polish Standard PN-EN 13200-1: 2005, *Spectator facilities - Part 1: Layout criteria for spectator viewing area – Specification.*

FIFA, 2005. *Equipment Regulations. Regulations Governing the Sports Equipment at FIFA Competitions*, Zurich: 13.

Neufert, E., 2005. *Podrecznik Projektowania architektoniczno-budowlanego*, wyd. III, Arkady, Warszawa: 37-38.

Nixdorf, S., 2008. *Stadium ATLAS. Technical Recommendations for Grandstands in Modern Stadiums*, Ernst & Sohn, Berlin: 130-137.

John, G. and Sheard, R., 1997. *Stadia. A Design and Development Guide*: 105-120

Lapaczewska K., 1986. *Pole widzenia. Badania modelowe*, Instytut Wzornictwa Przemyslowego, Prace i Materialy, Zeszyt 102, Warszawa : 7-16.

Pelczarski, Z., 2009. *Widownie współczesnych stadionów. Determinanty i problemy projektowe*, Oficyna Wydawnicza Politechniki Bialostockiej, Bialystok: 90, 152-154.

Rosner J., 1985. *Ergonmia*, Palstwowe Wydawnictwo Ekonomiczne, Warszawa: 164-170.

Ziobro E., 1989. *Ergonomia; wybrane zagadnienia*, Politechnika Wrocławska, Wroclaw: 144-45.

# CHAPTER 10

# The Application of Optical Illusions in Interior Design in Order to Improve the Visual Size and Proportions of the Rooms

*Anna Jaglarz*

Wroclaw University of Technology, Department of Architecture
St. Prusa 53/55, 50-317 Wroclaw, Poland
anna.jaglarz@pwr.wroc.pl

## ABSTRACT

Modern architects and interior designers are mostly used to large, bright space in which it is easy to demonstrate their design skills. However, it is still not enough products and services for large group of interior users, mainly flats users, with a small size of the area. Most of them are not completely satisfied with their rooms. Usually, spaces seem to be too small, too tight, or simply they have the wrong proportions. There are a many design problems in these small interiors, and they constitute the greatest challenge for designers.

In addition to the selection of equipment during the formation of a newly designed interiors, and repaired, or adapted rooms, designer should consider several important factors that affect the optical change in the size and proportions of rooms. It's possible to correct interior without redesigning or destructions with the help of optical illusions that changed the visual perception of interior space. A room can be visually enlarged, heightened, deepened, broadened, narrowed, or lowered. Things like color, lighting, materials, pattern, texture, forms, shapes, placement interior elements, horizontal or vertical lines, mirrors and even styles of interior equipment and finishing materials can correct the interior space because they can have a significant impact on the size and proportion of a space.

Color selection of both walls, floors, items of equipments and accessories is one way to change the visual size and proportions of the rooms. Using the visual effect of colors and appropriately selecting and comparing cold and warm, light and dark, complementary and contrasting colors, designer can easily manipulate the length, width and height of painted interiors.

Another factor, lighting, a decisive influence on the final image of the rooms. Suitably manipulating the color, intensity, location of light and number, type and size of points of light, we can improve the appearance of the interior optically.

Trying to "trick the eye" of the observer we should pay attention to the nature, diversity and the appropriate combination of finishing materials and materials of equipment. The effect can be a result of the material texture, pattern, color, gloss, etc.. Similarly, we can use to put the interior decorative designs and patterns. One more thing that also contributes to the change and correction of the interior space is the style of applied finishing materials and interior articles.

Spectacular changes with little effort can be achieved by the applied mirrors. Thanks reflections and optical illusions they provide an opportunity to significant visual enlarging of room space.

But the basic principle helpful in organizing small spaces is to care for the maintenance of order, not just architectural. Glut of different shapes and forms of elements, excessive amount of additives and the introduction of different levels on the walls and among the items of equipment may disturb the order and spoil even the best planned actions directed toward the arrangement of optical improve the appearance of the interior.

**Keywords**: interior design, visual perception, optical illusion

# 1    INTRODUCTION

Color, light, texture and pattern are effective means by which space may be articulated or defined. The surfaces treatment of walls, floors, ceilings and other elements of equipment and furnishings articulates the spatial boundaries of a room. Colors, light, texture and pattern are real factors which affect our perception of relative position in space and, therefore, our awareness and feeling of a room's size, dimension, scale and proportion. A room can be visually enlarged, heightened, deepened, broadened, narrowed, or lowered. Spaces may be made to appear larger than they are by unifying them with color and light that blend surfaces rather than fragment them.

Colors in combination with light and also things like materials, patterns, textures, forms, shapes, placement interior elements, horizontal or vertical lines, mirrors, windows and even styles of interior equipment and finishing materials can redefine our perception of living space.

The effect of color and light on the perception of space (the apparent, versus the actual, size and distance of objects from a viewer) will vary among individuals; however, the following are some general guidelines of how color and light may be used in the design of a space.

# 2 PROPER LIGHTING - THE SIMPLY WAY TO THE IMPROVEMENT IN THE SIZE AND PROPORTION OF THE INTERIOR

The most us, even if we just moved into quite new house, isn't fully satisfied with the inside in which we live. Usually the rooms seem too small for us, or simply have the wrong proportions. During the renovation or the adaptation we most often focus on materials and furniture, forgetting about one important factor. Illumination - after all an attractive appearance of the inside depends on lighting and lighting determines how we feel in the room.

Designers can shape and modify the visual experience of room by manipulating the perceptual role of lighting to define the visual boundaries and hierarchy of a space. Through the careful manipulation of light or correct using of light stresses designers can shape and modify the visual experience of interior space. Appropriately using the lighting they can optically improve its appearance. Below is a list of common problems and ways of how to deal with them.

- **The room is too small**

There's no doubt that cramped or dark spaces cause discomfort of users. Designer can create the perception of a larger space with the use of lighting. If the room is too small resignation from the central lighting in the ceiling to the number of lamps highlighting the room shape and focus the light on the walls allows for impressive interior magnification. Even and regular illumination of the interior by adding multiple light sources powered at the same time makes the interior is more spacious. A small room seems larger, if we turn the light on design elements placed in different locations. Directing lighting on the ceiling makes the interior seem enlarged. The effect of inside spaciousness can also be obtained by using the light reflected in the mirrors. Therefore strategically placing a mirror, or even a collection of mirrors, in the relevant, appropriate part of the room is an easy way to use the reflection of light to advantage in decorating a small space.

Designer can also create the perception of a larger space with the use of natural lighting. Maximizing the natural light will emphasize the space and make the room appear larger and brighter. Just as a lightened ceiling creates the sense of openness and spaciousness, so too, can large bright windows and natural light. By using floor to ceiling windows it gives the room or space a grander feel as well as allows natural sun light to come in which improves by giving the illusion of a larger scaled space. The usage of floor to ceiling windows makes any room seem larger and more spacious. The perspective to the user eye creates the feeling and physical reaction of enhanced space. Framing the window by hanging drapery panels at each edge of the window - but instead of starting at the edges and moving in, we should start at the edges and move outward along the wall - creates the optical illusion that the window is bigger. But ordinary blinds and window covers minimize window exposure and moderate natural light, so the window coverings should be used sparingly in smaller rooms.

- **The room is too low**

The effect of increasing the visual height of a room is made with the help of strongly illuminated ceiling apparently increases interior. Ceiling decoration with integrated lamps apparently increases interior. Besides, designers can gain this effect with the use of the 3-D structure of suspended ceilings, mirror glass, illuminating in the top part of a room, or by creation of a shiny surface.

Light cold colors and shiny textures visually level ceiling up. Illumination of ceiling with integrated lamps, can create interesting effect – ceiling "dangling" above a room. Compounded the impression of height of room can be obtained by darkening bottom ends of walls while simultaneously their upper parties are strongly illuminated. Light stressing vertical divisions (e.g. of doorways between rooms or of pillars) causes the impression of raising the ceiling. Placing the main lighting point above the line dividing colors of walls and the ceiling will cause the room seem higher.

Appropriate usage of drapery and blinds can give an optical illusion of the increased height. Long, light curtains hanged above the window close to the ceiling flowing onto the floor optically increase the height of the room.

- **The room is too narrow**

Lighting points creating cross stripes on the ceiling, in combination with the longer wall firmly lighted with the diffused light cause that the inside seems wider. Bright lighting crosswise walls with a few lighting points, at comparatively poor lighting longer walls optically correct proportion. Darkening zone above the main centre of attention causes the room seem wider. Backlighting horizontal divisions on shorter walls allows to get the effect of widening the interior. Putting on the longer wall the vertical elements of light, e.g. highlighted slits, can optically expand the inside.

Evenly illumination of walls and lower corners of the room with small lamps makes room look wider visually. If the space, on the contrary, must be narrowed, we should illuminate upper corners of the room. To make a rectangular room look more square, to a monotonous and equally illumination of the three walls designer should add an expressed light stress on the forth, the farthest wall.

- **The room is disproportionately high**

The ceiling illumination directed into the bottom visually makes the ceiling seems down. Directing the illumination onto the floor gives the impression of darkening and reducing the room. Backlighting horizontal shelves, oblong niches and plains allows to get the effect of lowering the inside. Lighting the inside mainly using wall lamps optically reduces interior. It is possible to optically make the room smaller by putting the main overhead light out at simultaneous lighting wall, floor, desk and table lamps placed very low. Putting the main lighting point below the line dividing colors of walls and the ceiling causes that the room seems lower than it is in fact. Curtains links suspended in the high interior at the eye level "stop eyesight" and visually reduce the room height.

Apart from mentioned above particular cases, designers can change visual proportions of the room through the experiments with the light underlying of some interior objects. Backlighting niches can seemingly enlarge them and cause that they will become more decorative. Light coming out from behind the screens and shields at simultaneous concealing behind them light fittings is a real decoration of the interior and creates the illusion of spacious depths. All the illusions of brightness-contrast in interior may be produced by lighting. Surfaces, objects, forms and details may appear larger or smaller, heavy or light thus also affect the perception of the whole space.

# 3 COLOR - AN EXTREMELY EFFECTIVE AND MANY-SIDES INSTRUMENT ENABLING THE IMPROVEMENT IN THE INTERIOR PROPORTIONS

Color is probably the most important factor to consider when trying to change room's proportions with interior design tricks and interior decorating information. Color has the ability to manipulate our sense of space, so it's possible to visually change the space in all three dimensions by correctly controlling ground-colors and color stresses. Color is a way to change the proportions of the rooms - of course it comes to visual change.

## 3.1 The effect of light and dark colors

Taking into account the different influences of light and dark colors it seems necessary to first examine the effect of light and dark colors within a small space. Light, pale or pastel colors make rooms and objects look larger than they are. Light colors visually broaden the space and a room seems wider and larger. The same applies to wall color. A brightly colored wall will appear larger than it actually is and using light wall colors with matching light trim and baseboards will also make space seem larger. Lighter colored ceilings and floors create the effect of a higher ceiling. Light ceilings will raise the apparent height of a room. Darker and deeper colors make rooms look smaller and tighter than they are in fact. Dark objects appear heavier. Dark color on the ceiling visually lowers it and makes a room seem lower. Combination of a dark floor and ceiling can greatly reduce the apparent height of a room and may seem oppressive.

## 3.2 The effect of contrast

Although the lighter and brighter colors help to create spacious and airy interiors, this does not mean that interior designers are limited to using lighter colors when they arrange and decorate a small room. In practice, dark colors contrasting with lighter colors can emphasize the effect of spaciousness. For example, a dark couch or armchair on light colored flooring and on light wall behind it, as a single dark element against a light background, will seem smaller than it actually is. By that means small room will appear more spacious. Therefore it is possible to use dark colors to create the sense of openness and spaciousness in a smaller space. The essential principle is to use darker colors in contrast with lighter colors to create the illusion of more space. It is also important to reduce of dark colors to a small surfaces and to a few small items.

## 3.3    The effect of warm and cool colors

Beside of the effect of light and dark colors and regardless of the use light and dark contrasts, designers can shape and modify the visual experience of interior space through the careful and correct manipulating the warm and cool colors.

Cool colors appear to move away. For example blues and greens recede and seem more distant. Warm colors approach. Colors in the red, yellow and orange family appear closer. This allows designer to easily manipulate the length, width and height of painted or colored interior. Reds and yellows advance a room. Walls covered in warm colors seem to move inward making a large room appear smaller. A strong, warm color on an end wall will shorten the apparent length of a room by drawing that wall forward. Cooler colors will cause the plane to recede, thereby expanding our perception of the room. Walls painted in cool colors appear more distant than they really are. Cool colors can make a small room seem larger. If a room is small, they can "broaden" its walls by using light cool colors in finishing – grey-blue, pearl, or pale green. If a room is larger, pastel or deep colors in a warm spectrum seem to be appropriate for it. Light, cool spaces are generally perceived as expansive, while the dark, warm spaces as diminishing.

Irrespective of the above conditions, there are colors that visually increase and broaden the space, and colors that create the effect of closeness of the space. To the colors that visually broaden the space the following ones belong: neutral white, light beige, warm beige-orange, different shades of yellow, cold light blue, blue, cold blue-green. To the colors that create the effect of restraint and closeness of the space the following ones belong: black, dark brown, blue-green, blue-violet, rich red, yellow-red, orange-red, warm reddish colors. There are also neutral colors that don't quite change the space perception. These are green, purple-red, violet and grey.

Application of light and warm colors in the main room makes the flat look wider, larger and more spacious. The effect of a bigger height is achieved by coloring the floor in rich and saturated colors. Red and all shades of red effectively accentuate horizontal lines and planes of the floor. Blue and all its shades on the floor are cold and unpleasant in the reception, especially on the large part of floor, but, at the same time, it seems big and wide. Warm yellow and all its shades on the floor make a room spare, bright and sunny; however, light color visually doesn't give a stable support for legs and heavy elements of interior.

## 3.4    The unify a space in order to enlarge the interior

In order to obtain a wider area should be avoided high contrast between the wall color and painted woodwork, furniture and decorations. The interior designers should also be avoided wall patterns, borders and anything that draws too much attention to the walls.

The broadening of visual living space can result from a common color - grade of a adjacent rooms, for example sitting room, a kitchen, and a dining room. Especially the color of floors and walls in this area should be the same as the main colors of major part that is usually living room. Strong-valued ceilings and floors may help to unify a space.

## 3.5    Use of color in special cases

- **Room or hallway is too long and too narrow**
In case we need to shorten the long and narrow room or long, narrow hallway, the end short walls should be painted in a deeper and warmer or darker color than the side long walls. The darkened walls will seem to come into the room, making them appear closer. The distance between the end walls will appear to decrease and therefore making the room look in proper proportion. Designer can turn the short end wall into an accent wall, using these decorative wall painting ideas. To widen a room or hallway, designer should provide light, pale colors on the walls, ceiling, and floor. Treatment of ceilings with a deeper color than the walls will create a more square shape.
- **Hallway or corridor is too short**
In the case of the short corridors we can paint the end wall with a lighter, cooler shade and elongate the space by painting stripes on walls (horizontal stripes on the side walls).
- **Floor surface is too large**
To make a large floor surface seem smaller, designer should select a floor covering that is darker than the room's walls. The floor color will define the boundaries of the room moving the eye downward.
- **Ceiling is too low**
To create the illusion of raising a low ceiling, designer should select a paint color that is lighter than the walls. Use of lighter ceilings and floors also create the effect of a higher ceiling.
- **Ceiling is too high**
Dark colors on the ceiling visually lowers it. A high ceiling can be lowered by not only selecting a tone that is darker than the walls, but painting it down to either picture or plate rail height.

## 4    FINISHING MATERIALS, TEXTURE, PATTERN, MIRRORS – ELEMENTS THAT ALLOW TO CREATE OPTICAL ILLUSION AND IMPROVE VISUAL SIZE AND PROPORTION OF THE ROOMS

A material's finish and paint's finish will also influence the proportions of a room. The surfaces of the same color, but with varying degrees of gloss are received by us in different ways. Matte or flat surfaces and finish absorb light and reflect the least amount of light making a room seem slightly smaller, but also more comfortable. Shiny surfaces and high gloss paints reflect the most light, so add more space to the interior and make a room look larger.

One of the best ways to enlarge a small room is the use of equipment with transparent and translucent materials. Transparent or translucent furniture, appliances and furniture will seem smaller and lighter than a comparable full-color.

To expand a room, we can use horizontal stripes on the walls. At the same time they cannot be too wide. Also, do not use too large divisions and elements on the walls. The smaller they are, the better. A small division, small parts, and various shades of them make a small space to grow. In the small room is better to avoid large patterns and materials with visible and express textures, because visually reduce the space. Interior decorating with stripes on walls, ceiling and floor dramatically transform inside, allowing to create various optical illusions and increase rooms dimensions. Stripes on walls or floor visually expands small spaces. Vertical striped wallpaper patterns or painting stripes on walls create optical illusions of higher ceiling and larger room, so they can be used in the very low room. Then vertical stripes on the walls make room look narrower, but airier. Horizontal striped wallpaper patterns or painted stripes on the wall stretch the space, making room look wider and more spacious. The use stripes on the floor, for example, by using stripes carpet, is also useful for creating optical illusion that visually increase living space.

The instrument that is usually used by interior designer to make room wider is a mirror. Mirror illusions are a specific way of proportion modification, in particular enlarging of room space. Mirror tricks can make room bigger (even twice). This result is created thanks to mirror reflection and optical illusions. Experiments with a mirror and mirror surfaces are justified especially in very small-scale rooms.

# 5    ARRANGEMENT, REORGANIZE AND DECLUTTER THE SPACE – BASIC PRINCIPLE HELPFUL IN ORGANIZING SMALL ROOMS

Excess of various forms and shapes of interior elements and the introduction of a large number of different levels generated by the elements in space and on the walls may spoil even the best planned actions directed toward the arrangement of optical improve the appearance of the interior. But the basic principle helpful in organizing a small spaces involving the care for the maintenance of order applies not only to architectural governance. While users of small flat may need a some necessary interior elements and fittings, for example, sofa and chair, a coffee table as an entertainment center in a small living room, there are other things that can clutter up the space, making it seem smaller than it actually is. To resolve this problem, we should get rid of the unnecessary features and save the maximum possible amount of open free space.

# 6    SUMMARY

Beside physical methods to modify the limits of interior space, there are also some design decisions and manipulations that can change the perception of a visual space. Correction of room's size and proportions without redesigning or destructions is possible with the help of optical illusions that change visual perception of interior space. Too small room can be visually enlarged, too low room - heightened, too narrow - broadened, disproportionately high room - lowered. Things like lighting, color, textures, patterns, finishing materials, mirrors and even interior elements placement can correct the interior space because they can have a significant impact on our perception of space and our awareness of a room's size, dimension, scale and proportion. With understanding the effect of these factors on the perception of space designers can shape and modify the visual experience of room. Proper planning, interior elements, furniture and mirrors placement can correct any, even very small, space.

## REFERENCES

Bielczyk, A. "Odpowiednie oswietlenie – prosty sposob na poprawe proporcji wnetrza". Accessed July 25, 2011, http://www.oswietlenie.pl/poprawa-proporcji-pomieszczen-c_23.html

"Design Guides for Interiors", Accessed October 14, 2011, http://www.wbdg.org

"Eyes For Design", Accessed December 29, 2011, http://eyesfordesigns.com/2011/09/

Flynn, J.E., Segil, A.W., Steffy G. R. 1988. *Architectural Interior Systems. Lighting, Acoustics, air Conditioning.* New York, Van Nostrand Reinhold.

"Interior Decorating Design Ideas for Using Stripes", Accessed December 29, 2011, http://homedecoratingmadeez.com/

"Interior Design optical illusion, or how to change the space". Accessed September 25, 2010, http://www.womanknows.com/decorating-repairs/news/422/

Jaglarz, A. 2011. *Perception and Illusion in Interior Design.* Lecture Notes in Computer Science. Vol. 6767: Universal Access in Human-Computer Interaction: context diversity - 6th International Conference, UAHCI 2011, held as part of HCI International 2011, Orlando, Fl, USA, July 9-14, 2011: proceedings.

"Oszukac oko – zludzenia optyczne w mieszkaniu", Accessed February 8, 2011, http://www.myfloor.pl/oszukac-oko-zludzenia-optyczne-w-mieszkaniu,7,artykul.html

"Panting Rooms to Manipulate the Space", Accessed September 11, 2011, http://www.housepaintingtutorials.com/painting-rooms.html

"Practical Interior Decorating Information. Room Proportion", Accessed September 10, 2011, http://www.interior-design-it-yourself.com/interior-decorating-information.html

Rossi, P. "Colors Can Affect the Size of a Room". Accessed November 15, 2011, http://nwrenovation.com/painting-articles/colors-can-affect-the-size-of-a-room/

# CHAPTER 11

# Large Scale Architecture. Design Human Factors and Ergonomics Aspects Based on State-of-the-Art Structures

## Does new architectural geometry require a reinspection of comfortable usage and the user's emotions?

*Nina Juzwa*

Technical University of Lodz
Faculty of Civil Engineering, Architecture and Environmental Engineering
Łódź, POLAND
nina.juzwa@polsl.pl

*Katarzyna Ujma-Wąsowicz*

Silesian University of Technology
Faculty of Architecture
Gliwice, POLAND
katarzyna.ujma-wasowicz@polsl.pl

## ABSTRACT

Nowadays, processes of novelty in architecture progress fast. Even if the world stopped turning after the fall of 2008, more than any other art form, architecture is linked to prosperity, innovation, to respect for design quality. As architecture is a costly thing it is made for the order of a given client. Thus the role of a client and an investor is of great importance. Therefore architecture must satisfy their requirements and utilitarian and aesthetic expectations.

Architecture, like art – could be divided into two simplified groups: the "great" architecture and the "casual" architecture. Usually we do not take casual buildings into consideration. They serve to make housing, working, commercial places and others. Currently the works of great architecture mean nothing if not varied and inventive. Even buildings that appear to rectilinear are full of details and elements that have not existed before. The society accepts the situation, especially when the architecture concerned is perceived as a part of the city landscape. The problem increases while talking about feelings of people who use interiors of the new architectural spaces. There are forms that assume various functions, same of them created to alarm or to make people afraid. Today, the architectural elements often follow towards abstraction, and our old friendly forms like "a window", "a door", "stairs" should be explained into the new language of architectural form. There follows an explanation to the users that provides information concerning the intended use of the architectural space of the building.

**Keywords**: human factors in large scale architecture, human feelings in architectural modern space, ergonomics of state-of-the-art buildings

# 1      INTRODUCTION

Architecture, similarly to art, may be considered within the categories of high art and universal art. "Universal" architecture is meant in discussions about our daily surroundings: places of living, trading, work, education etc. Its aim is to shape the environment built for the convenience and satisfaction of the residents. The other category, similarly to that in art, creates architecture that is highly original and innovative, equally in its form, as in the application of modern material technologies and construction engineering.

This is particularly clearly visible in urban architecture of postmodern cities of the end of the 20th century, which is dominated by a smooth, reflecting or half translucent curtain wall. The individual buildings become less visible. The overlapping images of glass and reflecting surfaces obscure the familiar surface roughness of old buildings and create an image adapted to the interplay of light and colour, allowing the creation of immateriality of the urban landscape.

Attempts to search a "language" of new expression that would show the interrelationship between the function of architectural objects and their form are visible in thinking about architecture and in constructions from the turn of the 21st century. These attempts stem from striving to break the 20th century idea of functionality. This new language should allow for a different than up to now attitude to architectural space. On the other hand, it would allow possibilities of practical realisation of innovative ideas. Modern architecture, particularly since the end of the last century, presents a continuous quest for novelty.

Dariusz Kozłowski talks about searching for novelty and weariness: "the recipient may be weary, but it is rather the impatience of a bored artist which is the driving force" of the difference in ever newer objects (Kozłowski 2011). Ewa

Rewers, on the other hand, talks about irony and ecstasy that rule over the imagination of architects and further, referring to the famous „less is more" of Mies van der Rohe and his Barcelona Pavilion presented at the 1929 International Exposition in Barcelona. She quotes Hans Poelzig's „We build modestly and it may cost as much as it has to" (Rewers 2005). Rewers shows the irony and loftiness of form, using the example of the Ming Pei's Louvre Pyramid. The clash of cultures and styles of the transparent building of the Pyramid with the historic façade of the Palace presents also a transgression from the past to the reality of tomorrow. The expression of this idea is the form of the glass structure filled with the spiral form of a staircase full of people visiting the Louvre and the Pyramid. This form was later repeated by Sir Normal Foster in the reconstructed interior of the Reichstag.

The idea to show in the regular geometry of a projection through the transparent façade a spiral of vertically rising crowds has a threefold meaning: aesthetical, metaphorical and demonstrating the "swirl" and irony of novelty. Before our eyes, a regular, Cartesian net of spatial arrangements disintegrates (Rewers 2005), showing in many modern examples aesthetical and technical possibilities of moving from a geometrical form to an organic one.

Modern architecture, and in particular, large scale architecture usually commissioned by a particular person or institution. For some customers the question of aesthetics remains of secondary importance, the future utility frequently is more important while the cost of realisation plays the most important role. In these cases, the aesthetical reception of the architectural solutions is important for an anonymous recipient or user.

Such approach raises many doubts. On the one hand, architecture which is testimony of its times should break its own codes (Eco 1990). On the other, it should make use of elements of the language of architecture understandable to generations (Kozłowski 2011), elements with which the recipient is familiar. The last characteristic gains a special weight if the object's functionality is the priority of the project – if the planned object is to serve a specific function.

Architecture is always a child of its times. Since Vitruvius, the idea is the basic material for creating architecture; however the thought itself does not constitute architecture. Architectural design in the form of a drawing or a mock-up is still only an idea, an element necessary to create a real object. The strict relationship between the thought and the material form of the created work depends on the times in which it is created.

The idea – form, quality and mode of use of architectural space initially appear in the architect's mind. Independently of the historical staffage, pragmatism and avant-garde aspirations architecture is imprinted with the world in which it is created. This takes place through the thought of the architect.

The formation of the object – the choice of construction, building technology, choice of materials, colours – takes place through the design in the form of a drawing or a spatial model. Their role is to transform a thought into a new reality which aims at creating new impressions in user's mind impressions similar to those that inspired the creator.

During the creation of an object the architect, similarly to an artist, not only struggles with matter, but also with ideas. Similarly to a manager of a large company, they have to react promptly and preferably without error to the possibilities, needs and dreams of the present because human dreams at the inception of the idea of a building change in a similar fashion to the materials available to architecture.

The process of design in architecture as well as in related fields has been dominated by computer tools – they replaced the classical drawing tools. Computer tools change the way architects work and some of the changes are perceived positively while others remain controversial. Undoubtedly, thank to computers, the process of design has been shortened, documentation is created more rapidly, the range and methods of collaboration with specialists have changed, as well as the way designs are presented. The models are also created differently – increasingly frequently the object being designed is created as a three-dimensional computer model, with the restriction (or with omission) of the use of sketches and intermediate artificial space models.

Computers are used in the production of building elements and are increasingly used in the realisation of architectural objects. The possibilities for the realisation of creative imagination increase. However, the most ground-breaking achievement in modern design is freeing up of the creator's imagination. In architecture, there are still unreal dreams, but as noted by Harbison the set of "not-to-build" objects (Harbison 2001) has been significantly reduced. Access to digital technology resulted in an increase of more than just quantity, but it also means that the users of new objects have to learn how to read the new architectural space.

Such method of design, despite making an impression of being easy, lays more responsibility on the architect – it concerns the prediction of the sensitivity of the future users to the impressions evoked by the designed architectural space.

Among the diversity of architectural ideas, most clearly visible are results of the creators' convictions as to the role of architecture in society which favours or disapproves of certain values and does not take notice of others. New possibilities of designing and building objects along with the appropriate intellectual climate for their creation influenced the architects' outlook on the world and often changed the concept of how architecture should be, as well as what can be considered as architecture. It can be expected that new ideas in architecture will be followed by new challenges for ergonomics and human factors in architectural design. To illustrate the problem we would like to present some selected trends of creative pursuits in architecture. They became a basis of consideration of the influence of human factors on new architecture.

These trends include:

- seeking to create an interior space between the old and new structure, a „box in a box", a space articulated very discreetly –so-called „in-between" space;
- fulfilment of the idea of softness of form as objects-organisms, objects-blobs, or a less radical striving for creating a bend surface;
- pursuit of energetic efficiency as a general idea, including articulation in the object of green, ecological architecture;
- pursuit of a new character of concepts in modern material technologies, in playing with materials for creating architecture using the familiar "language" of form.

## LE FRESNOY ART CENTER (TOURCOING – FRANCE) – arch. BERNARD TSHUMI

A modern search for creativity in factors forming or supporting the creation of innovative architecture is very well supplemented by the concept of a city of events. Its basic element is the introduction of a new factor that changes the current stereotype of perceiving given space. The idea of the competition aimed at positively changing the perception of the reality by the inhabitants of a small industrial town on the belgian-french border came from Alen Fischer. A writer, director, photographer envisaged the fulfilment of his dreams of creating a school of modern art in a complex of industrial halls that had been in the process of decay.

The organisers of the competition for the concept of a new object gave the participants full freedom. The participants were allowed to do everything with the halls present on the plot. The competition was won by Bernard Tschumi who covered all the objects by a large roof in one, courageous gesture creating a new space for various events. A large roof equipped with electronic technology covered the old objects as well as new spaces. The author refers to his project as an "encounter of an umbrella with a sewing machine, which generates a whole variety of difficult, incalculable and poetical events" (Tschumi 1994). The object, important for two reasons, introduces a cultural factor and digital world into the reality of a small town while in itself creating a new culture of architectural space. The most fascinating part is the creation of a totally new space allowing new and unexpected forms of use, creation of a new perception of architecture, and at the same time remaining in the climate of tradition and place. In particular, the latter is a tribute to the history and small-town community. However, in the light of lack of appropriate research, we do not know the answer to the question how strongly this new, possibly controversial, architectural form integrated with the social life of this small provincial town.

## MADRID AIPORT BARAJAS (SPAIN) – arch. ROBERT ROGERS

The airport is located north of Madrid and connected with the city by an underground and a motorway. Its huge scale is dissolved in the surrounding dessert-like landscape. Upon arrival there we immediately move to a completely different scale leaving the buildings and streets of Madrid in another reality. Here, we are surrounded by faded sand and rocks as far as we can see – our sight rests on the façade of the airport gentle in silhouette and colour.

The author of the new object is Robert Rogers. The work does not disappoint us – on the contrary, it has the best features characteristic for the work of this author. It surprises with the huge scale which is not overwhelming. It fascinates with its lightness and clarity of construction, space filled with light and invisible sky.

The new airport consists of two halls; the main object, terminal T4 and a satellite terminal TS4, which are located 2.5 km apart and are connected by an underground railway. The highly formally and aesthetically sophisticated geometry of the main hall of the terminal exhibits all possibilities of modern material engineering and all advantages of modern technology of model design. The use of computer technology to realise the thoughts and dreams of the architect allowed the creation of a multidimensional space, a hall with a subtle outline, enclosing a complex machinery of a modern airport.

Despite the large scale, the space aspires to "befriend" the user. This is achieved by the subdued colour scheme and translucency of the housing that blends the object into the Iberian landscape and thank to the lively colour scheme of the construction which enlivens the landscape from the inside. These functions are strongly supported by materials used for the finish of the interior, which alleviate the characteristic for airports atmosphere of domination of technology over humans.

Describing the impressions/emotions from the standpoint of the user I have to mention the railway connecting both terminals. It transports passengers from one terminal to another. We reach the carriage by moving along an indicated way, enter, move, exit and we are surprised to find ourselves in an identical place. IDENTICAL OR THE SAME PLACE? The surprise is the result of a similar aesthetics of the interior of both halls. They differ in colour, but the difference is apparent only after a little moment. The colour scheme, although different is similar which aesthetically is an almost perfect concept, but sometimes it leads to problems of understanding a place as it requires remembering small details.

### LIBRARY OF THE WARSAW UNIVERSITY – arch. MAREK BUDZYŃSKI

The building of the Library is one of the most frequently visited buildings that have been built in Warsaw in recent years. Its popularity is in part the result of the localisation in a place with a large number of young people studying in Warsaw. The interest is also the result of the large number of Warsaw inhabitants who enjoy visiting the object. The friendly atmosphere is created by the non-standard form of the open library space, the shape of the space of the interiors, access of light and shade, silence and peacefulness that are a result not only of the building's function but also of its interior design. Returning to the place is a source of pleasure. While fully accepting the architectural solution of the object it would be interesting to ask to user-reader: "how they receive the atmosphere of the interior of the halls of the library space?", or "does the height, spaciousness of the interiors allow being fully satisfied with the atmosphere of the space assigned to be the library reading room?".

To the visitor the building seems to be friendly. Green areas in urban architecture play an important role as stated by the architect – and it also attracts the inhabitants of Warsaw. A botanical garden open to the public is located on the roof of the building. The interest is heightened by a terrace with a view on the river banks and the adjacent to the library centre for science and art integration called "Kopernik".

In the roof garden, the outside of the green areas is dominated by solar installations stressing the important challenge of our times.

More sensitive observers also notice the simplicity of the beauty of the object whose architecture harmonises with the urban surroundings. This beauty has been recognised by the users: in the national open public competition of the magazine "Polityka" 2012 for the best and most beautiful building, the building of the Library reached the 14th position.

### PRITZKER 2009 PRIZE FOR PETER ZUMTHOR

The Pritzker Prize for Peter Zumthor in 2009, also called „architectural Nobel", was justified in the following way:

*These buildings have a commanding presence, yet, they prove the power of judicious intervention showing us again that modesty in approach and boldness in overall result are not mutually exclusive humility resides alongside strength. While some have called his architecture quiet, his buildings masterfully assert their presence, engaging many of our senses, not just our sight, but also our senses of touch, hearing and of smell.* (www.pritzkerprize... 2009)

Talking about the creation of Peter Zumthor we do not mean one particular, honoured object it is rather the philosophy of thinking about architecture. It is clearly visible in some of the selected objects described below.

- 2007 Brother Klaus Field Chapel (Eifel, Germany) - dedicated to Saint Nicolas, known as Brother Klaus; it is a spiritual place – tiny, built in the fields. The chapel was formed out of tree trunks arranged like a tent. That tent structure was surrounded with a concrete shell, with a little grid in the front.

The chapel was built by the owners of the plot on which it is located, with the help of the inhabitants of the village. It is a sacred place that is often visited as testified by many statements. The author stresses the difficulty of creating a small, closed space devoted to contemplation. It seems that the beauty of the place is created by the material – the wooden trunks that were used to create an effect of age of this sacred place, a discreet opening leading into this small holy interior. As usual for Zamthor, it is important how light is led inside through a small opening at the top of the triangular form; similarly discreet light inlet can be found in Kunsthaus Bregenz (1997) or Thermae in Vals (1996). Light and matter seem to be the main allies of the architect.

- 2000 Swiss Pavilion at World Exhibition EXPO in Hannover – the idea was to create an object, an architectural creation called in literature Gesamtkwerk. The Pavilion was conceived to talk about Switzerland through its form, material, smell, light and music. The form of the Pavilion consisted of unseasoned wooden board made of pine originating from Swiss forests. The boards connected by steel cables formed the sole stable structure of the Pavilion (ca. 2800 cubic meters). After the EXPO, the wooden elements were sold as timber.

1996 –Thermae in Vals were created as an answer of the inhabitants of a small village and a hotel to the need for an object that would increase the economic value of the place. Every year they create an offer for 40000 tourists visiting the town. The object is not large on the scale of modern architecture of place. Small objects connected by their function almost completely blend into the Alpine landscape.

The advanced technology of the interior furnishing along with traditional materials excellently respond to Zumthor's philosophy and his idea of creating architecture of place whom it serves. He says that the quality of architectural space does not always refer to novelty. The atmosphere created by light and shadow, by appropriate use of material, through its characteristics creates a space in which we should feel well and in which we would like to spend time. Some elements of architectural material sometimes move to the foreground, others remain in the background, as if in suspension, says the architect and … to those material elements adds the architect's talent…

When talking about the quality of architectural space Zumthor states: *"I think it's not such an intellectual, academic thing, quality in architecture. Atmosphere. Everybody can feel it… …… If an ordinary person thinks it is good, there it is good. Because people are not stupid. …… "* (www.pritzkerprize... 2009)

## 2      THE CURRENT STUDY

The undertaken study has been conducted from the standpoint of the architect profession. We put a particular emphasis on the problem of perception of the object by the user and their interaction with the surroundings. This interaction is understood as a need for contact with other human beings while also permitting to retain privacy.

At this point it needs to be stressed, that the analysed state-of-the-art architecture independently of its scale prefers space with glass dividing walls. So called "panoptical architecture" is expected to deliver a sense of security (I see everything), spaciousness, as well as to allow daylight to enter the internal structure of the building. In addition, state-of-the-art architecture favours a model of design with an important goal of creating unexpected impressions and spatial surprises described as "unexpected discoveries". This trend is considered justified as long as the user perceives it positively. In this approach particularly endangered are feelings of the user connected with the need for privacy, familiarity with the environment and freedom of movement. Invariably, an important question is also the role of the colour of the architectural environment, lighting and thermal comfort.

The aim of the conducted research was to answer the question:

"does the user accept the above trends and expects solutions of this type in all state-of-the-art buildings?"

The building selected it is not an object of a spectacular size, or usual in its function. It is a new building in its internal spatial division, however it is enclosed in old walls. The building, opened in 2011, used for teaching, belongs to the Silesian University of Technology. In this building teaching of students of the course "Interior design" takes place. The analysed object (previously being a cinema and a student club) is located in the centre of the city and with adjacent buildings it creates a frontage of a busy street. It consists of four storeys (the area of each of them does not exceed 400 m$^2$) connected by three stairways and two lifts. The interior design is characterised by a large amount of glazing and bold colour schemes in generally accessible spaces.

For the authors of the current study, it seemed interesting to investigate what kind of feelings would dominate among people who have predispositions for judging the functional and aesthetical solutions of others; who spend a large amount of time in the object; and who would like to create unusual architecture themselves.

A survey was conducted in January 2012. The interviewees were students of the course "Interior Design" of the Silesian University of Technology. The selected students were divided into two groups. The first group comprised of full-time students who spend in the building a few hours almost daily (Gr. 1). The second group comprised of extramural students who spend in the building weekends, staying for two days, from the morning till the evening (Gr. 2).

The survey form had the character of a set of open questions conceived by the current authors. The survey was returned by 51 persons: 41 full-time students (Gr. 1) and 10 extramural students (Gr. 2).

The students answered 14 questions directly relating to human factors and ergonomics. The questions concerned impressions of spaciousness and interior design, type of used internal walls, lighting and colour scheme. In the summary of the results we highlighted nine important problems.

The interviewees in both groups had similar, positive opinions in three areas: lightning, colour scheme of the interiors and general choice of the used materials of the walls and floors (problems 5, 6, 7, 8 and 9).

In the remaining areas, the assessment differed dramatically. In areas 1 and 2 the differences occur due to the number of students (the number of full-time students is much higher). In others, the differences seem to be explicable only by the type of days students spend in the building. The differences in reception of architectural solutions were particularly strongly visible in the type of used internal walls separating places of work and study from the areas of rest and communication. Full-time students who spend in the building 5 days a week (from Monday to Friday) made a larger number of critical remarks. For them, the all-present glass walls creating a panoptical space are unacceptable. Extramural students, who spend only weekends in the building, seem to concentrate more on functional imperfections (like the lack of a canteen or sitting spaces in the lobby where they could correct they projects between classes or interact socially) and they worry less about being "under surveillance" (problems 3 and 4).

## 3    CONCLUSIONS

It is important to create fine buildings; created objects should be good to the place where they are erected and to the user… The simplicity of the "prescription" of Peter Zumthor on impact with creative finesse present in his architectural solutions deserves attention. It seems that his thoughts advance well into the future. Creation of architecture that would take into account the role of the human factors in design seems to be an obvious element of assessing a work of architecture. Good emotions, positive impressions evoked in us by good architecture are important.

The results of the conducted investigations may be summarized by the following thoughts:
- it is possible to experiment with architecture in places where people spend time occasionally,
- the designer who wants to create "high" architecture should be particularly careful when creates space for work, study and rest. It is important for them to realise the need for maintaining the intimacy of the users;
- the interiors of state-of-the-art architecture that is characterised by open spaces, a large number of dividing glass walls where the user may feel under constant surveillance are received positively only when they are located in large scale buildings.

The concept and detailed solutions employed in objects which won international competitions and will become reference objects, usually defend themselves. Ergonomics and human factors in buildings of lower prestige should be under more control.

The obtained results are very encouraging and indicate the need for continuation of this research.

Figure 1. The interior of Library of the Warsaw University (from K. Budzyński archives)

# 4    REFERENCES

Budzyński M. 2011. Transforming space into city of nature and culture coexistence. In: Proceedings of 6th International Conference ULAR – Urban Landscape Renewal "City – a people friendly place." (*in print*), Gliwice, Poland

Eco U. 1990. Innowacja i powtórzenie: pomiędzy modernistyczną i postmodernistyczną estetyką. *Przekazy i Opinie 1/2:* 12-36

Frampton K. 1985. Modern Architecture. A Critical History. *Thames and Hudson* London

Horbison R 2001. Zbudowane, niezbudowane i nie-do-zbudowania. *Architektura Murator* Warszawa, Poland

Kozłowski D. 2010. Durability and Fleetness of Architecture In: *Defining Architectural Space* Kraków, Poland: 161-170

Rewers E. 2005. Post – Polis. *Universitas* Kraków, Poland

Tschumi B. 1994. Event Cities. Praxis. *The MIT Press* Cambridge, Mass, London

Zumthor P. 2010. Myślenie architekturą. *Karakter*, Kraków, Poland

http://www.pritzkerprize.com/laureates/2009: *Peter Zumthor*

CHAPTER 12

# Impact of Historically Grounded Social Acceptability on Ergonomics in Shaping Urban Space and Structures in European Cultural Circle

*Pawel Horn*

Department of Housing Design
Institute of Architecture and Urban Planning
Faculty of Architecture, Wroclaw University of Technology
B. Prusa Street 53/55, 50-317 Wroclaw, Poland
horn@hornarchitekci.pl

## ABSTRACT

The author argues that the current living conditions in big cities are affected by social acceptance of many adverse health and spatial parameters. The aim is to satisfy the basic need of living in a city, and the smaller the financial capabilities the greater flexibility in accepting compromise. The attractiveness of living in a city, especially in a city centre, slowly ceases to be the most important, historically conditioned value. Good, modern architectural projects in the area of the author's own city, where he is a practising architect set new standards for quality of residential space, and these new standards are becoming an increasing part of social expectations.

**Keywords**: contemporary urban design, residential housing standards, social expectations, impact of historical grounds in design of urban spaces

## 1. THE HISTORICAL GENESIS OF URBAN PLANNING AND LOCATION EUROPEAN TOWNS.

Cultural and historical diagram of a medieval European town came from the need to ensure the security and defence of the town and the neighbourhood residents. This factor was more important than the spatial qualities. The element that arranged structure within the city walls was a characteristic pattern with a formation of a town centre (market place) of various shapes, with the dominant structure associated with the secular power - the town hall or castle, and placed in close proximity to a sacred building. This centre, along with the development of towns and the importance of economics and commerce took on the characteristics of a shopping centre with the fairs places and the frontages around the market place were formed by houses with many services on the ground floor belonging to the wealthiest burghers. The centre of a European town served as the principal function in the whole of the town, focusing secular and religious authorities, crafts, services, commerce and entertainment. At the same time concentration on a relatively small area of so many functions, including housing, was associated with very high density of buildings. This gradual process of compacting buildings over the centuries contributed to the consolidation of spatial habits of residents. These habits remained, despite the changes of civilization. Location of buildings within the walls, forcing the inevitability of their large proximity to each other, did not guarantee the security any more. But despite this, cramped urban structure was not abandoned at once and town dwellers did not start building straight away more comfortable estates next to existing towns, but existing structures were slowly surrounded by successive rings of the new buildings, created then with the spirit of the era. The town centre still remained the most attractive place, bringing together the core of urban life. This attractiveness was often compensation for substandard living conditions, accepted in exchange for the opportunity to participate in city life. Most European towns have just such an evolutionary history of the development. Very rarely, the city were formed based on idealistic principles. However, they were created more as a self-contained units similar to today's districts, as a response to the problem of accommodating a large number of people in the historic urban centres.

## 2. THE VALUE OF THE URBAN CENTRE IN EUROPEAN CULTURE

Over the centuries, from the Middle Ages through the Renaissance urban development was a continuous process that, if there were no urban tissue losses in connection with military operations, consisted of continuation of building areas, supplementing of complexes of buildings, extensions and expansion of town limits. The breakthrough was the industrial era, when there was a rapid influx of population into the cities and the problem with the need to rapidly provide housing in a limited, existing structure of the city. Thus began a rapid annexing territories closest to the overcrowded towns what became the beginning of modern town planning and architecture. For the first time problems were noticed which concerned the conditions of residence and density, distance to the workplace, and other needs such as leisure, education, contact with the verdure. Simultaneously, the effects of the nineteenth-century urban planning objectives do not meet the needs of the modern city, as those structures do not fit for today's living conditions and the pace of modern civilization. This is related to the intensity of communication and transit within cities, the movement of the masses within the city in the daily cycle, increasing distances between home and work or study and the lack of time and space for recreation. Modern life fits no longer even in the urban structures of the early twentieth century, what causes the need to adapt, at the expense of quality of life. This is a critical moment in which the modern European city needs a new strategy for development and revitalization, as in the case of Wroclaw (Poland) – the author's dear city.

## 3. ERGONOMICS OF HOUSING CONDITIONS AND COMMUNICATION BETWEEN PLACES OF IMPLEMENTATION OF THE NECESSITIES OF LIFE VS SOCIAL ACCEPTABILITY OF DISTANCE, TIME TRAVEL AND THE QUALITY OF SPACE. COMPROMISES IN THE USE AND DESIGN OF URBAN SPACE, RELATED TO ECONOMIC AND SPATIAL BARRIERS

The parameters of public and residential space in the modern European city, with particular emphasis on Polish cities, are determined by historical and contemporary determinants of planning and construction standard values. At the same time we observe as architects discrepancies between the planning guidelines and the expectations of social and economic conditions. These guidelines often do not fit for the challenges of a rapidly changing economic and demographic situation in the city, which implies the very social acceptability of transient or substandard solutions. **Ergonomics of residential space in contemporary European city on the example of Wroclaw is the subject of a presentation.**

**In Wroclaw, single-family houses are located mostly outside the city centre, and down town in the enclaves with good spatial parameters. Currently, single-family housing is no longer localized down town, and outside of this zone it is being raised on basis of other determinants of city planning, legal and economic, thus a clear comparison of the two kinds is difficult.**

We analyse two zones of residential space (city centre with down town vs at the edge and outside of it but still in the city limits) in terms of lighting, acoustics, parking, recreation in the green, commuting time, access to basic services and above the basic price of housing. These factors determine the quality of life

## 3.1 Living in the city centre and down town

Wroclaw is a city with typical in Europe urban structure, which arose in an evolutionary way for centuries. The city had to cope with contemporary challenges of civilization and demographic within the existing urban fabric, and grew by attaching new settlements in situations of critical overgrowth of its population. However, the distinguishing feature of Wroclaw and other Polish cities are numerous wastages and losses in the urban structure as a result of hostilities in World War II. Paradoxically, these empty spaces now allow us to invest in new buildings and structures close to down town and owing to this those new projects are very popular due to its attractive location. These are mostly single buildings located in frontages of the streets or free-standing ones in quarters of new housing in those areas of the city, lacking historical tissue or where it is very scattered. Despite those lacks in urban structures, new building in those areas are still located within a dense network of transport and urban infrastructure.

Polish construction law contains a more liberal requirements for insolation of flats in supplementary building (so-called "fillings") in urban down town residential developments. This is mainly because of the distances between the existing buildings and density of structure. Even though the intention of Polish building regulations is to ensure good parameters of insolation of flats and ventilation (what is achieved by precise records) without this above-mentioned liberalization of rules in dense down town areas completion of buildings would not be at all possible. (This is not the subject of this essay, but it is worth mentioning that the current standards of the size of flats in Poland is about 300-430 ft$^2$ one bedroom apartment, two bedroom apartment of about 430-590 ft$^2$ , and over 590 ft$^2$ three-room apartment and more. Requirements of building regulations for insolation of rooms in flats in multifamily buildings depend on the size of the windows in relation to floor area and are specific over the year. Density results in poor insolation. Without taking into consideration the character of the city centre any development would not be possible, even at the lowest standards.) Similarly, building regulations are more liberal in terms of requirements for the parking of vehicles by residents in the city centre or down town, allowing street parking in the lane. But even this is often not sufficient, in cases where narrow streets often found in historic old quarters. The city authorities then bring the traffic limitations to allow residents to access their buildings and leave the car at least from the rear. Another problem is constantly increasing traffic volume and number of cars. This makes the interiors of quarters (courtyards) of down town buildings are occupied by parked cars and there is no green space or recreation. It is usually the result of spontaneous development of nobody's space. Only the actions of the city, is sponsoring the funds for the revitalization of such areas bear fruit in the form of an orderly and aesthetic development.

116

Figure 1. Typical quarter of historical, 19th century urban structure in Wroclaw. (photo: Google Earth content)

Figure 2. Example of extremely close proximity of 19th century tenement buildings, Wroclaw. Front (the left) and rear (the right) view. (photo: author)

The narrow streets' routes between 18/19century buildings which are usually 5-6-storey high are not currently able to accommodate traffic transport, often with a tram line, causing traffic jams, with no parking places for residents and people staying for a shorter period of time (consequences – mentioned courtyards packed with cars). They become dangerous, because of the very tight stops and walkways, a huge noise and exhaust polluted air, lack of separation of lanes for trams and buses, what also is a reason that public transportation does not exempt from getting stuck in traffic - discomfort and stress of everyday commuting to the inhabitants of the entire city. It is surprising that despite these disadvantages and difficulties in terms of ergonomics of the city, and despite high housing prices, one can observe that flats (generally speaking, living) in a central location are very popular. This is due to the above-described acceptance and the value the inhabitants assign to the city centre

Figure 3. Example of new multi-family building (so-called "filling") , down town of Wroclaw. (photo: author)

Figure 4. Example of new multi-family/office buildings (free-standing), located in an old district of Wroclaw within down town (currently under construction february 2012. (photo: author)

This follows from the fact that in European culture city centres of large, non-metropolitan areas do not have a character of single function office district, desolated and depopulated after working hours. It is due to the saturation of a mixture of residential and public services functions. City centres are attractive and live round the clock - such as Wroclaw's Market Square around Town Hall. Restaurants and cafés on the ground floors of the buildings around the Market Square operate late into the night, and on the upper floors are mostly flats. Here, parking problems, lack of recreational opportunities in the green and the noise of catering services on the ground floors and large events in the Market area are the main inconveniences - despite this the inhabitants do not want to move out from there, appreciating the value of their place of living which is the location. Most of the streets of the city centre and down town are designated by the residential buildings, that stood here since at least the 19th century. According to the acoustic map of Wroclaw (Wroclaw City Office 2008) noise level at the main arteries which are often surrounded by residential buildings in the city centre of Wroclaw is between 85-95 dB.

But for many people, especially young, childless and learning in higher education, living in the city centre means a shorter travel time to work or school, easy access to basic or cultural services – shopping centres, cinemas, theatres, opera houses, restaurants, galleries, city parks, and leisure or academic centres. Despite all the above-discussed shortcomings in the city centre apartments are very popular because historically conditioned cultural value of the city centre dominates over all the disadvantages. This shows that the above-described system: man - ambient conditions can be considered adequate and effective in terms of meeting the needs, and the biological cost of noise, air pollution and poor insolation is compensated by cultural factors.

## 3.2 Living outside of down town but inside the city limits

Reservation "in the city limits" is related to the fact that outside these boundaries there is a low-intensity residential development, associated with other types of employment than in the city and the lack of need for travel to work (employment in agriculture and local services), lower land prices, the other spatial planning in this type of zone. (In Wroclaw down town is an area and the official name of a district close to the city centre together with old town district. Other districts within a city are not called down town, though some of them are close to the city centre as well. From the point of view of a resident of far districts of the city, all the central districts of the city are often called 'city centre' rather than 'down town')

However, outside the down town area in Wroclaw there were raised in recent years many new multi-family housing estates mixed with single family housing estates and individually built houses. Residents of apartment buildings in multi-family housing estates, located in the down town area perimeter far from the centre can be divided into those:

- not possessing the means to build or buy one's own house and not wanting to live in a nuisance of the centre of the big city
- not wanting to live in a detached house intentionally and choosing apartments in apartment buildings to a standard equivalent to one's financial possibilities

⅄    for whom prices of flats outside the centre are the only means to obtain
    locale

A large number of new housing estates located in various parts of the city offers
a variety of conditions for residents. The more advantages and the more prestigious
location the higher the price. Settlements or individual buildings in locations with
more attractive parameters offer their residents a good or very good conditions
inside the housing and further their advantage may be the quality of common areas
in the settlement (the distance between the buildings, organization of inner space),
proximity to recreational areas and public spaces and services or schools and good
communication the city centre In such situations, the conditions of lighting,
acoustics, communication are usually pretty good and people choosing to live in
such a place do not have to make tough choices and balancing the pros and cons.
Acceptance of these conditions does not necessarily refer to either the cultural or
economic aspects. Having provided good access to down town and good spatial
conditions in the settlement, residents enjoy the advantages of living in the city
without the cost of biological nuisances, unless their estate is not in the lane
thoroughfares and noise, are not forced to travel to work or school spending a lot of
time in traffic jams, suffering from stress and long commuting times. What can be
seen as a great inconvenience is especially commuting time: spending up to two
hours on reaching one's work place in a distance of about 6-13 mi within a city
equals a journey of about 62 mi out of its limits.

Figure 5. New multi-family housing estates, located in west districts of Wroclaw. A good (the left)
and bad (the right) example of shaping of urban space, in respect of distance between buildings
and insolation dependent on it. (photo: author)

Settlements capable of using up the maximum of area designated for
development, in places which are less attractive – along the arteries, located far
from public transport and basic services - are the solution for people looking for
apartments available at the lowest prices. While making the decision to buy such a
flat people are often forced to accept high density development, a long time to get to
work or school, lack of nearby shop or other services, the lack of what to expect
while living far from the centre . This follows from the fact that the low price of
housing has been linked to the low cost of purchase of land in the unattractive
location. It may be puzzling why the economic factors outweigh quite easily

biological cost, that is a nuisance of long commuting and the high density of buildings, and the above-mentioned lack of access to basic services and recreation in the vicinity. In the opinion of writing this article, historically conditioned living habits relegate these factors to the background, causing them to skip in the face of strong economic argument.

These shortcomings in living conditions are effectively balanced by satisfaction from having one's own home and living in a new building. Regardless of the surrounding, even the lowest standard of new buildings has to meet the requirements of building codes, providing an inhabitant with a decent minimum of health conditions, and this makes satisfaction from from having one's own home in the described situation possible. The mechanism of compromise, originally associated with the consent to the adverse conditions for the advantages of living in the centre is of a form of consent to a different set of unfavourable conditions for the benefits of what is financial affordability.

## 4. CONCLUSIONS

Perhaps, centuries-old tradition of living in cities with high population density and the necessity of working out compromises in the balance of cultural or financial advantages and biological disadvantages helps to make trade-offs today. This follows from the basic fact that living in the city, even in worse conditions, and even far from the centre, makes it is easier to get a job and, despite inconveniences, to have access to a wider range of services than in a small town, even though in a small town housing prices are lower, the same as the noise level or air pollution. So the city as a whole can be seen as an area of attractive features, such as once the nucleus of a city within the walls.

Thus, if we consider the choices of residents in relation to historical factors, we may conclude that the social acceptability of adverse biological parameters in terms of place of inhabitance shall be disclosed in situations where they are offset by the benefits in a sphere of culture. Someone, for whom is not a value to live in the city and use of what it offers in terms of job and education possibilities, a variety of proposals for cultural, recreational and sporting activities is likely to incur high biological (mental and physical)  costs of such unfavourable parameters as discussed in the article. A frustration and discontent will not allow to determine their housing conditions as comfortable and rewarding.

On the other hand, a high level of social acceptability for the hardships of city life may not be determinant for the design of urban space. In addition to providing the minimum required by building regulations, urban planners and designers make their contributions through creative transcends social expectations. Functional planning of space arrangements, ensuring access of sunlight and the presence of greenery, aesthetics of architecture, good acoustics and communication service, proposed by the designers in the more expensive investments, slowly become in the eyes of the inhabitants of entire Wroclaw the value they desire as a primary standard.

Convenience and health, and aesthetic values of the surrounding environment slowly take place in a row of the most important aspects of living in the city, which is consistent with the spirit of our times. In Wroclaw it is fostered by the enormous potential of still undeveloped areas that are likely to become a place of shaping new attitudes in relation to what is the general principle of social acceptability in relation to the conditions of living in a big city.

**Note: The author's thesis and arguments are based on his own experience, derived from practice of an architect in Wroclaw, Poland.**

# REFERENCES

1. Wroclaw City Office, LEMITOR Ochrona Srodowiska Sp. z o.o., Geomatic Sp. z o.o., 2008 *Emission map, roads – Ldwn.,* viewed 28 February 2012
http://wrosystem.um.wroc.pl/beta_4/webdisk/6de093aa-ed68-41b5-af84-a7072d3e962d/LDWN_Drogowy_Emisja.pdf

# Facades and Multimedia Screens in Contemporary Architecture – Ergonomics of Use

*Agata Bonenberg*

Poznan University of Technology
Poznan, Poland
agata.bonenberg@put.poznan.pl

## ABSTRACT

Creation of attractive public spaces in cities increasingly involves the use of multimedia LED screens as an integral element of architecture. As a result, there is an increasing demand for illumination control, during which lighting should be subject to assessment of safety and comfort of use. The article illustrates positive and negative impact of LED screens in public spaces, proposing new standards which shall describe the functioning of such elements in line with the basic ergonomic principles.

**Keywords**: public spaces, multimedia screens, mesopic vision

## INTRODUCTION

The search for contemporary, attractive style in architecture and public spaces in cities increasingly involves the use of multimedia communication media as one of the elements of architecture. Many clients, recipients and users assess positively the aesthetics inspired by advanced elements of communication media infrastructure: multimedia screens, high-tech details, use of plastics, steel and glass. The use of motion, change of colours, reflection, and dynamics of the constantly changing communications increase attractiveness of architectural forms – and create a night panorama of urban spaces. On the other hand, light and sound are these factors which may cause users' discomfort. Ergonomics of technical infrastructure elements, their operating parameters and adjustment to the capabilities of human body start to be important problems of ergonomics within the urban space. Improper adjustment of luminous conditions on facades and multimedia screens may lead to blinding, car accidents, cause discomfort of pedestrians and the people living in the immediate vicinity. The use of multimedia screens and elevations also deepen the negative phenomenon of luminous chaos.

## 1. LIGHT IN URBAN PUBLIC SPACES

Observation of the contemporary urban public spaces shows that there is a demand for comprehensive research of the spatial and lighting phenomena, especially under conditions of night or mesopic vision. Establishment of a dynamic and variable picture is the result of changes in perception of the environment, temporary vanishing or exposure of respective elements of the urban panorama. Under certain conditions, this picture may interfere with identification and orienteering within the space. The phenomena related to luminous chaos and spaces exposure should be backed up with thorough research. Measurements of quantitative (photometric) and qualitative (colorimetric) features of light in correlation with the criteria of space evaluation shall be the basis for analysis of the quality of the luminous image of a city.

Figure 1. LED screens in the panorama of Shanghai, photo A. Bonenberg

From the point of view of spatial, social and cultural issues, public spaces are the main outlines of the functional and spatial structure. Their types include division depending on their intended use and form. In the context of the role which is played by them, we can distinguish public spaces with communication, trade and service, cultural, representative and recreational function. Each of these types requires proper luminous conditions in order to highlight the functional, aesthetical and compositional values. In each situation, the location of architecture being the source of light shall be subject to different conditions. However, it should always serve improvement of the social character of the public space.

## 2. ARCHITECTURE AND MULTIMEDIA

Integration of multimedia technology and architecture is a significant extension of the creative expression of a designer. The quality of formal solutions is influenced by the purpose and contents of multimedia communications. The possibility to link architecture to multimedia is a multi-thread tool for communication with the environment. Such communication may be dedicated to different types of information related to the city, society or culture. However, it usually concerns commercial issues, more rarely artistic ones.

### 2.1. Commercial function

Due to the omnipresence and the number of advertising stimuli, it is difficult to win the interest and attention of consumers. A movable picture with commercial communication offers more possibilities to influence the clients. It catches one's attention more easily: behavioral psychology explains it with the use of unconditioned responses. Static pictures instead, vanish from human consciousness in a short time. 1

### 2.2. Artistic, conceptual and aesthetical meaning

Artistic, conceptual and aesthetical meaning is based on the willingness to include urban or architectonic space in the flow of global information. With regard to shaping an informative society, this goal may be considered as one of the steps towards deepening of social relations built through media. This allows for consideration of such issues as architecture as an interface, and media communication in urban space as the means of communication of the society. An example of such action is the installation of the Media Wall in Beijing by Simone Giostra & Partners, in which a delicate, elegant picture displayed on multimedia surface dominates the surroundings of the object in the night landscape of the city.

---

[1] Kronhagel C., Műler R.,Tabel S., Teltenkoetter K., ah4 media facades, daab, Köln, 2006, p. 12

Similarly, one can interpret the facade of contemporary arts museum in Gratz, Austria, designed by Peter Cook where the light spots placed on elevations provide for the possibility of signalling and communication with the environment. Multimedia interiors of the Spanish pavilion for EXPO2010 in Shanghai play similar role.

Figure 2. LED screens constituting exhibition of the Spanish pavilion EXPO2010 in Shanghai, photo A. Bonenberg

## 3. AVAILABLE TECHNOLOGIES AND ERGONOMICS ISSUES

Lighting of public spaces and communication space in a city is necessary due to safety reasons. Under the conditions of night vision it is used to maintain visual comfort of public spaces which, in turn, is the prerequisite of an effective social integration. Lighting should be distributed in a regular manner to minimize contrasts – this method is used in designing of the present technical infrastructure of cities – a system of street lanterns.

Multimedia screens and elevations are an additive and supplement to this system. The technology used to produce them is Tricolour LED – a diode with structures to generate three basic colours (red, green, blue - RGB), giving possibility to mix them and to create any colour. The screen has also specific resolution, i.e. the number of horizontal and vertical pixels. The higher the resolution is, the more details the screen may contain. Refreshing frequency defines how many times during one second the picture is displayed on the screen. This value is responsible for blinking of the picture, and for observer's eyes fatigue.[2]

The basic parameters listed above provide for the possibility to adjust respective features of the picture, but designers and advertisers often choose to apply the solutions which may cause visual discomfort, especially at night. The basic problems include:

- Inappropriate adjustment of screen brightness and contrast,
- frequent use of blinding white colour, which is particularly onerous during night,
- designing of a communication in very contrasting colours and pictures changing at high speed,
- screen reflection on shiny walls, glass and glass elevations enhances the blinding effect,
- legibility of irregular, bizarre graphic layouts may cause distraction of the observer.

Frequent staying in such environment due to living or work in immediate vicinity of the same may cause inconveniences, especially under the conditions of night and mesopic vision. These include:

- Quick fatigue of eyesight,
- possible blinding, especially due to projection of white colour,
- distraction,
- eyes irritation i and blurred vision,
- headaches.

## 4. CORRECTIVE ACTIONS AND NEW STANDARDS

The basic preventive action concerning the effects of the common, mass use of multimedia screens and elevations is to develop standards within the scope of:

- placement and location of light element in the building elevation,
- direction of a light screen relatively to the communication tract (walking or driving),
- specification in the lighting standards of not only the minimum values for public spaces but also contrasts under the conditions of night vision,
- recommended use of infrastructure allowing for adjustment of luminous intensity depending on weather conditions and time of a day: rain, fog and other. Luminous effects are dependent on the environmental conditions.

---

[2] source: commercial materials http://www.projektgroup.pl/

There is an important demand for control and inventorying of illumination, during which lighting should be subject to assessment of safety and comfort of use. In 2011 a team of the employees of the Faculty of Architecture at the Poznań University of Technology performed research of the topic: *Method of analysis of public spaces in Poznań under the conditions of mesopic and scotopic vision*[3], which refers to the issues presented above.

The analyses are carried out on selected instances of the space of Poznań. The holistic approach to the issues of light in public spaces takes into account the architectonic, urban and luminous factors    . The purpose of this method is to provide multi-thread assessment of attractiveness of public spaces of a city with special regard to spatial links with the issues of quality and changeability of lighting. Deliberate use of artificial light, through exposition or hiding of respective parts of urban tissue, luminous combination, unification or separation, have significant effect on the social assessment of attractiveness of urban spaces.

## SUMMARY

To sum up, creation of attractive public spaces in cities increasingly involves the use of multimedia communication media as an integral element of architecture. Solutions using LED screens are promising means of artistic expression in construction of architectonic composition. Well-designed installations have positive effect on the users, they create meeting spaces: as it takes place in the above mentioned museum in Gratz designed by Peter Cook. At the same time, it is important that some adverse effects related to, in particular, commercial screens should not degrade the spaces in which they are placed. There should be standards in place which shall describe the functioning of such elements in line with the basic ergonomic principles.

## REFERENCES

Bańka A., Architektura psychologicznej przestrzeni życia, Gemini S.C., Poznań, 1997

Imielinski T., Nath B., Wireless Graffiti – Data, data everywhere, Proceedings of the 28th VLDB Conference, Hong Kong, China, 2002

Kronhagel C., Műler R.,Tabel S., Teltenkoetter K., ah4 media facades, daab, Köln, 2006, p. 12

---

[3] Pazder D., Bonenberg A., Nawrowski A., Kaźmierczak B., Innowacyjna metoda analizy przestrzeni publicznych Poznania w warunkach widzenia mezopowego i skotopowego

# Section II

*Ergonomics in Industrial Quality*

# CHAPTER 14

# Improvement of Lighting Quality in Advanced Main Control Rooms

*Chang-Fu Chuang[a], Sheue-Ling Hwang[b],Ju-Ling Chen[b], Chun-Hung Yang[b]*
[a]Atomic Energy Council 234,
[b]National Tsing Hua University, Hsinchu, Taiwan 300
[a]chuang@aec.gov.tw
[b]slhwang@ie.nthu.edu.tw
[b]go27@hotmail.com
[b]s9934808@m99.nthu.edu.tw

## ABSTRACT

The purpose of this research is to solve the glare problems by changing lighting environment configuration in nuclear power plant. Firstly, we interviewed the operators in order to understand current screen reflection and glare impact on operators. Then, we conducted an experiment and developed two strategies, 1) attaching capacitive touch screen films to touch screen, and 2) the subjects wearing polarized glasses in current nuclear power plant's lighting configuration environment. The dependent variables in this experiment included efficiency of operating, glare degree, visual fatigue and subjective evaluation. Thirdly, we simulated light paths in the current MCR and to find out the illumination level and the best lighting configuration based on principles of optics ($\theta i = \theta r$), and then verified the effect by an experiment. Additionally, we run a field experiment in nuclear power plant to verify the best improved design of the MCR.

**Keywords**: Nuclear Power Plant, Glare, Touch screen, Principles of optics, Visual fatigue.

## 1    INTRODUCTION

The advanced main control room （MCR） in the nuclear power plant widely used digital image displays on video display unit （VDU） as mainly human-

131

computer interface that provides operators highly integrated information to monitor and control system states directly on the screen. Because of the mirror material of touch screen, the glare caused by reflected light exists in the current MCR. The glare may affect the speed and accuracy for operators to read the important information on the screen that probably causes error, and may lead to a severe accident in the nuclear power plant. Since lighting sources in nuclear power plant are different between each screens and ceiling in MCR, if the relative lighting intensity or the position are not appropriate, it may cause glare and affect visual performance of operators, such as visual fatigue, blurred vision and so on （Hedge, 2010）, and it may also cause risks of operation in a nuclear power plant. Therefore, in order to enhance performance of operators and to avoid the operation errors, it is necessary to explore the issue of improvements of reflective problems.

## 2    LITERATURE REVIEW

### 2.1    GLARE

Glare is produced by brightness within field of vision that is sufficiently greater than the luminance to which the eyes are rightness within adapted so as to cause annoyance, discomfort, or reduce in visual performance and visibility. Direct glare is caused by light sources in the field of view, and reflected glare is caused by light being reflected by a surface in field of view (Sanders and McCormick, 2000).Glare is also classified according to its effects on the observer. Three types are recognized （Sanders and. McCormick, 2000):

(1) Discomfort glare: it produces discomfort, but does not necessarily interfere with visual performance or visibility;

(2) Disability glare: it reduces visual performance and visibility and is often accompanied by discomfort;

(3) Blinding glare: it is so intense that for an appreciable length of time after it has been removed, no object can be seen.

Glare is caused by the reflection of light off of surfaces and is a primary cause of eye strain .To avoid such problems there are some ways to eliminate glare:
(1) To reduce direct glare from luminaries.

(2) To reduce direct glare from windows.

(3) To reduce reflected glare.

(4) Consider materials, quantity and position of lights when designing lighting configuration environment (Hsieh,2004).

### 2.2    MAIN CONTROL ROOM LIGHTING ISSUES

According to NUREG-0700 of American Nuclear Regulatory Commission(NRC), the luminance level in primary operating area should be 50 foot candles (fc). Due to 1fc ≒ 10.76 lux, the luminance level should keep 500 lux.

Types of workstations include sit-stand workstations, stand-up consoles, sit-down consoles in Nuclear power plant MCR. The operators' performance may be affected by design characteristics that affect reach, vision, and comfort.

## 2.3 VISUAL FATIGUEMEASUREMENT

There are many methods to measure visual fatigue. Megaw (1990) pointed out that visual fatigue measurement methods can be divided into five categories:
(1) Eye movement function: includes accommodation, convergence, etc.;
(2) Visual acuity: includes Critical Fusion Frequency (CFF), visual sensitivity, etc.;
(3) Objective visual fatigue; 、
(4) Subjective visual fatigue;
(5) Other measurements, such as changes in viewing distance.
We applied on CFF, visual acuity and subjective visual fatigue in this research.

## 2.4 MENTAL WORKLOAD

Mental workload is one of important factors in human performance in complex systems. The mental workload is defined as "a measurable quantity of the information processing demands placed on an individual by a task"(Sanders et al.,1992). Mental workload may affect human performance. The increasing mental activities may result in latent human errors, but the low mental workload may decrease operators 'arousal, what we called it "Yerkes-Dodson law" (Yerkes and Dodson, 1908).In this study, the subjects' mental workload was measured by NASA-TLX. The NASA Task Load Index (TLX) is one of the popular methods for measuring subjective mental workload. The NASA-TLX is a multi-dimensional rating procedure developed by Hart et al., (1988). It derives an overall workload score based on a weighted average of ratings on six subscales. The following are the six components of mental workload : 1) Mental Demand 2) Physical Demand 3) Temporal Demand 4) Performance 5) Effort 6) Frustration Level.

## 3 RESEARCH METHOD

## 3.1 PRE-TESTING

To make sure the procedure and the feasibility of the formal experiments, we conducted two pre-test and according to the results of this test, Both of two participants are assigned to carry on the experiments during the lighting environment before improved (Environment I)and the lighting environment is improved (Environment II) in simulated WDP area. The independent variable is lighting environment (before and after improved), and the illumination at WDP area is 461.42 lux.The two dependent variables were CFF and operating performance included operating time and operating errors. The results of the pre-testing analyses show that two subjects were assigned to different order, the first subject operated in environment II first, and the second subject operated in the environment I first. The operation time of the second order was significantly lower than the first order (A: 329.61seconds, 177.26seconds;B: 647seconds, 306.157seconds), and the subjects were affected by different degree of glare that CFF was no significant difference, so we needed to improve the experiment design.

The improved method are as follows:

1) Fix the height of watching the screen to prevent the subjects to adjust viewing angle themselves.

2) The subjects can practice the task before the formal experiment to reduce the effect of learning.

## 3.2 EXPERIMENT I

### 3.2.1 EXPERIMENTAL DESIGN

The experiment tasks included three standard operating procedures for nuclear power plant, and used the standard pictures to simulated the scenario more authentic. The two strategies were developed, 1) attaching capacitive touch screen films to touch screen, and 2) the subjects wearing polarized glasses in MCC area where the glare is most in current nuclear power plant's lighting configuration environment. The research design of the study consisted of no capacitive touch screen films/ attach capacitive touch screen films and no polarized glasses/ wear polarized glasses .A 2X2 repeated measures design was used.

The dependent variables in this experiment included efficiency of operating, glare degree, visual fatigue and subjective evaluation:

i. Efficiency of operating: operating time and errors;

ii. Glare degree: glare illumination and glare area;

iii. Visual fatigue: visual acuity (vision) and CFF;

iv. Subjective evaluation: NASA-TLX and subjective visual fatigue questionnaires.

### 3.2.2 PARTICIPANTS AND EXPERIMENTAL APPARATUS

A total of eight students at the NTHU were randomly selected to participate in the study. The experimental apparatus consisted of Egg-crate, anti-glare acrylic, Note books , touch screen, Capacitive touch screen films , polarized glasses, V8/ MP3 recorder and Screen recording program.

The Measure apparatus were as follows:

i. Flicker Fusion System

ii. OPTECTM 2000 VISION TESTER

iii. Timer.

To simulate the current lighting configuration environment in MCR, the egg-crate and anti-glare acrylic which borrowed from nuclear power plant were used.

## 3.3    EXPERIMENT II

### 3.3.1  PARTICIPANTS AND EXPERIMENTAL APPARATUS

A total of thirty students from NTHU (20~65years old) were randomly selected to participate in the study. The experimental sequence of the study took approximately 30 minutes.

### 3.3.2  EXPERIMENTAL DESIGN

The treatment consisted of independent variables in this experiment were ambient illumination and light source position. Each treatment was follow by a posttest. The levels of the independent variables were as follows:
1) Ambient illumination:

Based on the lighting design guidelinesNUREG-0700of NRC, the standard ambient illumination in primary operating area should be 500 lux.
2) Light source position:
i. Environment I

We simulated light paths in the current MCR and to find out the best lighting configuration based on principles of optics ($\theta$ incident= $\theta$ reflection).
ii. Environment II

Based on anthropometric data from Council of Labor Affairs (CLA), the sitting eye height of males is 118.98 cm, and the standing eye height is 157 cm; the sitting eye height of females is 110.79 cm, and the standing eye height is 145 cm. According to these data, we simulated the light paths for males and females, and then we found the suitable positions for placing the lamps that would not produce the glares on the screens for both males and females in MCR. The best lighting configuration environment is to place the lamps 253.32cmand 563.65cm~ 647.5cmaway from the wall.

We simulated the improved lighting configuration environment according to WDP, MCC and SSC three areas in nuclear power plant to set up the experimental environment: The dependent variables in this experiment included efficiency of operating, glare degree, visual fatigue and subjective evaluation:
v. Efficiency of operating: operating time and errors;
vi. Glare degree: glare illumination and glare area;

## 4    RESULTS

## 4.1    THE RESULTS OF THE EXPERIMENT I

The experiment I used capacitive touch screen films and polarized glasses by four different conditions, 1) No capacitive touch screen films and no polarized glasses; 2) No capacitive touch screen films but wear polarized glasses; 3) Attach capacitive touch screen films but no polarized glasses; 4) Attach capacitive touch screen films and wear polarized glasses, to find out the better improvement in the current environment. The results indicated that the means showed that attach capacitive touch screen films and wear polarized glasses had decreased operation time and errors and reduced the glare illumination, area and CFF (Figure 1)

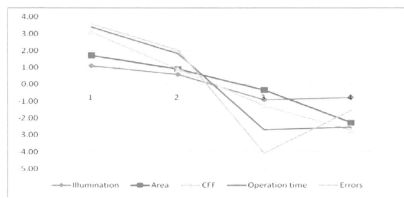

Fig. 1 Comparison of variables in four different conditions

Among the four conditions, we can find that attaching capacitive touch screen films decreased the errors better than wearing polarized glasses, and using two tools together can reduce visual fatigue more than with no tools. Therefore, we recommended attaching capacitive touch screen films to the screens in MCR in current lighting configuration.

## 4.2 THE RESULTS OF THE EXPERIMENT II

In this section, we considered the relative positions between operators' eyes and lamps. According to report by Institute Of Occupational Safety And Health (2008), in addition to consider the level of overall ambient illumination, lighting planning has to consider the lighting distribution, visual comfort and performance. The results was compiled as shown in Table 1.

Table 1 Comparison of two environment (before and after improved )

| Environment / Dependent variables | | WDP area | | MCC area | | SSC area | |
|---|---|---|---|---|---|---|---|
| | | I | II | I | II | I | II |
| Efficiency of operating | Operation time | no significant | | no significant | | no significant | |
| | Errors | more | less | more | less | no significant | |
| Glare degree | Illumination | higher | lower | higher | lower | higher | lower |
| | Area | bigger | smaller | bigger | smaller | bigger | smaller |
| | Head shift (side) | more | less | more | less | more | less |
| | Head shift (back) | more | less | more | less | more | less |
| Visual fatigue | Vision | no significant | | no significant | | no significant | |
| | CFF | no significant | | no significant | | no significant | |

| Subjective evaluation | NASA-TLX | no significant | | no significant | | no significant | |
|---|---|---|---|---|---|---|---|
| | Subjective questionnaire | worse | better | worse | better | worse | better |

## 5.   DISCUSSION

Several research questions were addressed in this study, and the principal findings suggested that (1) In the environment after improved, the lights caused no glare on the screens, there were no glare illumination and area on the screens, and the participants had no head shift, too ; (2) The statistic analysis by the average of vision and CFF still decreased in the environment after improved. This may explain that in the environment after improved could reduce the visual fatigue ; (3) The subjective visual fatigue evaluation was significant in WDP, MCC and SSC three areas between two environments.

Based on these results, we drawn the lighting configuration compared the current lighting configuration with the suggestive lighting configuration in Figure 2.

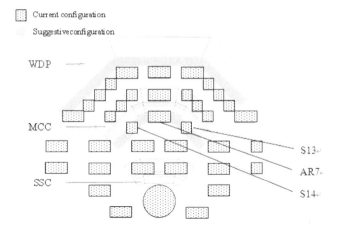

Fig. 2 Comparion of the current lighting configuration with suggestive lighting configuration

Three lamps S13, AR7 and S14 were marked out that the positions of these lamps would not cause any glare on the screens by previous research. These findings are consistent with Lin et al. (2008) study that the brightness of lamps which placed on top of workstations especially at 60~85 degree from the floor should be limited, and with Illuminating Engineering Society of North America (IESNA) findings that the closer the horizontal line of sight the smaller brightness can be received. We decreased the brightness of the lamps AR6 in the MCC area, and found that the glare reduced. The results of this experiment were consistent with our suggestive configuration.

# 6. CONCLUSION

Conclusions for the glare problems caused by lighting configuration in main control room are follows:

1) According to the results of experiment I, the nuclear power plant can attach capacitive touch screen films to the screens in main control room in current lighting configuration environment to get better operation performance and reduce visual fatigue.

2) The results of experiment II showed that the best lighting configuration environment is to place the lamps 253.32(cm) and 563.65(cm) ~ 647.5(cm) away from the wall, that will prevent the glare produced on the screen and reduce not only objective but also subjective visual fatigue.

3) Owing to lots of screens in MCR, too many lights would cause the glare easily. We recommend using several low-intensity luminaries instead of a few very bright ones, and avoiding placing lamps directly upon the operation plane.

# ACKNOWLEDGMENTS

This research was supported by Atomic Energy Council & National Science Council under the project # NSC 100-NU-E-007-008 –NU.

# REFERENCES

McCormick, E.J., & Sanders, M.S. (1982). Human factors in engineering and design. (5th ed.). New York: McGraw-Hill.

Hedge, A. (2010). Glare and Visual Performance, Cornell University, January 2010.

Megaw, T. (1990), The definition and measurement of visual fatigue. In Evaluation of human work, Wilson, J. R. and Corlett, E. N. (eds.), Taylor and Francis, London, pp.683-702.

Sanders, M.S., Mccormick, E.J., 1992. Human factors in engineering and design, 7th Edition. McCormick-Hill Inc., New York.

Yerkes, R. M. and Dodson, J. D. (1908), The relation of strength of stimulus to rapidity of habit-formation, Journal of Comparative Neurology and Psychology, 18, 459-482.

Hart, S.C., Staveland, L.E., 1988. Development of a multidimensional workload rating scale: results of empirical and theoretical research. In: Hancock, P.A., Meshkati, N. (Eds.), Human Mental Workload. Elsevier, Amsterdam. The Netherlands.

Illuminating Engineering Society of North America. IES lighting handbook (6th ed.): 1981 application volume. New York: Author.

Cegarra, J., and Chevalier, A., 2008. The use of Tholos software for combining measures of mental workload: Toward theoretical and methodological improvements. Psychonomic Society, Inc., 40, 988-1000.

Chun, T.-W., Tran, Q.-V., Choi, U.-D., & Kim, H.-G. (2008). Development of Monitoring System for Series HEV Bus with Touch Panel, 13th International Power Electronics and Motion Control Conference (EPE-PEMC 2008), pp. 1421-1425.

Tao, C., Ahang, X., & Wang, X. (2008). Research of environmental control systems for disabled people, APCMBE 2008, IFMBE Proceedings 19, pp. 476-479.

Institute of occupational safety& health(2008, November 10). Message posted to http://www.iosh.gov.tw/Publish.aspx?cnid=26&P=812

CHAPTER 15

# Development of Virtual Instructors for Enhancing Nuclear Power Plant Personnel Training Quality

*Chih-Wei Yang 1, Tsung-Ling Hsieh 2*
*Li-Chen Yang 3, and Tsung-Chieh Cheng 4*

Institute of Nuclear Energy Research
Taoyuan, Taiwan (R.O.C.)
yangcw@iner.gov.tw

## ABSTRACT

This study develops a virtual instructor and investigates strengths, weaknesses, opportunities, and threats of developing virtual instructors for enhancing nuclear power plant personnel training quality. A virtual instructor is an artificially intelligent agent that may be represented in the form of a three-dimensional graphical character with capabilities to provide a personalized learning experience tailored to human learning needs. In order for virtual instructors to improve and accelerate human learning performance, they must possess expertise in specific academic/knowledge domains and effectively guide learners through complex concepts/task and clarify misunderstanding. This study finds that an effective virtual instructor should be able to identify ways to adapt and guide training on an individual basis, according to the operator's cognitive characteristics and learning behavior. A virtual instructor should be able to monitor learning paths among the learning material, and provide emotional and instructional advices to the potential operator. The instructor should act similarly to a distinguished coach who is capable not only to provide the appropriate technical advices, but also to encourage the operators by providing emotional motivations, and taking advantage of the personal characteristics of each operator in order to gain maximum performance.

**Keywords**: virtual instructor, nuclear power plant, three-dimensional

# 1    INTRODUCTION

The quality of nuclear power plant (NPP) personnel training is strongly dependent on the availability of competent instructors (IAEA, 2004). Instructors should have the appropriate knowledge, skills and attitudes (KSAs) in their assigned areas of responsibility. There are five main phases of the systematic approach to training (SAT) process to outline typical responsibilities of instructors (IAEA, 1998), they are analysis phase, design phase, development phase, implementation phase, and evaluation phase. For example, in the implementation phase, instructors should deliver effective training, using proven techniques, including regular questioning of trainees to check understanding, reviewing information given and proposing case studies, as appropriate, to test the trainees' ability to apply the knowledge and skills.

Based on existing training policy, the NPP training organization should decide on the approach to instructor staffing, including the ratio between full time and occasional instructors, and the development of instructors for different training settings (e.g. classroom, simulator, on the job training). Simulator instructors are individuals who usually provide training mainly in a simulator setting. A simulator setting requires an instructor to have wider competence than that needed for classroom training. For example, a simulator instructor must be able to provide both classroom (pre-simulator) training and on the job (hands-on) training.

The training environment may need to be adapted to the availability of NPP personnel. For this reason, self-study solutions utilizing advanced training tools may be an alternative method in order to improve the knowledge of personnel. Some examples are as follows:

(1)    Distance learning: The trainee receives materials to study and learns the content and does assigned exercises. After completing the exercises, they are sent to an instructor for evaluation. At different steps of the training, an assessment should be done in order to reinforce the efficiency of the training.

(2)    Web-based, computer-based: The principle is the same as in the previous example. The difference consists in taking advantage of the web or the computer technology to improve the exchange of information and the interactivity of training (update of documents).

(3)    Virtual technology: This technology consists in designing virtual equipments or plant systems that will be operated from a computer. It should be used especially for parts of the plant which are not available during normal operating conditions (high level of radioactivity, identified dangers, etc.). It also should be used to train field operators in normal or emergency operating conditions.

Following the above descriptions, a virtual instructor should be able to monitor learning paths among the learning material, and provide emotional and instructional advices to the potential operator. This study develops a virtual instructor and investigates strengths, weaknesses, opportunities, and threats of developing virtual instructors for enhancing nuclear power plant personnel training quality.

## 2    STRENGTHS, WEAKNESSES, OPPORTUNITIES, AND THREATS OF DEVELOPED NPP VIRTUAL INSTRUCTORS

This section describes the developing process of NPP virtual instructors and discusses the strengths, weaknesses, opportunities, and threats of them.

## 2.1    Developing Process of NPP Virtual Instructors

Virtual reality (VR) is a way of visualizing, interacting with and navigating through an environment described by a three-dimensional (3D) computer model. A commonly accepted definition is "A computer system to create an artificial world in which the user has an impression of being in that world and with the ability to navigate through the world and manipulate objects in the world" (Louka, 1999). Training is currently one of the most popular areas of research in the VR field. A safe virtual environment can be used to simulate a real or planned one that is too dangerous, complex, or expensive to train in. VR-based training is particularly well suited to situations where cognitive and spatial skills are important.

A virtual instructor, defined in this paper, is an artificially intelligent agent that may be represented in the form of a 3D graphical character with capabilities to provide a personalized learning experience tailored to human learning needs. In order for virtual instructors to improve and accelerate human training quality, they must possess expertise in specific knowledge domains and match human pedagogical capabilities required to effectively guide learners through complex concepts/task and clarify misunderstanding, while at the same time, become intimately involved understanding the learner and their knowledge being learned.

(1)    VR system design: The first step is to design a virtual reality system. Using VR devices is not required if the purpose of the system was non-spatial. On the other hand, VR as a technology will have a unique value in providing strong spatial context for those applications that require it such as many training and educational systems.

(2)    Apparatus: The second step is to develop a large rear-project stereoscopic display for presentation on virtual environments (See Figure 1). The optical output of the display is approximately 5000 lumen.

(3)    Virtual environment constructing: In the third step, the 3D scene of the virtual nuclear control room, which contains the geometry and materials of the virtual objects, are constructed using Cinema 4D (http://www.maxon.net/), a 3D modeling and animation tool. The user interactions of the virtual objects and virtual system's responses by the user's interactions are programmed using a game authoring tool called Unity 3D (http://unity3d.com/). Any interaction scripts in Unity3D could be programmed using java script or C#.

Figure 1 Large rear-project stereoscopic display for presentation on virtual environments

(4) VR communicate with simulative system: Developing an operation training system of nuclear control room is very difficult. Because this operation environment has many controllers and display devices. And relations between controllers and display devices are very complicated. The VR programmers are very difficult to understand this operational logic. So, VR environment communicating with simulation system of nuclear control room is becoming very important. This study uses Windows API method which is a dynamic linking library included in Windows system. This study also uses the mapping program to map and translate the data between VR environment and nuclear simulation system. Thus, the VR training environment has the operation logics of nuclear control room. The information will calculate, share and transport automatically by Windows API and mapping program.

(5) Scenario design: Owing to the real control operation in nuclear control room have three key persons, including supervisor, operator and assist operator. They are communicating and operating at the same time. Supervisor will assign the task for operator. When operator received a mandate he will check the all conditions and talk to assistant operator to check or operate some controllers. So, in our scenario design, we designed some scenarios according to real operations of nuclear plant control room. Operators must operate the controllers or check some information while receiving the assignment from the supervisor.

(6) Realism evaluation: Witmer and Singer (1998) proposed presence questionnaire to evaluate a virtual training system. The scholars have suggested four evaluative faces to evaluate the presence in a virtual environment; they are control factors, sensory factors, distraction factors and realism factors respectively.

## 2.2　Strengths, Weaknesses, Opportunities, and Threats of Developed NPP Virtual Instructors

The NPP virtual instructors developed in this study (See Figure 2) can be used either as a passive instructor-led training or as an active plant operations self-training. Implementing passive instructor-led lessons with the instructors' guidance is an effective way to reduce the workload on the instructor without decreasing training quality. Because the NPP virtual instructors' training is consistent, operators follow instructor-guided training with a common set of knowledge and skills. Thus, the real instructors can teach the more complex principles of integrated unit operation.

The NPP virtual instructors can also be used to improve general knowledge of power plant theory and operating principles, helping users to develop cognitive skills that are essential for understanding complex operating procedures. Plant documents such as system descriptions, training manuals, plant drawings, and technical manuals are an ideal resource for general knowledge information, and are easily integrated into a NPP virtual instructor's lesson.

To clearly analyze advantages of NPP virtual instructors, this study investigates strengths, weaknesses, opportunities, and threats (SWOT) of developing virtual instructors for enhancing nuclear power plant personnel training quality (see Table 1). SWOT is actually an acronym that stands for Strengths, Weaknesses, Opportunities and Threats and is a commonly employed framework in the business world for analyzing the factors that influence a company's competitive position in the marketplace with an eye to the future. This structured examination of the factors relevant to the current status and future of the NPP virtual instructors will unlikely produce a final answer that members of the audience will consensually agree upon.

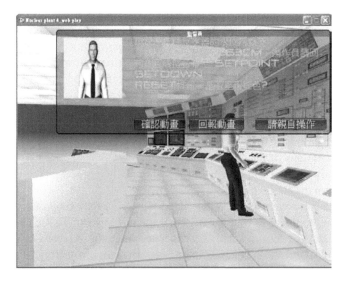

Figure 2　The NPP virtual instructors developed in this study

Table 1 SWOT of developed NPP virtual instructors

| SWOT | Descriptions |
|---|---|
| Strengths | The NPP virtual instructors provide effective power plant training through a "virtual" instructor. |
| | The NPP virtual instructors can be used to improve general knowledge of power plant theory and operating principles, helping users develop cognitive skills that are essential for understanding complex operating procedures. |
| | The NPP virtual instructors' training is consistent; operators follow instructor-guided training with a common set of knowledge and skills. Thus, the real instructors can teach the more complex principles of integrated unit operation. |
| | The NPP virtual instructors can be used especially for parts of the plant which are not available during normal operating conditions (high level of radioactivity, identified dangers, etc.). |
| | The NPP virtual instructors can be used in several applications, ranging from trainers in simulated worlds to non player characters for virtual main control room. |
| Weaknesses | NPP management will need to provide resources for the continuing training of instructors and professional development on the use of these new methods. |
| Opportunities | Continuing advances in VR technology along with concomitant system cost reductions have supported the development of more usable, useful, and accessible VR systems that can uniquely target a wide range of physical, psychological, and cognitive rehabilitation concerns and research questions. |
| | Tremendous growth in the interactive digital gaming area has driven development of the high quality, yet low cost graphics cards needed to make VR deliverable on a basic PC. |
| Threats | The NPP virtual instructors often influenced by such factors as one's faith in technology, economic concerns, frustration with the existing limitations of traditional tools, fear of technology, popular media influences, pragmatic awareness of current hardware limitations, curiosity and healthy skepticism. |
| | Administrators and financial officers believe that VR equipment is too expensive to incorporate into mainstream practice. |

# 3 ENHANCING NUCLEAR POWER PLANT PERSONNEL TRAINING QUALITY BY DEVELOPING VIRTUAL INSTRUCTORS

## 3.1 Nuclear Power Plant Personnel Training Quality

Training of plant personnel is an important factor in ensuring safe and reliable operation of nuclear power plants. Training programs help to provide reasonable assurance plant personnel have the knowledge, skills, and abilities to properly perform their roles and responsibilities. Training design should be based on the systematic analysis of job and task requirements. Therefore, training program development should be coordinated with the other elements of the human factors engineering (HFE) design process (USNRC, 2004).

The training program needs to include the requirement to monitor, evaluate, control and report on training program performance/quality and to identify organizational responsibilities for this. SAT is an ideal management tool for monitoring and controlling the quality of training and other human performance activities, because inherent in the SAT process is continual evaluation of the training programs as well as assessment of the performance of the trainees and job incumbents. Therefore, the study intends to enhance NPP personnel training quality by developing virtual instructors based on SAT (IAEA, 1996).

To achieve acceptable training quality, the training development should include the following five activities: (1) a systematic analysis of tasks and jobs to be performed; (2) development of learning objectives derived from an analysis of desired performance following training; (3) design and implementation of training based on the learning objectives; (4) evaluation of trainee mastery of the objectives during training; (5) evaluation and revision of the training based on the performance of trained personnel in the job setting. The following section will introduce the NPP personnel training quality enhancement by developing virtual instructors.

## 3.2 Nuclear Power Plant Personnel Training Quality Enhancement by Developing Virtual Instructors

The training program of NPP virtual instructors follows a systematic approach to training and includes classroom, simulator, and on-the-job training, as shown in Figure 3. And these measures of training quality due to HFE program include:

(1) Personnel task measurement: Primary and secondary personnel tasks should be measured. Primary tasks are those involved in performing the functional role of the operator to supervise the plant. Secondary tasks are those personnel must perform when interfacing with the plant, but which are not directed to the primary task, such as navigation and HSI configuration. NPP virtual instructors execute training and evaluations following plant procedures for conducting effective task performance evaluations.

(2) Situation awareness: Personnel situation awareness should be assessed. Besides using personnel task measurement, learning can be fairly and objectively evaluated while the NPP virtual instructor is being conducted. For example, individual performance follows a role-playing situation can be used as an evaluation technique by the NPP virtual instructor.

(3) Cognitive workload: Almost any evaluation of plant activities has potential relevance to training. For example, an inspection of reactor protection system maintenance might conclude that technicians performing calibration checks of the equipment were damaging the detectors when they removed their test equipment. This deficiency may be a result of inadequate training, but it might also be a result of other factors, such as excessive workload causing technicians to rush the work, or because of poor work habits due to inadequate supervision or attitudes. Therefore, personnel workload should be assessed.

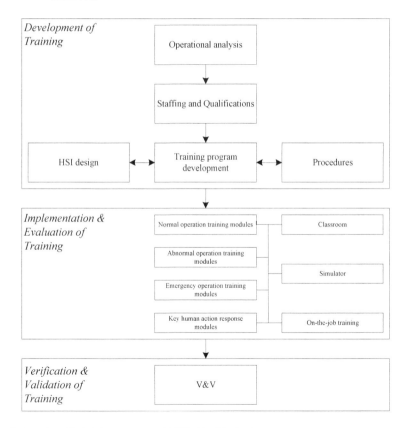

Figure 3  A systematic training program of NPP virtual instructors

(4)     Anthropometric and physiological factors: Anthropometric and physiological factors include such concerns as visibility of indications, accessibility of control devices, and ease of control device manipulation that should be measured where appropriate. Attention should be focused on those aspects of the design that can only be addressed during testing of the integrated system, e.g., the ability of personnel to effectively use the various controls, displays, workstations, or consoles in an integrated manner.

## 4     CONCLUSIONS AND RECOMMENDATIONS

In the past, much of the focus of formal NPP training and development programs was on the technical skills of NPP personnel, particularly those of control room operators. The environment in which NPPs operate is continually changing; placing new demands on NPP personnel to work more efficiently and effectively while continuing to maintain the high levels of safety required of NPPs. In this paper, a NPP virtual instructor that considers training along with other ways to achieve desired levels of training quality is suggested. The NPP virtual instructor proposed in this study follows SAT process to outline typical responsibilities of instructors which intends to enhance human performance (such as personnel task measurement, situation awareness, cognitive workload, and anthropometric and physiological factors. Besides it, this section would like to discuss a list of competence categories related to NPP training quality (IAEA, 2001), include:

(1)     Open communication not only reduces errors of NPP personnel but is a key point to changes in the communication structure and in reducing the barriers to achieving these changes.

(2)     Teamwork will build a climate of confidence between the team members, implement methods to work as a team, deal with conflict management, and encourage experience feedback.

(3)     Leadership will help to achieve the goals in an effective way through planning and organization, observation, facilitation.

(4)     Adaptability will facilitate the implementation of technical skills in encouraging flexibility, continuous learning and ability to transfer skills to new working situations.

(5)     Safety consciousness will encourage conservative decision making and use of risk analysis, and will help to identify weak points in organization and human performance.

(6)     Business focus will enhance employee's sense of responsibility for NPP performance, and develop an understanding of customer needs.

(7)     Professionalism will help the individual to understand his role in the organization, the most effective ways to carry out this role and why he should do it in a proper manner.

(8)     Problem resolution will help to identify potential problems and enhance the quality of decision making using team resources.

The above eight competence categories may not be achieved only by NPP virtual instructors. This study proposes strengths, weaknesses, opportunities, and threats of NPP virtual instructors. From SWOT analysis, plant managers can use advantages and avoid disadvantages of the NPP virtual instructor to enhance the training quality. This study finds that an effective virtual instructor should be able to identify ways to adapt and guide training on an individual basis, according to the operator's cognitive characteristics and learning behavior. In the near future we plan to build a new version of the NPP virtual instructor that improves based on HFE design process.

## REFERENCES

International Atomic Energy Agency, 1996, *Nuclear Power Plant Personnel Training and its Evaluation: A Guidebook*, Technical Reports Series No. 380 Vienna, Austria: International Atomic Energy Agency.

International Atomic Energy Agency, 1998, *Experience in the Use of Systematic Approach to Training (SAT) for Nuclear Power Plant Personnel.* Vienna, Austria: International Atomic Energy Agency.

International Atomic Energy Agency, 2001, *A systematic approach to human performance improvement in nuclear power plants: Training solutions.* Vienna, Austria: International Atomic Energy Agency.

International Atomic Energy Agency, 2004, *Development of Instructors for Nuclear Power Plant Personnel Training.* Vienna, Austria: International Atomic Energy Agency.

Louka, M. N., 1999, An introduction to Virtual Reality Technology, HWR-0588, Halden, Norway, OECD Halden Reactor Project.

Witmer, B. G., Singer, M. J., 1998, Measuring presence in virtual environments: A presence questionnaire, Presence: Teleoperators and Virtual Environments, Vol. 7, pp. 225-240.

U.S. Nuclear Regulatory Commission, 2004, Human Factors Engineering Program Review Model NUREG-0711 (Revision 2). Washington, D.C.: U.S. Nuclear Regulatory Commission.

# CHAPTER 16

# Applications of Ecological Interface Design on an Advanced Main Control Room

*Tsung-Ling Hsieh[1], Min-Chih Hsieh[2*], Sheue-Ling Hwang[2]*

Institute of Nuclear Energy Research
National Tsing Hua University
Hsinchu, Taiwan
[*]g9674019@cycu.org.tw

## ABSTRACT

Several studies showed that ecological interfaces improve monitoring and control performance in comparison to conventional interfaces. This study focuses on the different interface design with the different levels of automation on human performance during anticipated and unanticipated system status. The results showed that modified interface maintain the same situation awareness as current interface and provide a more effective way on monitoring tasks in different levels of automation.

**Keywords**: interface design, ecological interface design, levels of automation

## 1    INTRODUCTION

Over the past few years, several studies indicated that a good design of human-machine interface can enhance human performance and decrease the probability of human errors (Wickens, 2000; Jou et al., 2009). Take the advance nuclear power plant as an example, the main control room is a complex process control system that contains a large number of data of nuclear power plant states. There are a lot of studies focusing on human-system interface, levels of automation, and alarm system

in the advanced nuclear power plant to increase plant safety (Huang et al., 2007；
Jou et al., 2008；Norros and Nuutinen, 2005).

However, some operators felt the interface complex and unobvious during the
period of monitoring and operating. For example, reactor water level is a very
important value, but operator has never known the trend of water level in the page
of the feed water system. Unless they calculate the afflux and flux catchment of
water from reactor or read the assistant display page of the feed water system on
another LCD panel display. Therefore, a modified interface where critical
information can be detected easily is necessary.

Ecological interface design (EID) is a theoretical framework for designing
human-computer interface for complex system (Vicente & Rasmussen, 1992; Burns
& Hajdukiewicz, 2004). The framework explicitly aims to support user adaptation,
especially for unanticipated events (Lau & Jamieson, 2008). Jamieson (2007)
developed a system interface by EID and evaluated it with licensed operators, and
the results showed that ecological interfaces improve monitoring and control
performance in comparison to conventional interfaces. An ecological interface
should support good situation awareness (situation awareness, SA) (Skraaning et al,
2007).

Regarding SA, other factors such as levels of automation may also affect degree
of SA. Endsley and Kaber (1999) proposed implementing intermediary levels of
automation to improve SA and avoid out of the loop performance problems. The
results indicated that levels of automation combine human generation of options
with computer implementation to produce superior overall performance during
normal operation. Kaber, Onal, and Endsley (2000) discussed levels of automation
for telerobots and evaluated the effect on performance, operators' situation
awareness, and subjective workload. The results showed that high levels of
automation enhance performance during normal operating conditions via computer
processing, and low levels of automation increase operator situation awareness and
enhance human manual performance during system failure modes.

## 2    METHODOLOGY

In this study, we used five computers to simulate the feed water system of the
Longmen nuclear power plant as the experimental platform. Figure 1 shows the
layout of the feed water system workstation.

Figure 1 the layout of the feed water system workstation.

Two interfaces were employed in this experiment, including the current interface and modified interface. The current displays used in the control room of the Longmen nuclear power plant, has been stipulated by NUREG-0700. In accordance with the operator's comments, a new interface was modified by EID framework to show the calculational processes graphically (Figure 2), and thus the change of the water flow can be detected easily.

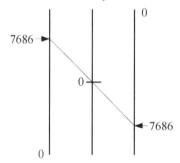

Figure 2 A sample of the EID framework.

According to the present situation of advanced control room in Longmen nuclear power plant, the operators execute tasks by automatic control or semi-automatic control most of the time. Therefore, automatic and semi-automatic control were considered in this experiment.

A few unanticipated emergency events were arranged in the experiment. During the emergency situation, the participant needs to execute emergency depressur-ization step by step, and monitor the state of the feed water system at the same time.

There are seven participants in this experiment, and each participant needs to monitor four LCD panel displays and perform a task on the fifth 19 inch LCD panel display. During each task, situation awareness (SART), error detection time, error detection rate were recorded. Figure 3 is the process of each task.

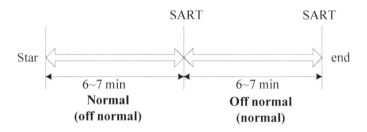

Figure 3 The process of this experiment.

There were several events occurred during the experiment, including water level increasing or decreasing and neutron flux. The participant has to report what kinds of the event were occurring, and the response time and correct rate were recorded. When each task carried on 7 minutes later, the LCD display was frozen and the participant was asked to rank the SA of the task performed before. Then after this task, participant ranked the SA of the task that performed during next 7 minutes. The Situation Awareness Rating Technique (SART) was used in the experiment, each rating scale of SART is given a numerical value from 1 to 10 (1, lowest SA; 10, highest SA).

## 3     RESULT

A general linear model was constructed to describe the data collected from the study. After the logarithmical transformation, the data distribution of the experiment are closer to the normal distribution.

Situation awareness, error detection time, and error detection rate were analyzed by ANOVA with three factors, interface type (current and modified by EID), levels of automation (automatic and semi-automatic), and scenarios (normal and unanticipated states).

There was no significant interaction effect between interface type and levels of automation ($P > 0.05$), between interface type and scenarios ($P > 0.05$), nor between levels of automation and scenarios ($P > 0.05$). There was no significant main effect of interface type ($P > 0.05$), levels of automation ($P > 0.05$), and scenario ($P > 0.05$) on situation awareness.

There was significant interaction effect between interfaces and scenarios on the error detection time ($F = 7.25$, $P < 0.05$) (Figure 4). The results showed that the error detection time were significantly different between the two interfaces ($F = 13.59$, $P < 0.05$). The effect of levels of automation was significant on the error detection time ($F = 4.22$, $P < 0.05$). The effect of scenarios was found significant on the error detection time ($F = 13.35$, $P < 0.05$).

The effect of interfaces was significant on the error detection rate ($F = 11.1$, $P < 0.05$). The correct rate of the modified interface is higher than that using current interface. There was no significant main effect of levels of automation ($P > 0.05$) and scenarios ($P > 0.05$) on the error detection rate.

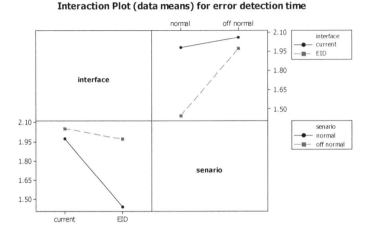

Figure 4 Interaction plot for error detection time.

# 4    DISCUSSION AND CONCLUSION

The results revealed the effect of interface design in the advanced control room of nuclear power plant. In this section, the effect of the situation awareness, error detection time, and error detection rate will be discussed.

The operator in complex and dynamic environment such as advanced control room in nuclear power plant has to maintain high SA. In addition, the interface design of control system is the important factor that can improve the situation awareness of the operator. The interface design of the advanced main control room in Longmen nuclear power plant may have been designed to maintain good situation awareness, therefore, the result indicated that no significant effect of interface type on situation awareness. In other words, modified interface maintain the same situation awareness as current interface on different levels of automation.

On the other hand, the error detection time using the modified interface was significant faster than that using the current interface, especially in unanticipated states. This result is consistent with the previous study of Lau and Jamieson et al. (2008). In their study, EID interface had an advantage in supporting operator performance during monitoring for unanticipated events. In the present study, under anticipated or unanticipated operation conditions, subjects performed better in terms of error detection time and detection rate when using modified interface than that using current interface. Besides, no matter under automatic or semi-automatic operation conditions, subject performed beter when using modified interface.

This study applied EID framework to modify some deficiencies that existed in the current interface. The experiment verified that the control system interface redesigned by EID framework could improve the operation performance and safety in the advanced control room.

# REFERENCES

Burns, C. M. and J. R. Hajdukiewicz,. 2004. Ecological interface design. Washington, D.C.: CRC press.

Endsley, MR. and DB. Kaber. 1999. Level of automation effects on performance, situation awareness and workload in a dynamic control task. *Ergonomics* 42(3): 462-492.

Huang, F.-H., Y.-L. Lee, S.-L. Hwang, T.-C. Yeen, Y.-C. Yu, C.-C. Hsu, and H.-w. Huang,. 2007. Experimental evaluation of human–system interaction on alarm design. *Nuclear Engineering and Design.* 237(3): 308-315.

Jou, Y.-T., T.-C. Yenn, C.-J. Lin, C.-W. Yang, and C.-C. Chiang,. 2009. Evaluation of operators' mental workload of human-system interface automation in the advanced nuclear power plants. *Nuclear Engineering and Design.* 239(11): 2537-2543.

Kaber, DB., E. Onal, and MR. Endsley. Design of automation for telerobots and the effect on performance, operator situation awareness, and subjective workload. *Human factors and ergonomics in manufacturing & service industries* 10(4): 409-430.

Lou, N., G. A. Jamieson, G. Skranning, Jr., and C. M. Burns,. 2008. Ecological interface design in the Nuclear Domain: An Empirical Evaluation of Ecological Displays for the Secondary Subsystems of a boiling Water Reactor Plant Simulator. IEEE transaction on nuclear science. 55(6): 3597-3610.

Lin, C. J., T.-C. Yenn, Y.-T. Jou, C.-W. Yang, and L.-Y. Cheng,. 2008. A model for ergonomic automation design of digitalized human-system interface in nuclear power plants. A model for ergonomic automation design of digitalized human-system interface in nuclear power plants, Las Vegas, USA.

Norros, L. and M. Nuutinen,. 2005. Performance-based usability evaluation of a safety information and alarm system. *International Journal of Human-Computer Studies.* 63(3): 328-361.

Skraaning, Jr. G., N. Lau, R. Welch, C. Nihlwing, G. Andresen, L. H. Brevig, Ø. Veland, G. A. Jamieson, C. M. Burns, and J. Kwok,. 2007. The Ecological Interface Design Experiment. University of Toronto, Toronto, ON, Canada. Available: http://www.mie.utoronto.ca/labs/cel/publications/files/tech_reports/CEL07-02.pdf.

Vicente, K. J. and J. Rasmussen,. 1992. Ecological interface design: Theoretical foundations. *IEEE Trans. Syst.* 22(4): 1-18.

Wickens, C. D. and J. G. Hollands,. 2000. *Engineering psychology and human performance.* NJ: Prentice-Hall Inc.

CHAPTER 17

# Safety Oriented Voice-based Interface for Vehicle's AV Systems: Talking Car System

*I-Cheng Leong[1], Sheue-Ling Hwang[1], Ming-Hsuan Wei[1], Pei-Ying Gu[1], Hsin-Chang Chang[2], Jian-Yung Hung[2] and Chih-Chung Kuo[2]*

[1]Department of Industrial Engineering and Engineering Management, The National Tsing Hua University, Hsinchu, 30013, Taiwan
jengleong_40000@hotmail.com; slhwang@ie.nthu.edu.tw

[2]Information and Communications Research Laboratories, Industrial Technology Research Institute, Chutung, Hsinchu, 310, Taiwan

## ABSTRACT

This research is aimed at developing an in-vehicle voice-based interface that is fault-tolerant, easy controlled, and with reasonable mental workload. Analyzing the requirements of users, user scenario, operational process (including the Voice Dialogue design) and panel interface, the voice-based interface called Talking Car has been designed. Users can control Talking Car with steering wheel control buttons and speech commands. Talking Car has concise graphical user interface (GUI) to let users glance the system status when necessary. Through experiment of driving simulation, subjects' behavior and response data when using voice-based interface were collected and analyzed. Besides, NASA-TLX is applied to evaluate subjects' mental workload.

**Keywords**: Voice-based interface, in-vehicle user interface, NASA-TLX, driving simulation, Talking Car

# 1    INTRODUCTION

## 1.1    Background and motivation

National Highway Traffic Safety Administration investigated 6,949 crashes that occurred between 2005 and 2007. This data indicate that distractions internal to the vehicle were a critical reason in about 11 percent of crashes studied (National Highway Traffic Safety Administration, 2010).

According the proposed structure of processing resources of (Klauer et al., 2006), the vertical modality dichotomy between auditory and visual resources can only be defined for perception, but that the code distinction between verbal and spatial process is relevant to all stages of processing. It is apparent that we can sometimes divide attention between the eye and ear better than between two auditory channels or two visual channels. Using audio-guided navigation is more suitable than visual display (Christopher & Justin, 2000). By appropriate use of speech recognition, drivers can effectively reduce the load on the visual tasks.

Since lots of car accidents caused by distracted driving, many car manufacturers have developed In-Vehicle Speech Systems by which drivers can use speech/commands to control the multimedia devices, such as dialing a phone, or play radio and CD. Drivers can focus on their driving while they are controlling the AV System. And several studies have shown that using Speech systems to operate multimedia device is more effective, less distractive and adjustable to the driver's cognitive capacity and driving situation (Pongtep et al., 2007; Carter & Graham, 2000; Forlines et al., 2005).

## 1.2    Objectives

Although In-Vehicle Speech Systems have been developed, there is only one (LUXGEN Think+) In-Vehicle Speech System in Taiwan. LUXGEN has a clear Chinese interface, and drivers can use the simple commands to dialing a phone, play music and CD, navigation and inquiry a contact person. However, LUXGEN exists some usability problems between the Speech commands and wheel steering buttons, and the Graphical User Interface of LUXGEN need drivers pay a lot of attention to operate.

However, speech recognition highly depends on the users' short term memory (Zhang & Wei, 2010). This study will focus on designing a less distracted In-Vehicle Speech Interface, and expect to find an optimal operation processing to control In-Vehicle Speech Interface so that drivers may focus on their driving when they are controlling the AV System in-vehicle and have a safe response to take care the road situations.

## 2    LITERATURE REVIEW

## 2.1    Driver and In-vehicle System

Most of the control systems in vehicle need drivers to use hands and eyes to control. Drivers have to operate the control system while driving that cause driving distraction easily. National Highway Traffic Safety Administration (2010) has defined the term distraction as a specific type of inattention that occurs when driver divert their attention away from the driving task to focus on another activity instead.

Glance Time is the quantitative term of visual distraction, and display glance time is defined as the driver's eyes began to shift the display from the forward view until the time they were fully focused on the forward view again (Itoh et al., 2004). Safer limitations on visual distraction and suggested that the total glance time in-vehicle telematics tasks must not require more than 10 seconds of visual attention to complete, of which every single glance shall not be longer than 1.5 seconds (Transport Canada, 2003).

## 2.2 Speech interaction systems

Operating Speech interaction systems can reduce the driving distraction but driver would increase the mental workload caused by thinking and listening the task. Driver involve even more complicated device operations that resulting in lack of spare capacity and influences driver behavior or operation of devices (Itoh et al., 2004).

'Flexible shortcuts' allows users to select any command by using 'continuous keyword input', which is a voice input method using a series of keywords related to the command (Teppei, 2008).

## 3. RESEARCH METHOD

## 3.1 Talking Car Interface Design

The main design goal is to minimize driver's mental workload when operating the driving control system. Therefore, we design the steering wheel control buttons to manipulate the speech interaction system with higher speech recognition. Drivers can use steering wheel control buttons and speech commands without hands off to manipulate the entire Talking Car system. The layout of steering wheel control buttons and the architecture between main page and others function of Talking Car is shown in Fig.1.

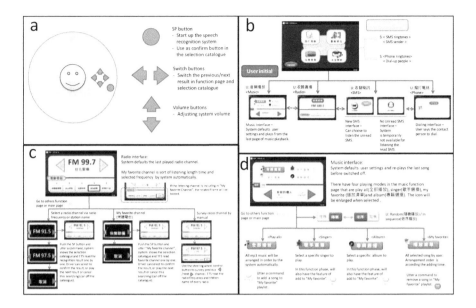

Fig.1 Interface of Talking Car System. a) Function of the steering wheel buttons. b) Architecture between main page and others function. c) Architecture and selection catalog in Radio functional page. d) Architecture of Music functional page.

## 3.2 Experimental design

LUXGEN Think+ System is the only speech interaction system in Taiwan. The operation procedure of simulated LUXGEN Think+ System is simplified in our driving simulator experiment.

By conducting a driving simulation experiment, simulated LUXGEN Think+ System as an experimental control group to compare the task completion time, we measure reaction time of users stepped on the brakes when "STOP!!!" message showed on the display, times of lane shift and task completion time while carrying out the driving task. Driving performance may be influenced by differences in the driving skill of the test subjects, among other factors [10]. So subjects of these two systems should be the same, and the driving performances of two systems are compared. Driving simulation test is arranged to divide into two parts. Subjects carry out the driving simulation of simulated LUXGEN Think+ System first and Talking Car System is finished after two weeks.

In the driving simulation experiment, two cameras are set to capture visual information of driver's side view and face region for driving performance analysis.

## 3.3    Experimental procedure

Fifteen participants, including 12 males and 3 females, are students from National Tsing Hua University in different major, and age range 20 to 35. Subjects have to hold the driving license at least 1 year and have a habit of listening to radio or music while driving. None of them has experience of using In-vehicle speech interaction system. The following sub-section describes the testing process.

The process of experiment took around 30 minutes, learning time took 5 minutes to introduce the system and practice to drive on the simulator, and complete tasks took 15 to 20 minutes. All tasks will be executed 2 times.

## 4      EXPERIMENTAL RESULTS

## 4.1    Task Completion Time

The completion time of related task was analyzed by Paired t-test. As the results of the Paired t-test, there are significant differences on the tasks completion time and shows the operation time of Talking Car System is longer than that of LUXGEN Think+ System. We speculate that LUXGEN Think+ is mostly used of visual recognition in operation process, and Talking Car System need user spend more time consumption to listen the system feedback than LUXGEN Think+ does.

## 4.2    Reaction Time

Subject's reaction time is recorded by simulated software. The calculation method of reaction time is started from the "STOP!!!" message showing on the display till subject stepping the brake to stop counting. Analyzing the average reaction time of related task by Paired t-test, Table 1 show that the reaction times of favorite radio channel selection (p=0.090) and of the specific album selection (p=0.19) are not significant enough but still exist the difference. Besides, the reaction time of the specific song selection (p=0.009) between Talking Car and LUXGEN are significantly different.

### Table 1 Paired t-test results of reaction time

| Task | Favorite radio channel | | Specific album | | Specific song | | Specific channel | Specific singer |
|---|---|---|---|---|---|---|---|---|
| (Unit:second) | TC | L | TC | L | TC | L | TC | TC |
| Mean | 0.73 | 0.82 | 0.73 | 0.78 | 0.76 | 1.03 | 0.74 | 0.74 |
| P-Value | 0.090 | | 0.19 | | 0.009 | | | |
| Significance | | | | | V | | | |

* TC: Talking Car, L: LUXGEN

## 4.3    Total Glance Time

In this experiment, we set two cameras to record subjects' eye movements. After the video analysis and calculating subject's every single glance time the total glance time in operating every single task can be obtained. Paired t-test was applied to verify the results indicated that total glance time of Talking Car System is significantly less than that of LUXGEN Think+ System in all related tasks (P-Value<0.001). It implies that Talking Car System is more suitable for driving use.

## 4.4    Times of Glance

Safer limitations on visual distraction and suggested that every single glance shall not be longer than 1.5 seconds. So we sort out the times of single glance time and the percentage of single glance time which is exceed 1.5 seconds. The single glance time exceeding 1.5 seconds of in Talking Car System are less than 10% and those in LUXGEN Think+ System are up to 35%. The results of Paired t-test indicated that single glance time exceeding 1.5 seconds in Talking Car System is significantly less than those in LUXGEN Think+ System in all related tasks (p≤0.026). That means drivers have less visual distraction when they are operating the Talking Car System.

## 4.5    Times of Displacement

The times of displacement is including lane shifting, craft and hunting. There exists some difference between the displacement times of two systems in this specific album selection (p=0.164) but no significant enough is shown in Table 2. It indicated that times of displacement of Talking Car System is significantly less than that of LUXGEN Think+ System in favorite radio channel selection (p=0.004) and specific song selection task (p=0.027). That implies the Talking Car System needs less spare capacity to manipulate, and the driver can focus on the road situation with a safer response.

Table 2 Paired t-test results of displacement

| Task | Favorite radio channel | | Specific album | | Specific song | | Specific channel | Specific singer |
|---|---|---|---|---|---|---|---|---|
| (Unit:second) | TC | L | TC | L | TC | L | TC | TC |
| Mean | 0.07 | 0.93 | 0.00 | 0.40 | 0.07 | 0.73 | 0.07 | 0 |
| P-Value | 0.004 | | 0.164 | | 0.027 | | | |
| Significance | V | | | | V | | | |

\* TC: Talking Car, L: LUXGEN

## 4.6    NASA-TLX

After the simulation experiment, subjects required to answer a NASA-TLX questionnaire. According to the score of each item, the difference of driver's mental workload between Talking Car System and LUXGEN Think+ System were evaluated. The average weighted score are shown in Table 3. The mental workload of LUXGEN Think+ System is higher than that of Talking Car System. The results of Paired t-test indicated that mental workload of Talking Car System is significantly lower than that of LUXGEN Think+ System (p=0.000), and it implied that Talking Car System provides a better user interface to the driver so that the driver can operate the Talking Car System with less mental workload.

Table 3 Average weighted score of NASA-TLX

| NASA-TLX weighted score | | | | | | | |
|---|---|---|---|---|---|---|---|
| Items | Mental Demands | Physical Demands | Temporal Demands | Own Performance | Effort | Frustration | Total |
| Talking Car | 11.47 | 7.64 | 6.00 | 6.89 | 11.20 | 6.00 | 49.20 |
| LUXGEN | 16.69 | 11.15 | 8.06 | 7.40 | 17.94 | 6.88 | 68.10 |

## 5    CONCLUSION

Regarding subjects' response and workload, show Talking Car system required less glance time, manual distraction and mental workload than that of using LUXGEN Think+ System. Therefore, drivers may focus on their driving when they are controlling the AV System in-vehicle and have a safer response to take care the road situations.

## REFERENCES

C. Carter and R. Graham, "Experimental Comparison of Manual and Voice Controls for the Operation of In-Vehicle Systems," *Proc. 14th Triennial Congress Int"l Ergonomics Assoc. and 44th Ann. Meeting Human Factors and Ergonomics Soc.* (IEA 2000/HFES 2000), Human Factors and Ergonomics Soc., 2000, pp. 286–289..

C. Forlines, B. Schmidt-Nielsen, B. Raj, P. Wittenburg, and P. Wolf (2005), "A Comparison between Spoken Queries and Menu-based interfaces for In-Car Digital Music Selection", TR2005-020, Cambridge, MA: Mitsubishi Electric Research Laboratories.

Christopher D. Wickens & Justin G. Hollands, "Attention, Time-sharing and Workload", Nancy Roberts, and Bill Webber, *Engineering Psychology and Human Performance (3rd edition)*, Prentice Hall, Upper Saddle River, New Jersey, pp.439-499, 2000.

K., Itoh, Y., Miki, N., Yoshitsugu, N., Kubo, and S. Mashimo. "Evaluation of a voice-Activated system using a driving simulator", SAE paper 2004-01-0232, 2004.

Klauer, S. G., Dingus, T. A., Neale, V. L., Sudweeks, J. D., Ramsey, D. J. (2006). The Impact of Driver Inattention on Near-Crash/Crash Risk: An Analysis Using the 100-Car Naturalistic Driving Study Data. DOT HS 810 594. Washington, DC: National Highway Traffic Safety Administration.

National Highway Traffic Safety Administration, "Overview of the NTHSA Distraction Plan", available at: http://www.nhtsa.gov/staticfiles/nti/distracted_driving/pdf/811299.pdf (accessed 26 January 2011), 2010

Pongtep Angkititrakul, Matteo Petracca, Amardeep Sathyanarayana, John H.L. Hansen, "UTDrive: Driver Behavior and Speech Interactive Systems for In-Vehicle Environments", IV 2007, Istanbul, Turkey, June 2007.

Teppei Nakano, "Flexible Shortcuts: Designing a New Speech User Interface for Command Execution", CHI Extended Abstracts, page 2621-2624, ACM, 2008.

Transport Canada, Standards Research and Development Branch Road Safety and Motor Vehicle Regulations Directorate, "Strategies for Reducing Driver Distraction from In-Vehicle Telematics Devices: A Discussion Document", available at: http://www.tc.gc.ca/eng/roadsafety/tp-tp14133-menu-147.htm (accessed 28 March 2011), 2003.

Zhang Hua and Wei Lieh Ng (2010). "Speech Recognition Interface Design for In-Vehicle System", Proceedings of the Second International Conference on Automotive User Interfaces and Interactive Vehicular Applications(AutomotiveUI 2010), November 11-12, 2010, Pittsburgh, Pennsylvania, USA.

CHAPTER 18

# Warning Message Design of LDWS

*Hsuan-Chih Cheng, Ying-Lien Lee\**
Department of Industrial Engineering and Management, Chaoyang University of
Technology, Taichung City, Taiwan

## ABSTRACT

Redundant coding can increase the bandwidth of communication and the robustness of channel, and in some cases, reduce response time. In this research, six designs of multidimensional warning of a lane deviation warning system (LDWS) are compared in a simulated driving environment: (1) directional auditory (DA): warning sound rendered to the left or right ear, (2) directional visual (DV): flashing light signal in the left or right visual field, (3) directional auditory and visual (DADV): directional warning sound and flashing light, (4) directional auditory and non-directional tactile (DANDT): directional warning sound and vibrating steering wheel, (5) directional visual and non-directional tactile (DVNDT): directional flashing light and vibrating steering wheel, and (6) directional auditory and visual and non-directional tactile (DADVNDT): directional warning sound and flashing light signal and vibrating steering wheel. Within-subject design is used to compare these designs. Response time (signal onset time to response) and error rate (percentage of time turning the steering wheel to the wrong direction) are recorded and compared. Correct response to the warning message of LDWS is almost as important as the primary driving task. The results of this research have important implications for the warning message design with directional dimension.

**Keywords**: multidimensional coding, lane deviation warning system (LDWS), driving task

## 1    INTRODUCTION

Information technology and car driving are tightly integrated in growing number of facets. At first, we only have small computers under the hood that control the operation of vehicles. Then, we start to see these computers being used to ensure the safety of driving. Then, we reach the stage of a full-blown computer in the central console of our car cabinet, which is capable of navigation, communication, entertainment, control and monitoring the ins and outs of the car, and whatnot. Driving in such kind of cabinet takes extra tolls of our sensory and cognitive systems to interact with these technologies. Although the driving task is the primary one, the secondary task of interacting with these technologies may interfere with the primary task by vying for the sensory channels (such as visual and auditory ones) or cognition resources. Therefore, the interfaces of these systems should take into account the resource competition for drivers' attention between the primary and secondary tasks. Priority should be set, for example, so that systems related to safety should be of higher priority when a competition occurs. Yet, prioritization among these systems is not easily achievable for after-market systems. For example, when a after-market navigation system is giving itinerary information, another system is also about to give an alarm of a potentially dangerous event. Due to the lack of integration, the sound of navigation instruction may make the alarm unnoticed. Therefore, the alarms of safety-related systems should be robust so that the message loss is minimized. One method to achieve this goal is redundant coding. Redundant coding also has the benefit of decreased response time. This research focuses on the alarm message design of safety systems that have directional signals, such as lane deviation warning system (LDWS), and blind spot monitoring system. Let's take LDWS for example.

A lane deviation warning system is capable of detecting situations when a vehicle is about to being steered out of the lane markings. The system then warns the driver about the potential danger to ensure that the driver is aware of it so that the necessary maneuvering can take place. Most systems warn the drivers via auditory and/or visual signals, which have some drawbacks. First, without encoding directional meaning about which side is about to go over the lane markings, the driver has to judge to which side one has to steer the car, a situation goes unnoticed. Second, our visual and auditory systems are sometimes overloaded by the various systems in the cabinet, such as the case mentioned above. We propose that the alarm should combine auditory, visual, and tactile channels to warn the driver to minimize information loss and that the directional meaning be encoded into the message so that the driver can take proper action immediately after receiving the message.

## 2    THEORETICAL BACKGROUND

Through practice, driving can become an automatic process that demands very little cognitive resource. Evaluating the road ahead, traffic condition around, and steering the vehicle can be orchestrated smoothly. But usually, there will be more than just one task while driving. One might have to interact with various systems in the cabinet, for example, interacting with a LDWS, which creates a dual task scenario. If the driver is engaged in other activities while driving, responding to safety-related systems may become third or fourth task, a situation in which the response time will be prolonged beyond the single task baseline. Any reduction in response time is invaluable. Auditory stimuli seems to be a good choice, since the response time to auditory stimuli is 30 to 50 milliseconds faster than to visual ones

(Woodworth & Schlossberg, 1965). Yet, there are situations when certain sensory channels are occupied so that information presented to them are lost. To overcome this situation, the signals of critical systems should use redundant coding combining adequate dimensions to minimize information lose (Eriksen & Hake, 1955). Responding to stimuli of multiple dimensions is also faster than to those of single dimension (Burke et al., 2006).

Redundant coding design also make the response time to secondary tasks faster by utilizing compatible dimensions. According to multiple resources theory (Wickens, 2002), cross-modal displays have advantages over intra-modal ones due to sensory channel competition, which is also true in driving. As mentioned earlier, directional information regarding which side is about to run over the lane markings is encoded in the stimuli. Although the stages of perception and cognition share the same resource, decoding binary information is easy.

# 3    RESEARCH METHOD

A driving simulation system is used to evaluate the proposed idea. The system is composed of a personal computer, a projector and projector screen, a Logitech steering wheel and pedals, a car seat, a headphone, and a USB I/O module. We use TORCS (http://torcs.sourceforge.net/), an open source car racing simulator, as the software platform, and write our own program to present the multimodal stimuli through means of vision (LEDs controlled by the USB I/O module), of audio (alarms rendered to one of the ears via the headphone), and of tactile (vibration of the steering wheel). Six conditions are compared in a within-subject experiment: (1) directional auditory (DA): warning sound rendered to the left or right ear, (2) directional visual (DV): flashing light signal in the left or right visual field, (3) directional auditory and visual (DADV): directional warning sound and flashing light, (4) directional auditory and non-directional tactile (DANDT): directional warning sound and vibrating steering wheel, (5) directional visual and non-directional tactile (DVNDT): directional flashing light and vibrating steering wheel, and (6) directional auditory and visual and non-directional tactile (DADVNDT): directional warning sound and flashing light signal and vibrating steering wheel. Response time (signal onset time to response) and error rate (percentage of time turning the steering wheel to the wrong direction) are recorded and compared.

## REFERENCES

Burke, J. L., Prewett, M. S., Gray, A. A., Yang, L., Stilson, F. R. B., Coovert, M. D., et al. (2006). *Comparing the effects of visual-auditory and visual-tactile feedback on user performance: a meta-analysis.* Paper presented at the Proceedings of the 8th international conference on Multimodal interfaces.

Eriksen, C. W., & Hake, H. N. (1955). Absolute judgement as a function of stimulus range and number of stimulus and response categories. *Journal of Experimental Psychology, 49,* 323-332.

Wickens, C. D. (2002). Multiple resources and performance prediction. *Theoretical Issues in Ergonomics Science, 3*(2), 159-177.

Woodworth, R. S., & Schlossberg, H. (1965). *Experimental psychology.* New York: Holt, Rinehart & Winston.

# CHAPTER 19

# Application of Taguchi Method on 3D Display Quality

*Ming-Hui Lin[a], Chin-Sen Chen[a], Yung-Sheng Chang[b], Sheue-Ling Hwang[b],*
*Pei-Chia Wang[b], Hsiao-Ting Huang[b], Chao-hua Wen[c]*
[a]Industrial Technology Research Institute, Hsinchu, Taiwan 310,
[b]National Tsing Hua University, Hsinchu, Taiwan 300
[c]National Taiwan University of Science and Technology, Taipei, Taiwan 106
Hsinchu, Taiwan
lmw@itri.org.tw, JIINSEN@itri.org.tw,
freshmango78@gmail.com, edting022@gmail.com,
d9734815@oz.nthu.edu.tw, slhwang@ie.nthu.edu.tw
chwen@mail.ntust.edu.tw

## ABSTRACT

In this study, an experiment by Taguchi Methods in the two-view autostereoscopic display was conducted. The independent variables were the display luminance, ambient illumination, range of disparity, brightness contrast ratio, and viewing time. The dependent variables were subjective image quality and perception evaluation, and depth perception measurement. The purpose of this study was to find out the best combination of independent variables in order the viewers to experience the best image quality and stereo effect while watching an autostereoscopic display.

Before the experiment started, participants were asked to measure their depth perception. During the experiment, 3D images were shown on the display every 15 seconds under different independent variables combination. At the last few minutes, participants needed to verbally answer the subjective image quality and perception evaluation. After the experiment, participants were asked to measure their depth perception again.

As the result showed, the ambient illumination statistically affected the subjective image perception evaluation while the range of disparity significantly affected the subjective perception evaluation. As for the display luminance, brightness contrast ration and viewing time, no effect was found on the dependent variables. The interaction between the range of disparity and viewing time significantly affected the subjective perception evaluation.

When the variables conditions were at medium ambient illumination (103lux), the subjective image quality was the best. As for subjective image perception, the viewers were able to experience the best image perception when the conditions were at high disparity range (0.85° ) and long viewing time (90 minutes).

**Keywords**: Autostereoscopic display, ambient illumination, display luminance, disparity, brightness contrast ratio, stereoscopic effect.

# 1    3D VISUAL EXPERIENCE

Since the introduction of television, much has been done to improve viewing experience. A logical next step is the introduction and improvement of 3D content (Seuntiëns et al., 2005).

In early research of the performance of a 3D television system, a 2D image quality model proposed by Engeldrum (2000, 2004) was often used as an evaluation method. Although the 2D image quality model doesn't well define the overall viewing experience of 3D display adequately, it still has its contribution and influence toward constructing a 3D image quality model.

There hasn't been a comprehensive 3D visual experience model formulated yet, however, it is likely to be a diverse set of image attributes contributes to the overall perceived quality of 3D-TV images (Seuntiëns et al., 2006).

Since image quality has proven to be a reliable way to assess the performance of 2D imaging systems, it is obvious that it also applied to 3D systems. On the other hand, the concept depth introduced in 3D-TVs clearly enhances the overall subjective performance of the TV. However, the added value depth in 3D display is not incorporated when subjects assess image quality. As a result, other concept for the overall performance of 3D display is needed. "Presence", "naturalness" and "viewing experience" have been mentioned as possibilities for such a concept. Until now, naturalness and viewing experience have been found to include the effect of image quality as well as the added value of depth. Moreover, naturalness is found to be a more promising concept than image quality to assess the overall performance of 3D displays (Kaptein, 2008).

Based on the concepts described above, Seuntiëns (2006) proposed a revised version of 3D visual experience model. The model proposed that naturalness, image quality, depth and visual comfort best described the overall 3D visual experience. The concept naturalness is a higher order concept, weighting both image quality and depth. As for visual comfort, there still need more research to confirm its role in the model.

## 2    FACTORS RELATED TO 3D VISUAL EXPERIENCE

### 2.1    Disparity

Disparity is the main causes for viewers to experience depth while watching stereoscopic images. Research has shown that depth has correlation with visual experience and could affect viewers' perceived image quality. Seuntiëns et al. (2007) used 3D video and found that increasing depth too much would decrease naturalness and viewing experience due to the artifacts that are created. Depth may also have a negative contribution to the image quality due to crosstalk, which leads to blurring and ghosting (Kaptein et al., 2008). Nevertheless, depth also has positive effect on visual experience. An increase in depth may lead to an enhanced sense of presence, provided depth is perceived as natural (IJsselsteijn et al., 1998).

### 2.2    Ambient illumination and display luminance

The combination of ambient illumination and display luminance would affect the viewers' perceived depth, naturalness and image quality when watching small-sized autostereoscopic display. As Pölönen et al. (2011) research shows, under lower ambient illumination (30 lux) and higher display luminance (102.87 cd/m$^2$), the perceived image quality would be better compared to other conditions. However, if one is watching an autostereoscopic display under high ambient illumination, adjusting the display luminance to higher condition would have better image quality.

Besides disparity, ambient illumination and display luminance, there are still other variables that could affect the viewers' 3D visual experience, such as image contrast ratio and viewing time. Hence, we would take them into consideration in our research.

## 3    METHOD

### 3.1    Participants

Altogether 28 subjects participated in the test (16 males and 12 females). 20 students were from National Tsing Hua University or National Chiao Tung University and 8 subjects who had experience with 3D application were from the Industrial Technology Research Institute of Taiwan. All subjects had 0.7 corrected visual acuity or better, normal color vision and normal stereo vision.

### 3.2    Experimental Design

The experiment was conducted by Taguchi Methods in $L_{27}(3^{13})$ orthogonal array. Each subject was randomly assigned to an experiment cell. Three two-factor interactions were also being assigned in the orthogonal array.

There were five independent variables in this experiment:

1) Display luminance: 24.85, 52.2, 109.1 cd/m$^2$
2) Ambient illumination: the color temperature of the LED bulbs was 2804K. Ambient illumination changed from 0,103 to 300 lux.
3) Range of disparity: -0.9°, 0°, 0.85°, all images' disparity were in normal distribution and had mean value at -0.9°, 0°, 0.85° respectively. The minus means crossed disparity, and plus means uncrossed disparity.
4) Range of brightness contrast ratio between the brightest point and the darkest point in images: 8.49, 18.87, 35.84.
5) Viewing time: 30, 60, 90 minutes.
   There were three dependent variables in this experiment:
1) Subjective image quality evaluation
2) Subjective image perception evaluation
3) Depth perception measurement

## 3.3    Experimental evnironment and apparatus

The experiment was conducted in an experimental room with 12 lamps to adjust ambient illumination. The apparatus included:

1)    24 inches autostereoscopic display
2)    OPTEC$^{TM}$ 2000 vision tester
3)    Depth measure instrument

## 3.4    Experimental procedure

The experiment aimed at analyzing different factors impact on participants' subjective image quality and perception evaluation and depth measurement while watching an autostereoscopic display.

The experimental procedure was as follows:

1) Before the experiment started, all participants were asked to take basic vision test, including visual acuity, color discrimination, and stereo depth perception, to ensure all participants have normal or corrected vision.
2) The process of the experiment was illustrated to the participants to make sure they clearly understand the procedure of the experiment.
3) Participants were asked to conduct a depth perception measurement before the experiment started.
4) Each participant was assigned to different experiment condition according to the orthogonal array. Participants were asked to answer the subjective image quality and perception evaluation verbally during the experiment.
5) Base on different experimental conditions, the experiment would last 30, 60, or 90 minutes.
6) After the experiment ended, the participants were asked to measure the depth perception again.

# 4    RESULT

## 4.1    The subjective image quality

The subjective image quality questionnaire includes two concepts: naturalness and overall image quality. The questionnaire is a 5-point scale, 1 stands for very poor and 5 stands for very good. Because both concepts' quality characteristic is the larger-the-better (LTB), data needs to be transformed into S/N ratio according to the formula as shown below:

Where $y_i$ is the performance value of the $i$th experiment trial and $n$ is the number of estimator groups in the outer orthogonal arrays.

ANOVA analysis of the naturalness S/N ratio shows that all factors and interactions F value is smaller than 2 hence have no significant effect on naturalness.

As for the ANOVA analysis of the overall image quality S/N ratio shows that the effect of ambient illumination is moderate (F(2,26) = 2.39, p = 0.208).

From the main effect of ambient illumination (Fig. 1) shows, at medium level the overall image quality will be the best due to its bigger S/N ratio.

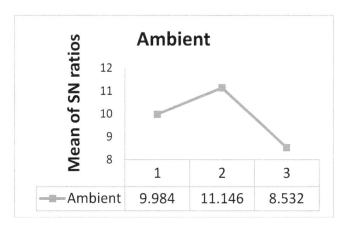

Figure 1 illustrate the main effec of ambient illumination on the overall image quality

## 4.2    The subjective image perception

The concept of subjective image perception, which is also the perception of depth, is also the larger-the-better.

ANOVA analysis of the perception of depth S/N ratio shows that the range of disparity (F(2,26) = 2.29, p = 0.217) and the interaction between range of disparity and viewing time (F(2,26) = 2.06, p = 0.251) had moderate effect.

From the interaction plot of the range of disparity and viewing time and main effect plot of the range of disparity shows, when the viewing condition is at high disparity range (images behind the display) and long viewing time (90 min.), the perception of depth will be the best.

## 4.3    Depth perception measurement

Before and after the experiment, subjects need to judge the depth (-21.3 cm, 0cm, 24.46cm) of three images with different disparities (-0.9 °, 0 °, 0.85°) in the same ambient illumination and display luminance of their experiment condition. The purpose of the depth perception measurement is to observe how the after-effect of the experiment affects the error and direction judgment of depth. The origin of the measurement of depth is located at the display, whereas negative value means that the depth of image is in front of the display, and vice versa.

### 4.3.1    Judgment error of depth measurement

When the image is in front of the display and on the display (2D image), the after-effect did not affect the judgment of depth measurement. However, when the image is behind the display, viewing time ($F(2,26) = 2.23$, $p = 0.223$) had moderate effect on the judgment of depth. From the main effect figure of the viewing time (Fig. 2) shows, when the viewing time is at high level condition (90 min), subjects would judge image that is behind the display more accurately.

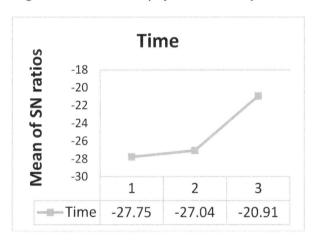

Figure 2 illistrate the effect of viewing time on the judgment

### 4.3.2    Judgment direction of depth measurement

The origin of the measurement of depth is set at where the display is located. If the measured value is minus, then the depth of the image is located in front of the display, and vice versa.

When the image is on or behind the display, the after-effect did not affect the judgment direction of depth. However, when the image is in front of the display, the range of disparity ($F(2,26) = 6.31$, $p = 0.058$) and the interaction of the range of disparity and viewing time ($F(4,26) = 4.38$, $p = 0.091$) had significant effect on the judgment of direction. Display (F(2,26) = 3.66, p = 0.125) and the range of brightness contrast ratio (F(2,26) = 2.91, p = 0.166) had moderate effect on the judgment of direction.

## 5    DISCUSSION

### 5.1    The interaction effect

The interaction of ambient illumination and the range of disparity and the interaction of ambient illumination and viewing time had no effect on either dependent variable. However, the interaction of the range of disparity and viewing time had moderate effect on subjective image perception. When the disparity range is high and the viewing time is long, subjects rated the images as much deeper. It also had significant effect on the judgment direction of depth when the image is in front of the display. When the disparity range is low and the viewing time is medium, images that are in front of the display are seen as less three-dimensional.

### 5.2    The main effects

Display luminance had moderate effect on the judgment direction of depth measurement. When the display luminance is high, images that are in front of the display are seen as less three-dimensional.

Ambient illumination had moderate affect on the overall image quality in the subjective image quality questionnaire when it is in medium level (103 lux).

The range of disparity had moderate affect on the subjective image perception. When the range of disparity is high (image is behind the display), subjects would have the best perception of depth.

The range of brightness contrast ratio had moderate effect on the judgment direction of depth measurement. When the range of brightness contrast ratio is high, images that are in front of the display are seen as less three-dimensional.

As for viewing time, it has moderate affect on the judgment error of depth measurement when the image is behind the display. When the viewing time is long for 90 minutes, subjects would judge the depth of image more accurately.

### 5.3    Independent variable optimal combination

The following Table 1 is the optimal combination of independent variable to each dependent variable. Under each optimal combination,

Table 1 Independent variable optimal combination

| Independent variable | | Optimal combination | Explanation |
|---|---|---|---|
| Sub. image quality | Overall image quality | B2 | Ambient illumination : 103 lux |
| Sub. Image perception | Depth perception | C3E3 | Range of disparity : 0.85o<br>Viewing time : 90 min |
| Judgment error of depth measurement | Image behind the display | E3 | Viewing time : 90 min |
| Judgment direction of depth measurement | Image in front of the display | A3C1D3E2 | Ambient illumination : 300 lux<br>Range of disparity : -0.9o<br>Range of brightness contrast ratio: 35.84<br>Viewing time : 60 min |

# 6    CONCLUSION

A Taguchi method was conducted to observe how display luminance, ambient illumination, range of disparity, range of brightness contrast ratio and viewing time affect the viewing image quality and the after-effect of depth measurement when watching an autostereoscopic display.

As the results showed, each dependent variable had its individual combination of independent variable optimal condition. And because there is no specific optimal conditions that can fit into each dependent variable, hence, depend on which dependent variable is in concern, the optimal combination could be selected as improvement.

## ACKNOWLEDGMENTS

We would like to acknowledge the research financial support by Industrial Technology Research Institute.

## REFERENCES

Engeldrum, P. 2000. Psychometric Scaling. Imcotek Press, Winchester, Massachusetts, USA.
Engeldrum, P. 2004. A theory of image quality: The image quality circle. *Journal of Imaging Science and Technology*, 48: 447-457.
IJsselsteijn, W., Ridder, H.D., Hamberg, R., Bouwhuis, D. and Freeman, J. 1998. Perceived depth and the feeling of presence in 3DTV. *Display,* vol. 18, 207-214.

Kaptin, R. G., Kuigsters, A., Lamnooij, M.T.M., IJsselsteijn, W.A., and Heynderickx, I. 2008. Performance evaluation of 3D-TV systems. *Proceeding SPIE Image Quality and System Performance V.* vol. 6808.

Pölönen, M., Salmimaa, M. and Häkkinen, J. 2011. Effect of ambient illumination level on perceived autostereoscopic display quality and depth perception. *Display,* vol. 32, issue3, 135-141

Seuntiens, P., Heynderickx, I and Ijsselsteijn, W. 2005. Viewing experience and naturalness of 3D images. *Proceedings of the SPIE,* 43-49.

Seuntiens, P., Vogels, I., and Keersop, A. van. 2007. Visual experience of 3D-TV with pixilated ambilight. *Proceeding of PRESENCE 2007.*

Seuntiens, P. 2006. Visual experience of 3D TV. PhD Thesis, Eindhoven University of Technology, The Netherlands.

CHAPTER 20

# Subjective Perceived Depth Measurement and Visual Comfort Evaluation for Viewing Stereoscopic Films

*Pei-Chia Wang*

National Tsing Hua University
Hsinchu, Taiwan
Email address: d9734815@oz.nthu.edu.tw

*Kuan-Yu Chen*

Compal Electronics Incorporated
Taipei, Taiwan
Email address: creailsator@gmail.com

*Sheue-Ling Hwang*

National Tsing Hua University
Hsinchu, Taiwan
Email address: slhwang@ie.nthu.edu.tw

*Chin-Sen Chen*

Industrial Technology Research Institute
Hsinchu, Taiwan
Email address: JIINSEN@itri.org.tw

## ABSTRACT

This study aimed at exploring the impact of subjective perceived depth distance and subjective comfort evaluation on the 3D viewing experience by means of ergonomics assessment. There were two independent variables. One variable was two levels of convergence distance, convex distance and concave distance. The other variable was two levels of ambient illumination, 0 and 300 Lux. The dependent variables were subjective perceived depth distance measurement and subjective visual comfort evaluation. This visual comfort was evaluated with a questionnaire with five-score Likert's scale consisted of six items, addressing dizziness, naturalness, depth, viewer crosstalk, slowed down focus change, and harmonization.

The result of this study revealed that convergence distance was more impactful than that of ambient illumination on perceived depth. However, ambient illumination slightly affected harmonization of subjective visual comfort when viewing 3D films. The regression equation showed that convergence distance was a predictor for perceived depth when viewing convex and concave images.

**Keywords**: stereoscopic, convergence distance, ambient illumination, perceived depth, and visual comfort.

# 1    INTRODUCTION

The film *Avatar* is an important milestone of the 3D movie industry and it introduces a new wave of stereoscopic technology. The major issues of viewing 3D films are visual comfort and image quality.

The perceived depth is a sensation of distance that is mostly determined by accommodation and convergence in the real world. The viewing condition would be the critical factor of the cause of visual fatigue. However, most researchers have observed the effect of perceived depth (Hoffman, Girshick, Akeley, et al., 2008) and ambient illumination environment (Lin and Huang, 2006; Chung and Lu, 2003) on the experience of watching CRT or TFT-LCD displays. There have been only a few relevant studies on 3D displays.

The subjective measurement of depth perception distance was adopted in this study. This research aimed at exploring the impact of ambient illumination and convergence distance on the 3D viewing experience by means of ergonomic assessment.

# 2    METHOD

## 2.1    Experimental Parameters

The experiment was a 2x2 mixed design with ambient illumination as a between-participants factor and convergence distance as a within-participant factor. One half of the subjects participated in low-level ambient illumination; the others participated in high-level ambient illumination.

There were two independent variables. One variable was convergence distance which consisted of two levels, concave and convex distances. Convex distance was the image of the distance from the camera to the target located in front of the TV screen, and consisted of 4 levels, 60 cm, 80 cm, 100 cm, and 120 cm. Concave distance was the image of the distance from the camera to the target located behind the TV screen, and consisted of 4 levels, 240 cm, 260 cm, 280 cm, and 300 cm. The other variable was two levels of ambient illumination, 0 and 300 Lux.

The dependent variables were perceived depth distance measurement and subjective visual comfort evaluation. A questionnaire evolved in visual comfort on a five-score Likert's scale. The questionnaire included six items, addressing dizziness, naturalness, depth, viewer crosstalk, slowed down focus change, and harmonization.

## 2.2    Apparatus and Experimental Environment

A 55-inch LED HDTV with 3D active shutter glasses was used in the experiment. Figure 1 shows the experimental environment. The surrounding was a non-reflective wall. There was seven yellowish bulbs hanging from the ceiling, and they can be adjusted the ambient illumination levels. The viewing distance was around two meters. Figure 2 illustrates the measurement of subjective perceived depth distance. There were a long ruler between the subject and the 3DTV, a mirror in front of the 3DTV, and an adjustable arrow on the top of the ruler. The device was used to find the proper location of the target for the subjects. If the subject perceived the object as in front of the 3DTV (i.e. concave distance), the subjects adjusted the head of the arrow to the exact location they saw. If the subject perceived the object as behind of the 3DTV (i.e. concave distance), they had to adjust the location of the arrow reflected in the mirror.

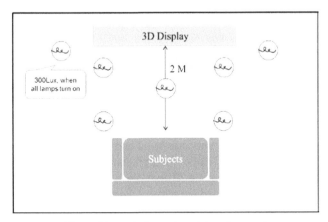

Figure 1 Layout of experimental environment

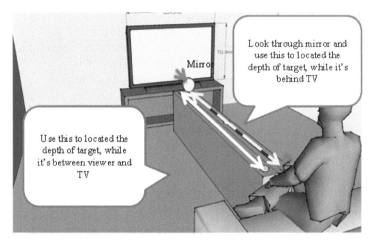

Figure 2 Measurement of perceived depth measurement

## 2.3 Stimuli

The task of the subjects was to finish two treatments, which were convex and concave 3D films taken by a 3D camera. The content included facial expression, gesture, people movement, flowers and trees in the garden, and traffic in the street scene. The subjects had to watch the four five-minute films in one treatment. The different levels of convergence distance of 3D films were randomly shown to eliminate the order effects.

## 2.4 Subjects

There were total of 16 subjects, including four engineers who worked in the field of 3D displays at the Industrial Technology Research Institute in Taiwan, and 12 graduate students from National Tsing Hua University in Taiwan who have never experienced 3D displays. All the subjects were paid for their participation.

## 2.5 Procedure

Only the subjects who were not color blind and possessed the correct or normal-to-correct visual acuity (equal to or higher than 0.8) and stereo acuity could participate in the formal experiment. At first, the subjects read the purpose and the procedure of the study, and signed the consent form. The subjects were assigned to different combinations of 3D films to eliminate the potential order effect. During one treatment, the subjects watched the four five-minute 3D films and measured the perceived depth distance in each film. They orally answered the subjective visual comfort evaluation questionnaire to the experimenter after 20 minutes when watching the films in one treatment. The subjects had to finish two treatments and there was a 10-minute interval for the subjects to take a rest.

# 3 RESULTS

## 3.1 Subjective Visual Comfort Evaluation

The data of visual comfort evaluation questionnaire was on an ordinal scale. The nature of the Likert's scale suggests applying nonparametric statistics for data analysis (Duffy and Chan, 2002).

The Wilcoxon Signed Ranks test showed that the main effect of convergence distance was not significant on dizziness ($Z = -0.083$, $p > 0.05$), naturalness ($Z = -0.171$, $p > 0.05$), depth ($Z = -0.021$, $p > 0.05$), viewer crosstalk ($Z = -0.502$, $p > 0.05$), slowed down focus change ($Z = -0.565$, $p > 0.05$), and harmonization ($Z = -0.478$, $p > 0.05$) of subjective visual comfort evaluation.

The Mann-Whitney U test revealed marginally significant effect of ambient illumination on subjective visual comfort evaluation in terms of harmonization ($Z = -1.890$, $p = 0.059 > 0.05$) , but not significant on dizziness ($Z = -0.145$, $p > 0.05$), naturalness ($Z = -0.535$, $p > 0.05$), depth ($Z = -1.740$, $p > 0.05$), viewer crosstalk ($Z = -0.963$, $p > 0.05$), and slowed down focus change ($Z = -1.022$, $p > 0.05$). The mean rank of harmonization in the high ambient illumination level, i.e. 300 Lux, was higher than those in the low ambient illumination, i.e. 0 Lux ($18.90 > 13.28$). It implied that the viewer felt visual comfort in the brighter environment.

## 2.1 Subjective Perceived Depth Distance

The raw data was split into two groups, which were convex distance and concave distance, and was analyzed by a repeated-measures ANOVA (Analysis of Variance) for mixed design. The result of convex distance group showed that the interaction effect between ambient illumination and convergence distance was not significant on subjective perceived depth distance ($F (3, 28) = 0.071$, $p > 0.05$), neither was the main effect of ambient illumination ($F (1, 28) = 1.029$, $p > 0.05$). However, the main effect of convergence distance attained statistical significance ($F (3, 28) = 33.483$, $p < 0.001$). The result of concave distance group did not yield a statistically significant interaction effect ambient illumination and convergence distance ($F (3, 28) = 1.085$, $p > 0.05$). The main effect of ambient illumination was marginally significant on perceived depth distance ($F (1, 28) = 3.844$, $p = 0.06 > 0.05$). Nevertheless, a significant main effect of convergence distance was found ($F (3, 28) = 12.025$, $p < 0.001$).

Furthermore, a significant correlation relationship between convergence distance and subjective perceived depth distance was attained ($r = 0.916$, $p < 0.001$), but it did not exist a significant correlation relationship between convergence distance and ambient illumination ($r = 0.022$, $p > 0.05$). The linear regression model was analyzed for two levels of convergence distance, convex distance and concave distance.

The statistical relation model between $X_i$ (convex distance of 60, 80, 100, and 120 cm) and $Y$ (perceived depth distance) could be expressed as Eq. (1), $r^2 = 0.600$.

$$Y = -0.013X_i^2 + 3.168X_i^2 - 72.333 \ (1)$$

The statistical relation model between $Xj$ (concave distance of 240, 260, 280, and 300 cm) and Y (perceived depth distance) could be expressed as Eq. (2), $r^2 = 0.183$.

$$Y = -0.00000874X_j^3 - 0.035X_j^2 + 1047.153 \ (2)$$

The result showed that Eq. (1) and (2) was significant (convex distance: $F = 45.761$, $p < 0.001$; concave: $F = 6.814$, $p < 0.05$). It implied that both two models were more reliably predictable.

# 4  CONCLUSIONS

Table 1 illustrates the effects of the two independent variables, where "V" means a significant variable, "○" means marginally significant, "X" means not significant, and "--" means without the interaction effect due to the data analyzed by a nonparametric method.

Table 1 The effects of convergence distance and ambient illumination on subjective comfort evaluation and perceived depth

| Independent variables | Dependent variables | |
|---|---|---|
| | Subjective comfort evaluation | Perceived depth |
| Convergence distance | X | V |
| Ambient illumination | ○ | X (for convex films), ○ (for concave films) |
| Convergence x Ambient | -- | X |

The main effect of ambient illumination was marginally significant on harmonization in terms of the subjective comfort evaluation. The subjects felt more visual comfort under the high level of ambient illumination (300 Lux). The result was similar to the previous study by Wang et al. (2012), who indicated that the viewer preferred the brighter ambient illumination of 300 lux when watching 3D films.

Convergence distance was significant on subjective perceived depth distance. However, the main effect of ambient illumination was slightly significant on perceived depth when watching concave 3D films. Two regression models each provided a predictor for the perceived depth distance when viewing the object with convex distance and concave distance. The two dissimilar models revealed that there might have other binocular and monocular cues that affect the perceived depth.

# ACKNOWLEDGEMENTS

This study was supported by National Science Council (NSC99-2221-E007-085-MY2), and the authors would like to thank the Electronics and Optoelectronics Research Laboratories of the Industrial Technology Research Institute (EOL/ITRI) for their technical support.

# REFERENCES

Chung, H. H. and S. Lu. 2003. Contrast-ratio analysis of sunlight-readable color LCDs for outdoor applications. *The Journal of the Society for Information Display* 11: 237-242.

Duffy, V. G. and A. H. S. Chan. 2002. Effects of virtual lighting on visual performance and eye fatigue. *Human Factors and Ergonomics in Manufacturing* 12: 193-209.

Hoffman, D. M., A. R. Girshick, K. Akeley, et al. 2008. Vergence-accommodation conflicts hinder visual performance and cause visual fatigue. *Journal of Vision* 8: 1–30.

Lin, C. C. and K. C. Huang. 2006. Effects of ambient illumination and screen luminance combination on character identification performance of desktop TFT-LCD monitors. *International Journal of Industrial Ergonomics* 36: 211-218.

Wang, P. C., S. L. Hwang, J. S. Chen, and K. Y. Chen. 2012 (In press). Luminance effects influencing perception of 3D TV imagery. *The Journal of the Society of Information Display*.

# CHAPTER 21

# Application of Integrated Score of Ergonomic Work Conditions (ErgQS) in Management of Ergonomic Hazards in Enterprise

*Wiesława M. Horst 1, Witold F. Horst 2, Maria K. Horst – Kończal 3*

Poznan University of Technology
Poznań, Poland
wieslawahorst@o2.pl

## ABSTRACT

The aim of the study was to define and use the quantitative index of ergonomic quality in management of WMSD symptoms and their causes resulting from occupational work. Questionnaires were distributed in four organizations in which were leading sampling survey: food production company, transportation company, supermarket, assembly company. Authors present an integrated index of ergonomic quality - ErgQS (Ergonomic Quality Score). This index may also be interpreted as the index of ergonomic quality of working conditions or else as an index of ergonomic quality of a work performance determined by these conditions. ErgQS was defined/ calculated based on the information about: WMSD symptoms, socio-economic consequences of WMSD symptoms and ergonomic risk factors.

The use of the index for assessment of work condition within one enterprise (in several departments) shows the experts the direction of particular interest, where correction activities should be undertaken in the technical, organizational and training aspect.

**Keywords**: ergonomic quality, ergonomic hazard, WMSD, musculoskeletal disorders

# 1    INTRODUCTION

The most frequently used method of defining the level of health protection at work is the assessment of compliance of existing work conditions with legal requirements, norms and ergonomic standards. The level of fulfillment of the above requirements is referred to ergonomic quality. This notion may pertain to individual technical objects, workstations and work processes (including the manner of performing work) as well as to the entire organization and its departments.

Ergonomic quality informs about the level of adjustment of objects and processes to the psychophysical (particularly somatic and physiological) abilities of workers and in practice it defines the level of health protection at work. This level directly influences the risk of occurrence of WMSD symptoms. WMSD generate losses both for the organization (lost revenues, passive presence at work, absenteeism, decrease of productivity and the quality of goods and services), for the entire society: medical treatment and rehabilitation costs, pensions, etc. and for the worker – loss of well-being and loss of incomes.

Various scientific approaches can be used to evaluate the ergonomic quality, depending on the purpose of the study. Information is acquired from subjective self-assessments made by workers/users (Horst W., Wachowiak F., 2010) or from assessments made by experts (i.a. Bruder R. 2006, Hassenzahl Mare, Axel Platz, Michael Burmester, 2000, Horst W. M., Dahlke G., 2008, NIOSH Publication No. 97-117, 1997). A result of these studies and analyses is the information about the number of people exposed to ergonomic risk factors and the sources of these risk factors, as well as, in case of subjective studies, the information about perceived of WMSD symptoms and their socio-economic consequences. Ergonomic quality is determined both by WMSD symptoms experienced by operators and by ergonomic risk factors (ERF) indicated by a worker or by an expert. Very often ERF and their health outcome - WMSD symptoms are analyzed separately. In the international literature however, there are more and more examples of presenting these two aspects combined, in the quantitative aspect (Venkatesh Balasubramanian, T.T. Narendran, V. Sai Praveen. 2011, prPN-N-8007:1988, Strasser H., 2011). Such approach is also the core of the ergonomic system managing WMSD symptoms and ERF.

In this paper is presented an attempt to combine knowledge of the ERF and their sources and health outcomes - the WMSD symptoms, in one integrated numerical index. Authors present an integrated index of ergonomic quality - ErgQS (Ergonomic Quality Score). This index may also be interpreted as the index of ergonomic quality of working conditions or else as an index of ergonomic quality of a work performance determined by these conditions. ErgQS was defined/ calculated based on the information about:

- WMSD symptoms,
- socio-economic consequences of WMSD symptoms and
- ERF.

This index allows acquiring quick preliminary assessment of ergonomics of working conditions at workstations in entire organization, indicating the areas of particular interest to the experts responsible for the management of health protection at work, and prevention of WMSD in particular. This article primarily indicates all possibilities offered by ErgQS in the area of WMSD prevention.

## 2     MATERIALS AND METHOD

The index was tested in years 2010-2011 (Horst - Kończal M.K, Horst W.F., Horst W.M, 2011). The data for calculating the index was acquired from the author's IDMS (identification of the musculoskeletal disorders and their causes, Horst W., 1976). Questionnaires were filling by the workers, used to identify of WMSD symptoms and their occupational conditions in organizations. The aim of the study was to define and use the quantitative index of ergonomic quality in management of WMSD symptoms and their causes resulting from occupational work.

Questionnaires were distributed in four organizations (the number of respondents is given in brackets) in which was leading sampling survey:
- food production company (FOOD), for all employees of both its divisions: VDT-workstation (20) and MH-manual handling (37),
- transportation company (TRANS), for all drivers of four types of trucks; Mercedes (8), Renault (27), Scania (13) and Volvo (12),
- supermarket (MARKET), for all cashiers in two stores of the same chain; market No-1 (9), market No-2 (9),
- assembly company (ASSEMBLY), for all employees of 4 departments; assembly department No-1 (29), assembly department No-2 (30), VDT-workstation (17), Store (8).

The ErgQS index was calculated based on the results of the survey done with a use of IDMS questionnaire in the above mentioned organizations, with a consideration of their specific conditions.

The IDMS questionnaire contains 9 WMSD symptoms: pain, trembling, tingling, numbness, loss of feeling, stiffness, burning, discomfort and coolness, experienced by workers in 9 segments of the musculoskeletal system: neck, shoulders/upper arms, upper back, elbows/lover arms, low back, wrists/hands, hips, knees/tights, ankles/feet.

Answers given in the questionnaire also inform about ERF including:
1. assuming awkward posture ($ERF_{AP}$) and
2. physical environmental conditions and organization of work and work breaks ($ERF_{ERFE\&O}$).

The questionnaire includes (e.g.) the following postures:
- head: leaning forward, sideways, backward; twisting, moving the get a better sight, resting on arms,
- torso: leaning forward, leaning sideways, leaning backward, twisting, sitting, without spine support,
- upper limbs: bent in elbows at an angle smaller than 90°; raised shoulders, raised elbows; twisted,
- forearms/wrists; bent/raised wrists,

  - lower limbs: sitting: on thighs; hips; without feet rest; on leg (foot); with crossed legs; with feet under the seat,
  - standing: supporting body weight on one leg; on toes; on both knees bent, on one knee bent,
  - kneeling or squatting.
  In regard to the physical environment conditions and organization of work and work breaks ($ERF_{ERFE\&O}$) the following were considered:
- possibility to change the body posture at one's will: from sitting to standing; from bent to straight and other,
- frequency of breaks: every hour, every two hours; less frequently,
- activity during breaks: lying on the back, on the side, smoking cigarettes; eating; watching TV, physical exercises,
- exposure to vibrations,
- coolness in: environment, objects or equipment, cool ventilation directed on the body,
- restraining of entire body movements,
- pressure on thighs (while sitting), forearms or other body segments.

**Table 1 Elements of IDMS questionnaire**

| Element | Number of questions/elements |
|---|---|
| 1. Symptoms of WMSD | 9 |
| 2. Body Segments | 9 |
| 3. Ergonomic Risk Factors postures | 23÷33 |
| 3.1. Manual Handling | 31 |
| 3.2. VDT workstation | 23 |
| 4. Ergonomic Risk Factors physical environment & work organization | 5+4 |

The questionnaire provides also a base for reasoning about the socio-economic results of symptoms of WMSD felt by workers, such as: sick absence, medical treatment, limiting activity at work and outside work, work resignation. A short statistic of elements considered in the IDMS questionnaire is presented in tab. 1.

# 3    RESULTS

To define an integrated ErgQS index, the following indexes were calculated separately for every group of surveyed workers/workstations (such as type of car, production department):

$TS_{WMSD}$ - a total index of the level of occurrence of all WMSD symptoms felt in selected segments of the musculoskeletal system (presented as a number of reported symptoms by all respondents compared to the possible number of occurrences).

$TS_{WMSDC}$ – a total index of reported cases of socio-economic results of musculoskeletal system disorders, (absence, treatment, presenteeism, etc.) caused by the occurrence of ERF at workstations, pertaining to all respondents.

$TS_{ERFAP}$ – a total index of identified ERF associated with an assumed awkward posture of entire body or its segments, reported by all respondents

$TS_{ERFE\&O}$ – a total index of identified ERF associated with a work organization and physical work environment, indicated by all respondents.

$TS_{all}$ – a total index of the number of ERF and their results defined as a sum of all the above defined indexes:

$$TS_{all} = TS_{WMSD} + TS_{WMSDC} + TS_{ERFE\&O} + TS_{ERFAP}$$

This index determines an aggregated level of occurrence of all the above defined ERF and their health related and socio-economic consequences in regard to the employees of the surveyed organization (department, location, workstation, etc.).

An integrated index of ergonomic quality (ErgQS) of workstations in entire organization (or its selected department) is defined as: ErgQS = 1 / $TS_{all}$.

The ErgQS index enables the quantitative assessment of ergonomic quality of all the workstations in an organization. This index has a specific numeric value and is defined on an absolute scale, which means that it also makes it possible to practically compare the level of implemented ergonomic solutions in different organizations. Higher values of the index indicate higher ergonomic quality.

## 3.1    A comparison of ergonomic quality of workstations and processes in the surveyed companies

The values of the score for four organization in which the survey was conducted are presented in tab. 2. It includes the indexes for all ERF in all the areas (col. 2, 3, 4, 5), as well as the total index of ERF and their consequences $TS_{all}$ (col. 6). The higher the level of the $TS_{all}$ index the higher the probability of WMSD occurrence. The last column (7) presents integrated values of ergonomic indexes ErgQS for every organization.

Table 2  Values of indexes or four surveyed organizations

| | Company | $TS_{WMSD}$ | $TS_{WMSDC}$ | $TS_{ERFE\&O}$ | $TS_{ERFAP}$ | $TS_{all}$ | ErgQS |
|---|---|---|---|---|---|---|---|
| | 1 | 2 | 3 | 4 | 5 | 6 | 7 |
| 1 | FOOD | 0,26 | 0,57 | 0,12 | 0,26 | 1,20 | 0,83 |
| 2 | MARKET | 0,40 | 0,70 | 0,20 | 0,65 | 1,94 | 0,51 |
| 3 | TRANSPORT | 0,36 | 0,52 | 0,39 | 0,48 | 1,74 | 0,57 |
| 4 | ASSEMBLY | 0,30 | 0,71 | 0,11 | 0,30 | 1,43 | 0,70 |

The highest value of ErgQS index was recorded in a highly automated food production company (FOOD - 0,83); lower levels of ergonomic quality were recorded for: assembly company (ASSEMBLY - 0,70), in which the manual assembly was dominating, transportation company (TRANSPORT - 0,57), and the lowest index was reached by a supermarket (MARKET - 0,51).

Primary factor determining the level of ergonomics quality (ErgQS) was the value of $TS_{MSDsC}$ index (tab. 2). This index reflects the socio-economic results, by already experienced ailments of the musculoskeletal system and their immediate results (necessity of medical treatment, sick absence, limited activity at work and outside work, changing employment). The highest value of a hazards level in this area was found in a assembly company (ASSEMBLY - 0,71) and the lowest in a transportation company (TRANPORT - 0,52).

The value of $TS_{MSDC}$ (0,71) in an assembly company was determined by ailments in the lumbar spine felt during last year and a week, which caused the necessity of medical treatment and sick absence.

The value of $TS_{MSDC}$ (0,70) index in a supermarket was mainly influenced by the ailments felt in the lumbar spine during last year and a week, and resulting limitation of activity at work and outside work.

The WMSD symptoms, as well as ERF (most of all $TS_{ERFAP}$) had relatively the highest index in supermarket (0,40 and 0,65) and in the transportation company (0,36 and 0,48), which contributed to their low ergonomic quality. Presented below is the analysis of the above mentioned indexes in the surveyed organization. The results of the study may serve as a recommendation for action for the employer.

## 3.2 Ergonomic quality of workstations and work processes in the food production company

Following the employer's decision, in the food production company, the ErgQs index was used for workstations and processes of the administration department, performing their occupational duties with a use of VDT for at least a half of the working time and for the workstations and processes in the production department. Conducted surveys (tab. 3) showed that the office work in this company entails a greater risk of WMSD symptoms occurrence than the work in production or in storage. ErgQS index in this department reached only 0,64 (tab. 3, col. 7).

Table 3   Value of indexes of ERF and their consequences for the food production company

| No | Company code | $TS_{WMSD}$ | $TS_{WMSDC}$ | $TS_{ERFE\&O}$ | $TS_{ERFAP}$ | $TS_{all}$ | ErgQS |
|----|--------------|-------------|--------------|----------------|--------------|------------|-------|
|    | 1 | 2 | 3 | 4 | 5 | 6 | 7 |
| 1 | FOOD-all | 0,26 | 0,57 | 0,12 | 0,26 | 1,20 | 0,83 |
| 2 | FOOD-VDT | 0,46 | 0,81 | 0,13 | 0,15 | 1,55 | 0,64 |
| 3 | FOOD-MH | 0,14 | 0,43 | 0,12 | 0,32 | 1,01 | 0,99 |

Table 4 Value of indexes presenting WMSD symptoms ($S_{WMSD}$) in selected segments of the musculoskeletal system

| No | Segment of musculoskeletal system | FOOD-VDT | FOOD-MH |
|---|---|---|---|
|  | 1 | 2 | 3 |
| 1 | Shoulders/arms | 0,10 | 0,04 |
| 2 | Thoracic spine | 0,09 | 0,05 |
| 3 | Lumbar spine | 0,25 | 0,14 |
| 4 | Knees/lower legs | 0,12 | 0,08 |
| $TS_{MSDC}$ |  | 0,81 | 0,43 |

Dominating influence on the value of the index had the level of already felt WMSD symptoms ($TS_{WMSDs} = 0,46$; tab. 3 col. 2; 3,5 times higher than for the MH – 0,14; tab. 3, col. 2) and their socio-economic consequences ($TS_{WMSDsC} = 0,81$; column 2, 2 times higher than for the MH – 0,43; tab. 3, col. 3), which resulted mainly from the ailments felt in the lumbar spine (0,25) and in knees and lower legs (0,12) - tab. 3. These consequences included sick absence, medical treatment, work changing and limiting activity outside work.

## 3.3 Ergonomic quality of workstation and work processes in the assembly company

Survey results show, that the highest chances of WMSD occurrence are in office work in this company – the ergonomic quality index in this department was relatively low (the lowest among all surveyed departments) and reached only 0,57; (tab. 5, col.7).

Table 5 Value of indexes of ERF and their consequences for the assembly company

| N o | Company code | $TS_{WMS}$ | $TS_{WMSD}$ | $TS_{ERFE\&}$ | $TS_{ERFA}$ | $TS_{al}$ | ErgQ |
|---|---|---|---|---|---|---|---|
|  | 1 | D | C | O | P | I | S |
|  |  | 2 | 3 | 4 | 5 | 6 | 7 |
| 1 | ASSEMBLY-all | 0,30 | 0,71 | 0,11 | 0,30 | 1,43 | 0,70 |
| 2 | ASSEMBLY-1 | 0,23 | 0,59 | 0,07 | 0,32 | 1,21 | 0,82 |
| 3 | ASSEMBLY-2 | 0,28 | 0,93 | 0,11 | 0,26 | 1,57 | 0,64 |
| 4 | ASSEMBLY-VDT | 0,54 | 0,80 | 0,19 | 0,23 | 1,76 | 0,57 |
| 5 | ASSEMBLY-store | 0,14 | 0,18 | 0,08 | 0,55 | 0,95 | 1,05 |

It was determined mainly by the high level of socio-economic consequences ($TS_{WMSDC} = 0,80$; tab. 5, col. 3) and the highest level of felt WMSD symptoms in all departments (TS$_{WMSDs}$ = 0,54; tab. 5 col. 2), which resulted from ailments felt in the cervical spine (0,12; tab. 6, col. 4) and lumbar spine (0,11; tab. 6, col. 4).

The factor which determines relatively high value of ErgQS index in assembly-1 department (0,82; tab. 5, col. 7) were also the socio-economic consequences of felt symptoms (0,59; tab. 5, col. 3).

Table 6  Value of indexes of WMSD symptoms in selected segments of the musculoskeletal system

| No | Segment 1 | ASSEMBLY-1 2 | ASSEMBLY-2 3 | ASSEMBLY-vdt 4 | ASSEMBLYstore 5 |
|---|---|---|---|---|---|
| 1 | Cervical spine | 0,03 | 0,05 | 0,12 | 0 |
| 2 | Lumbar spine | 0,04 | 0,03 | 0,11 | 0,03 |
| 3 | Hands /wrists | 0,05 | 0,07 | 0,07 | 0,02 |
| 4 | Ankles /feet | 0,02 | 0,02 | 0,08 | 0,05 |
| TS$_{MSD}$ | | 0,23 | 0,28 | 0,54 | 0,14 |

These consequences included: termination of employment, medical treatment, changing employment, limitation of activity outside work, caused mainly by ailments in hands and wrists (0,05; tab.5, col. 2).

## 3.4    Ergonomic    quality    of    workstations    and    work processes in the transportation company

Study results show, that the highest chances of WMSD occurrence in a transportation company are for Renault drivers, for whom the ergonomic quality index had the lowest value of 0,51 (tab.7, col. 7). Besides socio-economic index (TS$_{WMSDC}$ = 0,59) and ERF related to awkward posture (TS$_{ERFAP}$ = 0,55; tab. 7, col. 5) it was influenced by the highest level of hazards related to work organization and physical work environment of all departments (TS$_{ERFE\&O}$ = 0,44; tab. 7, col. 4), which resulted from high values (tab. 8).

Table 7  Value of indexes of ERF and their consequences for the transportation company

| No | Company code 1 | TS$_{WMSD}$ 2 | TS$_{WMSDC}$ 3 | TS$_{ERFE\&O}$ 4 | TS$_{ERFAP}$ 5 | TS$_{all}$ 6 | ErgQS 7 |
|---|---|---|---|---|---|---|---|
| 1 | TRANSPORT-all | 0,36 | 0,52 | 0,39 | 0,48 | 1,74 | 0,57 |
| 2 | Mercedes | 0,40 | 0,54 | 0,38 | 0,40 | 1,72 | 0,58 |

| 3 | Renault | 0,37 | 0,59 | 0,44 | 0,55 | 1,96 | 0,51 |
| 4 | Scania | 0,33 | 0,49 | 0,29 | 0,37 | 1,48 | 0,67 |
| 5 | Volvo | 0,31 | 0,36 | 0,37 | 0,52 | 1,55 | 0,64 |

A factor which influenced a relatively high value of ErgQS index for Scania vehicle (0,67) were the lowest of calculated values of $TS_{WMSDsC}$ (0,49), $TS_{ERFAP}$ (0,37) and $TS_{ERFE\&O}$ (0,29).

**Table 8 Value of indexes $TS_{E\&O}$ for the transportation company**

| No | Factor 1 | Mercedes 2 | Renault 3 | Scania 4 | Volvo 5 |
|----|----------|------------|-----------|----------|---------|
| 1 | Vibrations | 0,18 | 0,13 | 0,14 | 0,17 |
| 2 | Coolness | 0,05 | 0,10 | 0,03 | 0,03 |
| 3 | Restrain | 0,03 | 0,04 | 0,00 | 0,02 |
| 4 | Pressure on tights | 0,08 | 0,09 | 0,05 | 0,08 |
| 5 | Pressure on lower arms | 0,05 | 0,08 | 0,08 | 0,07 |
| $TS_{E\&O}$ | | 0,38 | 0,44 | 0,29 | 0,37 |

Low value of risk factors related to work organization and physical work environment was mainly a result of low values of indexes: coolness, restrain and pressure on tights (tab. 8).

## 4    DISCUSSION

In the analysis of the survey results, the authors (Horst Kończal M.K, Horst W.F., Horst W.M., 2011)] used their own Ergonomic Quality Score – ErgQS, which takes into account the reported in the survey:
- WMSD symptoms
- socio-economic consequences
- ergonomic risk factors ($ERF_{ERFE\&O}$): organization & environment
- ergonomic risk factors ($ERF_{AP}$): awkward postures.

Achieved results may serve as a base for comparison of ergonomic quality of specific products (such as vehicle brands - with regard to ERF for drivers).

The use of the index for assessment of work condition within one enterprise (in several departments) shows the experts the direction of particular interest, where correction activities should be undertaken in the technical, organizational and training aspect.

The article presents the indexes regarding organizations or their departments. Achieved results will indicate which departments should be taken care of in the first place. At this point it is possible to use the indexes regarding particular workstations, and their values will indicate with workstations require particular interest of ergonomists and/or occupation health departments.

Repeating the survey after a chosen period of time may serve as a base for evaluation of correctness and effectiveness of actions undertaken by an employer, including those suggested by the designers, whose responsibility was to redesign the workstations indicated in the previous survey.

Achieved ErgQS results indicate the areas of particular interest for the department of occupational safety and health at work.

Particularly important for the occupational health departments (also physiotherapy) are the total values of $TS_{WMDSs}$ and $TS_{WMSDC}$; the values of $TS_{ERFE\&O}$ $TS_{ERFAP}$ are useful for ergonomists. They will indicate the departments, which should be taken care of in the first place.

# REFERENCES

Bruder R. Ergonomic quality of Eppendorf piston –stroke pipette. Expertize Eppendorf reference. 2006 www.eppendorf.de/script/binres.php?RID.

Hassenzahl Mare, Axel Platz, Michael Burmester, Katrin Lehner, Hedonic and Ergonomic quality aspects determine a software's appeal. · Proceeding Chi '00 Proceedings Of The Sigchi Conference On Human Factors In Computing Systems Acm New York, NY, USA ©2000.

Horst W., Ankieta IDMS (identyfikacji dolegliwości mięśniowo-szkieletowych i ich przyczyn. UAM Poznań). Poznań 1976.

Horst W. M., Dahlke G., Kryteria oceny bezpieczeństwa i ergonomiczności pojazdu : wybrane zagadnienia. W: Bezpieczeństwo na drogach : edukacja i diagnostyka kierujących pojazdami : Monografia / W. M. Horst, G. Dahlke [red.]. Wyd. PP, Poznań 2008. - S. 91-106.

Horst W., Wachowiak F., Ergonomic requirements for city buses cabins - drivers' subjective feelings and comfort. Case study. In: Driver occupational safety: shaping well-being of drivers and passengers / ed. by W. Horst, G. Dahlke. Poznan: Poznan University of Technology, 2010. - S. 7-22.

Horst Kończal M.K, Horst W.F., Horst W.M. Diagnosis of the ergonomic quality of the truck driver workplace. w: Work-Related Musculoskeletal Disorders. Participatory Ergonomics. Wyd. PP. Poznań 2011. pp. 45 – 65 [Accepted for publication in Wyd. PP].

Jałosiński A., Horst W.M., Ergonomiczne kryteria oceny i zasady projektowania siedzisk do pracy na stanowisku VDT. W: Prewencja ergonomiczna schorzeń uwarunkowanych pracą. Wyd. PP, Poznań 2009. - s. 37-62.

NIOSH Elements of Ergonomics Programs A Primer Based on Workplace Evaluations of Musculoskeletal Disorders. NIOSH Publication No. 97-117, 1997.

prPN-N-8007:1988 Atestacja ergonomiczna maszyn i urządzeń. Podstawy metodyczne.

Strasser H., (Author, Editor) Assessment of the Ergonomic Quality of Hand-Held Tools and Computer Input Devices: Volume 1 Ergonomics, Human Factors and Safety. 2011.

Venkatesh Balasubramanian, T.T. Narendran, V. Sai Praveen. RBG risk scale: an integrated tool for ergonomic risk assessments. International Journal of Industrial and Systems Engineering 2011 - Vol. 8, No.1 pp. 104 - 116.

# CHAPTER 22

# Influence of Application Safety and Health Management Systems on Manufacturing Process Formation. A Case Study

*Piotr Kończal 1, Maria K. Horst-Kończal 2, Wiesława M. Horst 3*

Poznan University of Technology
Poznan, Poland
piotr.konczal@gmail.com

## ABSTRACT

The authors present an assessment of Health and Safety at work Management System - OHSAS 18001 done with use of ISRS method. The aim of the study was to show the influence of implementing the Management System on the manufacturing process in a company that produces goods from polyethylene using rotational molding process.

Research was done in the area of Health and Safety at work management before and after the implementation of the certified health and safety at work management system according to OHSAS 18001. The analysis was conducted in 2010-2011. For the assessment of health and safety authors used ISRS method on $5^{th}$ level. The evaluation was performed based on 324 questions. ISRS method is used to measure the effectiveness and quality of the management systems in enterprises in way of reducing losses caused by accidents, explosions, fires and other danger occurrences. The results of H&S management system audits conducted by ISRS gives measurable, spot result.

There was done an analysis of retrospective and current status of Health and Safety at work management system by ISRS. Prior to the implementation of the management system there was identified many lacks in the health and safety at work way of managing that could influence directly on the work process.

Implementation of the management system has enabled work process improvement through organizational and engineering changes, and by raising awareness of employees at the level of production workers and management.

Following the implementation of management system according to OHSAS 18001 the improvement of the health and safety in the following area was observed:

- Management and Administration – 38%,
- Planned H&S audits and maintenance – 64.8%,
- Accidents investigations – 31.7%,
- H&S rules and work permit – 55.6%,
- Analysis of accidents – 12.5%,
- Personal protective equipment – 0%,
- Health and safety at work – 15.9%,
- Promotion of safety issues – 33.4%.

The analysis of health and safety at work conditions in chosen organization identified the improvement of health and safety management after implementation of H&S management system. The expected result of the implementation of this management system is to reduce the occupational risks associated with work processes in the organization. Risk reduction should take place through the standardization of reporting by employees of the potentially accident events - nearmissess.

**Keywords:** health and safety at work, health and safety management system, OHSAS, ISRS

# 1 BRIEF DESCRIPTION OF CHOSEN COMAPNY

The object of this study was to company X as part of an international corporation engaged in the production of products for the construction industry. The company produces rotomoulded tanks from polyethylene. The products are designed for storage of various chemicals, oils and fuel as well as rainwater and drinking water. There are three management systems in company X. The first, quality management system (ISO 9001) is working since the beginning of business on the Polish market. In 2010 the company introduced two other management systems:

- ISO 14001 – Environmental Management System,
- OHSAS 18001 – Health and Safety Management System,

that have been certified in March 2011.

# 2 OHSAS 18001

The standard OHSAS 18001 (Occupational Health and Safety Assessment Series) is a guideline for the construction and implementation of the health and safety at work management system. The standard was established in response to the demand for such a system from the market. OHSAS Standard has been developed in such a way that was fully compatible with ISO 9001 and ISO 14001 (BS OHSAS 18001:2007).

## 3   ISRS

In 1978, in the Loss Control Institute in the USA International Safety Rating System ISRS was made. Basis of the system created Frank Bird (Top W. N. 1991,1993).

During the audit authors used a program called EVIS-ISRS developed by DNV (Det Norske Veritas) and Polish Labor Inspection (PIP). The idea of this program is to measure the effectiveness and quality of functioning in the company management systems, mostly to reduce losses caused by accidents, breakdowns, etc. Assessment of the management system carried out by ISRS gives specific measurable result, unlike the typical audit which more or less provides a confirmation of the requirements. ISRS covers the requirements of standards such as BS 8800, OHSAS 18001 and PN-N 18001. Moreover, it can be used as a manual of good practices in OHS. In ISRS entire business management system is divided into 20 key sub-areas, which were identified several strategic issues. For those issues have been developed specific question, scored on an appropriate scale (EVISA, 2004).

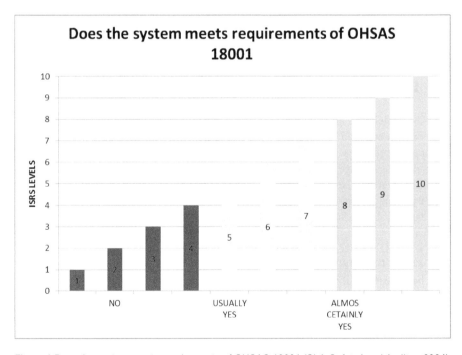

Figure 1 Does the system meets requirements of OHSAS 18001 (SLA-Safety Lead Auditor, 2004)

## 4    EFFECTIVENESS STUDY

Study the effectiveness of Management System in the company X was made with ISRS OES (Optional Element Score) method at $5^{th}$ level. The choice of such a method and level of assessment influenced by the fact that the company's Environmental and H&S System are young and not fully developed. Assessment was conducted on implemented management system according to standards ISO 14001 and OHSAS 18001 and also the H&S conditions before the implementation of the system - a retrospective study.
The audit was conducted in six areas required by the ISRS:

- leadership and administration,
- planned inspections and maintenance,
- accident/incident analysis,
- rules and work permits,
- personal protective equipment,

and two additional selected by the author:

- health and hygiene control,
- general promotion.

To achieve the $5^{th}$ level of ISRS in each of the evaluated areas company had to obtain at least 20% correct answers, with a minimum cumulative assessment of 40% and a minimum of Physical Condition Score at 65 to 100.

## 5    THE EFFECTS OF IMPLEMENTATION OF HEALTH AND SAFETY AT WORK MANAGEMENT SYSTEM

Different companies have different reason to implement H&S management system. In Europe mainly these are:

- Improvement of the image and competitiveness in the market,
- upgrading of health and safety,
- reducing sickness absenteeism and associated costs,
- corporate requirements.

In chosen entrepreneurship the decision about implementing the H&S system was made because the board wanted to formalized the issues of H&S and also it was a corporate requirement. Introduction of selected management systems led to huge improvement of H&S conditions what reflects in audit results. Audit was conducted in 8 areas, 6 obligatory and 2 selected by authors.

Table 1 Audit results before and after implementation of management systems

| Area | | Score before implementation of systems | | Score after implementation of systems | | Max score in area |
|---|---|---|---|---|---|---|
| | | Points | % | Points | % | Points |
| 1 | leadership and administration | 305,5 | 23,9 | 802,8 | 61,3 | 1310 |
| 3 | planned inspections and maintenance | 161 | 23,3 | 608,3 | 88,2 | 690 |
| 5 | accident/incident analysis | 164 | 27,1 | 356 | 58,8 | 605 |
| 8 | rules and work permits | 145 | 23,6 | 486,8 | 79,2 | 615 |
| 9 | accident/incident analysis | 151 | 27,5 | 220 | 40,0 | 550 |
| 11 | personal protective equipment | 198,3 | 52,2 | 197,1 | 78,2 | 380 |
| 12 | health and hygiene control | 276 | 39,4 | 386 | 55,8 | 692 |
| 17 | general promotion | 64 | 16,8 | 191 | 34,5 | 380 |

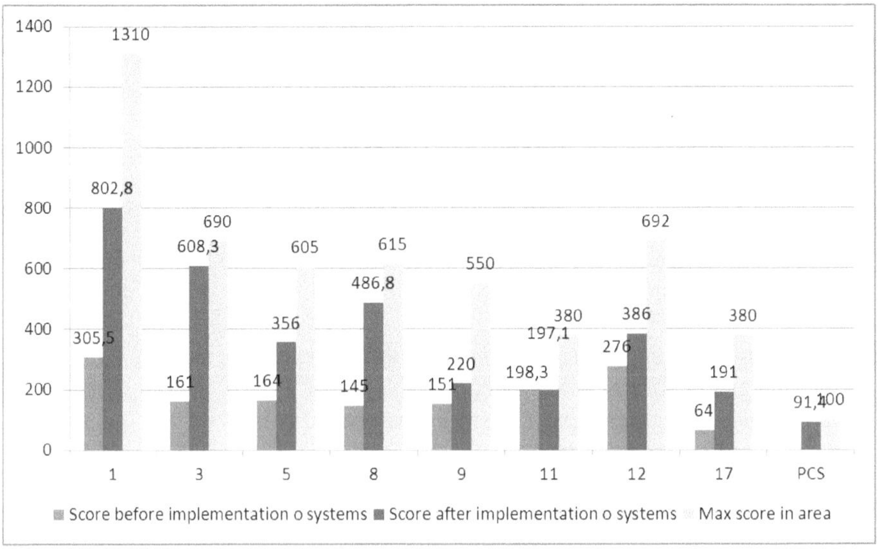

Figure 2 Audit score before and after implementation of management system in selected areas. (developed by authors)

Table 1 and figure 2 presents comparison of results of audits before and after implementing ISO 14001 and OHSAS 18001. The biggest growth was recorded in:

- leadership and administration (1),
- planned inspections and maintenance (3),
- accident/incident analysis (5),
- rules and work permits (8),
- general promotion (17).

The highest rank before implementation of selected systems company achieved in area 11. personal protective equipment – 52,2%. After it the best rank was in area 3 planned inspections and maintenance – 88,2%.

Before implementation of the system company achieved 1465 points from 5222 which is 28,2%. That means that the way of managing the H&S was not complying with requirements of the 18001 standard.

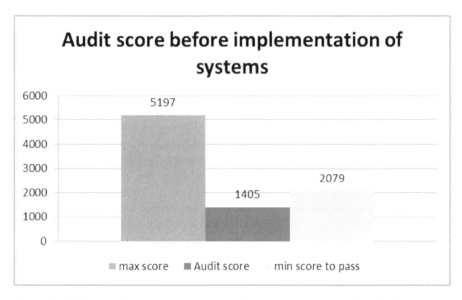

Figure 3 Audit score before implementation of management system for whole company. (developed by authors)

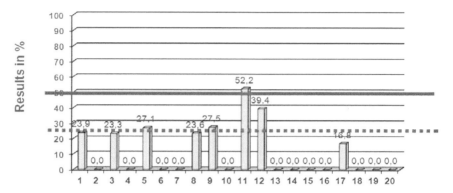

Figure 4 Results of audit in selected areas before implementation of systems (developed by authors)

In most of inspected areas before implementing OHSAS 18001 and ISO 14001 results were on lowest minimum to pass the audit in selected area (Figure 4). In Area 17. general promotion company achieved result 64 points (16,8%) which means that the company didn't pass audit in it. The dotted line on chart shows minimum level of points need to pass in an area. The red continuous line is minimum to pass a whole audit.

After the OHSAS 18001 and ISO 14001 was implemented company gained 3288 points which is 63% of total (figure 5). This allows to make a statement that the implemented system generally meets the requirements of ISO 14001 and OHSAS 18001. On the chart below there are results from all areas selected to audit. The highest rank was achieved in area 3 planned inspections and maintenance – 88,2%. Lowest ranks were in:

- general promotion – 34,5%,
- accident/incident analysis – 40%.

In all areas company exceeded the minimum for passing the audit.

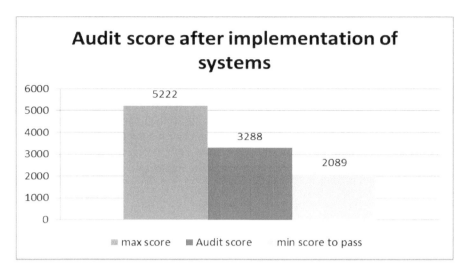

Figure 5 Audit score after implementation of management system for whole company. (developed by authors)

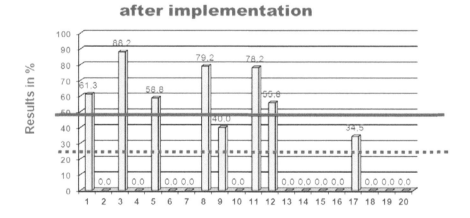

Figure 6 Results of audit in selected areas after implementation of systems (developed by authors)

When implementing the system in a procedural way company regulated environmental and H&S management. There was a formalization of hazard identification and risk assessment, hazardous waste management etc. Following the new risk assessment procedure, there was identified more risks than before. Many of them have been resolved by QuickFix or during 3 months of their identification. The Risk Assessment was reduced of one point. During the implementation of the system it was also identified training needs due to the specific of the work.

Figure 7 shows the accidents statistics in the first half of 2010 (before implementation of the ISO 14001 and OHSAS 18001) and 2011 (after the implementation of systems). As the graph shows the number of accidents decreased slightly from 5 to 4. Number of days lost due to accidents increases. However, in 2011, employment has increased significantly in chosen company from about 140 to 200 people (full-time employees). Therefore, the accident frequency rate is calculated as the ratio of accidents number to the number of employees, relative to 1000 workers declined from 34 in 2010 to 20 in 2011 (Figure 8) (Kończal P., 2011).

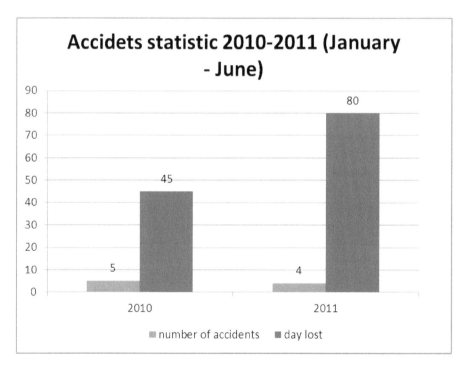

Figure 7 Accidents statistic January – June 2010-2011 in selected company (developed by authors)

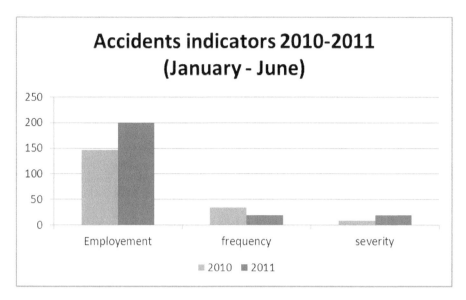

Figure 7 Accidents indicators January – June 2010-2011 in selected company (developed by authors)

The increase in severity of accidents after the implementation of the system seems to be disturbing, but previous studies in companies, especially big ones, showed that any change: implementation of new management methods, new machines, production lines, etc., in the short term, cause an increase in accidents (Pacholski L, 2000).

## 6 RESULTS

ISRS results that are achieved by company X are prove that its H&S management system (OHSAS 18001) is working properly. As it is said on (RYSUNEK 1) 5th level of ISRS means that system is mainly dealing with OHSAS 18001 demands. According to the fact that analyzed system is young there are few areas of management needs to be improve at first. These are:
- accident/incident analysis,
- accident/incident analysis,
- health and hygiene control,
- general promotion.

The key to success is using the Deming Cycle and systems approach to Environmental management and H&S at work.
The weakest areas were:
- general promotion – 34,5% (131 from 380 points),
- accident/incident analysis – 40% (220 from 550 points).

# REFERENCES

BS OHSAS 18001:2007 – Systemy zarządzania bezpieczeństwem i higieną pracy – wymagania.

EVISA Evaluation and Improvement of Safety – Program oceny i doskonalenia systemu zarządzania bezpieczeństwem i higieną pracy, DNV, Wydanie 3,1, Warszawa 2004.

Kończal P., Ocena efektywności Systemu Zarządzania Bezpieczeństwem i Higieną w Pracy w przedsiębiorstwie produkcyjnym na podstawie ISRS, Poznan 2011.

Pacholski L., red., Analiza wpływu polskiej transformacji gospodarczej na stan warunków pracy w przedsiębiorstwie produkcyjnym, Politechnika Poznańska, Poznań 2000.

SLA-Safety Lead Auditor, Kurs Auditora Wiodącego Systemu Zarządzania Bezpieczeństwem i Higieną Pracy wg: EVISA/ISRS oraz OHSAS 18001, Materiały szkoleniowe, Bukowina Tatrzańska, 15-19 listopada 2004.

Top W. N., International Safety Rating (part 1-4), International Loss Control Institute.

Top W. N., Safety & Loss Control Management and the International Safety Rating System, 1991.

Top W. N., Safety In Europe: Past - Present – Future, Insights Into Management National Safety Management Society, 1993.

Service for Managing Risk, Accessed February 2, 2012 www.dnv.pl.

Topves – Success in risk management, safety and loss control, Accessed February 2, 2012, www.topves.nl.

CHAPTER 23

# Basic Technical – Organizational Criteria Forming Labour Safety in the Medium Size Production – Services Companies

*Aleksandra Jasiak, Anna Berkowska*

Poznan University of Technology
Department of Ergonomics and Quality Engineering
aleksandra.jasiak@put.poznan.pl, bhpberkowska@gmail.com

## ABSTRACT

Purpose of this article is to verify criteria that form labour safety in the medium size production – services companies.

Content of this article consists of three parts: review of the literature, analysis of technical – organizational criteria and creating a concept of labour safety management model in medium size production – services companies.

Subject of the first part is the review of professional literature covering Polish Standards concerning labour safety, basic regulations concerning labour safety and occupational health binding in Poland as well as other laws and regulations which contain information governing rules of safety and health of labour in the company. Items of literature have been also analysed from the point of view of labour safety management.

Second part is the research part in which result of the studies have been presented concerning technical – organizational criteria forming labour safety in selected production – services companies. Presented results aim to compare criteria applied in selected companies.

Third part presents labour safety management model based on results obtained from the research. An algorithm of the proceedings has been formulated which essence is the improvement of the labour safety management system in production – services companies.

**Keywords**: labour safety, company, technical – organizational criteria

# 1    INTRODUCTION

Suitable working conditions improve productivity and reduce number of days of inability to work caused by poor working conditions as well as accidents at work. Improvement of working conditions influences activity results both of the companies  as well as the whole economy.

In Poland education in safety and related actions aiming on improvement of the labour safety quality in companies find larger and larger group of followers. More and more people and institutions recognize the benefits that may be achieved by rational approach to the problems of safety. But in spite of this positive phenomenon still too often present in our economy is the method of safety management directed more on the elimination of consequences than on formation of safe working conditions. After all in many companies management, employers and employees do not treat labour safety problems in proper way and do not spend adequate time and attention for this subject.

Purpose of this paper is the verification of criteria that shape labour safety in medium size production – service companies as well as creation of the model of labour safety management using Polish Standard PN-N-18001 because modern labour safety management is the activity oriented at  limitation of risk of an accident occurrence or loss of health or life and then keeping it at acceptable or even lower level. Aiming at quick change of safety management effectiveness a number of different activities have been taken and are still taken in our country (Poland) directed on creation and appropriate use of organizational tools. Safe working conditions and safe behaviour at work do not arise spontaneously. They have to be shaped first according to preconceived model. Therefore within the limits of integration with European Union Poland have to adapt themselves to European standards. Acting according to above principle basic directions of development in the field of technical – organizational progress and labour safety should be accepted. Management of labour safety should take place on all managerial levels.

# 2    THEORETICAL BASIS

Medium size companies have key importance for Polish economic development. They are indispensable for balanced functioning of the economy and contribute to the acceleration of economic growth. Main characteristics of this sector are as follows:
— ability to very quick reaction for market needs,
— openness for technical and organizational progress,
— potential ability for creating of new working places,

— low cost of working place,
— easy adaptation to the place, time and resources,
— appearance in all sectors of the economy.

Medium size companies constitute integral part of internal strengths of the region determining its development. Companies of this sector are mainly of local and regional nature. Local market is their basic source of supply of labour resources, materials etc. as well as basic sales market for the majority of the companies. Speaking about medium size companies generally taken into account are companies that:

— have low capital (assets) and employ small number of employees;
— have low market share and often their owners are managing person;
— are independent from other economic units (Dach Z. 2001).

Management of medium size companies is the process of obtaining and using of several different essential resources for supporting the goals of the organization (Bittel L. R., 1994). Basic factors influencing directly the extend of losses related to the human resources are as follows:

— demand for labour,
— labour productivity in poor conditions,
— minimum wage,
— awareness of the employees in the sphere of risks occurring in the workplace.

Based on the above mentioned factors it is possible to indicate many mechanisms that will lead to the improvement of safety and protection of health of the employees. It is however necessary that monitoring of the parameters is done on different levels. Management definitions as well as economic goals of the companies lead in the direction of improvement of working conditions with the task of minimizing of company's loses connected to poor labour safety. However most of those loses will be determined by external factors which include among others education on labour safety and health so that the changing needs of the society are satisfied. Labour safety management is a set of activities taken by persons belonging to the management staff with the purpose to reduce the risk of human life or heath loss to the level accepted from the point of view of law in force, economy and ethics (Pietrzak L., 2001). Effective execution of these and other resulting from the law requirements may facilitate introduction of labour safety management system which model as well as elements are described in Polish standard PN-N-18001:1999 "Systems of management of labour safety and health". Presented in this standard system of labour safety and health management consists of:

— engagement of the management as well as labour safety and health policy,
— planning,
— implementation and functioning,
— monitoring, auditing as well as corrective actions,
— review executed by the management and continuous improvement of the labour safety and health system management.

Implementation of the labour safety and health system management in the company using PN-N-18001 standard may not only facilitate fulfilment of the requirements of existing law but also enable achievement of tangible benefits connected to the improvement of labour safety and health level.

# 3    ANALYSIS OF TECHNICAL – ORGANIZATIONAL CRITERIA

To identify and analyse technical – organizational criteria research have been carried out on the group of twenty medium size production – service companies. Size of employment of surveyed companies is presented on fig. 1.

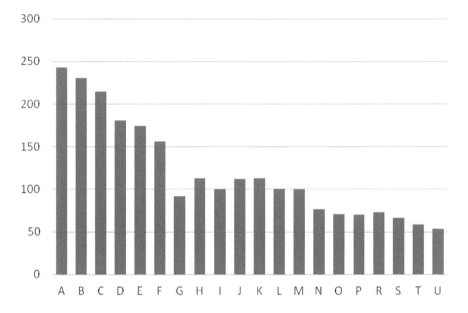

Figure 1. Employment size in surveyed companies (own work)

First survey in above companies has been carried out concerning technical criteria which include facilities, work spaces. From the analyse of studies that have been carried out it comes out that for twenty companies only one does not fulfil this criteria in 100% because facility used is in not proper technical condition. The reason of existing situation is the lack of possibility of interference in the facility as well as its surroundings because it is not the property of the company and the final location is under construction.

Next technical criteria taken into consideration is equipment, outfitting of above workplaces. Study of this criteria have been particularly focused on machinery that according to existing regulations of labour safety should be marked CE or be adapted to minimal requirements for the machines covered by Regulation of the Minister of Economy dated 30th October 2002 concerning minimal requirements concerning labour safety and health in the scope of machines use by employees during their work (Official Gazette no. 191, pos. 1596, changed y Official Gazette from 2003, no. 178, pos. 1745).

According to executed studies machinery in analysed companies in majority (39%) is new or adapted to minimal requirements (54%) that should be met, only small percentage (7%) of the machines is in the process of adaptation to existing regulations.

Whereas taking into account organizational criteria, after initial analyse few most important have been presented below, those which have particular influence on labour safety management, Table 1.

Table 1. Selected organizational criteria (own work)

| Question concerning problem studied | Percentage of answers | |
|---|---|---|
| | YES | NO |
| Whether the organizational structure is developed in the company? | 60% | 40% |
| Whether improvement of qualifications by the employees include their involvement in the activity for labour safety and health | 35% | 65% |
| Whether employees have required additional qualification certificates? | 94% | 6% |
| Whether in the companies there is an attitude towards continuous improvement of activities for labour safety and health? | 75% | 25% |
| Whether employees participate in trainings Concerning labour safety? | 100% | 0% |
| Whether procedures concerning labour safety management have been introduced in the company? | 45% | 55% |
| Whether procedures concerning avoiding accidents and occupational diseases have been introduced? | 75% | 25% |
| Whether employer secured for the employees proper working and protecting clothing as well as personal protecting equipment on required working place? | 90% | 10% |
| Whether relations between employees are recognized as positive? | 70% | 30% |
| Whether employer fulfils requirements covered in the legal regulations as well as other requirements concerning companies? | 65% | 35% |

Analysis of organizational criteria have shown us general picture of labour safety management. In nine companies that study has been carried out the system of labour safety management PN-N-18001 is already introduced and it organized many problems concerning management and the aspiration to the improvement of labour safety and health condition. But there are also companies that have no model of labour safety management introduced and only some of the requirements are fulfilled fully. In six companies the organizational structure is introduced which is intended to facilitate process of communication in the company as well as to faster execute the tasks. Employees in the majority do not have the education adequate to the job position held but mostly it concerns workers posts. To retrain the employee or to adjust his qualifications to the working place employers provide trainings, courses to gain adequate qualifications. Next criteria taken under consideration was the professional risk which has been evaluated in all companies under research however in many cases this documentation was not transferred to the employees and nobody informed them about the threats occurring on the job positions held. From the above analysis it appears also that not always the employer provides adequate working and protective clothes. However the biggest surprise is the evaluation of collaboration of co-workers which has been rated positively only in 70% what has been proved by a questionnaire conducted. Unfortunately bad interpersonal relations have negative influence on labour safety management. Probable reason of the situation occurred is increasing unemployment as well as continuous crisis causing lack of stabilization of companies as well as lack of possibilities of implementation of all required by existing regulations rules of labour safety and health.

# 4    MODEL OF LABOUR SAFETY MANAGEMENT

Creation of safe working condition as well as incentives for safe activity should be treated not only as fulfilment of the obligation imposed by law on the employer but also as the source of return of invested in labour safety resources. Therefore on the basis of:

— own experiences,
— substantive potential and experience of scientific centres as well as practitioners,
— knowledge of the existing structure of the companies under research,
— knowledge of the condition of medium size production – service companies in respect to technique and organization,
— knowledge of the financial condition of companies,
— knowledge of social conditions,

selection should be done of most favourable elements occurring in the systems of safety management, congruent to the companies under research conditions and structures and work out own system of safety management including characteristics of each activity, what is shown on fig. 2.

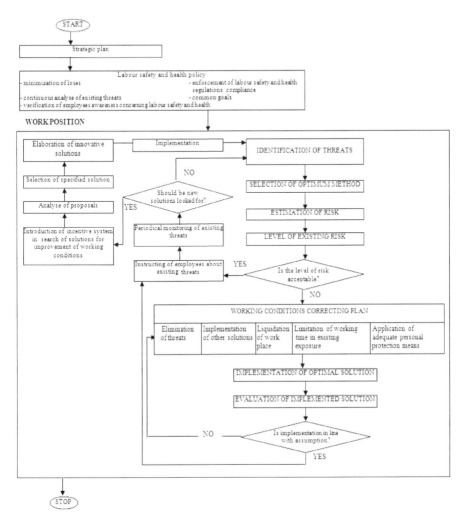

Figure 2. Model of labour safety management in small and medium size companies (own elaboration using: (OHSAS 18001, Guidelines 2001)).

## 5 SUMMARY

Initiated in some medium size production – service companies implementation of systems or elements of systems of safety management is advisable. It creates thereby the possibility of evaluation of their usefulness as well as allows, based on pilot conclusions and results, to make substantive and formal remarks to selected system of safety management. After refinement a uniform labour safety management system should be rather implemented as it allows the possibility of easy comparison of expenditures and effects.

Elaboration of a definite system is quite a long process therefore it should be propagated to immediately implement such elements of already existing systems which are common or congruent to each other.

System of safety management in companies under research is:

— reformable and open for new reasonable proposals,
— available to all their employees.

Implementation of the system should take place first in the companies which declare themselves the wish of its implementation. Safety management system includes rules of economic analysis of the issue to allow comparable planning and evaluation of the effects of the implemented elements of the system against the costs of these implementations and overheads of the company.

For possible for such implementation should be accepted:

— raising the rank of issues of labour safety and health of the employees,

— enlargement, and in some fields implementation, of risk analysis to define actions, methods and resources as well as costs of decreasing the level of risk,

— necessity of conducting trainings for persons form the management and supervision in the scope of foreseen directions of safety management as well as improvement of the effectiveness of professional education and trainings,

— if possible introduction based on existing structures and materials of a "Book of Safety Management" and the function of coordinator for safety management,

— expanding or reorganization of IT systems in the companies enabling comprehensive information about the results of the analysis carried on, implementation of safety system management as well as all matters related with accidents, professional diseases, their reasons as well as their effects for the company.

## REFERENCES

Berkowska A., Dangers, harfulnesses and nuisances determining hazards in the work of a community nurse, [in:] Charytonowicz J. (ed) Zastosowanie ergonomii: Wybrane kierunki badań ergonomicznych w 2012 roku, Polskie Towarzystwo Ergonomiczne PTErg, Wrocław, 2012.

Berkowska A., Drzewiecka M., The choice of optimal risk assessment method for an exemplary post, [in:] Salomon Sz. (ed.), Safety of the system: Human-Technical Object-Environment. Work health and safety in production, operation and maintenance, Częstochowa, 2011,

Boszko J., Struktura organizacyjna i drogi jej optymalizacji, Wydawnictwo Naukowo Techniczne, Warszawa, 1973.

Brown H. S., Angel D., Broszkiewicz R., Krzyśków B.: Occupational Safety and Health System in Poland in the 1990: a Regulatory System Adapting to Societal Transformation. Policy Sciences, 2001,

Czarnecki M., Ekonomiczno- społeczne kryteria oceny jakości, PWN, Warszawa, 1977.

Dach Z., Sektor małych i średnich przedsiębiorstw w perspektywie przystąpienia do Unii Europejskiej, Zeszyty Naukowe AE, Kraków, 2001,

Gierasimiuk J., Bezpieczeństwo pracy i ergonomia. Maszyny stanowiska pracy. Zestawienie tematyczne aktów prawnych i norm, tom I, CIOP, Warszawa, 1984.

Jasiak A., Misztal A., Makroergonomia i projektowanie makroergonomiczne. Materiały pomocnicze, Wydawnictwo Politechniki Poznańskiej, Poznań, 2004.

Koradecka D., Zarządzanie bezpieczeństwem i higieną pracy, CIOP, Warszawa, 2001.

Kowal E. , Ekonomiczno- społeczne aspekty ergonomii, Wydawnictwo naukowe PWN, Warszawa- Poznań, 2002.

Małysz F., Wypadki przy pracy i choroby zawodowe, Poradnik 153, Wydawnictwo „Biblioteczka Pracowniczka", Warszawa, 2003.

Martyniak Z., Organizacja i zarządzanie. 50 problemów teorii i praktyki, wyd. III, Książka i wiedza, Warszawa, 1986.

Olszewski J., Podstawy ergonomii i fizjologii pracy, Wydawnictwo Akademii Ekonomicznej w Poznaniu, 1997.

Pacholski L., Cybernetics and Systems: An International Journal, Taylor & Francis, 1998.

Pacholski L., Trzcieliński S., Przedsiębiorstwo konkurencyjne, Wydawnictwo Politechniki Poznańskiej, 2005.

PN-EN 60529:2003 Stopnie ochrony zapewnianej przez obudowy (Kod IP).

Podgórski D., Miareczko B., Pleban: Ocena zgodności maszyn oraz środków ochrony zbiorowej i indywidualnej z D. wymaganiami bezpieczeństwa pracy i ochrony zdrowia., (W: Seria: Bezpieczeństwo i Ochrona Człowieka w Środowisku Pracy, T.15) CIOP, Warszawa, 2001

Podgórski D., Pawłowska Z.: Ocena ryzyka zawodowego jako element systemu zarządzania bezpieczeństwem i higieną pracy, (W: Ocena ryzyka zawodowego. T.1 Podstawy metodyczne, Seria: Zarządzaniem Bezpieczeństwem i Higieną Pracy CIOP) , Warszawa, 2001

Podgórski, D. Pęciło M., Dudka G.: Procesowe zarządzanie bezpieczeństwem i higieną pracy, (W: Materiały Międzynarodowej Konferencji Bezpieczeństwa i Niezawodności (Safety and Reliability International Conference) KONBIN 2001) , Szczyrk, 2001

Polska Norma PN-80/Z-08052. Ochrona pracy. Niebezpieczne i szkodliwe czynniki występujące w procesie pracy. Klasyfikacja.

Polska Norma PN-N-18001:1999. Systemy zarządzania bezpieczeństwem i higieną pracy. Wymagania.

Polska Norma PN-N-18002:2000. Systemy zarządzania bezpieczeństwem i higieną pracy. Ogólne wytyczne do oceny ryzyka zawodowego.

Polska Norma PN-N-18004:2001. Systemy zarządzania bezpieczeństwem i higieną pracy. Wytyczne.

Polska Norma PN-N-18004:2001/Ap1:2002 Systemy zarządzania bezpieczeństwem i higieną pracy. Wytyczne.

Rączkowski B., BHP w praktyce, ODDK Sp.z o.o., Gdańsk, 2009.

Romanowska-Słomka I., Słomka A.,. Zarządzanie ryzykiem zawodowym. Tarnobus Sp. z o.o. Tarnobrzeg 2001 r.

Rozporządzenie Ministra Gospodarki z dnia 30 października 2002 r. w sprawie minimalnych wymagań dotyczących bezpieczeństwa i higieny pracy w zakresie użytkowania maszyn przez pracowników podczas pracy (Dz.U. Nr 191, poz. 1596, zm. Dz.U. z 2003 r. Nr 178, poz. 1745).

214

Rozporządzenie Ministra Pracy i Polityki Socjalnej z dnia 26.09.1997 r. w sprawie ogólnych przepisów bezpieczeństwa i higieny pracy (Tekst jednolity Dz. U. z 2003 r. Nr 169, poz. 1650 z późniejszymi zmianami).

Rozporządzenie Ministra Pracy i Polityki Społecznej z dnia 26 września 1997 r. w sprawie ogólnych przepisów bezpieczeństwa i higieny pracy (jednolity tekst Dz.U. z 2003 r. Nr 169, poz. 1650 z późn. zmianami),

Skuza L., Wypadki przy pracy od A do Z wypadki osób pod opieką szkoły i placówki, Wydanie VI rozszerzone, ODDK Sp.z o.o., Gdańsk, 2005.

Szymanek T., Zarychta W., Wypadki przy pracy i dochodzenie roszczeń, Wydawnictwo zrzeszenia prawników polskich, Warszawa, 2004.

Ustawa z dnia 26 czerwca 1974 r. Kodeks pracy (jednolity tekst Dz.U. z 1998 r. Nr 21, poz. 94 z późn. zmianami),

# A Method to Plan Human Error Prevention Strategy by Analyzing PSF Tendency – Case Study of a Pharmacy Factory and a Medical Center

*Masanao Ishihara, Kasumi Nishimura, Yusaku Okada*

Keio University
Yokohama, Japan
massa@z6.keio.jp

## ABSTRACT

In some industries, incident cases have been collected positively to grasp the seeds of future accidents. However, as the number of incident reports becomes larger, the safety manager will find it difficult to continue analyzing all seeds. In order to solve the problem, we intend to propose a way of evaluating the tendency of the human error seeds, which are called PSF (Performance Shaping Factors). Firstly, PSF Check Sheet is used to collect incident cases with PSF around them. Secondly, the book of the PSF Check Sheets is summarized through Comment Database into an Assessment Sheet, which contains a graph of the PSF tendency and comments about estimated latent problems in the work environment. As a result of our case studies, the validity of the method was confirmed, and advantages and disadvantages are found out. It can be installed effectively and promote safety activities.

**Keywords**: Human Error, Human Reliability, Performance Shaping Factors

# 1    INTRODUCTION

In some industries, incident cases have been collected positively to grasp seeds of future accidents. However, as the number of incident reports becomes larger, the safety manager will find it difficult to continue analyzing all seeds. Additionally, collected seeds have to be PSF (Performance Shaping Factors). Some seeds are just reported as self-reference but they cannot be effective to the other persons. As a result, only some PSF are grasped to take some precautions, and the other many seeds are merely recorded to be ignored. In order to solve the problem, we intend to propose a method to evaluating the PSF tendency.

This study will explain how the method collects PSF and how analyze the PSF tendency. Furthermore, it will show the result of some case studies and a subject for a further study.

# 2    PROPOSAL OF PSF CHECK SHEET

Firstly, we chose 15 principal PSF (Performance Shaping Factors) by analyzing past incident cases with Root Causes Analysis (RCA) and picking out latent factors of incidents. Each 15 PSF could be grouped into 8 categories (workplace, scheduling, manual, indication, instruction, the number of workers, handling the equipments, confirmation of the operation), and each of which is divided into some factors, such as workplace has 4 factors (lighting, noise, arrangement, working space).

Next, we summarized the study of choosing 15 PSF mentioned above and designed the PSF Check Sheet. We use this Sheet to see and point out the state of each factor. For example, with this sheet the worker could easily see and check if the lighting is too dark, too blight, or with no problem. It is helpful for analyzing the factors of the incidents. See the example shown in Figure 1.

# 3    ASSESSMENT SHEET

After PSF Check Sheets are used for collecting incident cases, the book of the sheets, namely certain amount of incident data, is summarized to figure out the overall situation of the workplace. The graph of the rates of each factor means PSF tendency in the work environment.

However, with merely the graph, it is difficult for the workers to understand the whole work condition. Also, because checking the work condition with PSF Check Sheets requires subjective viewpoints, if the workers consider their work environment is good enough and these PSF don't related to the incidents, they would not make a check in certain boxes on the sheets. In this situation, it is difficult to recognize the important PSF to improve the work condition.

## PSF Check-Sheet

ID

Date

Case Name

Place

Summary

Point to Check
- Check from the viewpoint to discover factors effecting performance in the work environment.
- Imagine an ideal work environment yourself, compare it and real environment.
- Check the item of "no problem" if the situation does not come under any item.

### workplace
see the matters objectively
the lighting was
- [ ] too dark
- [ ] too bright
- [ ] with no problem

the noise was
- [ ] made a lot around
- [ ] made by some people around
- [ ] with no problem

the arrangement was
- [ ] too untidy
- [ ] untidy
- [ ] with no problem

the working space was
- [ ] insufficient because of the structure
- [ ] a little mussed
- [ ] with no problem

### scheduling
Think about a sequence of the works seemed to be related to the incident
the time to spare was
- [ ] no allowance
- [ ] a little allowance
- [ ] some allowance
- [ ] enough allowance

work flow
- [ ] working parallely
- [ ] chanceled for some reasons
- [ ] with no problem

the preparation was
- [ ] not so enough that you became upset
- [ ] not so bad
- [ ] with no problem

### manual
the written procedure was
- [ ] lack of contents
- [ ] wrong contents
- [ ] able to be interpreted in several way
- [ ] with no problem

### indication
think about the clearity of labels, goods or format.
the indication was
- [ ] insufficient. There was no indication.
- [ ] difficlut to understand
- [ ] similar to another
- [ ] with no problem

### ordering
think whether there was a problem with communication
the content was
- [ ] insufficient
- [ ] wrog
- [ ] too much
- [ ] with no problem

the way of instructing was
- [ ] ambiguous
- [ ] urgent
- [ ] with no problem

the confirmation of the instruction was
- [ ] not done at all
- [ ] done by repeating the instruction
- [ ] done by pointing
- [ ] done by another person
- [ ] with no problem

### the number of the workers
think about the flow of the work which seemingly related to the incident
te number of the workers was
- [ ] aloways too much or too little
- [ ] too much or too little at that time
- [ ] with no problem

### handling of the equipments
think whether the operation of the equipments have influence on the incudent
you couldn't operate them well because
- [ ] you had no experience of it
- [ ] you had no manual of it
- [ ] you hadn't read the manual
- [ ] the manual was difficult to understar
- [ ] you are unaccustomed to it
- [ ] you hadn't confirmed the operation
- [ ] there was anther factor
- [ ] with no problem

### confirmation of the operation
the confirmation of the operation was
- [ ] not don at all (or not remember)
- [ ] merely checked for yourself
- [ ] checked by anothor person

Figure 1 PSF Check Sheet

Therefore, we derived some comments from the graph and made Assessment Sheet. It contains a graph of the PSF tendency and comments; the point to consider, the factors in the department, and the latent problems estimated from the PSF Tendency. The comments are based on the items of each factor (e.g. the item of "too dark", "too bright", or "with no problem" in the factor of "lighting") and the relations between them. It totally shows the considerable conditions of work environment. You can see the example shown in Figure 3. We handed it in the workers to help them to understand the condition of their workplace and improve it.

## 4 COMMENT DATABASE

To connect a graph of PSF Tendency and the comments in an Assessment Sheet automatically, Comment Database was designed. Table 1 is the outline of it.

Table 1 outline of the Comment Database

| Level | Condition | Pattern Explanation | Comments |
|-------|-----------|---------------------|----------|
| 0 | all factors are under 30% | The PSF collecting activity is not working. Give comments about how to introduce the activity of collecting incident cases with PSF Check Sheet | Table 2 |
| 1 | less than or equal to 3 factors are over 30% | The viewpoints of collecting PSF are biased. Give comments about the specific factors and for collecting more various viewpoints | Table 3 |
| 2 | more than 3 factors are over 30% | Various PSF are collected from various viewpoints. Give comments about how the Assessment Sheet can be used. | Table 4 |

The database has certain comments according to the "Level" of the summary of PSF Check Sheets. Because these PSF are related each others, if few factors are rated over 30% and the summary results in Level 0 or 1, it means the workers in the section could not improve their work environment sufficiently for themselves.

Additionally, each PSF listed in the Check Sheet has each comment. In a case of Level 1 or Level 2, the assessment sheet has the comments related to the level and the PSF rated over 30%.

# 5    CASE STUDIES

The mechanism of the method is totally explained above. Figure 2 shows the frame work of the method. In order to confirm the effect of the method, we carried out case studies in medical centers and pharmacy factories. A sample of the result is shown as Figure 3. The graph came from a department. According to the condition of the PSF tendency, the set of the comments was composed by Table 3, Table 5 and Table 6. Because the graph rates 2 PSF as over 30%, the pattern of the PSF tendency is Level 1 and the comments of Table 3 was chose. The PSF rate over 30% are "workflow" and "checking", so related comments were picked up from Comment Database: Table 5 and Table 6.

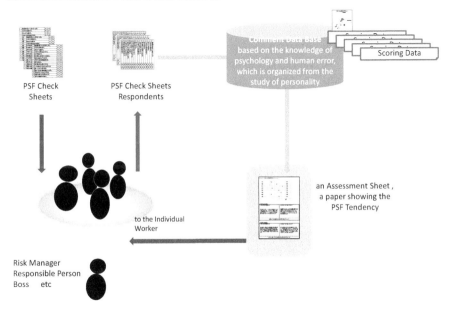

Figure 2 Framework of the Method

From the result of case studies, the validity of the proposed method was confirmed and some advantages and disadvantages are found out.

Advantages are below.

- Summarizing the incident cases is good to grasp total condition of the work environment. It supports safety managers to plan their safety strategy with long term and various view points to prevent human errors in the work environment.

Disadvantages are below.

- The analysis was not persuasive in some sections, because they collected few cases.
- This method doesn't search into each incident case. This makes it difficult by workers to understand the strategy and discuss it.

Given the results written above, we found it important to give the workers a fundamental education of safety management on human factors. Before PSF Check Sheets collection, we executed the learning program about human factors and human reliability for the workers. As a result, it was obtained that the workers positively collect incident cases and agree with the comments on the Assessment Sheet.

## 6    CONCLUSION

This study has proposed PSF Tendency Analysis for analyzing the tendency of the human error seeds. We have designed a check-sheet type of incident report. Then, we have constructed comment data base to evaluate the PSF tendency, and suggested Assessment Sheet. According to the result of the case studies, although we found the importance of getting consent with the workers about fundamental ideas about safety management, we confirmed that it will encourage various organizations to improve incident collecting activities.

## REFERENCES

A.D. Swain, and H.E. Guttman, 1983, Handbook of human reliability analysis with emphasis on nuclear power plant applications, NUREG/CR-1278, Washington D.C.

T. Yukimachi and M. Nagata, 2004., Reference List on Basis of GAP-W Concept and a Case Study, Human Factors in Japan, Vol.9, No.1, pp46-62,

T. Yukimachi and M. Nagata, 2004, Study of Performance Shaping Factors in Industrial Plant Operation and GAP-W Concept, Human Factors in Japan, Vol.9, No.1, pp7-14,

N. Sagawa, R. Fujitsuka, A. Furukawa, Y. Okada, 2008, A Study on Usability of a Reference List in Nursing Duties, Proceedings of the 38th Annual Meeting of Kanto-Branch, Japan Ergonomics Society, pp47-48

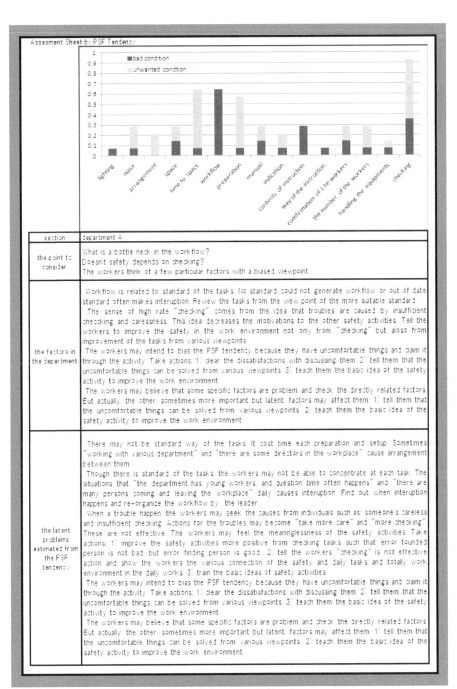

Figure 3 Assessment Sheet

Table 2 the comments of Level 1 PSF Tendency

| the point to consider | Have you got consensus on the activity in the department? |
|---|---|
| the factors in the department | There is a gap between the staff promoting this activity and the workers operating the check activity. The listed check items in the check sheet are considered as factors affecting the workers in your work environment. The fact that the incidents were happen means that the factors affected the workers. However, most factors have not been checked. It comes from some cases like that the workers don't understand the sense of the activity, don't accept, aren't trained enough and so on. Think of what the gap is and take an action for it. |
| the latent problems estimated from the PSF tendency | The workers may not accept the safety activity in their mind. "Our department is safe enough", "We are just doing because the staff told us to do it", "Such kind of the activity could not prevent any accident" etc. These kinds of dissatisfaction may change this safety activity into one of the error factors. Take actions; 1) Set meetings with the workers, 2) Have the workers to tell how they think of the activity, 3) Show them the real goal image. In order to motivate them to the activity, connect the motivations of the workers and the activity. "For costumer value", "for comfortable work environment", and so on. The word of "safety" isn't important but what makes sense is important. |
| | The workers may not understand or not be trained how to think and check against the incident cases. Of course they are doing because they are told, but they don't understand the basic idea of the activity and could not check the factors. Take actions; 1) have a time to teach them the basic idea of the safety activity to improve the work environment from various viewpoints, 2) tell them the point of the activity on the daily works, 3) when it is difficult to spread the idea in the workers, train a few workers intensively and spread through them with trials and errors in long term. |

Table 3 the comments of Level 2 PSF Tendency

| the point to consider | The workers think of a few particular factors with a biased viewpoint |
|---|---|
| the factors in the department | The factors in the work environment are affected each others. However, only a few factors are checked. It means that their viewpoints of collecting factors are very biased. Find out how biased and why biased their viewpoints are, and take actions for it. |
| the latent problems estimated from the PSF tendency | The workers may intend to bias the PSF tendency because they have uncomfortable things and claim it through the activity. Take actions; 1) clear the dissatisfactions with discussing them, 2) tell them that the uncomfortable things can be solved from various viewpoints, 3) teach them the basic idea of the safety activity to improve the work environment. |
| | The workers may believe that some specific factors are problem and check the directly related factors. But actually, the other, sometimes more important but latent, factors may affect them, 1) tell them that the uncomfortable things can be solved from various viewpoints, 2) teach them the basic idea of the safety activity to improve the work environment. |

Table 4 the comments of Level 3 PSF Tendency

| the point to consider | Do you take suitable actions to the work environment with the workers? |
|---|---|
| the factors in the department | The workers think of their work environment from various points to improve it, but the safety activities can be messed because of the variety. Find out what the most suitable actions are and what priorities are. |
| the latent problems estimated from the PSF tendency | The worker may have various ideas to improve their work environment but not be able to organize them. Take actions; 1) discuss the workers and the manager about the safety activities, 2) referring to this assessment sheet, and set priorities, 3) make a strategy for safety and execute it. |

Table 5 the comments for "workflow"

| the point to consider | What is a bottle neck in the workflow? |
|---|---|
| the factors in the department | Workflow is related to standard of the tasks. No standard, no workflow or out of date standard often makes interruption. Review the tasks from the view point of standard. |
| the latent problems estimated from the PSF tendency | There may not be standard way of the tasks. It cost time each preparation and setup. Sometimes "working with various department" and "there are some directors in the workplace" cause arrangement between them.<br>Though there is standard of the tasks, the workers may not be able to concentrate at each task. The situations that "the department has young workers, and question time often happens" and "there are many persons coming and leaving the workplace" daily causes interruption. Find out when interruption happens and re-organize the workflow by the leader. |

Table 6 the comments for "checking"

| the point to consider | Doesn't safety depend on checking? |
|---|---|
| the factors in the department | the sense of high rate "checking" comes from the idea that troubles are caused by insufficient checking and carelessness. This idea decreases the motivations to the other safety activities. Tell the workers to improve the safety in the work environment not only from "checking" but also from improvement of the tasks from various viewpoints. |
| the latent problems estimated from the PSF tendency | When a trouble happens, the workers may seek the causes from individuals such as: someone's careless and insufficient checking. Actions for the troubles may become "take more care" and "more checking". These are not effective. The workers may feel the meaninglessness of the safety activities. Take actions; 1) improve the safety activities more positive from checking tasks such that error founded person is not bad, but error finding person is good., 2) tell the workers "checking" is not effective action and show the workers the various connection of the safety and daily tasks and totally work environment in the daily works, 3) train the basic ideas of safety activities |

# CHAPTER 25

# Application of Advanced Reverse Engineering and Motion Capture Techniques in Motion Modeling of Human Lower Limbs

*Rychlik M., Stanowski A.*

Poznan University of Technology
Poznan, Poland
rychlik.michal@poczta.fm

## ABSTRACT

Every day thousands of people are getting hurt or damage of their health. Even a simple fracture or dislocation, is cause of exclude such person from performing the full duties. Statistics are showing, how serious the problem is. In 2006, musculoskeletal symptoms were the number 2 reason for physician visits. Musculoskeletal symptoms include pain, ache, soreness, discomfort, cramps, contractures, spasms, limitation of movement, stiffness, weakness, swelling, lump, mass, and tumors to the musculoskeletal system. There were more than 132 million physician visits for musculoskeletal symptoms in 2006 (National Ambulatory Medical Care Survey, 1998-2006). So it is important, to making fast and accurate diagnosis and the most efficient recovery. Presented paper is a part of research work, which are related to the motion capture systems (MOCAP) and computer modeling systems (CAx and Reverse Engineering). The authors of the article describe the procedure for obtaining of body movements data from the exoskeleton Gypsy5 (mechanical MOCAP system) and combine them with computer model of human skeleton (individual geometry of the bones). For better presentation of procedure the two cases of acquired body movement's sequences are presented. The first of them was done for patient with hereditary spastic paraplegia (HSP) during walking on flat floor and stairs. The second measurement was done for

226

health person in exact the same sequence of movements. The possibilities of using MOCAP systems and computer analysis of obtained results for diagnostic and rehabilitation procedures are discussed in further paragraphs.

**Keywords**: 3D geometry reconstruction, human lower limb anatomy, Motion Capture, Reverse Engineering

# 1 INTRODUCTION

One of the most frequent reasons for visits to the doctor, are the symptoms of musculoskeletal system. Musculoskeletal disorders cost the United States nearly $850 billion yearly. Currently employed workers in the United States miss nearly 440 million days of work because of musculoskeletal injuries (United States Bone and Joint Decade, 2008). Therefore, important and constantly developed elements are various systems for diagnosis and monitoring the treatment process.

The CAD systems are very well known by designers in their every day practice and numerical analysis. Computer models of real objects with advanced numerical tools, significantly improves the quality and reduce time of design process. In addition to the three-dimensional modeling systems, there are many other tools and techniques (such as Reverse Engineering techniques, motion capture systems, Rapid Prototyping machines), which will further enhance the capabilities of engineers. There are no obstacles that these tools could not be applied in medical practice and specifically in the field of bioengineering.

One of the dynamically developing and relatively new techniques which have been used in medical diagnostic, rehabilitation and Computer Aided Engineering (CAE) is the technique connected with virtual reality (VR) and Motion Capture systems (Chung S., 2000). The MOCAP systems altogether with the applications of computer analysis allow for full interaction between the user and the computer model in virtual space.

In this article the author's present application of the 3D computer models and Motion Capture systems for modeling and analysis of motion of the human lower limbs. Results of that computer analysis can be the basis for the diagnosis and evaluation of the treatment process.

# 2 METHODS AND MATERIALS

The first step in computer analysis of biological/medical objects is to obtain the correct and high accurate 3D model. Depending on the type of input data, there are two ways to reconstruct the geometry. The first uses medical imaging techniques such as CT and specialized software to process obtained DICOM images (Digital Imaging and Communications in Medicine). The second uses 3D scanning systems and specialized Reverse Engineering software to process obtained points clouds (coordinates of measuring points). For both ways, the end result is to obtain a fully parametric 3D model, which is compatible with CAD systems.

After obtaining a geometric model of the bones (individual model for the examined person) it is possible proceed to the second step: measuring the dynamic data (motion of the legs). For this process the Motion Capture system (Gypsy 5) was used.

## 2.1 Geometry Reconstruction

CT images of the lower limbs were subjected to digital processing in the program's ScanIP (SimpleWare). The most important part of data processing (next to the filter and remove the distortion) is the segmentation process. The main goal of segmentation is separation of anatomical structures from CT images. This process is extremely important step, since the correctness of its implementation depends on the accuracy of the resulting 3D model. In this example, the structures were wanted bone structures.

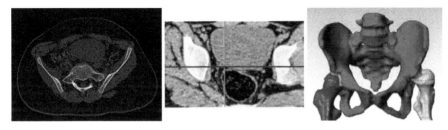

Figure 1  Geometry reconstruction process for the CT input data (from left): DICOM image, segmentation process, final 3D model (triangle surface grid)

Procedure for the second type of input data is following (Figure 2). The individual bones of the lower limbs were subjected to a 3D scanning process (the structural light scanner). The final result of the scanning is a points cloud, which describing the object's surface. For processing of the points cloud the special Reverse Engineering software Geomagic Studio was used.

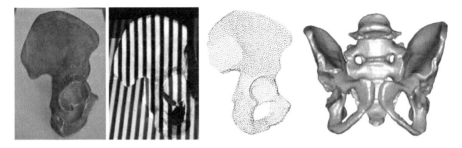

Figure 2  3D scanning procedure (from left): input geometry - bone, the measurement process (structural light), points cloud, final 3D model (triangle surface grid)

In this paper the process of geometry reconstruction was performed for all lower limb bones, beginning from the pelvis and ending with the toes of the foot. The final result of reconstruction is CAD model of the lower limb skeleton structure (Figure 3).

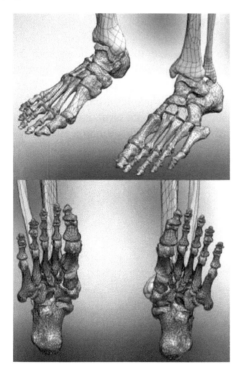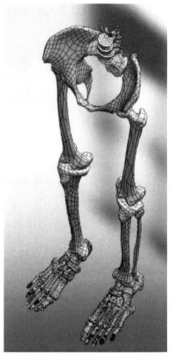

Figure 3  The final CAD model of the lower limb (skeleton structure)

## 2.2    Motion Data – Motion Capture System

The Motion Capture systems (also known as Motion Tracking or MOCAP), they provide digital recording of the motion parameters of a real person (figure or real-world objects) and then necessary conversion and processing of the acquired data aiming at making appropriate modifications in virtual reality (Kucharski T., 2005). Four basic equipment groups are distinguished in the MOCAP technique: mechanical, optical, magnetic and inertial. The main difference between them is the data acquisition method, complexity level of the equipment, accuracy and data acquisition speed.

In presented work the mechanical motion capture system (rigid exoskeleton) was used. The rigid elements are connected by joints allowing for a desirable freedom of movements. The joints are equipped with angular shift sensors. Such an arrangement (Figure 4) constitutes a specific external skeleton attached to the body of the user. The joints allow for a limited freedom of movements and enable angular shift measurement in all required axes. System used in this research (named Gypsy 5), includes 37 angle sensors (for measuring of limb movements) and 2 gyrocompasses (used for measurement of the position and rotational displacement between the hip and the chest). The resolution of the sensors is $0.125^0$.

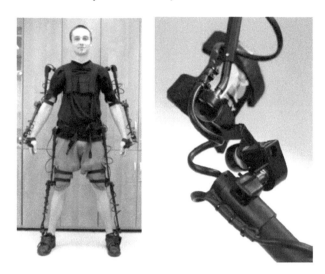

Figure 4  The mechanical MOCAP system (from left): rigid exoskeleton (Gypsy 5), example of the joint

## 3.    RESULTS

The measurement and analysis process of the motion is realized in three basic steps: 1) preparation of measurement, 2) recording data of motion sequences, 3) integration of the 3D models, with motion data and analysis of measurement results.

## 3.1    Preparation of Measurement

In order to prepare the system adequately, it is necessary to make calibration photos of person with special calibration cube (Figure 5a). They enable to load anthropometric parameters of the patient to the measurement system. Calibration process is needed to determining of localization of all kinematics nodes between each bone. After calibration the measuring system is wearing by the patient (Figure 5b). Sensors of the exoskeleton recording individual positions of kinematics nodes and sending them to the motion capture software (Figure 5c). For better visualization of results is possible to used full three dimensional CAD model of bottom limbs bones (Figure 5d). After these preparations the sequence of movements can be recorded.

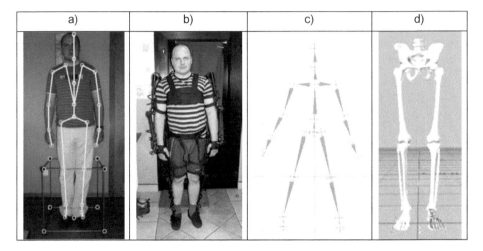

| a) | b) | c) | d) |

Figure 5 Calibration of measuring system: a) calibration photo (determining of localization of kinematics nodes), b) patient wearing the movement measuring system Gypsy 5, c) the view of the skeleton in the motion capture software, d) virtual model of bones of bottom limbs

## 3.2 Recording Data of Motion Sequence

After the calibration process we can proceed to perform the measurement of the motion sequences. In presented example the sequence of movements of the person during walking on a flat floor was performed. As a research object the patient (40 years old) with hareditary spastic paraplegia (HSP) was measured (Figure 6). For better understanding results of measurements done for person with HSP, are compared with measurements of healthy man (age: 22 years) without pathologic degenerations of lower limb system.

Figure 6 Movement sequence of patient with hareditary spastic paraplegia (few frames of one step)

Recorded movements, give possibility to precise analyze trajectory of each part of body (kinematics nodes). Also speed and angels of bones can be analyzed (refreshment of data registration 60frames/sec). Obtained data are saved, as digital signal and they can be analyzed with using of various numerical methods. The one of possibility is trajectory analysis of leg joints.

## 3.3 Integration 3D Models with Motion data and Analysis of Mesurement Results

The measurements results of limbs movement of the patient's was combined with derived in previously step the individual CAD models of the bones. The integration process was performed using the Motion Builder software. It consisted of attribution the results of measurements obtained to each joint with the corresponding nodes in kinematics model of skeleton (characterization process). The result of the integration process is the CAD model of lower limbs containing complete information with mutual position of the joints (in fact also the bones) in the performance of motion (Figure 7).

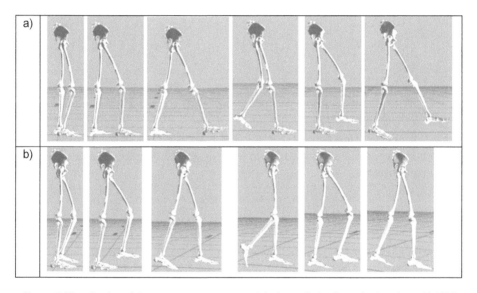

Figure 7 Visualization of the movement sequence of the lower limbs (bones): a) patient with HSP, b) healthy man

On Figure 8 is presented graphs of trajectory of main joints of leg: 1) femur joint, 2) knee joint, 3) ankle joint. Movement of each joint is described by set of three curves. Each curve corresponds with changes of joint angel in axes X, Y, Z in three orthogonal planes XY, YZ and ZX.

The main difference in movement between patient with HSP and healthy man is visible in dissimilarity of angle in axis X. For knee joint (Figure 8.2) we can observe flat area in comparison for analogical area of health patient.. In this case patient has not correct response movement system of knee and this joint is to

232

stiffness. Also steps are slower then normal speed. Basis on this knowledge the treatment and rehabilitation procedure can be planed.

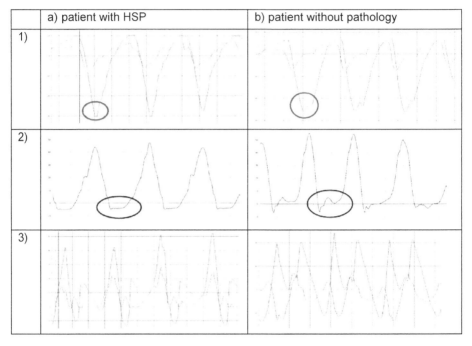

| | a) patient with HSP | b) patient without pathology |
|---|---|---|
| 1) | | |
| 2) | | |
| 3) | | |

Figure 8  Analysis of trajectory of legs joints: 1) femur joint, 2) knee joint, 3) ankle joint, a) patient with hareditary spastic paraplegia HSP, b) healthy man - no diseases, (in circle areas of big differences in motion)

## 4    CONCLUSIONS

Motion Captures systems enable measuring of body movements in full 3D space. Results of that measuring process can be applied in medical diagnosis, rehabilitation or Computer Aided Engineering (CAE). Obtained data contain lot of information, such as: positions of joints, speed, accelerations. This knowledge, as a digital data, can be used in many different computer analysis.

One of application of that recorded data is analysis of posture and joints kinematics of person with diseases in movement functions. Analyze of trajectories can be used in planning of rehabilitation procedure. Also during treatment the control of rehabilitation progress can be done. Possibility of result visualization in slow motion is very useful for precise describing of limitations of movements.

Presented in this paper the applicability of MOCAP systems are just a small part of the real possibilities of using this technique in the medical or rehabilitation.

Existing systems used in medicine such as BTS Smart (Lovecchio N at all., 2008), although provide possibility of data registrations about the patient's limbs (optical systems), but does not used the data on the internal structure (skeleton, bones) and their mutual relations during movement (set in space). Presented in this article system, allows for full integration of the individual geometric model of lower limb bones of the patient with information about its motion derived from MOCAP systems.

Parallel to the author's work presented in this article the other teams work on similar systems. A team of researchers from the University of Duisburg-Essen has performed its reconstructions on the basis of the optical system with the use of markers (D. Raab, Z. Tang, Paula J., and Kecskeméthy A., 2011). This system is characterized by other limitations than the mechanical system. Limitations are associated with elements such as: placing markers on the patient's skin, a complex system of image analysis and processing of video data on the position of the limbs. It would be purposeful and very interesting to compare results from various systems, optical and mechanical, executed for the same patient

The area of further research and development should be a system of reducing measurement errors due to the shift axis of rotation of the joints with the axis of rotation sensors on exoskeleton. In addition, a significant challenge is to develop a system for measuring motion of individual elements of the foot. Previous studies and visualizations, define the foot as a element of the rigidly connected bones.

Generally the possibilities offered by the application of measurement techniques MOCAP and possibilities of the computer analysis in CAx systems are very extensive. The information obtained may give rise to new forms of rehabilitation and the impact on an objective assessment of health status as well as a faster treatment process.

## ACKNOWLEDGMENTS

The part of this work was supported under research grant no: N518 496039 from the Polish Ministry of Science 2010-2012.

## REFERENCES

Chung S. 2000. Interactively responsive animation of human walking in virtual environments. The Department of Computer Science of The George Washington University. May 21.

Kucharski T. 2005. Modeling shape and motion of animated characters, Nonpublished PhD thesis. Wroclaw University of Technology.

Lovecchio N, Galante D., Turci M., Sforza Ch. and Ferrario Virgilio F., 2008, Quantitative Analysis of Rotational Movements of Knee in Healthy Subjects During Treadmill Barefoot Walking, The Open Sports Medicine Journal, 2, pp 28-33.

National Ambulatory Medical Care Survey, 1998-2006, U.S. Department of Health and Human Services, Centers Disease Control and Prevention, National Center for Health Statistics.

Raab D., Tang Z., Pauli J., Kecskeméthy A., 2011, MobileBody: An Integrated Gait Motion Analysis Tool Including Data-Fusion with Patient-Specific Bone Geometry, In Proceedings of the First International Symposium on Digital Human Modeling (DHM2011), ID2190, Lyon, France.

United States Bone and Joint Decade. 2008. The Burden of Musculoskeletal Diseases in the United State. Rosemont. IL American Academy of Orthopedic Surgeons.

# Section III

**Human Factors in Terrorism**

# CHAPTER 26

# Human Factors in Counter-terrorism

*A.W. Stedmon, G. Lawson, R. Saikayasit, C. White, C. Howard*

Human Factors Research Group, Faculty of Engineering
The University Nottingham, Nottingham, UK
alex.stedmon@nottingham.ac.uk

## ABSTRACT

This chapter introduces the timeliness and context of human factors research in counter-terrorism by revisiting a number of themes identified a decade ago and updating the research base based on recent findings. Specific attention is given to surveillance (e.g. traditional CCTV and semi- or full-automation solutions), mass transit hubs (e.g. aviation security), emergency responders (e.g. teamworking and issues of vulnerability) and virtual reality (e.g. developing scenarios to support training). From this a systems approach is offered as a way to understand the complex interactions and wider systems thinking behind the process of security from the viewpoint of those who try to uphold it (e.g. security staff), those who use it (the wellbeing and positive experiences of the general public) and those who aim to undermine it (e.g. the terrorists).

**Keywords**: counter-terrorism, human factors, systems approach, resilience

## 1    THE COUNTER-TERRORISM DOMAIN

After the 9/11 attacks in 2001 an important paper was published (Hancock & Hart, 2002) outlining aspects of counter-terrorism from a human factors perspective. A key focus was the discussion of issues in aviation security and military research that could be applied to other areas of public safety. From an overview of intelligence analysis, military and emergency response, and airport security, new directions were prescribed where human factors could provide a valuable contribution including: performance assessment, signal detection/monitoring, operator perception and alertness, team performance, selection/training, modelling/simulation, and data mining/visualisation (Hancock & Hart, 2002).

Whilst this paper did not contain new research, it detailed the existing knowledge base that could be communicated to stakeholders in support of future human factors involvement. However, a decade later, human factors in counter-terrorism is still an under-developed field where the literature is limited. For example, in the Ergonomics Abstracts database (http://tandf.informaworld.com) that holds 141,818 records (accessed 29/02/12) only 12 publications have the word 'terrorism' and just two have 'counter-terrorism' in their title. This of course excludes ergonomics/human factors research published in related areas (although only 260 publications mention of 'security' in their titles) but it does highlight a gap in the profile of human factors and counter-terrorism.

Over the same timeframe terrorist campaigns have shifted their modus operandi toward indiscriminate, mass impact activities, often with coordinated and targeted secondary attacks aimed at inflicting maximum casualties. Several incidents (e.g. the Bali bombings and Mumbai attack) focused on crowded public spaces in and around mass transit hubs and social venues. In these situations it can be very difficult to conduct counter-terrorism initiatives but it remains relatively easy for terrorists to plot their attacks given the open access of these areas. In the UK this was illustrated when the 7/7 bombers gained access to the London Underground network during the morning rush-hour (Pape, 2005).

In response to this the UK Government has produced a report outlining their 'CONTEST' counter-terrorism strategy identifying Al Qa'ida and associated militant groups as the main terrorist threat to the UK (HM Government, 2009). The CONTEST strategy has been constructed to address four key areas:

- **Pursue** - to stop terrorist attacks
- **Prevent** - to stop people becoming terrorists/supporting violent extremism
- **Protect** - to strengthen our protection against terrorist attack
- **Prepare** - where an attack cannot be stopped, to mitigate its impact

Terrorist activities in crowded public spaces are usually preceded by a period of 'hostile reconaissance' in order to gather essential field intelligence (Linett, 2005). In these situations terrorists are at their most vulnerable, both to detection and disruption however a major human factors challenge is how to monitor such areas whilst also protecting people going about their daily lives, free from fear. Thus, solutions are required, based on the integration of science and fundamental principles of human behavior. From research and practitioner perspectives, the ergonomics discipline has much to offer this endeavor including:

- an applied knowledge of psychology which can be used to analyze suspicious behaviors
- a systems-approach to consider not only hostile intent but the groups and wider-social networks in which terrorists might operate
- methods and techniques to support the development of new initiatives and approaches for detecting terrorist activities.

To address these issues a strategic UK security consortium supported by the Engineering and Physical Sciences Research Council (EPSRC) responded to stakeholder requests for novel interventions that elicit robust, reliable and usable indicators of criminal and terrorist activities. The 'Shades of Grey' consortium draws upon expertise from applied social and physical sciences, computing and engineering to define, design, and deliver a science of interventions. From a human factors perspective, this translates into a range of activities including: user requirements elicitation (Saikayasit et al., in print) and fundamental empirical research (Lawson et al., in print). The following sections revisit some of the themes from Hancock & Hart's paper (e.g. surveillance, mass transit hubs, and emergency response) and present more recent human factors research within, and relevant to, these areas as well as highlighting research needs.

## 2. RESEARCH AREAS

### 2.1 Surveillance

Surveillance is a critical aspect of counter-terrorism initiatives and encompasses activities such as intercepting communication data, identifying suspicious behaviors, monitoring individuals and safeguarding crowded public spaces. Video surveillance has received a lot of research attention in relation to operator workload, human performance and the development of full- or semi-automated systems (Troscianko et al., 2004; Blechko, Darker & Gale, 2008; Dadashi, Pridmore & Stedmon, 2009). However, at a fundamental level, little is still known about how operators perform observation tasks such as person recognition or object identification (Keval and Sasse, 2010). In addition, the monotony of vigilance tasks can induce declining task engagement associated with loss of attentional resources and objective performance (Warm, Matthews & Finomore, 2008). Automated surveillance systems are emerging to support the human operator and a range of guidance is available for the design of control rooms, such as the number of monitors to be viewed, operator viewing distances and image presentations (Wood and Clarke, 2006). However, many of these recommendations lack empirical research and there is little evidence of their effectiveness in real contexts of use (Dadashi, Pridmore & Stedmon, 2009). Furthermore, operators often perform a range of security tasks across a variety of surveillance and communication technologies within the wider security system and seamlessly tracking suspects in crowded public spaces is often beyond the abilities of both human operators and current automated solutions (Dee & Velastin, 2008).

More research is required to investigate the wider systemic issues of surveillance from the complexity of tasks operators conduct through to the 'process of security' from target identification and communications, to active surveillance and onward handover of intelligence in a timely and effective manner without compromising system integrity.

## 2.2    Mass transit hubs

Mass transit hubs are typical of many transport and social venues. Within airport security, the detection of dangerous items in baggage relies on a human operator's ability to interpret a two dimensional X-ray image of a three dimensional object (Beecroft, McDonald & Voge, 2007). False negative and false positive errors are commonplace due to the ambiguous appearances of potential targets (Gale, Purdy & Wooding, 2005). A potential threat may be missed during baggage screening or a false positive might be identified due to the unusual orientation of an object, the target being obscured by other objects, lack of relevant operator knowledge (especially if the target object is not commonly seen or a novel/improvised explosive device) and limitations in available scanning time (von Bastien, Schwaninger & Michel, 2008). From recent research a number of solutions might improve detection performance such as specific training to improve recognition abilities as opposed to improving visual scanning techniques and multi-view screening to improve the potential of accurate identification (von Bastien, Schwaninger & Michel, 2008).

Screening tasks should also be designed to encourage a cautious criterion however, whilst fewer threats are likely to be missed, an increased false alarm rate could lead to delays that might aggravate some members of the public (Beecroft, McDonald & Voge, 2007). This reinforces the notion of a total systems perspective for security where both the interventions put in place are designed to identify potential hostile activities without compromising the user experience of the general public. One of the innovative approaches within the Shades of Grey consortium is to look at supporting the general user experience whilst also investigating active interventions to support crime detection.

## 2.3    Emergency response

Response to a large scale crisis such as a terrorist attack is likely to be a multi-agent operation, involving representatives from the immediate emergency services (e.g. police, paramedics, fire and rescue), military organizations, local and national government bodies, and volunteer groups, ad-hoc teams and the media. Each of these should communicate accurately, share pertinent information and work collaboratively in order to respond to the emergency effectively (Balasubramanian et al., 2006). Often it is the fire and rescue services that are able to mobilize personnel at short notice who are trained not just in fire-fighting but search and rescue, flood response and more recently in dealing with incidents involving hazardous substances (Albores & Shaw, 2008).

Since the attack on the World Trade Centre the dangers that emergency responders face when confronting terrorist attacks have become clear. With the change in the nature of terrorist campaigns there is the danger of coordinated secondary attacks or the consequences of buildings damaged by blasts (e.g. glass and debris or their destruction as with the twin towers). With this in mind, there needs to be a coordinated understanding of the overall system requirements in mitigating danger to the emergency responders whilst also maximizing the chances of saving those caught up in an initial attack. Treatment in the first hour after a serious injury is critical to the overall chances of victim survival, however this also places the emergency responders in the greatest danger (Karasová & Lawson, 2008). The staff involved in emergency response must master situation assessment, decision-making and accurate communication, all under extreme situations, to execute their roles effectively.

There is an important difference between well-defined teams and ad-hoc teams formed in crisis scenarios in relation to increased cognitive demands of shared understanding and awareness of the situation. In civilian aviation, teams of crew-members are typically allocated by roster and may not have worked together before (Smith et al., 2008). To overcome this, teams and individuals who may be involved in emergency response should undergo similar training where possible to ensure that knowledge and procedures are based on shared understandings. There is evidence to suggest that shared mental models directly affect team performance (Fiore et al., 2003) and that through training, teams can develop a shared understanding of a given situation without which members will not utilize the available resources efficiently and may even have different goals.

## 2.4    Virtual environments for training

In recent years, reduced costs and increasing processor power have made computer simulation and virtual reality (VR) a feasible form of training in many fields (Smith & Ericson, 2009). Such computer simulations have been used by researchers in a number of areas relating to terrorist attacks such as examining human behavior during evacuations, organizing mass vaccinations, communication infrastructures for emergency services and even predicting the effects of bomb blasts on building structures (Albores & Shaw 2008).

Security staff and emergency responders must know how to execute safety procedures in highly stressful situations and simulations present an opportunity to train and measure the efficiency of emergency protocols (Zarraonandia, et al., 2009). In addition, VR can be used to continuously develop new and more effective training systems to match the constantly moving threat of terrorism (Swift et al., 2005). VR solutions also lend themselves to interdisciplinary training of distributed teams and information obtained from these simulations can then be used to support emergency managers in planning and coordinating response activities (Fiedrich & Burghardt 2007).

In order to train and prepare for terrorist attacks, emergency response personnel using virtual environments (VEs) are able to conduct tasks safely and without risk to the trainee or members of the public (Meador, McClure & Brett, 2006). In addition to this, not only do trainees display equivalent levels of learning in VEs when compared to real-world training, they enjoy the experience more and retain the learned knowledge for longer (Smith & Ericson 2009).

Agent-based systems can be used to simulate human behavior during large-scale, complex disasters and the inclusion of human behavior simulation in these VEs allows for training in complex situations that might assist teamwork and decision-making (Meador, McClure & Brett, 2006). One such simulation is the 'Virtual Terrorist' which has been developed as an intelligent autonomous agent in a VE to simulate terrorist behavior (Meador, McClure & Brett, 2006). Based on cognitive task analyses that model a typical terrorist behavior this agent can then be used to help train security staff and emergency responders dealing with terrorist activities in different scenarios.

It has been proposed that the use of VR simulations could also be deployed in real-time when emergencies occur to help predict post-attack effects and inform decision makers (Boukerche, Zhand & Pazzi, 2009). This would require a combination of existing simulations including building and fire, traffic and information flow, and hospital intake scenarios so that projections and forecasts could then be used to coordinate response teams and ensure the most effective strategies are used (Jain & McLean, 2003). In this way, modeling emergency responses as a real incident unfolds could support the effective allocation of resources during an initial attack and also help support resource allocation for any secondary attacks (Albores & Shaw, 2008).

## 3.   A SYSTEMS APPROACH TO SECURITY & RESILIENCE

Whilst there has been progress in the last decade in various areas of security, there is still a lack of an integrated approach to understanding the domain. As such, it is now necessary to take a macro-ergonomics systems approach focusing on the system as a whole and how different factors interact with other people, technologies and/or other artefacts (Kleiner, 2006). From an organizational context these factors represent socio-technical entities that influence systemic performance in terms of integrity, credibility and performance. By understanding macro-level ergonomics, the principle is that the organizational level can be devolved to the micro level (Kleiner, 2006). Macro-ergonomic work system models have been used to describe factors influencing the performance of the overall system of security screening and inspection (SSI) systems of cargo and passengers (Kraemer et al., 2009).

These SSI systems are used to identify possible threats on passenger and cargo transits across US borders by air, rail and sea (Kraemer et al., 2009). In the case of airport security, since the 9/11 attacks, airports have adopted more technologies and procedures to detect potential threats (Hancock and Hart, 2002). Many aspects of security and threat identification still depend on the performance of frontline security personnel who are often low paid, poorly motivated and lack higher levels of education and training (Hancock and Hart, 2002). In various contexts, security personnel are often required to monitor multiple channels over extended periods of time (e.g. border control desks, baggage handling, CCTV control rooms) whilst detecting low probability events (Kraemer et al., 2009). Monitoring tasks require focused attention and vigilance often in poor work design conditions (e.g. long hours, low status, low responsibility) which affect overall performance (Hancock &

Hart, 2002; Dadashi, Pridmore & Stedmon, 2009). Returning to the SSI system approach, in order to understand individual performance a five-factor framework has been proposed that contributes to the 'stress load' of the front line security worker (Figure 1).

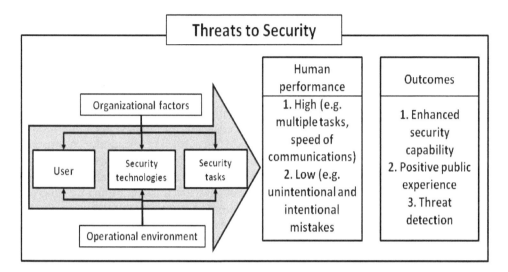

Figure 1: Macro-ergonomic framework for SSI systems (adapted from: Kraemer *et al.*, 2009)

Taking a user-centred approach, the central factor of the framework is the frontline security worker (e.g. user) who has specific skills, unique characteristics, perceptions, behaviors and needs. Within the SSI system, the security worker is able to use technologies and tools to perform specific security screening tasks. The technologies and tools support the security task but are influenced by human factors such as task and workload factors (overload/underload task monotony/repetition). In addition, organizational factors (e.g. training, management support, culture and organizational structures) as well as the operational environment (e.g. noise, climate, temperature) also interact in delivering the overall human performance contribution to the total system and desired work outcomes. Within this macro-ergonomic approach, the complexity of the task and resulting human performance in the SSI system may include errors (e.g. missing a threat signal) or violations (e.g. protocol violations in order to adapt to the dynamic operational conditions) (Kraemer et al., 2009). It can also be used to assess aspects of teamwork (using multiple inputs) and different levels of stakeholder participation and perspectives.

The macro-ergonomic framework has been used to form a basis for understanding user requirements within the Shades of Grey consortium, focusing on the interacting factors that contribute to the overall performance outcomes Saikayasit et al., in print). In this way, the aim is to better understand the needs of different stakeholders and use these to promote a greater awareness of a systems approach and system resilience. One example of this is that syndromic surveillance systems can be developed to understand the inter-relationships of different agents in the total system. For example, by gathering and analyzing data from hospitals, doctors, chemist sales, agricultural supplies, and even absenteeism from work and social networks it may be possible to assist security personnel in their tasks of identifying suspicious patterns of behavior, networks or group behaviours and potential targets of hostile intent (Stoto et al., 2004).

## 4. CONCLUSION

This chapter took an important paper from 2002 that defined the application of human factors to counter-terrorism in the aftermath of the 9/11 attacks. Since then research publications in the specific area of counter-terrorism have been limited but related research exists which has been used to revisit and update the knowledge base of key themes. What emerges however, is that whilst there is focused attention on specific areas (e.g. CCTV, aviation security, teamworking, virtual reality) there is little work being done to consider the wider systems thinking behind the process of security from the viewpoint of those who try to uphold it (e.g. security staff), those who use it (the wellbeing and positive experiences of the general public) and those who aim to undermine it (e.g. the terrorists). In this way, hostile intent is embedded within the wider system balancing different user needs against system capabilities which sets a foundation for future research exploring overall system effectiveness and resilience.

## ACKNOWLEDGMENT

The authors acknowledge the EPSRC for funding this work as part of the Shades of Grey EPSRC research project (EP/H02302X/1).

## REFERENCES

Albores, P., and Shaw, D., 2008. Government preparedness: Using simulation to prepare for a terrorist attack. *Computers and Operations Research.* 35(6), 1924-1943.

Balasubramanian, V., Massaguer, D., Mehrotra, S., et al. 2006. DrillSim: A simulation framework for emergency response drills. *IEEE International Conference on Intelligence and Security Informatics.* San Diego, CA, USA, 23-24 May, 237-248.

Beecroft, M., McDonald, M., and Voge, T., 2007. Achieving personal security in future domestic travel - technology and user needs. *IET Intelligent Transport Systems.* 1(2) 69-74.

Blechko, A., Darker, I., and Gale, A., 2008. Skills in detecting gun carrying from CCTV. *42nd Annual IEEE International Carnahan Conference on Security Technology,* 13-16 October, 265-271.

Boukerche, A., Zhang, M., and Pazzi, R.W., 2009. An adaptive virtual simulation and real-time emergency response system. *Proceedings from the International Conference on Virtual Environments, Human-Computer Interfaces and Measurements Systems,* Hong Kong, China May 11-13.

Dadashi, N., Pridmore, T., and Stedmon, A.W., 2009. Automatic components of integrated CCTV surveillance systems: Functionality, accuracy and confidence, *Sixth IEEE International Conference on Advanced Video and Signal Based Surveillance*, Genova, Italy 2-4 September, 376-381.

Dee, H., and Velastin, S., 2008. How close are we to solving the problem of automated visual surveillance? *Machine Vision and Applications.* 19(5) 329-343.

Fiore, S.M., Jentsch, F., Bowers, C.A., et al. 2003. Shared mental models at the intra- and inter-team level: Applications to counter-terrorism and crisis response for homeland security. *Human Factors and Ergonomics Society Annual Meeting Proceedings, Cognitive Engineering and Decision Making.* 439-443.

Fiedrich, F., and Burghardt, P., 2007. Agent-based systems for disaster management. *Communications of the ACM Archive.* 50(3), 41-42.

Gale, A.G., Purdy, K., and Wooding, D., 2005. Designing out terrorism: Human factors issues in airport baggage inspection. In: D. de Waard, K.R. Brookhuis, R. van Egmond, and T. Boersema (Eds) *Human Factors Design, Safety, and Management.* Maastricht: Shaker.

Hancock, P. A., and Hart, S. G., 2002. Defeating terrorism: what can human factors/ergonomics offer? *Ergonomics in Design.* 10, 6-16.

HM Government., 2009. *Pursue, Prevent, Protect, Prepare: The United Kingdom's Strategy for Countering International Terrorism.* Her Majesty's Stationary Office, ID 6072968 03/09 (Crown Copyright).

Jain, S., and McLean, C., 2003. A framework for modelling and simulation for emergency response. *Proceedings of the 35th Conference on Winter Simulation.*1068-1076.

Karasová, V., and Lawson, G., 2008. Methods for predicting human behaviour in emergencies: An analysis of scientific literature. *Proceedings of the GIS Research UK 16th Annual conference GISRUK 2008*, Manchester Metropolitan University, 2-4 April, 217-221.

Keval, H.U., and Sasse, M.A., 2010. Not the usual suspects: A study of factors reducing the effectiveness of CCTV. *Security Journal.* 23(2), 134–154.

Kleiner, B. M., 2006. Macroergonomics: Analysis and design of work systems. *Applied Ergonomics.* 37, 81-89.

Kraemer, S., Carayon, P., and Sanquist, T. F., 2009. Human and organisational factors in security screening and inspection systems: Conceptual framework and key research needs. *Cognition, Technology and Work.* 11, 29-41.

Lawson, G., Stedmon, A.W., Saikayasit, R., et al. In Print. Interpreting deceptive behaviors. *Proceedings of 4th International Conference on Applied Human Factors and Ergonomics (AHFE 2012).* 21-25 July, San Francisco.

Linett, H., 2005. *Living with Terrorism: Survival Lessons from the Streets of Jerusalem,* Boulder, Colarado: Paladin Press.

Meador, D.P., McClure, R.S., and Brett, B.E., 2006. Development of a virtual terrorist. Proceedings of the *Human Factors and Ergonomics Society Annual Meeting.* San Francisco, 16-20 October, 2255-2258.

Pape, R.A., 2005. *Dying to Win. The Strategic Logic of Suicide Terrorism.* Random House Inc.

Saikayasit, R., Stedmon, A.W., Lawson, G., and Fussey, P. In Print. User requirements for security and counter-terrorism initiatives. *Proceedings of 4th International Conference on Applied Human Factors and Ergonomics (AHFE 2012).* 21-25 July, San Francisco.

Smith, P.A., Baber, C., Hunter, J., et al. 2008. Measuring team skills in crime scene investigation: Exploring ad-hoc teams. *Ergonomics,* 51(10), 1463-1488.

Smith, S., and Ericson, E., 2009. Using immersive game-based virtual reality to teach fire-safety skills to children. *Virtual Reality.* 13, 87-99.

Stoto, M.A., Schonlau, M., and Mariano, L.T., 2004. Syndromic surveillance: Is it worth the effort? *Chance.* 17(1), 19–24.

Swift, C., Rosen, J.M., Boezer, G., et al. 2005. Homeland security and virtual reality: Building a strategic adaptive response system (STARS). In, J.D. Westwood, R.S. Haluck, H.M. Hoffman, et al. (Eds.) *Medicine Meets Virtual Reality: 13*. IOS Press, 549-555.

Troscianko, T., Holmes, A., Stillman, J., et al. 2004. What happens next? The predictability of natural behaviour viewed through CCTV cameras. *Perception.* 33, 87-101.

von Bastian, C.C., Schwaninger, A., and Michel, S., 2008. Do multi-view X-ray systems improve X-ray image interpretation in airport security screening? *Zeitschrift für Arbeitswissenschaft.* 3, 165-173.

Warm, J.S., Matthews, G. and Finomore, V.S. (2008) Vigilance, workload, and stress. In, P.A. Hancock, and J.L. Szalma (Eds) *Performance under Stress*. Abingdon, Oxon, GBR: Ashgate Publishing, Limited. 115-141.

Wood, J., and Clarke, T., 2006. Practical guidelines for CCTV ergonomics. In, R.N. Pikaar, E.A.P. Koningsveld and P.J.M. Settels (Eds) *Proceedings of IEA 2006 Congress, International Ergonomics Association 16th Triennial Congress, Maastricht*. The Netherlands, July 9-14, 2006, Elsevier, ISSN: 0003-6870, CD-ROM.

Zarraonandia, T., Vargas, M.R.R., Díaz, P., et al. 2009. A virtual environment for learning airport emergency management protocols. In, J.A. Jacko (Ed) *Human-Computer interaction. Ambient, Ubiquitous and Intelligent Interaction*. Springer Berlin/Heidelberg. 228-235.

# CHAPTER 27

# Interpreting Deceptive Behaviors

*G. Lawson, A.W. Stedmon, R. Saikayasit, E. Crundall*

Human Factors Research Group, Faculty of Engineering, The University of Nottingham, Nottingham, UK

glyn.lawson@nottingham.ac.uk

## ABSTRACT

This chapter reports recent research into observable behaviors as indicators of deceptive intent. In an initial study (Lawson et al., 2011), participants were either told to lie or tell the truth during an interview. They were covertly recorded while waiting for the interview. Furthermore, half of the participants faced a mirror during the waiting period; the others faced a blank wall. Analysis of their behavior revealed that without a mirror, participants expecting to lie spent less time moving their hands than those expecting to tell the truth; the opposite was seen in the presence of a mirror. This finding was used as the basis for further analyses of specific hand movements which showed liars spent significantly less time touching objects (other than their own bodies/hands) than truth tellers. In accordance with previous literature, the effect sizes seen were small, yet contribute to an understanding of observable indicators of deceptive behavior.

**Keywords**: deceptive behavior, body movement

## 1    INTRODUCTION

This paper presents a study conducted as part of a major UK security consortium (Shades of Grey: EP/H02302X/1) that aims to develop a suite of interventions for identifying terrorist activities. The consortium was funded in response to an increase in acts associated with terrorism (Home Office Statistical Bulletin, 2010).

Prior research has demonstrated that there is no single, reliable cue that can be used to identify people who are acting deceptively (Vrij, 2004). However, liars may experience one or more of three interlinking processes: emotion, cognitive effort and attempted behavioral control, which evoke verbal, non-verbal or physiological responses that differ to those made by truth-tellers (Zuckerman et al., 1981; Vrij, 2004). Several non-verbal behaviors have previously been associated with the processes mentioned above (e.g. smiling, Memon et al., 2003, gaze aversion, Vrij, 2008), although the effect sizes for these behaviors are often small and therefore can at best provide weak cues to deception (e.g. DePaulo et al., 2003).

Research has demonstrated that one approach to improve observers' ability to detect deceit is to increase the cognitive load of suspects, such as asking them to report events in reverse order (Vrij et al., 2008). However, Lawson et al. (2011) identified that monitoring the suspect *prior* to an interview may yield further clues to deception. Most previous work has focused on deceptive behavior demonstrated *during* the interview itself.

The analysis presented in this chapter extends the original study conducted by Lawson et al. (2011). The methodology and main findings of the original study are summarized below; the further analysis conducted for this chapter is reported in section 2.

For the original study (Lawson et al., 2011), 80 participants were recruited from the undergraduate student population at the University of Nottingham. They were invited to take part in a study to understand deception skills in interview; few other details were given about the purpose of the study during the recruitment stage. Participants were randomly assigned to one of two deception levels: either truth-telling or lying. The truth-tellers were told that they would be interviewed about their degree courses and should give only truthful answers; the liars were told that they should give no truthful answers. While waiting for the interview, half of the participants in each group faced a full-length mirror; the other half faced a blank wall. This intervention was studied as raised self-awareness has previously been shown to affect study participants' performance through a focus and evaluation of oneself (Wicklund and Duval, 1971). However, the effects of raised self-awareness had not previously been investigated on those intending to deceive.

The researcher left the participants under the pretence of going to find the interviewer. The participants were discretely recorded by a hidden camcorder for five minutes until the researcher returned and debriefed them on the true purpose of the study, which was an investigation of observable behavior while waiting to be interviewed.

The participants' behavior captured on the covert video camera was subsequently analyzed. Their body movements were recorded according to a high-level taxonomy. This included the duration and frequency of movement for hand/arm (i.e. any movement on either hand or arm); foot or leg (i.e. any movement on either foot or leg); or whole body. Gaze direction was also recorded. The taxonomy was based on prior research into indicators of deceptive behavior (Ekman, 1997; DePaulo et al., 2003; Memon et al., 2003; Vrij 2004, 2008; Ekman and O'Sullivan, 2006). Analysis of the behavior obtained from the video footage showed a significant interaction for the duration of hand/arm movements between deception level and self-awareness. Participants expecting to lie spent more time moving their hands in the presence of a mirror; the opposite was seen for those expecting to tell the truth (Lawson et al., 2011).

The significant interaction suggested that monitoring behavior while manipulating self-awareness during the pre-interview setting may have potential for identifying intention to deceive. However, it was recognized that the taxonomy used for coding the behaviors looked only at the duration of time spent moving hands/arms, while participants actually displayed a range of different actions (e.g. tapping, scratching face) which may provide more reliable indicators of deceptive behavior.

## 2. ANALYSIS OF TYPES OF HAND MOVEMENTS

Thus, based on the original study (Lawson et al., (2011), further investigation was made into the types of hand movements participants demonstrated while waiting to be interviewed. The same video footage was used for the analysis and participants were analyzed according to one of four conditions:
1. Expecting to tell the truth/ high self-awareness (truth/mirror)
2. Expecting to tell the truth/low self-awareness (truth/no-mirror)
3. Expecting to lie/high self-awareness (lying/mirror)
4. Expecting to lie/low self-awareness (lying/no mirror)

## 2.1 Method

Initially, the video footage was reviewed to identify all different categories of hand movements. This review identified the following sub-categories:

- **Hand-Hand** – this was when the participants' hands were touching each other, for example, drumming fingers together, picking nails, wringing hands, stroking one hand with the other.
- **Hands Only and Separately** – this category included drumming fingers or repetitive tapping of one or both hands separately, clenching fists, stretching fingers, rubbing fingers on same hand(s) together, flexing arm muscles and stretching arms. The first criterion for this category was that the hands were not in contact with one another. Second, the hands were not in contact with anything else, except for when the participant was drumming fingers or repetitively tapping another object or their own body.
- **Adjusting Clothes** – including fastening or unfastening straps or zips, pulling down top, turning up trousers, picking off fluff
- **Touching Face** – rubbing, picking, scratching or stroking the face, excluding repetitive finger movements such as drumming or tapping. This excluded bringing the hand up to the face as a result of an involuntary movement such as a sneeze.
- **Touching Hair** – as with touching face, but the moving hand was in contact with the hair.
- **Touching Body, Arms or Legs** – as with touching face, but the

moving hand was in contact with the body, arms or legs. The hands could be in contact with clothes rather than skin, but this category differed from 'Adjusting Clothes' in that the state of the clothing was not altered in any way as a result. For example, rubbing a sleeved arm with the opposing hand would be classified as 'Touching Body' whereas pushing up a sleeve would be classified as 'Adjusting Clothes'.

- **Biting Nails**
- **Touching Other Object** – touching an object rather than the body, such as a mobile phone, handbag, switches. This excluded repetitive finger movements such as drumming or tapping. It also excluded touching the chair to aid standing or a general shift of body position.
- **Folding/Unfolding Arms**
- **Other** – all other hand or arm movements. These were generally arm movements made as a result of the participant making a general shift of body position. It also included involuntary movements such as sneezing.

The original video footage was re-coded using the above hand sub-categories. The list of sub-categories encompassed all possible hand movements (exhaustive) but did not overlap in any way (mutually exclusive).

The total amount of time spent carrying out each type of hand-movement (in seconds) was analyzed using a series of 2 x 2 ANOVAs, each with self-awareness (mirror vs. no mirror) and deception level (liars vs. truth tellers) as between groups variables. Thus, the analysis aimed to identify whether type of hand movement differed between liars or truth tellers, or between the high and low self-awareness conditions, or whether there were any significant interactions.

## 2.2    Results

The results are presented below, categorized by each hand sub-category.

### Hand-Hand

There were no effects of deception or self-awareness, and no interaction.

**Table 1  Hand-hand behavior**

| Effect | F | df | p | Eta$^2$ |
|---|---|---|---|---|
| Deception level | 0.293 | 1,76 | NS | 0.004 |
| Self-awareness | 1.176 | 1,76 | NS | 0.015 |
| Deception*self-awareness | 3.177 | 1,76 | NS | 0.040 |

## Hands Only and Separately

The effect of deception level approached significance with liars (mean=26.197s) demonstrating more of this type of hand movement than truth-tellers (mean=13.681s).

**Table 2  Hands only and separately behavior**

| Effect | F | df | p | Eta$^2$ |
|---|---|---|---|---|
| Deception level | 3.642 | 1,76 | 0.060 | 0.046 |
| Self-awareness | 0.719 | 1,76 | NS | 0.009 |
| Deception*self-awareness | 0.228 | 1,76 | NS | 0.003 |

## Adjusting Clothes

There were no effects of deception or self-awareness, and no interaction.

**Table 3  Adjusting clothes behavior**

| Effect | F | df | p | Eta$^2$ |
|---|---|---|---|---|
| Deception level | 0.102 | 1,76 | NS | 0.001 |
| Self-awareness | 0.396 | 1,76 | NS | 0.005 |
| Deception*self-awareness | 0.069 | 1,76 | NS | 0.001 |

## Touching Face

There were no effects of deception or self-awareness, and no interaction.

**Table 4  Touching face behavior**

| Effect | F | df | p | Eta$^2$ |
|---|---|---|---|---|
| Deception level | 0.340 | 1,76 | NS | 0.004 |
| Self-awareness | 0.077 | 1,76 | NS | 0.001 |
| Deception*self-awareness | 0.214 | 1,76 | NS | 0.003 |

## Touching Hair

As might be expected, participants spent more time touching or playing with their hair more in the presence of a mirror (mean=27.168s) than in the no mirror condition (mean=10.246s). No other effects reached statistical significance.

Table 5 Touching hair behavior

| Effect | F | df | p | Eta$^2$ |
|---|---|---|---|---|
| Deception level | 0.022 | 1,76 | NS | 0.000 |
| Self-awareness | 6.473 | 1,76 | 0.013 | 0.078 |
| Deception*self-awareness | 0.173 | 1,76 | NS | 0.002 |

## Touching Body, Arms or Legs

There were no effects of deception or self-awareness, and no interaction.

Table 6 Touching body, arms or legs behavior

| Effect | F | df | p | Eta$^2$ |
|---|---|---|---|---|
| Deception level | 0.702 | 1,76 | NS | 0.002 |
| Self-awareness | 0.251 | 1,76 | NS | 0.017 |
| Deception*self-awareness | 0.379 | 1,76 | NS | 0.010 |

## Biting Nails

The ANOVA revealed no statistically significant effects, but participants bit their nails slightly more in the no mirror condition (mean = 6.761s) than in the mirror condition (mean = 1.468). Counter to expectations, truth-tellers (mean = 6.644s) bit their nails slightly more than the liars (mean = 1.585s). However, the reliability of this analysis may have been reduced by the large number of participants who did not engage in any nail-biting, resulting in a large number of zeros in the dataset.

Table 7 Biting nails

| Effect | F | df | p | Eta$^2$ |
|---|---|---|---|---|
| Deception level | 3.512 | 1,76 | 0.065 | 0.044 |
| Self-awareness | 3.844 | 1,76 | 0.054 | 0.048 |
| Deception*self-awareness | 2.649 | 1,76 | NS | 0.034 |

## Touching Other Object

There was a significant effect of deception level, with liars touching other objects (mean=7.066s) less than truth-tellers (mean=26.438s). Self-awareness did not reach significance, and there was no interaction.

Table 8  Touching other object behavior

| Effect | F | df | p | Eta$^2$ |
|---|---|---|---|---|
| Deception level | 4.819 | 1,76 | 0.031 | 0.060 |
| Self-awareness | 0.215 | 1,76 | NS | 0.003 |
| Deception*self-awareness | 0.144 | 1,76 | NS | 0.002 |

## Folding/Unfolding Arms

There were no effects of deception or self-awareness, and no interaction.

Table 9  Folding/unfolding arms  behavior

| Effect | F | df | p | Eta$^2$ |
|---|---|---|---|---|
| Deception level | 0.690 | 1,76 | NS | 0.009 |
| Self-awareness | 1.555 | 1,76 | NS | 0.020 |
| Deception*self-awareness | 0.029 | 1,76 | NS | 0.000 |

## Other

There were no effects of deception or self-awareness, and no interaction for the *other* category.

Table 10  Other  behavior

| Effect | F | df | p | Eta$^2$ |
|---|---|---|---|---|
| Deception level | 0.917 | 1,76 | NS | 0.012 |
| Self-awareness | 1.717 | 1,76 | NS | 0.022 |
| Deception*self-awareness | 0.521 | 1,76 | NS | 0.007 |

## 3.    Discussion

The most notable behavior was for *touching other object* (F=4.819; df=1,76; p=0.031; Eta²=0.06) for which liars spent significantly less time moving their hands than for truth tellers. Thus, this category may have potential for identifying those intending to deceive. *Hands only and separately* (F=3.642; df=1,76; p=0.06; Eta²=0.046) approached significance, with liars spending more time moving their hands than truth tellers. However, supporting previous research in this area, the effect sizes in both categories are relatively small, and therefore this approach is likely to have limited use for detecting deception during a live observation.

*Biting nails* approached significance for both deception level (F=3.512; df=1.76; p=0.065; Eta²=0.044) and self-awareness (F=3.844; df=1.76; p=0.054; Eta²=0.048) although the fact that several participants spent no time biting their nails reduced the reliability of this data, and therefore conclusions cannot be drawn about the usefulness of this measure for detecting deceptive behavior. Touching hair demonstrated a significant main effect of self-awareness (F=6.473; df=1,76; p=0.013; Eta²=0.078), with those in front of a mirror spending more time touching or playing with their hair. This result would be expected and in itself might not useful for detecting deceptive behavior.

It is noteworthy that the primary finding of the original study (Lawson et al., 2011) was an interaction between deception level and self-awareness for duration of hand movements and yet no significant interactions were found in this study. Thus, the differences between liars and truth-tellers found in this study (*touching other object*) did not differ by self-awareness level. Future work aims to investigate whether categorizing the behaviors by the emotion of which they are indicative (e.g. anxiety, boredom) reveals any main effects or interactions to identify whether raised self-awareness is a useful intervention for amplifying the differences between truth-tellers and liars. However, this approach introduces greater levels of subjectivity (e.g. whether finger tapping is indicative of anxiety or boredom) and therefore it is important that a reliable coding scheme is developed. Other future work also aims to evaluate the instinctual response of an observer, for example whether people are able to determine simply by watching the video whether a participant intends to lie.

## 4.    CONCLUSIONS

The analysis conducted for this investigation revealed that participants who are expecting to lie in an interview spend less time touching objects other than their own bodies than participants who are expecting to tell the truth. Moving hands only (and separately) approached significance. However, and in support of previous literature, the effect sizes seen were small. Differences in time spent biting nails between truth tellers and liars cannot reliability be concluded as indicative of deception. Further work is needed to understand the accuracy of subjective judgments made by observers of the video footage.

## ACKNOWLEDGMENTS

The authors acknowledge the EPSRC for funding this work as part of the Shades of Grey EPSRC research project (EP/H02302X/1).

## REFERENCES

DePaulo, B. M., Lindsay, J. J., Malone, B. E., et al. 2003. Cues to deception. *Psychological Bulletin.* 129: 74-118.

Ekman, P., 1997. Deception, lying and demeanor, in: *States of mind: American and post-soviet perspectives on contemporary issues in psychology,* eds. D.F. Halpern, A Voiskunskii. Oxford: Oxford University Press, 93-105.

Ekman, P., and O'Sullivan, M., 2006. From flawed self-assessment to blatant whoppers: the utility of voluntary and involuntary behavior in detecting deception. *Behavioral Sciences & the Law,* 24: 673-686.

Home office statistical bulletin. 2010. Accessed February 2010, www.statistics.gov.uk

Lawson, G., Stedmon, A., Zhang, C., et al. 2011. Deception and Self-Awareness. *Engineering Psychology and Cognitive Ergonomics, Proceedings of the 9th International Conference, EPCE 2011, Held as Part of HCI International 2011, Orlando 9-14 July 2011.* Heidelberg: Springer.

Memon, A., Vrij, A., and Bull, R., 2003. *Psychology and law: truthfulness, accuracy and credibility* (2nd ed.). Wiley, Chichester, 1-55.

Vrij, A., 2004. Why professionals fail to catch liars and how they can improve. *Legal and Criminological* Psychology. 9: 159-181.

Vrij, A., 2008. *Detecting lies and deceit: pitfalls and opportunities* (2nd ed.). West Sussex: Wiley, 1-188.

Vrij, A., Mann, S.A., Fisher, R.P., et al. 2008. Increasing cognitive load to facilitate lie detection: the benefit of recalling an event in reverse order. *Law and Human Behavior,* 32: 253-265.

Wicklund, R.A., Duval, S. 1971. Opinion change and performance facilitation as a result of objective self awareness. *Journal of Experimental Social Psychology* 7: 319–342.

Zuckerman, M., DePaulo, B. M., and Rosenthal, R., 1981. Verbal and nonverbal communication of deception. In: Advances in Experimental Social Psychology (Vol. 14), ed L. Berkowitz, New York: Academic Press, 1-57.

# CHAPTER 28

# User Requirements for Security and Counter-terrorism Initiatives

*R. Saikayasit[1], A. Stedmon[1], G. Lawson[1] and P. Fussey[2]*

[1]Human Factors Research Group, Faculty of Engineering,
The University of Nottingham
Nottingham, UK
[2]University of Essex, Essex, UK
Rose.saikayasit@nottingham.ac.uk

## ABSTRACT

The involvement of end-users and stakeholders in the development of a product or process is considered essential as their reactions, interactions, use and acceptance dictate the effectiveness, success and the overall performance of any implementation. End-user and stakeholder involvement has been adopted as part of user-centered design methodology in many fields such as healthcare, health and safety, product design and human-computer interaction. This paper describes the application of user requirements gathering methods from existing human factors techniques and discusses the specific issues and challenges of requirements generation in the sensitive domain of security and counter-terrorism. The knowledge was gained during the development of design specifications for a suite of interventions for two distinct end-user groups: front-line security personnel and top-level security agencies. A case study is presented of the user requirements elicitation conducted at a major entertainment venue. Furthermore, key emergent themes were identified and are reported.

**Keywords**: user requirements, security and counter-terrorism, end-user and resource management, sensitive domains

# 1    INTRODUCTION TO USER REQUIREMENTS

This paper presents aspects of a user requirements elicitation exercise conducted as part of the UK 'Shades of Grey' security consortium. This research aims to design and develop a suite of interventions that can be applied in crowded public spaces to amplify the signal-to-noise ratio of suspicious behaviors in order to improve the rate of real-time detection of terrorist activities.

User requirements elicitation is an important approach forming part of the user-centered design methodology. This approach ensures that both end-users and stakeholders are fully involved throughout the design and development process so that their needs, limitations, capabilities and preferences are understood and designed for in the system or equipment, with consideration of the context of use (Wilson, 1995).

While user requirements elicitation is generally considered essential to the development of products and processes (Wilson, 1995), involving end-users and stakeholders is often a challenge to researchers, especially in domains where they are difficult to access, recruit and manage (Lawson and D'Cruz, 2011) and information is generally confidential and restricted to outsiders (Lawson *et al.*, 2009). These challenges are expanded upon below, after an overview of user requirements methodologies.

## 1.1    User requirements elicitation methodology

There is a growing interest in user-centered design, which enables end-users to participate in the development process (Salvo, 2001). Traditionally, expert-centered approaches focus the design work on the developers' observations and interpretations of the context of use along with a vision of what the products should be and how it should be used by the end-users (Salvo, 2001). Human factors methods such as questionnaires, interviews, focus groups, observations, tasks analysis and self-recorded diaries can all be used as part of the user requirements elicitation (Preece *et al.*, 2007). These methods provide different levels and types of data. Thus a range of methods are often selected to complement each other with consideration of the resources available (e.g. time, number of researchers and cost), and the preferences and capabilities of end-users.

Interviews can be used to provide insight into thought processes or highlight problems from the perspectives of end-users. However this process takes time and may require end-users to neglect their primary work tasks which can affect their willingness to participate. In contrast, questionnaires can be completed in a relatively short period of time and can be administered to a larger number of respondents, with lower costs, however the data gathered can lack detailed insight and can be inefficient for probing complicated issues (Sinclair, 2005). Focus groups can also be used to gather requirements where the researcher acts as facilitator and prompter to encourage participants to discuss issues around pre-defined themes. However this method can be difficult to apply in sensitive domains where the degree of anonymity and confidentiality may prevent end-users from discussing issues in front of other participants.

Other challenges in sensitive domains can include: difficulty in gaining access to information that is generally confidential and restricted to outsiders (Lawson *et al.*, 2009); the necessity to involve vulnerable participants (e.g. young children, differently-abled patients and elderly users) to reflect the breadth of the target population (Salvo, 2001); and dealing with unpleasant topics with potential to distress participants, such as studies of survivors of the World Trade Center (e.g. Gershon *et al.*, 2007) and posttraumatic stress disorder in police officers after an emergency (e.g. Carlier *et al.*, 1997).

Specific methods for identifying and gathering user needs in the security domain are under-developed which presents complications and challenges in the design process of security solutions. Furthermore, existing methods such as observational techniques including ethnography can be considered too intrusive, disruptive and inappropriate (Crabtree *et al.*, 2003). For example, when security personnel are dealing with members of the public, it would be inappropriate and unethical to video their interaction without informed consent. In such contexts the aim is often to observe natural behaviors and the presence of the researchers and video cameras could alter behavior, thus changing the nature of the interaction altogether.

## *1.2*   Security from a human factors perspective

In order to understand the full context of use in security applications, a macro-ergonomics system approach was adopted within the Shades of Grey project, as part of the user requirements gathering and analysis. This approach focuses on how different factors within a system as a whole interact as well as influence one another, such as when more than one individual interacts with other people, technologies and/or other artifacts (Kleiner, 2006). Macro-ergonomic work system models have previously been adopted to elucidate factors influencing the performance of security screening and inspection (SSI) systems of cargo and passengers (Kraemer *et al.*, 2009). These SSI systems are used to identify possible chemical, biological, radiological and nuclear threats on passenger and cargo transits across US borders by air, rail and sea (Kraemer *et al.*, 2009). In various contexts, security personnel are often required to monitor multiple channels over extended periods of time (e.g. border control desks, baggage handling, CCTV control rooms) whilst detecting low probability events (Kraemer *et al.*, 2009).

Many aspects of security and threat identification still depend on the performance of frontline security personnel who can be influenced by aspects such as organizational factors (e.g. training, management support, culture and structures), technologies and tools (whether they help improve the work or are seen as added tasks), operational environment (noise, climate, temperature) and workload (e.g. underload/overload, repetition/monotony) shown in Figure 1.

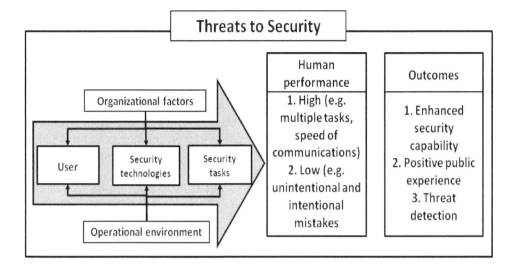

Figure 1: Macro-ergonomic conceptual framework for SSI systems (adapted from Kraemer *et al.*, 2009).

Following the user-centered approach, the central factor in the work system is the front-line security worker whose skills, tacit knowledge, unique characteristics, perceptions, behaviors and needs all contribute to the performance outcomes of the overall system. The macro-ergonomic framework has been used to form a basis for understanding user requirements within the Shades of Grey project, focusing on the interacting factors and their influence of overall performance.

## 2   SHADES OF GREY – USER REQUIREMENTS CASE STUDY

A series of user requirements gathering exercises were conducted at an entertainment venue (referred to as 'Venue X'). To ensure anonymity and confidentiality, the full nature of the work, staffing, location and other sensitive information are not fully disclosed.

'Venue X' is an organization that owns and runs an indoor venue, hosting large-scale national and international events for up to 10,000 attendees at any one time. They also have facilities available for conferences, classes and activities for a range of age groups. The venue is open to the general public throughout the day and often late into the evenings, even if a large-scale event is taking place at the same time. The organization has in-house employees but also employs a security company to supervise large-scale events.

## 2.1 Method

The user requirements process began with the development of a set of questions required to develop the researchers' understanding of the overall work process at 'Venue X'. These questions were designed to assess factors as outlined by the macro-ergonomic framework such as organizational, operational, technologies and tools, tasks/workload and human security personnel. 'Venue X' was recruited using a snowball sampling method, where researchers from the Shades of Grey project were introduced to the venue by another agency who participated in an earlier user requirements gathering exercise. This sampling method was used as security agencies and organizations are often reluctant to share confidential and sensitive information to outsiders. As a referral was made by a security agency 'Venue X' agreed to take part in the exercise.

## 2.2 Semi-structured interviews

Informal semi-structured interviews were conducted at an initial meeting with managers to gain an understanding of their day-to-day tasks and procedures in dealing with different aspects of security. This process enabled participants to discuss their work while the researchers assumed the roles of 'learners' rather than 'hypothesis testers' (McNeese et al., 1995). This approach also enabled the researchers to gather background information regarding the organizational culture, the work schedules, and the way different teams within the organization interact and communicate with one another. The initial interviews were transcribed and analyzed for emergent themes, categorizing the specific focus of the work at 'Venue X'.

## 2.3 Field observation

An eight hour field observation took place at 'Venue X' during an evening event where approximately 10,000 people attended. The purpose of this field observation was to allow researchers to gain a deeper understanding of the existing security activities in a live setting and seeing emergent themes gathered from the informal interviews in a real context of work. The field observation focused on communication networks, tasks, contexts and protocols, some of which were designed specifically for the event.

Prior to the event, several management and supervisor meetings took place as different teams organizing the event worked together to ensure a smooth running of the evening. The event manager in charge of the organization for the event had been working on arrangements leading up to that evening several months in advance, and was thus able to pass over information and inform managers of different departments of decisions which have been made prior to the day of the event.

The first pre-event meeting took place between managers and supervisors, who later held meetings within their teams to disseminate information to their staff. Smaller teams included security staff (including door supervisors, front-line personal supervising the venue, CCTV operators and supervisors patrolling the venue).

Based on past experience of similar events, and coupled with event intelligence from other venues 'Venue X' was able to predict the crowd demographics and expected behaviors for the event. This enabled them to implement coping strategies and safety precautions for the event to support security personnel in the venue. This was also an important aspect of supporting the general public's experience of the event so that their user needs were taken into account as well.

Informal interviews with staff on duty also took place during the observation on an ad-hoc basis. Information gathered from these interviews, the field observation and the initial information interviews was used to contribute to the framework of emerging themes.

## 2.4 Results

The data gathered from interviews and the field observation was used to tabulate a summary of results which illustrated the factors influencing the overall performance of 'Venue X' for different groups of staff within the organization. Factors reported by Kraemer *et al.* (2009) which influence human performance and the overall work system were used to categorize results.

It became apparent that 'Venue X' relied on several teams to work closely with each other to ensure safety and customer satisfaction during an event. These teams included staff in the two control rooms, responsible for different areas of the venue, a security team overseeing entries and exits, security personnel on the floor during the event, and the managers on duty who oversaw the venue throughout. One of the main priorities was to provide good customer service and ensure the safety of attendees, artists (i.e. performers) and staff. The factors affecting the human performance of each team are summarized in Table 1.

Table 1 Summary of results from interviews and observation at 'Venue X'

| Factors | Control rooms | Security Team |
|---|---|---|
| Organizational factors | • Control rooms are central to the organization<br>• Operators coordinate activities around the venue during an event<br>• Managers often leave staff to respond to incidences and will only advise on request or in an emergency<br>• Management emphasizes the importance of communication and encourages staff to communicate incidents to supervisors | • Security at 'Venue X' is outsourced to a professional event management company<br>• Tasks include patrolling perimeter, crowd control, access control and screening patrons<br>• The head of the team joins the management meeting with staff at 'Venue X' then passes on instructions and protocols to his/her team |
| Technologies/tools | • CCTV cameras are positioned over the venue<br>• Cameras can be viewed, controlled (i.e. zoom and reposition) from the control rooms.<br>• Incidents are reported to the control rooms so that CCTV cameras can be positioned to capture them. | • Supervisors have radios which can be used to contact the head of security and other staff within their team<br>• Supervisors and head of security contact control room if any incidents need coordinating |
| Operator/user characteristics | • Being very integral to the whole operation, staff Security Industry Authority licensed to monitor CCTV and are highly experienced and highly trained for large scale events<br>• 'Venue X' ensures that the most experienced staff are on duty during large events. | • Supervisors and head of security are all Security Industry Authority (SIA) licensed<br>• Less experienced staff are often paired with those with more experience |

| Factors | Control rooms | Security Team |
|---|---|---|
| Tasks/workload | • The Workload is not evenly distributed among control rooms.<br>• 'Control Room A' is responsible for constantly coordinating all event related activities (e.g. between floor supervisors, paramedics and housekeeping). | • As security staff work as a team and have been assigned specific tasks which they are responsible for within the team, the workload is less, however they have to stay vigilant and those stationed on the floor, where people were able to stand and watch the performance, were required to interact with clients throughout to ensure safety. |
| Operational environment | • Control Room A' is located near the performance area, therefore the noise level is extremely high, causing constant vibrations in the room. The glass panel overlooking the venue also meant all lights had to be switched off, so staff work in the dark during an event. | • Most security staff are stationed in the floor during performances<br>• This means the noise level (combined with ear protection) makes communication<br>• The standing floor often gets overheated with hundreds of people and heat exhaustion is not uncommon. |

## 3  DISCUSSION

It can be seen from Table 1 that the main security work during an event, or even on a day-to-day basis at 'Venue X' is conducted by different teams of staff, with different needs, priorities and backgrounds. Management operate a 'hands-on' philosophy with the day-to-day running of the venue and are present at management and supervisors meetings leading up to an event. Two managers are always on duty during an event and are contactable via the telephone and radio communication networks. One of the managers will be assigned the task of the 'decision maker' during an event and will be seated in one of the control rooms. The decision maker will have the ultimate control in case of an emergency (i.e. whether to evacuate, call in emergency responders). The other manager on duty during an event will patrol the venue performing several safety checks (e.g. fire exits) and dealing with localized incidents when requested by staff.

All teams work together and collaborate during an event to achieve the same shared goals. Each smaller team has their own goals that they need to accomplish which contribute to the overall performance of 'Venue X' as a total system.

Interviews with management suggested that they are keen on encouraging front-line staff to understand the thinking behind different sets of protocols and safety measures, so that staff can assess the situation, and understand how protocols can be implemented or applied. They encourage personnel to communicate anything they see which might be 'out of the ordinary' and to report these to the control rooms. The control room operators will then decide whether to contact management and get advice on the situation.

From this initial exploration into the user needs of 'Venue X' it is important to consider that different end-user groups have different needs, capabilities, priorities and experience when designing a series of interventions. The researchers now understand how security, as a process, operates within the current protocols that have been designed for different situations. This has led to the development of the next stage of user requirements elicitation and a key focus for the Shades of Grey consortium, how future interventions might be incorporated into established practices.

Interventions implemented to aid existing security practices should ensure that they are not perceived as extra work for staff. With the example of 'Venue X' control rooms staff can have extremely high workload during a busy event, keeping a log of all communications and incidents, which is necessary for any post-incident analysis but which is also regarded as additional work. Introducing additional steps or work to the control room might delay the coordination and communication of activities in an emergency. It is therefore vital and integral to the whole organization, particularly during a large scale event that such potential barriers to effective organization do not compromise the overall system integrity.

In order for a new system of interventions to be effectively used and successfully implemented, the various needs of different end-user groups as well as those of the stakeholders (i.e. decision and policy makers) need to be understood and addressed in the overall design and implementation of solutions. Failure to address needs of the end-users may lead to the lack of acceptability, lack of use of the system and compromise the overall effectiveness.

## 4    CONCLUSION

This case study has shown the importance in understanding the context of work and other factors contributing to the overall performance of a system of security through user requirements gathering exercises. Without interviews and observation work in situ, the lack of understanding and appreciation of the work context can lead to inappropriate design of interventions. This case study also highlighted the presence of different end-user groups within the same organization whose needs and requirements all need to be addressed in the design and implementation of future security solutions.

## ACKNOWLEDGMENTS

The authors would like to acknowledge the EPSRC for funding this work as part of the shades of Grey EPSRC research project (EP/H02302X/1). The authors would also like to thank all the stakeholder and end-users who took part in the user requirements elicitation.

## REFERENCES

Carlier, I. V. E., Lamberts, R. D. and Gersons, B. P. R. 1997. Risk factors of post-traumatic stress symptomatology in police officers: a prospective analysis. In *Journal of Nervous & Mental Disease* 185(8): 498-506

Crabtree, A., Hemmings, T., Rodden, T., Cheverst, K., Clarke, K., Dewsbury, G., Huges, J. And Rouncefield, M. 2003. Designing with care: adapting cultural probes to inform design in sensitive settings. *Proceedings of OzCHI 2003*: 4-13

Gershon, R. R. M., Qureshi, K. A., Rubin, M.S. and Raveis, V. H. 2007. Factors associated with high-rise evacuation: qualitative results from the World Trade Centre evaluation study. In *Prehospital and Disaster Medicine* 22(3): 165-173

Kleiner, B. M. 2006. Macro-ergonomics: analysis and design of work systems. *Applied Ergonomics* 37: 81-89

Kraemer, S., Carayon, P. and Sanquist, T. F. 2009. Human and organisational factors in security screening and inspection systems: conceptual framework and key research needs. *Cognition, Technology and Work* 11: 29-41

Lawson, G., Sharples, S., Cobb, S. and Clarke, D. 2009. Predicting the human response to an emergency. In *Contemporary ergonomics* 2009, ed. P. D. Bust. Proceedings of the Ergonomics Society's Annual conference, London, UK, Taylor and Francis: 525-532

Lawson, G. and D'Cruz, M. 2011. Ergonomics methods and the digital factory. In *Intelligent Manufacturing system DiFac*, eds L. Canetta, C. Redaelli and M. Flores, London, Springer

McNeese, M. C., Zaff, B. S., Citera, M., Brown, C. E. and Whitaker, R. 1995. AKADAM: Eliciting user knowledge to support participatory ergonomics. *International Journal of Industrial Ergonomics* 15: 345-363

Preece, J., Rogers, Y. and Sharp, H. 2007. *Interaction design: beyond human-computer interaction*, 2nd edition, John Wiley & Son Ltd. Hoboken, NJ

Salvo, M. J. 2001. Ethics of engagement: user-centred design and rhetorical methodology. In *Technical Communication Quarterly* 10(3): 273-290

Sinclair, M. A. 2005. Participative assessment. In *Evaluation of Human Work: A Practical Ergonomics Methodology*, 2nd and Revised Edition, eds J. R. Wilson and E. N. Corlett, CRC Press, Taylor and Francis Group: 83-112

Wilson, J. R. 1995. Ergonomics and participation. In *Evaluation of Human Work: A Practical Ergonomics Methodology*, 2nd and Revised Edition, eds J. R. Wilson and E. N. Corlett, CRC Press, Taylor and Francis Group: 932-963

# Section IV

---

*Enterprise ICT and Work*

# CHAPTER 29

# The Influence of Enterprise Resource Planning in Role Management ☐ Case Study in a Given Portuguese Industry

*Ana Patrícia Pessoa Brito, Gilson Ludmer*

Federal University of Pernambuco, Recife - Brazil

## ABSTRACT

The Enterprise Resource Planning (ERP) is a technology increasingly present and represents big changes in organizations, causing changes in business practices, work processes, behaviors and functions. The central scope of this work was to understand the influence of Enterprise Resource Planning in managerial roles in the post-implementation in a given industrial company. Thus, was performed a case study using interpretative approach in a given timber industry and its products installed in Portugal. The interpretations of the results indicated a predominance of transactional use of ERP by a technical management, imprisoned for operating functions. The role of ERP is primarily perceived as an organizer and centralizer. The skills required by the ERP were basically associated with its use in direct data feed and imply disciplinary behaviors and attitudes regarding the need for organization, concentration and method, eliminating any possibility of improvisation and creation by the user and causing the loss of managerial autonomy. Concerning strategic issues the use of the ERP was perceived as insignificant.

**Keywords:** Enterprise Resource Planning. Managerial functions. Interpretive Case Study

## INTRODUCTION

Over the past few years, companies around the world have implemented the ERP which are considered a lever for organizations to gain competitive advantage. The use of an ERP system affects the organization's structure and actions of its users (Askenaes & Westelius 2000). Yeh and OuYang (2010) affirms that users are forced to learn a new way of working. For the authors, ERP is a significant factor that affects the tasks of managerial work and creates a new form of organization that challenges the traditional management model. Taking into account, the importance of ERP in the generation of changes in organizations and the impact that this technology can cause in the people, the general objective of this paper is to contribute to understanding the influence of Enterprise Resource Planning in managerial roles in post-implementation. The relevance of this academic study is to contribute to better understand how these technologies affect these management activities and interpretation of how these influences occur and are perceived in different organizational processes. More specifically, this study sought to investigate the occurrence of changes in managerial roles, analyze the perceptions of managers about the skills required; the policy issues associated with the use of enterprise resource planning and also analyze the impacts of ERP on the autonomy of managers. This article is based on a case study in a given manufacturing industry installed in Portugal where an interpretive approach was used. Subsequently, the theoretical elements used, the methodology, the results, and finally, conclusions are presented.

## 2 THEORETICAL REVIEW

Not limited exclusively to a specific model, this work was supported in studies of Mintzberg (2010) to understand what the roles of a manager in the organization are. According to Mintzberg (2010) managers need to act on three fronts: should be geared to people, to information and to actions, because this is the essential balance to the practice of management, which can lead to one side and to another, depending on the pressures of the moment. Classical literature presents the managerial roles as being very dynamic. Mintzberg (2010) gives the management the following characteristics: the relentless pace, the brevity and variety of its activities, fragmentation and discontinuity of work, the guidance for action. Weick (1983) confirms this dynamic nature of the managerial function stating that managerial work is not a discrete process, static, one decision at a time, because in fact there is an "flux and reflux" of meetings, requests, pressures and negotiations. It is also possible to identify in the literature focus on the strategic management function. The responsibility of leadership, decision-making power, the focus on enterprise results, making plans, uniting efforts and developing strategies to achieve them. Barnard (1968) stands out among other management functions to formulate and define the organizational objectives. Barnard (1968) says that the function of formulating grand purposes and to provide for a redefinition is sensitive system that requires managers, communication skills, experience in interpretation, imagination, and the delegation of responsibility.

The use of an ERP system affects the organization's structure and actions of its users (Askenaes & Westelius 2000). The level of influence of the ERP in the activities of stakeholders in organization is commented by Ludmer and Rodrigues (2003) who says that in the workplace, the ERP model seeks to provide a set of interconnected business processes that use the least amount of work possible. According to the author, since the ERP and its construction are based on the principles of reengineering, this goal is achieved through: redefinition of job roles in order that employees have multi-skills and can take action in several interrelated processes, and maintenance of power work under constant pressure, since the processes are highly interdependent and activities take place "just in time⬜ In his analysis of the case study, Yeh and OuYang (2010) reveals the process of mutual influence between the ERP system and its context, exploring the cultural, political and power issues. According to him, ERP users are forced to learn a new way of working and changing the way they think about their work and their organization. Beheshti (2006) states that the purpose of ERP is to increase the competitiveness of the company by improving its ability to generate accurate and timely information for management decision-making. Bahrami and Jordan (2009) however warns that despite the general view has been widely recognized about decision support functionality of ERP systems, managers do not always take advantage of information from the software in their decision-making processes. To Bahrami and Jordan (2009) the difficulty of achieving a reasonable level of integration between business processes and data that in many cases, leads to an ERP implementation coupled or semi-integrated is a major factor that prevents the use of ERP more strategic sense. The author states that the ERP modules coupled with a centralized mindset are contributing factors to transform the data that could be used for decision making at a strategic level in departmental data useful only for the operational use of day-to-day. Lim et al (2005) also warned that not all organizations have been able to put the potential of ERP systems in their strategic use. The authors point to research conducted by Ross and Vitale (2000) where it was observed that the use of ERP systems often remain in very fundamental level applications, namely ERP packages are often used to perform only simple business operations. Ranganathan and Kannabiran (2004) identify the approaches to management as a critical factor affecting the performance and strategic use of information systems in organizations. For the authors, the profile management and their attitude towards the ERP can be crucial to the field of technology, focusing on use-oriented strategic functions and extracting the best benefits to organizational management. Peng and Nunes (2009) also warns to the fact that possibility of managers fail to retrieve relevant and necessary information from the ERP system and thus have affected his decision-making process. Peng and Nunes (2009) indicate that business managers needs to have different information according to their needs, their styles of personal decision, their experience and real contexts from situations. For this purpose, format and content of reports generated by ERP systems should be changed in a flexible and customized according to the real needs of managers. However, as the author states, not all ERP packages available in the market can be flexible enough to meet this user requirement. Furthermore, structures, formats and content of reports generated in a specific national context can`t be easily used or even translated into other national contexts, foreign ERP systems can`t meet the needs of local businesses due to cultural and political differences. In this respect Ludmer and Rodrigues (2003) and Soh et al (2000) also argue that adaptation of ERP facilities are highly debatable because there may be considerable mismatches between the country, industry and specific business practices of the company and the reference models contained ERP systems.

According to Govers and Amelsvoort (2007) the introduction of ERP software indicates a change process, which is not always successful. The alleged benefits expected as a control, competitive advantage, efficiency and flexibility do not always appear. The image seems to have changed the ERP, a highly promising technology was seen as a highly demanding technology, and it is this socio-organizational ERP and increasingly demanding, which may have an influence on the bureaucracy. Govers (2003) states that the practice has shown that many ERP projects have serious difficulties with regard to customization, and what actually happens is that the company eventually become adapted to the ERP instead of this software being adapted to the company. Consequently the various organizational functions are required to build a new social reality within the limits of ERP, which in turn brings a uniform approach in order to impose normative and programmable behaviors. The ERP analysis by Govers (2003), led to a spiral of formalization and standardization exerting additional pressure on people to adapt to the rules on how to implement and manage business processes, as determined by the software. Govers (2006a) warns that in fact, the ERP consultants are subtly forcing organizations to adapt as much as possible to the ERP and intentionally, pushing companies to a more bureaucratic and away from the originally desired flexibility. The ERP does not change the fundamentals of bureaucratic organization, but it might involve the emergence of a new figure, which lives in a new kind of iron cage: the hummus erpius, the worker who lives in a bureaucracy in which the administration of the rules changed the place. (Ludmer & Rodrigues Filho 2003)

According to Govers (2006b) the bureaucratization of the ERP is therefore also a political management which clearly shows that the ERP can be used as an instrument of power. Govers (2006b) also warns that because of the idea that the behavior is normative and programmable, managers came to believe that people and processes can be organized efficiently. In other words, it is believed that the behavior can be calculated by applying as standard of obedience. However, this idea that the behavior is programmable is a mistake. In different contexts, often, makes no sense and is not feasible uniform behavior. Some circumstances has its own variety and dynamics that need localized and particular treatise.

Boersma and Kingma (2005) states that the standard procedures of ERP contains a kind of script to users, informing them of what actions should be taken, when, where and how. This script also implies system instructions at work on organizing activities between co-workers, with management and with other elements of the business cycle. Making an analogy Boersma and Kingma (2005) says that these normative scripts can operate like a highway in a traffic system, directing drivers to their destinations. In their work, Ludmer and Falk (2007) identified the ERP as associated with centralization and loss of autonomy, seen as playing a role as technocrat. In the company studied, Ludmer and Falk (2007) state that the ERP is seen as an organizer, but in such a way that seems to influence significantly the autonomy, thinking and acting of managers and employees. Moreover, in all areas of the company, the author found managers resent about the lack of flexibility in their jobs and often embarrassed by the excess of discipline imposed. ERP involves large organizational transformation processes whose implications are significant for the model, structure and style of management (Yeh & OuYang 2010). The same author asserts that some studies indicate that ERP seems to affect employment, the allocation of power and the perceived value of the stakeholders. What is also consistent with the finding of Davenport (2000), who argues that ERP tends to impose its own logic on the company's strategy, culture and organization and that this logic may or may not conform to existing organizational arrangements. Govers (2003) says that ERP can bring a greater degree of centralization and an additional energy management, serving as an instrument of power, because it offers them the opportunity to gain knowledge anonymously, and perception and control in respect to work processes and the performers of the process. Permeating the implementation of information systems there are processes of control and domination. (Walsham & Waema 1994)

## 3 METHODOLOGICAL PROCEDURES

According Haguette (1987) the problem under investigation is that dictates the method of investigation. Thus, considering the nature of the problem under study was chosen to qualitative research. As stated by Gray (2009) qualitative data can be a powerful source for analysis. Qualitative research is highly contextual, being collected in the natural setting of "real life." Regarding the epistemological view, was chosen the interpretive research. According to Orlikowski and Baroudi (1991) a lens through which the research in information systems are conducted and that influence the methods used is interpretivism, where the studies assume that people create and associate their own subjective and inter subjective meanings as interact with the world around them.

The research strategy adopted for this study is a case study in depth which according to Walsham (1995) is a vehicle often used for interpretive research, and implies a search that involves frequent visits to the field site over a prolonged period time. Ever desiring to study the managerial roles related to the use of Enterprise Resource Planning, the choice is justified by case study that, second says Myers (1997) is the appropriate method for understanding the interactions between innovations related to information technology and organizational contexts. For this study, a case study using interpretative approach in a given manufacturing industry located in Portugal was executed as well as in-depth organizational analysis of areas of the same organization. The selection of the study areas was intended for the use of the ERP system and the availability of management.

Permission to conduct this research was performed under the condition of maintaining confidentiality of the identity of the company. Interviews were conducted semi-structured, without following standardization, based on a list of issues and questions addressed in accordance with the development of each interview. Responses was transcribed with the rigor and care necessary for the avoidance of any loss of data.

According to Bauer and Gaskell (2002) there is no method for selection of respondents in qualitative research, given that no amount of interviews leads to a more detailed understanding of the problem. In this study were selected to be interviewed actors belonging to the top level of the company, directors and managers, who deal directly with the ERP. In a period of approximately two months, the first interviews were conducted, which were repeated in a second phase, after a period of approximately twelve months, according to the evaluation of the results that were achieved in relation to the objectives of the study. To observe the natural conditions of work, the interviews were conducted in the respective departments of each company manager. Direct observation, field notes, reports of activities and official documents provided by the company, were also used as secondary sources. Subsequently, all information gathered was submitted to an interpretive process of analysis, ensuring its accuracy and range.

## 4 SEARCH RESULT

The manufacturing industry object of the present study is called Gama Industry. It s been the market leader in Portugal for 50 years. Gama Industry exports about 80% of production for the world and is part of a holding company Gama Group, which has 1100 employees, of which 580 are part of the industry. The business group Gama consists of companies of various branches of business. The business of manufacturing is the highest expression in the group with the Gama Industry assuming greater role. The Gama has implemented SAP solutions in the areas of Finance, Logistics, Sales, Inventory Management, Production, Maintenance and Human Resources. With the implementation of SAP was the restructuring of departments with the intention to centralize the information of the entire holding in the industry Gama. The commercial area of was designated Gama industry to another holding company, Global distribution.

In the Gama, the use of SAP is very restricted to organize and centralize information in the appropriate departments. The adviser to the president said "the company is not taking the return that could". The information generated by SAP are not often used. Joining this, the integration, classic feature inherent in the ERP, is not checked in the Gama in its fullness. The marketing was out of the system. In the sales area, there were operations with SAP have been suspended. In the financial and accounting services to use is basically done by assistants. In the area of production, workers only feed the system and in the area of human resource, management also points to a low use of analytical reports obtained by the SAP.

According to the respondents ERP brought a greater bureaucracy to Gama. This bureaucracy was brought to the obligation to formalize and standardize that is inherent in these systems and at the same time with the imposition of the industry have to adapt to the ERP instead of this software be adapted to the industry. The department follows rules, is our means in terms of department. This scheme was designed by a team of experts." (Director of Information Systems and Management) Associated with this is the strategic centralization in the presidency of the enterprise and the consequent loss of managerial autonomy observed in the Gama. "Some managers had the knowledge and were employed by disclosure, now reports are seen, treated and analyzed by the president who decides, no longer depends on a single person. This took the power of some, who did not adapt had to leave the enterprise" (Advisor to the President). ⬜What is best for the company is already defined in the system and departments only have to comply with the determination. Thus, the company's goals will be achieved.⬜(Accounting Manager)

## 5 CONCLUSIONS

### 5.1 The ERP states behaviors and may threaten the management role.

As the use of ERP in the areas studied is very operational, the skills required, mentioned by the leaders are basically associated with its use in direct data feed. Besides being "required skills" the need of organization, method, and concentration are "imposed behavior" by ERP forcing the disciplinary action eliminating any possibility of improvisation and its creation by the user. The ERP serve as objects and tools to reinforce the Weberian Iron Cage Rationalistic spirit that trap people in bureaucratic organizations (Ludmer & Rodrigues Filho 2003) (Gosain et al. 2005).

### 5.2 Prevalence of the use of the ERP transactional functions contributing to the formation of a managerial profile technical and imprisoned operational functions

The managers interviewed have extremely technical and operational profile, lined on one side, a centralized vision and dictatorial attitude of the company and secondly, the pre-established processes. The dynamic nature of the managerial role found in classical literature (Mintzberg 2010) (Weick 1983) was not perceived in the managers interviewed in this company. On the other hand, functions that essentially would not be in a managerial role, are mentioned by the leaders of the Gama to translate their role in the department: sender reports, helpdesk, facilitating the tasks of the ERP, the presidential adviser, computer support ... The use of ERP by management occurs in a limited way, with a tendency much accounting / financial and focused mainly on transactional and operational aspects. Moreover, managers do not directly use the system; only coordinate the activities of his aides in the feeding of data. The displays of strategic and analytical activities are very rare. The industry Gama revealed very technical managers and very caught the attention of the ERP responses. SAP is seen as a major technical tool. A manager is defined as "someone in the service of the company that generates reports useful for analysis of the head⬜ for that, SAP helps a lot. Some managers proved to be bothered arguing feel very attached to technical issues and confessing that they can not "stop and think".

## 5.3 The ERP is used as an instrument of presidential power by reducing the managerial autonomy

In reviewing the literature it was found that ERP requires a kind of discipline, rather than management, and therefore can be classified as an instrument of powerful force that has a great influence on the balance of power in companies, can be used to maintain or develop the structure of the desired power (Govers 2003). This behavior is confirmed by the Gama's president with a centralized and dictatorial profile that controls and limits the processes and people in the company. The Gama industry seems to have a highly centralized decision-making. The department heads meet specific functions to feed the machine and only for this purpose, the ERP seems to be very useful. The bureaucratization brought with ERP leads to a formalization and standardization which limits individual behavior. In addition to providing guidelines and parameterized processes, SAP makes the job easy to control, because it logs data at the same time provides information for those who the presidency wish and blocks for the other, as declared by the managers, "information from other departments were available where it was relevant by criteria of the CEO. What also confirms the reasoning of Govers (2003) that the ERP leads to a virtual supervision where people would become "manageable distance". In Gama, ERP takes a strategic centralization in the presidency of the company and the consequent loss of managerial autonomy, "by SAP all processes are already established in advance by the company." One use of the ERP also aligned to the idea of Boersma and Kingma (2005) who says that the standard procedures of ERP contain a kind of "script" to users, informing them of what actions should be taken, when, where and how.

In conclusion, ERP affects the structure, culture and power relations in a company. Because of its omnipotence, the ERP has a great influence on the values and organizational standards. The ERP can create a technical isomorphism where companies tend to rely on the application of the rules determined by ERP software. (2006b):

## REFERENCES

Askenaes, L. & Westelius, A., 2000. Five roles of an information system: a social constructionist approach to analyzing the use of ERP systems. *ICIS 2000 Proceedings*, pp.426-434. Available at: http://aisel.aisnet.org/icis2000/4.

Bahrami, A.B. & Jordan, E., 2009. IMPACTS OF ENTERPRISE RESOURCE PLANNING IMPLEMENTATION ON DECISION MAKING PROCESSES IN AUSTRALIAN ORGANISATIONS. *PACIS 2009 Proceedings*. Available at: http://aisel.aisnet.org/pacis2009/30/.

Barnard, C.I., 1968. *Functions of the executive*, Cambridge: Harvard University Press.

Bauer, M.W. & Gaskell, G., 2002. *Pesquisa qualitativa com texto, imagem e som: um manual prático* Vozes, ed., Editora Vozes. Available at: http://www.scielo.br /scielo.php?script=sci_arttext&pid=S1415-65552004000200016&lng=pt&nrm =iso&tlng=pt.

Beheshti, H.M., 2006. What managers should know about ERP/ERP II. *Management Research News*, 29(4), pp.184□193. Available at: http://www.emeraldinsight.com /Insight/viewContentItem.do?contentId=1554343&contentType=Article.

Boersma, K. & Kingma, S., 2005. Developing a cultural perspective on ERP. *Business Process Management Journal*, 11(2), pp.123-136. Available at: http://www.emeraldinsight.com/10.1108/14637150510591138.

Davenport, T.H., 2000. *Mission Critical: Realizing the Promise of Enterprise Systems*, Harvard Business School Press. Available at: http://books.google.com /books?hl=en&lr=&id=p2kwgnefpQEC&oi=fnd&pg=PR7&dq=Mission+critical:+realizing+the+promise+of+enterprise+systems&ots =slnN4A3-pO&sig=FmuZ68MTdh832DP_BxfrdrUDg1Y.

Gosain, S., Lee, Z. & Kim, Y., 2005. The management of cross-functional inter-dependencies in ERP implementations: emergent coordination patterns. *European Journal of Information Systems*, 14(4), pp.371-387. Available at: http://www.palgrave-journals.com/doifinder/10.1057/palgrave.ejis.3000549.

Govers, M.J.G., 2006a. ERP software undermines organisational flexibility. *Automatisering Gids*, 10, p.17.

Govers, M.J.G., 2006b. Enterprise IT: socio-effects, and an alternative approach to deal with it. Available at: http://archypel.com/pages/download/socio-effects_IT.pdf [Accessed January 5, 2011].

Govers, M.J.G., 2003. *Setting out for modern bureaucraties with ERP systems?* Radboud University Nijmegen.

Govers, M.J.G. & Amelsvoort, P.V., 2007. ERP computerisation and limiting bureaucracy: reality or illusion? Available at: http://www.archypel.com/pages /download/ERP_and_limiting_bureaucracy.pdf [Accessed December 15, 2010].

Gray, D.E., 2009. *Doing Research in the Real World*, England: Sage Publications. Available at: http://scholar.google.com/scholar?hl=en&btnG=Search &q=intitle:Doing+Research+in+the+real+world#0.

Haguette, T.M.F., 1987. *Metodologias qualitativas na sociologia*, Petrópolis: vozes.

Lim, E.T.K., Pan, S.L. & Tan, C.W., 2005. Managing user acceptance towards enterprise resource planning (ERP) systems □ understanding the dissonance between user expectations and managerial policies. *European Journal of Information Systems*, 14(2), pp.135-149. Available at: http://www.palgrave-journals.com/doifinder /10.1057/palgrave.ejis.3000531.

Ludmer, G. & Falk J.A, 2007. Sistemas Integrados de Gestão e Conhecimento Organizacional: dinâmica das interações na pós-implementação em uma regional de uma empresa de serviços de telecomunicações. *Journal of Information Systems and Technology Management*, 4(2), pp.151-174.

Ludmer, G. & Rodrigues Filho, J., 2003. ERP e teoria crítica: alertas para os riscos de uma disparada para um novo tipo de Iron Cage. In *Iberoamerican Academy of Management. Third International Conference*. São Paulo.

Mintzberg, H., 2010. *Managing. Desvendando o dia a dia da gestão*, Porto Alegre: Bookman.

Myers, M.D., 1997. Qualitative Research in Information Systems. *MIS Quarterly*, 21(2), pp.241-242. Available at: http://www.qual.auckland.ac.nz/.

Orlikowski, W.J. & Baroudi, J.J., 1991. Studying Information Technology in Organizations: Research Approaches and Assumptions. *Information Systems Research*, 2(1), pp.1-28. Available at: http://isr.journal.informs.org/cgi/doi /10.1287/isre.2.1.1.

Peng, G.C. & Nunes, J.M.B., 2009. Identification and Assessment of Risks Associated with ERP Post-Implementation in China. *Journal of Enterprise Information Management*, 22(5), pp.587-614. Available at: http://dx.doi.org/10.1108 /17410390910993554.

278

Ranganathan, C. & Kannabiran, G., 2004. Effective management of information systems function: an exploratory study of Indian organizations. *International Journal of Information Management*, 24(3), pp.247-266. Available at: http://linkinghub.elsevier.com/retrieve/pii/S0268401204000295.

Ross, J.W. & Vitale, M.R., 2000. The ERP revolution: surviving vs thriving. *Information Systems Frontier*, 2(2), pp.233-241.

Soh, C., Kien, S.S. & Tay-Yap, J., 2000. Enterprise resource planning: cultural fits and misfits: is ERP a universal solution? *Communications of the ACM*, 43(4), pp.47-51. Available at: http://dl.acm.org/ft_gateway.cfm?id=332070&type=html [Accessed January 24, 2012].

Walsham, G., 1995. Interpretive case studies in IS research: nature and method. *European Journal of Information Systems*, 4(2), pp.74-81. Available at: http://www.palgrave-journals.com/doifinder/10.1057/ejis.1995.9.

Walsham, G. & Waema, T., 1994. Information systems strategy and implementation: a case study of a building society. *ACM Transactions on Information Systems*, 12(2), pp.150-173. Available at: http://portal.acm.org/citation.cfm?doid=196734.196744.

Weick, K.E., 1983. The presumption of logic in Executive Thought and Action. In *symposium on the Functioning of the Executive Mind*. Case Western Reserve University.

Yeh, J.Y. & OuYang, Y.-C., 2010. How an organization changes in ERP implementation: a Taiwan semiconductor case study. *Business Process Management Journal*, 16(2), pp.209-225. Available at: http://www.emeraldinsight.com/10.1108/14637151011035561.

# CHAPTER 30

# Ergonomic Characteristic of Software for Enterprise Management Systems

*Krzysztof Hankiewicz*

Poznan University of Technology
Faculty of Engineering Management
Poznań, POLAND
E-mail: krzysztof.hankiewicz@put.poznan.pl

## ABSTRACT

The paper discusses the ergonomic features incorporated into software as designed for enterprise management systems. The ergonomic side of designing such software entails ensuring it is user-friendly, decreases stress and workloads and accounts for the sensory perception capabilities of humans. In addition to assessing the user interface, the author describes the system's functionalities which are key to ensuring suitable working conditions. Rather than replacing human operators in their intellectual work, the system helps them perform repetitive tasks such as data processing and standard report preparation.

The paper relies on literature describing management system design and on research done in selected enterprises. The enterprises surveyed by the author applied management software having various levels of complexity. They were examined by the case study method. The surveys involved interviewing staff and inspecting software.

The paper shows that to best meet ergonomic requirements, management software needs more than a proper user interface and/or software whose features are confined to a particular computer workstation. Instead, one should consider the IT system as a whole and study information and decision flows as well as data locations and data flows. One should also note that the system in question comprises more than a man and a technical object as it is made up of multiple human operators and multiple technical objects.

The study findings suggest ways to improve software in the selected field. They help identify the features of management software which influence ergonomics and potentially contribute to better enterprise management.

**Keywords**: ergonomics, software, management systems

# 1    SOFTWARE ERGONOMICS

The ergonomic focus in software development is on interactions between man and computer which rely on software. Software plays the role of translating human commands given to a computer into a language that a computer can understand and presenting a computer's output in a form comprehensible to the human mind.

In line with the orientation on users and their needs, such considerations are essential and critically impact on the final shape of an IT product. A great number of subjective and objective factors affect the satisfaction with software use.

The key principles of ergonomic software design are to [Preece, et. al, 1994]:

- know the target population of users
- facilitate the cognitive process
- anticipate and plan for errors
- maintain cohesion and lucidity.

All of these very general requirements boil down to satisfying the user.

Customer satisfaction, in its turn, is critically influenced by [Hankiewicz, Prussak 2005]:

- the graphical interface,
- the legibility of symbols and the clarity of dialogue language,
- control over access to system information,
- the ease of undoing operations,
- the ease of correcting erroneously entered data.

User involvement in software design may also be invaluable. Some of the benefits of involving persons who will potentially use a future system have been studied by Preece, among others. (Preece, et. al, 2002)

The primary factors for the quality of system interaction are:

- user communication with the system,
- user interface,
- system prompts and feedback,
- system response time,
- the help feature.

An IT management system is not only a tool but also a conveyer of tasks. As users interact with the system's database, they may generate tasks by applying predefined algorithms. Possible interactions between tasks, tools and users may produce work outcomes, organizational results and changes in human welfare.

The user and the computer are linked by a set of actions collectively referred to as communication. The two basic forms of communication are entry and dialogue. In most applications, entry communication has been replaced with dialogue which allows the user to intervene into the processing of previously entered data. In the case of indirect communication, dialogue should be designed with proper account taken of user profile which comprises user skills, prior training and experience in computer use.

System designers should recognize the needs of every potential user. They should also remember that users who work with a system grow more advanced and effectively change their profiles. A system should be adjustable to meet the individual needs of a given user and expand user capacities. On the other hand, a system should be somewhat "distrustful" of user actions, particularly those which are irreversible, hard to reverse or which involve the loss of sensitive data. Such "distrust" should be expressed by prompting users to confirm their choices and preventing the entry of invalid data. The system should force informed decisions, i.e. prevent automatic acceptance at confirmation prompts. It should also avoid excessive confirmation requests not to disturb workflows and concentration or cause fatigue.

Many people find working with computers stressful although they rarely experience similar problems when using other devices, including those whose operation is highly complex. A major role in the process is played by the user interface. As it turns out, the interfaces of other devices have been better developed. Yet, user interface improvements can also be seen in computers, especially personal computers. For a number of years now, work has been under way to develop a graphical user interface (GUI). The progress has been significant helping to popularize personal computers. Yet, today's products are still short of perfect as the interfaces still need to become fully intuitive. On the other hand, many of the tasks performed by software are highly complex – in fact, their complexity increases as development continues.

In designing an interface, one should differentiate between character-based and graphical interfaces. The purpose behind replacing the character-based interface with a graphical equivalent is to extend the bandwidth, i.e. the range of useful information visible to a user at a given time, and increase the speed of performing specific tasks. This can be achieved by presenting information graphically and additionally by applying animation. It is also essential to use the right colors. Even when added to plain text, colors make the message easier to convey. Further enhancements result from text bolding and changes in font size and text layout.

An interface can be made substantially more intuitive by allowing direct manipulation with a pointing device.

In designing an interface, one should recognize the importance of screen item layout, command phrasing, the understandability of prompts, the presentation of non-textual information, highlighting and information encoding.

When laying out screen items, decisions are made as to where to place individual information items, forms and dialogue boxes, how to present them graphically and which color codes and schemes to apply. It may be advisable to set out a portion of the screen to display the current status of a program. The screen may need to be divided into sections dedicated to displaying information of various types. The division may be used to separate input and output information and system notifications. When deciding about the layout of user screen, care must be taken to avoid arrangements which run counter to operator habits.

The names selected for program items should be easily identifiable; the prompts and questions should be easy to understand. Such unambiguity should be ensured in particular for abbreviations. One should avoid symbols, especially those known only to a limited number of users.

Care should also be taken to ensure that computer prompts and feedback are fully comprehensible to the user. The information units which need particular attention from the user should be displayed in a different font (of different type, size and/or color). At best, such information should be placed in the center of the screen (especially for error messages). Thematically related and frequently showed information should be displayed in the same place and form.

Non-textual messages which are nonverbal and given in an alphanumeric form, such as telephone or identification numbers, should be provided in a way which facilitates their easy entry, searching and memorization.

Alphanumeric characters should be grouped in sets of two, three or four, separated with spaces. This is of particular importance for strings of more than six characters. Larger and frequently searched data sets are best placed in tables or sorted.

Texts made up of letters should be adjusted to the left whereas those made up exclusively of digits, especially when recognized as numbers, should be adjusted to the right. Similar data should be presented in the same way. Use should be made of traditional data entry methods – this refers to data order and standard formats.

Messages, in particular those used to convey information, confirmations and error alerts, should be phrased in a uniform manner, especially when they referring to similar circumstances occurring in different parts of the system. The messages should not require that the user look up the related code in software documentation. Verbal messages replacing codes should be clear and unambiguous as well as easy to understand without referring to the documentation.

Software designers should always figure on operator errors. Errors require time as needed for cause identification and correction. It is therefore desirable that the system reduces the number of errors which may be made. It is easy to protect a system from errors if the operator is not permitted to modify data. The task is much more challenging when the operator is allowed to enter and remove data from databases. In such a case, the system should support fast detection and should immediately notify the operator.

Dialogue box design is a critical factor for error occurrence. One should recognize the effects of user psychology. A program should neither bore nor overwhelm a user. If immediately alerted to an error, a user will be in a better position to remember the situation, learn his lesson and avoid similar mistakes in the future.

Some errors are due to the overburdening of an operator's memory. This is why one should rely on information which requires remembering or at least reduce the amount of such information. Much help to users comes from generalization, i.e. the option to repeat the same choices for items of various types. An example are such editing commands as cut, copy and paste which perform the same task on various objects. The principle of using such tasks for editable text was expanded in

Windows 95 which first applied it to files. The function is so obvious today that few people realize that earlier systems did not support it. A particularly effective generalization is in graphical interfaces where individual actions are performed by pointing to an icon (e.g. with a mouse). It is essential that the icon can easily be associated with the action it triggers. Moreover, system messages which catch the user's attention and provide error alerts should be carefully worded and informative. They should be free of abbreviations and codes which confuse users and force them to refer back to the documentation to check their meaning. Being forced to do this would stress out the user and additionally disrupt the workflow. It is best to match message wording and user skill levels. More advanced users should be informed with short concise messages which nevertheless offer access to more detailed information. Beginners require more wordy messages which describe the actions to be taken at any given time in precise step-by-step instructions.

It is essential for the user to receive system feedback at the very moment it is expected. Usually such time is short so the user does not have to wait for the system to respond. It is difficult to define the response time acceptable to users as this depends on the situation. What is certain, however, is that if the system response time is kept short, the computer and the user are capable of exchanging more information which makes the workflow more efficient. Yet, despite progress in the development of IT equipment, short response times prove very expensive to achieve. The difficulty often lies in the internal organization of the system and the method and location of data storage. Users may be willing to tolerate longer response times when they realize that large volumes of data are being processed or that data are being retrieved from remote databases. A user convinced that an operation is not complex will not accept a longer response time. It is therefore essential that the user be kept informed on system status and the progress in task completion. Users should be aware that the system is in operation and that they need to wait. A message to this effect may be given graphically or verbally in the status bar or a separate dialogue box. For prolonged waiting times, it is advisable to display a progress bar.

Finally, note that overly short response times are not always desirable as they may make the operator anxious about not being able to keep up with the system. To prevent that, system responses may be intentionally deferred.

When faced with a problem, an operator may use a more experienced user as a source of information. This is commonly an option with systems which have been in place for an extended duration as a relatively large number of prior users would then be available. It may be advisable to have an advanced user coach new users on system operation. It is particularly valuable to have the advanced users point out issues which they have found to be challenging. The drawback of using such help is that one preoccupies persons who may have other duties to perform.

At any rate, complete information on a system can be found in system documentation. Such documentation should be helpful in solving operating problems. System documentation comes in various forms. The three most common ones are:

- the user guide – user guides explain how to perform each task with the software at hand; a user guide should enumerate all available system functions; users who wish to rely on such documentation on their own need certain experience;
- the quick reference guide – quick reference guides contain the basic system information provided in a nutshell; such documentation is intended for advanced users who have previously worked with similar systems;
- the full reference manual – such manuals provide in-depth information on each system function; the full reference manual is designed for advanced users who wish to explore and use all system functions;
- the tutorial – tutorials provide step-by-step instructions on how to use the system; tutorials are intended for inexperienced users who want to learn how to operate the system.

The type of documentation included with the system depends on system type, target users and, in some circumstances, on marketing considerations. It is essential that the text of the documentation is supplemented with graphical symbols, tables, graphs and photographs to make it easier to follow. It is also critical to provide a sufficient number of examples to make documentation content easier to understand, in particular to beginner users.

Faster access to the information needed to solve a given problem is offered in the program help facility available directly from the system. An additional advantage of program documentation is the ease of updating by way of file replacement. Extended program help should include context-sensitive help which refers one straight to the help topic relevant for the task at hand.

Program help should contain a keyword index for easy search. When frequent help updates are needed, it is beneficial to use program help linked to a web-based equivalent.

An efficient way to bring a user to operate a system is to use a training program. Training programs resemble system guides. They explain the system's main functions by means of demonstrations and exercises. These help the user learn to operate the system. Ideally, a training program should include tests to check user knowledge.

All in all, a help system should offer users the choice of using either program help or system documentation. In either case, help should be sufficiently complete and the search option sufficiently easy for the user not to have to seek the assistance of more advanced operators. It is best to ensure that users are given a chance to go over a system's basic functions with the help of a training program.

## 2 THE ERGONOMIC FEATURES OF A COMPUTER SYSTEM

While ergonomic studies of a single computer terminal focus on the user interface, the ergonomics of an entire system also involve the cooperation of multiple persons. Such cooperation takes place through and with the help of an IT system. The ergonomic features of a system are therefore tied to the method of accessing data and documents and to documentation and information flows in the system. Access to data and documents can be defined in two ways. Firstly, before documents are available in a system, they need to be entered and saved in a particular location defined in prescribed procedures. Secondly, access to documents must be controlled with access rights. Such rights may concern the entry, viewing, modification and removal of documents and data as well as document approval. The placement of a document in a particular location does not mean it will reach the right person at the right time. For that to happen, the system should be fitted with mechanisms for notifying given employees of the arrival of new information. Employees should not need to keep checking whether a document has shown up. This task should be performed by the system which should then send notifications to proper workers. The system should keep track of a document's author, the time of its drafting as well as the authors and the time of successive modifications. Once an approved document has been modified, it should lose its approved status and be submitted for re-approval.

Having to enter all of the necessary information into the system is rather strenuous. While high-capacity scanners make an easy task of entering hard-copy correspondence, the entry of oral information provided in face-to-face or telephone conversations requires greater discipline and more effort on the part of the employees. The documents generated in a company may be entered directly into the system. Those received electronically (by email or facsimile) can be entered automatically without human involvement.

One of the reasons why all documents and information should be available in the system rather than as hard copy is to reduce retrieval time. The time put into searching for information and communicating with other employees to retrieve all of the necessary data may exceed 30% of the total time used to operate a system. Interestingly, such time is comparable to the amount of time needed to produce documents. The employees use the remaining time on data administration, communication and analysis. The reason information searches turn out to be so time-consuming is that they require going through a great number of physical folders, many kept in separate rooms, sometimes even in other departments of the company. Such searches also involve other workers as they are approached and asked to reply to inquiries. One should not underestimate the time required to draw up documents. Having to prepare documents under the stress of working against the clock under short deadlines is particularly strenuous. Yet, such situations are common when studies and reports are being compiled for the management. All the more valuable are systems which automatically generate summaries and reports. They do the work faster, make the documents available at any time and put them together without the errors often precipitated by haste. The stress reduction and efficiency gains achieved in preparing reports and summaries for other company units and the management are very significant, especially that easy data searches and automatic summary generation will substantially reduce the time needed for report preparation. Thus the cost of improving a computer management system will soon pay off with the additional benefit of better ergonomics for the workers. The benefit of being able to provide the management with the most up-to-date information cannot be overestimated as it may ultimately make a difference between enterprise survival or failure in a competitive environment.

Equally large amounts of time are spent by executives as they search for information. In their case, however, much of the burden is passed on to various assistants and other subordinates who are charged with answering inquiries and retrieving information. This increases the proportion of time spent on interpersonal communication in total working time. Communication time also includes meetings at which employees exchange information in face-to-face contacts. The time spent to perform such activities can also be significantly reduced by applying an IT system which generates summaries and studies based on information provided in databases.

The benefits of adopting an IT system go beyond the mere time savings for employees and executives and the related stress and effort reduction and efficiency gains. Such systems additionally offer a range of other advantages. The immediate entry of payment documents into the system helps rationally manage cash flows and plan payments ahead of time thanks, for instance, to having invoices available in the system immediately after their entry and even before they have been approved.

Another benefit is the access to cross-data in a way not supported in traditional systems. When first establishing contact with a customer (a client, supplier or vendor), the subject matter of the contacts is referred to as the case (or task, as in some systems). In further contacts with the customer or with any other company, the case is accompanied by a reference name or number. In effect, retrieving information on a given case is not only possible but also very simple even though such information may involve a number of different companies. An essential part of a system is the module which notifies employees of the arrival of new documents, prompting them to approve the documents, enter them in books of account, prepare offerings or proposals, fulfill orders, etc. A very important task is to alert employees of any uncompleted tasks.

An enterprise's document and information flow management system may be a module of a comprehensive IT enterprise management system or constitute an independent system which communicates with the enterprise's other systems. Regardless of the degree of integration, such systems should cover various areas of management such as document management, document and information flow management, case management and workflow management.

The drawback of such extended and automated systems is that a user's activities are constantly monitored, which may lead to a sense of being cornered. For that reason, employees should be allowed to, e.g. use the company telephone to attend to urgent personal matters without the need to record the call in the system. If this would be undesirable due to software design, one should seek to provide a separate line not controlled by the system.

## 3    CONCLUSIONS

Ergonomic designers of IT management systems focus on items typically associated with the man-computer relationship, i.e. user communication with the system, user interface, system feedback and prompts, system response time and system help. To ensure that the IT system better responds to human capacities, this range should be extended by additionally including actions which seem to have very little to do with ergonomics. These include studies of document and information entry methods and document and information flows in the system. One should also recognize that many of today's IT enterprise management systems act as intermediaries in interpersonal communication. This is true for relations within the enterprise in which the system is installed as well as external contacts with business partners and customers. Such an extended scope of ergonomic focus can be expected to reduce the strain of working with the IT system. One could also expect a reduction in error rates. An ability to respond faster to customer needs and market change is also likely to improve enterprise performance. In addition, the employees themselves will certainly benefit by being more satisfied with their jobs.

## REFERENCES

Hankiewicz, K.; Prussak, W., 2005, Usability Estimation of Quality Management System Software, In *HCI International. 11th International Conference on Human-Computer Interaction* G. Salvendy. ed., Vol. 4. Theories, Models and Processes in HCI, MIRA Digital Publ.

ISO/IEC 9241-11, 1998. Ergonomic requirements for office work with visual display terminals (VDTs). Part 11. Guidance on usability.

Nielsen, J., 1993. Usability Engineering. London: Academic Press.

Preece, J., Rogers, Y., Sharp, H., Benyon, D., Holland, S. & Carey, T., 1994 Human-Computer Interaction. Essex, England: Addison-Wesley Longman Limited.

Preece, J., Rogers, Y., & Sharp, H., 2002. Interaction design: Beyond human-computer interaction. New York, NY: John Wiley & Sons.

Stephenson N., 2003, In the Beginning Was the Command Line, HarperCollins Publishers.

# CHAPTER 31

# Human Factors and Well-balanced Improvement of Engineering

*Leszek Pacholski*

Poznan University of Technology
Poznan, Poland
leszek.pacholski@put.poznan.pl

## ABSTRACT

Professional competences from the area of *Knowledge Engineering* have contemporary become the basis for engineering design and operation of intelligent constructions, systems and processes. Artifacts and processes are characteristics for the stage of transition from the pre-industrial era to the era of creativity and empathy. Contemporary engineer innovations characterize a more clear articulation of human needs, natural limitations and possibilities of man. It means that became to impose to engineering the direction of development through the articulation of his needs: mobility, solving ecological problems, sense of security, communication and exchange of information and knowledge, health care, creativity and empathy. There have been gradually occurring following categories of efficiency (regulators) in the development of *Engineering*, which have a form of economical, biological and social costs. In particular cases, determined right, can even mean, that we are legitimate to impose a trenchant barrier to the unlimited development of *Engineering*. Taking into account all arguments presented before, the case study of a potential limitation of improvement of the modern *Computer Engineering* is analyzed. The case concerns *Artificial Intelligence* as basic innovation of *the Sixth Kondratieff Business Cycle*. Facing the disputable question of potential domination of the *Artificial Intelligence* over the natural one, we should ask whether the improvement of this area of *Engineering* should be limited. It is worth to be aware that modern *Computer Engineering* has limitations of its improvement, which are important from the human-centric point of view.

**Keywords**: Human Factors, improvement of engineering, Artificial Intelligence

# 1     INTRODUCTION: NATURE AND IMPROVEMENT OF MODERN ENGINEERING

The subject of my deliberation: improvement of engineering, requires defining following terms: *science* and *technology*. *Science* answers the question: why a determined problem takes place? It searches the solution to this problem; it tries to explain and describe it. *Technology* focuses on questions, like: how to solve the problem? Or even: how to implement the problem's solution into practice? In parallel to the distinction between science and technology; there exist two term that relate to those areas: *theoretical science* and *practical science*. The idea of *practical science (engineering)* reflects a specific complex that *technology*, as *imperfect knowledge*, has in face of "pure science" and its scientific aspirations. Determined *imperfect knowledge* constitutes the basis for such statement. Like: *science* deals with knowledge about, what already exists and *technology* creates things that don't exist yet. However, in practice *science* and *technology* do often cooperate on the same field. *Science* enters the process of practical use of own inventions – and so it enters the area of *technology*. The process of technological progress enables *technology* to discover new phenomena, for which it finds solutions – and so it automatically becomes equal with *science*. However, the character of scientific and engineering research is different than the profile of theoretical study. Scientific research in the range of technology focus on creating methods for solving a problem that results from an already discovered phenomenon, which is however too complicated to find an accurate solution for its questions. *Engineering science* often use quasi-empirical methods, which are not applied in the *science* in its classical interpretation. Terms: *engineering* and *technology* often function as synonyms. In old French the word: *engineer* was equivocal with the term: *war machinery constructor*. In turn, the English word *engineering* ostensibly doesn't come from *engine*, but from ingenious; its genesis is Latin – from words: *ingeniosus*, which was connected with human genius.

Contemporary engineering deals with designing and exploiting technical creations (in form of *artifacts* or *processes*) with use of knowledge, experience and intuition. *The art of engineering* constitutes an interesting combination of knowledge and skills. Pragmatism is often its attribute. It is caused by the necessity of finding solutions, even before *science* fully explains certain phenomena.

# 2     RESOURCE FOUNDATIONS OF ENGINEERING CREATIVITY

Modern engineering work includes three resource foundations of the world surrounding us: matter, energy and human knowledge and skills (intelligence). Foundations are the subject of engineering design and operation in form of *artifacts* (constructions and systems) or *processes* (technological sequences and organizational solutions). View to limited number of resources and their exhaust character (matter and energy), knowledge and skills became lately resources of particular rank. *Knowledge engineering* exposes four extraordinary characteristics of this asset:

- dominativeness, resulting directly from the fact that efficient of use of knowledge is a crucial factor of competitive position of a company,
- inexhaustibleness – it means that knowledge, unlike other assets, can not be consumed; its value rises while it is exploited with more intensity,
- symultanivity – the knowledge itself might be used in many (enterprises) in the same time and in many places (traditional assets can't be distributed in such way),
- non-linearity, which means that a small scope of knowledge might cause unimaginable big consequences or vice versa: enormous knowledge might become useless.

Professional competences from the area of *knowledge engineering* have contemporary become the basis for engineering design and operation of intelligent constructions, systems and processes. However, in combination with a so-called *social intelligence* of the company they have formed so-called *intelligent organizations*. Artifacts and processes of this sort are characteristics for the stage of transition from the *pre-industrial era* to the *era of creativity and empathy*.

Modern forms of *engineering* usually were built as an answer to a so-called *social imperative*, i.e. *a need of a group*. The picture (Fig. 1) shows an exemplary set of such imperatives and so-called *basis innovation*, which represent the engineer answer to those imperatives (Pacholski, 2006). It is named *Kondratieff's business cycles*.

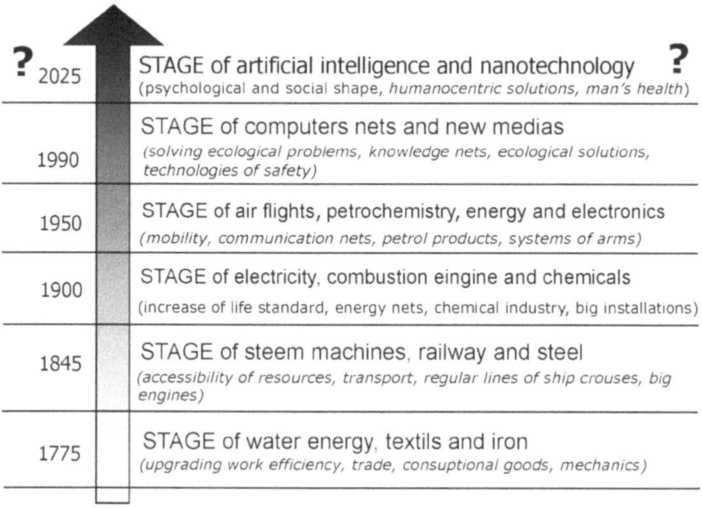

Fig. 1. The social imperatives and basis innovations (the *Kondratieff's business cycles*)

# 3    HUMAN FACTORS IN ENGINEERING

Engineer innovations of the industrial era characterize with the fact that they constituted a more clear articulation of human needs, natural limitations and possibilities of man. It is man that became to impose to engineering the direction of development through the articulation of his needs: mobility, solving ecological problems, sense of security, communication and exchange of information and knowledge, health care, creativity and empathy. An industrial and information era man, have simultaneously begun the trial of reducing the unlimited development of engineering taking as the main the human aspect. There have been gradually occurring following categories of efficiency (regulators) in the development of engineering, which have a form of economical, biological and social costs.

In the final part of the Second World War Americans have discovered that 400 aerial disasters of plains, that were fully airworthy and which happened already after the realization of the task, were caused by pilots' errors, who confused steering devices in the process of landing. A new area of knowledge arises: *Human Factors Engineering*, along with anthropo-technical standards.

In the beginning of the fifties of the 20th century, British discover the Long Wall Case. Perfectly modernized (from the point of view of the classical engineering) process of coal mining doesn't bring expected results because of reasons that, today, we call *macroergonomics* (structure and size of working teams, content of work, company identity, culture of the organization). Within frames of the new discipline of *engineering* called: *Human Factors / Ergonomics*; as well as facing a strong pressure of ecological groups, designers and explorers of new creations of techniques, discovers that modern engineering cannot be focused strictly on technical components.

In modern techniques, oriented on human as a central aspect, an engineer can't design and explore an aliened technical creation. He must form a system entity, which includes: man and technique, taking under account the subjective role of man.

The historical sequence of paradigms of engineering must be replaced by a *humanocentric* sequence (Pacholski, 2006) within which man has a subjective role (Fig. 2). And so, in the second part of the 20th century engineering has discovered that a *technocentric* world, in which man plays in manufacturing processes a role of a marginal factor in a realization system, with the central role of technical creations, would be an absurd. We have discovered that the order created in Paradise had the *anthropocentric* character. God place man in the central position and gave him rights to decide about names of other creations. Such situation means that we have the right to subordinate the scientific and technological improvement, to limitations and abilities of man. In particular cases, determined right, can even mean that we are legitimate to impose a trenchant barrier to the unlimited development of engineering.

The historical sequence of paradigms of engineering

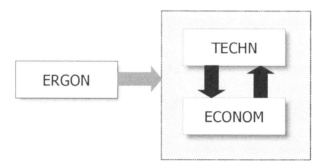

The humanocentric sequence of paradigms of contemporary engineering

Fig. 2. Historical and contemporary sequences of paradigms of engineering

It seems that these are clear examples of such barriers:
- world limitations of carbon dioxide emission,
- non-proliferation of atomic arms,
- concept of limits of genetic engineering,
- control of production of determined goods within frames of Economic Communities,
- *Intelligent Systems* of individual distribution of the electricity,
- systems of preliminary segregation of waste and garbage.

## 4     LIMITATIONS OF IMPROVEMENT – CASE STUDY

Taking into account all arguments presented before, we analyze further the case study of a potential limitation of improvement of the modern *computer engineering*. The case concerns *artificial intelligence* as *basic innovation* of the sixth *Kondratieff business cycle*.

From all intelligent abilities of human mind, the special place belongs to the ability of concluding. The simplest, almost mechanical forms of concluding are related with the ability of counting, calculating and combining. This has directed first efforts of smoothing through technique natural limitations (concerning mostly the speed of operating) and abilities of human intelligence by building mechanical counters. Work on upgrading calculating machines (Fig. 3) has led to constructing electromechanical calculating machines. First electromechanical computers were built during the Second World War in USA and in Germany. In November 1945 the description of ENIAC, the first electronic digital machine, has initiated the development of modern computers.

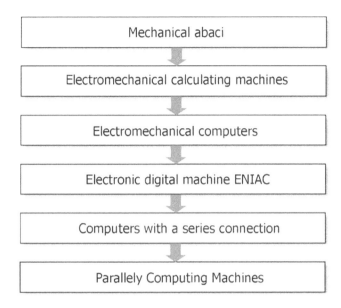

Fig. 3. Upgrading of calculating machines

It is possible to distinguish two groups of technological solutions in the brief history of contemporary computers (Zielinski, 2000; Pacholski, 2009):

a) machines with a so-called *serial processing*, in which the limit of abilities of increasing the speed of operation is determined by the speed of the electromagnetic wave. The architecture of action in this group was based in turn on following elements:

- electromechanical transmitters,
- valves (first generation computers),
- transistors,
- integrated circuit (*third generation computers*),
- so-called *Large Scale Integration* circuits,
- LSI systems,

b) machines with a so-called *parallel processing*; the fourth generation computers were built with use of *Very Large Scale Integration* circuits (VLSI).

On the basis of VLSI circuits began work on the structure of systems processing knowledge. Processing knowledge is characteristic for so-called *Expert Systems*. There is a common opinion that work concerning questions mentioned above is determined by the name of fifth generation computers. They could initiate the course of *Artificial Intelligence* within the sphere of modern technologies of computers. Processing knowledge can be realized by next two generations of computer technologies. The sixth generation is built with use of so-called *neurocomputers*, which are funded on optic and hybrid technologies. Those devices are material mathematic models of human's neural network. Main centers of technology have begun work on realization of so-called *molecular computers* that are included into the seventh generation computers. The development and improvement of tools of *Artificial Intelligence* enable practical eliminating the maladjustment of bivalent logic to the description of real phenomena and of the opacity of database. The differences between traditional software and A*rtificial Intelligence* programs are as follows:

- Digital processing vs. Symbolic processing,
- Algorithmic supply of action vs. Declarative method of recording knowledge,
- Batch or interactive processing vs. Interactive software environment,
- Possibility of validation of the program vs. Lack of practical possibility of full control of correctness of functioning of the program,
- Development of the program based on specification vs. Development of programs based on creation of prototypes and on their improvement,
- Presentation and application of data vs. Presentation and application of knowledge,
- Using databases vs. Using basis of knowledge.

It seems, that because of instrumental and technological questions the idea of common use of so-called *intelligent computers* must be postponed at least to the beginning of the six *Konratieff business cycle*, that is to the thirties of the 21st century.

However, transitional solutions from this sphere, including the fourth generation of computers, appear already at present. For example: in the domain of automatics, robotics and management (Casals and Fernandez-Caballero, 2007; Gao, Shang and Kokossis, 2009; Mirhosseyni and Webb, 2009; Patterson, 2007; Schelsinger, 2007). Although it seems hard to believe, perhaps the contemporary young generation will live to times, in which visions of science fiction writers, predicting entrance in our life of so-called, thinking, humanoid robots will find fulfillment. Building an *artificial man* i.e., an intelligent machine has; always absorbed minds of inventors. Although it seems surprising, contemporary theoreticians and philosophers argue already about the range and rights, that in future would be possible to grant to so-called *intelligent automats*.

The interest in biological sciences recalled at the beginning of this study started investigation of functioning of neural mechanisms of the human mind. These analysis in the eighties, XX century, came back to research concerning *artificial neural networks*. Period of the 90-ies was dominated by a course determined with name of *connexionism*, which in practical applications combined advantages and demerits of *expert systems* and *neural networks* and was supplemented with new methodological instruments, like e.g. *the fuzzy logic* and *genetic algorithms*. This has led to creating the idea of so-called *hybrid systems*, consisting *expert systems* and *neural networks*, but also a supplement in form of *fuzzy logic, genetic algorithms* and application of the theory of chaos within the area of *artificial intelligence*. Robots exploiting first programs based on *artificial intelligence* are already settled, among others in the USA, Europe and countries of the Far East, assembled in production lines in factories. Human Factors/Ergonomics determinants are still the main premise of their application. In industry robots appeared coming from military programs.

The army and police in the USA are using since the certain time robots, which one could call intelligent, for conducting battle action, in anti-terrorist and rescue actions and during cataclysms. In the end of 2008, in Iraq and Afghanistan fields of battle appeared 12 thousands of sapper's robots, called "mines hovers" accompanied by about 22 types of other intelligent terrestrial machines, including ones for offensive operations. American military program called *Future Combat Systems* assumes that the first robot-soldier would appear on a regular battlefield until the year 2025. Among non-military solutions, the biggest sensation is the *android CB2*, created to the pattern of a small child. The intelligent software manipulating this robot is interpreting external stimuli and it is learning appropriate reaction on the principle of a neural network, similarly to a case of a learning person or evolving child. In the first decade of the 21$^{st}$ century we face a rather dynamic development of so-called *domestic robots, robots - companions for the elderly, disables and lonely,* as well as *toy-robots.* According to estimations, until the end of 2010 the global value of the market of robots mentioned above reaches 25 billions of US dollars.

A document concerning development in United Kingdom to the year 2056, prepared by the British government assumes a possibility that robots will think and feel by this date. They will have to have the same rights (voting right, fear treatment, right to work) and obligations (taxes) as human citizens. The government of South Korea has ordered a preparation of a similar paper.

# 5    SUMMARY

In summary, we return to the issue of limits of improvement of the modern computer engineering. Facing the disputable question of potential domination of the artificial intelligence over the natural one, we should ask whether the improvement of this area of engineering should be limited – similarly to limitations mentioned before, i.e. concerning: carbon dioxide emission, concept of development of the genetic engineering or non-proliferation of the atomic arms? However, for pure *intellectual pleasure*, in this study there has been signaled the existence of a theory of so-called *Parallel Multiuniverses* (Pacholski, 2009). According to this theory, the essence of human being is creating and processing information. Man is a *Genetic Information*, which exists in a determined space (Universe) of time (Fig. 4).

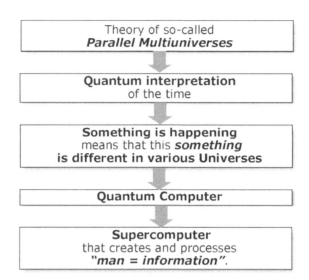

Fig. 4. The theory of parallel multiuniverses leading to the idea of Quantum Computer

The theory is based on *Quantum interpretation* of the time, which claims that different moments of the time constitute different universes. So, time doesn't flow like it has been presented in classical physics. The fact, that something is happening means only that this "something" is different in various universes. There are some physical experiments that seem to confirm presented theory. The phenomenon of *Quantum Interference* consisting in getting qualitatively different images at further letting photons through one and then two cracks (the fleck of light and the interference fringe) is a classic example. Shortly speaking, the *theory of parallel multiuniverses* can lead to the idea of *Quantum Computer*. An escalation of such philosophy directs us to the idea of a *Supercomputer* that creates and processes "man-information". Fortunately, supporters of the theory of so-called *Parallel Multiuniverses* deny the possibility of potential existence of such device within limits of our Universe.

However, (in the perspective of one of dominating fundamental innovations of the sixth *Kondratieff's business cycle*) it is worth to be aware that modern computer engineering has limitations of its improvement, which are important from the humanocetric (ergonomic) point of view.

## REFERENCES

Casals, A., and A. Fernandez-Caballero. 2007. Robotics and autonomous systems in the 50th anniversary of artificial intelligence. *Robotics and Autonomous Systems* 55(12): 837–839.

Gao, Y., Z. Shang, and A. Kokossis. 2009. Agent-based intelligent systems development for decision support in chemical process industry. *Expert Systems with Applications* 36(8): 11099–11107.

Mirhosseyni, S.H.L. and P. Webb. 2009. A hybrid fuzzy knowledge-based expert systems and genetic algorithm for efficient selection and assignment of material handling equipment. *Expert Systems with Applications* 36(9): 11875–11887.

Pacholski, L. 2006. Ergonomic issues of the neural integrated human-computer interaction. *Cybernetics and Systems* 37(2–3): 219–228.

Pacholski, L. 2009. Human factor in robotical and managerial applications of artificial intelligence. In. *Macroergonomics vs. Social Ergonomics,* ed. L. Pacholski. Publishing House of Poznan University of Technology. 43–60.

Patterson, D.J. 2007. Involving intelligent assistants in active human communication. AAAI Spring Symposium. Technical Report SS-07-04: 98–99.

Schelsinger, B., et al. 2007. A semi-autonomous interactive robot. AAAI Workshop. Technical Report WS-06-15: 40–44.

Zielinski, J.S. 2000. *Intelligent Systems in management.* Warszawa: PWN. (in Polish).

# CHAPTER 32

# The Assessment Criteria of the Ergonomic Quality of Anthropotechnical Mega-systems

*Marcin Butlewski & Edwin Tytyk*

Poznan University of Technology
Poznan, Poland
e-mail: edwin.tytyk@put.poznan.pl
e-mail: marcin.butlewski@put.poznan.pl

## ABSTRACT

Modern societies increasingly depend on the technical measures that are the essential tools of any intentional action. It is hard to imagine any activity performed on a large scale, not supported by computers, machines, or even tools, which are the technical factor in anthropotechnical systems. These systems, in conjunction with the technical advancements in the modern world, are also becoming more sophisticated, and globalization makes them increasingly linked to other systems. For this reason, we can talk about anthropotechnical mega-systems – that is, complex systems with many people and many different technical objects, joined by a number of dependencies, of which only some are visible at first glance. Such mega-systems are, for example, industrial companies, airports and seaports, railway stations, shopping malls, residential buildings, and cities. The functioning of individuals and social groups is dependent on the ergonomic (and exactly macroergonomic) quality of complex technical systems, which are the technical members of anthropotechnical mega-systems. What is important is that quality is not, as previously thought, dependent on one or several factors that a user can affect, but results from a whole range of completely independent elements of the work and the recreational environment.

An important methodological problem is to assess the ergonomic quality of such anthropotechnical mega-systems. This article presents the concept of the macroergonomical method of diagnosis, which is based on sets of criteria that describe the properties of systems in meeting the requirements of different user groups. This approach allows for seeing the diversity of roles performed by the man in anthropotechnical mega-systems and the resulting consequences.

**Keywords**: anthropotechnical mega-systems, ergonomic quality

## ERGONOMIC QUALITY

Ergonomic quality is a multicomponent criterion, which is often difficult to determine, even by those on which it has a direct impact. For its precise definition it is necessary to examine both the degree of fulfillment of the requirements for the object, as well as other factors, independent of the met requirements, which actually have a significant influence on the interaction between the user and the technical environment. Ergonomic quality is also a highly subjective feature, which is based on the adaptation of the object to the psycho-physical, variable, and differently understood human needs, that is, based on their perception of contact with the technical object, which is difficult to assess, and creates equally serious problems of methodology and classification (Słowikowski 2001). Often the ergonomic quality is understood by users of technical systems as a form of functionality, but this is also not a one-dimensional criterion. Due to the limited capacity of human perception, the criterion of functionality is also limited by the number of provided product features, above which the functionality of the actual object starts to decrease (Rust, Thompson, Hamilton 2006).

We are accustomed to showing the ergonomic system for human interaction with the technical object as in Figure 1. However, the relationship as shown is incomplete if one takes into account the increasing complexity of the present situations, where the dependencies must be considered in a multidimensional way.

300

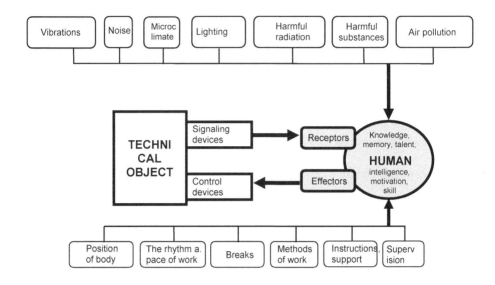

Fig. 1. Ergonomic man – technical object system (Adapted from (Tytyk 2001))

The model presented in Figure 1 has a fundamental flaw, and a simultaneous advantage, of simplifying the role taken by man in work processes. This simplification has been accepted because of the need for a simple way to show the interaction between man and the technical equipment and their surrounding environment. Further on into this article the authors present another model, which takes into account the human presence increasingly common in mega-systems.

## ANTHROPOTECHNICAL MEGA-SYSTEM

An anthropotechnical mega-system is the world we live in. But to attempt to systematize the term, we must refer to the definition of the system, which defines it as a group or groups of inter-related elements in systems, performing the superior function or set of such functions as a whole (Gadomski, 2006). However, in each system groups may form subsystems, while systems are located within larger super-systems. An anthropotechnical mega-system will therefore be a system composed of many different man – technical object subsystems, in which internal relations between the subsystems will take place. Anthropotechnical mega-systems may be places like cities, railway stations, companies, etc.

Such an approach showing the need for super-system thinking emerged in the 1980s and 90s, when the concept of complex studies of man – technical object systems appeared, called macroergonomics (Pacholski, 1995, 2011). Relations between "the human subsystem," consisting of many interacting people, and the "technical subsystem" consisting of many interacting machines, devices, tools, equipment, vehicles, buildings, etc., have a different character, that is relevant to sociology, economics, politics, culture, and ecology. The complexity of such systems makes them immensely difficult to study, in particular – prospective engineering activities (Tytyk, Butlewski, 2011).

Technical mega-systems have at least one thing in common: their functioning requires a structured interaction of many hundreds or thousands of people. The scale of this cooperation means that in the management of such technical ecosystems, problems arise that were previously unknown or not found so frequently. Some of these problems have a source in the complexity of technical mega-systems: if they consist of thousands of interdependent parts, the probability of breakdown or unpredictable functioning of the whole increases exponentially with the increase of unreliability of these parts and their number. Technical mega-systems have little immunity to the extreme forces of nature, such as earthquakes, typhoons, floods, or fires. This demonstrates their very fragile homeostasis. At the same time their action alters the course of natural phenomena (the greenhouse effect, ozone depletion), and produces natural disasters (typhoons, floods, or fires). Mega-system breakdown is also due to their low resistance to the destructive actions aimed deliberately at selected elements or actions inconsistent with the procedures, and due to human error. The consequences of breakdown of mega-systems are disastrous – it is enough to recall the effects of the disaster at Chernobyl nuclear power plant in Ukraine and the damage done to the Fukushima nuclear power plant, the explosion of space shuttle Columbia, or the terrorist attacks on the World Trade Center in New York.

Furthermore, it is important to realize that during the construction and operation of such technological giants as skyscrapers, ships, ports, oil rigs, airports, highways, underwater or outer space work – extreme working conditions and specific ergonomic issues exist. There, issues appear such as the question of the principles on which to base the management of human systems in such mega-systems and the design principles of technical ecosystems, since it is usually not possible to build and test a prototype, and the models developed for the purpose of computer simulation are too simplistic and far from reality. Here, interdisciplinary and holistic knowledge is necessary, which includes most notably the fields of science dealing with the functioning of human groups: social psychology, sociology, cultural studies, religious studies, philosophy, and knowledge about the natural environment.

## ROLE OF THE HUMAN FACTOR IN ANTHROPOTECHNICAL MEGA-SYSTEMS

Since the rapid development of industry started to require the building of ever more complex technical organizational and production structures, the role of man cooperating with such structures underwent far-reaching changes. Man as a user of technology was increasingly moved away (both literally, in space and in the intellectual sense) from the executed technological process. Man had already ceased to be a source of strength and energy required for a technological task, and for some time has lost meaning even as the controller of the process.

The introduction of automated machines has caused that the man – operator has become an unimportant addition to the "self-reliant," "wise," complicated, and expensive machine. His role is usually to perform auxiliary work such as: turning on the machines, loading the machines with material for processing, emptying the container with the finished products, disposing of waste (e.g. chips), observing the indicators reporting of correct technological process, and switching off the machine in case of an accident (sometimes this step proceeds independently of man) – in the case when we are dealing with production systems. In service mega-systems, man's functions are even more limited. The distinctions regarding the superiority of the human being in, among other things: the reception of weak or very weak stimuli (e.g. detection of small amounts of light and voice), the receiving and creation of information by means of light and of voice, the memorization of large amounts of information over a long period of time and the ability to recall the relevant facts at the right time, formulated so far seem to slowly lose their relevance (cf. McCormick 1964).

Production unassisted by the work of a number of machines with a high degree of automaticity is slowly becoming a domain only for artists. Automated machines usually do not cause risk of accident, are equipped with numerous sensors and safeguards to protect the individual against intrusion into the zone of technological operations. However, their advantage in terms of technology is a disadvantage in terms of ergonomics, as they are both "unergonomic" – forcing a man to perform the same set of simple actions, forcing a static posture, generating noise. The job of the operator is monotonous work, poor intellectually, boring, and does not require thinking. This situation lasts as long as such a machine or computer system is operating normally. If the machine happens to break down, the situation changes dramatically. The operator usually does not have the qualifications and powers to remove the causes of failure and has to call a service technician (setter, mechanic, renovator, etc.), and often – an entire service team consisting of specialists: electronic engineers, electricians, mechanics, plumbers, programmers, etc. These are often people with a university degree and extensive professional experience. They must carry out an accurate diagnosis of the condition of the machine or any other technical system, make the right decision, and quickly eliminate the fault – quickly, because the downtime usually causes large losses. Time pressure enhances mental tension and additional stress is caused by the infrequent occurrence of emergency situations and the inability to predict when the event necessitating immediate action will occur. Similar characteristic features are in work in dispatcher positions, where the human role is to supervise the technological process carried out by the complex, automated, or robotic machine systems. A dispatcher's work is responsible, and yet – apart from emergency situations – very monotonous and tedious, forcing prolonged sitting, and tracking control devices. These are only some, but not the only problems, associated with the so-called human factor in technical mega-systems.

Immeasurable in terms of depth and complexity are the social aspects that are present in technical mega-systems. Social systems are associated with technical mega-systems by very strong and very close bonds of internal dependencies. This is necessary for the smooth functioning of mega-systems, but at the same time a natural function of human communities that have evolved for thousands of centuries in the ability to build relationships between people. New anthropotechnical conditions also cause changes in the social character, whose trends and dynamics are difficult to predict. An example is the collapse of authority in the field of technology adoption and acceptance of certain patterns of behavior, which once happened as a result of permission for local elites. This trend disappeared with the increasing education (Butlewski 2011). Hence it is difficult to estimate the future effects of actions taken today.

Currently we are witnessing a situation in which the need to maintain close ties of information often goes hand in hand with the territorial limitation of the field of operation of the social system and individual people. Restriction of freedom of information and territory leads to a sense of a lack of freedom, which gives birth to aggression, irrational behavior, or apathy and disease. Excessive stress is even the cause of suicide. The social system becomes a crowd prone to manipulation, psychosis, panic and other disorderly conduct that is unreasonable and unpredictable. Therefore depressive disorders are now a major health issue in the U.S. workplace (Burton, Conti W.N., D.J. 2008).

## ERGONOMIC QUALITY CRITERIA OF A MEGA-SYSTEM

Ergonomic quality modeling allows one to specify groups of criteria which compose it. It is noteworthy that the ergonomic conditions for adjustment of technical objects may be different for different audiences and their needs increase with the enlargement of the mega-system. Figure 2 presents a model of ergonomic quality in the man – technical mega-system arrangement. Purposefully only three groups of criteria are presented, which the authors believe include the other requirements, widely reported in literature in the field of ergonomics. It seems that this description does not limit the consideration of individual components of the criteria. We will therefore have to deal with not only optimization between the technical possibilities and their costs, as in the case of microsystems, but we must also contend with the necessity of quantifying and evaluating groups of audiences. It is difficult to imagine that you can meet the expectations of all.

The mega-criterion of ergonomic quality in terms of meeting the psychological and physical needs of man leads to the conclusion that the basic criteria to be met by all subsystems of an anthropometric mega-system are:

- Provision of an optimum influence, tailored to individual, articulated as well as hidden, human needs, from the physical environment both at work and at all other dimensions of human functioning,
- Provision of an adequate (individually tailored) level of physical and mental load, which will be mutually offsetting in excess of the limit of optimal range for any of the subsystems,

– Meeting the needs of people, both obvious (self-realization) and hidden and nonstandard.

Of course, criteria presented in this way have a wide scope, however, for the complexity of mega-systems these criteria seem appropriate.

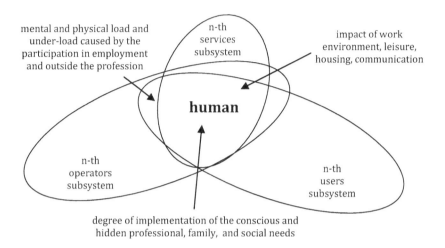

Fig. 2. Ergonomic criteria on the background of the man – anthropotechnical system arrangement

When defining the ergonomic model of man – anthropotechnical system, indicated also is the existence of three types of possible technical subsystems from the ergonomic and functional point of view. The operator subsystem is a system in which a person acts as the operator, and thus has a direct impact on the behavior of objects while at the same time also being influenced by this subsystem, for example, when driving a car. In the service system there is a much stronger bond, as it occurs when a person is dependent on the correctness or the possibility of operation of the technical elements, such as repair of the damaged subsystem. Such a relationship also causes a large impact on a person, among others, in relation to the proximity of technology and the previously mentioned stress and time pressure. In the case of a user subsystem, it is assumed that the user has no effect on the technical objects included in the system in a manner likely to cause a significant change in their functionality, though he is the recipient of technical and organizational conditions provided by the system.

The ergonomic quality criteria also apply to the effects of the other systems included in the ergonomic mega-system, that cannot be described as technical – for example, an ecosystem, and which affect the ergonomic quality of the conditions of human functioning.

# SUMMARY

Anthropometric mega-systems will increasingly make up the world around us. Of course, their nature will be diverse. In the production processes they will be associated with an increasing dehumanization (automatic machines replacing the work of many people) or an increasing integration (human settlements connected by computer-infrastructure networks). However, such systems will continue to require support, maintenance, monitoring, redesign and utilization, and this will require an increasingly more in depth knowledge of ergonomics.

Hence it is important that the ergonomic quality of solutions used is adjusted to the needs of users. There is a possibility to achieve an increase in productivity by 10-38% thanks to a higher ergonomic quality of simple products like hand tools (Butlewski, Tytyk 2008). If we take under consideration the effect of synergy, we can assume that the more complicated the systems are, the higher productivity we can achieve, because of the ergonomic quality of those systems. Another need of ergonomics in anthropotechnical mega-systems is the adjustment for all types of users, especially those who are disabled. Because of the demographical changes, it is necessary to provide work-related solutions for the elderly and the handicapped. The lengthening of the life span, caused by the advances in medical care and improved living conditions, produces the necessity of elongating the time of occupational activity. In addition, during the next 25 years the number of people able to work in the EU will decrease by over 20 million to 280 million people (an over 7% decrease), which means that people theoretically unable to work will have to perform increasingly more labor (Tytyk, Butlewski 2011).

Safety, ergonomics, and the operating conditions of people in technical mega-systems have a decisive impact on the environmental quality of life for many people (sometimes hundreds of thousands or millions) living in the acting area of the technical mega-system. There is an urgent need for the accumulation of knowledge concerning the functioning of technical ecosystems. For this knowledge not to have a "general theory on everything" nature, diagnostic tests are necessary for particular technical ecosystems, operating on the basis of specific technical mega-systems, computer simulations, allowing for the analysis of virtual critical states and for the development of a "philosophy of technical ecosystems." It should be noted that the technical aspects are intertwined with the social and some issues have a philosophical nature, but with the consequences being very real. Perhaps in the future we will begin to answer the question of how to estimate a human life, which is at stake in these types of systems, and whether what we usually recognize as moral (respect for property and the opportunity to own by no means limited) does not conflict with the overall good of the people in anthropotechnical systems. This approach might be able to help in the future a larger number of people to function in the even more technical world, without excessive health and social costs.

# REFERENCES

Burton, W.N. Conti, D.J. Depression in the workplace: The role of the corporate medical director, Journal of Occupational and Environmental Medicine, Volume 50, Issue 4, April 2008, Pages 476-481

Butlewski M., Uwarunkowania historyczne rozwoju subregionów Wielkopolski raport FORESIGHT „Sieci Gospodarcze Wielkopolski" - Uwarunkowania rozwoju gospodarczego w subregionach Wielkopolski i tożsamość subregionów 2011

Butlewski, M., The Method Of Matching Ergonomic Nonpowered Hand Tools To Maintenance Tasks For The Handicapped 2nd International Conference On Applied Human Factors And Ergonomics Jointly With: 12th International Conference On Human Aspects Of Advanced Manufacturing (HAAMAHA) 2008,

Ergonomia. (Red.) L. Pacholski. Poznań, Wyd. Politechniki Poznańskiej 1986.

Gadomski An approach to an Unified Engineering Meta-Ontology: Universal Domain Paradigm: SPG Representation of a Domain-of-Activity. e-paper, the ENEA Agency MKEM Server (It), http://erg4146.casaccia.enea.it/wwwerg26701/gad-diag0.htm, last updating 15 Aug. 2006.

Jabłoński J. (red.), Ergonomia produktu. Ergonomiczne zasady projektowa-nia produktów. Wydawnictwo Politechniki Poznańskiej, Poznań 2006

Kowal E., Ekonomiczno-społeczne aspekty ergonomii. Wydawnictwo Naukowe PWN, Warszawa – Poznań, 2002

McCormick E. J., Antropotechnika, przystosowanie konstrukcji maszyn i urządzeń do człowieka. 1964. Wydawnictwo, WNT

Pacholski L., Macroergonomic evaluation of the work process quality of the multiagent manufacturing system; (in): Ergonomics Design: Interfaces – Products – Information; p. 445-448, ABEGRO, Rio de Janeiro, 1995

Pacholski, L.. Jasiak A. Makroergonomia, Wydawnictwo Politechniki Poznańskiej, Poznań 2011.

Rust R., Thompson D., Hamilton R., Nie dodawaj kolejnych funkcji.., Haward Business Review Polska, kwiecień 2006

Słowikowski J., Przesłanki ergonomiczne „wyczuwania" maszyny przez człowieka, Bezpieczeństwo Pracy 7-8/2001

Tytyk E, Podstawy metodologii projektowania ergonomicznego, (w): Ergonomia produktu. Ergonomiczne zasady projektowania produktów (red. J. Jabłoński), WPP, Poznań 2006

Tytyk E., Butlewski M., Ergonomia w technice, Wydawnictwo Politechniki Poznańskiej, ISBN 978-83-7775-048-3, Poznań 2011

Tytyk E., Butlewski M., Ergonomiczna jakość zmechanizowanych narzędzi ręcznych jako przesłanka ich doboru do prac obsługowych, (w) Tytyk E. (red.) Inżynieria ergonomiczna. Teoria. Wydawnictwo Politechniki Poznańskiej, Poznań 2011

Tytyk E., Projektowanie ergonomiczne. Wyd. Naukowe PWN, Warszawa – Poznań, 2001

CHAPTER 33

# Ergonomic Engineering of Anthropotechnical Mega-Systems

*Edwin Tytyk*

Poznan University of Technology
Poznan, Poland
e-mail: edwin.tytyk@put.poznan.pl

## ABSTRACT

In the paper, defined is the concept of ergonomic engineering as a type of prospective engineering measures, which change the world of technical objects (artifacts) in the direction of increased ergonomic quality, safety for humans, and care for the environment. To effectively achieve this goal, it is necessary to make changes in thinking about the proposed system and the structure of the design process. Important elements of the new structure of the design process of anthropotechnical systems are decision criteria, in terms of content belonging to the area of ergonomics. The paper presents the formulation of the principles of ergonomic decision criteria, tailored to the specific steps of design procedure and expressed in the engineering language.

The level of difficulty of the diagnostic and design work grows with the increase in the complexity of the examined or designed systems and reaches the highest level in the design of complex anthropotechnical mega-systems made up of many people and many technical objects (e.g., industrial plants, airports, railway stations, shopping malls, hospitals and schools, etc.) This type of measures can be called macroergonomic engineering.

Methods of ergonomic diagnosis of anthropotechnical mega-systems are based on the concept of checklists, usually modified to reflect the special characteristics of the examined mega-systems.

Designing of anthropotechnical mega-systems encounters major methodological difficulties due to the multi-interdisciplinary nature of problems to be solved, incompleteness and uncertainty of knowledge of a prospective nature, and the risk of the appearance of unforeseen, adverse effects of engineering measures, which may become apparent in the long term.

The paper presents a concept of the structure of decision criteria that are relevant in the design process of anthropotechnical mega-systems of an industrial nature.

## 1. INTRODUCTION

Despite more than a 60-year period of development of practical and theoretical (in that order) ergonomic studies, still there are questions about the nature, scope and purpose of these studies. One reason for this is probably the undefined formal status of ergonomics in the Polish system of scientific areas and disciplines.

No doubts were had by prof. Wojciech Bogumił Jastrzębowski, who is internationally recognized as the author of the name and forerunner of ergonomics, who in 1857 wrote, that "the name Ergonomics ... means the <u>Science</u> of Labor ..." (7, p. 193). Another internationally known Pole had no doubts, the founder of the philosophical trend of praxeology, prof. Tadeusz Kotarbiński when defining "Ergological <u>Sciences</u> (from the Greek *ergon*, in English work or to work) – one way or another are interested in human activity" (8, p. 468). The community of scientists and practitioners gathered in the Polish Ergonomic Society did not have any doubts either when in the year 1983 in the Statute PTErg it formulated the definition: "Ergonomics is an applied <u>science</u>, aiming at the optimal adaptation of tools, machinery, equipment, technology, organization, and the physical working environment, as well as objects of everyday use for the physiological, psychological, and social requirements and needs of man" (underline ET).

In countries where there is a high economic and civilization level, the dilemma of the status of ergonomics is not present. Simply, it is a useful and necessary knowledge, which allows for the manufacture of industrial products with high quality, measured by, among others, the commercially very valuable "friendliness" to the user (so-called user friendly products) – and this is enough for ergonomics to attract attention and support from policy-makers in the areas of industry, education, and politics. There, the ergonomist profession exists, one can get a degree from this area of knowledge, one can undertake studying in this field.

However, we can see a chance to break the existing deadlock in this matter. We must extensively develop ergonomic engineering, a causative, practical trend of ergonomics, supported simultaneously with the built up theoretical knowledge, and consequently – definitely strive toward the fields of technical sciences and develop "technical thinking" in ergonomics.

## 2. THE TERM: ERGONOMIC ENGINEERING

An important distinguishing feature of ergonomics is the practical aspect, which is known as applied or engineering, and these terms are used interchangeably. This feature is emphasized in the definition of ergonomics, as adopted by the Polish Ergonomic Society in 1983. This fact can justify the use of the name "ergonomic engineering."

Tadeusz Pszczołowski (1987, p. 90) defines engineering as a "modernly expanding section of practical sciences, which transforms a selected fragment of reality, including inorganic and organic matter, plants, animals, and humans." According to this source, the term "engineering" is used in different senses, which are defined by the predicates added to it, such as social or social engineering, human or ergonomic, systems or systems engineering, but also: land, hydro, materials, chemical, biochemical, military, domestic. In English language literature there are still other terms used for the word engineering: job, industrial, sales, values, products, management. Serious concerns can be raised when calling ergonomics "human engineering," but this may be the result of that it was done over thirty years ago.

Thus we can say that engineering is the practical, efficient side of various disciplines and branches of knowledge. It also includes the knowledge (science) of ergonomics, and therefore it can legitimately be called "ergonomic engineering." It can be defined as the science and ability to perform engineering (design and implementation), whose aim is to produce technical objects with a high ergonomic quality and safe, healthy, and favorable conditions for interaction between humans and technical objects. The ergonomic quality of a technical object (product, tool, machine, apparatus, device, equipment, vehicle) is the degree to which it meets certain assessment criteria, formulated in the form of ergonomic requirements contained in standards and other mandatory records (e.g. ministerial regulations), literature, and also expressed in the opinions of the users of these objects (cf. Tytyk, 2009, pp. 88-90). A model set of ergonomic quality features aimed at hand tools is shown in Figure 1.

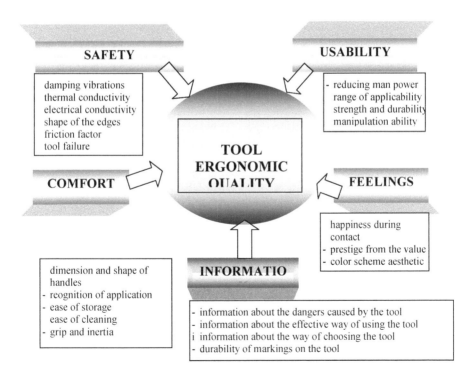

Fig. 1. Model of the ergonomic quality of nonpowered hand tools (Adapted from (Butlewski, 2008))

Ergonomic quality is an important category that consists of a comprehensive product quality, also known as the ergonomics of the product. It is mostly invested in the areas of usability and usefulness. With the advance of civilization and the increase in the standard of living, ergonomics is increasingly important in the overall quality features of different products. Support for this thesis is A. Maslow's known concept of the hierarchy of human needs – after the satisfaction of basic needs (existence, security, fulfillment of the principal functions) higher needs emerge: convenience, satisfaction, physical and psychological comfort, and beauty.

**Table 1. The types of activities belonging to the ergonomic engineering field (own elaborating)**

| No. | Criterion for the division of activities from the ergonomic engineering field | Name of engineering activities in the field of ergonomics: "... ergonomics..." |
|---|---|---|
| 1 | Phase of the existence of the man – technical object system | - correctional<br>- conceptual |
| 2 | Specificity of the technical component of the man - technical object system | - working conditions<br>- means of work<br>- hand tool<br>- sporting equipment<br>- consumer goods<br>- vehicle<br>- "little objects"<br>- technical mega-system<br>- ... |
| 3 | Specificity of the human component of the man - technical object system | - for people with disabilities<br>- for seniors<br>- for preschool children<br>- for students<br>- cognitive<br>- ... |
| 4 | Specificity of the man – technical object system as a whole | - production hall<br>- public building<br>- office<br>- home<br>- architecture<br>- transport (land, water, air)<br>- anthropotechnical mega-system (macroergonomics)<br>- ... |

| No. | Criterion for the division of activities from the ergonomic engineering field | Name of engineering activities in the field of ergonomics: "… ergonomics…" |
|-----|---|---|
| 5 | Specificity of intentional actions | - physical work<br>- creative work<br>- cognitive<br>- artistic work<br>- professional sport<br>- leisure and recreation<br>- treatment and rehabilitation<br>- … |
| 6 | Specificity of the industry | - material processing technology<br>- chemical technology<br>- agriculture<br>- forestry<br>- aviation<br>- construction<br>- mining<br>- metallurgy<br>- military |

Within ergonomic engineering we can distinguish various activities, which are commonly called "… ergonomics" with a complementary word to give the meaning (Table 1). This is not correct in terms of language, because ergonomics as a science, and probably in the near future, as a scientific discipline, is consistent in terms of paradigm and specificity of the object of study. Therefore there are no different types of ergonomics – there are only the various activities undertaken in this emerging discipline. It should be noted that the divisions shown in Table 1 are not separable.

The engineering character of ergonomics manifests itself particularly clearly in the division criteria of ergonomic engineering tasks numbered 1, 2, 4, and 6 (Table 1). The criterion of phase of the existence of the system (No. 1) allows for the separation of conceptual and correctional actions, which have an already established methodology (ergonomic diagnosis: questionnaire and instrumental methods, ergonomic design: methods with a various "power of interference" into the design process – cf. (Tytyk, 2001; 2009; Słowikowski, 2000)). The criterion specificity of the technical subsystem (No. 2) as well as the specificity of the whole man – technical object system (No. 4) allow for the distinction of engineering activities, widely described in numerous literature sources. The criterion of specificity of the industry (No. 6) clearly attributes ergonomic activities to various "branches" – sectors of industries, manufacturing, or the national economy and is associated with all the other division criteria of ergonomic engineering. In this case, literature has an "island," specialist character; the subject of ergonomics is not always visible in the titles of books (cf. (Giefing, 1999, pp. 147-155)).

Ergonomic knowledge at the moment is so well established and advanced that it can be seen as an emerging scientific discipline. Although opinions on this subject are divided (cf. (Chapanis, 1979; Franus, 1992; 2001; Rosner, 1978; Jabłoński, 2005; 2011)), this does not mean that the practical usefulness and exploratory attractiveness of this interdisciplinary science is exhausted. On the contrary – we constantly observe the beginning of new areas of exploration of this informal science.

## 3.    ANTHROPOTECHNICAL MEGA-SYSTEMS (ATMS)

It is easy to imagine the structure of the systems of a higher complexity, composed of many elementary man – technical object systems, (Fig. 2). They can be called "anthropotechnical mega-systems" (ATMS). In practice, these can be for example a production department, office, hospital, school, neighborhood, and even a mega-system called a technical civilization. Here we enter into the area of scientific study called macroergonomics (Pacholski, 2000).

The environment of systems with a higher degree of complexity is different than for the elementary system: here factors come into play that belong to economics, sociology, ecology, politics, culture.

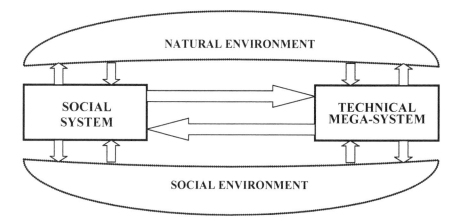

Fig. 2. Anthropotechnical mega-system  (Adapted from (Tytyk, 2009, p. 91))

To make a macroergonomic evaluation of ATMS, one must take into account the criteria belonging to many different areas of meaning. Conditions for the formulation of criteria are provided in Table 2.

Based on the assumptions presented in Table 2, one can formulate ATMS evaluation criteria or the decision criteria used for making design decisions. The lean framework of this article does not permit for the further development of this problem. However, one can assume that the individual criteria areas (column 2 in Table 2) will be evaluated on a scale of relative values in the range (0 - 1), where zero means an extremely poor score, and one – an extremely good score.

A problem will arise for the aggregation of these indicators. Aggregation models based on averaging (algebraic, geometric, weighted) can be considered inappropriate, because one cannot find a logical justification for the hypothesis of one-way interaction of the criteria belonging to different areas. Therefore, a graphical interpretation of the aggregation of criteria belonging to different areas of meaning is proposed. For the example shown in Table 2, which highlights eight areas of ATMS criteria, a graphical illustration of the aggregation model is shown in Figure 3. The individual "rays" (1 - 8) represent the values of criteria areas on a scale of (0 - 1).

**Table 2. Principles for the formulation of macroergonomic evaluation criteria**

| No. | Criteria area of meaning | Substantive grounds for the formulation of criteria |
|---|---|---|
| 1 | 2 | 3 |
| 1 | Ergonomics | psychology, physiology, anthropometry, medicine, technology, work organization, management, work economy, occupational risk, qualitology |
| 2 | Ecology | ATMS' impact as a relatively isolated whole and its components on ecosystems (human, animal, plant) in various stages of ATMS and its components |
| 3 | Economy | ratio of output to input, efficiency, productivity, costs of reliability and resistance to changing operating conditions |
| 4 | Sociology | professional qualifications and education of people, professional ethos, gratuities for work, market demand for labor in the profession, worker mobility, interpersonal relationships, social behavior |
| 5 | Culture | moral principles, ethical and religious beliefs, family ties, traditions, compliance of corporate principles with ones own system of values |
| 6 | Well-being | subjective feeling of "ones place," an understanding of the world, acceptance of the rules of functioning in society, health, stability, perspective of personal and professional development, social interaction |
| 7 | Social safety | salary and other gratuities, job security, health care, social security, pensions, pro-family policy |
| 8 | Public safety | lack of danger to themselves and their families from other social, political, or religious groups, protection against natural disasters and technological, biological, chemical, or energy disasters |

The maximum aggregate score is equal to the surface area of a regular octagon with radius equal to unity (R = 1), namely:

$$S_{\bullet 8} = \tfrac{1}{2} \cdot 8\, R^2 \sin \text{☉} = 2\,,2 \tag{1}$$

The aggregate measure of the macroergonomic quality of ATMS can be defined as the relation of surface area of an inequilateral octagon, formed from the assessed criteria areas in order of decreasing values of score indicators, to the surface area of an equilateral octagon with sides equal to unity (maximum).

After simple mathematical transformations, for any number of areas of evaluation criteria (*i*) it could be in the formula:

$$Q_{ATMS} = \tfrac{1}{8} \cdot \bullet\, x_i \cdot x_{i+1} \qquad\qquad \text{if:} \quad x_i \,\text{¤}\, x_{i+1} \tag{2}$$

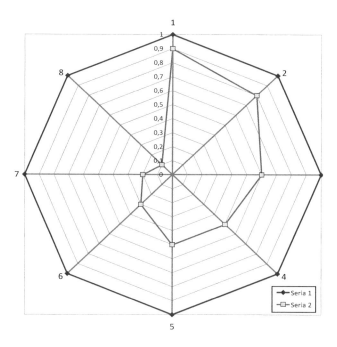

Figure 3. Graphical interpretation of the aggregation model of macroergonomic quality assessments

Using the formula (2), for the eight sub-points, in order of decreasing value: 0.9 – 0.8 – 0.6 – 0.5 – 0.5 – 0.3 – 0.2 – 0.1 we receive the aggregate scores of marcoergonomic quality with the value $Q_{ATMS} = 0,25875$ (Fig. 3).

# 4. CORRECTIONAL AND CONCEPTUAL ERGONOMIC ENGINEERING

Methods and techniques of ergonomic engineering are usually addressed to diagnostic and design activities. General diagnostic methods are based on questionnaire techniques, in particular – surveys and checklists, while detailed methods of diagnosis are various techniques for apparatus measurement and evaluation, consisting of the measurement results being compared with the values of maximum intensity or concentration (NDN or NDS) adopted as acceptable or recommended (Górska, 1988; Rybarczyk, 2000). Ergonomic design methods need to be more sophisticated, because they focus on changes in creative thinking and evaluation of design concepts (Tytyk, 2001; Słowikowski, 2000). Currently, very extensive amounts of literature exist about methods of diagnosis and less numerous – on ergonomic development, but its discussion would exceed considerably the volume of this work. Table 3 shows only the stages of development of design methodology and gives their characteristic features. Each of these steps has its "own" literature.

**Table 3. The developmental stages of ergonomic design methods (Adapted from (Tytyk, 2009, p. 86))**

| Stage | | Characteristic features |
|---|---|---|
| No. | Nature of changes | |
| I | Noticing the need for change | Perception of needs and opportunities of inclusion of human features in the design and revision of projects |
| II | Surface changes – formal | Enrichment of the structure of the technical design process – formal and procedural changes |
| III | Deep changes – methodological | Modification of the structure of the design process – methodological changes, the creation of rules and theories |
| IV | Assistance of heuristic processes | Computer-aided ergonomic design, automation of routine design |

Methodological reflections on ergonomic design lead to the following conclusions:
- The design goal should be to develop a project (project documentation) of not only an efficiently working machine (usage artifact), but above all a functioning **system** consisting of operators and technical objects, cooperating under optimal environmental conditions in the workplace (or other intentional action) of man.
- The good of the people as components of this system should be prioritized in relation to the technical requirements.
- Design is an activity carried out **BY the people and FOR the people** – it should therefore be of a humanocentric nature.

In conclusion – if we believe that ergonomics deserves the status of a scientific discipline, then the chance for it should be seen in the development of ergonomic engineering. This development in turn depends on the quality and universality of an education of technical thinking that is humanocentrically oriented and on positivist "work on the basics" carried out in the direction of educating faculty and gaining next degrees within the existing scientific disciplines, but in specialties referring to ergonomic engineering.

## REFERENCES

Butlewski M., *Metodyka ergonomicznego projektowania oraz doboru niezmechanizowanych narzędzi ręcznych*. Rozprawa doktorska, maszynopis niepublikowany, Wydział Informatyki i Inżynierii Zarządzania Politechniki Poznańskiej, 2008

Chapanis A., 1979, *Quo vadis, Ergonomia?* Ergonomia, tom 2, nr 2, Zakład Narodowy im. Ossolińskich, Wrocław, s. 109-122

Franus E., 1992, *Struktura i ogólna metodologia nauki ergonomii.* Towarzystwo Autorów i Wydawców Prac Naukowych Universitas, Kraków

Franus E., *Myślenie techniczne.* Zakład Narodowy im. Ossolińskich, Wyd. PAN, Kraków, 1978

Giefing D.F, 1999, *Podkrzesywanie drzew w lesie.* Wyd. Akademii Rolniczej, Poznań

Górska E., 1988, *Diagnoza ergonomiczna stanowisk pracy.* Oficyna Wydawnicza Politechniki Warszawskiej, Warszawa

Jabłoński J., 2011, *Obszar i przedmiot poznania ergonomii. Odniesienia poznawcze i ontologiczne.* Wyd. Politechniki Poznańskiej

Jabłoński J., *Czy ergonomia jest nauką?* 2005, Wyd. Politechniki Poznańskiej, Poznań

Jastrzębowski W.B., *Rys ergonomii, czyli nauki o pracy.* Ergonomia, tom 20, nr 2, 1997

Karwowski W., Kantola J., Rodrick D., Salvendy G., *Macroergonomic Aspects of Manufacturing*; (in): W. Hendrick at all (Eds.), Macroergonomics, Human Factors and Ergonomice; Lawrence Erlbaum Associates, Inc. Publishers, Mahwah, New Jersey, 2002

Kotarbiński T., *Traktat o dobrej robocie.* Wyd. 3, Wrocław, Zakład im. Ossolińskich, 1965

Pacholski L., 2000, *Macroergonomic Aspects of Total Quality Management*; (in): D. Koradecka, W. Karwowski, B. Das (Eds.), Ergonomics and Safety for Business Quality and Productivity; Warszawa

Pochanke H., *Dydaktyka techniki.* Warszawa, 1985

Pszczołowski T., *Mała encyklopedia prakseologii i teorii organizacji.* Wyd. Zakład Prakseologii Instytutu Filozofii i Socjologii PAN, Zakład im. Ossolińskich, Wrocław, 1978

Rosner J., 1978, *Rozwój i stan ergonomii jako nauki.* Ergonomia, tom 1, nr 1, Zakład Narodowy im. Ossolińskich, Wrocław, s. 17-33

Rybarczyk W., 2000, *Rozważania o ergonomii w gospodarce.* Wyd. Centrum Zastosowań Ergonomii, Zielona Góra

Słowikowski J., *Metodologiczne problemy projektowania ergonomicznego w budowie maszyn.* Wyd. CIOP, Warszawa, 2000

Talejko E., *Zagadnienia psychologii twórczości technicznej.* Poznań, 1971

Tytyk E., 2009, *Inżynieria ergonomiczna jako komponent inżynierii zarządzania jakością warunków pracy.* (w): Ergonomia – Technika i Technologia – Zarządzanie, red. M. Fertsch, Wyd. Politechniki Poznańskiej, Poznań, s. 81-97

Tytyk E., *Methods of ergonomic design of human-machine systems.* International Encyclopedia of Ergonomics and Human Factors, 2-nd Edition, Edited by W. Karwowski, Chapter 330, Louisville, USA, 2005

Tytyk E., *Projektowanie ergonomiczne*, 2001, Wyd. Naukowe PWN, Poznań – Warszawa

CHAPTER 34

# Factors Adversely Affected the Productivity of Software Designers Applying CASE Tools

*Andrzej Borucki*

Institute of Management Engineering
Poznan University of Technology, Poland

## ABSTRACT

A key role in developing software is played by the intellectual resources of project teams, i.e. their knowledge. The knowledge used to produce software may be either open or hidden. In order to manage software development effectively, advantage needs to be taken of the knowledge held by each design team member. This is to ensure that the resulting software meets the requirements defined by its future user and arrives on time and on budget. Today's managers responsible for planning and organizing software development projects ask themselves the fundamental questions of whether CASE (Computer Aided Software Engineering) tools boost the efficiency of software designers, and whether such tools help managers in managing IT projects.

**Keywords:** Project Management,Software Engineering, Reguirement Engineering

# INTRODUCTION

According to a long-held view, today's software design requires the use of CASE tools and, in particular, a high-level language. As knowledge grows, it is placed in repositories available to designers and ready for use on a wide range of projects. Multiple use of the knowledge stored in CASE tool repositories is believed to be critical for rapid software development and expected to streamline the design process. The knowledge held in such repositories concerns software design, implementation and testing. The options for its use are numerous and depend on the knowledge management strategy adopted for a given design project. Two distinctive knowledge management strategies are available for managing software development. These are the knowledge codification strategy and the knowledge personalization strategy (Jashapara A.,2004).

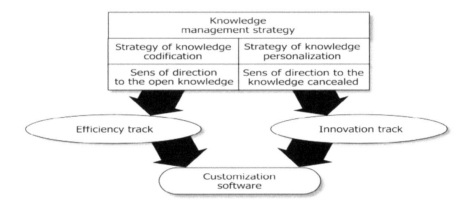

Figure 1. Alternative approaches to the knowledge management strategy.

As proposed by the knowledge codification strategy, project management involves rational gathering of knowledge on technical issues encountered in the design process and inserting it into databases, procedures and the documentation of logical and physical models developed during a project. The use of a knowledge codification strategy in IT project management resembles approaches to repetitive production where the technological process of producing a product follows a prescribed structure and comprises such individual components as operations, measures and items. Each component of the technological process employed to produce a machine part is described in great detail in a database (of standards) in terms of the depletion of physical and financial resources, machine and equipment uptime and labor intensity. Similarly in software development, the primary purpose of the information held in CASE tool databases is to build logical and physical software models suited for repeated use on various projects. It is difficult today to imagine the design of a system context model or the drafting of data flow charts without the use of CASE tools. The knowledge codification strategy is intended to make the software development process repeatable and base it on well-tested components and development methods. Such components may be used at any stage of software development, i.e. during requirement analysis, implementation, software testing and maintenance. In some cases, one may generate a component-based software source code in the form of charts depicting the system's logical model. Amidst all such components, particular attention should be given to COTS (Commercial Off-The-Shelf) components which may be delivered by an external software supplier and as autonomous ready-made software items. These include database systems, software supporting the common functionalities of a web-based store, software used for the remote control of system operation, etc. The use of components and tested software development methods shortens lead times, reduces deployment uncertainty, enhances software reliability, increases compliance with standards and reduces costs. The intention therefore is to make software development ever more efficient and effective. The knowledge repository contains data dictionaries, chart and form preparation tools as well as report and software language generators. The knowledge codification strategy is the most common approach in today's project management. It generates a huge demand for software supporting CASE project management.

## CASE tools and software designer productivity

Software development productivity can be defined in many ways. Its most popular measure is the number of source code lines (such as code lines written per month) or the number of completed program functions delivered within a specified time, as proposed by Albrecht (Katonye G.,et al, I.,1998 ). The latter measure of software designer productivity expressed per person per month is applied regardless of the language selected for programming purposes. The number of functions is typically multiplied by a relevant weight reflecting software complexity. Software complexity depends chiefly on the complexity of the developed algorithms and interface structure.

The productivity of software designers depends on a number of factors, the most critical of which is the knowledge of the particular field in which the software is to be applied. Over many years of observing software development (Borucki A.,2008 ), we have found that the greater a designers' knowledge of a given field, the shorter it takes to develop software, regardless of the degree of reliance on various CASE tools in software development. Designers who intimately understand the business processes which the future software is intended to support tend to be more productive. In many of the projects carried out under our management, team members varied greatly in their productivity.

Some software units produced by designers knowledgeable in a given business area emerged more than a dozen times faster than those developed by less productive designers. In some cases, it made business sense to refer the designers to additional courses to fill their knowledge gaps and help them better understand the business processes involved in a given corporate function such as warehousing, marketing or logistics. Differences in designer productivity due to knowledge gaps in specific fields come across as the most conspicuous in small design teams. Comparisons of the productivity of individual designers are only sensible on the assumption that the software they produce is of comparable quality. Our experience in managing IT projects shows that a strategy of knowledge personalization in software development helps improve designer knowledge in a given business field and boosts their productivity.

The key benefits of using CASE tools in software development to streamline processes lie in allowing for:

1. system modeling at various abstraction levels;
2. recording project outcomes in uniform formats at each stage of design;
3. easier communication among project contractors and the future software user;
4. easier horizontal as well as vertical communication among project designers;
5. the use of a prior project knowledge base;
6. the use of standard methods of developing various systems;
7. the application of uniform methods of assessing project quality;
8. the setting up of a knowledge repository for the needs of future projects – such repositories are made up of databases of specifications concerning labor intensity, contractors, the developed objects, charts, algorithms, etc.;
9. the standardization of input and output data formats;

All of the above CASE functions support the gathering of knowledge on a current project and help develop a model of a given enterprise with the use of either the Gantt chart or the PERT method. The knowledge codification strategy helps gather the knowledge deemed to be of value at various levels of application. The highest level incorporates knowledge of strategic importance for individual projects and knowledge seen as elite. A lower level may include knowledge on design techniques and methodology, design templates and documentation drafting support tools. A further level may hold knowledge on standards and guidelines applied in project implementation.

The knowledge codification strategy requires the use of expensive technology to apply CASE tools. It entails high costs of strict compliance with the design and documentation principles stipulated in CASE manuals. The knowledge codification strategy makes it necessary to use CASE tools.

Based on two decades of experience in managing IT projects, we have grown somewhat critical of the use of CASE tools. Despite extensive reliance on such tools, most projects run over time and budgets. The reason for this lies in misguided knowledge management strategies and excessive reliance on CASE tools for software design. This is particularly visible in innovative projects designed in a non-linear fashion and those which follow sequences up to a certain point in the project only to move into a feedback loop which was not envisioned in the original project model. The non-linearity of the design process also means that the structure of a given programming project carried out by a design team may not be fully known at project launch. Negative impacts of excessive reliance on CASE tools could be seen in application projects which make use of a number of different IT technologies.

On the down side, CASE tools undermine designer productivity by contributing to:
1. limiting the creativity of software designers;
2. large costs of CASE tool acquisition and deployment;
3. making the design process more rigid by strict adherence to the design methodology imposed by CASE tools;
4. limiting project management following changes in the requirements model;
5. imposing limits on the roles assigned to design team members;
6. preventing osmotic communication among members of the same design team and across multiple design teams assigned to a single project;
7. inadequate mapping of business processes in the requirements model caused by excessive reliance on templates and knowledge derived from other projects;
8. compounding project management problems wherever the design process becomes non-linear and random;
9. increasing confusion in project management where a certain degree of confusion already exists caused by poorly selected project scope, inadequate work plans and erroneous estimates of working time and resource needs;
10. the need for indicators to assess the quality of project stages and the entire project whenever such tools are used;
11. "thoughtless" work resulting from strict reliance on knowledge provided in the CASE tools database;
12. reducing project managers to the role of planners rather than project leaders;
13. mistakenly treating designers as "expandable" in the belief that project documentation offers complete security against designers dropping out from projects;
14. defensive project management;
15. an emphasis on details while losing sight of the big picture (entire project).

In order for the knowledge personalization strategy to work successfully, a number of preconditions need to be met concerning employee remuneration for sharing their knowledge with other workers and creating a workplace atmosphere conducive to initiating osmotic communication. One way to do this is by physically designating in an office a common area shared by all project participants to encourage informal exchanges of information among the personnel assigned to the same "executive level" of the project's different modules (De Marco T.,Lister T.,2002). Such a mutual exchange of information on problem issues encountered in implementing various project components promotes innovation and mitigates failure risk. Managing projects in keeping with the knowledge personalization strategy does not entail any additional expenses.

# Knowledge management strategy in requirements engineering

The most daunting part of software design is to analyze the requirements. To this end, designers and future software users need to recognize and understand the services and limitations known as system requirements.

As is commonly believed, critical problems in software development which may potentially cause failures are likely to arise in:

- managing changes in software logic to achieve conformity with an enterprise's business model,
- managing changes in the project schedule.

To ensure that software design complies with the expectations of future users, the design process must reflect software requirements. Software requirement specifications (SRS) form a document describing the requirements expressed by the future user and guidelines for ensuring compliance with such requirements in the process of software development and after allocation to the IT infrastructure in which the software will operate( Berenbach B.,et al .,2009).

The solutions expressed as software functionality should be geared mainly to ensuring the enterprise achieves a competitive advantage in its market.

Long-time experience in managing IT projects shows that an increasing number of corporate IT projects are conducted in a very volatile business environment sensitive to rapidly changing markets affected by the recent crises. Many companies are forced to redefine their market positions and develop new competitiveness strategies. Such new circumstances call for new approaches to software design so as to respond faster to new demands and be able to provide software users with new services within the required time. The requirements put to designers are ever more demanding in terms of the time allowed to deliver new software services. Requirement changes are managed in the four stages of change identification, change analysis, change monitoring and change planning. All requirement modifications need to be examined not only for implementation issues but also in terms of the time and cost of their adoption. In a rapidly changing environment, requirements management becomes an essential skill of software engineers. Due to highly changeable business models and shorter software modification time, application developers are forced to resort to adaptive rather than predictive and non-linear methods which allow for the structure of a given project to be partially unknown at project launch as the design project includes feedback loops decisive for the definitions and solutions adopted further into the design process(Laplant P.,2009). Such an approach to software development is only possible when knowledge management in the project team follows the knowledge personalization strategy. The strategy aims at osmotic communication which allows exchanges of information among project team members on the basis of equal access. Project team members are granted access to all project information regardless of whether such information is of direct relevance for the design tasks they perform or concerns another set of responsibilities. The assumption is that one can never be sure if a given piece of information will not be needed further in the project and turn out to be helpful in solving a problem which cannot yet be foreseen. If the knowledge codification strategy is applied in software design management resulting in extensive use of CASE tools, volatile requirements will wreak havoc with painstakingly composed project schedules and prevent clients, i.e. future software users, from intervening on an ongoing basis. This is because to intervene in such a manner, the clients would first have to learn the language and detailed rules of using specific CASE tools. The benefits of replacing CASE tools with the knowledge personalization strategy become obvious on very unique and challenging projects conducted to highly changeable software design requirements.

## Project management and the Goldratt theory of constraints

An examination of the effectiveness of using alternative knowledge management strategies in software design suggests the questions of whether the knowledge personalization strategy may support the knowledge codification strategy which relies heavily on CASE tools; and whether another third knowledge management option exists in software design which would allow the two other strategies to coexist.

It seems that as of late, the Goldratt theory of constrains has contributed to a pronounced breakthrough in project management. According to the theory of constraints, project management involves two key constraints which adversely affect project time and budget. These concern:

- the capacity to perform certain tasks, i.e. bottlenecks which affect project time and budget,
- human resource management.

Under the theory of constrains, the effectiveness of an entire process depends on the weakest link in a project, i.e. the completion of critical chain tasks. A definition of the critical chain could be based on the net theory proposed by Petri

(Blikle A.,2010). The Petri net is a directed graph with two types of nodes: one used to define elementary project tasks, i.e. "transitions", and another used to define the conditions which need to be met for a transition to fire – these are referred to as Petri net "places". Each graph arc connects a place with a transition and may run from a place to a transition or reverse. If a graph arc arrow points to a transition, the transition is said to have an input. If the arrow points from a transition to a place, it is said to have an output. Places in a net which serve as transition inputs contain dots which mark the available resources necessary to perform a given project task. The resources may be product-based and constitute project items produced at earlier project stages which are subject to "wear and tear" or be developed by a given transition, or tool-based and used by a given transition: these are not subject to wear-and-tear and can be used by other transitions. The tool resources of a software project are people, computers and CASE tools.

In order for an elementary project task to be considered as completed, i.e. in order for "a transition to fire", to use the language of Petri nets, all of the resources necessary for the completion of a given task, i.e. the so called marks or tokens, must be found at the transition's input (the places must contain dots indicating the availability of individual resource categories) whereas no tokens should be found at outputs, meaning that output places are ready to accept resources, as shown in Figures 2 and 3.

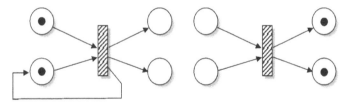

Figure 2. A transition before it fires      Figure 3. A transition after it has fired

In his work, Blikle points out problems with unambiguously defined notion of the critical chain in the Goldratt theory of constraints. By modeling a given project with the use of Petri nets, one can define the critical chain as a set of transitions having the following properties:
1. The acceleration of any transition which belongs to a critical chain accelerates the completion of the entire project.
2. The delay of any transition which belongs to a critical chain delays the completion of the entire project.
3. Many transition sets usually come as candidates to become a critical chain.
4. The ultimate choice of a critical chain from among the transition sets
   Seen as candidates to become a critical chain, is made by project leader.

The first step in the theory-of-constraints procedure is to identify the critical chain, i.e. the bottleneck. IT projects comprise a complex network of links between transitions and lack the determinism whose manifestations include unforeseen cycles, i.e. reversals to tasks which were once thought to be completed. The resources used in transitions have achievability specified in time. At many project steps, the allocation of resources depends on a manager's decision, as taken at a particular time, and has the nature of an intervention.

According to the theory of constraints, further project management steps involve the assignment to critical chain tasks of additional resources used to make transitions fire, if and where possible. If a project manager is unable to mobilize resources immediately ("hot" resources), he should have resources delivered as soon as possible ("cold" resources). Such a delivery would concern resources used in other parallel projects or resources which need to be acquired outside the project organization. The fundamental challenge for managers of projects managed by the theory of constraints is to subordinate the work of their whole team to an on-time on-budget completion of tasks, i.e. transitions belonging to the critical chain. Allocating additional resources to transitions which belong to the critical chain and mobilizing the team to complete the transitions depends on the intensity of links between the individual goals pursued by the designers in a given project and the needs of the entire team, i.e. the completion of those software design tasks which constitute a bottleneck by increasing the throughput of the design process and restricting the "rigidity" of project schedules and budgets. One of the basic drawbacks of the knowledge codification strategy is its heavy reliance on CASE tools for improving the economic effectiveness of the design process. Such reliance loosens interpersonal relationships within teams and breaks them up. The resulting subgroups of project teams are more loyal towards themselves then towards the team as a whole. A fragmented project team may be more prone to internal competition. This may stimulate creative energies among its members but also weaken the team and disrupt its effort to pursue common goals.

# CONCLUSIONS

IT companies develop ever more complex software systems ever more rapidly thanks to continuing advances in software engineering. Software design relies on highly efficient methods and techniques to aid designers and manage IT projects. Design firms have long searched for a way to boost the productivity of individual designers and entire design teams. One of the options considered was to improve process efficiency – this approach entails the knowledge codification strategy. Another was to develop innovative design solutions, which can be achieved by pursuing the knowledge personalization strategy. The choice of the

326

codification strategy implies the use of CASE tools to build knowledge repositories which support software design. Regrettably, excessive use of such tools has caused a number of issues which have ultimately contributed to reducing software design innovation and designer efficiency. Each software development project is the result of teamwork. To achieve the objective of producing a good design, an effort is needed to subordinate the individual goals of each designer and designer sub-team to such an overarching objective. Once managers become aware of the common objective and realize the importance of individual goals, designers tend to be more efficient. By pursuing the knowledge personalization strategy, design team members gain unrestricted equal access to knowledge resources and a strong sense of contributing to the common objective and belonging to a community. It seems that a rational way to manage software development projects is to apply the theory of constraints which approves moderate reliance on CASE tools coupled with psychological and sociological considerations in personnel management. All in all, project management by the principles set out in the theory of constrains allows for the knowledge codification strategy to coexist with the knowledge personalization strategy.

# REFERENCES

Berenbach B., Paulish D., Katzmeir J., Rodorfer A.,(2009)Software and Systems Requirements Engineering : In Practice. New York: MCGraw-Hill Professional.

Blikle A.,(2010) Dwa paradygmaty w zarządzaniu projektami : teoria ograniczeń i teoria Petriego , WWW.firmyrodzinne.pl.

Borucki A., (2008)Use of Customization to Enounce the Ergonomic Qualities of Software, Monograph red.L.M.Pacholski, J.M.Marcinkowski, W.M.Horst, Employee Wellness Ergonomics and Occupation Safety, Poznań.

Cushman W.H.,Rosenberg D.J., (1991)Human Factors in Product Design, Elsevier, Amsterdam.

DeMarco T, Lister T.,(2002),Czynnik ludzki,skuteczne przedsięwzięcie i wydajne zespoły.WNT,Warszawa.

Goldratt Eliyahu M.,(2009) Critical Chain, MINT Books ,Warszawa.

Jashapara A., (2004), Knowledge Management :An Integrated Approach, Pearson Education Limited.

Kotonye G., Sommervile J.,(1998) Requirement Engineering:Process and Techinques, John Wiley and Sous.

Laplante P.,(2009),Requirement Engineering for Software and Systems(1st.ed) Redmond, WA:CRC Press.

Proctor T., (2002) Twórcze rozwiązywanie problemów, GWP ,Gdańsk.

Sokołowska J.,(2005) Psychologia decyzji ryzykownych, WSWPS Academica.

Stallman A., Greene J.,(2005) Applied Software Project Management. Cambridge, MA: O'Reilly Media.

# Section V

---

*Learning and Training*

# CHAPTER 35

# Optimization of Gifted and Talented Students' Activity: Cognitive and Organizational View

*Irina PLAKSENKOVA, Oleksandr BUROV,*
*Volodymyr KAMYSHYN, Mykhailo PERTSEV*
Institute of Gifted Child
Kyiv, Ukraine
ipdakini86@gmail.com

## ABSTRACT

It's a well known fact that the cognitive abilities of every person (especially a gifted teenager student) are not stable during the school year and even at shorter time intervals (months, weeks, days). Such oscillations aren't taken into account by educational programs, teachers and parents' requirements for the teenager. It can decrease the effectiveness of learning and cause the deterioration of health by accumulation of fatigue and neuro emotional stress.

Furthermore, statistics indicate that most children's health significantly deteriorated over the period of schooling, and children receive a certain number of chronic diseases at the age of 17 [1]. The Main Health Administration of Kyiv City State Administration informed that the level of school-age children chronic diseases increased nearly till 70% per 1,000 students (Kyiv, 2010).

At the same time it is known that the manifestation of the disease is preceded by the appearance of functional abnormalities resulting from nonoptimal work and rest schedule, non-optimality of physical and mental workload, inadequate training requirements and cognitive opportunities to relevant psycho-physiological resources of the child.

The theoretical and applied models, concepts, and structures developed to optimize gifted/talented learning/education activity system as sociotechnical one, including its organizational structure, policies, and processes are discussed.

Important parts of the system (sub-systems) include students' abilities assessment, selection of those who are gifted in science and research, health monitoring.

The aim of our research was to test the effectiveness of computerized methods developed for daily monitoring of the adequacy level of the teenager's functional and cognitive abilities to training load.

**Keywords**: students' abilities assessment, daily monitoring, learning activity optimization.

# 1    RESEARCH METHOD

Organization of the research involves gifted teenager students in a daily (except weekends) task performance after the 6th lesson. Duration: 2 months. Results presented below were obtained in a pilot research for practicing the technique and logistics of research in Ukrainian gymnasium 'Obolon' (December, 2011). One of the obligatory conditions to make such kind of research is an agreement that School signs with all stakeholders (parents, students) in accordance with the requirements of bioethics, acting in Ukraine. Another peculiarity of this type of research is automated test performance and researcher participation only in the preparation stage of the research (system setup and preparation of the subject).

The structure of each daily experimental testing activity consists of several phases (Figure 1).

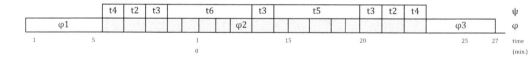

Fig. 1. Time chart of daily experimental research in schools.

Notations: ti - tests of SPFR (ψ);

φj - physiological measurements: φ1 (background) - heart rate, blood pressure, EPD; φ2 - activity modeling - ECG; φ3 (renewal) - heart rate, blood pressure, EPD.

Daily experimental research is based on the use of a computer system SPFR (system of psycho-physiological research) to monitor the cognitive activity of students [2]. The survey includes test performance with the parallel registration of the ECG RR-intervals duration (without interruption through the use of "Solveig" equipment), blood pressure and heart rate before and after the test performance. Electropuncture diagnostics (EPD) [3] was conducted twice a week (on Monday and Friday) after a test session for each subject.

The research includes:

- primary examination which meets the requirements for career guidance procedure [4];
- daily research (monitoring of students' cognitive activity) includes the implementation of logical-combinatorial tasks (5 test tasks), the cardiovascular system control (ECG registration) and blood pressure (before and after test performance), EPD. Age group of subjects: boys 15 - 17 years. Duration of daily research 1 month.

The test block includes:

- Short-term memory test T2. There is presented a table of 12 random numbers from 11 to 99. The number of correctly reproduced numbers is considered as a result.
- Time perception test T3. Where the subject is proposed to press any key on he keyboard after the sound signal through 60 seconds (calculation of time is carried out without the use of wristwatch, etc).
- Activity and mood self-assessment test T4. It's an abridged version of Health-Activity-Mood test. The subject is proposed to give a subjective assessment of his/her state in 7-point scale, giving answers on five pairs of questions-characteristics.
- Numbers permutation test (combinatorial) in ascending order T5. It consists of a sequence of numbers (from 0 to 9) which are not repeated and placed in a random order. Test performance time is fixed and calculated individually for each subject according to the results of the training test.
- Numbers permutation test (combinatorial) in ascending order T6. The tasks are of the same type as in T5 but the test performance time is free.

## 2    RESULTS

Methodical support of the research includes the system of legal, organizational, informational and software tools.

Previous studies using this system demonstrated its effectiveness as well as ease of use. At the same time, it should be noted that the process of 'adjusting' of a subject is sufficient even after the training session, which, in our opinion, is related to the automation of test performance and the formation of specific psychomotor response. But after a few days the improvement of the individual average time spent on the test performance begins to show up, it also can vary during the research period (Figure 3). Meanwhile, the dynamics of heart rate has a different character (Figure 4).

The average values of subjects' physiological parameters during the testing period indicate the individual character of their dynamics in the initial state (after classes) and after the test execution (Table 1). Comparison of the physiological parameters variation as a reaction to stress (cognitive tests are simple and meet the requirements of the logical skills of first grade pupils in general secondary school) indicates that even these activities can serve as a functional test of the fatigue occurrence, which was established in previous studies [5].

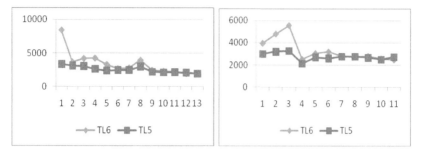

Fig. 3. Daily test performance dynamics of two subjects (10th grade students). On the axis of abscissas - days of testing, ordinates - the average time of test performance in tests T6 and T5.

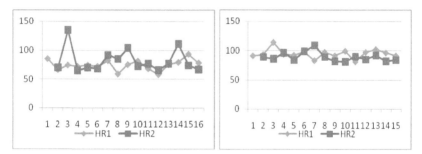

Fig. 4. Daily heart rate dynamics of the same subjects. On the axis of abscissas - days of testing, ordinates - heart rate before and after test performance.

We should notice the heart rate increase of the first subject, while another has a stable average myocardial activity. Herewith the subjective assessment of mood under the influence of the testing activities after its completion is more stable in comparison with the original state, just after the last lesson (Figure 5).

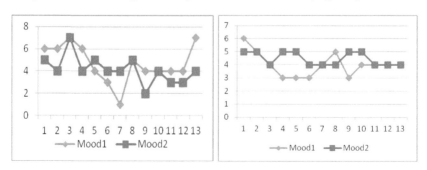

Fig. 5.

Table 1 Average values of subjects' physiological parameters during testing days

| Subject | ATS$_1$ | ATS$_2$ | ATD$_1$ | ATD$_2$ | HR$_1$ | HR$_2$ |
|---------|---------|---------|---------|---------|--------|--------|
| SVV | 123,83 | 125,18 | 71,67 | 73,09 | 75,5 | 81,27 |
| SMR | 131,80 | 130,60 | 81,47 | 77,87 | 90,73 | 92,53 |
| SMM | 134,27 | 131,36 | 81,93 | 73,64 | 94,6 | 89,14 |
| SBK | 116,93 | 110,54 | 71,57 | 71,08 | 66 | 67,38 |
| MOR | 128,36 | 123,1 | 72,73 | 76,40 | 73,45 | 83,90 |
| RYI | 114,17 | 109,33 | 84,50 | 119,80 | 72,00 | 69,80 |

Note: **ATS1** and **ATS2** – blood pressure, systole, before and after the test, mm Hg., **ATD1** and **ATD2** – blood pressure, diastole, mm Hg., **HR1, HR2** – heart rate, before and after testing activities.

## 3    DISCUSSION

The main problem to provide the student with adequate information is what should be measured, which indices of a human state (psychological, physiological) can be valid for this and can secure a high level of prediction accuracy.

The peculiarity of learning activity is that information processes are more involved in regulation of cognitive activity than energetic ones, and conceptual model of learning activity is a result of psycho-physiological adaptation to this activity. It is relatively close to an operator activity and, from substantial viewpoint, consists in discreet comparison of information received from the education system with the conceptual learning model used, in carrying out a particular task. In other words, in forming and using an „information contour" that exists in active state in the process of purposeful activity [6]. This contour includes afferent inputs, decision making block, action acceptor and the program of action, as well as the object of action (conceptual model of the knowledge area).

To avoid a negative impact of learning in the e-environment on health of the user and to increase a learning efficiency, it can be useful to monitor a child mental status, his/her ability to act in education process. As a result of such an investigation just before to start current day's education exercises, it can be useful for a student to get information about what type of learning is more useful at the current moment. Special recommendations can be provided to this student immediately after a short procedure of investigation. These recommendations deal with individualization of current type of learning to increase its efficiency. The appropriate research has demonstrated high accuracy of prediction of human cognitive work productivity.

We should not forget to mention that student-computer interaction is not always effective: one day a student can work a lot, another day he/she is not so effective. It is a well known fact that operator's functional state fluctuations can result the fitness-for-work deviations from an individual "norm" and he/she cannot meet requirements of the occupation. A student is an operator as well because of the mental nature of his/her work. It is important to give students and teachers the knowledge of the way they can adapt the education process to an individual and vice versa.

## 4     CONCLUSIONS

The system to assess a student's preferable domain (the highest achieved area of cognitive abilities) and to optimize their work (learning activity) depending on their current functional state was developed.

Schoolchildren's individual psycho-physiological peculiarities of the cognitive activity dynamics were studied.

An identification accuracy of individual and common influence occurrence of the learning process on the health and effectiveness of student learning throughout the sub-semester intervals was demonstrated.

## REFERENCES

1.   Shudro S.A. Workload in high school as a risk factor for students' health. - *Medical Perspectives.* ¬- *2007.* - t.XII.-№4.-Pp. 108-114. (in Ukrainian)
2.   Burov O.Yu., Plaksenkova I.O., Pertsev M.A. Technique for psycho-physiological study of cognitive   activity dynamic. *Seminar Proceedings Psycho-physiological aspects of giftedness: theory and practice. 3 February 2012, Kyiv, Ukraine.* Pp. 11-17. (in Ukrainian)
3.   Rudenko S.A. Study of school children chronic fatigue with method of acupuncture diagnostic by Nakatani. *Seminar Proceedings Psycho-physiological aspects of giftedness: theory and practice. 3 February 2012, Kyiv, Ukraine.* Pp.92-97. (in Ukrainian)
4.   Automation of the experiment during psychophysiological research in the physiology of labor (guidelines) // Reshetyuk A.L., Burov O.Yu., Polyakov O.A, Chetvernya U.V., Korobeinykov V.G., Burova O.O., Zorina T.V. (compiler – Burov O.Yu.) - *MOZ Ukrainy, Resp. Tsentr naukovoi medychnoi informatsii. - Kyiv: 1993.-* Pp. 12. (in Ukrainian)

5.  Burov O., Filatova I., Burova K. Cognitive work and individual response to external factors. *2008 AHFE International Conference, 14-17 July 2008, Ceasar APlace, Las Vegas, Nevada USA. Conference Proceedings. Ed. By Waldemar Karwowski and Gavriel Salvendy.*

6.  Burov O.Yu., Kamyshin V.V., Burova O.O. Student ability status' identification when working in learning environment. *Proceedings of the ECHA 2010 Conference, July 7-9, in Paris.*

# CHAPTER 36

# Hybrid Design Model of E-learning Course at Education Institution Based on SECI Model

*Satria Hutomo Jihan, Jussi Kantola*

Department of Knowledge Service Engineering, KAIST (Korea Advanced Institute of Science and Technology), Daejeon, Republic of Korea
Correspondence to: satriahj@kaist.ac.kr, jussi@kaist.edu

## ABSTRACT

Nowadays, the Internet has changed the way people learn in their daily life. Electronic learning (E-learning) has gradually been taking promising place in education institutions due to its technological advantages. Utilizing Nonaka's four knowledge creation basic processes (Socialization, Externalization, Combination and Internalization) and the combination of asynchronous & synchronous methods in the form of Absorb-Do-Connect-Create activities, this study proposes a hybrid design model of E-learning. The learning activities will follow either online method or with face-to-face meeting. The new hybrid E-learning design is expected to enhance students' practical knowledge and competencies related to domain topics.

**Keywords:** SECI model, hybrid E-learning, competency, knowledge creation

## 1. INTRODUCTION

With rapid technology development nowadays, the Internet has become almost inseparable element enhancing every aspects of human life, including the way how people learn in their life. Using technological tools including the Internet, electronic learning (E-learning) now is actively used in education institutions. There are many benefits of using E-learning, such as time and location flexibility, cost and time saving, self-paced and just-for me learning and unlimited access and use of learning material (Hiltz and Wellman, 1997).

However, E-learning participants may face time consuming demand from their environments (family, work, etc.) that may make it difficult to focus on the undergoing learning process (Fung, 2004). They also may not feel the teaching presence - the sense of belonging to the classroom. Moreover, the online communication tools can't accommodate tacit knowledge transfer which usually can be transferred during face-to-face meeting.

To address these challenges, some attempts have been done to combine the face-to-face meeting and E-learning methods into blended E-learning (Garrison and Kanuka, 2004; Rossett, 2003; Oh and Park, 2009). The hybrid E-learning combines the benefits of traditional class room and E-learning. It doesn't just add online contents to the traditional class room, but blended E-learning restructures and rethinks about the teaching and the learning relationship (Garrison and Kanuka, 2004).

This study proposes a new hybrid design model of E-learning combining asynchronous and synchronous method which can supervise students' learning and interaction using CMS (Course Management System) platform. The proposed hybrid model is an open E-learning model, which encompasses knowledge transfer within or outside the formal system. The goal of the proposed model is to improve the practical knowledge and competencies gained by the learner and also to support new knowledge creation.

## 2. RELATED WORKS

### 2.1 E-learning

There are many definitions and similar concepts as E-learning that vary according to the researchers and study objectives. Gunasekaran, McNeil, and Shaul (2002) define E-learning as Internet-enabled learning of which components include content delivery, the management of the learning experience and the networked community of learners, content developers, and experts. Nichols (2003) defines online learning as an education method that occurs only in the web, the sole medium for student learning and contact. Qwaider (2011) extends the E-learning definition to the purpose of E-learning itself as learning using electronic means & technology to not only learn facts (know-what) for a specific subject but also having practical skills (know-how) and developing competencies in the given domain.

### 2.2 SECI Model

Knowledge creation is the result of continuous interaction between tacit knowledge and explicit knowledge within or beyond organizational boundaries (Nonaka, Toyama, and Hirata, 2008). Tacit knowledge is intellectual property in peoples' minds that is based on their experiences, wisdom, insight and their internalized information about a specific context which largely acquired through association with other people while explicit knowledge refers to knowledge that is transmittable in formal, systematic language. In Nonaka's SECI model (figure 1), there are 4 knowledge conversion processes, which lead to new knowledge creation: socialization, externalization, combination, and internalization.

- Socialization: In this stage, tacit knowledge is transferred through shared experience in daily social interaction, specifically emphasizing to live/experience in shared time and space.
- Externalization: In this stage, the tacit knowledge is made explicit through languages, images, models, or others modes of expression, and then shared with in the group.
- Combination: In this stage, the explicit knowledge is combined, edited, and processed to more complex and systematic sets of explicit knowledge. Combination process may include "breakdown" concept, which mean that the existing concept is broken into more specific concepts and further systematic explicit knowledge.
- Internalization: During this stage, explicit knowledge is converted to personalized intrinsic knowledge through practicing with a conscious mind, such as reflection, action, experiment, simulation, etc.

Figure 1 SECI model (adapted from Nonaka, 1994)

The knowledge transfer processes can be done inside the formal system or outside the formal system. The knowledge creation environment includes both of these systems. Similarly, E-learning 2.0 approach suggests the openness between learners and knowledge providers in the frame of formal and informal learning opportunities (Ivanova, 2008). Social interaction in informal learning is an important issue to make learning more useful and to give meaningful context for learners (Lave and Wenger, 1990).

## 2.3 Knowledge Management System

In higher education courses where the course members change every period, the growing community can be facilitated through KMS (Knowledge Management System).

KMS is defined as a distributed system for managing knowledge in an organization, supporting creation, capture, storage and dissemination of expertise and knowledge with goal to get the right information from the right people to the right people at the right time (Tiwana, 2000). Integrating KMS with E-learning will bring some benefits (Qwaider, 2011) as following:

- Improving learning efficiency through knowledge management and learning feedback.
- Using knowledge management activity can cost less time in making teaching materials for teachers.

Lee (2011) proposes the E-learning model, which is supported by KMS to support knowledge creation and innovation. In his model, KMS supports the SECI processes in learning interaction by using various KM functions such as knowledge capture, knowledge validation, knowledge storage, knowledge sharing and transferring and knowledge utilizing.

## 3. PROPOSED HYBRID E-LEARNING MODEL

Hybrid learning is a learning approach which integrates the benefits of multiple delivery methods, commonly a mixture of online and face-to-face in classroom methods in order to optimize learning outcome (Delialioglu and Yildirim, 2007). Because hybrid learning depends on various specific contextual needs and contingencies (e.g. discipline, developmental level, and resource) the design of hybrid learning can be very complex and unique (Garrison and Kanuka, 2004; Rossett et al., 2003).

### 3.1 Hybrid E-learning Framework

The guideline/framework of hybrid E-learning is proposed in this paper, Figure 2. The first stage of hybrid E-learning activities is an assessment test. The purpose of this stage is to understand students' current related competencies and skills and to have a base to understand the improvement of the competencies in the end of the E-learning process. The assessment result can also determine the students' class/batch.

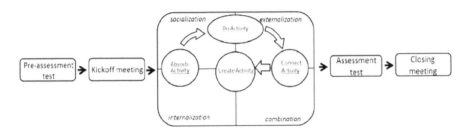

Figure 2 Learning stages in hybrid e-learning

Each class/batch can have different composition of learning activities. After class enrollment has been established, face-to-face meeting should be done when the class starts at the first time. The kickoff meeting is important for students to know each other and to deepen the learners' commitment to the course. Recommended activities in this stage are ice breaker, introduction to class syllabus and class subject, and the establishment of class agreement.

The latter stages in learning in hybrid E-learning encompass 4 kinds of main activity: Absorb activity, Do activity, Connect activity and Create Activity. These activities are extensions from learning activities, which are suggested by Horton (2006). The instances of these 4 kinds of activities are explained in table 1.

**Table 1 Hybrid E-learning activities**

| Learning Activities | Purpose | Knowledge Transfer process | Activity Examples |
|---|---|---|---|
| Absorb | Learners read, listen, watch learning materials or observe the instructor's presentation. | Internalization, Socialization | Lecture video, online paper / article, lecture demonstration |
| Do | Learners do something with what they are learning, to deepen the understanding of the concept and procedures. | Socialization, Externalization | teamwork activity, case study / role-playing scenarios |
| Connect | Leads learners to connect current learning to their work, their lives or their prior learning | Externalization, Combination | Citing exploration (to search example of the theory in website and summarize it) and group critique. |
| Create | Create the new knowledge from the existing one to solve the existing problem or improve knowledge state of art. | Internalization, Socialization, Externalization, Combination | Research paper assignment, term project. |

These learning activities also follow cognitive process of Bloom's Revised Taxonomy (Krathwohl, 2002). In Absorb activities, students internalize factual and conceptual knowledge (know-about) by remembering and understanding the learning documents and learning demonstration. In Do activities, students practice and apply what they have learned so that they may gain procedural knowledge (know-how) and deepened conceptual knowledge (know-why). In Connect activities, students analyze and evaluate their learning result in order to have conditional knowledge (know-when) and relation knowledge (know-with). In Create activities, students are expected to be able to generate/produce new knowledge produce by putting elements together or integrate learning materials from outside course.

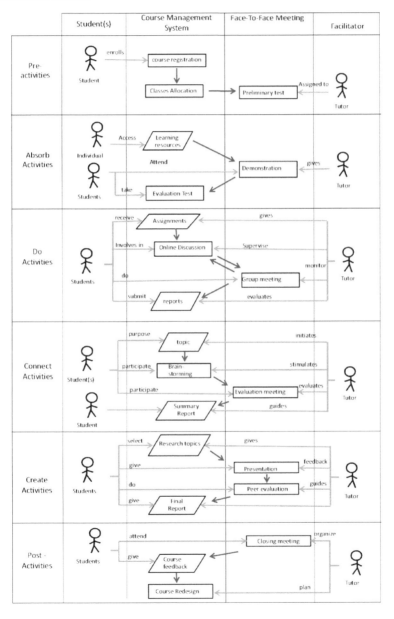

Figure 3 Details of hybrid E-learning activities

Figure 3 exhibits the details of hybrid E-learning activities. There are two kinds of activities based on the place where it is conducted: Online learning activities within CMS and offline learning activities via face-to face meeting. Offline learning activities can be categorized as synchronous learning whereas online learning activities can be categorized as either synchronous or asynchronous learning.

## 3.2. User roles analysis

### 3.2.1 Students

As an individual, a student has specific roles in hybrid E-learning activities. First, student will take pre-assessment test, where the students who have similar test result will be grouped in same class. This is to ease facilitator/tutor defining the suitable activities/delivery approach for that class. Second, a student may adjust his/her learning pace in Absorb activity because learning materials are delivered using asynchronous method. Third, a student may use a topic/case that is interesting to him/her or to his/her experience for further exploration in Connect activity or Create activity, because one of the purposes of hybrid E-learning is to develop self-awareness (meta-knowledge) for applying related knowledge in familiar situation first.

As a group, students' role in hybrid E-learning is to undergo constructive learning where social interaction plays important roles (Oliver, 2001). Since the hybrid course design is student centered, students are expected to actively work with others, to share their experiences, to give review, and to reflect on the result.

Student's personal experience / knowledge is important to contribute in creating new knowledge. Evaluation in every learning stage, either by written test or report, may help to monitor the students' learning progress as well as to capture their personal knowledge in explicit form. Combining students' result will provide useful references for the facilitator to redesign the course in a more effective way and to develop richer teaching material.

### 3.2.2 Facilitator

Facilitators' involvement in hybrid E-learning activity is not the same as their involvement in traditional class meetings, because hybrid E-learning involves learner-centered activities and some activities are based on online platforms so that the facilitator can have more free time to do other activities.

Although the instructor/facilitator role doesn't become the center in hybrid E-learning, their presence is important to fulfill interaction needs, to increase learning satisfaction and learning experience (Garrison and Cleveland-Innes, 2005). The facilitator's presence in face-to-face meeting is also important in the socialization process where students can observe the facilitator directly and gain insight/tacit knowledge.

Facilitators have responsibility in designing the course contents and the learning activities with the consideration of student's interaction, collaboration, and learning pace while utilizing available online tools and face-to-face meetings.

Facilitators should do evaluation process not only based on the score obtained by students in the test, but also based on the students' interaction in online discussion or in class participation. Giving feedback and advices to students may help to catalyze learning process and to sharpen the student's meta-knowledge.

## 4. FINDINGS

A questionnaire survey was conducted during Oct 2011-Nov 2011 to explore students' experiences and expectations for E-learning. The participants were 22 KAIST students who had taken or were taking the Ethics and Safety course, a mandatory E-learning course for students. The course itself was conventional E-learning course without any face-to-face meeting. The questionnaire had 2 main sections: evaluation of the current system and expectation of future system.

The result exhibits that more than half of the respondents (68%) agreed that face-to-face meeting is needed to complement E-learning and most of the respondents (91%) supported that interaction with other students would help them to understand more the course material. Moreover, 78% of the respondents agreed that explicit documents from students' personal experience and knowledge would help them to understand better the subject domain and 68% of the respondents agreed that those kinds of contributions from each student would help to nurture development of the overall course material. These results indicate that the utilization of the SECI process will help students to have better learning process. While the respondents agreed that the conventional E-learning has the biggest impact on the improvement of their personal knowledge, it neglects the improvement of students' competencies and practical skills, such as communication skill, to achieve better result in their study field.

## 5. DISCUSSION

## 5.1 Limitations of Hybrid E-learning

Hybrid E-learning has a limitation in student's geographical coverage, because it requires students to be in nearby places so that they can have a face to face meeting. Learners who live in very far areas or even in different countries may not attend due to transportation/ cost issue. The alternative solution for this issue is utilizing the synchronous online communication tools, such as VoIP, webinar, Video conferencing, etc. However these tools are lack of sensory senses, so that learners might not perceive learning process effectively and learners might not be able to grasp tacit knowledge from facilitator.

## 5.2 The successfulness of Hybrid E-learning

The activity details and the composition between online learning and classroom meeting may depend on the nature of learning domain itself, for example social domain may need more direct meeting than science domain. However, we can measure whether the hybrid E-learning is successful or not from students' perspective because hybrid E-learning is students centric.

The students are expected to have practical skills to meet society and industry demand (Yordanova, 2007). This is in line-with survey results which exhibit that the students have expectation to have real-world examples which they could practice specific skills (Siritongthaworn and Krairit, 2006).

We argue that hybrid E-learning approach can improve students' competencies in terms of collaborative works with others, cognitive-awareness to apply the related knowledge, and in other self-improvements. Collaborative works with others, such as brain storming, giving feedback to others, allows students to increase their interpersonal skills and to have better learning results. In collaborative works, every student shares their personal experience and knowledge related to learning domain, so other students can have wider view about the application of knowledge they learn.

The successful hybrid learning course will bring benefits to students, to facilitators and also to educational institutions. It will increase students' satisfaction, course completion rate, students' self-achievement, and students' expense saving due to reduced commuting time (Heterick and Twigg, 2003 cited in Garrison and Kanuka, 2004). For instructors, the blended learning course provides them with more time to spend with the students individually and improve quality of interaction with the students. With hybrid learning approaches, institutions can increase flexibility in scheduling courses and improve the usage of their limited resources such as class rooms and parking lot (Oh and Park, 2009).

## 6. CLOSING & FUTURE WORKS

Hybrid E-learning is not merely adding the technology aspect to teaching activities, but it encompasses transformation of overall course design itself by considering students' interaction, teaching presence, students' learning pace, and collaborative learning. The SECI model in hybrid E-learning is utilized through 4 kinds of learning activities: Absorb-Do-Connect-Create activities. All of these activities can help students to have higher cognitive processes, from remembering descriptive knowledge to creating a novel knowledge product.

This study can be extended to how to redesign hybrid E-learning course using the knowledge products from students so that the course has continuous improvement over the time. The proposed approach can also be extended to the implementation of hybrid E-learning in companies, where the hybrid E-learning itself may support organizational learning in the companies to reach their common goals based on a human-centric approach. Therefore, the approach proposed in this paper may suit well for vocational on-the-job training in different work roles.

# 7. REFERENCE

Delialioglu, O. and Yildirim, Z. 2007. Students' perceptions on effective dimensions of interactive learning in a blended learning environment. *Educational Technology & Society* 10(2): 133-146.

Fung, Y.Y.H. 2004. Collaborative online learning: interaction patterns and limiting factors. *Open Learning* 19(2): 135-149.

Garrison, D. and Kanuka, H. 2004. Blended learning: Uncovering its transformative potential in higher education. *The Internet and Higher Education* 7(2): 95-105.

Garrison, D.R. and Cleveland-Innes, M. 2005. Online Learning : Interaction Is Not Enough. *American Journal of Distance Education* 19(3): 133-148.

Gunasekaran, A., McNeil, R.D, and Shaul, D. 2002. E-learning: research and applications. *Industrial and Commercial Training* 34(2): 44-53.

Heterick, B. and Twigg, C. 2003. The Learning MarketSpace. Online, retrieved on December 5, 2003 from http://www.center.rpi.edu/LForum/LM/Feb03.html

Hiltz, S. R. and Wellman, B. 1997. Asynchronous learning networks as a virtual classroom. *Communications of the ACM* 40(9): 44-49.

Horton, W. K. 2006. *E-learning by Design.* Pfeiffer, San Francisco: John Wiley & Sons.

Ivanova, M. 2008. Knowledge Building and Competence Development in eLearning 2.0 Systems. In: *Proceedings of International Conferences on Knowledge Management and New Media Technology.* Graz, Austria, 3-5 September 2008, pp. 84–91.

Krathwohl, D. 2002. A revision of Bloom's taxonomy: An overview. *Theory into Practice* 41(4): 212–218.

Lave, J. and Wenger, E. 1990. *Situated learning: Legitimate Peripheral Participation.* Cambridge, UK: Cambridge University Press.

Lee, D. Y. 2011. *Design of an E-learning system based on SECI model.* Ms. KAIST

Nichols, M. 2003. A theory for eLearning. *Journal of Educational Technology and Society* 6(2):1-10.

Nonaka, I. Toyama, R., & Hirata, T. 2008. *Managing Flow : A process theory of the knowledge-based firm.* Basingstroke: Palgrave Macmillan.

Nonaka, I. 1994. A Dynamic Theory of Organizational Knowledge Creation. *Organization Science* 5(1):14-37.

Oh, E. and Park, S. 2009. How are universities involved in blended instruction. *Journal of Educational Technology & Society* 12(3): 327-342.

Oliver, R. 2001. Developing E-learning environments that support knowledge construction in higher education. In: *Working for excellence in the e-economy,* S. Stoney, J. Burn, Ed. Churchlands, Australia, 2001, pp. 407-413.

Qwaider, W. Q. 2011. Integrated of Knowledge Management and E- Learning System. *International Journal of Hybrid Information Technology* 4(4)

Rossett, B.A. et al. 2003. Strategies for Building Blended Learning. *Learning Circuits* 4(7): 4-9.

Siritongthaworn, S. and Krairit, D. 2006. Satisfaction in E-learning: The context of supplementary instruction. *Campus-Wide Information Systems* 23(2): 76–91.

Tiwana, A. 2000. *The Knowledge Management Toolkit.* New Jersey: Prentice Hall.

Yordanova, K. 2007. Integration of Knowledge management and E-learning – common features. *CompSysTech 07 Proceedings of the 2007 international conference on Computer systems and technologies* 1: 1-6.

# CHAPTER 37

# Careers and Further Education of Engineers against the Background of Globalization

*Martin Kröll*
Ruhr-University Bochum
Bochum, Germany
martin.kroell@rub.de

## ABSTRACT

Increasing globalization, demographical change and competitive rivalry lead to more complex and intransparent factors, which influence the occupational vita of employees, in particular engineers. The traditional career understanding, which is based on climbing the career ladder through working on different hierarchical positions in one company (i.e. management career), breaks up and is complemented by new career paths. In the occupational area of engineers, an equivalent implementation of more expert and project careers besides the traditional management career is discussed The present study analyzes career plans and further education activities of 544 engineers and discusses to which extent employment biographies of engineers can still be planned in advance. Furthermore, it is examined whether engineers orientate their career planning and their further education activities towards specific career paths and ideas. The central research question can be summarized as follows:

What are the factors used by engineers to plan their employment biography and which implications does this have for their further education behavior?

**Keywords:** further education, management, expert or project career

## 1    THEORETICAL BACKGROUND, SUBSTANTIATION OF HYPOTHESES

For a long time a management career was regarded as the ideal career model for engineers which should be used as an orientation to plan career paths and competence development activities. This term denotes a "traditional career in the form of a continuous, hierarchical advancement accompanied by respective qualifications"(Thom, 2008, p. 11). The realization that not every engineer is capable of filling a management position is not surprising. In addition more and more companies prefer the concept of flat hierarchies and therefore less management positions are available in firms. This fact leads to a situation, in which the demand of management positions cannot be satisfied by the companies. While mainly dual career systems (= management and expert careers; Kunz, 2005) were discussed for a long time in career research, Majer and Mayrhofer (2007) pointed to triadic career systems recently. These consist of the three career areas of management career, expert career and project career. According to Thom (2008, p. 252), expert careers are defined as career paths that comprise an advancement across different positions which require specialist knowledge. In contrast to management and project careers, specific knowledge has priority and personnel responsibility is not adopted or limited to a low degree. Furthermore, supervision tasks play a minor role in this career. In contrast, the project career includes project-related leadership and management tasks and can therefore be defined as a mixture of classic management and specific expert career. The positions in the context of a project career are usually temporally determined, which implies an increased flexibility of career development. Over the last years, this path grew in popularity in career research and was even described by some researchers as a separate profession. However, recent research indicates a lack of comparability of management and project careers. Corria-Simpson, Langer and Handor (2008) mention in particular the lack of a defined career path and of formal mentoring programs as disadvantages of a project career. Also, status and influence possibilities do not seem to be available in the project career to a degree desired by engineers (Hodgson, Paton & Cicmil, 2011). Similar disadvantages can be observed regarding a comparison of expert and management career. This imbalance may lead to a situation where employees decide to pursue a management career even if this does not match their competencies and interests. It can be assumed that lacking support through companies and the disorientation with regard to career perspectives lead to dissatisfaction of those engineers who choose an expert or project career. Based on this assumption the following hypothesis is developed:

**H 1:** Engineers who chose a management career are (1) more satisfied with their occupational situation than engineers who chose an expert or project career and (2) rate their future occupational perspectives most positively.

In times of increasing skilled worker shortages in technical occupations, it is of great interest for companies to retain engineers in the long term. It is therefore of interest to analyze for what reasons engineers change their career path or leave the company. On the basis of equity theory (Adams, 1965; Skiba & Rosenberg, 2011), which assumes that individuals expect fair compensation in exchange for their labor input, it can be assumed that under-challenged engineers do not feel promoted according to their potential and thus a sense of injustice is created. It is conceivable that they feel overqualified and underpaid and strive for a position with more challenges (and therefore also a higher salary).

In addition, the argumentation can be supported by the stimulus-contribution theory (March & Simon, 1958). Accordingly, employees receive incentives from the organization and provide contributions in return. Dissatisfaction and ultimately termination of the employment relationship can occur when the incentives granted fall below the contributions rendered. In the case of highly qualified, underchallenged employees, this seems likely: They can easily have the perception that their contributions (in the form of competences and qualifications) exceed the incentives received (in the form of an interesting, challenging task or appropriate payment). In line with these considerations, Gerstenfeld and Rosica (1970) could show that the main reasons for an engineer's job change included interesting tasks and the opportunity for personal development.

H 2: Underchallenged engineers are more likely to change their job than overchallenged engineers.

Especially for engineers, the importance of lifelong learning is stressed repeatedly since they contribute a lot to the technical progress and are especially concerned with the threat of outmoded knowledge). However, this poses the question of the way learning should happen and the kind of knowledge that should be acquired in the process.

Scientists distinguish basically between explicit and tacit knowledge (Nonaka & Takeuchi, 1997; Lam, 2000). Explicit knowledge is formally verbalizable knowledge, which is based on theories and can be described as "intellectual knowledge" (Nonaka & Takeuchi 1997). In contrast, tacit knowledge is not easily verbalizable, context-specific and often the owner does not consciously know about it. For instance it can consistof "experience-knowledge" or learned routines and behavior. Tacit knowledge, which is influenced by skills, techniques, know-how and routines, occupies a central role regarding organizations' competitiveness and maintenance of product and service quality. According to Lam (2000), innovation capacity and learning in an organization depend largely on the generation and sharing of tacit knowledge. However, the acquisition of tacit knowledge can only occur through practical experience in the respective relevant work context („Learning by Doing", c.f. Nonaka & Takeuchi, p. 82). The transfer of this kind of knowledge is considered as experience-based, action-oriented and dependent on social interaction. It may be assumed that engineers are aware of the importance of tacit knowledge for the generation of competitive advantages for the company and hence for their own careers. Thus, they should prefer forms of further education that enable the acquisition of new tacit knowledge.

H 3: Regardless of their chosen career path, engineers attribute greater importance to workplace-based forms of further education than to external further education activities.

Against the background of theories of nonlinear dynamic systems, which assume that in the long term the acting of systems is influenced by non-predictable interdependencies, Strunk (2009) states that careers are becoming more complex. This development affects mainly younger engineers who have not decided about their career planning yet and are particularly challenged to choose alternative career paths. In this context, the project career can be interpreted as an alternative Career Pathway in terms of Strunk (2009) as it exhibits greater flexibility and offers the opportunity to opt for a management or expert career later on, or to continue following the project career. Against the background of the increasing complexity of career development, it appears useful for younger engineers to increasingly enter into the project career, as this career type offers significant advantages - more diverse development options and a possible change into an expert or management career - in the case of undetermined career planning.

**H 4:** Regarding their future career planning, particular younger engineers are oriented towards alternative career designs (here operationalized through the project career).

Women are significantly underrepresented in the engineering profession and often have to contend with prejudice. According to Herrmann (2004, p. 262), female engineers are more frequently assigned to project activities in companies. In an inquiry of 151 men and 130 women from middle management and human resource management, Funken (quoted after Nitsche, 2011) could additionally show that female managers engage less frequently in management careers, but more often in project careers. For this reason it is stated:

**H 5:** Female engineers choose project careers more often than male engineers and also aim for this type of career more frequently as a future career path.

# 2    METHOD

The present study is based on an online survey (42 closed questions) of engineers (88% males), which has been carried out in 2008 through the Association of German Engineers (Lienert, 2010). The data are obtained from 544 completed questionnaires (response rate 17%). Six-point Likert-scales were used for measurement of variables, ranging from "I totally agree" to "Does not apply at all". The answers were coded with numbers from 1 to 6, with low values representing a high agreement. Data analysis was conducted using the statistic software SPSS. (general linear model [t-tests, ANOVA], $\chi^2$-tests, factor analysis (principal component analysis, Varimax rotation), cross-classified tablets, correlation coefficients according to Pearson). Table 1 provides an overview of demographic characteristics of respondents and is complemented by Table 2, which refers to the career paths of engineers surveyed.

**Table 1:**    Demographic characteristics of sample participants

| Variable | | N | Proportion |
|---|---|---|---|
| Age | 21-30 | 145 | 27% |
| | 31-40 | 186 | 35% |
| | 41-50 | 130 | 24% |
| | 51-60 | 55 | 10% |
| | über 60 | 18 | 3% |
| Highest educational attainment | Diploma, university of applied sciences | 239 | 45 % |
| | Diploma, university | 190 | 36% |
| | Doctor's degree | 61 | 11% |
| | Miscellaneous (e. g. Bachelor, Master) | 54 | 8% |
| Scope of work | Development | 126 | 24% |
| | Sales and marketing | 61 | 12% |
| | Manufacturing and production | 53 | 10% |
| | Construction | 45 | 9% |
| | Research | 43 | 8% |
| | Miscellaneous (e. g. administration, consulting, logistics, management) | 216 | 37% |

**Table 2:** Career path chosen / desired for the future / desired by the organization (N = 534)

| Career path | Chosen | Desired for the future | Desired by the organization |
|---|---|---|---|
| Expert career | 36,7% | 21,8% | 38,4% |
| Management career | 33,0% | 56,6% | 23,3% |
| Project career | 27,5% | 19,5% | 36,3% |

# 3 RESULTS

Though there are criteria to categorize occupational biographies to one of the three career paths (management, expert or project career), the present study asked the engineers to categorize themselves into one of the three categories (1) at the present time and (2) as planned for the future. This procedural method is advantageous because the self-assessment and the self-concept of the respondents, which influence the engineers' behavior in a crucial way, are taken as a foundation for subsequent analysis. Additionally it was found that the career types differ with regards to career indicators as well as preferred further education topics. For instance engineers with a management career attach more importance to personnel responsibility as an indicator of a successful career compared to the two other groups ($F(2, 507) = 13,58$, $p<.001$, Post hoc Bonferroni tests: $p<.01$ for both groups). Furthermore engineers with a management career attach more value to achieving high positions in the company than engineers with an expert career ($(F(2, 505) = 4,67$, $p<.01$, Post hoc Bonferroni test $p<.01$ for both groups). In contrast for expert engineers it is significantly more important to be an accepted expert in their field ($(F(2, 509) = 6,93$, $p<.001$). Management career engineers think that communication behavior is especially relevant for their career ($M=1,43$, $SD=0,58$), followed by management competences ($M=1,44$, $SD=0,64$) and team development ($M=1,47$, $SD=0,65$). For those who chose an expert career, technical expert knowledge is most important ($M=1,41$; $SD=0,66$). Engineers with a project career attach high importance to the topic project management ($M=1,41$, $SD=0,66$). Post-hoc Scheffé tests show that the three groups differ significantly with regards to the topics expert and economic knowledge, team development as well as management knowledge ($p<.05$). In reference to communication competencies and self-management, project and management career engineers rate these topics significantly higher than expert engineers.

The previous results support the assumption, that a differentiation between the three career paths is reasonable and well-founded, so that this classification can be used for further analysis.

With respect to H1, it appears that engineers who have chosen a management career ($N=172$, $M=2,76$; $SD=1,00$) are significantly more satisfied with their occupational situation than those engineers who chose an expert career ($N=186$, $M=3,15$, $SD=1,01$) or a project career ($N=137$, $M=3,13$, $SD=1,00$). A univariant comparison of means confirmed significant group differences ($F [3, 505] = 5,31$; $p < 0,01$). Bonferroni Post hoc tests showed that management career engineers are significantly more satisfied than the two other groups ($p < .01$ for both groups).

Furthermore, the question "How do you rate your occupational perspectives?" was analyzed (1=very good, 6=insufficient). Engineers who followed a management career rated their occupational perspectives as positive (M=2,06, SD=0,95). Those who chose an expert or project career evaluate their occupational perspectives in the range of good and moderate (expert career: M=2,56, SD=1,09; project career: M=2,37, SD=0,95). As variance homogeneity was not given, the non-parametric Kruskall-Wallis-test was used. The analysis confirmed significant group differences (H[2] = 26,27, p <.001). Mann-Whitney tests were calculated to find out if management engineers differ significantly from the two other groups (assignment of Bonferroni-correction, all results are significant on the level .0025). It could be shown that engineers with a management career differ significantly from those who have chosen an expert (U = 12375, r = -.25) or project career (U = 9872, r = -.19).

To investigate H2, answers to the question „All in all, how do you feel at work?" were analyzed (six-point Likert-scale, 1= highly overchallenged, 6=highly underchallenged). Engineers were divided into two groups: those who tend to be overchallenged (answers 1 to 3) and engineers who were rather underchallenged (answers 4-6). These groups were analyzed with regards to a planned job change (two answer possibilities: job change planned vs. no job change planned). In the first instance, it was assured that engineers of the three career paths (management, expert or project career) did not differ in their disposition to change jobs ($\chi2$ = 5,29, p>0,05) or in their assessment to the over-vs. underchallenged group ($\chi2$ = 3,43, p>0,05). Cross tablets were used to investigate the disposition to change jobs (see table 3). It was found that a greater proportion of underchallenged than of overchallenged employees planned a job change. Particularly management engineers differ in their disposition to a change: 69 % belonging to this group with the willingness to change their job were underchallenged. In the groups of expert and project careers, 62% of engineers planning to change jobs are underchallenged.

| | All career paths | | Management career | | Expert career | | Project career | |
|---|---|---|---|---|---|---|---|---|
| | Over-challenged | Under-challenged | Over-challenged | Under-challenged | Over-challenged | Under-challenged | Over-challenged | Under-challenged |
| Job change planned | 98 (36 %) | 177 (64 %) | 27 (31 %) | 60 (69 %) | 37 (38 %) | 60 (62 %) | 33 (38 %) | 53 (62 %) |
| No job change planned | 71 (55 %) | 87 (45 %) | 26 (49 %) | 27 (51 %) | 26 (38 %) | 43 (62 %) | 19 (56 %) | 15 (44 %) |
| Chi-Quadrat | $\chi^2$ = 3,65, p=.06 | | $\chi^2$ = 4,55, p<.05* | | $\chi^2$ = 0,00, p=.95 | | $\chi^2$ = 3,04, p=.08 | |

**Table 3:** Over- and underchallenge of respondents and their disposition to change their job, depending on career path chosen (percent values relating to a planned job change)

To address H3, the question „Which further education offers do you use for your career?" was analyzed (6-point Likert-scale, 1=very relevant, 6=not relevant). Table 4 shows means and standard deviations for the whole sample as well as for the three different career paths. Furthermore, significant group differences were investigated using one-way ANOVAs.

| Item | Mean | | | | Standard deviation | | | | Group differences | |
|------|------|--|--|--|--------------------|--|--|--|------------------|--|
| | Total | Expert. | Manage | Project | Total | Expert. | Manage | Project | F | Sigifi-cance |
| Learning by Doing | 1,65 | 1,68 | 1,66 | 1,58 | 0,81 | 0,88 | 0,71 | 0,81 | 0,74 | ,480 |
| Courses/seminars at external educational institutions | 2,51 | 2,52 | 2,32 | 2,67 | 1,17 | 1,18 | 1,05 | 1,27 | 3,67 | ,026* |
| Visitation of trade fairs/ expert conferences | 2,81 | 2,88 | 2,72 | 2,86 | 1,26 | 1,24 | 1,21 | 1,34 | 0,80 | ,451 |
| Tuiton at working place | 3,41 | 3,26 | 3,51 | 3,47 | 1,44 | 1,42 | 1,41 | 1,49 | 1,54 | ,216 |
| Quality cycles, workshops, work groups | 3,47 | 3,60 | 3,20 | 3,50 | 1,54 | 1,49 | 1,51 | 1,57 | 3,20 | ,042* |
| E-Learning | 3,81 | 3,73 | 3,91 | 3,79 | 1,55 | 1,62 | 1,52 | 1,49 | 0,62 | ,537 |
| PhD | 4,90 | 4,90 | 4,87 | 4,97 | 1,83 | 1,80 | 1,86 | 1,85 | 0,09 | ,916 |
| Master of Science/of Engineering | 5,16 | 5,10 | 5,09 | 5,27 | 1,61 | 1,62 | 1,67 | 1,55 | 0,49 | ,614 |
| MBA | 5,38 | 5,44 | 5,36 | 5,27 | 1,34 | 1,16 | 1,45 | 1,47 | 0,57 | ,565 |

**Table 4:** Means and standard deviations of nine further education offers and group differences between the three career paths

For future career development, learning by doing was rated as most important (M=1,65; SD=0,81). A MBA, Master or PhD were evaluated as not very relevant for further education. Significant group differences could be found with regards to "courses / seminars at external education institutions" ($F[2,507]= 3,67$, $p<.05$) as well as „quality cycles, workshops, work groups" ($F[2,487]= 3,20$, $p<.05$). Post-hoc Scheffé-Tests showed that "courses / seminars at external education institutions" are regarded as more relevant by managers than by engineers who chose a project career. Likewise, quality cycles, workshops and workgroups are regarded as more important by management engineers.

With regard to H4, the question "What is your planned career path in future?" was considered (see table 5). Results show that the wish for a management career is most prominent among younger engineers (65 % of engineers between 21and30 years, compared to 47% in the group of 51 to 60 and above). In terms of project careers a similar distribution can be found. In contrast, there is a development in the opposite direction with regard to the wish to choose an expert career: The older the engineers are, the more often they prefer an expert career (11% of the 21-30-year-old engineers, 33 % of over 50year-old engineers).

| Age group | Expert career | Management career | Project career | Total |
|-----------|---------------|-------------------|----------------|-------|
| 21-30 | 16 (11%) | 92 (65%) | 33 (24%) | 141 |
| 31-40 | 43 (24%) | 108 (59%) | 31 (17%) | 182 |
| 41-50 | 33 (27%) | 69 (55%) | 23 (18%) | 125 |
| 51- >60 | 23 (33%) | 28 (47%) | 15 (20%) | 66 |

**Table 5:** Planned career paths of engineers, classified according to age

With regard to H5, table 6 shows chosen and aspired career paths of male and female engineers. It could be found that women chose a management career less often than their male counterparts, but decided for a project career more often. In contrast, as many women as men prefer a management career for their future career planning (58% men vs. 56% women). More women than men strive for a project career in future (19% m vs. 26% w) while men focus more strongly on an expert career (23% m vs. 17% w).

| Chosen career path | Male | Female | | Planned career path | Male | Female |
|---|---|---|---|---|---|---|
| Expert career | 170 (37%) | 22 (39%) | | Expert career | 102 (23%) | 10 (18%) |
| Management career | 158 (35%) | 16 (28%) | | Management career | 262 (58%) | 32 (56%) |
| Project career | 125 (28%) | 19 (33%) | | Project career | 87 (19%) | 15 (26%) |
| Total | 453 | 57 | | Total | 451 | 57 |

**Table 6:** Chosen and planned career path of male and female engineers

# 4 DISCUSSION

The present study detected interesting developments in the career planning of engineers and demonstrated the need for action with regard to a stronger establishment of alternative career paths.

H1 could be confirmed based on the study data. Engineers who have chosen a management career are more satisfied with their current job situation and evaluate their future career more positively than expert or project career engineers. If companies want to use diverse potentials in an optimal way, it is advisable to institutionalize the expert or project career path and connect it to attractive incentive systems. A lack of advancement opportunities and career prospects can lead to demotivation and the migration of high-performance engineers. Therefore, the equivalent establishment of alternative career paths is a necessary precondition to retain engineers in companies.

H2, which assumed that a key reason for changing jobs was a feeling of being underchallenged, could be confirmed. It was found that a feeling of underchallenge leads to an increase.in willingness to change, especially in the group of executives. This result provides evidence for the argument outlined above, that underchallenged employees have the impression that their high qualifications do not meet with sufficient consideration of the company in the form of challenging activities or attractive remuneration. In addition, a change is usually associated with additional efforts that increase still further if the change requires the acquisition of new competences. It is likely that organization members who are already feeling overwhelmed tend to see themselves as not able to take on this extra effort. In contrast, those who feel underchallenged should be prone to accept the extra effort associated with a change more willingly. It is therefore recommended for companies to establish regular competence evaluations, systematic performance reviews and structured employee interviews in order to identify underchallenged employees and promote them by higher challenges or career development programs. In this manner the potentials of this group can be used and a job change to another employer can be avoided.

With respect to H3, it could be confirmed that engineers attach greater importance to workplace-based forms of further education than to external further education offers. One explanation for the importance of learning by doing among engineers could be that they have already experienced the difficulties associated with further education (e.g. the transfer problem, i.e. the lacking implementation of theoretical knowledge into praxis), and therefore assess particularly workplace-based further education activities as beneficial to their future career path. Practical implications emerge mainly for the providers of external further education activitites (e. g, universities) who are facing the challenge to attract and retain this target group. The results of the present study point to the danger that institutions of higher education may miss the demand in terms of the living and working conditions and respective wishes of engineers when developing their further education offers.

With respect to H4, it was shown that particularly the management career is chosen frequently as a future career path by younger engineers. Such a clear trend cannot be identified regarding project careers. Therefore, H4 cannot be confirmed. It is surprising that the younger engineers stick to the "old" career pattern of a management career. One reason may be that the engineers recognize that support and promotion measures by the company and the further education providers are developed most comprehensively for the management career in comparison to other careers such as the expert and project career (see H1).

The development of career aspirations as a function of age can also be interpreted to mean that a large part of the engineers has the desire for a management career early in their working life, but fail to realize this wish because of limited available jobs at leading positions. Accordingly, the wish is adjusted to reality as time passes. In this context, there is the risk of unfulfilled expectations among this group in the course of their career. These can have a negative effect on work behavior and performance. The establishment of a project career as an alternative guiding career would counteract the problem and provide the connection and further development of technical and managerial skills.

Regarding H5, it is confirmed that female engineers choose and plan a project career more frequently than male engineers. However, the percentage is significantly below that concerning management careers planned for the future. The large discrepancy between chosen management careers and desired future management careers especially concerning female engineers indicates that companies do not integrate female employees in management careers in a sufficient way. Especially against the background of the introduction of quotas to increase women's share in management positions, the present results do indeed point to the existence of a "glass ceiling" that female engineers seem to encounter.

The limits of the present study can be seen with regards to the following aspects: (1) The dimensions (e.g. satisfaction, overchallenge, future perspectives) could have been measured with more than one item. (2) Due to the low number of women in our sample, the detected gender differences should be investigated with a larger sample to increase explanatory power. (3) Domain specific artifacts could have biased the study results. For instance it can be assumed that the specialization of the engineers (e.g. industrial, construction or civil engineer) has an influence on the career opportunities in the three different career paths (management, expert, project career). Furthermore it is possible that engineers differ in their needs for further education based on their former education (i.e. university of applied sciences, university, doctor's degree). Even if explorative analysis within the framework of the present study indicates that differences are marginal (if at all existent), further research is needed to reveal possible domain specific differences.

# 5    PROSPECT

The preceding comments have made clear that the unilateral orientation of engineers' career paths towards the management career is ultimately only helpful in a limited way. The increase of support for the development and establishment of expert and project careers, without neglecting the management career, seems urgently necessary. Provided the potentials of the engineers are to be used, all actors are challenged in this context: Education and training providers, enterprises, human resources departments, managers in their role as staff developers as well as the engineers themselves. A key result of this paper is that a differentiation at least between expert, management and project careers is meaningful and that the different wishes regarding further education as well as orientation points for career development are to be considered. Based on the differences between engineers who have opted for a management, expert or project career which were outlined above, there is the possibility to align the further education of engineers more target group specifically in the future and to use the types of further education more adequately.

If engineers are involved heavily in innovative development processes and play a central role in them and if innovations are based on the combination of two or more subject areas in many cases, then it is not surprising that particularly expert career engineers want to acquire further expertise and keep their professional skills up-to-date. They need this to link current knowledge and professional competencies to develop innovative ideas. If, regarding the further education of engineers, primarily programs for soft skills development are expanded and disproportionately offered by companies as well as by further education providers, this will ultimately lead to a discrimination of expert career engineers.

# 6    REFERENCES

Adams, J.S. (1965).    Inequity in social exchange.    In L. Berkowitz (Ed.),    Advances in experimental social psychology (Vol. 2, pp. 267-299). New York:.

Corria-Simpson, N.-D. A., Langer, B. R. & Handor, P. (2008). Project Controls: a Roadmap to a Choice Career. *AACE International Transactions*, 1-10.

Gerstenfeld, A.& Rosica, G. (1970). Why engineers transfer: Survey pinpoints reasons for job changes. Business Horizons, 13(2), 43.

Hermann, A. (2004): Karrieremuster im Management. Wiesbaden.

Hodgson, D., Paton, S. & Cicmil, S. (2011). Great expectations and hard times: The paradoxical experience of the engineer as project manager. Journal of Project Management, doi: 10.1016/j.ijproman.2011.01.005

Lam, A. (2000). Tacit Knowledge, Organizational Learning and Societal Institutions: An Integrated Framework. Organization Studies, 21(3), 487-513.

Lienert, A. (2010): Neuen Karrieren bei Ingenieurinnen und Ingenieuren? Bochum (unveröffentlichte Masterarbeit)

Majer, C. & Mayrhofer (2007). Konsequent Karriere machen. Personal 11/2007, 36-39.

March, J. G. & Simon, H. A. (1958). Organizations. New York, John Wiley.

Nitsche, S. (2011). Kooperation kontra Konkurrenz. Medieninformation Nr. 219/2011. http://www.pressestelle.tu-berlin.de/medieninformationen/2011/juli_2011/medien information_nr_2192011/ (Abruf am 10.01.2012).

Nonaka, I. & Takeuchi, H.: Die Organisation des Wissens. Frankfurt: Campus.

Skiba, M., & Rosenberg, S. (2011). The Disutility of Equity Theory in Contemporary Management Practice. Journal Of Business & Economic Studies, 17(2), 1-19.

Strunk, G. (2009): Die Komplexitätshypothese der Karriereforschung. Frankfurt 2009.

Thom, N. (2008). Moderne Personalentwicklung: Mitarbeiterpotenziale erkennen, entwickeln und fördern. Wiesbaden.

# CHAPTER 38

# The Occupational Health and Safety Training Outline for the Managers

*Tappura, S.[1], Hämäläinen, P.[2]*

1 Tampere University of Technology, Tampere, Finland
2 VTT Technical Research Centre of Finland, Tampere, Finland
sari.tappura@tut.fi

## ABSTRACT

Safety research generally addresses a manager's key role in promoting occupational health and safety (OHS). Managers are responsible for conforming to the OHS regulations, and they have an essential role in implementing the OHS policy and procedures (OHS management) in their area of responsibility. To be able to comply with these requirements, managers need training on OHS issues. The OHS training provides managers with knowledge of their responsibilities and the tools to deal with them. Furthermore, it influences their perceptions and attitudes toward occupational safety.

The objective of this article is to propose an outline for line managers' OHS training program. The theoretical framework of the paper is based on OHS regulations and appropriate literature. The empirical framework is based on two training programs carried out in two Finnish case organizations.

The experiences and feedback from the two managers' OHS training programs were positive. The training helped the participants to better outline their duties and gave them tools to emphasize OHS in their area of responsibility. The proposed OHS training program can be used as an outline for the future programs. However, every training program shall be designed in close co-operation with the respective organization based on the managers' needs and the level of their safety competence.

**Keywords**: occupational health and safety (OHS), management safety training, management safety competence

# 1    INTRODUCTION

## 1.1    Managers' essential role in promoting occupational health and safety

Safety research and literature generally addresses managers' key role in promoting occupational health and safety. Managers are commonly considered to be key players in safety improvements and safety implementation, since they have the capacity and power to make decisions about safety investments, and can influence the safety culture (e.g. Dedobbeleer and Béland, 1998; DeJoy, et al., 2004; Schaffer, et al., 2004; Fernández-Muñiz, Montes-Peón, and Vázques-Ordás, 2009; Flin, O'Connor, and Crichton, 2000; Flin, 2003; Hofmann and Stetzer, 1996; Maurino, et al., 1995; Rundmo, 1996; Rundmo and Hale, 2003; Zohar, 1980). In Guldenmund's (2000) study, he found that management's safety activity was one of the frequently appearing aspects of safety culture. According to Bentley and Haslam (2001), supervisors' impact on safety arises both from their attitudes and actions. According to Simola (2005), the safety awareness, competence, and commitment of managers are important in achieving positive results to promote occupational health and safety. In his study, Zohar (2002) found that training supervisors to increase their monitoring and rewarding behaviors positively affected injury rates and safety climate scores.

Despite the managers' essential role, some managers might show only a weak commitment to safety, and prioritize production criteria rather than safety. Thus, the level of implementation of safety management procedures may be quite low, allocation of resources to preventive actions could be limited, and the managers may only seek to avoid legal responsibilities in formal compliance with regulations (Fernández-Muñiz, et al., 2009).

Top and line managers are responsible for implementing the safety policy and procedures, as well as safe working conditions, at the workplace. Safety information and education is often given only to the top managers, and this is not adequate for establishing good safety conditions. Also, the middle and line managers need safety knowledge in order to be able to systematically improve safety (Seppala, 1995). However, in many cases, the safety competence of line managers is insufficient given that they may be missing out on safety training (Kletz, 1991; Palukka and Salminen, 2003; Simola, 2005). Therefore, the managers may not be aware of their responsibilities or the company's safety policy. The safety procedures may not be known and implemented properly, or they may vary among departments within the company (Carder and Ragan, 2005).

In addition to the lack of OHS training and knowledge, managers often lack other training as well. There may be several weaknesses in the management skills and supervisory process that hinder rather than promote safety, and this might impede the improvement actions. Besides OHS competence, managers need leadership skills to maintain safe performance in the workplace. (Carder and Ragan, 2005; Hofmann and Morgeson, 2004; Flin, et al., 2008)

## 1.2    The OHS competence requirements for the managers

In this article, managers' OHS competencies refer to the skills and knowledge needed to fulfil the OHS legislative and company-specific requirements. Managers at all levels should be aware of these requirements. The continual improvement of their knowledge should also be encouraged (Australian Government, 2000).

The European OSH Framework Directive (D 89/391/ETY), as well as further OHS directives, are the fundamentals of European safety and health legislation. In Finland, the Framework Directive has been transposed into the Finnish Occupational Safety Act (L 23.8.2002/738), the Occupational Health Care Act (L 21.12.2001/1383) and the supplementary lower degree regulations. These regulations define the responsibilities of employers and employees. In this article, the focus is on the employers' responsibilities and on the managers' role representing the employer.

In addition to the regulations, OHS management system specifications like OHSAS 18001 (2007) provide a framework and guidelines for managing OHS issues. These guidelines help the employer to proactively improve safety performance and promote employees' health and safety.

Managers represent their organization and employer, and OHS duties are thus part of their operational responsibilities. The managers should be aware of and obey their duties, and the safety policy and procedures of the organization in order to promote occupational safety in the workplace. Thus, they require knowledge and tools to manage these responsibilities (Simola, 2005).

## 1.3    OHS training themes and effectiveness

Different studies have shown that safety training is one of the factors that affect safety culture and attitudes toward safety (Williamson, et al., 1997). The attitudinal factors can be influenced by training and development interventions. Changing attitudes may change behavior (e.g. risk-taking behavior), and thus directly and indirectly affect safety culture and accident rates. Safety training should be designed based on safety objectives and evaluated against them (Harvey, et al., 2001; Morrow and Crum, 1998). In a Finnish study (Hämäläinen and Anttila, 2009), interviewed organizations considered safety competence and training as one of the main methods to improve safety and health at workplaces.

To establish a unified and strong safety culture, designing and implementing training and education programs for the managers is crucial. Training provides managers with knowledge of their responsibilities and company-wide OHS policy and procedures. Furthermore, training provides the managers with competence and proper tools for promoting safety, and it also influences their perceptions and attitudes toward safety (Carder and Ragan, 2005).

Employee safety training has been investigated extensively (e.g. Sinclair, et al., 2003; Burke, et al., 2006; Dong, et al., 2004; Weidner, et al., 1998), while investigations regarding the managers' and supervisors' safety training have been rather limited. According to Finnish studies (Salminen and Palukka, 2007; Palukka and Salminen, 2003), there are differences in safety training between educational levels. Managers often study at higher levels, where the safety training may be insufficient or even absent from the curriculum. Inadequate safety training leaves the managers with a limited understanding of their legal and corporate responsibilities (Osmundsen, et al., 2008).

OHS training will help the managers to manage safety more systematically and to commit themselves to the unified safety policy and procedures. According to Seppala (1995), the OHS training should emphasize the foreline managers' potential possibilities and duties with respect to carrying out OHS activities in their area of responsibility. The safety policy, goals and the safety programs of the company should be reviewed and discussed in the training, and the integration of OHS and other goals should be emphasized. Other topics of the OHS training mentioned were, for example, hazard identification, safety inspection rounds, accident investigations, job organization and instruction, work psychology, and work environment. In the OHS training, the daily work issues should be handled rather than general and overly theoretical issues. According to Tappura and Hämäläinen (2011), e.g. management responsibilities, continual improvement, communication, control, and measurement should be included in managers' OHS training.

According to Tappura and Hämäläinen (2011), effective training consists of joint discussions constructing a shared understanding of safety issues and promoting commitment to safety procedures. Thus, it is important to use company-specific cases and procedures to link theory and practice. The effectiveness of the training depends also on the willingness of the managers to learn new competencies and to change their behaviors.

## 1.4    Background of the two training cases

This article is based on experiences from two safety-training programs. The first was realized in a large Finnish manufacturing company (Tappura and Hämäläinen, 2011) and the second as part of the Master of Business Administration (MBA) program. In the first training case, the participants were the line managers at the same company, but from different departments such as manufacturing, quality, maintenance, and safety. However, they did not have very unified safety processes and certain safety problems occurred at the company. Consequently, the top management launched a safety development program. The training of the line managers was seen as very important for catalyzing further development. As a result, managers' OHS training was planned to raise their safety awareness, competence and commitment to safety.

The participants of the second training case came from different kinds of organizations such as the municipal sector, private construction sector, training organizations, and the public health sector. Their main responsibility areas were construction, maintenance, vocational education and IT-specialist organization. The participants were quite positively oriented to safety, because they participated in the course voluntarily.

## 2    OBJECTIVES

The objective of this article is to propose an outline for the line managers' OHS training program. The training helps the managers to better outline their duties and to develop their OHS competence in order to promote occupational health and safety within their areas of responsibility. The theoretical framework of the article is based on the OHS regulations and appropriate literature. The empirical framework is based on two training programs carried out by the researchers/authors. The article presents the main issues and experiences of the two training programs to be used as a basis for developing future company-specific or general management OHS training programs.

## 3    METHODS

This study is based on two OHS training programs of line managers carried out by the researchers (authors) in 2009 and 2010. The preliminary OHS training program was designed by the researchers and was based on the employer's OHS responsibilities according to the Finnish occupational health and safety regulations (L 23.8.2002/738; L 21.12.2001/1383; L 20.1.2006/44). The safety training program was further developed in close collaboration with the company in question, based on its needs. There were three meetings between the company safety management and the researchers, where the needs, safety policy, and safety practices of the company were evaluated. The program was presented to the top managers of the company, who accepted and committed to it. After the training, feedback was requested from the participants regarding the content and success of the training. Also, a feedback opportunity was provided to the top management and line managers in the company.

The second training program was based on the experiences and feedback of the first training. The outline was further developed by the researchers according to the initial feedback. The developed training program was presented to the participants at the beginning of the second training, and they were asked about their conception and needs for the training and prioritization of the themes. The program was further fine-tuned based on this discussion. After the second training, feedback was requested and taken into account in developing the outline for the training.

## 4    RESULTS

The preliminary training program was designed based on the OHS regulations and employer responsibilities. It consisted of following main issues:

- OHS regulations and managers' role and responsibilities
- Continuous monitoring and improvement of the working environment
- Monitoring of the working environment, the work community, and work practices

- Hazard identification and risk assessment
- Occupational injuries and near-miss reporting and investigation
- Intervention in a case of violating occupational health and safety
- Safety-performance measuring and reporting
- Corrective actions control

The preliminary OHS training program was further developed with the company safety manager by adding several company-specific examples. In addition, more information was needed regarding safety motivation, statistical information on occupational injuries and sickness absences in Finland, machine safety regulations, and examples of good OHS procedures as well as other comparison material from other organizations. The first OHS training for the line managers was arranged in November 2009. There were nineteen line managers (including the safety manager) participating and representing different departments of the company. The training lasted two days. During the first training, the matters that promote safety in the company were discussed. According to the participants, the most important issues to emphasize in OHS management are managers' commitment to OHS, unified OHS procedures and clear goals, good OHS competence, training and motivation of managers, effective safety-performance measurement and control, and open communication concerning OHS issues. The managers' tools for OHS management were also discussed. According to the participants, managers' competence is especially important in the areas of:

- Communication (safety meetings and discussions)
- Accident investigation (an investigation model and information sharing between departments)
- Risk assessment (information sharing between departments and corrective-actions control).

During the safety training, some weaknesses in management practices were found. Competing priorities and safety issues were sometimes ignored, for example, the PPEs were sometimes considered unnecessary and too expensive taking into account the risks. The managers did not know all the safety procedures and their safety responsibilities, and were thus unsure about their safety authority. In addition, the managers were unsure about whether and how they should interfere with workers' intentional and continual safety misconduct. Some examples of interference were discussed to give the managers unified guidelines to interfere in cases of misconduct.

The feedback from the training program was collected at the end of the training, and was predominantly positive. The training program gave the participants an overview of their OHS responsibilities and tools to emphasize and promote safety in their areas of responsibility. The training helped the managers to better outline their duties, to commit themselves to the unified safety practices of the company, and to further develop them. Peer communication during the training was seen as very important and necessary to continue after the training as well. However, participants felt that they needed even more information about legal aspects and a more practical approach to the safety issues.

The second training was arranged in 2010 as two sessions of two days each. There were eight line managers participating from different organizations. At the beginning of the training, participants agreed on the suggested contents of the training with minor adjustments. After the training, they were requested to give feedback. According to the feedback, they found the training very advantageous to their work. The training helped them to clarify their legal responsibilities. They got knowledge and tools for managing safety issues in their own organizations. They emphasized systematic safety management, especially risk assessment, accident investigation, analytic approach, and continuous improvement of safety issues.

# 5    DISCUSSION

This article proposes an outline for the managers' occupational health and safety training. The outline is based on the European and Finnish occupational safety regulations (D 89/391/ETY; L 23.8.2002/738; L 21.12.2001/1383; L 20.1.2006/44) and appropriate literature. It is further developed based on the experiences and feedback of the two management safety training programs; the experiences of the training programs were mainly positive. According to the participants, the training helped them to better outline their duties and gave them competence and tools to emphasize OHS in their area of responsibility. They needed even more information regarding the OHS requirements, and their responsibilities and rights. They also required peer-discussion with other managers later on to cope with their duties and to determine good OHS practices.

The jointly-developed training program can be used as an outline for future OHS training programs in different kinds of organizations. There are certain OHS issues that should be included in a managers' safety training, and based on the literature and experiences from the two management safety training sessions, the management OHS training should be composed of the following themes:

- OHS regulations and their mandatory requirements
- Managers' role, responsibilities and authority to intervene in violations of OHS
- Motivation and justification of OHS from economic and ethical perspectives
- OHS policy, goals, programs, and procedures of the organization in question
- Continuous monitoring and improvement procedures of the working environment, the work community, and work practices
- Hazard identification, risk assessment, and information sharing to prevent risks from being actualized
- OHS orientation and training
- Occupational injuries and near-miss reporting, investigation, and subsequent learning

- Work-related health problems in working community and psychosocial work environment
- Safety performance measurement and reporting
- Corrective actions control
- OHS communication (meetings, inspections rounds and discussions)
- OHS cooperation, supporting organizations and specialists

According to the feedback of the training cases, the legal responsibilities of the managers should be emphasized, and interference in cases of misconduct should be advised. Understanding the economic aspects of safety would help the managers to prioritize competing goals, and motivate them to improve occupational health and safety.

Emphasis on these issues depends on the OHS education and competence of the managers. However, every training program should be designed in close co-operation with the respective organization based on its needs and objectives (Harvey, et al., 2001; Tappura and Hämäläinen, 2011).

Managers' competence and commitment to OHS does not solely affect the safety performance of the organization. The employees' competence and commitment is also important and they should be trained as well to motivate and inspire them to change their attitudes and behavior towards safety. Otherwise, the effects of the managers' training might become obsolete (Tappura and Hämäläinen, 2011). The managers have the possibility and power to enable the employees' OHS training. By so doing, managers also demonstrate appropriate attitudes and values, which play a significant part in employees' attitudes and commitment (Carder and Ragan, 2005).

According to Reason (1997), safety culture should be socially engineered since it arises from shared practices. Effective training should support collective learning and taking the proper steps to promote occupational health and safety. During OHS training, these practices should be discussed and improved in order to be suitable and consistent throughout the company.

Managers often lack OHS training, as well as other management training (Simola, 2005; Palukka and Salminen, 2003; Carder and Ragan, 2005). Lack of management and leadership skills may impede the overall improvement actions and safety performance. Thus, alongside the managers' OHS competence, the management and leadership skills should also be improved in order to promote employees' health and safety.

## ACKNOWLEDGEMENTS

The authors would like to acknowledge the Finnish Foundation for Economic Education and Tampere University of Technology for funding this study. The authors also thank the managers who actively participated in the two pilot training programs.

# REFERENCES

Australian Government 2000. SRC Commission. The Management of Occupational Health and Safety in Commonwealth Agencies. Australia.

Bentley, T. and R. Haslam. 2001. A comparison of safety practices used by managers of high and low accident rate post offices. Safety Science 37: 19-37.

Burke, M. J., S. A. Sarpy, K. Smith-Crowe, S. Chan-Serafin, R O. Salvador, and G. Islam. 2006. Relative effectiveness of worker safety and health training methods. American Journal of Public Health 96, 2: 315-324.

Carder, B. and P. Ragan. 2005. Measurement matters. How effective assessment drives business and safety performance. ASQ Quality Press, Milwaukee, Wisconsin.

D 89/391/EEC. Council directive of 12 June 1989 on the introduction of measures to encourage improvements in the safety and health of workers at work.

Dedobbeleer, N. and F. Béland. 1998. Is risk perception one of the dimensions of safety climate? In: Feyer, A., Williamson, A. (Eds.), Occupational Injury: Risk, Prevention and Intervention. Taylor and Francis, London.

DeJoy, D.M., B. S. Schaffer, M. G. Wilson, et al. 2004. Creating safer workplaces: assessing the determinants and role of safety climate. Journal of Safety Research 35: 81–90.

Dong, X, P. Entzel, Y. Men, R. Chowdhury, and S. Schneider, S. 2004. Effects of safety and health training on work related injury among construction laborers. Journal of Occupational Environment Medicine 46, 12: 12222-1228.

Fernández-Muñiz, B., J. M. Montes-Peón, and C. J. Vázques-Ordás. 2009. Relation between occupational safety management and firm performance. Safety Science 47: 980-991.

Flin, R. 2003. "Danger - men at work": Management influence on safety. Human Factors and Ergonomics in Manufacturing 13: 261-268.

Flin, R., P. O'Connor, and M. Crichton. 2008. Safety at the Sharp End. A Guide to Non-Technical Skills. Ashgate Publishin Limited.

Flin, R., K. Mearns, P. O'Connor, and R. Bryden. 2000. Measuring safety climate: identifying the common features. Safety Science 34: 177-192.

Guldenmund, F. 2000. The nature of safety culture: A review of theory and research. Safety Science 34: 215-257.

Harvey, J., H. Bolam, D. Gregory, and G. Erdos. 2001. The effectiveness of training to change safety culture and attitudes within a highly regulated environment. Personnel Review 30, 6: 615-636.

Hofmann, D. and F. Morgeson. 2004. The role of leadership in safety. In J. Barling and M. Frone (eds.) The Psychology of Workplace Safety. APA Books, Washington.

Hofmann, D.A.and A. Stetzer. 1996. A cross-level investigation of factors influencing unsafe behaviours and accidents. Personnel Psychology 49: 307–339.

Hämäläinen, P. and S. Anttila. 2009. Successful occupational safety and health work in Finland. Proceedings of the 17th World Congress on Ergonomics IEA 2009, CD-ROM. International Ergonomics Association.

Kletz, T. 1991. Process safety –an engineering achievement. Proceedings of the Institution of Mechanical Engineers, Part E: Journal of Process Mechanical Engineering, v 205, n E1, p 11-15.

L 23.8.2002/738. Työturvallisuuslaki. The Finnish Occupational Health and Safety Act. (in Finnish)

L 21.12.2001/1383. Työterveyshuoltolaki. The Finnish Occupational Health Care Act. (in Finnish)

L 20.1.2006/44. Laki työsuojelun valvonnasta ja työpaikan työsuojeluyhteistoiminnasta. The Finnish Act of Occupational Safety Supervision and Collaboration at Work. (in Finnish)

Maurino, D., J. Reason, N. Johnson, and R. Lee. 1995. Beyond aviation human factors. Aldershot: Ashgate.

Morrow, P. C.and M. R. Crum. 1998. The effects of perceived and objective safety risk on employee outcomes. Journal of Vocational Behaviour 53, 2: 300-313.

Osmundsen, P., T. Aven, and J E. Vinnem. 2008. Safety, economic incentives and insurance in the Norwegian petroleum industry. Reliability Engineering and System Safety 93: 137–143.

OHSAS 18001:2007. Occupational health and safety management system. Requirements.

Palukka, P. and S. Salminen. 2003. Työturvallisuuskoulutuksen valtakunnallinen selvitys (The national study of occupational safety training). (In Finnish)

Reason, J. 1997. Managing the risks of organizational accidents. Ashgate Publishing Limited.

Rundmo, T. 1996. Associations between risk perception and safety. Safety Science 24: 197–209.

Rundmo, T. and A. Hale. 2003. Managers' attitudes towards safety and accident prevention. Safety Science 41: 557–574.

Salminen, S. and P. Palukka. 2007. Occupational Safety training in the Finnish education system. The Journal of Occupational Health and Safety Australia and New Zealand 23, 4: 383-389

Seppala, A. 1995. Promoting safety by training supervisors and safety representatives for daily safety work. Safety Science 20: 317-322.

Simola, A. 2005. Turvallisuuden johtaminen esimiestyönä. Tapaustutkimus pitkäkestoisen kehittämishankkeen läpiviennistä teräksen jatkojalostustehtaassa (Safety leadership as a line supervisor's task. A case study of the implementation of a long-term development project at a steel works). Doctoral dissertation, University of Oulu, Department of Industrial Engineering and Management, Work Science, Finland. (In Finnish)

Sinclair, R. C., R. Smith, M. Colligan, M. Prince, T. Nguyen, and L. Stayner. 2003. Evaluation of a safety training program in three food service companies. Journal of Safety Research 34, 5: 547-558.

Tappura, S. and P. Hämäläinen. 2011. Promoting occupational health, safety and well-being by training line managers. In: J. Lindfors, M. Savolainen, and S. Väyrynen (Eds.) Proceedings of the 43th Annual Nordic Ergonomics Society Conference NES 2011, Oulu, Finland, September 18-21, 2011, 295-300.

Williamson, A. M., A.-M. Feyer, D. Cairns, and D. Biancotti. 1997. The development of a measure of safety climate: the roles of safety perceptions and attitudes. Safety Science 25: 15-27.

Weidner, B. L., A. R. Gotsch, C. D. Delnevo, J. B. Newman, and B. McDonald, B. 1998. Worker health and safety training: Assessing impact among responders. American Journal of Industrial Medicine 33: 241–246.

Zohar, D. 1980. Safety climate in industrial organizations: theoretical and applied implications. Journal of Applied Psychology 65: 95–102.

Zohar, D. 2002. Modifying supervisory practices to improve sub-unit safety. A leadership-based intervention model. Journal of Applied Psychology 87: 156-163.

# CHAPTER 39

# Benefits of Combining Social Influence and Ergonomics to Improve the Use of Personal Protective Equipment (PPE)

*Eric BRANGIER & Javier BARCENILLA*

Université de Lorraine. INTERPSY-ETIC, EA 4165.
UFR Sciences Humaines et Arts - BP 30309 - Île du Saulcy 57006 Metz (France).
Brangier@univ-metz.fr

## ABSTRACT

The aim of this study was to try different techniques to see in which manner we can lead workers to use safety protections. A sample of 100 operators was divided into 5 groups. The first was the control group and the other four were submitted each one to a different procedure to encourage them to use PPE: (1) the simple distribution of a safety booklet; (2) the distribution of the same booklet followed by questions about operators' work and security; (3) the distribution of the booklet followed by injunctive questions; and (4) the distribution of the booklet supplemented by a social influence (based on a free submitted compliance). Results show that the main factor determining a better integration of the safety instructions by the operators (measured by observing the use or the lack of use of safety protections) concerns the free and public commitment of the operators in relation to safety. Others procedures did not produce any effect. Furthermore, the group of operators who simply received the safety booklet does not differ from the control group. Finally, results underline the idea that the use or non-use of PPE depends also on work situations, i.e. real activity and work constraints, and on the ergonomics characteristics of the workplaces.

**Key words.** Personal protective equipment, Protection devices, Evaluation, Free submitted compliance/free-accepted submission, Safety and Security.

## 1.    INTRODUCTION

This article summarizes a research about the factors that influence the wear of personal protective equipments (helmet, gloves, shoes) by operators in a factory. It aims at presenting the results of an intervention in the field of safety and prevention of professional risks.

Our research focuses on the ergonomics aspects of written safety instructions, on the factors related to the distribution of a safety booklet to workers and their approach of it, and on the safety policy of a company in the industrial sector. After we had written the booklet applying ergonomics guidelines, we tried (1) to measure the consequences of this safety passport on the operators' real behaviour, and (2) to see whether security-based behaviour was improved by several methods of distribution of the instructions to the operators, especially by techniques of social influence. In other words, this research tries to measure the relative impact of the psychosocial factors (distribution methods of the instructions) and the cognitive factors (reading and understanding of the instructions) in the enforcement of safety-based behaviours at work.

In order to understand the effects of such intervention, we will present first issues and problematics related with compliance to safety instructions at work, then we will present the methodology and the results of the research. Finally, we will discuss about implications of possible manners to encourage workers to observe safety instructions.

## 2.    PROBLEM AND METHODOLOGY

## 2. 1. Problem

The non-respect of safety instructions in industry is a recurrent problem. Safety instructions are sometimes considered as useless recommendations while they actually define imperative rules that no individual is allowed to disobey. Their objective is to optimize people's safety by dictating pre-established procedures. Apart from a number of exceptional cases of contradictory or erroneous instructions, their ill-management exposes the operators, among others, to risks of cuts (gloves), falls (shoes), deafness (ear protections) and more widely to professional diseases, witch consequences, can be catastrophic for the persons concerned and damageable for the company. In the particular case of personal protective equipment, the operator's security depends mainly on whether the protection gear is worn or not.

Among all the factors mentioned, this research tried to check the efficiency of several methods of distribution of a safety booklet in a company on the wear of personal protective equipment. The point was to know which options would be the most effective among the following:

- a procedure with no distribution at all of the booklets (control group);

- a procedure with written communication, i.e. the booklets were given without any further precision (broadcasting of knowledge);
- a procedure with a sharpening of the operator's attention: the booklets were given and then there was a verbalization phase in order to repeat some instructions (short memory brush-up);
- a procedure with directive communication: The booklets were given, then the operators were asked to fill in a questionnaire in order to check whether they exactly followed the instructions. This step reminds us of the traditional method of hierarchical communication in which one demands the operators to wear their safety protections;
- a procedure of freely granted submission. This kind of submission makes a person to behave quite unusually: he gets the feeling that he performs freely his own will when a task is requested from him. The "freely granted submission" strategy, called also "the foot in the door" strategy, is the one used often by a salesman. It means bringing a person to do what is expected from him by means of a first harmless approach.

In other words, our problem is to measure the respective consequences of four methods of communication in security behaviour:
- distribution of security information;
- distribution and validation of the acquisition of security knowledge;
- distribution and induction of security knowledge;
- distribution and involvement in security acts.

The general hypothesis is that security-based behaviour is improved by the distribution of the instructions, slightly more by the knowledge of these instructions, and still more by a commitment in the act.

## 2.2.    Description of the firm.

The company, from the industrial sector, manufactures furniture of average range. It is highly automated and employs about 450 persons in 6 sectors of activity: tasks on special machines (cuts, press), machining, assembling, carpentry, unloading and loading. Safety policy is rather old-fashioned and more than 13% of the workers owned only their first-aid diploma. Incidentally, there was no document concerning safety instructions in this company, and the person in charge of security told us he would fain let us prepare a first draft for a booklet called "prevention and safety", in order to encourage the wearing of personal protective equipment.

## 2.3.    The safety booklet.

This 20-pages booklet is of a small and convenient size and gives instructions about: grip of station, protections to be worn by the workers, hygiene and security policies inside the firm, task-related postures, movements in the factory, and finally instructions to be followed in case of fire. The booklet gathers also information about the safety rules and important telephone numbers. Once put back, a detachable sheet of the booklet must be filled in and signed by the operator. On this sheet, he must write that he has been informed of the rules and must promise he will follow them. This booklet was meant for all the employees and so it was written with a constant concern that every worker could understand it (Hartley, 1994).

## 2.4    The sample

Our sample is made of 100 operators dispatched in 5 groups of 20 persons each. Each group is formed according to the worker's distribution in 6 workshops of the factory so as to have a sample representative of the whole staff. Within this sample group, the workers were chosen at random. So, each group includes: 3 operators from the loading zone/department, 3 from the unloading zone, 3 from the manufacturing zone, 3 from the woodwork zone, 3 from the machine manufacturing zone and 5 from the assembling zone. It is of course obvious that the nature of the job should determine the kind of body protections to be worn (tab. 1).

Table 1.  Making-up of the 5 groups for the experience

| Sectors of activity | Manpower by group | Total staff complements | Required protections |
| --- | --- | --- | --- |
| Unloading Dpt. | 3 | 15 | Shoes |
| Loading Dpt. | 3 | 15 | Shoes + gloves |
| Assembling Dpt. | 3 | 15 | Shoes + gloves |
| Manufacturing Dpt. | 3 | 15 | Shoes + gloves + helmets |
| Carpentry Dpt. | 3 | 15 | Shoes + gloves + helmets |
| Woodwork Dpt. | 5 | 25 | Shoes + gloves + helmets |
| Total | 20 | 100 | |

## 2.5    Procedures

In order to measure the consequences of the distribution methods of the booklets about the wear of PPEs, we applied a specific method of distribution to each group:

- Group 1: control group. This is the group without the booklet. As the distribution of the booklet was progressive, we intentionally postponed the distribution of the booklet to the workers in this first group.
- Group 2: this group received the booklet with no further information.
- Group 3: this group was asked to remember the instructions orally (a shift in the workers' attention). Three days after the booklet had been given to them, the operators in this group had to answer to open-ended questions about topics tackled in the booklet. For example: "Can you tell me about the moment you start your job? Can you tell me about your movements within the company? What must be done in case of accident? What must be done in case of fire?" This forced the workers to remember the topics developed in the booklet.
- Group 4: subjected group. Three days after the booklet had been given, the operators were asked closed-up questions that can be considered as safety commands. Some examples: "When you are starting your job every day, are

you wearing your gloves? When you walk inside the company, do you pay special attention to whether a trolley can be behind you or not? Do you immediately call the first-aid workers for help in case an accident happens?" And so on.

- Group 5: freely granted submission. Three days after being given the booklet, the operators were asked to answer a questionnaire that serves as a salesman's method to sell. This questionnaire directs their answers and leads the subjects to say they were in favour of the development of safety at work. The questions were like these: According to you, is safety in France a priority in companies? According to you, is safety a priority in your company? Do you think that safety must be a priority in a company? Do you consider it fair for a company to develop safety-based behaviours? Would you agree to develop safety in your job? 20 subjects answered on a scale in 10 points. After this first engaging act was made, the workers were then asked to gather in groups of 5 persons each to participate to a meeting about safety in the factory. This meeting time was held outside working hours so as to increase the psychological impact of their act: the workers could not justify their attending it by the fact they were paid or by the fact it could be considered as a break in their working day. The operator had thus no external reasons to justify their being at this meeting. We asked them to take part in it by telling them they were free to accept or not, that no one was willing to influence them, and that no one but themselves could decide whether or not they wished to attend. As for us, we were introduced as being part of the university staff, and we said we were carrying out a research on safety. To this purpose, it was perfectly coherent to gather information about safety in their working tasks. Of course, each worker was free to take part or not in this thirty-minute meeting during which the subjects had to express themselves on the last question in the questionnaire (would you agree on developing safety in your job?). Five minutes before the end of the meeting, we asked them what they would be ready to do about this purpose. Their opinion served as a conclusion of the meeting. Thus, the operators left the meeting declaring freely and publicly - i.e. in front of their colleagues – what they were going to do with respect to safety.

Following these incentives to safety, we have observed the use of personal protective equipments (shoes, gloves, helmets) over two weeks - 10 working days. The 100 operators were being observed without knowing it, once a day at a precise time and at random, in order to see whether protections were worn or not, and were not informed of the process being carried out. Furthermore, to preserve privacy, we kept the names of the employees secret.

## 3. RESULTS ANALYSIS

Results (tab. 2) should be analysed collectively and individually. On the collective level, we measured the global effects of incentives on the use of personal protective equipments and we drew up a wearing rate for each group. On the individual level, we identified the number of operators who always used their protection (called "in safety") or those who forgot them at least once (caught "at fault").

Table 2. Indicators of the use of protections for each group.

| | Use of protections over 10 days | | | Total percentage of wear | Nb. of operators observed | |
|---|---|---|---|---|---|---|
| | Gloves | Helmet | Shoes | | At fault | In safety |
| Group 1 (control) | 81/170 | 82/90 | 173/200 | 74.3% | 14 | 6 |
| Group 2 | 96/170 | 85/90 | 180/200 | 80.9% | 16 | 4 |
| Group 3 | 90/170 | 90/90 | 164/200 | 76.6% | 15 | 5 |
| Group 4 | 79/170 | 75/90 | 180/200 | 77.9% | 15 | 5 |
| Group 5 | 122/170 | 90/90 | 190/200 | 87.5% | 7 | 13 |

Nb: the results of the use of protections are reported to the number of days of observation and to the number of operators concerned with a given equipment. Example: in group 1, 17 operators must wear gloves as requested for their job. Over 10 days, we should in theory have 170 observations (workers) that observe this safety instruction, whereas we actually get 81.

Group 1 (control) is not very representative of the enforcement of the safety instructions. Only 6 subjects out of 20 follow the instructions about the use of PPEs, and the wearing rate is 74.32%. This control group has not been influenced and serves as a comparison group.

Group 2 has a percentage of 80.91% for the wear of safety equipment. Operators do not respect the use of gloves in the manufacturing zone, carpentry zone and assembling zone (only 56.4% of the operators wear their gloves while helmets are use by 94% of operators and shoes by 90.5%). No significant difference is noticed between the first group and the second one. This tends to prove that having a safety booklet does not imply any improvement in the wear of protections. Security-based behaviour is not developed by the only distribution of the safety instructions. As a matter of fact, the booklet presents instructions already known by the operators: wear yours protections, move along the indicated paths, remain calm in case of accident, etc. It does not give additional and new information, but only presents in a simple, pleasant and easy way what people very often already know.

Results in groups 3 and 4 show that the wearing average of safety equipment is respectively 76.6% and 77.9%. In both groups, 15 operators out of 20 are caught at fault. These results tend to prove that there was no significant effect due to a shift of attention - whether it is no inductive or injunctive - on the security-based behaviour. The fact of being aware of the problems of safety at work and in his own workplace does not redirect the operator's security-based behaviour. Furthermore, it appears that the wear of personal protective equipment is not affected by the availability of knew knowledge. In other words, the fact of knowing and having this knowledge in one's consciousness does not imply a rise in the wearing rate. Perhaps this is about a lack of conviction or motivation. This interpretation is even confirmed by the analysis of verbal data obtained from operators' from open-ended questions. Indeed, when asked about their arrival to the workstation at the beginning of their work, only 3 operators out of 20 knew how to describe this very moment by giving details about the body protections they had to wear. Similarly, when asked about their movements inside the factory, only 6 out of 20 mentioned safety and the 14 others gave various and very personal answers. On the other hand, all the operators questioned on the topics of accident and fire gave exact answers about the attitude to take in these cases. For these topics, the answers developed were even beyond the scope of the booklet, since new elements in the attitude to adopt appeared like, activating the alarm, using the extinguishers, limiting accidents, protecting the equipment, etc.). It seems that when risk is high, instructions are clearly identified and verbalizations match the requirements.

However, the group 5 results show a new tendency in relation to other groups. They show a positive effect both at the level of attitudes and behaviour. Indeed, the answers to the questionnaire - considered as a salesman's method (foot-in-the-door)- emphasize a beginning of worker's involvement in the safety field (table 3).

Table 3. Submitted questions in group 5 and answers obtained. (N=20)

| Questions submitted, scaled from 0 to 10 | Average scores |
| --- | --- |
| 1. According to you, is safety in France a priority in companies? | 6.35 |
| 2. According to you, is safety a priority in your company? | 6.1 |
| 3. Do you think that safety must be a priority in a company? | 9,4. |
| 4. Do you consider legitimate that a company should develop safety-linked behaviours? | 9.4 |
| 5. Do you agree on developing safety at work? | 9.4 |

The last question reaches the heart of the subject. The operators are straightforwardly asked whether they are ready to develop safety-linked behaviours in their job. Average answer is 9.4 out of 10, with 14 operators answering " yes, completely ", or 10 out of 10 to this question. This first engagement thus obtained, the operators were to confirm their acceptance of security-based behaviours by publicly declaring (in front of their colleagues), in a free way (as nothing forced them to do so) and free of charge (as they came on their leisure time) that they felt ready to get involved in those behaviours. It must also be pointed out that this engagement could not be attributed/claimed to/by any other person than itself.

After answering the questionnaire and participating to the subsequent meeting, the results clearly show that the whole process influenced the individual's behaviour. Indeed, results show that operators got involved in safety as a result of the meeting (87.5% against 74.3%, i.e. an average gain of 13.2%). While 14 operators were caught up with no protection in group 1, 16 in group 2, 15 in groups 3 and 4, they were only 7 in group 5; i.e. a Khi2 (4, N=100) = 12.03; p =.017.

The results of table 2 can also be distributed according to the nature of workshop involved. Table 4 shows the relation between the type of activity in every workshop, the need of personal protective equipment and the observed behaviour.

The results show that the workers in the unloading zone (all groups taken together) have the best percentage in protection wear (15/15 in safety). This is the activity sector with the weakest constraints, with only shoes to be worn. The results in the assembling zone (19 against 6) and in the loading zone (9 against 6) show that although they require only 2 kinds of protections (shoes and gloves), work and safety instructions are interpreted in different ways. Work in the loading zone is physically more difficult, and the wear of gloves and shoes added to that of a back-supporting belt, is necessary for an efficient protection. In this activity sector operators are relatively well protected during their work period. However, in the activity sector related with the assembling tasks, we can notice a certain degree of carelessness from the operators who sometimes work a whole day with their complete protection equipment and then forget it the next day. The results in the woodwork zone (13 against 2), carpentry zone (13 against 2) and manufacturing zone (13 against 2), in which the 3 kinds of protections are necessary (gloves, ear-protection, shoes) clearly show that operators use personal protective equipment on a rather irregular basis. In short, the analysis of the differences - in use of PPEs - between the five working groups gives: Khi 2 (5, N = 100) = 39.57; p <.0001.

Table 4. Relations between the nature of the operators' activity, the need of personal protective equipment and the rate of individual protection wear, all groups taken together.

| Sectors of activity | Compulsory protections | No. of operators observed over | |
|---|---|---|---|
| | | Faulty | In safety |
| Unloading zone | Shoes | 0 | 15 |
| Loading  zone | Shoes + gloves | 9 | 6 |
| Assembling zone | Shoes + gloves | 19 | 6 |
| Manufacturing zone | Shoes + gloves + helmets | 13 | 2 |
| Carpentry zone | Shoes + gloves + helmets | 13 | 2 |
| Woodwork zone | Shoes + gloves + helmets | 13 | 2 |

From these results we can notice that the type of activity and the number of protections needed to be in safety, both cause workers not to follow the safety instructions: the more numerous and appropriate the protections are, the lower the wearing rate and vice versa. We can observe in the operator's activity that they tend to make a digression about safety so as to arrange their activity as a function of constraints of their work: the complexity of procedures, of postures, of coordination of different methods to accomplish their activity. The non-respect of safety instructions is a sort of regulation to fit to the constraints and the reality of tasks.

## 4.    DISCUSSION - CONCLUSION.

Considered as a whole, the results help us to evaluate the kind of intervention needed to encourage workers to use personal protective equipment. The most powerful experimental group was that in which the foot-in-the-door method was applied to a meeting group, which brought the operators to strengthen their opinions on safety and to publicly promise they would develop their security-linked attitudes. According to the results, the main factor for a better acceptance and enforcement of the safety instructions by the operators is linked to the commitment of the people in charge of safety.

The shift in attention (by open-ended questions or injunctive/yes-no questions) had no consequences at all car we do not observe a relation between the contents of the verbalizations and the subjects' behaviours related with the use of personal protective equipments. In groups 3 and 4, the verbalizations integrated the instructions written in the safety booklet pretty well, but the behaviours did not give a concrete expression to them.

Furthermore, the group that only received the safety passport does not differ from the control group (without instruction). Although we constantly had in mind to improve ergonomics aspects of written instructions: to write easy-to-understand instructions with a respect for syntactic and semantic criteria, to use basic words and explicit drawings, etc., these ergonomics factors were not sufficient to lead or encourage workers to use safety protections. It was just as expected!

The results also point out that such a research in industrial psychology proves to be insufficient: the characteristics of the operators' actual tasks explain for a great deal the use -or not- of protections.

As a conclusion, this action research raises eventually four issues:

First, while a number of companies still tackle the issue of body protections with posters or a video film watched by the operators on the very first day of his joining the firm, it would be advisable to seriously question the efficiency of these methods of communication. It appears obvious to us that they do not achieve their purpose! Have these films the role to clear through customs the company or to legitimate afterward sanctions? Either it is a sign of belief according to which the knowledge of a problem induces a behavioural change? In psychology, every one knows the insufficiency of this assumption. Being aware of a safety instruction does not always imply people will follow them. Equation is much too easy. It is therefore not fair for those in charge of security to just write socially acceptable recommendations (which are often well known) while they should really work on the employees' behaviours and help them to get more comfort, safety and efficiency. With safety at work, this intervention is also dealing with action and the results of this action must be a priority.

Secondly, by focussing on the action, this type of psychosocial intervention matches the ergonomic intervention but it also shows its own limits. Indeed, the process of the freely granted submission method enables us to make the operators behave according to the safety instructions, but they do not question the instructions themselves. Instructions are always relevant? Can they be applied unanimously to all kinds of working situations? Obviously, the differences in the use of personal protections - in the five workshops - are closely linked to the differences in the real activity. Work analysis will indeed show that some security regulations are now used to satisfy the operator's individual or collective objectives: to fit out the comfort, to work faster, to grant himself more autonomy. However, the evaluation of five intervention methods enables us to connect the results obtained to procedures implemented to encourage workers to use personal protective equipment.

Third, it is possible to decide on the consequences of the intervention because we now know a procedure/method will help to encourage the use of protections. Let us remember that 450 people are employed in the company; it seems now interesting to apply the fifth procedure to each of the 350 other operators. In the same idea, it becomes necessary to take into account the ergonomic characteristics of protections and to evaluate their adaptation to several real working situations.

Last and fourth, the security department must not only improve the reliability of the technical system by limiting most risks and developing the workers' level of knowledge - thanks to working procedures and training (Bahr, 1997, Brangier & Barcenilla, 2002) - but it must also add methods of psychosocial interventions which aim at the internalization of security-based behaviours at work.

## BIBLIOGRAPHY

Bahr, N., J. *System safety engineering and risk assessment : a practical approach*. London : Taylor & Francis. 1997.

Bradley, G., L. Group influences upon preferences for personal protection : a simulation study. *Journal of safety research, 25*, 2, 99-105. 1995.

Brangier, E., Barcenilla, J. Ergonomie des aides textuelles au travail : analyse de la compatibilité homme-tâche-document dans dix entreprises, *Bulletin de Psychologie*, 55, 2, 193-204. 2002.

Hartley, J. *Designing instructional Text*. London : Kogan Page. 1994.

Joule, R-V., Beauvois J-L. *La soumission librement consentie*, Paris : PUF. 1998.

Kiesler, C. A. *The psychology of commitment. Experiments linking behavior to belief*. New-York : Academic Press. 1971.

# Section VI

*Flexible Work Force and Work Schedule*

# CHAPTER 40

# Social and Organizational Ergonomics and Temporary Work in Industrial Firms

*Manfred Bornewasser*

University of Greifswald
Greifswald, Germany
bornewas@uni-greifswald.de

## ABSTRACT

Traditionally, ergonomics is defined as the study of designing work equipment and devices that fit the human body, its movements and its cognitive capabilities. Thus ergonomics describes systemic states which are constituted by people in different work places, by technical instruments, and by structural features which regulate relations between people, people and machines and between machines. This conception has been enlarged and elaborated by aspects of the design of social conditions like team work or job shops and even organizational structures like work time arrangements and modern employment systems. The present paper concerns an important aspect of modern organizational ergonomics, namely temporary work in industrial firms.

**Keywords**: ergonomics, temporary agency work, commitment, triadic employment arrangement

## 1    DIMENSIONS OF ERGONOMICS

Ergonomics has its roots in physical interactions, e.g. movements and postures between bodies and technical objects. However, work cannot be reduced to a fit of bodies and devices, but it also includes broader aspects of work like perception and decision making and of the work place or work station (e.g. light, smells), of the social arrangements (e.g. leadership, teamwork), and of organizational features (e.g. staffing arrangements, design of working time). This implies that the individual worker is part of different subsystems, which leads to a differentiation of micro- and macro-levels of ergonomics, with physical ergonomics at the microlevel and organizational ergonomics as the macrolevel. In between are to be seen cognitive and social ergonomics (Dul & Weerdmeester, 2008). The final goal of macroergonomics is a completely efficient work system at both the macro- and the micro-level which - for an employer - results in improved productivity, and –for an employee – leads to satisfaction, health, safety, and commitment. Consequently, from a methodological point of view, ergonomics has to consider different levels of analysis (organization, team, workers) and different relations between levels.

The present paper concerns an important aspect of modern organizational ergonomics, namely temporary work in industrial firms. Temporary work is seen as a special staffing arrangement of industrial enterprises, leading to temporary or time-limited employments during finite peak periods of demand or during finite work projects. In Germany, this staffing arrangement has become a legitimate instrument of flexibility and has found its regulation in a special law called Arbeitnehmerüberlassungsgesetz (AÜG), which implies that a service agency by itself employs workers who, for a limited duration, are rented, lent or leased to firms. Round about 40 percent of these low-skilled temporary workers are categorized as helpers, often showing deficient or obsolete qualifications. These helpers are in the centre oft he present paper.

The implementation of such a flexible staffing arrangement has strong implications for the social structure of employees, as it leads to a differentiation of an internal and an external workforce. This differentiation implies a lot of consequences for both, internal and external workers. Externals are confronted with discriminations concerning pay and treatment, internals with additional work and feelings of threat.

Finally, at the individual level, being a temporary worker and experiencing the differentiation of internal and external workers has consequences for the development of affective relations to work and organization. That is the cause of the general assumption of lower work satisfaction and commitment with temporary workers. This general assumption has its reason in the exchange conditions in client firms which often lead to manifest or latent disadvantages for temporary workers.

## 2    EROSIONS OF TRADITIONAL FORMS OF EMPLOYMENT AND RISE OF TEMPORARY WORK

Social and economic changes have underscored the need of organizations for greater flexibility in their employment systems. Traditional HRM all over the western world is challenged to develop new practices that enable organizations to adapt quickly to rapid developments in production technology, volatility of output demands and greater diversity in labor markets. A popular characterization of this concern is expressed by the approach of the "flexible firm" (Atkinson 1984). This term describes different ways of adaptation, especially to changing labor markets, and is often closely associated with the assumption of eroding standard employment relationships with its elements of unlimited and full-time employment.

There are two distinct strategies of flexible labor utilization in the centre of the flexible firm approach: The first strategy concerns the enhancement of the firms' ability to perform a variety of tasks and to participate in decision-making, while the second strategy aims at a limitation of workers' involvement and – in the case of demand – an extension of the labor force via external labor markets. These two strategies have been differently referred to as functional versus numerical flexibility, internal versus external flexibility or as organization-focused versus job-focused employment relations (Tsui, Pearce, Porter & Hite 1995). This idea implies three basic assumptions:

- First, establishment of a high performance work system via internal labour markets vs. externalization of contracting out workers via external labour markets. This leads to the formation of two different work systems, the difference of which could be described by reference to high and low skills, involvement, and commitment.

- Second, standard employment relationship vs. non-standard contingent employment relationship. Unlimited and full-time employment constitutes an internal labor market of the firm, whereas part-time, temporary work, and contract work of unsure duration constitutes an external market of the firm.

- Third, segmentation of a core work force vs. a peripheral workforce within the organization, leading to a differentiation of fixed and variable parts of the workforce. Connected to this differentiation is the idea that the core workers are highly trained, skilled and committed, that they are of higher value to the organization, and that they should be buffered and protected from overload and from extreme fluctuations of demand. In contrast, the peripheral workers are mostly seen as low skilled and low committed, as less valuable and expendable.

Thus, there is a theory-driven tendency to combine on the one hand high performance work systems, standard employment relationships and core work force as a basis of an internal market strategy of contracting in, while externalization, contingent employment relationships and periphery work force constitute the basis of an external market strategy of contracting out. Of course, this clear-cut differentiation has instigated a lot of critical comments (Kalleberg, 2011). Independent of this theoretical controversy, reality in Germany shows a clear tendency for organizations to rely more and more on external labor markets, to use services of external help agencies and to accept more or less moderate forms of segmentation of their work forces. This option is motivated primarily by economic reasons: First, by highly volatile output demands, leading to mismatches of work force and demand capacities especially in peak periods, which are no longer to be compensated by internal flexibilization; second, by avoidance of high costs in the course of cancellation of workers with regular employment contracts, third by avoidance of long-term overload of the internal work force. These motives are legitimized and supported by institutional permissions of contingent work as a means of bridging between the first and the second labor market and by the establishment of nearly 17.000 service help agencies. At the same time, the liberalization or even elimination of limitations concerning the duration of limited employment could be seen as inducements to run more and more external HRM strategies. In a representative study, Bellmann and Kühl (2007) show, that the proportion of industrial firms which use temporary work arrangements has increased up to nearly 50% and that this increase is the result of an intensified use especially by larger organizations. Accordingly, in Germany the number of temporary workers has increased up to nearly one million cases in 2011.

## 3    TEMPORARY WORK DESIGNED IN A TRIADIC EMPLOYMENT ARRANGEMENT

The German labor market is regarded as conservative and regimented. This creates certainty for all parties and is at the centre of the so-called typical or normal employment relation in Germany.  Its security is derived primarily from the underlying contract. It regulates work content, working hours, working conditions and pay, and makes a statement about the intended duration of the contract. Atypical employment however, is characterized by the fact, that there exists insecurity about the duration of employment (Polivka & Nardone, 1996). Temporary work in Germany shows both elements: A temporary worker has an open-ended contract with an agency, however, the time to work within each client firm is open. Temporary work implies, therefore, always aspects of uncertainty.

Temporary work occurs when a service help agency (lender) gives its own employees to a client firm  (borrower) in such a way, that the client hands over the authority to issue directives to his lent temporary worker    (§1, Arbeitnehmerüberlassungsgesetz, AÜG).  By that, the typical triangular design of temporary work relations is constituted (see Figure 1).

Figure 1 Design of triangular employment of temporary work

The triadic relationship leads initially to the fact that explicit key design parameters such as pay or job content can not be negotiated directly between the client firm and the temporary worker. Instead, both are negotiated between agency and client. Similarly, temporary workers rarely have access to career and developmental opportunities in the client firm. This means that the bargaining chips by both partners are reduced to rather implicit or social factors such as promises of equal treatment or appreciation. This might lead to a paradoxical situation for the client firm: On the one hand, it relies on integrating the new employees in order to stimulate their achievement motivation, on the other hand, the company has only little means to induce workers' willingness and preparedness to do their jobs. Thus, active relationship building in the client firm is of tremendous importance.

To assess the relationship building process conceptually, reference could be taken to the model of the "Area of Acceptance" by Simon (1951). It describes some kind of a mentally negotiated space between two negotiators which contains all the explicit and implicit obligations, both partner can rely on in further exchange situations. Let us imagine that a contractor (in this case, the client firm) shall exercise his authority over an employee (temporary employee) in order to move him to provide a certain performance. Whether the temporary employee shows compliance, depends on three factors: (a) what the advantages and disadvantages are that the employee expects when he shows compliance (b) how good or bad he is compensated for any disadvantages, which might occur in the run of his employment with the client, and (c) whether there are alternative employments. The sum of the values of all three features (expected costs, compensation and alternatives) determines what might be expected by the superiors of the client without inducing resistance. The mental space that includes all those instructions, which will be followed either willingly or grudgingly, constitute the "Area of Acceptance" or some kind of an acceptable range. This range of acceptance probably differs for normal and contingent workers.

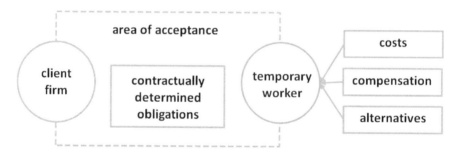

Figure 2 Schematic representation of the area of acceptance and its influencing factors (own illustration based on Simon, 1951)

Temporary employees in the helper segment practice due to their low entry qualifications only minor and largely standardized activities. Their knowledge is of minor importance fort he client firm and they can always be procured at the external labor market if necessary. Helpers often come out of unemployment and have little or no alternative employment opportunities. The resulting asymmetry of power leads to a strong dependence of the helpers of the decision makers on the part of the client firm, which is expressed in a feeling of "marketization" of one's own person. For the client firm, this leads to a relatively broad area of acceptance, implying the possibility to assign even extremely unpleasant tasks to the low-skilled temporary worker. For members of the core work force as well as high-skilled temporal workers the situation is quite opposite.

# 4    INFLUENCES OF TEMPORARY WORK DESIGN ON MOTIVATION AND COMMITMENT

Now the question is, which kind of emotional and behavioral consequences this employment situation has for temporal workers in client firms. Temporal workers are continuously confronted with this unusual design of employment. Uncertainty, associated risks in life planning and a different bonding and integration into hierarchical structures are permanently connected with this kind of temporary work. It can be assumed that these aspects have a direct impact on work attitudes and performance of temporary employees. This concerns in particular the affective bonding or commitment to the client firm.

Commitment is a construct mainly investigated within the framework of standard employment. However, commitment research within the last fifteen years has increasingly paid attention to contingent work forms, such as temporary work (e.g. Gallagher & McLean Parks, 2001). Most of the studies within this research field consider commitment as an outcome of the fulfillment or breach of psychological contracts or as a reciprocal contribution in an acceptable exchange relation.   These results support the evidence from research on exchange relationships that reciprocal exchanges and successfully performed negotiations are fundamental for commitment to the other exchange party involved. The general interest in commitment is due to its beneficial outcomes to the organization, which have been shown for permanent employees as well as for temps (Felfe et al., 2008). Concerning temporary work, client organizations should additionally consider the temp workers' exchange perceptions if they plan to frequently hire the same person or even to take over a temporary worker into the permanent staff in the future. Therefore, commitment reflects the establishment of well-functioning social exchange relationships which are the bases for successful future collaborations. Till now, there are only quite a few studies on the commitment of German temporary workers with emphasis on health and performance outcomes and different commitment foci like the agency or client organizations (Galais &Moser 2009).

## 5    RESULTS OF AN EMPIRICAL STUDY

In an empirical study of 190 core and peripheral workers two central questions were investigated: Do both groups of workers show differences in terms of affective commitment and which factors determine the amount of affective commitment to the service agency and the client firm. Based on theory-based difference hypotheses,  participating companies launched a questionnaire to their personnel, which contained several standard questions from established forms  containing issues affecting work and organizational commitment (e.g. job insecurity, stress experience, job satisfaction, perceived organizational support).  Statistical analysis was performed using the software packages SPSS and AMOS.  Only a choice of results is presented here (for more details see Bornewasser, 2011).

Data showed, that core workers and temporary workers differ significantly in terms of affective commitment to the company where they do their job (3.96 vs. 3.02, t = -5.89, p <.000.). Core workers thus have a significantly higher affective commitment. Thus, it might be assumed that temporary workers have to spend more efforts to realize their productivity. Temporary workers did not differ in their commitment to the agency or the client firm (3.03 vs. 3:17; T = -0.67, p <.20). This result leads to the assumption that there is a so-called simultaneous, dual commitment, compared with the core workers, however, to a much lesser level. This finding confirms the results of Felfe et al. (2008).

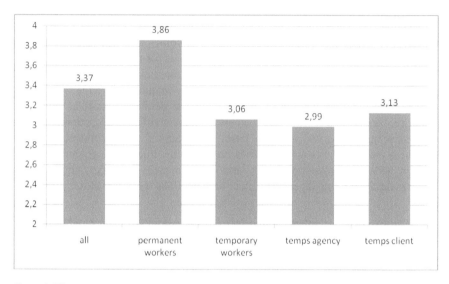

Figure 3 Affective commitment of permanent and temporary workers

This result was analyzed for moderating effects of the length of employment, perceived job insecurity and perceived organizational support, by application of structural equation models. There are three main results:

(1) The affective commitment of the temporary workers varies along the duration of the temporary employment relationship in the client firm. It rises in the initial period (1-7 months) and after that falls off again (8-20 months; 3.24 vs. 2.49). Thus, it seems as if temporary workers use their affective commitment in the beginning probably as an instrument to be included in the core workforce of the client firm. If these efforts of takeover fail, temporary workers start to increase their emotional distance to the client firm. This might also lead to an impairment of health or an increase of sick days (Galais & Moser, 2009).

(2) Job insecurity and perceived organizational support are two key causal factors for the affective commitment of temporal employees to the service agency and the client firm. Here, the perceived benefits play a much more prominent role and the expected negative effect of job insecurity on affective commitment is reversed, that means, uncertainty leads to increased affective commitment. This may be due to the fact, that temporary workers in client firms express a high level of affective commitment and motivation in the hope of a takeover. Another reason is that those temporal employees who already worked for a very long time in the client firm, do no longer realize differences of insecurity and thus show high commitment levels comparable to those of core workers (although each temporary workers exactly knows, that it is he who has to leave the client firm in a case of capacity crisis).

(3) There are clear moderating effects of the factor duration of employment. With increasing duration of the temporary employment the relationship between job insecurity and commitment is reversed. Job insecurity then correlates, as expected, more negative with affective commitment (short duration: $r = .35$; long duration $r = - .25$). Furthermore, there was a significant interaction between job insecurity and duration of employment ($\beta = - .27$, $p < .03$), signalling a decrease of expectancy concerning a takeover with increasing intervals of employment with the client firm. However, as there is no increase of affective commitment to the service agency, this result could be indicative of some kind of resignation und alienation.

# 6    CONSEQUENCES FOR A HUMAN RESOURCE MANAGEMENT CONCERNING TEMPORARY EMPLOYMENT

There is a lot of evidence showing that external flexibilization leads to a segmentation of work forces in the client organization. This segmentation creates two social groups with some kind of a borderline, which could be seen as a line of separation and mutual discrimination. Lau and Murnighan (1998) call it a faultline. Such a faultline denotes a group composition where multiple diversity dimensions are aligned in such a way that a group can split into antagonistic subgroups. Client firms must adjust to the changed situation resulting from volatility of markets and the possibility of external flexibility. The made-up border between unlimited internal and limited external workforce needs concrete integration measures, to increase the willingness to cooperate even with predictable non-acceptance and to reduce various forms of opportunism. This requires a quality management system in which on the one hand the standardization of recruitment, training, performance evaluation and also the cancellation of temporary workers is established. On the hand there are programs to be installed, which prepare especially superiors for their active integration job.

Temporary work cannot be seen as a homogeneous phenomenon. It has a lot of appearances. Differentiation is necessary. Although the classical model of temporary work as a low-skilled, marginal workforce continues to exist, there is also a growing number of high-skilled employees, who temporarily work as professionals and specialists. Thus, there are at least two segments of temporary work. Both segments are fundamentally different from each other in terms of the nature of the exchange ratios between temporary workers and client firms. While helpers continue to be regarded as largely interchangeable, workers in the skilled segment of the uniqueness of their knowledge have a relatively high bargaining power. The gap between both segments probably will be broadened in the coming years.

In Germany, temporary agency work has already for a long time functioned to provide temporary staff as a short term cover for someone who was temporarily absent or to accommodate seasonal or short-term business fluctuations. It was easy to find candidates in the segment of low skilled workers, which often are young and even long-term unemployed. The unemployed - this was at least expected - may acquire new skills and gain some working experience that increase his productivity and his employability and thus improve his future labor market prospects. Temporary work was and still is seen as a bridge, a path, a springboard or a stepping-stone into a permanent employment with a client firm. However, reality shows that only a small proportion of 7% of temporary agency helpers gets a chance of a transition and to become a member of the internal core work force (Lehmer & Ziegler, 2010). Actually in western Europe there is a discussion which factors „make transitions pay" (EMCO, 2010).

The external numerical flexibility, notably in the helper segment, creates a severe training problem. Generally, the employer is responsible for corporate training. Due to the splitting of the role of employer (agency and client), there is a specific situation concerning training. As the service agency feels always threatened by a takeover of better qualified workers, it has no strong interest in qualification programs. This is true for the client firm, too, because low-skilled temporary workers are only needed for simple tasks or because they are employed only for a short term. In light of this conflict, it is necessary to find alternative ways to install training programs, which might advance prospects of temporary workers. One possibility could be that agencies and clients come to arrange financial transfer payments in cases of takeover. Alternatively there could be some kind of dual education, which combines phases of work with the client and phases of certificated training in a special academy. Otherwise there is the problem of so called „low pay no pay careers" (unemployed – temporary work – unemployed).

Independent of labor market policy concerning temporary agency work, personnel managers have to find ways to better integrate employees of both core and periphery segments for the duration of the assignment. In the low-skilled segment appear equality of treatment, fairness and a level of participation as an important instrument for increasing motivation, commitment, and performance, all are important conditions of maintaining the health of temporary workers.

## ACKNOWLEDGMENTS

The research work is being carried out as part of the research project entitled „Flex4Work – Integration and implementation of flexibility strategies concerning permanent and temporary staff". This project is supported jointly by the German Federal Ministry of Education and Research and the European Social Fund (Support Code 01FH09127).

## REFERENCES

Atkinson, J. 1984. Flexibility, uncertainty and manpower management. IMS report No. 89. Brighton.
Bellmann, L. and A. Kühl. 2007. Weitere Expansion der Leiharbeit? Eine Bestandsaufnahme auf der Basis des IAB-Betriebspanels. Hans Böckler-Stiftung. Berlin.

Bornewasser, M. 2011. Psychologische Aspekte der Zeitarbeit. In. *Beiträge zur Flexibilisierung Band 1 – Schwerpunkt Zeitarbeit*, eds. R.B. Bouncken and M. Bornewasser. 7-32. Bayreuth/Greifswald.

Dul, J. and B. Weerdmeester. 2008. Ergonomics for Beginners: A Quick Reference Guide. CRC Press.

The Employment Committee 2010. Making transitions pay – EMCO Opinion. Council of the European Union. Bruxelles.

Felfe, J., R. Schmook, B. Schyns and B. Six. 2008. Does the form of employment make a difference? Commitment of traditional, temporary, and self-employed workers. *Journal of Vocational Behavior,* 72: 81-94.

Galais, N. and K. Moser. 2009. Organizational commitment and the well-being of temporary agency workers. *Human Relations*, 62: 589-620.

Gallagher, D. G. and J. McLean Parks. 2001. I pledge thee my troth … contingently commitment and the contingent work relationship. *Human Resource Management Review,* 11: 181-208.

Kalleberg, A.L. 2011. Good jobs, bad jobs. The rise of polarized and precarious employment systems in the United States, 1970s to 2000s. Russell Sage Foundation, New York.

Lau, D. C. and J. K. Murnighan. 1998. Demographic diversity and faultlines: The compositional dynamics of organizational groups. *Academy of Management Review*, 23: 325-340.

Lehmer, F. and K. Ziegler. 2010. Brückenfunktion der Leiharbeit: Zumindest ein schmaler Steg. IAB Kurzbericht, July 2010.

Polivka, A. E. and T. Nardone. 1996. On the definition of contingent work. *Monthly Labor Revue*, 112: 9-16.

Simon, H. A. 1951. A formal theory of the employment relationship. *Econometrica* 19: 293-305.

Tsui, A. S., J. L. Pearce, L. W. Porter and J. P. Hite. 1995. Choice of employee-organisation relationship: Influence of external and internal organisational factors. In *Research in personnel and human resources management,* ed. G. R. Ferris. 117–151. Greenwich.

CHAPTER 41

# Two Roads of Ergonomics

*Stanisław Janik, Andrzej Niemiec*

Poznan University of Technology
Poznan, Poland
stanislaw.janik@put.poznan.pl, andrzej.niemiec@gazeta.pl

## ABSTRACT

Nowadays, almost everywhere meaning of the ergonomics is being emphasized in the daily living, but first of all in the business. Companies are investing in devices meeting more and more excessive ergonomic criteria. Additionally regulations are imposing an obligation to organize positions this way on the employer so that they meet the to say the least minimal demands for the health and safety at work and the ergonomics in the scope of the amount of area falling to one employee, office furniture, illumination, level of noise and the heating and the ventilation. This road is solving the number of issues associated with safety and comfort, the quality affecting the improvement and the work output. Solutions from the scope of the ergonomics on the workstation are minimizing coming into existence of illnesses of the layout of the movement, teams of muscle tensions, teams of the surcharge in the range bar, wrist and many other associated with pursued disappointment. This trend in the improvement of the working conditions, state not only the indicator of the development, but also the fashion for the ergonomic business which unfortunately isn't being moved to small single enterprises, with the registered office at home or the flat of the employee. Development of the information technology and simultaneously reality stewards, limiting expenses and are supporting coming into existence of workstations in living quarters. In private places, where the employee is his employer simultaneously the regulations concerning the ergonomic organization of a place of employment aren't applicable. With substantial amount of the competition which in its flat performs the career designers of websites, architects, computer specialists, financial advisers, translators, land stewards, computer graphic designers are deciding and so on forth. They perform their work in rooms of home everyday use, so as living rooms or bedrooms. Nobody specially is designing these rooms and is suiting including the pursued profession. They didn't fit in order to into the interior and the design of the family house. In the process very often they aren't accomplishing primary requirements of the work study. Yes rally technology and economics forced progress into coming into existence of two outermost roads. First is marking out the development dispatched to the improvement in conditions and qualities of work, imposing an obligation to employers of planning ergonomic posts for employed employees. The second road, it direction of the return to conditions in which the work was apart from the influence of the ergonomics. Which the balance was upset between the performed work in conditions of the full safety, in and with broad development of the ergonomics. In the course of the performance of work however the aspiration to the welfare is a superior aspiration.

**Keywords**: ergonomics, homeworking, welfare, job security.

## 1.  CURRENT TRENDS OF ERGONOMICS

Nowadays, almost everywhere, the meaning of ergonomics is being emphasized in the daily living, but first of all in business. Companies are investing in devices meeting more and more excessive ergonomic criteria. Additionally, regulations are imposing on the employer an obligation to organize workstations in a way to meet at least minimal demands for the health and safety at work. Whereas in the sphere of ergonomics in general they provide the required amount of the area for one employee, the office furniture, the illumination, the level of noise and the heating and the ventilation. (Polish Minister of Labor and Social Policy of 26 September 1997) Such direction of action is solving the number of issues associated with safety and comfort, influencing the improvement, the quality and the work output. In this scope an optimum forming of the entire system of the activity of the man is a basic aim of the ergonomics which comprises: possibilities of the man, the organization and measures used at the work and the product which is a result of this activity. The ergonomic optimization of working conditions is needed and applied in all branches of industry, trade, communication, and in the clerical work (Studenski, 1996). Principles of ergonomics are more rarely used in the household, and at present more and more smaller companies are coming into existence, of which an own flat is a registered office and a place of the performance of work for the entrepreneur. Contemporary trends and technical solutions being in their favor allow forecasting that with time more and more people will work virtually. In developed countries, in which this kind of employment already has its rooted tradition, about 15% of employees is choosing this form of work In Poland, as well as in other post-communist countries, the post-transition period has not developed this form of employment in a satisfying degree. At present, employment of virtual employees in Poland reaches 3% (Herbst, 2008). But it is being forecasted that in the future decade more and more employees will be choosing that contemporary employment status. The group of virtual contractors will not only constitute the specialists living in the country, but also the ones who are abroad or represent different nationalities. This tendency is appearing with transformations, which are occurring in proceeding reality because telecommuting which enables the performance of professional tasks at home has many pluses. First of them is deciding the hour of starting work, breaks, and also lacking the constant scrutiny from the side an employer or a principal. Smaller running costs of employees or employing the most talented people are next advantages, irrespective of the place of inhabiting them. A frugality is a next plus of the homeworking. The employee doesn't worry for spending money for the everyday commuting to work. Moreover, he has not to attach importance to his workwear clothes. There is no direct contact with the customer, so he can economize on the purchase of business clothes. But by numerous positives there are also elements weighting to the disadvantage of this employment status. The homeworking requires a great self-control and discipline above all because the employee alone is choosing working hours for himself, and it is not easy to force oneself to perform professional activities, if we drink our morning coffee or finish breakfast. In addition, in private house, we rarely have selected rooms for work, and the profession is being carried out in rooms of everyday home use. It is necessary to consider the fact that the average area of a flat occupied by the entire household in Poland in 2010 was 70 square meters (Ulman 2010). Therefore, in living conditions of the entire family the homeworking is bringing a lot of inconvenience with itself. Still, despite all impediments, the most important objective of such solution is to realize tasks determined in the contract and present results in agreed time is often the sole criterion which the employee should fulfill. There are no regulations, which exact safety rules and ergonomics of the health at home. Paradoxically, the law is forbidding everyone who isn't holding the prosecutor's order even to go on the premises of the house. So, it is limiting the inspection of services of the health and safety at work and simultaneously it gives the sense of lack of necessity for carrying about the appliance of standards determining the equipment of a workstation, so that it would be safe and ergonomic. The employer or the principal do not control, what hours the order was fulfilled in, because exclusively an effect of the work is counting. The homeworking is attractive particularly for women who are nursing small children, because such an employment status enables to combine the work with maternal duties. It requires a great self-control and discipline, but how is showing the practice this task is not unfeasible. The homeworking has also another defect. It is above all about a type of the signed agreement with the employer. It is mainly fee-for-task agreement and specific-task contract; it means that the employee is not entitled to full entitlements resulting from the contract of employment. Thus, it is possible to assume that the development of the ergonomics reached a crossroads, in which one direction is a road of a rapid development into which the knowledge was harnessed from many fields of science. Such direction of the development is solving the number of issues connected with the safety influencing the improvement of the quality and work outputs. This trend in the improvement of working conditions, states not only the indicator of development, but also the fashion for ergonomic business. In the same time development of the information technology and simultaneously reality of economics are conducive to come into existence small, often one-man companies cooperating with larger companies. Those micro-enterprises, with their seats in employee's house or flat, do not meet all ergonomic standards. In private spaces, where the employee is one's own principal, regulations concerning ergonomics of work aren't applicable. The biggest number of professions that can be performed in own flat constitute websites designers, architects, computer specialists, financial advisers, translators, property managers, computer graphic designers, etc. They perform their work in rooms of everyday living, so as bedrooms or living rooms. Nobody designs nor equips these rooms with consideration of the performed occupation. They didn't fit to the inside and the decor of the family house. Therefore, very often they aren't accomplishing primary requirements of the ergonomics. This way the progress of technology and economics forced coming into existence two extreme roads. The first one is marking out the development dispatched to the improvement in conditions and qualities of work imposing an obligation to employers of the ergonomic design of working conditions. The second road is directed to the return to conditions in which the work was apart from the influence of the ergonomics. A balance between working conditions in large enterprises and small companies was disturbed. The second direction of the ergonomics in a sense is moving the mankind back to the period of the early phase of the industrial era, when man was only a second-rate element of the manufacturing process; where workers paid costs of the produced good with their torment during the performance of work, and in the end wasted their health.

## 2    GENESIS OF THE WORK ERGONOMICS

The road of the ergonomics has its beginning already in distant times where in prehistorical homesteads the first tools for performing essential activities arose. There were mainly tools for defense, hunt, setting skins and preparing meals. This at first cohesive direction evaluated to modern times and only in the 19th century pioneers of the contemporary ergonomics became interested in the issue of improvement of the working conditions in companies and enterprises. Thereby ergonomics was closely appointed direction of its development. It occurred that small solutions and improvements could upgrade the efficiency of work and, most of all, reduce fatigue and range of accidents (Rosner, 1985). Additionally apart from humanistic premises, a main reason of interest in man and his possibilities was a conclusion, which the first observers of manufacturing process reached. It said of the fact that the man in spite of the great adaptability remained most brittle and unpredictable component of work system. So an awareness of designers and employers was more and more universal, that man would be able to achieve great work output under the condition of harmonious cooperating with technical means of production and the financial and social environment. It is only possible when the work and measures to make it are designed up to the standards of psychosocial possibilities of man. Directions of civilization and technological changes in processes and means of production appoint the new formula of the ergonomics. These changes, since the 19th century, take place in more and more rapid pace, determined by revolutions: first was industrial, and then information one (Bańka,

1976). During theme, the objective was to make technique a helpful tool for man and mean for achieving his goals, instead of being the objective of activity itself; so that the production was for the man, rather than man for the production. As one of the first research works in the ergonomics, in contemporary of this word for meaning, it is possible to determine examinations from 1908, which F. W. Taylor established the relation between the size of shovels, mass of charge, muscle strength of the worker and the work output with this tool. In result of conducted attempts, Taylor drew recommendations up as for optimum sizes of shovels. In order to ensure always the optimum weight of the cargo, which he described to be 8 - 9 kg, he recommended using shovels with a large surface to light materials, and smaller ones to heavier materials. From his surname arose a system called "taylorism" (scientific management), acknowledged as first system optimizing the work. In these times, it significantly enabled to increase the efficiency of production. Such an organization, called the stream working system (assembly line), however reduced the man for the role of a robot performing mechanically straight activities, depriving him of creative satisfaction from the work. Ignoring psychological and social factors of motivation at work, it aroused the opposition of employees and turned out as a result to be a barrier of further progress of the productivity in the production. Most efficient work organization turned out to be a so-called nest organization, in which a production cycle is being carried out by a group of employees being able to perform all action in this cycle and exchanging tasks. Such an organization is releasing the creative relationship to the work, is giving a sense of satisfaction and is heightening a responsibility for the work (James, 1997). All changes which occurred in the work organization are good demonstrative history lessons proving the profitability of considering demands of the ergonomics in practice. The term "ergonomics" was presented and proposed for science concerning work in 1857 by a Polish scientist - Wojciech Jastrzębowski. We wrote in his paper "Ergonomic draft, i.e. science based on rights taken from Natural Science" in „Przyroda i Przemysł" magazine: "the name of Ergonomics is taken from Greek words (ergon - work, nomos = natural laws) and it means Science on Work, i.e. on how man uses strength, force, skills and abilities given to him by God". Unfortunately, the newly appointed field of science not at once developed so as it effects transferred into the improvement of work conditions in factories, workshops and other places of the performance of work. Only eighty years later, first reliable ergonomic researches were initiated in USA and in UK. During the Second World War, planes and other war machines achieved the top degree of complexity. The man was forced to cope with new challenges and duties, resulting from the use of such complicated equipment. It extorted additional mental strain on him while processing a lot of information and taking many various decisions in a very short time. The result of these observations was an attempt to determine limits of mental, physical and environmental strain of the man. Experiences in the area of ergonomic research got in the time of war were transferred in the area of the civil industry after finishing it; it resulted in the distinct increase of work output and qualities of products. In post-war years, a spontaneous development of ergonomics started as a field of activity and science applied in the industry, caused by seeking new means of increasing the work output. The main reason for initiating study on efficiency and effectiveness in the man – to – technical object system was observing striking disproportions between possibilities of the efficient effect of technical means and possibilities of the man in the field of his professional occupation. Concluded, that the best quality of technical means is not able to guarantee of high production effects, if applied machines and processes will be dehumanized. This way one started putting the rally great weight to the meaning of ergonomics in work, initially calling this field of science the humanism of the work. Very quickly, one started introduce new technical solutions to production; they took under consideration biological requirements of human organism. Automation of manufacturing processes, maximum use of recommendations of physiology, hygiene, psychology and sociology of work were spread in creating financial and psychological-social work conditions (Kotarbiński, 1965). It was acknowledged that a humanization is also passing the initiative on the employee in creating working conditions so that he considers himself to be the subject and the creator of his work process. This course manifested itself also through turnabouts in working systems, making it rich, widening the scope of responsibilities and implementing the flextime. The name "ergonomics" for this discipline of science has been popularized since 1949, after forming Scientific Ergonomic Society in Great Britain, and measure of the development of ergonomic ideas in the world was coming into existence in 1962 International Ergonomics Association. In Poland, a Polish Ergonomics Association (PTErg) was established in 1977, which is an IEA member. Earlier a Committee of Ergonomics already existed by Polish Academy of Sciences (1974) and Committee of Ergonomics and Health and Safety at Work at Polish Federation of Engineering Associations established in sixties, when they started leading large scale ergonomic examinations in a few biggest industrial plants in Poland. For the first time in Poland universal explaining the password ergonomics turned up at Large Popular Encyclopedia in 1970 as the supplement in the group of new terms, not included in twelve previous volumes. This fact confirms that the term obtained a full linguistic right to exist only in the seventies in Poland. In world literature it is possible to find a lot of definition of the ergonomics. These definitions do not compete with each other, they mutually complete each other; they add certain accents, depending on the field of science represented by their authors. International Labour Organization (ILO) in 1961 determined ergonomics as: "(...) joint application of some biological and technical sciences for providing optimization of conditions of mutual increase of work efficiency in relations between man and work and for contribution to the prosperity of the worker". In articles of International Ergonomics Association (IEA) was included in 1967 following definition this new discipline of the knowledge: "Ergonomics is about relations occurring between man and his occupation, equipment and environment (material), in its widest interpretation, including work, leisure, situations at home and on the road". The Committee of the Ergonomics of Polish Academy of Sciences in 1982 approved such an expression: "(...) the main task of ergonomics is optimum adaptation of material products of man and conditions of using them to psychical and physical features of man, taking under consideration factor of his material and social environment. Ergonomics' objective is to provide man a state of wellness in the system man – to technique, both in his professional and private activity". This way so it is possible to assume that the ergonomics has a lot of meanings and applications. That's why it should count it among the group of theories about the work. The ergonomics is joining three groups of detailed sciences to itself: the theory about the man, the theory about the labor organization and the theory about the technique. It means that it has multidisciplinary character. The ergonomics is using the knowledge of many sciences for the basic, pragmatic target of improving working conditions, rest of the man - through adapting technical devices to his needs and the education of the user. The mentioned below approximate list of fields of study is not complete; moreover scientists as part of individual specializations are developing detailed new solutions optimizing work mainly from humanistic premises, and acknowledging that man is a subject and most important element with it. Therefore, according to this idea the production and the product should be shaped for man as the measure to achieve his purposes, above all for his better quality of life and health.

## 3  CONTEMPORARY DIRECTIONS OF ERGONOMICS

As a result of the permanent progress of information, telecommunications and multimedia technology it is observed that the environment we live in is subjected to constant changes. At present, the civilization entered into a new phase of development; and as the contemporary civilization we are becoming an information society. The newly created quality of the communication results from its accessibility, massive and globalization nature. Along with it, a way of the performance of work is being changed. A work performed exclusively using hands is vanishing, and in difficult and onerous duties man is more and more often replaced with machines, or performs his work using them. At present, equally with disseminating the knowledge about possibilities of modern technologies, the quality of the occupational activity influences on organization of work space. Concept of ergonomic environment isn't already only a comfort but a standard guaranteeing the pleasant and healthy surroundings. The ergonomic optimization of working conditions is needed and applied in all branches of industry, trade, communication, in the office, in the household (Tytyk, 1998). Occurring relations between man, the technical structure and the type of performed activities are an object of the ergonomic design. Man is the most sensitive element of this system, and his psychophysical possibilities are limiting the productivity. Therefore, ergonomics recommends adapting the space, the environment, and the labor organization to the man. In order to solve specific issues ergonomics is using different sciences and disciplines; therefore it is being ranked among comprehensive studies. These are achievements of the learning and contemporary solutions from the perspective of the ergonomics on the workstation serve man in minimizing coming into existence of illnesses of the layout of the movement, syndrome of muscle tensions, syndrome of the tension within shoulder, wrist and many other associated with the performed work. At present, the scope of the effect of the ergonomics is widening on other, apart from the work, fields of man's activities; it is talking about the ergonomic for living conditions, sport and rest, ergonomic flats, furniture, cars, toys. The developing in last years ergonomics of the product is aimed at increasing the functionality also of consumer goods. One of contemporary definitions of the widely comprehended ergonomics and containing the aim for it acting is as follows: "the ergonomics is a field of science and practice, for which the forming of the human activity is a purpose - in it above all of work - appropriately to his physiological and psychological properties". These definitions lead to a statement that a basic aim of the ergonomics is an optimal forming of the entire system of man's activity, in which all elements are composed of possibilities of man, organization and measures used at work, as well as the product which is a result of this activity. The ergonomic optimization of working conditions is needed and applied in all branches of industry, trade, communication, in the office, but not only because principles of the ergonomics are often also used e.g. in the household. The optimization of work is following above all of humanistic premises. Man is a subject and most important element of it, the production and the product should be shaped for man as the measure to achieve his purposes, above all of better quality of life. Such an attitude is opposing the view from the period of the early phase of the industrial era, when the man was only an imperfect element of the manufacturing process. Apart from humanistic premises, the reason of main interest in the man and his possibilities of manufacturing process was a conclusion which observers reached; it said that man, in spite of his great adaptability remained the most brittle and unpredictable system component. An awareness of designers and employers was more and more universal on the fact that the man would be able to achieve the great work output under the condition of harmonious cooperating with technical measures of the work and with the material and social environment. This trend in the improvement of the working conditions, states not only the indicator of development, but also the fashion for the ergonomic business, which unfortunately isn't being moved to small single enterprises, with the office registered at home or flat of the employee. The development of information technology and simultaneously a reality of economics have influence on curtailing expenses favor coming into existence of workstations in living quarters. In private apartments, where the employee is simultaneously his own employer, regulations concerning the ergonomics of work aren't applicable. Substantial amount of the professions, which are being performed in their flats constitute: websites designers, architects, computer specialists, financial advisers, translators, property managers, computer graphic designers etc. Since implementing the ergonomics is usually connected with a smaller or bigger economic cost, with reference to a specific workstation, we usually associate so-called ergonomic demands with the minimal requirements. Ensuring such minimal requirements should allow usually preserving health through the entire period of the career and a smooth functioning after finishing it. For everyone the contemporary ergonomics offers much more than so far. It is presenting both structural, technological and organizational solutions which will let getting much more benefits in a form of greater productivity of work and its smaller nuisance, e.g. through reducing tiredness, reducing sick leave, numbers of occupational diseases or accidents. Task of the ergonomics is an optimal forming of the working practice, both individual of his elements as well as the relation between them. It means ensuring of the work output carried out in conditions of not only full safety, but letting the broad development of the employee in his intellectual, psychological and social sphere (Janik, 2008). These optimal working conditions are creating possibilities and motivation at the employee for improving efficiency. According to the contemporary knowledge, it is assumed that the essential condition in work projection is its safety. The forming of such working conditions requires knowledge about the reliability of action of not only technical system, but also man - of his physical and psychological possibilities, as well as knowledge about differences between possibilities of individual people. Because not every person can work on any workstation. After all, there are such positions, on which are needed e.g. the above average physical strength or the ability to the logical reasoning (Kowal, 2002). It is necessary to remember also about the fact that an optimization of working conditions is not only a good design, but also unceasing correcting and taking into account happening changes in the organization, or equipping the work post. It should be a process of constant analysis and evaluation of working conditions, made on the basis of the system criteria. Only then, when this process has constant character, it will be possible to create an ergonomic place of employment. All three elements appearing in the arrangement man - machine - environment, are important for the correct forming of working conditions, on account of the interaction. However, a knowledge of man is a crucial issue. The base of creating principles of the forming of optimal working conditions there is an acquaintance of psychophysical possibilities of the man. Especially, knowledge of restrictions of these abilities is important. When work exceeds abilities of the adaptation, an intellectual and physical tiredness appears, complaints and occupational diseases are unfolding. The work output is losing on it and when this phenomenon is not connected with one person, it means that the workplace was designed badly and requires an immediate reorganization. About 40% of people, out of everyone working in the European Union, perform their everyday tasks by the computer (Górska, 2002). The environment of the office work is posing many threats to its users. It does not concern the only possibility of the accident at work, but for the long-term influence of particular elements of equipping the office for bone structure of the employee, his eyesight, frame of mind etc. Effects of this influence are not visible at once, but slowly, systematically are leading to declining of the spine, impairing the eyesight, impairing the hearing, increasing allergy, increasing pain and degenerative states within almost of whole body. Only years later we observe consequences of the improper organization of a work place in the form of illnesses and irreparable degenerative changes, and even then not always we associate these complaints with the applied mode of the working life.

## 4.  ERGONOMICS WORK AT HOME

The growing importance of ergonomics in the office started in the sixties of 20th century. However, it gained momentum in the next decade. Then an equipment and decor of offices started changing. The research on the influence of the sitting mode on the human health extorted changes in the approach of designers. Designers began to seek of new forms and solutions, which would eliminate the detrimental effect of the seat, which is an unnatural position of for human body. In order to facilitate the typing, one developed various stands and supports of legs, loins or adjust the desk's height. All in order to maintain an upright position and to prevent back and neck pain. Then opened office became popular, where the worker is mobile and has the ability to move. But since the 80s of last century, the computer started to play an increasingly more important role in the work of man. At the very beginning however these devices took up much space on the desk what led to the development of a completely new dimension to the work area. A role of shelves, additional tables, or monitor tables increased. A model of the group and team work has more and more often been applied in the consecutive decade what indicated the growth in importance of space intended to meetings. At present, the computer so much reduced its dimensions, that is not taking up much space. Notebooks became common which put on the table when are needed for the work, and fold and put off into the any place, when they are not needed (Lewandowski, 2000). It is usually such portable computers are equipment of home offices of the work. The computer as a basic tool to help in the practicing a profession changes the standard approach to work. For the majority of people, the career is related with getting up early morning and with a trip with car or public transport to the workplace, where one should botch its eight hours. Then it is possible to return to the cozy apartment or house. However, many people do not have to leave their four angles in order to work, because they perform their job duties directly at home. For ages, among others, writers and poets were working this way, nowadays thanks to possibilities offered by the Internet, also people associated with IT industry, graphics and it similar. All of them are called freelancers. These are people who are conducting the telecommuting, mainly by performing a single job or are employed on a permanent basis they can however, perform their duties at home. Such a work has great advantages, among others we can freely dispose of time, we can divide our duties of discretion and so on. On the other hand, the downside is that private space, turns simultaneously in space in which we work, what should be separated. Working at home is also connected with an unlimited time spent above professional duties and frequently can very much protract. In the legal field, the contemporary ergonomics has quite a lot to do. Just for example to compare the current regulation forbidding men to pick objects up weighing more than 50 kg and 30 kg more than 4 times per hour with results of researches and recommendations of the International Labor Office and the U.S. NIOSH. These recommendations assume that 90% of healthy men without risk may increase only about 23 kg. A major task for the Polish ergonomics should be also (and as soon as possible - in the face of an ageing population) to make with children and teenagers under 16 years health and safety regulations were also respected, in particular in the area of transport manual and of work on the computer. Creating a friendly environment to disabled people is an effect of action undertaken by the contemporary ergonomics. It is regarding not only the better structure and the organization of workstations, but also the entire infrastructure. In the recent time in Poland, it was done quite a lot in this field, although there is still a lot to do. And even though in many scopes of work studies, legal regulations are being implemented, in the issue of the practiced profession in house rules are dead, and appointed institutions for spreading the ergonomics in this area do not deal with the growing problem. Perhaps in this case it is necessary to spread information, training and arousing awareness of ergonomics. The matter seems to be very serious because when we will accept that the majority of works of micro-entrepreneurs, for whom an own flat is a registered office, they execute orders working with the computer, it is forming it with the risk of coming into existence of many illnesses. Conducted findings amongst 75 respondents working on workstations with the computer showed, that only about 17% of respondents did not identify any problems related to work behind a desk. However at the 83% of respondents were found to have a number of health complaints. Over 35 respondents (approximately 50% of respondents) reported complaints of pain: neck, back, complaints of eyes, discomfort seat. Moreover, 27% granted that they suffer from headaches and exaggerated mental strains (Jóźwiak, 2001). Generally, it is possible to say that these complaints are caused mainly by: incorrect placing the computer position and its equipment (chair, table, monitor, keyboard, etc.) which forces the user to adopt the disadvantageous position during the work, overload of particular group of muscles and the skeletal system. The reason is an inappropriate organization of a workstation and techniques of the work and assuming the incorrect position during performed duties what extorts additional effort on the organism. In addition, one must be taken into account, the existence of the overload of particular group of muscles and the skeletal system, as well as repeating of the same activities. At home, when deadlines are short, and a lot of work to do often you give up micro breaks and relaxation exercises which allow the regeneration of tired parts of the body. Therefore, on a basis of these observations it is possible brightly to state that a computer work place requires an ergonomic design, of what at home is not actually applying. Everyone, who carries out his work at home should be aware that placing the computer, the monitor and the keyboard, as well as the appropriate, shaped chair will actually enable the work with the computer physically and mentally less burdensome. Unpleasant consequences of the performance of work in the wrong extorted position can also be menstrual complaints, miscarriages, and even impotence. Some of the most frequent complaints waited in their place in the List of Occupational Diseases, which is attached to the Regulation of the Council of Ministers from 30 July 2002 on the list of occupational diseases, the detailed rules of conduct in the reporting of suspected cases, the identification and determination of occupational diseases and the relevant entities in the matters. The most common diseases related to the way of making office work are carpal tunnel syndrome, pains and degenerative states of the spine, tennis elbow, epicondylitis of the humerus, chronic inflammation of the tendon and its sheath, periarticular shoulder inflammation, ocular diseases and allergies. Simultaneously proved that the rise in the amount of working hours in a sitting position had increased the risk of the appearance of muscle-skeletal disorders. However even exclusive holiday armchairs are not able entirely to eliminate risks of the appearance of pain. To go against these inconveniences, it began practically to exploit the concept of dynamic seat. It consists on forcing changes into arranging the spine, thereby changing the pressure on its individual parts (Janik, Grygiel, 2010). Many scientific studies conducted laboratory analyzes of the influence of different sitting positions on the activity of muscles and assumed posture. However, most of them did not report any significant results. It is possible to think, why this is so that at home ergonomic seats are not being used at the performance of work, having at its disposal ordinary chairs, and a comfortable designed ergonomic armchair, after all diametrically different comfort is feeling seats. Fitness of the man, ahead of everything physical, we judge generally based on the achievements of top-class athletes ("this is what man is capable"), forgetting that the average man in productive age is represents far lower level. And then there is more and more clearly observed phenomenon of ageing of the population. So ergonomics "here and now" and the ergonomics of tomorrow must face adapting work processes not only to less efficient, but also more and more old employee (Lewandowski, 2000). Offices, more than private flats, houses or cities, are showing, in what way a society is reasoning, how perceives the future and appearing innovations. We spend most of the day in the workplace, there we collect gadgets, photographs which confirms our status, but also can strengthen or weaken our willingness to work. In addition to behaviors common to all societies, such as using of technical equipment, the need of an exchange of ideas, the approach towards the privacy, social relations and attitude to work, differ depending on the country, and sometimes, depending on the region. Many factors influence, so the atmosphere and working conditions. The quality of our workplace has a key importance for achieving business purposes and of the maximization of results of the work of employed employees. They are after all a fundamental value of the company. Caring for emotional employing involvement of individual people and their sense of well-being at work should be never pushed off to the background. However, methods of the work are changing over the years. Thus, the way of arrangement of the offices is changing. All the time the technological development is taking place. In consequence, our offices are subject for changes and metamorphosis. Perhaps in the future workplace will never change so far in order to be completely remote. On the contrary, in the recent time direct exchange of experience and work in groups gaining in popularity. Of course, a manner of labor organization, communication and human relationships are changing in order to cope with more and more increasing demands of the market. For the example, in Italy the corporate culture is dividing employees into the boss and the rest of the team (Pokorska, 2004). Such approach creates barriers between the boss (usually man) and members of his team. Here is a very strong orientation towards individual success. Situation is different in the Anglo-Saxon model. There, the team constantly evaluates and performs the continuous exchange of ideas between the manager and the rest of the team. Currently designing an office space so as to promote the exchange of ideas, freedom of contact and communication, it is rather a rarity. To put it in one sentence, however, office is designed, the greatest emphasis should be placed on the aesthetics and what emotions it is arousing. The workplace should have a soul and a clear character. Unfortunately, our offices are far from this ideal. We have more to do with places that do not say anything about the character and values of the company, and sometimes communicate wrong values and block the natural dynamics of the team. Preparing the workplace that stimulates creativity, requires a custom approach. A friendly design, which will create the unofficial, family atmosphere, is needed through appropriate furniture and their innovative placing. The point is to replace the sterile-looking office space something what associates with warm and well. Of course the innovation of such place of work should go hand in hand with the friendliness of materials for the environment, the energy saving and the possibility of recycling particular elements (Jasińska, Janik, Jasiński, 2009). Today, while creating places of employment, we must remember, what kind of people are going the exactly labor market up. They are representatives of the new generation. The possibility of reconciling work and family life is one of the most cherished values for them. Therefore, flexible statuses of employment are already not only possibility which employers have, but also a need, when they want to pull, hold and build the loyalty of employees. A contemporary generation encloses mobile people, who expect more and more flexible approach to methods of the work and more and more diversified offices. Therefore, there is a need to offer new products, made of new materials; products, which will come out beyond what at that particular moment we are calling the good taste or the fashion even. The office space has a potential to develop the productivity and the creativity, however nothing can happen without investment. Planning the office space and hence the office-residential space is coming from the need of rational using the area (Pokorska, 2004). Independent consultants should be employed for its planning. Currently, the area to work is a valuable asset. Therefore, it is important to identify the real needs of the company and how best to answer for her purposes with appropriate project. It is good, so that it takes place before choosing of the area to the office, and even before the design of the building. In some cases the wrong building affects the functioning of the entire company negatively, and even to run its business. A well-planned area contributes to the success of the company, strengthens its brand and internal sense of belonging. The office of the future in the spotlight sees a man who works as best he can, feeling in addition relaxed and motivated. This man is not already interested in static work and feels better in the new place of employment, where can be himself. People from the contemporary generation appreciate the flexibility of time and place of work and its equipment and decor. Furniture modeled on the ones which we have at home, will ensure the convenience, nice colors and the diversity. Certainly we live in times at which radical changes are turning up. They will influence the nature and the appearance of office customs and for perceiving by employees the new reality. A rapidly growing virtual world implies new needs and opportunities. It is forcing coming into existence new ways and directions of the development. It is good as it is a good way supporting the improvement of the working conditions. Unfortunately, quoted examples show that the development of the technology opened new field of the employment, where the mankind in a way is going back to the past. But it is possible to suppose the theory about the ergonomics will find a remedy so that the work at home will be still attractive, comfortable and safe. Then the universality of such a form of employment will remain natural.

# REFERENCES

Bańka, J., 1976. *"Humanizacja techniki, Główne zagadnienia i kierunki eutyfroniki"* Wydawnictwo Śląsk, Katowice

Górska, E., 2002. *"Ergonomia: projektowanie, diagnoza, eksperymenty."* Polskie Wydawnictwo Naukowe. Warszawa

Herbst, J., 2008. *"Od trzeciego sektora do przedsiębiorczości społecznej – wyniki badań ekonomii społecznej w Polsce"* Stowarzyszenie Klon, Jawor

James, P., and N. Thorpe, 1997. *"Dawne wynalazki"* Wydawnictwo Świat Książki. Warszawa

Janik, S., and D., Grygiel, 2010. *"Ecological aspects of macroergonomics Advences In Occupational, Social, and Organizational Ergonomice."* Edid by Peter Vink and Jussi Kantola. CRS Press Taylor& Francis Group, Boca Raton London, New York

Janik, S., 2008. *"Macroergonomic aspects of ecological production."* Paper presented at AHFE International Conference, jointly with 11th International Conference on Human Aspects of Advanced Manufacturing. Las Vegas

Jasińska, E., S. Janik, and M. Jasiński, 2009. *"Oddziaływanie lidera na otoczenie."* Międzynarodowy Kongres Górnictwa Rud Miedzi – Perspektywy i wyzwania, Lubin

Jóźwiak, Z., 2001.*"Stanowiska pracy z monitorami ekranowymi - wymagania ergonomiczne."* Instytut Medycyny Pracy, Łódź

Kotarbiński, T., 1965. *"Traktat o dobrej robocie"* Zakład im. Ossolińskich, Wrocław

Kowal, E., 2002. *"Ekonomiczno-społeczne aspekty ergonomii."* Wydawnictwo PWN, Warszawa-Poznań

Lewandowski, J., 2000. *"Zarządzanie bezpieczeństwem pracy w przedsiębiorstwie"*, Politechnika Łódzka, Łódź

Minister of Labour and Social Policy of 26 September 1997. Poland

Pokorska, B. 2004. *"Przedsiębiorca w systemie franczyzowym"*, Polska Agencja Rozwoju Przedsiębiorczości, Warszawa

Rosner, J., 1985. *"Ergonomia"*, Państwowe Wydawnictwo Ekonomiczne, Warszawa

Studenski, R., 1996, *"Organizacja bezpieczeństwa pracy w przedsiębiorstwie"*, Wydawnictwo Politechniki Śląskiej, Gliwice

Tytyk, E., 1998. *"Dobre i złe tradycje w kształtowaniu środowiska pracy i życia człowieka"*, Wydawnictwo Ergonomiczne, Zielona Góra

Ulman, P., 2010. *"Sytuacja mieszkaniowa polskich rodzin w świetle danych z badania budżetów gospodarstw domowych."*, Katedra Statystyki, Uniwersytet Ekonomiczny w Krakowie, Kraków

# CHAPTER 42

# Working Time Configuration in a Call Center Using a Simulation Approach

*Patricia Stock, Michael Leupold, Gert Zülch*

Karlsruhe Institute of Technology
Karlsruhe, GERMANY
patricia.stock@kit.edu

## ABSTRACT

Working time configuration is highly complex in practice due to the fact that the organization's operations have to be taken into account in addition to aspects of law and collective negotiations. In the services sector in particular, dynamic factors have to be taken into account that stem from the stochastic nature of arrival intervals and processing times. This results in such a wide range of possible configurations for a company's working time system that operations planners are barely able to take them all into account without the help of suitable methods of planning and evaluation. However, there are few tools that currently offer this kind of dynamic assessment. Only personnel-based simulation offers the opportunity to carry out a dynamic and prospective evaluation, although previously only work-related stresses on employees could be modeled. This was the situation in which the OSim-GAM simulation tool was refined in order to allow prospective working time configuration that takes the work-life balance of employees into account. This paper presents the developed simulation tool using a call center as a case study.

**Keywords**: working time configuration, work-life balance, stress, call center

# 1    COMPLEXITY OF WORKING TIME CONFIGURATION

## 1.1    Factors influencing working time configuration

A useful working time model must take into account a wide range of framework conditions, above all the relevant provisions of statutory and collective bargaining law, goals relating to operations and employees' interests, and ergonomic recommendations (see Hornberger and Knauth, 2000, p. 25). Some of these underlying conditions are counteractive, which means that not all goals can usually be realized to the same extent. An expert from France has shown that taking into account all legal regulations governing working time configuration would make a permissible working time model impossible. Working time configuration becomes even more complex when attempting to find not only a suitable working time model, but a working time system. A working time system is a composition of various working time models for a single enterprise or department. For example, there may be different working time models according to profession or full-time/part-time employees (Bogus 2002, p. 29).

Employee-related goals have also recently come to the fore. Longer or even atypical working time mean that many employees are faced with the prospect of working in the evening at times usually reserved for family life or leisure, or even the weekends (Stock, Bogus, Stowasser 2004, p. 32). The resulting conflicts with family life, leisure time, the fulfillment of voluntary duties etc. can have negative consequences for employees that in turn impact their performance at work. The literature mentions side-effects such as mental stress, poor health, low satisfaction with work and private lives, and increased fluctuation to name but a few (e.g. Frone, Russel, and Cooper, 1997, pp. 330; Greenhaus, Collins, and Shaw, 2003, pp. 525). For this reason, in February 2011, the German Federal Government, business associations and the German Confederation of Trade Unions (DGB) launched the "Family-Friendly Working Times" initiative aimed at improving work-life balance (BMFSFJ 2011).

## 1.2    The need for the dynamic assessment of working time models

Conventional, static methods of planning and assessment (such as the benefit-in-use analysis, the balance of arguments or checklists) can only give quantified statements regarding the anticipated consequences of using different working time systems to a limited extent (see Bogus 2002, pp. 25). Only a dynamic assessment using a computer-based model that takes into account the chronological work process within a company makes it possible to gain a comprehensive overview of the potential (positive or negative) consequences of planned working time models. When attempting to do so, simple analytical or numerical solutions (for example based on queueing theory) very quickly reach their limits (Bogus 2002, p. 53).

On the one hand, dynamic influences on working time can result from stochastic operation times, which are typical for the field of services. Although average customer service times are often known, actual operation times can vary, and this variance must be represented using suitable probability distributions. On the other hand, customer numbers are also subject to fluctuation due to the fact that the intervals between arrivals are usually also stochastic. This makes it more difficult to adjust available capacity to meet stochastic capacity requirements (see for example Zülch, Stock 2011, p. 1).

## 1.3  Need for a tool to prospectively assess working time configuration taking work-life balance into account

There are currently no tools available that allow decision-makers at a company to assess the efficacy of revising the working time system prospectively (i.e. before realization) while taking dynamic aspects as well as the work-life balance of the employees into account. At most, existing processes for designing working time models take working time preferences into account but not the stress on the individual (see Zülch and Stock 2011 for an overview of methods of dynamically assessing working time models). Such forecasts are however an essential part of contemporary HR assignment planning. This makes it more difficult for the decision-maker to view the problem in its entirety and take appropriate restructuring measures for their own company, explicitly taking into account the interactions between employees' work and private lives.

The ARBWOL project (German acronym for "Working time configuration whilst considering the work-life balance") is dedicated to this subject. The leading research hypothesis states that different employees have differing social roles which are distinguished from each other based on the workload they have to deal with in their private life (see Zülch, Schmidt, Stock in this proceedings). Appropriate working time models may compensate for this strain in line with the concept of work-life balance.

## 2  SIMULATION-BASED ASSESSMENT OF WORKING TIME MODELS

The tool used to assess working time models is the object-oriented simulation tool *OSim-GAM* (German acronym for "Object Simulator for the Design of Working Time Models"; Bogus 2002, pp. 160) developed at the ifab-Institute of Human and Industrial Engineering of the Karlsruhe Institute of Technology (KIT, formerly University of Karlsruhe). This tool allows for the simulation both of production and service companies and has already been used successfully in numerous simulation studies for configuring working time models (e.g. Zülch, Stock, Hrdina 2008; Zülch, Stock, Leupold 2011).

The various orders/customers are modeled using activity networks that show the chronological interdependencies of activities during processing in the form of a network diagram. An activity network can have internal or external triggers, for example the arrival of a customer. Both the activities and the triggers for the flowcharts can be modeled stochastically using different distributions in order to adequately depict dynamic aspects. As an outcome the simulation provides various operational (e.g. throughput time, employee workload), monetary (e.g. costs per customer) and employee-related (e.g. time stress, physical stress) key figures for analysis. For more information, please refer to Zülch, Fischer, Jonsson 2000 or Zülch, Stock, Hrdina 2008, for example.

However, in order to configure and assess working time models which are able to improve the work-life balance, *OSim-GAM* had to be supplemented by additional concepts for modeling and assessing conflicts between professional and private life. The social roles in *OSim-GAM* therefore serve, on the one hand, as a category of strains and conflicts in terms of specific roles. On the other hand, they can be used to interpret the simulation findings with relation to specific social roles (Zülch, Stock, Leupold 2011, pp. 2164). The social roles are currently represented using restrictions on working time, i.e. by specifying the periods of time in which potential conflicts with employees' private lives may occur. Once an employee survey (which is currently ongoing) has been concluded, a role/strain model will be implemented that will describe the causal relationships between social role on the one hand and strain and working time preferences on the other.

The following describes simulation-based working time configuration using the configuration of working time at a call center.

# 3       WORKING TIME CONFIGURATION AT A CALL CENTER

## 3.1     Modeling the initial situation

The call center functions as a service provider for a number of industries, i.e. technical services, banking or taking orders for retail and tourism. A high level of flexibility in terms of time and subject matter is therefore expected of the employees. The simulation study described here looked at one department of the call center in which employees mainly served as providers of information.

### 3.1.1  Modeling the different types of inquiries

The department examined can be reached under several telephone numbers that are structured according to the types of inquiry received. The processing of a customer generally comprises two separate activities: the telephone call itself and post-processing on the computer. An inquiry type can therefore be modeled as the sequence of these two activities, each of which is assigned a distributed processing time. The average duration of the call and the post-processing depends on the nature of the inquiry. A conversation to request technical services lasts an average of 2.9 minutes (standard deviation sd= 0.7 minutes), and the post-processing in order to arrange a service representative 0.7 minutes (sd = 0.3 minutes). Taking a report of the loss of service, on the other hand, takes just 1.1 minutes (sd = 0.7 minutes) and requires 0.4 minutes of post-processing (sd = 0.6 minutes). As the processing times are subject to several influences like the time of day or the operator's level of training, no random distribution could be derived for modelling them. Instead, all recorded processing times were entered into the simulation tool which then draws one of them randomly every time a call arrives.

### 3.1.2   Modeling the frequency of calls

The volume of calls varies with regard to the frequency of customer inquiries. Figure 1 shows fluctuations in the volume of calls to request technical services over the course of the day. 79.3% of calls are received between 8.00 a.m. and 8.00 p.m., although calls continue to come in through the evening and into the night making a minimum level of staffing necessary. The daily fluctuations in other inquiries follow a similar distribution.

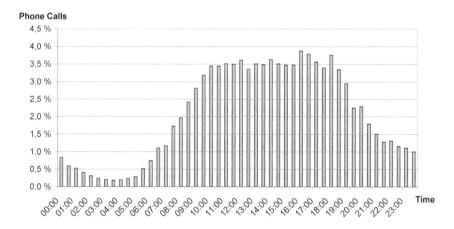

Figure 1  Fluctuations over the course of the day in the volume of calls for technical services

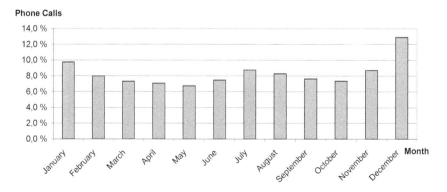

Figure 2  Fluctuations over the course of the year in the volume of calls for technical services

There is also a characteristic pattern in the volume of calls over the course of the year, as shown in Figure 2. At the end of 2010, the total volume of calls for 2011 was estimated at around 1.6 million. Experience shows that this estimate has to be adjusted as the year progresses, with more calls than planned usually being received.

In the simulation model, the calls' interarrival times were modelled using exponential distributions. To account for fluctuations, the simulation period was divided into time-slots of 30 minutes and the parameters of the exponential distribution for each of those slots were set individually.

### 3.1.3    Modeling staff

According to REFA (1991, p. 224), the theoretical personnel level consists of the available and the non-exploitable personnel capacity. However, this simulation study cannot use this theoretical value as it would require forecasting the absentee rate of employees (e.g. as a result of illness or vacation) resulting in another influential factor that would have to be taken into account. Instead, the attendance rate based on the times when the employees were logged into the call center's telephone system is used as a parameter in the simulation.

Every employee is modeled for this purpose, with data stored on their weekly working hours and their minimum and maximum shift durations. The employees and activities are then linked using the definition of qualifications. Employees specialize in three different types of inquiry. While all employees can field inquiries for technical services, only some can accept reports of lost service or provide status information. These specializations are relevant for working time configuration as a minimum level of staffing must be maintained for all types of inquiry.

There are usually 151 people working at the call center. In 2010, on average 20.0% of the employees were full-time and 7.2% were temporary. The other employees worked part-time for a varying number of hours a week, with 24.2% working 30 hours and 12.6% working 27 hours a week. 29.8% of employees were students who also only work part-time, but on the other hand are highly flexible and are mainly assigned to peak times.

### 3.1.4    Modeling the working time system

The working time system used for obtaining the results in this paper was the initial working time system used by the call center. The employees' working times were exported from the call center's planning tool and imported into the simulation model as is.

## 3.2    Modeling the social roles of employees

The social roles for the entire call center were recorded using an employee survey. This was done in order to guarantee the anonymity of respondents, which could not have been done if only the department being examined here had been surveyed. As all of the call center's departments are comparable in terms of their personnel structure, it is safe to assume that the social roles are similar.

Cluster analysis was used to identify seven different social roles that differ in terms of their working time preferences. It is apparent that the employees are not equally distributed between the defined clusters. Only 15.3% of employees are almost always available to work and would only experience a slight scheduling conflict of they were required to work nights. On the other hand, 45.9% of employees experience constant conflicts with their private lives, although these are described as slight during the day and moderate or severe at other times (for more information please refer to Zülch, Schmidt, Stock in this proceedings).

In order to analyze the effects of different distributions of social roles among the employees, 40 different scenarios were constructed randomly. Each of those scenarios uses the same absolute frequency of social roles corresponding to the frequency observed in the employee survey.

## 3.3 Simulation results

In order to eliminate stochastic effects, the simulation was repeated 8 times using different random number seeds. The simulation week chosen consisted of 5 weeks with a moderate number of calls. Each simulation run took about three hours on a desktop PC.

Table 1 shows selected key figures as well as their coefficients of variation throughout the simulation runs. The coefficients of variation are relatively low which means that the stochastic system behaves relatively stable. As can be seen utilization of the employees is relatively low. On one the hand this can be attributed to the fact that even during times of low call volumes, a reserve capacity of operators has to be present in order to meet certain service levels. On the other hand only direct functions have been integrated into the simulation model.

| Key figure | Mean value | Coefficient of variation |
|---|---|---|
| Number of calls | 171,197 | 0.26 % |
| Service rate | 98 % | 0.00 % |
| Employee utilization | 52 % | 0.34 % |

Table 1  Selected key figures of the simulation model

Figure 3 depicts the mean number of conflicts per employee arising in the simulated period structured by severity of conflict and role distibution scenario. It is obvious that the number of conflicts varies depending on which employee in the working system belongs to which of the roles. Still the variation coefficients throughout the role distribution scenarios are low at 5.8 % for severe conflicts and 4.6 % for moderate and minor conflicts. Thus it may be feasible to use the mean number of conflicts of the 40 role distribution scenarios for describing the potential for conflicts of the working time system as a whole.

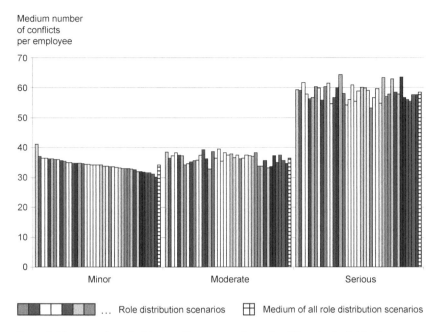

Figure 3  Mean number of arising conflicts per employee for different role distributions

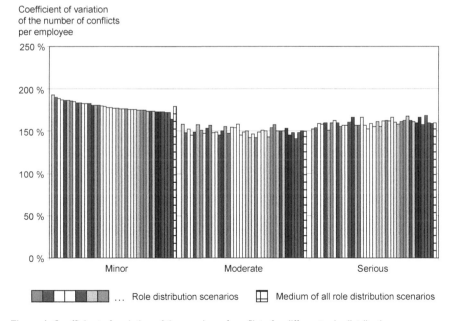

Figure 4  Coefficient of variation of the number of conflicts for different role distributions

As can be seen in Figure 4, the coefficient of variation for the number of conflicts among the employees is rather high at over 100 %. This means that some employees experience conflicts by far more often than others. This as well could be used as a key figure for measuring the impact of different working time systems and describing what could be called "fairness of conflict distribution" of a working time system.

## 4    OUTLOOK

In order to improve the simulation tool and the methodology used for assessing working time systems, subsequent simulation studies have to be performed. These simulation studies should go one step further and should not only be concerned with evaluating an existing working time system but should help to evaluate different working time systems taking into account operational and personnel-related key figures. A systematic study could then reveal if and how the two new key figures drafted above could be used to rate the potential of conflict and the fairness of conflict distribution for a specific working time system regarding the social roles of the employees.

It would also be interesting to find out how different working time systems perform under different call volume scenarios. Those can be easily constructed by adapting the parameters of the exponential distributions underlying the calls' interarrival times.

The presentation held at the conference will feature additional current results beyond what has been presented in this paper.

## ACKNOWLEDGMENTS

The research work is being carried out as part of the research project entitled "Working time configuration considering the work-life balance with the help of computer-based simulation" (Arbeitszeitgestaltung unter Berücksichtigung der Work-Life-Balance mit Hilfe der rechnerunterstützten Simulation – ARBWOL). This project is supported jointly by the German Federal Ministry of Education and Research and the European Social Fund (Support Code 01FH09046).

## REFERENCES

BMFSFJ – Bundesministerium für Familie, Senioren, Frauen und Jugend (ed.) "Charta für familienbewusste Arbeitszeiten." Berlin, 8. February 2011. Accessed February 15, 2012, http://www.erfolgsfaktor-familie.de/data/downloads/webseiten/Charta_%20 Initiative_Familienbewusste_Arbeitszeiten.pdf.

Bogus, T. 2002. *Simulationsbasierte Gestaltung von Arbeitszeitmodellen in Dienstleistungsbetrieben mit kundenfrequenzabhängigem Arbeitszeitbedarf.* Aachen: Shaker Verlag. (ifab-Forschungsberichte aus dem Institut für Arbeitswissenschaft und Betriebsorganisation der Universität Karlsruhe, Band 31)

Frone, M. R., M. Russel, and M. L. Cooper. 1998. Relation of work-family conflict to health outcomes: A four-year longitudinal study of employed parents. *Journal of Occupational and Organizational Psychology* 70(4): 325–335.

Greenhaus, J. H., K. M. Collins, and J. Shaw. 2003. The relation between work-family balance and quality of life. *Journal of Vocational Behavior* 63(3), 510–531.

Hornberger, S. and P. Knauth. 2000. Innovative Flexibilisierung der Arbeitszeit. In. *Innovatives Arbeitszeitmanagement*, eds. P. Knauth and G. Zülch. Aachen: Shaker Verlag. (ifab-Forschungsberichte aus dem Institut für Arbeitswissenschaft und Betriebsorganisation der Universität Karlsruhe, Band 22)

Stock, P., T. Bogus, and S. Stowasser. 2004. *Auswirkungen flexibler Arbeitszeitmodelle auf den Personaleinsatz und die Belastung des Personals*. Aachen: Shaker Verlag. (ifab-Forschungsberichte aus dem Institut für Arbeitswissenschaft und Betriebsorganisation der Universität Karlsruhe, Band 32)

Zülch, G., J. Fischer, and U. Jonsson. 2000. An integrated object model for activity network based simulation. In. *WSC'00 Proceedings of the 2000 Winter Simulation Conference*, eds. J. A. Joines, R. R. Barton, K. Kang et al. Compact disk of the WSC'00, pp. 371–380. Accessed February 15, 2012, http://informs-sim.org/wsc00papers/053.PDF.

Zülch, G., D. Schmidt and P. Stock. 2012. "Influence of the Social Role of an Employee on Working Time Configuration". Paper presented at 4th AHFE International Conference, San Francisco, CA, 2012.

Zülch, G. and P. Stock. 2011. Dynamic Assessment Of Working Time Models. In. *Innovation in Product and Production*, eds D. Spath, R. Ilg, and T. Krause. Conference Proceedings of the 21st International Conference on Production Research ICPR 21. (CD-ROM, 157_Zuelch.pdf)

Zülch, G., P. Stock, and J. Hrdina. 2008. Working Time Configuration in Hospitals Using Personnel-oriented Simulation. In.*Lean Business Systems and Beyond*, ed. T. Koch. Boston: Springer-Verlag, pp. 493–501.

Zülch, G., P. Stock, and M. Leupold. 2011. Simulation-aided design and evaluation of flexible working times. In. Proceedings of the 2011 Winter Simulation Conference, eds. S. Jain, R. R. Creasey, J. Himmelspach et al. Compact disk of the WSC'11, pp. 2159-2170. Accessed February 15, 2012, http://www.informs-sim.org/wsc11papers/194.pdf.

# CHAPTER 43

# Influence of the Social Role of an Employee on Working Time Configuration

*Gert Zülch, Daniel Schmidt, Patricia Stock*

Karlsruhe Institute of Technology
Karlsruhe, GERMANY
gert.zuelch@kit.edu

## ABSTRACT

The issue of flexible working hours has become increasingly important in recent years. Flexible working hours may have various advantages both for the company and for its employees but they can also lead to conflicts between work and personal needs of employees. These arising conflicts usually affect family, hobbies or honorary posts held by the employee. They can also lead to individual negative consequences, e.g. to psychological stress, or low satisfaction with work or even private life. The aim of the German ARBWOL project is to analyze the effects of flexible working hours on the work-life balance of employees. For this purpose, social roles reflecting the employee's position in his or her private environment and the resulting obligations and expectations have to be defined.

In order to identify these social roles, a questionnaire was created in which the employees made time-based statements and gave preferences about their time use in their private lives. From this information, social roles of employees were defined using cluster analysis. The paper presents first results of this analysis taking employees at a call center as an example. The results serve the comprehensive study of the relationship between the social roles of employees, the stress from their work and private lives and the resulting potential for conflicts. From these results, recommendations for the design of stress-reducing flexible working hours shall be derived.

**Keywords**: working time configuration, work-life-balance, stress, call center

# 1    INTRODUCTION

The rapid development of communications and information technologies, rising demand for service (especially in after-sales) and the changing demands of consumers are giving rise to new work tasks and new organizational forms. Many aspects of customer service are handled by call centers. Many new jobs have been created at call centers in recent years. Be they in-house (company-operated call center) or legally distinct service providers handling outsourced orders on behalf of companies, these new forms of employment present new challenges for operators, employees and the law on occupational health and safety. One important aspect is configuring working time to take work-life balance into account. The ARBWOL project (German acronym for "Working time configuration whilst considering the work-life balance") was launched to address this subject. The leading research hypothesis states that different employees have differing social roles which are distinguished from each other based on work load they have to deal with in their private life. Appropriate working time models may compensate for this strain in line with the concept of work-life balance.

# 2    DETERMINING THE STRESS ON EMPLOYEES AT A CALL CENTER

## 2.1    Conducting the employee survey

As part of the ARBWOL project, an employee survey was conducted at a service department in order to identify the social roles that occur and investigate the relationship between work-related and private levels of stress. The aim of this survey was therefore to record the relevant demographic and behavioral characteristics required to assign a specific social role to a person, as well as to gain a subjective impression of the work-related and private stress situations of employees, particularly as a result of conflicts between roles.

As no instrument existed that was suitable for collecting all of the information required, a new questionnaire had to be developed. The questionnaire is based on the existing SALSA questionnaire (German acronym for "Salutogenetic Subjective Work Analysis"; Rimann, Udris, 1999). The primary focus of SALSA is the work sphere; the instrument dedicates only six questions to the private life. Therefore, the questionnaire was expanded by adding questions concerning the interplay between professional and private life following Carlson et al. (2000). Furthermore, questions concerning the practiced working time models as well as the employees' preferences for working time and their activities in their private lives were added. The expanded questionnaire comprised 295 items.

Two hospitals, a call center and two retail chains agreed to participate in the employee survey, and about 2700 questionnaires were distributed in total. The response rate was 41.9%, which can be taken as an indication of the general significance of the issue. The industries involved exhibit particularly high demand for flexible working hours, and are characterized by their atypical working time patterns.

## 2.2 Stress situation for employees at a call center

The call center acts as a service provider for a number of industries, i.e. technical services, banking or taking orders for retail and tourism. A high level of flexibility in terms of time and subject matter is therefore expected of the employees. A highly-flexible shift system is operated, and a high number of part-time employees are used to cover peaks in demand. Just under 500 questionnaires were distributed, of which 137 were returned (response rate of 28%). 79% of respondents were women. The average age was 33.9 years.

The perceived requirements in terms of the task identities, required qualifications and degree of responsibility tend to be below the standard SALSA values for the service industry. A qualitative underload and overload as well as a quantitative overload are perceived in the medium level, the values are however significantly higher than the standard SALSA values for the service industry. The stresses from work that are typical for call centers (noise, remaining seated for extended periods of time, working on a computer screen and the air conditioning) are very strongly represented. The management of the call center is already attempting to remedy the situation in this regard using appropriate measures such as baffle boards and humidifiers.

The demands and stresses are offset by resources aimed at promoting or restoring health. A high level of representation is therefore desired in this regard. In light of this, the fact that almost all indicators for organizational and social resources fall substantially short of the standard SALSA values for the service industry is considered negative. Only the perceived social atmosphere and social support from colleagues are at a similar level to the standard SALSA values or higher. There is therefore potential for improvement in this regard.

Carlson et al. (2000, p. 251) distinguish between two general directions of conflict ("work interference with family" and "family interference with work") and between three forms of conflict (time based, stress based and behavior based). This results in six different key figures. Generally, the key indicators in the survey results are at a low level. Interference of work with private life is, however, significantly more pronounced than the interference of private life with work.

## 2.3 Satisfaction with working time configuration

44.5% of respondents work full-time, 42.3% part-time and 12.4% in minimal employment, with the part-time and minimally employed workers working an average of 24.2 hours a week. When asked about their working hours, 17.2% claimed to work fixed hours. 7.0% work in a flexitime arrangement, and 36.7% in shifts. The other respondents have flexible working arrangements with working hours that cannot be planned. However, the concrete working time configuration also depends on the manager or the working atmosphere in the relevant department.

In this context, 65.5% of employees stated that their working time preferences are often or almost always taken into account. 58.1% of respondents are often or almost always required to work on a Saturday (47.1% on a Sunday). One positive finding is that only 14.7% report frequent, short-term changes to their working hours. However, such changes are perceived as highly disruptive by 44.5% of employees.

Figure 1 shows employees' level of satisfaction with their own working time model and working hours. While the various working time models used only tend to have an impact on employees' satisfaction with their working time configurations, there are substantial differences with regard to satisfaction with the timing of employees' own working hours in relation to their partners'. As is to be expected, satisfaction is highest among those who work the most concurrently with their partner.

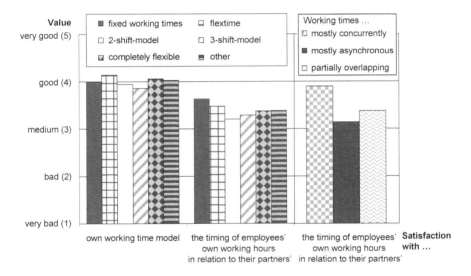

Figure 1 Satisfaction with working time configuration

# 3    CLUSTER ANALYSIS

The data for the aforementioned call center was used in the cluster analysis. In order to collect information on the use of private time and potential conflicts between private and professional lives at the call center using the aforementioned questionnaire, respondents were presented with a series of items in which they were asked to provide information on how they use their private time. The data was also broken down according to daily and weekly activities. Respondents were also asked to give their time preferences outside of work.

For example, they were asked how many hours a day they spend time on the following activities: partnership, childcare, looking after a dependent, taking care of animals, traveling to work, consuming media and sleeping. They were also asked how many hours a week they spend time on the following activities: secondary employment, honorary positions, housework, home improvement, meeting friends or keeping in touch, adult education, sports/fitness, hobbies, wellness and religious/spiritual interests.

Additionally, respondents were able to enter their conflicts into a weekly schedule broken down by day and individual hours showing when they had to work during a particular period of time. The week was subdivided into 168 units of time, which corresponds to the number of hours per week. It was possible to assign potential conflicts three levels of priority. The following priorities were used: 1 – slight conflict; 2 – moderate conflict; 3 – severe conflict. This relatively detailed record of potential conflicts was required to make quantified statements regarding conflicts between the private and professional lives of individual employees.

The k-means method (Bacher et al. 2010, pp. 299; Everitt et al. 2001) was used to classify conflicts between work and private lives. This relatively common deterministic cluster analysis method was chosen over other methods. One reason for choosing this method was that it can cluster large number of cases. The k-means method calculates the cluster centers using a three-stage iterative algorithm: (1) Based on a predetermined initial partition for the starting values of the cluster centers, the observations are assigned to cluster centers in a way that (2) minimizes the total cluster variance sum of squares, on the basis of which (3) the cluster centers are recalculated. This iteration ends when modifying the cluster centers no longer changes the assignment of the observations (Bacher et al. 2010, p. 300). However, there are three preconditions for using the k-means method:

- Firstly, the variables used for classification must have scales that are at the same level and the same length (see Bacher et al. 2010, pp. 175). In order to harmonize the scales they must be standardized using a z-transformation. This raises the question of which measure of similarity or distance to use for classification. These investigations used the square of the Euclidean distance as a measure of (dis)similarity.

- Secondly, the number of clusters must be known in advance. Test statistics were used to find the ideal number of clusters as described by Bacher et al. (2010, pp. 305) and Schendera (2010, pp. 129). The criterion of "explained variance $(\eta^2)$", the criterion of the "relative improvement of the explanation of the variance (PRE coefficients)" and the criterion of the "best variance ratio (Fmax statistic)" were applied.

- Thirdly, the $k$ observations used in the initial partitions of the cluster centers must be specified. The initial value process implemented in this software was applied using *SPSS*. While relatively simple, this algorithm depends on the sequence in which the observations were entered (Bacher et al. 2010, p. 336). The process was therefore run several times using different sequences for the observations.

## 3.1 Results of the cluster analysis

The cluster analysis presented in this paper is based on a call center. As already discussed in chapter 2.2, the questionnaire was returned by 137 respondents. The figures for returned questionnaires were further revised downwards after checking the questionnaires for completeness. Of the total of 137 questionnaires returned, 111 unique cases could be used for the cluster analysis. The lower number of cases for the cluster analysis was mainly due to the exclusion of questionnaires that had not been completed in full. The weekly schedule in which respondents were asked to enter any conflicts was particularly time-consuming as they had to carefully consider where conflicts arise between their private lives and their work, and how serious they judge these conflicts to be.

Respondents at the call center were guaranteed anonymity. This means that no analyses will be carried out that could allow responses to be traced back to individual employees. For the cluster analysis, this means that clusters with a frequency of less than 10 are not described in order to protect the anonymity of the individual respondents.

The results of the cluster analysis are described in detail below. When classifying respondents, a solution with seven clusters is preferred based on the aforementioned criteria for determining the number of clusters. First of all, it is apparent that the employees are not equally distributed between the defined clusters. Cluster 6 is by far the largest cluster. The seven types of employee working at the call center can be described as follows in terms of the conflicts between their private and working lives (see also Figure 2):

- Cluster 1 (n=17): This group of employees have almost no conflicts at any point during the week, including the weekend, and if they do occur they are described by respondents as slight. These slight conflicts mainly arise from Monday to Friday in the early evening between 5.00 p.m. and 8.00 p.m. On Saturday these conflicts arise between 1.00 p.m. and 8.00 p.m.

- Cluster 2 (n=11): This employee group does not experience any conflicts from Monday to Friday between the hours of 6.00 a.m. and 4.00 p.m. However, there are slight conflicts between their working and private lives between 4.00 p.m. and 6.00 a.m. At the weekend, moderately severe conflicts prevail on Saturdays before 12.00 noon. Otherwise, conflicts at the weekend are severe.

- Cluster 3 (n=16): This category of employee did not indicate any conflicts between their working and private lives on any day of the week, including the weekend, between 8.00 a.m. and 10.00 p.m. After that time there were slight conflicts between 10.00 p.m. and midnight, and moderately severe conflicts between midnight and 7.00 a.m.

- Cluster 4 (n=14): This category of employee did not indicate any conflicts on any day of the week between 7.00 a.m. and 7.00 p.m., with slight conflicts between 7.00 p.m. and 9.00 p.m. as well as between 6.00 a.m. and 7.00 a.m., and moderately severe conflicts until 10.00 p.m. and between 5.00 a.m. and 6.00 a.m. If these respondents were required to work between 10.00 p.m. and 5.00 a.m., this would constitute a severe conflict.

- Cluster 5 (n=20): The employee category always indicates conflicts between their working and private lives. Conflicts are seen as slight from Monday to Friday, between 8.00 a.m. and 9.00 p.m. Between 6.00 a.m. and 8.00 a.m., and between 9.00 p.m. and 10.00 p.m., conflicts are classified as moderately severe. Conflicts are described as severe from 10.00 p.m. to 6.00 a.m. The classification of conflicts on Saturdays stays the same, with the exception of moderately severe conflicts from 6.00 p.m. Slight conflicts are experienced on Sundays between 11.00 a.m. and 1.00 p.m. Conflicts are either moderately severe or severe for the rest of the day.

- Cluster 6 (n=31): This group (also the largest) described constant conflicts between their private and working lives, as cluster 5 did. Conflicts are seen as slight from Monday to Friday, between 7.00 a.m. and 4.00 p.m. Between 6.00 a.m. and 7.00 a.m., and between 4.00 p.m. and 7.00 p.m., conflicts are classified as moderately severe. Conflicts are described as severe from 7.00 p.m. to 6.00 a.m. Conflicts on Saturdays between 6.00 a.m. and 6.00 p.m. and on Sunday between 12 noon and 4.00 p.m. are seen as moderately severe, and severe for the rest of the weekend.

- Cluster 7 (n=2): This cluster, already described as an atypical group of people before the detailed description, indicates severe conflicts between their work and private lives between 6.00 a.m. and 10.00 a.m. The group also indicates moderately severe conflicts between 4.00 a.m. and 6.00 a.m., as well as between 10.00 a.m. and 3.00 p.m. This is why this group is described as atypical in terms of the conflicts it describes when compared with the other clusters.

These clusters can be described in even more detail based on the information provided by respondents on how they use their time and how they prioritize their spare time activities. This kind of detailed description is given below, omitting cluster 7 in order to preserve respondents' privacy.

The description is based on the relative importance of various leisure activities. Respondents were asked to prioritize their private activities by level of importance. The biggest priority for all respondents was their partner, followed by sleep. These were followed by contact with friends and housework. Hobbies came in fifth place, then education, media consumption, childcare and sports. Wellness placed above looking after a dependent. These were followed by looking after animals and home improvement. At the bottom of the list came secondary employment, then honorary positions and finally religious/spiritual interests.

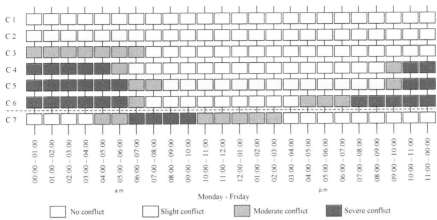

Figure 2 Conflicts between the work and private lives of different employee types at a call center

The following categories of employee are described in more detail based on these results and the data on the use of time in private lives. The following activities of private life, listed in the order of their importance, characterize the seven clusters at the company:

- Cluster 1 is characterized by the high priority it assigns to caring for a family member in comparison to the other clusters. Housework is also seen as very important. Hobbies, on the other hand, are assigned less importance than in the other groups. These priorities are also reflected in the use of time, with this cluster dedicating the most time to housework and secondary employment.

- Cluster 2 assigns less importance to certain private activities. This group considers secondary employment to be of least importance. It takes a similar view of the importance of caring for a family member, honorary positions and home improvement. On average, this group is slightly younger and does a below-average amount of housework. It also devotes the least time to sports on average.

- Cluster 3 attaches above-average priority to honorary positions, education and hobbies. Taking care of an animal, on the other hand, is not seen as very important. In terms of the use of time, this group spends the least time with children if there even are any. On the other hand, it indicates the most amount of time for home improvement compared to the other groups.

- Cluster 4: This category of employee most highly values caring for children and animals in comparison to the other categories. Hobbies, sports/fitness and wellness are also assigned the most importance in comparison to the average. This group assigns the least importance to home improvement. In terms of the use of time, it can also be said that this group dedicates the least amount of time to relationships compared to the other groups.

- Cluster 5 is characterized by the low priority it assigns to childcare, housework and wellness. Secondary employment, on the other hand, is seen as more important. An analysis of this group's use of spare time shows that a very high value is given to secondary employment, just like cluster 1.

- Cluster 6 considers childcare to be very important, as cluster 4 does. Education, hobbies and sports are assigned the least importance in comparison to the other employee categories. The percentage of group members with their own children is the highest in this group.

- Cluster 7 is not described in more detail because of the low number of people.

As already mentioned, this cluster analysis used data on time-based conflicts between respondents' private and professional lives. The clusters were described using both the data on the use of time and the employees' biographical data. Each cluster can be clearly and plausibly interpreted without contradiction, provided it does not fall short of the aforementioned anonymity threshold.

The analysis of the call center essentially shows that normal working hours (mornings and afternoons) are preferred by most employees. However, there are groups of people who do not indicate any conflict if they were required to work at night, as well as groups who could work at weekends without suffering any serious conflict with their private lives.

Because of the low number of participants it is not possible to apply these results to any industry or even the entire population of Germany. However, there are first signs that it is possible to break down the heterogeneous multitude of the workforce into a number of more manageable groups with regard to their working time preferences. According to these results, conflicts between working and private lives can in fact serve an investigation into the relationship between the social roles and stresses from people's work and private lives, as well as the resulting potential for conflict.

## 4    CONCLUSIONS

The working hypothesis of this paper is that there are different types of employee with social roles that vary in terms of the stresses resulting from their private lives, and for whom a balance can be struck between their work and private lives using specific working time models. The use of time was recorded by presenting respondents with a series of items in which they were asked to provide information on their use of time outside of work, and any conflicts arising. Respondents were also asked to give their time preferences outside of work.

A cluster analysis using the k-means method was used to define the different types of employee in terms of the scheduling conflicts between their work and private lives. The results showed that it is possible to break down the heterogeneous multitude of the workforce into a number of more manageable groups that can be described in terms of their characteristics. Further investigations will be carried out into conflicts between employees' work and private lives. In addition to the analysis of employee surveys in other service areas, a simulation study is currently in preparation that will look at the effects of various working time models on work-life balance in relation to the social roles determined in the course of the cluster analysis (for details refer to Stock, Leupold and Zülch in the same proceedings).

## ACKNOWLEDGMENTS

The research work is being carried out as part of the research project entitled "Working time configuration considering the work-life balance with the help of computer-based simulation" (Arbeitszeitgestaltung unter Berücksichtigung der Work-Life-Balance mit Hilfe der rechnerunterstützten Simulation – ARBWOL). This project is supported jointly by the German Federal Ministry of Education and Research and the European Social Fund (Support Code 01FH09046).

## REFERENCES

Bacher, J., et al. 2010. *Clusteranalyse*. München: Oldenbourg.

Carlson, D. S., K. M. Kacmar, and L. J. Williams. 2002. Construction and Initial Validation of a Multidimensional Measure of Work-Family Conflict. *Journal of Vocational Behavior* 56(2): 249-276.

Everitt, B. S., S. Landau, and M. Leese. 2001. *Cluster analysis*. London: Edward Arnold.

Rimann, M. and I. Udris. 1999. Fragebogen "Salutogenetische Subjektive Arbeitsanalyse" (SALSA). In. *Handbuch psychologischer Arbeitsanalysen*, ed. H. Dunckel. Zürich: vdf Hochschulverlag an der ETH, pp.404-419.

Schendera, C. 2010. *Clusteranalyse mit SPSS*. München: Oldenbourg.

SPSS, Version 18, IBM Deutschland GmbH, Ehningen.

Zülch, G., P. Stock, and D. Schmidt. 2012. Analysis of the strain on employees in the retail sector considering work-life balance. *Work: A Journal of Prevention, Assessment and Rehabilitation* 41(Supplement 1): 2675-2682. Accessed February 29, 2012. http://iospress.metapress.com/content/w9716708601264kv/fulltext.pdf.

Zülch, G., P. Stock, D. Schmidt, and M. Leupold. 2010. Conflicts between professional and private life in the different life stages. In: *1st European FEES Conference on Ergonomics, 10-12 October 2010, Bruges, Belgium*, ed. Federation of the European Ergonomics Societies (FEES). Assenede: Medicongress, 2010. Accessed December 18, 2010. http://www.ece2010.be/papers/download.php?f=./poster_papers/P14%20-%20 Zuelch.pdf.

Zülch, G., P. Stock, D. Schmidt, and M. Leupold. 2011. Conflicts between Work and Private Life Caused by Working Times. In. *Human Factors in Organisational Design and Management – X, Volume 1*, eds. M. Göbel, et al. Santa Monica, CA: IEA Press, 2011, pp. I-159 – I-164.

# CHAPTER 44

# Agent-based Planning and Simulation-based Assessment to Improve the Work-Life Balance of Hospital Staff

*Thilo GAMBER, Gert ZÜLCH*

Karlsruhe Institute of Technology
Karlsruhe, GERMANY
gert.zuelch@kit.edu

## ABSTRACT

This paper presents an agent-based planning system for use in hospitals aimed at achieving a better work-life balance for employees within the same job category. In doing so, it attempts to better take into account employees' individual preferences and the balancing of working time preferences within a job category by improving the rostering. A dynamic assessment of work processes is carried out using simulations with regard to the operating requirements and the degree to which they are fulfilled by the planned working times of the employees.

**Keywords**: healthcare, nurse rostering, working time preferences

## 1    INDIVIDUAL WORKING TIME PREFERENCES OF HOSPITAL STAFF

Personnel requirements in hospitals are subject to a high level of fluctuation, and cannot be planned using traditional methods on account of the highly stochastic nature of patient arrivals. Furthermore, the work tasks arising in connection with the treatment of the patient and the time required depend on the nature and extent of treatment necessary. In addition, running a hospital generally requires round-the-clock staffing (see Zülch, Stock and Hrdina 2008). It is important to make sure that minimum staffing levels for each job category (physicians, nursing staff) are maintained at all times. This means that staff has to be on duty outside normal working hours, i.e. late and night shifts, for example.

However, only taking operating requirements into account can result in scheduling conflicts between employees' work and private lives. This often causes problems with finding a work-life balance.

The paper presents the results of a research project. In the course of the project, the *ProSis* (*Pr*eference-*o*riented Planning and *Si*mulation of Work *S*chedules) software tool was developed to help resolve this problem. During its development, a number of regulatory mechanisms were investigated to ascertain whether they can support the aforementioned goal of rostering based on individual preferences that also achieves a fair balance of interests. One reason for developing a software-based planning tool is that new regulatory mechanisms for social systems cannot simply be tested in real-life studies. It is therefore necessary to develop suitable methods for this purpose and to evaluate potential regulatory mechanisms in a modeled environment with the aim of achieving predetermined targets.

## 2 AGENT-BASED PLANNING APPROACH TO WORKING HOURS SCHEDULING

As part of the research, an agent-based planning procedure was developed to improve an existing roster using virtual swap negotiations (see Gamber and Börkircher 2008; Gamber, Börkircher and Zülch 2010; Gamber and Zülch 2010; Zülch, Gamber and Freytag 2011). In addition to the individual working hours preferences, this also takes into account the balance of the extent to which working hours preferences are met within a job category. The planning system therefore models an agent or electronic representative for each employee who only pursues the goal of maximizing the benefit to itself, in this case the degree to which its working hours preferences are put into effect.

In order to represent the employee as an agent within this planning procedure, a range of data is required for use in modeling the various attributes of the agent. The relevant working hours from the starting scenario are stored for each agent as a working hours calendar, while the individual working hours preferences are saved in a preference calendar (see Fig. 1).

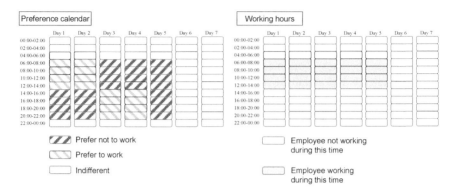

Figure 1: Example of an agent's working hours and preference calendars

The working hours preferences shown in Figure 1 can also be weighted: Preferences with more weighting are given priority in the planning procedure. The employee's qualifications can also be modeled as attributes of the agent. For the sake of simplicity, the following assumes that all employees within a job category (e.g. physicians, nursing staff or more specialized job categories) have the same level of qualification and are therefore interchangeable. The procedure incorporates the modeling and specification of the swap using qualification profiles. However, it has become apparent that modeling this kind of system would be much too time-consuming in connection with this project. Further attributes that are relevant to the performance of the procedure can be assigned to the agents, such as individual working hours regulations, in order to reflect the applicable of employment contracts.

The developed concepts were implemented in the form of prototypes. The *ProSis* planning tool for working hours scheduling was designed for use on a PC, but can essentially be used on any platform. The implementation was tested on a computer with the *Microsoft Windows XP* operating system. The *ProSis* tool was developed in the *Eclipse* development environment using the *JAVA* programming language. The *JADE* (*Java Agent Development* Framework) agent platform was used, which complies with *FIPA* (*Foundation for Intelligent Physical Agents*) standards.

Figure 2 shows the scheme of the planning tool. First the fundamental parameters are set (process step 0) depending on the planning task (e.g. entire shifts or smaller units of time). Then, process step 1 involves preparing templates for preference and working hours patterns. In process step 2, these templates are assigned to individual agents. The templates (working hours and preferences) are then adjusted for the individual agents (see e.g. Figure 1).

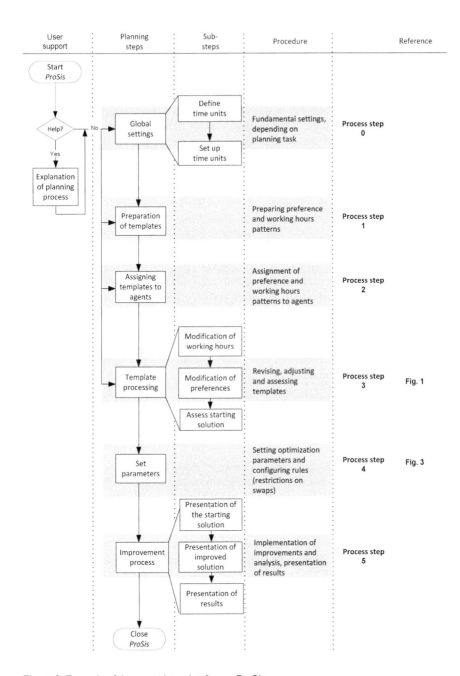

Figure 2: Example of the agent-based software *ProSis*

Depending on the template, the starting scenario can be introduced in process step 2 or 3. Potential starting scenarios can thus either be derived using an existing working hours model or entered separately by the user.

In the process step 4, the optimization parameters (choice of negotiating strategies) are set and the rules for determining improvement (restrictions on swaps) are configured. A range of negotiation strategies have been incorporated into the procedure for this purpose, some of which are familiar from the literature (e.g. Chiaramonte and Chiaramonte 2008, pp. 679) while others were developed especially. The process benefits in this regard from its generic structure, as further negotiation strategies can be added relatively easily. The same applies with regard to the generic nature of the configuration of rules (restrictions on changes). A selection of important ergonomic and labor law regulations have been incorporated into the current version of the procedure. For example, regulations governing daily working hours and average weekly working hours have been stored in the system (pursuant to the German Working Hours Act ArbZG 1994; Figure 3).

This agent-based planning results in a modified individual roster for every employee. Key ratios can be used to assess the improvement between the original working hours schedule and the modified version (e.g. average number of swaps, average increase in benefit, level of satisfaction). The level of satisfaction, for example, takes into account the initial benefit to the agent. In so doing, the agent's individual situation with regard to its preferences is standardized using a uniform scale. Satisfaction level 1 holds when the agent achieves the maximum possible benefit. The fairness of the swaps in particular is assessed. Other key figures are calculated by modeling the work system in question using a simulation tools. This mainly concerns key figures relating to operations, personnel and patients.

## 4 ILLUSTRATIVE APPLICATION OF THE DEVELOPED PLANNING PROCEDURE

The newly developed procedure is illustrated using the rostering of the nursing staff on a hospital ward. For this pilot application a survey was used to identify the working hours preferred by individual hospital employees. Then, the existing working hours schedules were stored in the *ProSis* system, and the employees were modeled together with their preferences. In addition to selecting an improvement strategy (e.g. focus on fulfilling swap requests or fairness), restrictions on swaps are also set at this juncture, as described above. The virtual agents give employees the opportunity to swap working hours. Swapping should not result in any changes to the operating figures as those involved in the swap have the same level of qualification.

A subset of strategies was selected for this pilot application from the range of swap strategies available. These swap strategies were used to generate improved rosters, which were then assessed using the specific key figures for the increase in benefit and fairness in *ProSis*. These improved rosters provide the basis for the simulation studies. The underlying work system is included in the evaluation of the planning outcome. The planning solutions from the planning procedure are transferred to the simulation procedure.

The *ProSis* proceedure is then initiated, resulting in a solution for improving each employee's individual roster. This solution is then evaluated using the key figures implemented in the *ProSis* system.

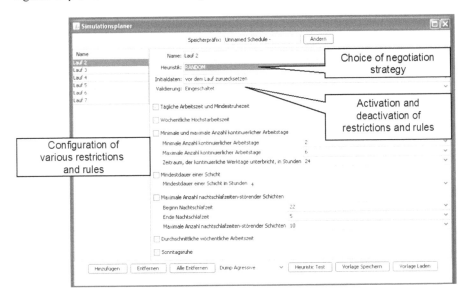

Figure 3: Defining optimization parameters and the configuration of rules in *ProSis*

In the pilot application, the improved planning solution was assessed using the existing simulation model for a hospital system (see Zülch, Stock, Hrdina and Gamber 2008). Figure 4 describes the parameters for the work system being assessed in this case. The work system under investigation is an intensive care unit with 20 beds that treats about 5,500 patients each year. With the exception of a few emergency cases, the admission of new patients follows the hospital's surgery schedule. A total of 52 nurses work on the investigated unit (30 full-time and 22 part-time). The nurses can be assigned universally, and work according to a traditional three-shift system, i.e. four full-time and part-time shift groups during the day and one separate night shift.

The intervals between patient arrivals were modeled stochastically. The treatment processes and associated work tasks for the employees are based on the different patient types (condition types), which are represented in the model using flowcharts similar to network diagrams in the form of clinical pathways (see Küttner and Roeder 2007, pp. 19, for example, for the definition; see Zülch, Stock and Hrdina 2007, 2008b on the method). Using this approach, the patients in the modeled unit were sub-divided into seven different types based on the relevant clinical pathways.

For the simulation studies, the results of the agent-based improvement were manually transferred to the simulation tool *OSim-GAM* (*O*bject *sim*ulator). This simulation tool, developed by Jonsson (2000) at the ifab-Institute of Human and

Industrial Engineering of the Karlsruhe Institute of Technology was expanded by Bogus (2002) for rostering in the services sector. The version used here, *OSim-GAM* (German acronym for*OSim* for the design of working time models), was firstly applied to problems in the hospitals sector by Zülch, Stock and Hrdina (2006a; 2006b; 2007; 2008).

| | Hospital | Restrictions on swaps |
|---|---|---|
| Number of employees | 52, 30 full-time | (stemming from the German Working Hours Act and ergonomic recommendations etc.): |
| Simulated period | 26 weeks | Compliance with maximum weekly working hours, compliance with average working hours, minimum shift duration: 4 hours |
| Shift system | 3 shifts, rotating | |
| Qualification | Generalists | |
| Qualification frequency | 100% | |
| Swap format | Whole shifts | |
| Number of cycle plans | 25 | |

Figure 4: Illustrative application of the developed method to a hospital work system

This pilot application shows that individual working hours preferences can be met without requiring additional human resources capacity. The swaps implemented change the key figures calculated for each individual employee on the basis of simulation results (like time stress, physical stress and utilization) depending on which tasks have to be performed at certain times of day and days of the week. It is therefore possible to contribute to improving individual employees' quality of life by satisfying swap requests without suffering disadvantages in terms of operations (e.g. increased personnel expenses as a result of increased human resources requirements).

The assessment criteria in *OSim-GAM* (particularly personnel capacity utilization, time stress, level of fatigue) were used to assess the planning results in addition to the criteria used in *ProSis* (which evaluate the benefit and fairness of the swaps). The key figures provided by the simulation procedure can be used to evaluate the dynamic impact on personnel and patients. The verification of both procedures was based on the nursing staff as this was also the area for which the most data could be collected. For the purposes of verifying the procedures potential swaps were not prevented by differing levels of qualification due to the assumption that everyone in this job category has the same level of qualification.

The results showed that the level of employee satisfaction was raised by taking their working hours preferences into account (statement based on the key figures provided by *ProSis*). The degree of goal achievement for the criterion "capacity utilization" is robust in the model. A degree of goal achievement can assume a value between 0 % and 100 %, where 100 % represents the ideal value for the specified assessment figure (for details refer to Zülch, Grobel and Jonsson 1995, pp. 315). The values for time stress were either slightly better or slightly worse depending on the negotiation strategy used (statement based on the key figures provided by *OSim-GAM*).

## 5 SUMMARY OF RESULTS AND LOOK AHEAD AT FUTURE RESEARCH

The research described here involved developing an agent-based planning procedure that modifies the rosters of employees within a job category by swapping their shifts or units of time. This swapping is aimed at generating rosters that are better suited to employees' private interests. Data was recorded to conduct a practical test of the procedure. The working hours preferences of hospital employees were determined. The *ProSis* agent-based procedure was then designed and implemented as a prototype.

The *Osim-GAM* simulation tool was used to test the functionality of the approach. The nursing staff work system already modeled in previous research was used (Zülch et al. 2008). Simulation studies verified the efficacy of the procedure.

The benefit of simulation for a comprehensive analysis was also made clear. Alongside the individual agent-based approaches to be found in the literature (e.g. Chiaramonte and Chiaramonte 2008; Wang and Wang 2009) for preparing and modifying rosters, the use of simulation in combination with an agent-based planning procedure is new. Also, thanks to its generic structure, the *ProSis* procedure offers high potential for expansion.

There are several conceivable directions that follow-up research to the work described here could take. For example, there is research potential with regard to the adjustment of time parameters in connection with personnel assignment management, as in real life it is often necessary to react quickly to staffing changes. Furthermore, the research described here has created the conditions for reacting quickly to a shortage of personnel, e.g. in the event that employees fall ill. Pertinent questions could include, for example, how many employees have to be kept available to compensate for a short-term lack of staff (on standby or on call depending on the working hours model used).

Based on the generic implementation in this case, the process has the potential to be applied to other areas and professions which face similar personnel assignment issues with regard to the systems used. For example, the following are potential areas of investigation: assignment planning for out-patient social services, retail shift planning or planning the assignment of air traffic controllers.

In the project documented here, the effects on patients are only considered in the form of operating performance figures. Since it is safe to assume that the performance of personnel varies according to the time of day, patients are at varying levels of risk of being the victims of treatment errors. The increasing frequency of errors on the part of employees with long or excessively long shifts (e.g. Knesebeck et al. 2010) should be mentioned in this regard as a recognized scientific fact. Further research could therefore give rise to a model that assesses the risk to the patient depending on the time of day and the nature and severity of their condition, as well as the shift length and qualification etc. of the treating personnel. This could give rise to an additional decision-making criterion in connection with working hours schedules that is applied in addition to the rules already implemented and could help improve the hospital work system in the long run.

The authors would like to thank the German Research Foundation (Deutsche Forschungsgemeinschaft DFG) for their support of the work discussed in this paper as part of the project "Balancing work and private life by planning socially acceptable working hours, illustrated using the example of hospitals".

## REFERENCES

ArbZG, Arbeitszeitgesetz. Released on June 6, 1994, last amended on July 15, 2009.

Chiaramonte, Michael V. and Chiaramonte, L. M. 2008. An agent-based nurse rostering system under minimal staffing conditions. International Journal of Production Economics 114: 679-713.

Eclipse IDE for Java Developers 2000, 2011. Version Indigo 3.7 (Build: 20110916-0149, Service Release 1). Ottawa: ON Eclipse Foundation Inc. Eclipse contributors and others. Accessed October 10, 2011, http://www.eclipse.org/.

FIPA-Standards 2002. The Foundation for Intelligent Physical Agents. Los Alamitos, CA. IEEE Computer Society. Accessed February 29, 2012, http://www.fipa.org/; http://ieee.org/.

Gamber, T. and M. Börkircher. 2008. Vereinbarkeit von Familien- und Berufsleben bei der Gestaltung flexibler Arbeitszeiten. In: Mittelpunkt Mensch, ed. Deutscher Studienpreis, 227-247. Wiesbaden: VS Verlag für Sozialwissenschaften.

Gamber, T., and G. Zülch. 2010. Agentenbasierte Planung und Simulation der Arbeits-zeitgestaltung in Krankenhäusern. In: Integrationsaspekte der Simulation: Technik, Organisation und Personal, eds. G. Zülch and P. Stock. 373-380. Karlsruhe: KIT Scientific Publishing.

Gamber, T., M. Börkircher, and G. Zülch. 2010. Vereinbarkeit von Berufs- und Privatleben im Krankenhausbereich durch Gestaltung sozialverträglicher Arbeitszeiten. In: Arbeits- und Lebenswelten gestalten, ed. Gesellschaft für Arbeitswissenschaft. 409-412. Dortmund: GfA-Press.

Jade 2010 (Java Agent DEvelopment Plattform). Version 4.0 (20.04.2010). Rom: Telecom Italia. Accessed June 30, 2010, http://jade.tilab.com/.

Java 2010. JDK Version 1.6.0_17, 2009; JRI Verion 6, 2010. Santa Clara CA: Sun Microsystems/Oracle, 2010. Accessed June 30, 2010, http://www.java.com/de/.

Jonsson, U. 2000. Ein integriertes Objektmodell zu durchlaufplanorientierten Simulation von Produktionssystemen. Aachen: Shaker Verlag.

Knesebeck, O., J. Klein, K. Grosse Frie, K. Blum and J. Siegrist. 2000. Psychosoziale Arbeitsbelastungen bei chirurgisch tätigen Krankenhausärzten. Deutsches Ärzteblatt, 107: 248-253.

Microsoft Windows XP 2007. Version 5.1 (Build 2600.xpsp_sp3_gdr.101209-1647: Service Pack 3). Redmond WA, Microsoft.

OSim-GAM 2010. Objektsimulator für die Gestaltung vor Arbeitszeitmodellen, Version 3.0. Karlsruhe: Institut für Arbeitswissenschaft und Betriebsorganisation, Karlsruher Institut für Technologie.

ProSis 2012. Präferenzorientierte Planung und Simulation von Arbeitsschichten. Karlsruhe: Institut für Arbeitswissenschaft und Betriebsorganisation, Karlsruher Institut für Technologie.

Zülch, G., T. Gamber, and R. Freytag. 2011. Improving the work life balance of hospital staff through an agent-based rostering procedure. In: Healthcare Systems Ergonomics and Patient Safety 2011. An Alliance between Professionals und Citizens for Patient Safety and Quality of Life, eds. S. Albolino, S. Bagnara, T. Bellandi, Javier Llaneza, G. Rosal and R. Tartaglia, 291-295. London: CRC Press, Taylor & Francis Group.

Zülch, G., T. Grobel, and U. Jonsson. 1995. Indicators for the Evaluation of Organizational Performance. In: Benchmarking – Theory and Practice, ed. Asbjørn Rolstadås, 311-321, London et al.: Chapmann & Hall.

Zülch, G., P. Stock, and J. Hrdina. 2006. Working Time Configuration in Hospitals Using Personnel-oriented Simulation. In: Lean Business Systems and Beyond, ed. T. Koch, 331-336, Wroclaw: University of Technology. (=2006a)

Zülch, G., P. Stock, and J. Hrdina. 2006. Simulationsbasierte Gestaltung flexibler Arbeitszeiten im Krankenhaus. In: Simulation in Produktion und Logistik 2006, ed. S. Wenzel, 183-192, Erlangen: SCS Publishing House. (=2006b)

Zülch, G., P. Stock, and J. Hrdina. 2007. Process Optimization and Efficient Personnel Employment in Hospitals. In: Operations Research Proceedings 2006, eds. K.-H. Waldmann and U.M. Stocker, 325-332. Berlin, Heidelberg: Springer.

Zülch, G., P. Stock, and J. Hrdina. 2008. Dynamic Analysis and Reorganisation Measures in Hospitals Using the Clinical Pathway Approach. In: Creating and designing the healthcare experience. International Conference, Healthcare Systems Ergonomics and Patient Safety 2008, HEPS, June 25-27, 2008, Strasbourg, Convention and Conference Centre.

Zülch, G., P. Stock, J. Hrdina, and T. Gamber. 2008. Arbeitszeitgestaltung mit Hilfe der Simulation. In: Arbeitsgestaltung für KMU, ed. Rektor der Technischen Universität Ilmenau, Peter Scharff. 51–58. Ilmenau: Verlag ISLE.

# Section VII

*Adapting for Special Groups*

# CHAPTER 45

# Optimizing Job Design for Older Adult Workers

*Martha Sanders*

Quinnipiac University
Hamden, CT, USA
Martha.sanders@quinnipiac.edu

## ABSTRACT

Older workers are the fastest growing sector of the labor force. Over 65% of adults aged 55 and over are working or actively seeking employment. This generation of older adults is seeking employment that provides both positive social experiences and financial gains. However, few studies have identified the unique psychosocial needs of older workers or the ways in which businesses can optimize work experiences to promote successful aging needs. The purpose of this study was to examine how job characteristics (decision-making, skill variety, coworker support, supervisor support) impact successful aging outcomes (social network, emotional support, personal sense of control, generativity) in older sales associates in the homebuilding industry. The broad research question was: How does job design influence successful aging in older workers in the home building industry? Results of four standardized tests including the Job Content Questionnaire, the Mirowsky-Ross Personal Sense of Control Scale, the MacArthur study Social Network and Emotional Support scales, and the Loyola Generativity Scale indicated that a healthy job design can also promote successful aging. Retail homebuilding jobs provided opportunities for older workers to use a variety of skills, make independent decisions, and receive coworker / supervisor support. Workplace job design variables contributed to 23% of the variance in generativity and 15.5% of the variance in personal sense of control, two key measures of successful aging. Recommendations are offered on ways managers can design jobs to reciprocally meet the needs of business and older workers.

**Keywords**: Older workers, job design, successful aging

# 1 INTRODUCTION

## 1.2 Background on Older Workers

The number of older adults in the United States will reach 70 million by the year 2030 and will comprise greater than 20% of the population (Bureau of Labor Statistics [BLS], 2010). Accordingly, one in five older adults are remaining or returning to the labor force to reap the financial and social rewards of paid employment (American Association of Retired Persons [AARP], 2003). Among adults of traditional retirement age, 28% and 38% of women and men, respectively, are still employed (BLS). While businesses recognize the value of older workers in terms of productivity, work ethic, and corporate knowledge, few studies have identified management approaches or job design strategies that best utilize older workers' generational strengths and promote their psychosocial needs related to aging successfully. If businesses value the continued labor force participation of older workers and if older adults seek positive work experiences that fulfill their psychosocial needs, then organizations need to understand how workplaces can be designed to optimize successful aging for older workers.

Ilmarinen (2006) coined the term "age management" to address management concerns and accommodations for aging workers in contemporary workplaces. Ilmarinen recommended modifications for physical tasks and opportunities for continued leadership and professional growth through work. Karasek and Theorell (1990) demonstrated that psychosocial characteristics of the job influence worker health; such characteristics include opportunities to use a variety of skills, make decisions, and experience coworker and supervisor support. For the older worker, health may also include indices of successful aging. The MacArthur studies equated successful aging with good physical and cognitive health in addition to meaningful involvement in life, such as paid employment (Erikson, 1997; Rowe & Kahn, 1998; Seeman, Lusignolo, Albert, & Berkman, 2001).

Previous studies have found that a healthy job design can impact older workers' experiences. Wahlstedt, Nygard, Kemmlert, Torgen, and Bjorksten (2000) found that jobs organized in a work team format increased older workers' perceptions of support from coworkers more so than jobs organized in a traditional hierarchical structure. Ross and Wright (1998) found that jobs providing opportunities for autonomy, decision-making, and task variety were associated with high levels of perceived personal control in workers of all ages. These studies suggest that the design of the job may influence the psychosocial health of older workers. However, the extent to which the job design characteristics also contribute to successful aging warrants further investigation.

This study examined the extent to which workplace job design characteristics influence successful aging in older sales associates in the home building industry. Job design characteristics were defined as skill variety, decision authority, coworker support, and supervisor support. Successful aging was defined according to the gerontology literature as perceived control over life events, passing along knowledge to younger generations, having a social network, and feeling emotional support from friends (Erikson, 1997; McAdams & de St. Aubin, 1992; Rowe & Kahn, 1998; Seeman et al., 2001).

## 1.2 Research Purpose

This study answered the following broad research question: How does psychosocial job design (opportunities for skill variety, decision-making, coworker support, and supervisor support) influence successful aging (personal control, generativity, social network, and emotional support) in older sales associates in the retail home building industry. Specific questions included the following:

1. What are the mean levels of job design characteristics and successful aging variables for this study sample as compared to industry norms?

2. What is the relationship between job design variables and successful aging variables?

3. How much variance does each job design variable contribute to successful aging outcomes?

## 2 Methods

### 2.1 Participants

The population for the study was workers aged 55 and older employed full or part time as sales associates in the home building industry. The sample was a convenience sample of older workers employed in nationwide, independently-owned, and local home building stores in Southern New England.

### 2.2 Instruments

Job design characteristics were measured using the job control and job support scales of the Job Content Questionnaire (Karsek, 1986). The four variables were skill discretion (the ability to use a variety of skills on the job), decision authority (the ability to make decisions), supervisor support (the perception of support from supervisors), and coworker support (perception of support from coworkers). Successful aging variables were measured using the MacArthur study Social Network (for social network) and Emotional Support scales (emotional support), the Mirowsky-Ross Personal Sense of Control scale (personal control), and the Loyola Generativity Scale (generativity). Each test was comprised of ordinal level, Likert-like questions that created an overall interval-level scale and raw score (Karasek, 1986; McAdams & de St. Aubin, 1992; Mirowsky & Ross, 1991; Seeman et al., 2001).

### 2.3 Research Design and Data Collection

A cross-sectional survey design was used to study older workers' perceptions of their job characteristics and their perceptions of successful aging.

Participants were recruited from over 75 homebuilding stores in Southern New England. In national chains, workers were approached individually and asked to participate (after permission granted from regional directors); in independently-owned stores, managers were approached prior to asking individual participants. Participants were introduced to the study, given an informed consent, a small coffee token as appreciation, and were asked to complete the surveys within one week. Participants completed four standardized assessments that took about 10 to 15 minutes to complete. The researcher returned to the store to retrieve the completed surveys at an agreed upon time.

## 2.4    Data Analysis

Data were analyzed using SPSS Version 17.0 for descriptive statistics and chi-square analysis for demographic information and measures of central tendency for each variable. Pearson product-moment correlations identified relationships among variables. Standard multiple regression equations determined the contribution of job design variables to successful aging outcomes.

## 3    Results

## 3.1    Participant Demographics

A total of 142 older workers were invited to participate in the study. One hundred and fifteen (115) older workers completed the surveys for a response rate of 81%. The final sample was 109 once surveys were examined for inclusion criteria and completeness.

*Personal background.* The *ages* of the participants ranged from 55 to 81 years old with a mean age of 64.03 years (SD = 6.26). The *gender distribution* of the sample consisted of 82.6% men and 17.4% women. The *marital status* of the majority of the sample was married (74.3%) or divorced (11.9%). Older male workers tended to be married and living with spouses, while older female workers tended to be non-married (single, divorced, or widowed) and living alone or with a family member. The greater majority of respondents *perceived their health* to be average, above average, or excellent (93.5%). The educational attainment for this sample was higher than the national average for completion of high school (99.1%), *some* college education (62.4%), and bachelor degree (23.9%).

*Work-related context.* Older workers in this sample had been working in the labor force a mean of 44.82 years (range 15 - 63 years; SD = 9.22). Older workers had worked in their present job from 2 months to 60 years with a mean of 9.54 years (SD = 10.40 years). When a comparison was made between older workers' current and past job experiences, it was found that over two-thirds of the sample (71%) had worked in previous jobs that were related to their current job position (including trade, organizational, or customer service experiences). Although individuals in the sample may have been healthier and better educated than national standards, they were representative of older sales associate positions in the region studied.

## 3.2 Job Design Characteristics and Successful Aging Variables

*Job Design Characteristics*. Scores on Skill Discretion and Decision Authority were significantly higher than normative values for similar occupational categories indicating that older workers perceived opportunities to use skills, learn new information, be creative, and make independent decisions in this job position. Scores for Supervisor Support and Coworker Support were comparable to occupational norms suggesting that workers perceived close relationships with supervisors and respect from coworkers. Older workers overwhelmingly felt they passed along knowledge to customers (99%) and coworkers (97%) while working. Overall, they were positive about psychosocial aspects of the job (see Table 1); the job was considered "healthy" job for this sample.

*Successful Aging Characteristics*. Scores on the Social Network scale and Emotional Support scale were both higher than age-related norms. The Personal Sense of Control Scores were also higher than community-based studies of older adults. Generativity scores were reported as similar to other age-related samples. Overall, successful aging outcomes indicated that this sample of older workers was a high functioning group with a social network size, emotional support, and a personal sense of control that was comparable to highly functioning older adults in the MacArthur studies (Seeman et al., 2001).

**Table 1**
**Descriptive Statistics for Variables under Study**

| Variable | Range | Scale Mean | SD | Mean 4-pt scale |
|---|---|---|---|---|
| Skill Discretion | 26.00- 44.00 | 34.89 | 4.35 | 2.91 |
| Decision Authority | 24.00- 48.00 | 34.53 | 6.68 | 2.87 |
| Coworker Support | 8.00 - 16.00 | 12.37 | 1.55 | 3.08 |
| Supervisor Support | 4.00 - 16.00 | 12.35 | 2.18 | 3.09 |
| Social Network | 1.00 - 54.00 | 14.92 | 8.67 | - |
| Emotional Support | 1.50 - 4.00 | 3.59 | .51 | 3.60 |
| Personal Control | 18.00- 32.00 | 24.74 | 2.89 | 3.09 |
| Loyola Generativity | 8.00 - 24.00 | 18.48 | 3.55 | 3.07 |

*Relationships between Job Design and Successful Aging*. Job design characteristics were related to the successful aging outcomes of Emotional Support,

Personal Sense of Control, and Generativity. A small, univariate relationship existed between Emotional Support and Decision Authority (r = .200, p<.05). Personal Sense of Control was related to Coworker Support (r = .339, p =.000) and Skill Discretion (*r* = .257, *p*=.007). Generativity was related to Skill Discretion (*r* = .438, *p* =.000), Decision Authority   (*r* =.276, *p* =.004), and Coworker Support (*r* =.219, *p* =.022). There were no relationships between Social Network and job design characteristics. (See Table 2.)

**Table 2**
**Correlation Matrix for Job Design and Successful Aging Variables**

| Variable | 1 | 2 | 3 | 4 | 5 | 6 | 7 | 8 |
|---|---|---|---|---|---|---|---|---|
| 1. Skill Discretion | 1 | .57** | .22* | .26** | -.13 | .19 | .26** | .44** |
| 2. Decision Authority | | 1 | .26** | .33** | -.08 | .20* | .13 | .28** |
| 3. Coworker Support | | | 1 | .41** | .07 | .19 | .34** | .22* |
| 4. Supervisor Support | | | | 1 | -.09 | .02 | .15 | .03 |
| 5. Social Network | | | | | 1 | .27** | -.08 | .14 |
| 6. Emotional Support | | | | | | 1 | .23* | .35** |
| 7. Personal Sense Of Control | | | | | | | 1 | .32** |
| 8. Generativity | | | | | | | | 1 |

*Note.* ** Correlation is significant at the 0.01 level (2-tailed).
* Correlation is significant at the 0.05 level (2-tailed).

## 3.3    Job Design Contributions to Successful Aging Outcomes

When job design variables were regressed on successful aging outcomes, job design variables collectively contributed 23% to the variance in Generativity and 15.5% of the variance in Personal Sense of Control.  Job design variables contributed to 7.6% of the variance in Emotional Support and only 3.8% of the variance in Social Network. The specific job design variables that contributed the most to successful aging variables were Skill Discretion and Coworker Support. Skill Discretion alone explained 11.4 % of the variance in Generativity suggesting that jobs designed with opportunities to develop and use a variety of skills contributed to their opportunities to pass along knowledge to younger generations. Decision Authority and Coworker support contributed minimally to variance in Generativity.

Job design characteristics collectively contributed 15.5% to the variance in Personal Sense of Control scores. Coworker Support and Skill Discretion contributed individually 7.9% and 3.8% respectively to the variance in Personal Sense of Control. These findings suggest that workplaces that foster teamwork and skill development may also be contributing to older adults' perceptions that they have control over their social environments, and more broadly, over the life events. Although workplace coworker support was strong, the lack of association between social network and workplace coworker support may indicate that the work relationships developed by this sample of older workers were not strong enough to be considered close personal friends upon whom older workers relied upon for emotional support.

# 4 Discussion and Recommendations

## 4.1 Summary

This study surveyed 109 older sales associates in the retail home building industry to determine their perceptions of job design characteristics and to examine how job design may contribute to healthy aging. Results indicated that these jobs offered older workers opportunities to make autonomous decisions, use a variety of skills, and receive support from workers and supervisors. Thus, the job design for a sales associate in retail homebuilding provided positive work experiences for these older workers. The job design also contributed to two measures of successful aging: feeling a sense of personal control over their lives and generativity. The most important job design features that promoted successful aging were the ability to use a variety of skills on the job and feeling support of coworkers. In fact, these characteristics were more important than opportunities to make independent decisions, typically important to job control and healthy work. Decision authority was not a strong contributor to successful aging outcomes, which suggests that, for older workers, their ability to use skills and interact successfully with coworkers may be more important than making authoritative decisions akin to climbing the corporate ladder. The ability to use previous skills, learn new skills, be creative, and pass along knowledge was important for the majority of the sample.

## 4.2 Recommendations for Managing Older Workers

A healthy job design for older workers may benefit older workers and businesses alike. For example, as older workers appreciate passing along knowledge to younger generations and customers, this process simultaneously improves customer relations and promotes knowledge management. The following recommendations are drawn from the results of the study:

1. *Design job tasks to include a variety of skills.* Managers should design jobs so that older workers have opportunities to use a variety of different skills

on the job. Skill discretion is associated with challenge, new learning, and creativity.

2. *Provide skill updating, niche expertise, and new skill training.* Skill expertise enabled older workers to pass along knowledge to customers and coworkers and to feel in control of one's life. Managers can provide ongoing retraining and updating of skills. They can develop older workers' expertise in specific product areas. While a BLS study found that over 50% of older workers are not trained on the job, professional development enables older workers to remain effective and generative. New skill learning is equally important for cognitive health.

3. *Promote coworker support.* Coworker support provided older workers with a sense of camaraderie, social engagement, and in control of their social environment. Managers can foster respect and support for older workers by developing work teams, and collaborative or cooperative work tasks with other employees.

4. *Make use of older workers' work and life experiences.* Over 70% of the workers utilized past technical, organizational and "people" skills on their jobs. Managers can place older workers in job positions that utilize their broad base of experience.

5. *Promote opportunities for generativity.* Managers can promote older workers mentoring and training younger workers or even customers. Managers can further develop intergenerational collaboration and project work teams in which older workers become the subject matter experts.

6. *Delegating style of supervision.* Older adults will benefit from a flexible style of leadership in which close supervision is not necessary once job tasks are learned. Older workers in this sample were positive about opportunities to make independent decisions without supervision. They sought variety, challenge, and creativity in their work. A situational style of leadership would be relevant to older workers in which supervisors manage according to the capabilities of the workers. Older workers have the commitment and confidence to be delegated tasks once content is learned.

## 5    CONCLUSIONS

Workplaces can be designed to reciprocally serve the needs of business and older workers alike. In this study, a healthy job design contributed to two measures of successful aging: generativity and personal sense of control. Managers can optimize older workers' job experiences by designing jobs with opportunities to use a variety of skills, learn new skills, and experience the support of coworkers.

## ACKNOWLEDGMENTS

The authors would like to acknowledge the older workers that made this possible.

## REFERENCES

American Association of Retired Persons (2003). *Staying ahead of the curve: The AARP working in retirement study.* AARP Knowledge Management: Washington, DC. Accessed on November 11, 2004. http://research.aarp.org/econ/multiwork_2003_1.pdf

Bureau of Labor Statistics (BLS). 2010. "Labor Force Statistics from the Current Population Survey" Retrieved at: http://www.bls.gov/cps/cpsaat3.pdf.

Erikson, E. (1997). *The life cycle completed: Extended version.* New York: W.W. Norton & Company.

Ilmarinen, J. (2006). "Towards a longer and better working life: A challenge of work force ageing." *Medicina del Lavoro, 97*: 143-147.

Karasek, R., and Theorell, T. (1990). *Healthy work: Stress, productivity, and the reconstruction of working life.* New York: Basic Books.

Karasek, R. A. (1986). *Job Content Questionnaire and user's guide.* New York: Job-Heart Project at Columbia University.

McAdams, D. P., and de St. Aubin, E. (1992). A theory of generativity and its assessment through self-report, behavioral acts, and narrative themes in autobiography. *Journal of Personality and Social Psychology, 62*(6), 1003 - 1015.

Mirowsky, J., and Ross, C. E. (1991). "Eliminating defense bias and agreement bias from measures of the sense of control: A 2 X 2 index". *Social Psychology Quarterly, 54:* 127-145.

Ross, C. E., and Wright, M. P. (1998). "Women's work, men's work, and the sense of control." *Work and Occupations, 25*: 333-356.

Rowe, J. W., and Kahn, R. L. (1998). *Successful aging.* New York: Dell Publishing, Random House.

Seeman, T., Lusignolo, T. M., Albert, M., and Berkman, L. (2001). "Social relationships, social support, and patterns of cognitive aging in healthy, high- functioning older adults: MacArthur studies of successful aging." *Health Psychology, 2*: 243-255.

Wahlstedt, K. G. I., Nygard, C. H., Kemmlert, K., Torgen, M., and Bjorksten, M. G. (2000).The effects of a change in work organization upon the work environment and musculoskeletal symptoms among letter carriers. *International Journal of Occupational Safety and Ergonomics, 6:* 237-255.

CHAPTER 46

# Design for All on Board: Boat Design in the Era of Access for (Almost) Everybody

*Silvia Piardi, Andrea Ratti, Sebastiano Ercoli*

Politecnico di Milano
Milano, Italy
silvia.piardi@polimi.it

## ABSTRACT

Sea adventures books count a long list of disabled characters, from the one-legged Captain Hook, who also lost a hand due to a crocodile bite, to the one-legged Melville's Captain Achab and the one-legged pirate Long John Silver of Treasury Island.

Not only literature, but also history and current events show excellent physically impaired sailors. Horatio Nelson lost an eye and a hand, though he continued to command the British Royal Navy; the Italian Counter admiral Agostino Straulino, born in 1914, won gold and silver Olympic medals on the Star class, even if elderly and almost blind; Gianmarco Borea crossed the Ocean more than once with only one arm; aged 65, Francis Chichester sailed around the world in solitary after diagnosed with cancer – and he broke all the records with only one docking in Sydney.

How should boats for the disabled be designed? Can the boats fitted for the physically impaired be used with the same relish and results by the supposed 'normal' people, as 'design for all' philosophy suggests? This paper answers to these questions starting from a cultural and historical point of view. Analyzing the most notable world experiences, we try to define design guidelines, criticalities and opportunities.

**Keywords**: disabled, boating, yachting

# 1 ACCESSIBLE BOAT HISTORY

The designer always minds of the relationship between the object he or she is designing and its user. This relationship is the object of the discipline of ergonomics. One of the pioneers in this field was the American industrial designer Henry Dreyfuss), the author of one of the first manuals of anthropometry for designers. In his book, man and women were measured in every part and in every posture. The wide range of references allowed designers to develop any kind of object with correct size and proportions. Though a great step in the improvement of usability of products, designs were characterized by an opposition between 'normal' and 'special' projects.

In the 1950s, the 'Barrier Free' movement rises in the US as an answer to the people affected by poliomyelitis and then to the veterans of the Vietnam war (Steffan and Tosi, 2010).

In the 1960s the name 'Design för Alle' ('Design for All') is coined in Sweden, during the maximum splendor of the Nordic 'universalist' welfare model (ibid.).

In the 1970s the quadriplegic American architect Ronald L. Mace coined the name 'Universal design' for the approach he developed. The approach consists in starting a project from the needs of the differently able users. Universal Design assumes that these needs are not different from the ones of supposed 'normal' users; they just change in strength. In effects, 'normal' is a relative concept: everyone is potentially disabled in relation to the characteristics of the environment where he or she is (i.e. everyone is blind in a dark room). The Universal Design approach helps to identify the critical aspects of the project. If the designer is able to satisfy the strongest requirements, the resulting object will be more usable for everyone. The approach is radically different from the traditional design for the disabled: not special objects for a special category of users, but universal objects for the widest audience possible.

Since a long time sailing is a way of educating, aggregating and helping people with different kinds and degrees of disabilities, but this idea has been connected to the philosophy of accessible design only in the last thirty years.

The history of accessible boating starts probably with the history of Les Glénans, the famous French sailing school. Le Glénans (Les Glénans, 2012) was founded at the end of World War II – in 1947, to be more precise – in the Glénan archipelago, off the southern Brittany coast, by a group of former French resistance members. The aim of the school was to give to war veterans, many of which had previously been displaced, the chance to regain zest for life. Based entirely on voluntary work, sea life at the *Archipél* of Le Glénan acquired a pedagogical function, forging men's and women's qualities to the values of responsibility, collective life, solidarity and autonomy. Thus the motto of Les Glénans: '*école de voile, école de mer, école de vie*' ('school of sail, school of sea, school of life').

The spirit of sailing demonstrated to be open to different abilities and capabilities. This is furthermore confirmed by a wide series of initiatives that match sport and disability. One of the most famous is Sailability World Inc., an international nonprofit organization that promotes "...the activity of sailing, enriches the lives of people with any type of disability, the elderly, the financially and socially disadvantaged" (Sailability, 2010).

An intense activity of this kind is managed in Italy by Circolo Vela Gargnano on the Lake Garda. Since 1996 the Gargnano yacht club started the 'Homerus' project (Gaoso, 2012). The aim of this project is to teach to visually impaired and blind people to sail by themselves. Unlike previous experiences, where blind sailors were directed by seeing skippers, Homerus allows blind people to race in complete autonomy, thanks to sounding buoys. To avoid collisions, each of the two sailing boats is equipped with an audio device which signals the position of the other. The combination of the two systems allows the two racers to know their own position and the position of the challenger.

The Lake Garda is home to two other initiatives. 'EOS la vela per tutti' ('EOS sailing for all') is an association wanted of the judge Michele Dusi, in collaboration with two local hospitals, which aim is the rehabilitation through the experience of sailing, united to the healthy atmosphere and to physical activity. The association owns and manages the multihull sailboat 'Emozioni' (length 14 ft, circa 4.26 m), designed to be operated by a single disabled sailor, possibly with the help of an additional sailor. The same-level collaboration of a disabled and a non-disable person can be very challenging for both of them. The second initiative is 'Hyak', a project for the rehabilitation of people affected by mental diseases. Autistic patients can indeed improve their condition with the practice of sailing, which requires and stimulates collaboration, solidarity and sharing of an aim.

The town of Gargnano has become the Italian focal point of sailing and disability: since 1997 it hosts the meeting 'Navigando nel grande mare della solidarietà' ('Navigating in the great sea of solidarity'), an occasion of meeting and comparison for all the different experiences of accessible sailing. All the Italian initiatives passed here: 'Nave di Carta' ('Papership') by Comunità Exodus for children and mentally disabled, the 15m 'Bamboo', 'Matti per la vela' ('Mad for sailing') and many others.

Every new experience is a won battle for increasing accessibility.

## 2    CASE STUDIES

The boat can be interpreted as a sport tool, as a prosthesis, as a house or as a mean of transport; or simply as the sum all of the above. During the last years, designers worked to change the boats, making them simpler and easier to use, in the effort of increasing the user base – that is, the market. Set the myth of harsh sailors, servo systems allowed everybody to be at the helm, independently on the physical strength; the slogan 'easy sailing' was coined. This trend is convergent to universal design. In this section we analyze quite a wide range of case studies – real projects and concepts, large motor yachts and small, fast skiffs – with the purpose of giving a bird's eye view over the application of these principles.

## 2.1 Class 2.4mR

The 2.4mR sailing class was introduced as the Mini Class by the Scandinavian Sailing Federation in 1982 and was adopted as an international class in 1993 (International Sailing Federation, 2010). A metric class, it consists in a single-user yacht with bulb keel, characterized by affordability and usability of a common dinghy. The class had a strong evolution in its first years and it is now a very reliable and performing boat. Being a metric class, it counts different competing projects. The project called 'Nolin Mark III' is the most famous: probably circa 90% of circulating 2.4mR boats is of this kind (Associazione Classe Italiana 2.4 S.I., 2012). Two other projects share the most of the remaining quotas; they are the Swedish Stradivari 2.4 and Eide 2.4. A small number of boats is then designed and produced at amateur level; they contribute to the 'health' of the class, introducing variations and ideas. The rules of the class are however very strict and constantly updated.

The 2.4mR boats are designed to be operated by a single sailor, who sits safely in the middle of the hull, in front of the riggings and near the rudder stock. Thus, navigation does not require a high physical effort and regattas can be disputed in a 'open' formula, that is people of both sexes, different ages and different physical conditions can race at the same level – and they share the same ranking. Thanks to this characteristic, the class was adopted in 2000 for the IX Sydney Paralympic Summer Games.

## 2.2. Other paralympic classes

Beyond the 2.4mR, the International Federation for Disabled Sailing (2012) recognizes other sailing classes:

- SKUD 18. The acronym stands for Skiff of Universal Design. Developed in more than twenty years of experience by the , it allows a wide versatility in relation with the installed equipment and rigging. In Paralympic Games it is operated by two people sitting in the centreline, but it can be navigated by 'normal' people even riding trapeze (Mitchell, 2006 cited in Access Sailing, 2006). The boat can be equipped with a manual joystick for its navigation.
- International Access 2.3 (lenght overall 2.3 m). About 700 boats of this class are divided in eighteen Countries.
- Access 303, a keelboat largely similar to Access 2.3. The main difference is that it can host two sailors, making it particularly suitable for coaching.

## 2.3 Concept works at the Politecnico di Milano

The theme of Universal Design in boating was a source of inspiration for many of our students. Here we present a selection of projects that were conceived as a Master's degree thesis in Industrial Design or as a Master work in Yacht Design at the Politecnico di Milano.

In 2005, Maurizio Redaelli develops the thesis project 'Tutti a bordo' ('Everybody on board'), supervisor Andrea Ratti. Thanks to a stage at the Mattia & Cecco shipyard, where the accessible boat 'Lo Spirito di Stella' was designed and built, the Master's degree thesis work explored the connection between easy sailing and design for all. The thesis project started from the Mattia 56 catamaran hull (total length 56 ft, circa 17 m). The two hulls were slightly modified to increase stability and liveability. Then, the interiors and exteriors were developed. Technical plants (engine, generator, electric wires, water tanks etc.) were not developed, but suitable space was calculated for their installation.

The catamaran is characterized by wide passages and corridors, dimensioned on the width of a wheelchair. Accessibility to the wharf and to the sea is allowed by the same element, a hydraulic flying bridge derived from the models produced by Besenzoni, Italy. The movement from one main to the lower deck is allowed by a 'elevator-platform' first developed by P-M-P for the yacht 'Lo Spirito di Stella'.

The project originally improves the usability of the two piloting stations thanks to a seat which can be hydraulically moved from one station to the other. Interiors (kitchen, dinette, bathroom, bedrooms) have all been studied to allow accessibility on a wheelchair. Safety has been improved with round edges. The earning of a pleasant and modern aspect does not give the feeling of a 'special' product, in accordance with the spirit of Universal Design.

'Supernormal' by Giulia Monacci is a Master's degree thesis project, supervisor Silvia Piardi. Inspired by Naoto Fukasawa's and Jasper Morrison's search for everyday 'normal' objects, the project aims to design an accessible 30 m (100 ft) motoryacht in which accessible solutions are not stressed or highlighted. The hearth of the project lies in the owner's cabin, which – thanks to a movable platform – becomes the element that connects the ship decks.

The theme of a large motoryacht for a tetraplegic buyer is explored also by the project 'Octopus' by Jacopo Bevilacqua, Alessandro Lo Iacono, Rafael Matsumura, Elena Pozzi, Roberta Ragni, Aurora Restucci, Fabrizio Uzzi e Michele Vandone, developed during the Master in Yacht Design in 2011-2012. The project originality resides mainly in the stern of the yacht, where a movable terrace can becomes the access point to the sea. The terrace can even be moved under the water level, helping the disabled to go up again onboard.

# 3    DESIGN GUIDELINES

The starting point for the formulation of our collection of guidelines for inclusive boat design are the famous seven principles of Universal Design developed at the North Carolina State University by a group of architects, designers and engineers led by Ronald Mace (Connell et al., 1997): equitable use, flexibility in use, simple and intuitive use, tolerance for error, low physical effort, size and space for approach and use.

These principles were adopted in the 2000s by Waypoint Yacht Charter Service, an American broker specialized in accessible travel (Backstrom, 2008). After years of experience in the field, the charter director Sherri Backstrom and her colleagues modified the Universal Design principles adapting them to "the changeable nature of water-borne environments" (ibid.). The five Waypoint-Backstrom Principles are:

1. Begin with Universal Design
2. Design for Self-Sufficiency
3. Design for Extraordinary Conditions
4. Design for Modularity and Revision
5. Design for Seamless Intermodal Transfer.

Starting from these principles, summed up with our experience in the field and on the case studies, we issued a series of guidelines that should characterize an accessible boat project. The keywords are interconnected and blurred: the boat should be considered in its unity, where all the parameters are interconnected.

- Safety. At the basic level, safety means that the boat must be unsinkable and fireproof. Then, safety is connected to performance: manoeuvrability and speed allow the avoidance of dangers (i.e. a collision or a storm). Often a compulsory safety equipment is determined by strict rules, but the training of the crew – especially of the skipper – is even more important: prevention is ideed the main way of safeguard.
- Reliability. It has both a functional and a psychological role. It consists in the ability of the boat of coming back to a safe harbor. The feeling of safety is confirmed by a high-quality design of the boat, by a high-quality construction and by a high-quality equipment and rigging.
- Ease of navigation. Manoeuvrability and ease of holding the course are more and more important as the trend of 'easy sailing' is rising. Hydraulic riggings are an example of solutions aimed at simplifying boating.
- Affordance. The purpose and functioning of each component should be simply understandable by everyone – the objects should *suggest* the users how to use them. This requirement is very important, especially in the field of interface design. Designers should consider the abilities of all kinds of users. Often little details can make the difference – the little bump on the key "5" on a PC or telephone keyboard makes the difference between 'usable' and 'unusable' for a blind person.

- Comfort. The boat and its parts should not only be simple to use, they should be comfortable too.
- Pleasantness. Though design should not be reduced to the mere aesthetics, after being functional, objects should be enjoyable and pleasing.

To effectively design an accessible boat, this should be considered also in its relationship with the environment – mainly the wharf and the sea.

## 4    CONCLUSIONS

Life expectancy is increasing and so does the average age of boating enthusiasts. In Western world, disability is strongly connected to age, i.e. as reported by Battisti (2003, p. 26) for Italy. Summing the two considerations, we gather that the improvement of accessibility is very important for the nautical industry of the future.

In Italy, 26% (increasing) of disabled people practice a sport and 40% of them is at least 44 years old (ibid., p. 31). What does it mean? It means that what is currently considered a 'special' solution for a limited target will become the standard in new designs, going further on the course of 'easy sailing'. This trend is already influencing the market: the introduction of auxiliary methods for navigation already expanded the target, both in respect of age and sailing capabilities.

A new, interesting design area is rising. It involves interiors layout, access, instrumentation and accessories. At the same time, it involves all that is around the boat: wharfs and dock equipment are often less accessible than boats themselves and should be designed with care (United States Access Board, 2003). Prejudices and worries are often the biggest obstacle to 'boating for all'; designers are summoned to surpass these obstacles through a careful and respectful work.

## ACKNOWLEDGMENTS

The authors would like to acknowledge all the teachers and all the students of the Master in Yacht Design of the Design School and of the Department INDACO (Industrial Design, Arts, Communication and Fashion) of the Politecnico di Milano, which they launched in 2001 and which they still foster – the Master is currently at its 11th edition. The Master, thanks to the great experience of the professionals who lead the lessons and to the energies and creativity of the students, was a steady stimulus for exploring, questioning, rethinking and redesigning the world of boating.

The authors thank also the colleagues and the students of the MS in Nautical and Naval Design, established in La Spezia since 2005, the two years program of the Università di Genova and the Politecnico di Milano.

# REFERENCES

Access Sailing, 2006. *Skud 18.* [online] Available at: <http://www.accesssailing.com/?Page= 19874&MenuID=Boats%2F13166%2F0> [Accessed 17 February 2012]

Associazione Classe Italiana 2.4 S.I., 2012, *Le origini del 2.4.* [online] Available at: <http://www.duepuntoquattro.it/index.php?option=com_content&task=view&id=1&Ite mid=9> [Accessed 7 February 2012]

Backstrom, S. 2008. *The Waypoint-Backstrom principles. Maritime Inclusive Environments and Practice (Human-Centered Seaworthiness).* [pdf] New Dehli: Design for All India. Available at: <http://www.designforall.in/newsletter_Nov2008.pdf> [Accessed 15 January 2012].

Battisti, D. ed., 2003. *Tecnologie per la disabilità: una società senza esclusi.* Roma: Commissione interministeriale sullo sviluppo e l'impiego delle tecnologie dell'informazione per le categorie deboli.

Connell B. R. et al., 1997. *Principles of Universal Design.* [online] Available at: <http://www.ncsu.edu/project/design-projects/udi/center-for-universal-design/the-principles-of-universal-design> [Accessed 10 February 2012]

Gaoso, A. 2012. *Chi è Homerus.* [online] Available at: <http://www.homerus.it/Vela-non-vedenti/chiehomerus.html> [Accessed 12 February 2012]

International Sailing Federation (ISAF), 2010. *International 2.4mR Class Rules 2010.* [pdf] Southampton, UK: ISAF. Available at: <http://www.sailing.org/tools/documents/ 24m2010CR150212-%5B12108%5D.pdf> [Accessed 20 February 2012].

International Federation for Disabled Sailing (IFDS), 2012. *Paralympic Games.* [online] Available at: <http://www.sailing.org/paralympics/index.php> [Accessed 20 February 2012].

Les Glénans, 2012. *Qui sommes-nous?* [online] Available at: <http://www.glenans.asso.fr/ FR/presentation/art_11462.php> [Accessed 10 January 2012]

Sailability, 2010. *Welcome to Sailability.* [online] Available at: <http://www.sailability.org/> [Accessed 6 February 2012]

Steffan I., Tosi F. 2010. Ergonomia per il progetto e Design For All. In: SIE (Società Italiana di Ergonomia), *Ergonomia, valore sociale e sostenibilità, Atti del IX Congresso nazionale SIE, Società italiana di ergonomia.* Roma, Italy 27-28-29 October 2010. Roma: Nuova Cultura.

United States Access Board, 2003. *Accessible boating facilities. A summary of accessible guidelines for recreation facilities.* [online] Available at: <http://www.access-board.gov/recreation/guides/pdfs/boating.pdf> [Accessed 5 February 2012]

CHAAPTER 47

# Indirect Estimation Method of Data for Ergonomic Design on the Base of Disability Research in Polish 2011 Census

*Marcin Butlewski*

Poznan University of Technology
Poznan, Poland
e-mail: marcin.butlewski@put.poznan.pl

## ABSTRACT

The following article presents issues connected with the shaping of ergonomic quality of the technical environment for people with disabilities, due to the difficulty of obtaining and verifying data on the needs of its users.

The problem of disablement concerns a substantial group of the whole society. It is estimated that over 500 million people in the world (10% of the global population) suffer from different kinds of disabilities. In Poland, according to official data from 2002 there are 14.3% of disabled people in the total population, it means nearly every sixth Pole is disabled.

However, official data on persons with disabilities is often limited to information that is too general to make it useful in the design process. In particular, when there is no potential for the development of technical objects from the start, solving problems of disabled people (but not only theirs) must happen as a compromise. This compromise must take into account not only the needs of different disability groups, which tend to be antagonistic towards each other, but also the needs of all other user groups, as well as an economic calculation of the proposed solutions. When designing or adjusting the technical environment it is therefore necessary to reach for the tools which allow for the representation of reality as closely as possible, while at the same time having a lack of data and small samples or regions. For this purpose, a type of estimation that allows us to estimate the parameter values and the distribution in the general population, based on observations obtained in the sample should be used. Particularly interesting seems to be indirect estimation, which allows to achieve a much smaller variance of the estimator in the case of restrictions on the sample or its territoriality (e.g., the value of the relative fraction estimation error is less than almost 75% when compared with direct estimation). The advantage of indirect estimation for determining the needs of disabled people in a given area is that it will be able to strengthen the so-called estimation by borrowing information, which will necessitate resorting to data "scattered" in other records (e.g. social security records).

Indirect estimation may therefore be a way to obtain the necessary level of ergonomic quality of technical creations, through their adaptation to the needs of users in a more perfect manner. Application of the methods in the field of indirect estimation should avoid generalizations that may distort the characteristics of local communities. The higher ergonomic quality obtained in this way will significantly contribute to the life improvement of people being so affected by fate.

**Keywords**: Indirect estimation, disability in Poland, ergonomic quality

## THE PROBLEM OF DISABLEMENT AND ERGONOMIC DESIGN

The official data show that 500-650 million people all around the world, that is, above 10% of the global population, suffer from different kinds of disabilities (Butlewski M., Tytyk E., 2009). In Poland, the number of disabled people in 2002 was surprisingly high, that is 5,456,700 people. It constitutes 14,3% of the whole Polish population, which means that almost every sixth Pole can be diagnosed as suffering from a certain degree of disability (Balcerzak-Paradowska B., 2002). One of the most numerous groups among the disabled is people with the dysfunction of a motor organ. A disabled person with the dysfunction of a motor organ has a limited dexterity of lower or upper limbs or spine due to a lasting health loss that is an outcome of injury, brain deficiency or disease, or bodily harm or deformity in skeletal, muscle, or nervous systems. A motor disability is thus a dysfunction of a motor organ, which can be caused by multiple reasons, but the consequence is always a decrease of motor dexterity (Butlewski M., Tytyk E., 2009). In turn, the most numerous group out of people with a motor disability is people with a locomotive disability, people who have problems with moving around. Injuries of a motor organ, next to circulatory system ailments, account for the main cause of disablement. Motor organ dysfunctions are most often caused by ailments (45.7%) and injuries (45.1%) and only in 6.4%, are an outcome of congenital diseases and in 2.4% of other causes (Szozda, 2010). These figures do not take into account elderly people who due to their age possess some limited dexterity (Thring, 1973). In the

456

year 1950, there were about 200 million people above the age of sixty, while currently, this group in the global population has reached 500 million people. It is assumed that by the year 2020, the number of elderly people all around the world will have rocketed to one billion and by 2025 will have reached 1.2 billion (Olszewski, 1997). This research has suggested that more and more activities will have to be performed by disabled people. Unfortunately, the phenomenon of a loss of dexterity is inextricably linked with age. As the years pass by, the number of disabled people is increasing steadily in both the biological and legal respect. The result of this is a considerable contribution of sixty-year old and above people to the population of disabled people. It is estimated that 36% of people aged 65 and over have problems with managing stairs (LaCroix, et al. 1993). The figures that have been presented in this study are just estimates and do not include an open admittance for the loss of dexterity among elderly people. It is often the reason why they are not diagnosed as having a disability. We know now, that elderly people in Poland will have to work even until they are 67 years old, so it is very possible that the problem of adjusting work to the abilities of less efficient people will be growing.

Also it is estimated that over 48% of companies do not provide disabled people with adequate conditions in their workplace. About 38% declare that the different job opportunities are directed at people who are not diagnosed as disabled (GUNB Report 2006). This situation contributes to the fact that disabled people are poorly educated, earn far less than "healthy" people, and feel isolated and unwanted by the society (Ochonczenko. 2000, p.2). When we also take under consideration that 63% of employers are afraid that a disabled person will be ineffective at work, and 30% will not hire disabled people because they are afraid that they will have more privileges, we cannot wonder that over 86% of people with disabilities of working age with a lower than average level of education were in 2010 professionally inactive. The situation is improving - in 2007 the activity rate of disabled people of working age in Poland was 22.6%, in 2008 - 23.9%, in 2009 it increased to 24.6% and in 2010 it reached 25.9%. (ONLINE - http://www.niepelnosprawni.gov.pl). But still it means that about 75% of disabled people are not labor active.

The above mentioned facts point to the need to undertake actions, which will contribute to an increased activity, also in employment, of disabled people. While we recognize the need to adapt the technical environment for the disabled, to do so on a large scale for all is of course impossible. Hence the way to optimize the processes of ergonomic design in this respect may be to determine the anticipated user groups and their needs, which is a prerequisite for creating a functional ergonomic system. Unfortunately, knowledge of the needs of users, especially for people with disabilities may be difficult due to their "nonstandardness" but also the lack of broad lines of communication with those responsible for the engineering design process. This situation makes the need for the widest possible information concerning persons with disabilities, particularly those that will contribute to the welfare of the largest possible audience. Hence, the interest in statistical methods which will allow for achieving a greater efficiency in the process of ergonomic design.

# DISABILITY IN POLISH CENSUSES

Already in the first Polish census in 1921 it was asked, with the typical bluntness of those times, about a disability (blindness, deafness, and lack of limbs). During the following four national censuses in Poland disabilities had not been studied. The return to disability issues came in 1974 only in a micro-census, when it was asked about any legal disability only. By contrast, from the 1978 census and on both categories of disability, legal and biological, were taken into account. The questions added to the census questionnaire at the time functioned in the following censuses, which provided the relative comparability of results. A similar concept was to study the scope of disability in the micro-census of 1995 (Kostubiec 1995). However, question 8 in Form A allowed only for the determination of a biological disability: whether because of a disability or a chronic illness is one of a completely or severely limited capacity to perform basic activities for ones age? As explained in the methodology of the census, "Those who declare a disability in this sense, often referred to as the biological, gave both the degree of reduction of capacity to perform basic life functions in two categories - a total restriction, severe restriction (partial disability). In both cases it was only for long-term states of disability, existing for at least 6 months."

The study of disability in the national census from 2002 differed from previous censuses. On the one hand it put more emphasis on the problems of the disabled from the previously unknown effects of the political and economic changes in 1989, while on the other hand it showed the chances and opportunities to defend their interests. In particular, the labor market for disabled people, health and social care, and new opportunities related to the study of the economic activity of the population and experience gained from the questionnaires about the population health status in 1996. The census questionnaire in the national census from 2002 had reserved questions 12 – 14 for the disability survey, which had to be answered – according to general census rules - without showing any evidence of legal status. In question 12. Do you have a completely or severely limited capacity to perform basic activities for your age (work, education, self-service, games, etc.) because of a disability or chronic illness? One had to choose one of three options: (1 yes, completely, 2 yes, severely, 3 no). Question 13. Do you have (issued by the appropriate adjudicating authority), the decision pronouncing the current inability to work, the degree of disability, the desirability of retraining, handicap, or (in the case of children under 16 years of age) entitlement to a care allowance? Choosing a response option 1, one should go to question 14, but option 2 not to go to Question 15. Question 14 set the legal level of disability: How was the inability to work / disability / handicap qualified?

- Group I disability or severe disability or total inability to work and dependent existence, or the inability to work on the farm with the entitlement to a care allowance
- Group II disability or moderate disability or total disability
- Group III disability or slight disability or partial disability to work or inability to work on a farm or the advisability of reclassification.

Based on data collected in 2002, the statistical offices of individual provinces developed their own copyright publications on disability in the region.

The full census ended on June 30, 2011 was a full direct questionnaire (computer type device with hand-held, mobile, Internet and self-registration via traditional mail) included only 15 features, of which only the question of nationality and language have had their coverage of administrative data and were spontaneous, independent sources with no basis in any documents. The list of questions in the full study, which were asked during the census in 2011 of all residents of Poland, included appendix 1. The list of questions to more specific topics, called modules, included appendix 2. Module 6 is dedicated to questions about disability. It contained the following questions:

**1. Do you have a limited ability to perform usual activities (school, job, home management, self-service) due to health problems (disability or chronic illness) lasting 6 months or longer?**

    1. yes, completely limited

    2. yes, severely limited

    3. yes, moderately limited

    4. no, I have no limitations

    5. I do not want to answer this question

**2. How long have you had a limited ability to perform usual activities?**

    1. from 6 months to 1 year

    2. from 1 year to 5 years

    3. from 5 years to 10 years

    4. 10 years or more

    5. I do not want to answer this question

**3. Please specify the reason why you feel a limited ability to perform usual activities (can select no more than 3 replies)**

    1. injuries and diseases of musculoskeletal system

    2. damage and disease of eye

    3. damage to hearing

    4. cardiovascular disease

    5. neurological disease

    6. other disorders not specified

    7. I do not want to answer this question

**4. Do you have the current decision pronouncing the disability, incapacity to work, or disability?**

    1. yes

    2. not

    3. I do not want to answer this question

**5. How was the disability / incapacity / handicap qualified?**

    1. significant degree of disability or total disability and dependent existence, or group I disability, or long-term inability to work on a farm with entitlement to the care allowance

    2. moderate degree of disability or total disability, or group II disability;

    3. slight degree of disability or partial disability, or the desirability of retraining,

or group III disability, or long-term inability to work on a farm with no entitlement to care allowance;

4. disability for people under 16 years old

5. I do not want to answer this question

All modules were implemented on the same large 20% random sample. Thanks to this, all disabled persons in the household, which has been drawn to participate in the survey, can be described by the set of features defined in other modules. For example, the level and direction of education and the profession learned by the disabled is described in module 2, the characteristics associated with economic activity (module 3), commuting to work (module 4), sources of income (module 5), fertility and reproductive plans of a disabled person (Module 7 - provided that a disabled person is a woman between the ages of 16 to 49 years), country of birth and citizenship (module 8), domestic and foreign migration (module 9), nationality, language, and religion (module 10). A 20% sample allows for estimation of statistical tables with sufficient detail if it is able to be supported by the data contained in the records of ZUS (Social Insurance), KRUS (Agricultural Social Insurance Fund), PFRON (State Fund for Rehabilitation of Persons with Disabilities) and EKSMON (National Electronic Monitoring System of Rulings of Disability) .

## ESTIMATION FOR DESIGN PURPOSES

Very often in the design process that includes identification of needs of disabled people there is a need to draw conclusions on the basis of incomplete or fragmentary information. What is more, there is a revealed need for an option that is more favorable for one group of users, which requires at least an estimate, which group will benefit more from a particular solution. There is thus a need for estimation. Estimation is a valuation of parameter values or a distribution in the general population, based on observations obtained in the sample. However, the estimation of people with disabilities on the background of the general population is not a simple problem. There is a problem of too small and too dispersed domains, and the representativeness of surveys based on samples.

A domain we understand as a subpopulation of interest in a survey for example: geographic domains such as a city district. Direct estimators use data only from the study units in the domain and time period of interest and include standard weighted survey estimators. Direct estimation also gives good design properties: unbiased estimators and valid confidence, but this is not reliable enough if the sample size is extremely small (Breidt, 2004). If we take as a goal to estimate needs of people with disabilities within districts of a city we can assume that we will have small domains. Small domains, or small areas, means that it may be zero in some domains and usually model-based inference is necessary to yield estimates of adequate precision. So this means that the definition depends on sampling resources and precision requirements (Breidt, 2004). To estimate small domains we need some additional source and that is why we need to use indirect estimators. Indirect estimators are using data from outside the domain and/or time period of interest and they can "borrow strength" across time or space (Rao, 2003). This means that the accuracy of estimation does not need to connect with the size of the random sample, you need to eliminate the sources of systematic errors or minimize their importance. (Paradysz, 2008). Indirect estimation is therefore applicable in identifying the needs of people with disabilities as evidenced by the work: (Report Small Area Estimation Models for Disability - Commonwealth Department of Australia 2004; Bizier V, Yong Y., Veilleux 2009)

However, small-area estimation, also known as small area statistics has certain conditions and restrictions (Paradysz, 2008), among which we can mention those related to:

1) Requirement to have a good information infrastructure (administrative records, surveys conducted regularly)
2) The requirement of the permanent borders of the administrative division, which contrary to appearances, does not appear an easy task - administrative changes make it necessary to merge the data of various common structures,
3) Are required to be made by a well-prepared team of statisticians,
4) Difficulty in assessing the precision of estimation for many of the estimators,
5) Difficulty in assessing the truthfulness of the assumptions a priori, which occurs more often in statistics of small areas than in the classical method of representation; which means that there is a greater probability of failure of assumptions regarding the distribution of the variables that are studied and complementary as well as the similarity of small and large areas,
6) The effect of closing the gap between the areas where they are in fact much more strongly differentiated, which is common in the outlying areas in terms of the highlighted features or exaggerating differences less common when the estimators take absurd values (when the distributions of the studied traits are extremely asymmetric),

These issues are the factors to be considered, but this does not change the fact that the use of indirect estimators will significantly increase the quality of design data to estimates of the disabled.

## WHAT DATA DO WE NEED TO ESTIMATE

The above considerations clearly show that, despite the high detail and large sample in statistical data there is a lack of strict design information, such as the functionality of the disabled. It therefore seems reasonable that future studies in addition to the disclosure of the territorial distribution would show information hitherto neglected and pertaining to a functional disability and not only a disability in the sense of law or general biology. Helpful here would be the classification of

Goldsmiths, which groups individuals according to their motor abilities (Branowski 2006):

1) For activities transferring from and to the wheelchair which requires the help of two accompanying persons
2) Move only with the assistance of a companion or with an electric drive
3) Move and are independent if they have a right and a specially adapted wheelchair architecture
4) Move in and enjoy the architecture, provided that there is an application of the principles of universal design
5) Have a limited physical ability, which causes difficulty in overcoming stairs (such as children's stroller)
6) Because of the failure of dimensional architecture feel inconvenienced during use of such devices as the toilet
7) Have an overall capacity of a normal human adult, for example, without lifting weights, and the possibility of climbing the stairs
8) Can perform any actions, such as jumping, climbing ladders, dancing, moving heavy items.

Of course, the next division would be, those associated with the ability to see, hear, and other cognitive, psychomotor or mental functions, which will enable to identify the needs of the disabled and elderly and, therefore, essential functions of the proposed equipment to be designed. Such studies, combined with statistics of small areas may allow us to capture dependencies, which we currently may not know.

## SUMMARY

As was shown in the article indirect estimation may be a way to obtain the necessary level of ergonomic quality of technical creations, through their adaptation to the needs of users in a more perfect manner. It is highly possible that the application of the methods in the field of indirect estimation will avoid generalizations that may distort the characteristics of local communities and that it will also be cheaper than direct ways of direct estimation.

## REFERENCES

Balcerzak-Paradowska B., (red.), Sytuacja osób niepełnosprawnych w Polsce. Raport Instytutu Pracy i Spraw Socjalnych, Warszawa 2002.

Bizier V, Yong Y., Veilleux Model-based Approach to Small Area Estimation of disability counts and rates using Data from the 2006 Participation and Activity Limitation Survey Section on Survey Research Methods – JSM 2009

Bolach E., Bolach B., Seidel W. Motywacja do uprawiania podnoszenia ciężarów „Powerlifting" przez osoby niepełnosprawne (w:) Problemy dymorfizmu płciowego w sporcie. – Wyd.AWF, Katowice 2000

Branowski B., Zabłocki M., Kreacja i kontaminacja zasad projektowania i zasad konstrukcji w projektowaniu dla osób niepełnosprawnych, (w): Ergonomia produktu.

462

Ergonomiczne zasady projektowania produktów, (red.) Jan Jabłoński, Wyd. Politechniki Poznańskiej, 2006, ISBN: 83-7143-238-0

Breidt F. J., Small Area Estimation for Natural Resource Survey, Monitoring Science & Technology Symposium, Denver, Colorado, September 23, 2004

Butlewski M., Tytyk E., "Ergonomic features of construction solutions to assist stair climbing by disabled people"; (in): The Ergonomics and Safety in Environment of Human Live. Monograph, p. 19-34; Eds. G.Dahlke and A.Gorny, Publishing House of Poznan University of Technology, Poznan, 2009; ISBN: 83-7143-848-6

Butlewski M., Tytyk E., Niezmechanizowane narzędzia ręczne dla osób o ograniczonej sprawności ruchowej. Zastosowania Ergonomii, nr 1-2, Wyd. PAN, PTErg, CZE, Poznań – Wrocław – Zielona Góra, 2007, ISSN 1232 – 7573 s. 129

Commonwealth Department of Australia 2004 Research Paper: Small Area Estimation Models for Disability (Methodology Advisory Committee), Nov 2003 Latest ISSUE Released at 11:30 AM (CANBERRA TIME) 26/05/2004 First Issue

Jasiak A., Misztal A.: Makroergonomia i projektowanie makroergonomiczne, Wydawnictwo Politechniki Poznańskiej, Poznań 2004.

Kabsch A. Potrzeby rehabilitacji w przewidywalnej przyszłości, w: Ergonomia niepełnosprawnym w przyszłości, red. J. Lewandowski, J. Lecewicz-Bartoszewska, M. Sekieta, Wyd. Politechniki Łódzkiej, Łódź2003,

Kostrubiec S. (Ed.), Osoby niepełnosprawne na rynku pracy. Studia i analizy statystyczne. GUS, Warszawa 1995.

Lacroix, A.Z.; Guralnik, J.M.; Berkman L.F.; Wallace, R.B.; Satterfield, S., Maintaining mobility in late life. II. Smoking, alcohol consumption, physical activity, and body mass index, American Journal of Epidemiology, Volume 137, Issue 8, 1993, Pages 858-869

Ochonczenko H., Problemy i potrzeby osób niepełnosprawnyc w opinii liderów środowiska lokalnego (na podstawie badań w byłym województwie zielonogórskim) Nasze Forum 2000

Olszewski J., 1997; Podstawy ergonomii i fizjologii pracy. Wydawnictwio Akademii Ekonomicznej, Poznań

ONLINE Niepełnosprawność w liczbach 2010 http://www.niepelnosprawni.gov.pl/niepelnosprawnosc-w-liczbach/

Paradysz J. (2008), Kryteria dobroci estymacji dla małych obszarów, konferencja naukowa z okazji jubileuszu 90-lecia GUS: Statystyka społeczna: dokonania — szanse — perspektywy, Kraków, 28—30 stycznia 2008

Piasecki M., Osoby z różnymi rodzajami niepełnosprawności – specyficzne zagadnienia w poradnictwie zawodowym, Warszawa, 22 maja 2007 r. za Polacy o niepełnosprawności TNS OBOP, 2004

Piasecki M., Osoby z różnymi rodzajami niepełnosprawności – specyficzne zagadnienia w poradnictwie zawodowym, Warszawa, 22 maja 2007 r. za Polacy o niepełnosprawności TNS OBOP, 2004

Rao, J.N.K. (2003), Small Area Estimation, Hoboken, NJ: Wiley

Szozda R. Fizjologia pracy, Atest 9/2010, s. 42

Thring, M.W, Mechanical aids for the disabled, Engineering in Medicine Volume 2, Issue 2, April 1973, Pages 32-40

Tytyk E., Butlewski M., Ergonomiczna jakość zmechanizowanych narzędzi ręcznych jako przesłanka ich doboru do prac obsługowych, (w) Tytyk E. (red.) Inżynieria ergonomiczna. Teoria. Wydawnictwo Politechniki Poznańskiej, Poznań 2011

Tytyk E., Projektowanie ergonomiczne. Wyd. Naukowe PWN, Warszawa – Poznań, 2001

CHAPTER 48

# Designing Wearable and Environmental Systems for Elderly Monitoring at Home

*Sabrina Muschiato, Maximiliano Romero, Paolo Perego, Fiammetta Costa,*
*Giuseppe Andreoni*
Indaco Department, Politecnico di Milano, Milano, Italy
sabrina.muschiato@polimi.it

## ABSTRACT

This work presents a research on assistive technology acceptability conducted in the framework of a wider public funded project called MAMMA (Multimodal Ageing Monitoring and Assistance) involving companies, university and a health care institution aimed at developing a telemonitoring system to support autonomously living elderly people.

MAMMA system is aimed to connect permanently elderly patients with their relatives and care givers. In this way, users can continue to live their everyday life with naturality but continuously monitored as in the hospital. In case of anomalous events or health conditions, the system will provide an alarm to care givers and relatives. The monitoring is based on the collection of biological data (ECG and blood pressure) and information related to users movements. The collected data are transmitted to a hub where they are processed, compared and evaluated. The hub is able to create patterns between biological condition and movements (if the user is sleeping, walking, falling...) and define anomalous situations. Eventual alarms are transmitted to the remote assistant & health management.

In our hypothesis, if it is possible to provide this technology to elderly people and if they accept it, then it will be easy to use it for social improvement goals. In the situation where technological issues are solved, acceptability is the key factor. For this reason we developed a multidisciplinary approach to design systems and services that can match the specific needs of elderly users..

## INTRODUCTION

Starting from the fact that the European population is aging, many applied research projects focused on ergonomics are developing different issues about the integration of home care services based on Information Communication Technology systems.

European population is growing, in 2025 the over-65s is predicted to reach more than 30%, which is equivalent to 211 million citizens. By 2050, this group could reach more than 42%, corresponding to almost 265 million people (Marsili et al 2002). One Italian of 5 is over-65 and the forecast for 2030 is 1 of 4 (CEIS 2009). In this condition it is easy to prevent a collapse of the National Sanitary System, and in consequence each country tries to find alternatives. In Italy the public healthcare system is autonomously managed by each region, and the policy can be largely different in many of them.

In any case, frequently, traditional hospital infrastructures are not economically sustainable and depend Regional governments' contributions. Lombardia Region is trying to implement strategies focused on de-hospitalization and telemonitoring in order to reduce costs. Thus many applied research projects focused on home care services based on Information Communication Technology systems have been carried out.

The application of Ambient Intelligence for health care purposes has been defined as Ambient Assisted Living. At this point, and in particular for elderly target population, other problems start to become important like acceptability issues. ICT based projects for telemonitoring or e-health more often do not pay sufficient attention to improvements in human-technology interaction and acceptability. Frequently, many technological well developed products are not implemented because final users don't accept them. European Community is trying to solve this gap with the so called "challenge 5" (Framework Program 7) funding research projects focused on home care for elderly, chronic diseases and disabled people: "This challenge addresses advanced ICT research for sustainable high-quality healthcare, demographic ageing, social and economic inclusion, and the governance of our societies"(http://cordis.europa.eu). Accessibility and acceptability issues of technological systems for healthcare are key-points for success. More often service provider do not apply ergonomic approaches for service design and development (Martin J.L., et.al, 2008)

Design and applied ergonomics can offer several tools to develop projects in which the house becomes an interactive environment for non-invasive support to users with special needs, related to health but also with "social needs".

An ergonomic approach to design domestic context integrated by technology can show that the interest on issues related to Ambient Intelligence. Ambient Intelligence can have a real application in the field of improving the quality of life for people whit special needs.

This approach can push forward proposals for the development of a new interaction between the physical world and the digital world, in order to lay the foundations for a new life style for elderly people like a "safe and pleasurable way of dwelling".

To this purpose the MAMMA (Multimodal Ageing Monitoring and Assistance) project, co-funded by Lombardy Region and involving companies, university and health care institution, develops a telemonitoring systems for elderly people to support autonomously living at their homes.

.

MAMMA system is aimed to connect permanently elderly people with their relatives and care givers. In this way, users can continue to live their everyday life with naturality but continuously monitored as in the hospital. In case of anomalous events or health conditions, the system will provide an alarm to care givers and relatives.

## METHODOLOGY

In our hypothesis, if it is possible to provide technologies to elderly people and if they accept it, then it will be easy to use it for environmental and social improvement goals. In the situation where technological issues are solved, acceptability is the key factor.

For this reason, an important part of the research is dedicated to usability (Rubin, J., 1984) and acceptability (Jordan, P.W., 2000) evaluation of the interface console and products.

Design strategy to improve acceptability of ICT and overcome elderly wariness is based on embedding cameras and sensors in furniture and garments. Early acceptability evaluation is conducted in real environment through Wizard of Oz techniques (Maudsley et al 1993): mock-up of the alternative products have been produced and provided to elderly people involved in the project who were asked to put the pieces of furniture in their home and to dress the wearable devices for one week. Elderly's acceptance and preferences are assessed through focus groups, interviews and diaries filled out by care givers. The final configuration of the monitoring products and devices is consequently designed according to this results. Final usability evaluation of the whole system is also handled through on site tests.

The devices, organized in two subsystems, were tested in two small apartments in a residence located in Monza, near Milan. In these apartments live two men aged 85 and 89 years. For five days the first subsystem has been tested by the first user and the second subsystem from the second user, then they were inverted and tested for another five days.

Figure 1: the preliminar and final focus group

The final focus group allowed users who tested the products, to describe their experiences with fellow guests by highlighting positive aspects of the experience and allowing their opinions of people who not having used the products. Also people who haven't used the systems could express their own opinion of the perception of the two systems in their daily lives and it had a negative opinion. The comments gathered from all trial participants and the focus groups, were considered important by the companies involved in the project and compiled by researchers to be released in the process of product design developed in recent months.

## THE SYSTEMS

The project involves different kind of products for home monitoring of elderly people. In order to increase the acceptance of technologies by the elderly it was considered to split the products in two main subsystems.

The first subsystem is composed of a patch that collect biomedical data (ECG and blood pressure) and cameras embedded in a lamp and a guard robe. The second subsystem is based on a wearable shirt with embedded sensors and infrared video-cameras placed in the room in order to collect information related to users movements .

The collected data are transmitted to a hub where they are processed, compared and evaluated. The hub is able to create patterns between biological condition and movements (if the user is sleeping, walking, falling...) and define anomalous situations. Eventual alarms are transmitted to the remote assistant & health management. The hub provides also access to data and sanitary services through an ad hoc console with different interaction levels according to the different users: elderly people, relatives, caregivers…

The technical part is devoted to the definition of the systems' architecture and user interface, to the hardware and software development of cameras, accelerometers, biosensors and hub and to the production of prototypes for the final experimentation with end users.

The parallel development of acceptability and technical issues permits to tune the telemonitoring system on elderly psychological and social needs according to User Centered Design principles. As Norman (Norman 1998) states we are now in a mature phase of technology where product development has to shift from a technology centered to a user centered perspective.

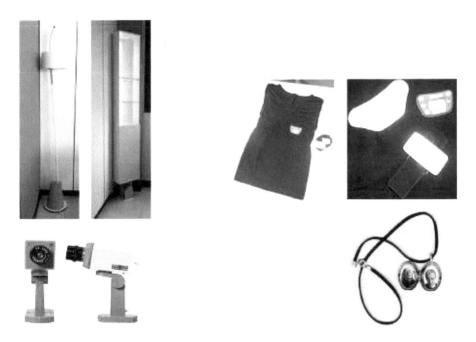

Figure 2: the two IT subsystems for monitoring elderly at home.

## RESULTS

At the end of the tests we can say that users have actively participated in the evaluation phase of the acceptability of IT systems. So in the second phases of the research, users preferences and suggestions can be integrated in the design process of the different products involved .

The comments gathered before, during and after the tests of the two subsystems, show a lot of qualitative data relating to the possibility that older people accept, or not, IT in their day living activities.

Users have declared that they can accept IT devices if they have a benefit for increase their security or monitoring specific pathological conditions.

It appeared evident that if users don't clearly link technology to a pathological condition, products are rejected by the majority of users. If the critical factor depends on safety from falls and trauma, the products are not easily accepted in homes by the elderly because they do not consider them so dangerous problems.

.

After the test, although the evidence of positive opinion of users who had tried devices, indirect participants didn't accept all the proposed products, but declared the need to choose a device customized it to their pathological needs.

The large sized objects with hidden cameras, such as furniture, are less welcome then lamps and normal video cameras. Wearable items are did not show any problem and just a few remarks about improving usability were highlighted.

| Focus groups pre test | Test in user's home | Focus groups post test |
|---|---|---|
| 20 users | User 1<br>Test systems devices for 10 day | 20 users |
| 18 don't accept systems devices | | 9 don't accept systems devices |
| 2 accept systems device | User 2<br>Test systems devices for 10 day | 2 accept systems device |
| | | 7 accept one device but only if it's personalized |

Figure 3: tests results.

## DISCUSSION AND CONCLUSION

The Mamma research show the achievement of ergonomics and ICT issues to satisfy end users.

Maybe it is better to design single devices that can be increased in relation with health user needs. A system can be seen like a foreign object and too complex to understand, so elderly users can be scared to insert this systems in their homes. IT system should have a high level of customization that elderly users can adapted to his state of health.

The integrated application of Ambient Intelligence to maximize life quality of life of inhabitant with special needs like elderly is a promising strategy requiring the collaboration of different stakeholders from research and practice community, such as experts in ethnography, physical and cognitive ergonomics, independent designers and architects, staff and managers from manufacturing companies and elderly/disabled care institutions.

Further researches will be needed also to face practical problem as the integration of ICT in existing homes, to develop new combination of communication technologies in the physical spaces for the domestic environment using opportunities that IT gives us.

## REFERENCES

CEIS, Rapporto Sanità 2009, Sanità e sviluppo economico, Fondazione Economia Tor Vergata, Health Communication srl, Roma

Cordis, Better Technology for Europe's silver surfers, Re-search UE results magazine, p.31. N°32, March 2011, Luxemburg

Giudice, F., La Rosa, G., Risitano, A. (2006) Product Design for the Environment. A Life Cycle Approach, CRC Press, Taylor and Francis Group, USA.

http://cordis.europa.eu/fp7/ict/programme/challenge5_en.html

Jordan, P.W. (2000) Pleasure with Products: Beyond Usabili-ty, Taylor and Francis

Marsili M., Sorvillo M.P. (2002), Previsioni della popolazione residente per sesso, età e regione dal 1.1.2001 al 1.1.2051, Istituto Nazionale di Statistica, ISTAT, Roma 2002

Martin J.L., et.al, (2008), Medical device development: The challenge for ergonomics, Applied Ergonomics 39, pp. 271–283

Maudsley, D., Greenberg, S. & Mander, R., (1993), Prototyp-ing an intelligent agent through Wizard of Oz. in INTERCHI '93 Conference Proceedings, pp. 277-284.

Melnik, D., et.al (2005), Environment and Human Well-being. A Practical Strategy, United Nations, Earthscan

Norman, D. (1998),The invisible computer, MIT Press

Rubin, J. (1984) Handbook of usability testing: how to plan, design and conduct effective tests, John Wiley & sons, New York.

CHAPTER 49

# Development of Risk Sensitivity of Workers

*Toru Nakata*

Research Institute of Secure Systems,
National Institute of Advanced Science and Technology (AIST)
toru-nakata@aist.go.jp

## ABSTRACT

Most of industrial accidents can be prevented by risk-sensitive workers who notice foretastes of the dangers and take measures against them. Therefore it is one of the most important tasks of the company management to develop the workers' sensitivity for risks in the workplace. For this purpose, we must find out the difference between safe workers and high risk workers. When we are investigating workers condition in various companies, we often find the difference: there is a characteristic group of workers who tends to repeat more mistakes than ordinary workers. Knowing the difference, we can detect who are the most dangerous workers in the company, and then we can educate and guide them to behave safer. We carried out survey on 157 workers in a construction company. By using questionnaire sheet, we asked them what and how they feel risk. In addition to this, the workers answered about their experience of injuries in the works. The result shows there are 4 types of workers, namely novices, intermediate workers, safe workers and dangerous workers. The difference among the 4 types is mainly by personal experience of the work and injuries. The dangerous workers tend to repeat minor injuries, so they are reluctant to care about risks. The experience of instructing and assisting junior workers helps such dangerous workers to behave safer.

**Keywords** Industrial Safety, User Interview, Risk Sensitiveness, Vector Quantization.

## 1. INTRODUCTION

If you know what are dangerous well, you can avoid accidents well. Risk-sensitiveness of the workers is the key factor on industrial safety.

Conventional researches are, however, somewhat reluctant to investigate what the workers think. Actually, some our papers in past were criticized by some reviewers who believe the priority of safety research is located at mechanical protections and management, not the workers' mindset. They regarded our thinking on workers' mindset as transferring responsibility from the safety managers to the workers.

In spite of such criticisms, this paper focuses on the difference of the workers' mindset and risk-sensitivity. Difference is difference. We must protect the workers by correcting each worker's negligence for the risks. The company management has to develop each worker to obtain correct risk sensitivity in the workplace. Famous "Hawthorne experiment" (Gillespie, 1993) is one of the researches to approach worker's consciousness for safety. Today, management program of Southwest Airlines to increase communication among the workers (Freiberg and Freiberg, 1996) is reputed as good practice for increase safety of their operation. To consider what the workers think will be better and more precise than dealing each worker equally as an existence that makes mistakes according to "Human Error Probabilities" determined by Gertman and Blackman (1994).

In general, it is difficult to measure workers' mindsets on safety. We have only limited ways to measure it. We adopt questionnaire sheet method, which is to collect opinions of each worker by using papers, not by direct conversations. Therefore we need some sophisticated techniques to analyze the data and to get meaningful results. We test the validity of classification methods for analysis of such opinion data.

## 2. METHODS

We carried out survey on 157 workers in a construction company in Japan.

By using questionnaire sheet shown in Fig.1, we asked them what and how they feel risk. The sheet has three parts of topics about general hazards, risk of using tools, and risky conditions. For each risk, the workers answered injury experience, estimation of amount of each risk, and practice to wear protectors (i.e. helmet and lifeline). The answers are transferred into numbers for statistical processing: we deal answer word "Strongly" as 2, "So-so" as 1, "Not especially" is 0, and so on. Then we recompose each data for 3 types as the following:

- Experiences of injuries on each hazard or risk.
- Opinions about how they attach importance for each hazard and risk.
- Practices of wearing protectors for each hazard or risk.

We also heard the workers about their experience of injuries in the works.

Now we get a huge matrix data connecting experience, understanding of risks, and practices of each worker.

472

**Circle word of your answer**

<< Questions about General Hazards >>

 About injuries by falling from height

| How many times have you been injured? | Never / 1—2 / 3—4 / 5— |
| Are you careful about this type of accidents? | Strongly / So-so / Not especially |
| Do you wear the protector for this accident? | Always / Sometimes / Never / Protector not exist |

This part has more questions about hazards of pinching, hyperthermia, cutting, slipping, and "anything else".

<< Questions about Risk around Tools >>

 Accidents when you are on stepladder or scaffold

| How many times have you been injured? | Never / 1—2 / 3—4 / 5— |
| Are you careful about this type of accidents? | Strongly / So-so / Not particular |
| Do you wear the protector for this accident? | Always / Sometimes / Never / Protector not exist |

This part has more questions about step differences of floors, obstacles over the head, falling of objects, handheld machines, big machines, and "anything else".

<< Questions about Risky Conditions >>

How often is the schedule of work changed with short notice?
Often / Sometimes / Rarely / Never
- - - - - - - - - - - - - - - - - - - - - - - - - - - - - - - - - - - - - - - - - - - - - - - - - - - -
Do you think that such situation increases the risk of accidents?
Yes / No / Do not know.

This part has more questions about shortage of personnel, harried works, unaccustomed works, fatigue, dim lights, slow colleagues, careless colleagues, novice colleagues, working alone, lack of communication to colleagues, censorious supervisors, and insufficient briefing.

Fig.1. The questionnaires sheet. Questions are planned for inspecting attitude against common hazards, attitude for correct use of tools, and attitude against dangerous situations.

Fig.2. Overview of the clustering workers into four groups by using Vector Quantization technique in respect to experience of the job, experience of injuries, personal practice for safety, and opinions on hazards in the workplace.

## 3. RESULTS AND DISCUSSION

### 3.1 Analysis of Naïve Way

Before using complex analysis, let us observe the result with basic methods.

Does the length of work experience explain data enough? The answer was no. The length of experience had very weak relationships between workers' answers like data shown in Fig.3.

This result means that understandings for risk do not simply depend on temporal length of experience. Even though the work length is same, mindset and practice for safety differs person by person. Therefore, we need other ways of analysis which can detect the differences.

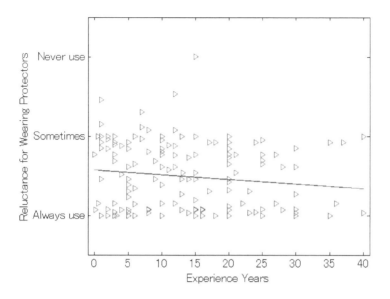

Fig3. Answers about custom of wearing protectors versus length of work experience. The work experience does not necessarily evoke practice of wearing protectors.

## 3.2 Classified Types of the Worker

Processing the data matrix by using clustering method, we can divide the workers into groups of the workers with similar mindset.

Each group will represent typical worker. In many cases, major difference among the worker is experience. So the division will be along with the experience, and the groups will represent novice type, intermediate type, and expert type.

We choose Vector-Quantization method as the clustering method (Gersho and Gray, 1991). Vector-Quantization method is especially good at clustering non-linear and high-dimensional data. But it has a limitation that the number of clusters must be power of two. We select 4 as the number of the clusters.

We classified the 157 workers into 4 groups, each of which represents typical characteristic of the workers. The distribution of experience lengths are weekly correlates with the difference of group (Table 1, Fig.4).

Table 1. Basic features of the classified group

| Group | All | Novice | Injury Repeater | Inter-mediate | Expert |
|---|---|---|---|---|---|
| Number of people | 157 | 17 | 36 | 61 | 43 |
| Median of experienced year | 12 | 5 | 10 | 12 | 15 |

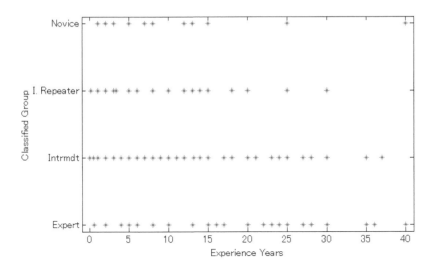

Fig.4. Distribution of lengths of work experience of the 4 groups. The correlation coefficient between the groups and years is 0.24.

We attach names for the groups as the followings:
1.　Expert Type
This cluster consists of 43 safety minded workers, who are very willing to wear the protectors and pay adequate attentions to actual risks in the workplace.
2.　Intermediate Type
This cluster consists of 61 workers, who have average amount of experience on the job and injuries. Their wills to wear protectors are stronger than that of novices.
3.　Injury Repeater Type
This cluster consists of 36 risky workers, who are rather reluctant to wear the protectors. They also pay less attention to dangers, so that they suffer more injuries. Since the injuries are minor, they do not realize the risks of critical accidents and do not reform their mind.
4.　Novice Type
This group consists of 17 workers, who have little experience of the job and injuries. They are reluctant to wear the protectors. They are also rather unwary against the risks, so they have less knowledge about actually patterns of accidents.

## 3.3 Differences among the Group

We found that practice of wearing the protector is good for preventing injuries (Fig.5). Especially the workers of Novice type think little of the protectors, and have been injured more frequently. Intermediate type covers the injury risk by their skill rather than using the protectors.

The mystery is that the people of Injury Repeater type answered they usually wear the protector, in spited of their frequent injuries experiences. We can

476

understand it by focusing on opinions of risk-sensitiveness (Fig. 6). Injury Repeaters are, in general, least careful for hazards. Even though they suffers injuries many times, they just wear the protectors and do not modify their carelessness. They suffer rather minor injuries that are not enough to reclaim their mind.

It is also interesting that being most careful against all hazards and risky conditions are not the best. Regarding the opinions about carefulness, Experts group answered less careful than Intermediate group (Fig.7). Magnitudes of wariness of Experts are at similar level of Novices. That means the experts have learned the difference between really dangerous factors and rather weaker risks, so they do not attach high wariness for all factors.

The people of Intermediate group have the highest wariness, but they less wear the protectors in practice. This gap between mind and practice prohibits the workers from becoming experts.

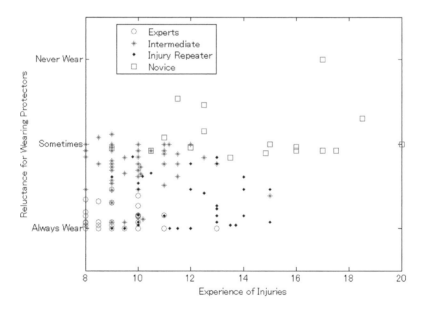

Fig.5. Distribution of the workers' answers on the plane of injuries experience vs. practice of wearing the protectors. The practice of wearing protectors has influence on injury prevention.

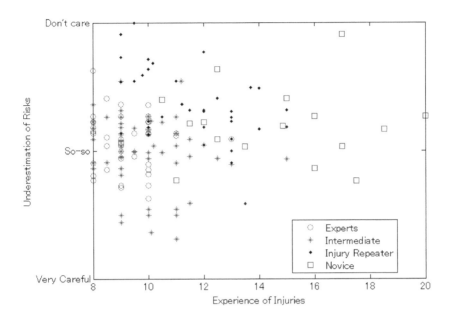

Fig.6. Distribution of the workers' answers on the plane of injuries experience vs. opinions of underestimation of the risks.

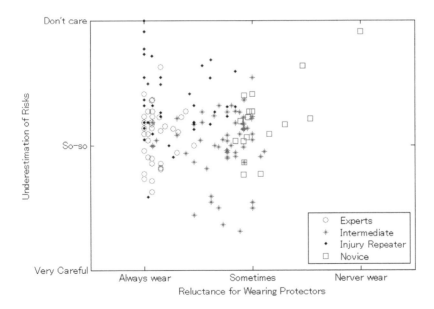

Fig.7. Distribution of the workers' answers on the plane of practice of wearing the protectors vs. opinions of underestimation of the risks.

## 4. CONCLUSION: DEVELOPMENT OF WORKER'S SAFETY MIND

In ordinary case, a worker has safety mind of the novice type at the start of the jobs (Fig.8). After taking experience of the jobs for several years, the worker develops to the intermediate type.

According to personal interview which was carried out along with the questionnaire, there are 2 types of trigger to reclaim the intermediate workers to esteem safe and to promote themselves to the trusty worker type. One is experience to instruct and to assist junior workers. The other is to suffer a severe injury.

Using the database made from the 157 workers' opinions, we can classify workers into the 4 group. If some worker is classified as immature type, we should approach them for development to experts. The points of the approach are 1) instruction to wear the protector, and 2) education of the difference of really dangerous risks and others.

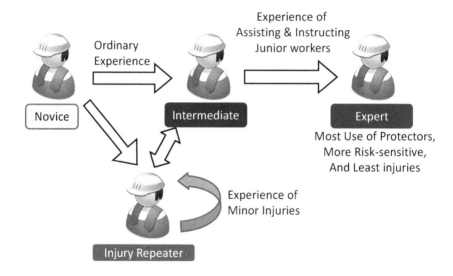

Fig.8. Developmental path of workers' risk sensitivity.

## REFERENCE

Freiberg, K., Freiberg, J. (1996). NUTS!: Southwest Airlines' Crazy Recipe for Business and Personal Success. Broadway Books.

Gersho, A., Gray, R.M. (1991). Vector Quantization and Signal Compression. Springer.

Gertman, D. I., Blackman, H.S., (1994). Human Reliability and Safety Analysis Data Handbook. Wiley-Interscience.

Gillespie, R. (1993). Manufacturing Knowledge: A History of the Hawthorne Experiments. Cambridge University Press.

CHAPTER 50

# Ergonomic Shaping of Learning Places for School Children

*Przemyslaw Nowakowski*

Wroclaw University of Technology, Department of Architecture
Wroclaw, Poland
przemyslaw.nowakowski@pwr.wroc.pl

## ABSTRACT

A proper development of human body, particularly in the childhood and adolescence period requires an intense movement activity. The young, however, are more and more overburdened with school responsibilities and they have to spend much more time at school and when doing homework at home than ever before. Furthermore, children frequently give up games and play involving movement and spend more time at a computer. The sitting lifestyle and use of furniture for sitting of poor ergonomic quality have had a bad influence on their psychophysical condition. Unfortunately, defects in posture, overweight and stress are common contemporary social phenomena, which seriously affect the young generation. To divert such trends, some system social, cultural, educational and designing actions in the relationship of man – technical product should be undertaken.

Organizing the education process and a selection of school equipment, including the furniture equipment, requires a new approach, which further may change the negative trends in the psychophysical development of young generation. The preventive actions aimed at maintaining a proper development of human body may be supported by specialized ergonomic furniture equipment adjusted to various age groups of children and teenagers of diversified psychophysical condition.

In the paper, the following issues will be discussed: evolution of a school system as well as evolution of school furniture, phenomena conditioning production of furniture for children, the facets of culture shaping work space for children and teenagers, postural defects and ergonomic aspects and criteria of shaping furniture for study.

**Keywords**: ergonomics, sitting position in school, school equipment

# 1    INTRODUCTION

A lifestyle of man considerably affects human psychophysical condition. Physical activity has always been deemed a way of keeping fit and slim. However, civilizational changes including prolonging study and work times, development and popularization of mass media, computerization and automation have had a significant influence on changing lifestyles, customs and values in highly developed countries. There is a strong tendency among both children and grown-up people to limit their kinesthetic and sensory activities for the sake of the static ones. Children nowadays spend more time playing at home unlike in the past, when they spent their playtime outdoors.

Children's young bodies require quite a lot of kinesthetic activity for their appropriate development. However, contemporary school curriculums are so extended that children stay longer at school and doing their homework takes them much more time at home. Common availability of the internet and computers make children give up sensory plays for the sake of static entertainment provided by electronic devices. A sitting lifestyle and using some sitting furniture of poor quality deteriorate children and youth's bodily conditions. Postural defects, overweight and stress are unfortunately common social phenomena which mostly affect the young generation. To counteract such negative trends systematic preventive actions in the relation of man and technical objects should be undertaken on a large social, cultural, educational and design scale.

The education system should be organized in a manner that promotes movement, kinesthetic activity and a healthy lifestyle. Additionally, a selection of furniture and equipment supporting studying should ensure comfort of use and eliminate unfavorable work conditions. Preventive actions at school may be supported by ergonomic furniture which allows maintaining an appropriate sitting position when studying.

# 2    EVOLUTION OF THE EDUCATION SYSTEM
# AND FURNITURE FOR PUPILS

The European education system was solely based upon the church education or private tutoring for the wealthy, generally with individual teaching one-to-one. Not until the education reforms in the nineteenth century began by the Swiss educator Johann Heinrich Pestalozzi contributed to emergence of teaching methodology, the education system and compulsory full-time education for all children.

Use of industrial technologies for mass furniture production and other school equipment forced some size standardization, particularly in terms of the sitting height and the height and size of the tabletop adjusted to body measurements of pupils at different ages. In 70. and 80. of the nineteenth century first professional books on designing school objects and furniture appeared (mainly in Germany and the USA). Simultaneously, many doctors dealt with the issue of „school illnesses" in the light of their medical and preventive activities, including in particular ergonomic shaping of furniture equipment. Orthopedic requirements were recognized and systematized quite early, however at that stage of technology preventing the spine illnesses was effective to an insignificant extent. It was observed that the most serious postural defects occurred when writing papers (Berquet, 1971).

At the turn of the nineteenth and twentieth century the education system was not fully centralized yet. However, the state administration already took a great interest in education. It was reflected in producing furniture systems already at an industrial scale in quite a lot of countries, which was coordinated by state and regional administration of various levels. Also, the first two decades of the twentieth century was a time of experimenting in the field of industrial production, design and effective matching with requirements of more and more unified school systems. Until the 60. the most famous architects such as Richard Riemerschmid, Eliel and Eero Saarinen, Jean Prouve and Arne Jacobsen also committed to designing not only school buildings but also to specific furniture for educational facilities (Mueller, 1998). It was the time when measurement standardization referring to adjusting pupils' desks and chairs to various age groups became popularized.

## 3    FACTORS INFLUENCING PUPIL'S FURNITURE PRODUCTION

Production of contemporary furniture for pupils and those designed for schools in particular depend on many social, technical and economic factors. There are **humanocentric** and **medical factors** (anthropometry, ergonomics, orthopedics, optics, ophthalmology), **pedagogy** (teaching systems, flexibility in using furniture systems, furniture typology and purpose and use of furniture), **designing** (spatial requirements, functionality, forms and aesthetic values, selection of colors), **construction** (durability, products life-span, juxtaposition of elements and systems, material costs and expenditures), **official rules and regulations** (norms, use safety, producers' warranties, the natural environment protection), **economic factors** (production technologies, production and distribution costs, technical progress, shopping and bargaining procedures).

The ergonomic quality, and aesthetic in particular, of school furniture commonly available is frequently poorer compared with home or office furniture. It results mainly from normative and usage requirements (durability and life-cycle), but mostly it is a consequence of purchasing furniture for schools through bargaining procedures in which the lowest price is the most often the only criterion for choosing an offer. Moreover, bargaining requirements are formulated by clerks and administration workers of various levels who usually do not have appropriate knowledge on ergonomics and orthopedics. Thus, such a situation is not favorable to a development of furniture branch and consequently a visual range of furniture for pupils has not altered for decades. Also, fulfilling all the mentioned requirements in a production process is not easy, and some of them are even mutually exclusive (aesthetic values – durability – production expenditures).

Furniture design, production and distribution for pupils is therefore a complicated process in which specialists such as teachers, designers, doctors, technologists, process engineers, tradesmen and clerks have to take active part. Effects of their actions are sometimes mutually exclusive. A role of clerks in particular combines lots of contradictions – on the one hand high safety, use and durability requirements and on the other hand low budgets for purchasing school furniture.

# 4    CONSEQUENCES OF WORKING LONG HOURS IN A SITTING POSITION

A negative consequence of the long hours sitting life style is (particularly with children and young people) is various ailments, which can be generally classified as postural defects. Moreover, members of contemporary society seem to be not aware of health hazards caused by assuming incorrect sitting positions on uncomfortable sitting furniture. It may result from the fact that an adaptive and defensive system of human body does not respond early enough to the pathogenic phenomena.

Going children to school and necessity of sitting in classes long hours coincides with their strong natural and biological urge to move. Contemporary school systems however care very little for stimulating kinesthetic activities of pupils. Therefore, they reinforce incorrect habits and psychophysical defects related to them.

Postural defects in childhood and adolescence may have different reasons. They can be caused by habit defects, sometimes called environmental defects, which result from disorders in correct bodily postures (Children postural defects). Factors which contribute to these postural defects among children and adolescents are muscular insufficiency maintaining the body, sitting for long hours, assuming incorrect sitting positions, and using seats and work stations of poor ergonomic quality. Incorrect sitting positions at school therefore lead to various spinal illnesses such as the sagittal and frontal curvatures of spine.

The correct human posture is characterized by the head kept straight, the physiological curvatures of spine in the sagittal plane, a well vaulted chest and rib cage, a pelvis well supported in a sitting position, straight legs and well vaulted feet. Postural defects in adolescence may be the reason of serious illnesses whose initial stage is usually imperceptible and unfortunately proceeds as insignificant. A defect and an ailment unrecognized early enough and not treated becomes reinforced and established, which in turn may lead to even more serious illnesses in the adulthood. One of the major factors preventing postural defects is movement. However, in the present educational conditions, sensory and kinesthetic activity is significantly limited. Preventive measures to promote and maintain correct sitting habits is therefore of much importance to maintain and stimulate correct postures. Childhood is the best period for adopting preventive measures against postural defects.

Polish statistical data referring to the health prevention among children are alarming. They indicate that approximately 60 % of school children have some postural defects and scoliosis is observed among over 35% of children population (Postural defects profile). The data include both children with confirmed clear and distinct postural defects and adolescents with considerable deficiency in kinesthetic activity, poor physical dexterity and fitness and various movement system disorders. Lack of kinesthetic activity is also one of major causes of overweight, which already refers to approximately 30% of children and adolescents in the European Union countries. To prevent this phenomena in Poland, the education administration, the health ministry and media promote and recommend health prevention programs. However, to improve the spinal conditions among children and adolescents, also including the grown up people, they should be activated kinesthetically and their lifestyles should be changed. In shaping the school environment and home work stations a considerable role may be played by furniture designers cooperating with orthopedists, parents and teachers.

# 5    ERGONOMIC ASPECTS OF SHAPING FURNITURE FOR STUDY

A conventional study position and place for a pupil is designed for reading and writing first. These activities are considered as little precise and clear, not requiring any special or advanced equipment and light. The exception is writing, which may be deemed a precise activity, in which pupils usually assume rather leaning position in order to minimize the sight distance to even 20 – 30 centimeters.

The study places for pupils comprise two major elements – a seat with a backrest and a desktop. They may additionally be equipped with a footrest, a hook for a backpack, containers, drawers, shelves for books, notebooks, and school accessories.

Ergonomics encompasses adjusting the entirety of environment for needs and capabilities of man. One of the areas of its interest is therefore relations man – technical product. In case of study places for pupils an important matter is adjusting their individual elements to requirements (also anthropometric) for maintaining possibly the best work and study comfort in a sitting position. The comfort should be ensured by eliminating various physical loads (muscular overloads) and psychical (such as stress).

In a sitting position a child should relieve their legs and back and reduce an energetic cost during an activity. However, a natural position of man is a straight standing position and a specific spinal arrangement related to it, with its three specific curvatures: lordosis (in the area of neck), kyphosis (in the area of shoulder baldes) and lordosis (in the area of loins). Sitting causes a change in the system of pelvis – spine and their mutual arrangement and deformation of a neutral muscular and skeletal system. In a sitting position the spine becomes deformed causing unequal and irregular overloading of its individual circles and intervertebral discs, in the area of loins particularly. Static overloading some back muscles overlap the spinal burden and contribute to development of various illnesses and postural defects. Sitting hunched over at a desk also causes constricting and failure of internal organs, and pressure upon the diaphragm make the breath less efficient, quicker and shallow. Maintaining a healthy sitting position poses therefore an ergonomic challenge to orthopedists, anthropologists, designers, teachers, parents and also children themselves.

Eliminating deformations of the neutral muscular and spinal system, static overloading of muscles and maintaining of correct sitting positions require appropriate body support in certain places, particularly in the area of thighs, pelvis and loins. Sometimes, a footrest is additionally required as well as support for elbows, forearms, hands, the neck and head, etc. Simultaneously, in designing a system of supporting points for parts of the body, a principle of **dynamic sitting** should be taken into consideration. It is connected with a necessity of changing a sitting position every few minutes in order to involuntarily exercise muscles overloaded as a result of a static position. Ergonomic seats should therefore be flexible and change positions of supporting points depending on a current position of body, however with a constant and correct position of the spine and pelvis. Dynamic sitting allows for a changing and symmetrical overloading individual parts of muscles and regeneration of those not overloaded. Therefore sitting long hours is not too fatiguing. Ability of dynamic sitting is favorable to such behaviors as easier perception of static overloading of muscles, searching for a more comfortable position, stimulation of mobile activity while sitting, maintaining a correct sitting position (relatively neutral for the spine), relief to the circulatory system and internal organs, relaxing of tense muscles, concentration and psychophysical improvement. The purpose of dynamic sitting is not only possibility to work longer hours in a sitting position, but also maintaining conditions for activities beyond work in the sitting position.

Ensuring dynamic sitting and as a result possibility of changing positions when sitting reduces static overloading of the muscular and skeletal system and allows for stimulating the intervertebral discs and their regeneration. Appropriate exercise when working and occasional work in a standing position are efficient preventive measures when sitting long, and also they provide opportunity for regeneration in a situation of static overloading. Physiologists and orthopedists claim that frequent changes of sitting positions and activation of the movement apparatus are necessary, particularly among children and adolescents, that is in the time of the body growth and development (Dynamic sitting).

It is supposed that when reading and writing a neutral sitting position is assumed, in which the angle of bends of parts: body – thighs – calves is 90°. A longer maintaining such a position requires however a convex prominent rest for the loins part, in order to maintain a natural arrangement of the spine. Such an arrangement is also optimal for the muscles work, since they support the entire muscular and skeletal system of the breast part. In fact, a longer maintaining of a straight sitting position is not possible at the norm pupils' desks. Since the desktop placed at the level of elbows is too low for writing, and the sight distance is too big. In such circumstances leaning forward and lowering the head is necessary. Such a spine position causes a considerable bending of the spine in the sagittal direction and deformation of the neck lordosis and the lumber lordosis in particular. Improving the sitting conditions at desktops for studying in a position leaning forward is possible mainly thanks to elevating and additionally slightly sloping the work desktop. In this case resting the forearm and the body itself is more effective and at the same time the sight distance is shortened. A shorter sight distance is particular significant when learning to write , which is a more precise work. By raising and sloping a desktop the sight distance gets shortened and should be approximately 30 centimeters. This slight distance at a possibly straight sitting position should be a criterion for defining a desktop height and a distance between the desktop and a seat. However, a desktop which is too high (reaching the armpits) causes considerable raising of hands placed upon it and raising the arms at the same time. Pulling up both arms is tiring and reduces dexterity and freedom of hand movements. A child then drops a free arm then and sits at a table in a non-symmetrical manner with on hand writing on a desktop and the other free put down. Writing at a desktop which is too high with the other hand lowered (not active at a writing task) may cause scoliosis. However, shortening the distance to the desktop level may lead to short-sightedness (Berquet, 1971).

Sitting people are usually leaning back for relaxation. When sitting on a stiff chair  one must then move themselves forward. When leaning backwards, the natural spinal system changes its shape and the lumbar curvature instead of being concave becomes convex in the sagittal direction. The head at the same time frequently tilted and lowered additionally overloads the neck deforming this spinal curvature.

Spinal curvatures considerably overload both spine circles and intervertebral discs and back muscles. Assuming such positions may lead to various illnesses such as hyperkyphosis (deformation of the spine in the sagittal area, where an excessive lumbar spinal curvature occurs).

Desktops selected for different age groups should be then approximately 10 – 15 centimeters higher than an elbows' level of a sitting pupil, so slightly higher than selected nowadays according to norms established by the ministry of education. For younger children, who find learning to write a more difficult and time consuming tasks,  sloped desktops are recommended.

## 6    ERGONOMIC CRITERIA FOR SHAPING FURNITURE FOR PUPILS

Kinesthetic activity for school children considerably decreased compared with the past. Physical plays are preferably displaced by the static ones, particularly with the use of a computer. Also, the school tasks and responsibilities are more complex and time consuming, requiring concentration when sitting for a long time. Pro-health prevention therefore encompasses meeting many ergonomic requirements including every day activities and furniture equipment. The everyday activities encompass:

- dynamic sitting and frequent changing positions,
- frequent breaks when working in a sitting position and short physical exercise,
- rest of the entire surface of feet when sitting,
- horizontal rest of the entire surface of thighs on a seat,
- maintaining the bend of thighs and calves at the angle of 90°,
- freedom of moving thighs on a seat.

As far as equipment is concerned the following should be taken into consideration:

- arrangement of adjusted to each other elements of the system of seat – backrest – desktop – footrest in terms of their size,
- adjusting a workstation to a size (and age) and kinesthetic abilities of a pupil,
- ensuring a principle of dynamic sitting,
- implementing sloping desktops,

- implementing anti-sliding mats or confining battens,
- size of desktops that enable placing both forearms and elbows,
- use of seats with backrests leaning backwards and forwards with an adjustable rest for loins,
- profiling the front part of a seat not pressing the thighs in the knee bend,
- supporting the lumbar part in both the straight position and when leaning forward or backward.

Meeting these criteria requires both individual awareness of pupils and their parents, systematic exercise and selection of relatively expensive equipment, well adjusted to the body size of growing children. In the scope of design and realization extending the offer of ergonomic school furniture requires cooperation of many specialists, and also promoting actions for common enhancing of pro-health awareness.

## 7    CONCLUSIONS

Frequent occurrence of postural defects reflects the contemporary cultural changes and reversal from kinesthetic activities towards the sitting life style. However, the sitting equipment commonly available for pupils to study is not suitable for proper bodily postures. The analysis of pupils' sitting positions at different desktops provides evidence for the fact that generally available standardized and horizontal desktops are not high enough for writing tasks. Therefore, the school table should be higher and provided with a possibility of regulating and sloping the tops, and chairs with an adjustable level of seats, with the functionality of sloping them forwards and backwards. Paradoxically, these postulates are not new, and the old school tables with mechanisms enabling arranging and moving their individual elements seem to be a particular model for meeting the above mentioned requirements. Besides, some modern systems of chairs and tables are clearly modeled upon the old-style school tables. However, they are not frequently used and mostly occur in the highly developed countries of the western world.

Designers of furniture for pupils may therefore make efforts to improve the ergonomic quality of the contemporary school furniture. The ergonomic changes encompass both furniture used in schools and at home. Durability, safety of use and the economic value comprise the basic criteria of choosing the school furniture and even often furniture for home (for doing homework and studying at home). These criteria however should not obscure the pragmatic aspects of furniture use (ergonomic, orthopedic) and its aesthetic qualities. Then comfort of work, aesthetic experience, health prevention should be the most important aim and inspiration for designing and manufacturing the contemporary furniture for school children.

## REFERENCES

Berquet, K. 1971. *Schulmoebel, Geschichte, Auswahl, Anpassung*. Ferdinand Duemmlers Verlag. Bonn, Germany.

Mueller, T. 1998. *Das Klassenzimmer, Schulmoebel im 20. Jahrhundert.* Prestel. Munich, Germany

„Children postural defects", Accessed November 15, 2010,
www.wady-postawy.blogspot.com/

„Dynamic sitting", Accessed November 15, 2010,
http://www.leuwico.com/web_d/ergonomics/dynamic_sitting/

„Postural defects profile", Accessed November 20, 2010
mz.gov.pl/wwwfiles/ma_struktura/docs/profil_wad_postawy_24022010.pdf

# Section VIII

## Ship Design

# CHAPTER 51

# Operating a Boat

*Massimo Musio-Sale*

Università di Genova
Genova, ITALY
musio-sale@arch.unige.it

## ABSTRACT

**Fundamentals of ergonomics applied to helm stations of boats.**

From the tiller hinged on the stern, up to the steering wheel introduced by the great Vessels, depending on the type of vessel to which it refers, may also profoundly change the requirements of good ergonomics of a helm station.

The development of marine engineering knowledge has led to the development of ever-increasing size of vessels; so, the solution of a traditional tiller would become inconsistent with the figure of the helmsman, therefore a single man would not have been able to control properly. To do this, the solution was introduced through chains and ropes, pulleys referrals through appropriate scaling, by leading the rudder to the steering wheel. In doing so, any vessel, however large, could be ruled by one man. From sea to air or land the steering wheel is an application derived from the marine "rudder wheel", which can be traced back to the sixteenth century.

Nowadays, however, the traditional tiller still exists and is highly appreciated for boats -usually small and light- designed following the priority for simple, reliable and efficient solutions. The spread of boats of larger size, has described the adoption of a fairly standard setting, designed to meet ergonomic requirements for the functional control of the boat, in the most direct and efficient way.

The introduction of modern technologies, for transfer the control signals and monitoring, have developed interesting application of these "drive by wire", totally innovative if compared to traditional mechanical technologies. Even more interesting are the "Wi-Fi" or "blue-tooth" solutions, now entering the market. Today times, to control the movement of a large vessel through a joystick is one of the most proven and reliable solutions. Moreover, the modern technology allows applications of remote controls able to drive a boat from anywhere, even from outside the boat itself, like in a videogame. This introduces new and different ergonomic problems.

**Keywords**: ergonomic, technology, command bridge

# 1 INTRODUCTION

We can say that there are two basic principles of operation to control the movements of a boat: acting on fluid dynamics -through the diversion of the fluid streamlines operated by the rudder- or acting on drives -through the action of the propellers, sails, or simply of rowing-. The first case, in order to be efficient, needs to have a good boat speed, while, if the boat is stationary, there are no fluid streamlines along the hull. The second case, however, can be efficient even applying the required effect on a stationary boat. This simple observation tells us that the rudder is an organ of direction inefficient while manoeuvring in tight spaces, stationary or nearly stationary, whereas the sensitivity is excellent for its response in high speed, both to make large evolutions, or to regulate small movements.

Just the need to feel the rudder response is the reason why -traditionally- the expert sailor prefers the use of the control stick, than any other remote control, specially in the tradition of sailing boats, because through the stick the skipper can "hear" gait and balance of the boat, even in relation to the wind direction.

The resistance of rudder on the rudder-stick is proportional to the mass of the boat. Although, every good boat should be balanced and shows almost no resistance on the tiller to stay en route, a simple-tiller must exert a force proportional to the mass of the boat to cause her turning. Especially, when the boat is big and heavy, the rudder surface should be wide as well, to move the amount of water that enough to obtain the evolution. The rudder on the sternpost, traditionally must have a lever arm proportionate to the effort required for its operation; the tiller -which is the lever arm- should therefore be proportionate to the size of the boat: longer for large and heavy boats, shorter for smaller and lighter crafts; the manoeuvring action is carried by one person regardless of the size of the vessel; the stick can not be too short, to require excessive effort, nor too long, to be developed kinematics too large, not related to the size of the anthropometric measurements of a single person.

When the rudder is related to a big boat, necessarily requires an effort of vigorous exercise; the adequate solution is the introduction of the steering wheel, already successfully applied in the construction of the great galleons, dating back to the sixteenth century.

When the government is exercised through the action of the boat propulsion, as mentioned, it may be on oars, sails or propellers. Ignoring the case of small rowing boats, for the treatment of which would require a separate ergonomic in-depth, we can observe that sails are traditionally controlled by a series of cables -halyards and sheets- manned by sailors in charge of the exercise of their functions.

The coordination of the operation of sails, by seafaring tradition, observes a strict military hierarchy, where the captain gives directives, reverberated through the boatswain and promptly executed by all the sailors, each closely to their skills. The chain of command does not change much in principles, in relation to the size of the boat. While a small sailing-boat may be conducted by a single person -able to control both the rudder, and the sheets of sails- gradually growing in the size of boats, we need more people to control every single part; this not only because of the distance here existing between the different tack of sails and the bridge, with the steering wheel. In modern sailing boats, as we shall see later, all material functions can be performed by servo-mechanisms: a single person may be able to control and exercise the entire operational process, allowing to have a small crew, even to drive large boats.

The control of propellers requires the presence of a propulsion engine. This can be endothermic or electric; it requires anyhow a proper device to modulate intensity and direction. We will see later that we can act on an inverter device, on a throttle or even on the same pitch of propeller to get the required operation, but, in all cases, it is necessary to have a remote control -mechanical, electrical or hydraulic- that transfers and actuates the order, given by the captain, through a lever, a valve or a button. To complete this application, it is interesting note how, in the tradition, the position of the helm station was mainly placed in the aft of the boat, while in our days, the same is generally moved to the area between the third and the first half of the hull, towards the bow.

In sailing ships and boats the oldest, the aft position of the station allows both the proximity of the rudder to the wheel -to reduce remote control cables- and the favourable position to control in sight the trim of sails, the arrangement of crew and the general behaviour of the boat, in relation to route and waves.

The recent implementation of forward-position of the helm, toward bow and amidships, is preferable for motorboats -generally faster than sailboats- since external visibility results facilitate so to allows better control and greater security for traffic situations, while arriving. The forward position is made easier by the adoption of remote controls -electric or hydraulic- more flexible and less bulky than conventional mechanical systems.

## 2   OBJECTIVES

With regard to the objectives related to the ergonomics of a helm station for boats, apart from the many different types and sizes available, we must remember two basic requirements: safety in the operation of the vessel and, especially in the case of recreational crafts, the pleasant feed-back, while driving the boat in its operations. In those circumstances, to better understand the design choices underlying the arrangement of a control station on board a boat, it helps to understand the interactions that the captain must exercise in order to have perfect control of the vessel.

A sailing boat and a motorboat -one slow and one fast- are cases that show very different requirements of ergonomics: in the first case the control is determinate by the balance between intensity and direction of the wind, sail trim and angle of the rudder in relation to the direction of the route pursued; in the second case, the control is caused by the boat operating-speed and the angle of surfing the wave series, to the destination arrival.

To define the objectives of the theme, we analyse hereinafter some archetypal cases.

Figure 1  Small, single-mast, sailing boats (Cutter or Sloop)

These are boats (semi-cab or cab), designed for short cruises. In recent times, this type is combined with the definition of *easy sailor*. In the light of looking at the simplification of sailing fitting, the aim is to concentrate the entire control in the cockpit of the boat. This is made possible with simple references of halyards and sheets of jibs that become operational near the mainsheet and the tiller stick. Under these conditions only two people (but even one, if expert) can easily sail the boat while reducing risk and effort in favour of fun.

Figure 2  Large size Sailing yachts with one or more masts (Cutter, Ketch and Schooner)

The control of these tall ships normally should refer to an experienced crew, close-knit and well controlled. Certainly there is to be considered the fact of having the entire length of the boat, as all masts and equipment, obstructing the visibility around the ship with regard to the outline of other boats or traffic, while manoeuvring and approaching. The aim of facilitating these functions has recently introduced solutions derived from motorboats typology, such as for the installation of the helm station in a high position on a true flying bridge.

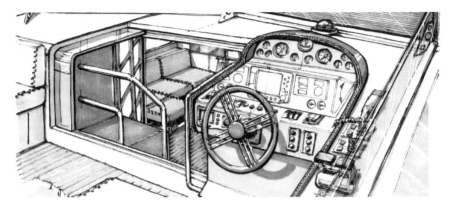

Figure 3  Small size motorboats.

Usually fast, sporty and easy to handle, they are open, semi-cab or cab, short cruises designed for couples or families. The driving position is usually placed amidships, in a definitely more advanced than in the sailboat cases. The steering station is located on-axis or on one side, preferably starboard side, to simplify the passage of the remote operated handcuffs with the right hand and to facilitate the visibility on the right side of the boat from which the priority must respect of those who's crosses. The commands closely reflects the aesthetics and ergonomics of car solutions, with a steering wheel and a cloche-stick for each engine, similar to the gearbox-stick, preferably to operate with the right hand, to control functions of the inverter and throttle.

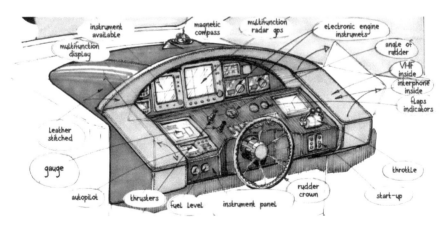

Figure 4  Large size motorboats

Usually slow, but in recent times also fast and very fast, they are commonly blue water yachts, cabins designed to accommodate many passengers on-board for long stays. Control is often available in more than one station, each time facilitating the specific functions related to the location of the various command posts. The main station, hosted on a high bridge (sometimes called bridge-deck) is usually between mid-ship and the first third of the bow, in a position of high visibility, often crowned by both sides of special terraces -called wings- to facilitate manoeuvres and approaches through outlets of specific commands.

Often there is a helm station placed on the top deck to enjoy the open-air driving; this arrangement is called the flying bridge and stylistically derived from the type of fisherman-boats. Sometimes larger boats are also equipped with a small helm for special operation, located in the cockpit, to facilitate the control of the boat during landing manoeuvres, stern to the dock.

## 3 METHODS

In all cases displayed hereinabove, apart from the different locations of the helm that characterize the various types, the common purpose of all solutions is always to try to concentrate in one location, interfaced with the size of a person's anthropometry, all functions of control and command of the boat.

Everything in modern boats is completely mechanized, to ability the control, decision and command of the entire vessel even only by single person. Just on board sailboats, the most sport ones, there is still a certain amount of manual skills for operation and adjustment of sails. Modern technology, mechanical, hydraulic, electrical and electronics has greatly changed the way of conducting a boat. Over time, the helm position from the stern, outside of the boat, have migrated into areas of greater visibility by introducing indoor and highly specialized solutions. The introduction of mechanical propulsion has caused the need to check the functioning of the management of the ship, not only for the direction of route, but also for the operation of engines. The design of a command post, being able to make use of modern technologies of translation data, does not fail to consider the different characteristics associated with the use of many different ways of going at sea. If staying on board a sailboat the contact with the open air looks very engaging, on a motorboat the speed sometimes is so high to need a protection for driver and guests behind a large windshield. The need to be able to move on board, even during navigation, still remains a condition usually appropriate in each vessel. To stay on car seat with safety belts buckled up, is a condition checked only on fast racing speedboats with survival capsule for pilot and crew. In all other cases, people can be free to move themselves, this in order to carry out their functions: either navigation, or simply to enjoy the life on board.

**The primary point of view.** The command position is therefore never determined in a fixed place. Man and commands -or controls- are in interface relation by moving the entire body -even by few steps- and not only by eyes or hands. An ergonomic condition is however almost always verified: this is the height of eyes of captain and the helmsman; it must be just the right position to see both the instruments and the boat outside around. It must be kept unchanged as much as possible either for the standing, or the sitting position. This justifies the presence of bridge chairs or sofas, raised in a high position and complemented by appropriate footrests. The seats, for this function, are so typical of a share of 65 to 75 cm from the ground, so that the eyes are always at a height between 160 and 170 cm from the ground.

**The compass.** The supreme device that reigns in the cockpit of a boat is the magnetic compass, as well as the main control is the helm wheel. Usually the compass is placed on the axis of the bridge, also coinciding with the axis of vision of the helmsman and the helm wheel. Its coexistence with other instruments on board is often difficult because the proper functioning of the compass is affected by the proximity of metal objects and effects caused by the operation of the rest of the electrical and electronic equipment. The functional logic would therefore dismiss the compass as possible from other sources of potential interference, but it takes to make the reading difficult, therefore, to adopt the values and compasses necessarily larger when located further away from the eyes.

**Other navigation devices.** Nowadays, the compass, while being a tool however fitted on board -because of the regulations prescribed by the principal Class Agencies- is not considered as specific and efficient as a modern GPS –Global Positioning System-, able to base tracking and direction of the ship on a satellite triangulation, certainly more accurate than the magnetic technology. The GPS instrument is therefore the principal reference in setting up a helm station of a boat, not intended for the sole coastal use. In addition to GPS, other useful devices for navigation are the autopilot, the depth sounder and the LOG. The first is an instrument that combines the control/command of the settled route, the automatic control helm, and often has a rudder-angle indicator to understand the exact position of the rudder from the keel line. The depth sounder is the instrument that measures the depth of water under the boat, while the LOG is the device that measures the instantaneous speed of the boat. The shape of these instruments may be available according to a modular series, or collection of instruments that combine multiple functions within integrated groups. In the case of sailing vessels, another device is considered essential: it is the indicator of the speed and apparent wind direction; it can be single or separated according to different indications. It is very useful to determine the ideal sail trim on the type of sailing chosen.

On the ergonomic layout logic, fingers, in order to make adjustments or programming, often should touch all these kits; so they must be related to the command of the helm-wheel and often they are arranged to crown the wheel in easily and accessible location.

**The instruments of engine control.** In addition to the navigation instruments there are other devices needed to check the functioning of propulsion. Primarily observing the rotation speed, and temperatures associated with the plant lubrication and cooling: these elements are monitored in each internal combustion engine. The lubrication system is also observed in terms of pressure, which the oil flows in the circuit. Equally helpful is revealed the capacity of engine to run the alternator to recharge the batteries and the charge state of the same. All this wealth of information means presence of many different instruments: revolution counter, usually larger and clear, crowned by the pressure gauges for water and oil temperature, oil pressure, ammeter and voltmeter; a big one gauge and 5 small ones, this for each engine on board. These are "instruments of reference" only, which should not be adjusted with the fingers. For this reason they can be located in an area of the dashboard, even away from the hands of the driver. In addition to all these devices -sometimes compacted into multifunction gauges- a series of warning lights and acoustic alarms to indicate malfunction of any of the above features are required; this means more visible presence of many colour lights, according to the gravity of alarm, and, in addition, the installation of alarm buzzers, usually discreetly hidden behind the dashboard, in a position not visible. In terms of ergonomic design, these instruments should be placed in relation to motor control. In aviation it is therefore usual to have these elements on the central panel between the pilot and co-pilot, adjacent to the controls power of the handcuffs. Even in the nautical field that provision would certainly appropriate, but not always the morphology of dashboards allows this rational solution.

**The board controls.** Depending on the boat size and general complexity, there are commands for many different functions. The main controls are those already described: the helm wheel and throttles of engines, including their ignitions; these are often associated with minor adjustments for other commands. They are ranging from command of the bow thrusters, than, for the control of surfing through flaps, and again, for the control of trimming by changing the drive angle with respect to the keel (the power trim command, present on the Z-drive or stern drive type Arneson), up to controls for the anti-roll fins or other kinds of devices to change the trim of the boat. These commands, useful for navigation, are joined by other commands, so-called auxiliary; present to give access to many service functions. They range, from the anchor winch, controls the switches for navigation lights, the horns and other services such as bilge pumps, engine room exhaust fans, controls, fuel levels, wipers, etc. Finally there are some service commands, for the comfort on board, which are still linked to the bridge. Among these are usually switches to allow the current various rooms: those who operate the autoclave for the distribution of water; those that allow the operation of on-board refrigerators and other major utilities. All these service switches, in larger vessels are placed in a special panel of electric utilities, where, in a centralized manner, throughout the body vessel can be controlled and managed. Now we can understand how the helm station of a boat is a complex place, and sometimes chaotic for the coexistence of many different functions to perform. A dashboard can include many elements of command and control that often constitute a veritable "forest", not only on the engineering, but also on perceptual point of view (either for mechanical or electrical cables and for wheels, throttles, switches, buttons, gauges, monitors, etc.). Scope of ergonomics applied to a helm-station project, is to try to simplify the perception of all functional objects, and make natural, instinctive and easy to use, the general interface, not to mention aspects related to the satisfaction associated with being able to exercise the command of the boat.

# 4     CONCLUSIONS

The research of ergonomics of the boats' helm station is oriented to establish priorities for command actions; this is obtained by simplifying the perception of a lot of information coming from the in-dash navigation systems, ever more refined and performing. In terms of perception, the modern video allows to reproduce any image and lend them to return information either in digital or analogical form. A display of the latest generation can play the illusion of a traditional pressure gauge, as well as an optoelectronic display or three-dimensional image, also as photography. It all depends on what is being transmitted to the instrument. With this solution a few monitors are suitable to replace many different instruments, ready to support information related to navigation, alternating with the operation of motors, or CCTV cameras, all on one screen, according to the request of the operator. The latest videos are also available in combination with touch-screen function; this allows to integrate the functions of command thanks to virtual switches appearing on display; in addition, this technology simply shift and change images on the screen by using hands; this could displays different information, moment to moment. The modern technology is therefore offering the possibility to make helm stations more compact and simplified in the provision, although reflecting increasingly large and powerful performances. It is difficult to introduce the dimension of "time" necessarily present in order to have on the bridge the various functions and controls, according to the program of different operations to play. Because the usability is likely to be more complex, having to play each instrument with many commands and control-functions at different times, is now according to a logic of computer science type.

The entire chain for the traditional control of the boat (rudder, handcuffs and etc.) in order to facilitate the manoeuvres is now available with a single element: thanks to the interface of a computer, the system can translate the impulses to the various devices with a single joystick, controlling every movement of the boat. Pressing once the propulsion engines, once the thrusters, through the joystick the boat can performs translations rotations diagonal movements and all sort of combined motion. And more: by interfacing the device with the GPS reception, we can program that the boat remains motionless on the geographical location where the joystick remains inactive in a central position, automatically opposing any movement caused by drift or leeway due to the presence of stream. Under these conditions, a boat mooring becomes easier as the parking of a car, and whoever, potentially, may seem an expert seaman. Other interesting possibilities to control a vessel are now available thanks to wireless technologies, Wi-Fi, Blue-tooth, infrared, or other data-transmission systems over the air. Thanks to these technologies, we can have portable control stations capable of driving a boat as if it were a toy model, being able to control the movements on board even from outside the boat, such as from shore or from a tender.

In the performance of the art of sailing boats the satisfaction of the command of the vessel must be enjoyed through systems that, in the complexity of the performance, provide an immediate feedback, able to "feel" the behaviour of the boat as physical perception, balancing touch and hear, as well as the view of the all-around atmosphere. In the case of sport boats, the skippers are accustomed to declare that the conduct of a good boat, as well as by sight, must perceive on the "lower back"; this goes to show that in ergonomics, the answer to the conduct of a boat must be seen as an experience of total involvement, and not just a response to aseptic implemented by reading data and gauges. This is the real goal to the best ergonomics of a helm station must be able to define, by borrowing innovative potentials and advanced technologies.

Figure 5  Vision of the future

Last, but not the least, we could be envisaged of a future situation already existing: We may imagine enjoying on-board the view from outside the boat, by returning on a HD screen. This image can be taken from a satellite camera or from a camera installed in an aircraft "drone" or on a radio-controlled blimp.

This solution could be used as a toy of the yacht to look others-around and how we are, at anytime. Our image, as it appears to others, may show a narcissistic spirit, but -in fact- it could be considered also serious and safety, useful for early recognitions of a landing or for sailing in dangerous situations.

## ACKNOWLEDGMENT

The authors would like to acknowledge Elisa Bassani and Elisabetta Pianta.

## REFERENCES

MUSIO-SALE M., YACHT DESIGN - dal concept alla rappresentazione - Editrice Tecniche Nuove MILANO: *figure 3 and figure 4, principles of ergonomy for boating ("progettare pensando all'uomo" pag. 101 -113)*
SCIARRELLI C. , LO YACHT – Mursia Editore, Milano: *principles of yachting*
ZIGNEGO M.I., CRUISE VESSELS DESIGN - 2009 DOGMA Editore, Savona: *principles of ship design*

CHAPTER 52

# 10 Meters Daysailer *"for All"*. Sustainable Technological Solutions for Easy Navigation

*Emilio Rossi, Giuseppe Di Bucchianico, Massimo Di Nicolantonio*

University of Chieti-Pescara
Pescara, Italy
e.rossi@unich.it

## ABSTRACT

In yacht design, the issue concerning the mobility of people on board represents one of principal problems to deal both from the technical point of view than functional. To this is added another factor: the consideration of characteristics and specificity of users when they move away from the definition of "standard user". On this side, the Design for All (DfA) Approach try to respond to the meaning of human diversity considering abilities and different psychophysical conditions of possible users as point of strength for the design of products and services usable by everyone.

This paper presents the results of a design experimentation conducted through the DfA Approach on a 10 meters daysailer that suggests sustainable solutions in order to facilitate the mobility on board and the navigation.

**Keywords**: Design for All, yacht design, users identification, easy navigation

## 1    INTRODUCTION: THE DESIGN FOR ALL APPROACH

According to the EIDD Stockholm Declaration (European Institute for Design and Disability), Design for All (DfA) is "design for human diversity, social inclusion and equality". This definition represents the fundamental principle underlying the design of environments, facilities, everyday objects and services, usable autonomously by individuals with diversified needs and abilities. Indeed, in DfA Approach, human diversity is meant as a "resource" to exploit rather than a "constrain" (Accolla, and Bandini Buti, 2005).

The main thesis of DfA Approach is that it extends its interests to a group of users that potentially embraces all humanity, with its psychophysical, cultural and social characteristics, both momentary than permanent. So not only people with disabilities, but also those that, at different levels, move away from the psychophysical or sociocultural conditions of user considered as "standard". These users, indeed, represent only 5% of entire population but for which we probably design more than 95% of products and services.

In this background also the meaning of "user" is extended from the "final one" to "All" the individuals in the entire product supply chain: we tend to pursue the satisfaction of needs and aspirations of all people who, for various reasons, want and have a reasonable chance to benefit the product in an independent and comfortable way (Accolla, 2009). This open-mindedness, however, may constitute a problem for the designer that, since the complexity of factors present, needs to know as much as possible characteristics and specific needs of target users in order to realize a product as inclusive as possible and not a priori discriminatory.

## 2    OBJECTIVES: THE MOBILITY "FOR ALL" ON SAILBOATS

Recent improvements of inclusive design have allowed to develop a key concept: handicap conditions are mainly attributable to environmental characteristics, maybe also as a result of a bad design process, than directly to the physical condition of users.

In sailboats, the review of literature relating to the activities to be performed, has underlined that most disadvantageous conditions, from the point of view of psychophysical conditions of people, are those concerning the issue of "mobility on board" (Larsson, and Eliasson, 2000), because conditioned by a diversified series of environmental and design conditions (the wave motion, the inclination of the hull in navigation, the lack of supports, etc.) which determine the level of comfortable usability of the boat.

The research here presented has studied a representative sample of sailboats (from 3,50 meters to 17,00 meters) and has confirmed that the problem of mobility on board showed in the literature has not yet been dealt with DfA Approach: many boats tend to "confine" the crew with possible disability in static positions that don't allow any type of movement and restricting, than, the pleasure resulting from the navigation experience. On this side the cockpit is the most critical spatial element because appears to be one of the areas more dynamically used of the whole boat.

From the comparison between principal problems related to management activities and psychophysical capabilities involved, arises the target of the research: purpose a design solution of a 10 meters daysailer that, while maintaining the possibility to perform tasks that normally take place on board, it reduces the level of possible "social exclusion" resulting from inappropriate morphology of the vehicle itself. The research has identified some technological and functional "for All" solutions to solving two typical areas of interest: the first one regards the spatial configuration of entire boat layout and the second one regards the steering area.

# 3    METHOD: THE DESIGN FOR ALL APPROACH IN YACHT DESIGN

In the study of "for All" mobility on sailboats, the meta-project process was articulated in three main phases. The first one (3.1) has analysed the various components of a sailboat that could be potential "disabling" elements. The second one (3.2) has defined the "limit" user for the project. The third one (3.3) has drafted the design requirements list useful for the definition of a 10 meter daysailer "for All" concept design.

## 3.1    Analysis of Yacht's Disabling Elements

The first meta-project phase has analysed the various sailboat elements the by form, dimension, and way of use could be potential "disabling" elements for users. Moreover, to this was added the consideration of the attitude of users to carry out in the most correct way as possible. The analysis has revelated the existance of four main group of disabling elements.

The first one is represented by system for conduct the boat and, in particuar, the steering wheel. This element, by form and characteristic, obligates to take particulars postures which, if sustained for a long time, can lead in situation of physical discomfort.

The area of access on board represents the second one. This element often is made with a narrow and rigid footbridge that much limits movements and makes precarious the balance; such solution makes almost impossible the access into the sea making them of the elements that affect the level of accessibility on board.

The third one is represented by the equipments used to conduct and control sails and mast (winches, ropes, etc.) for which, because the specificity and the necessary experience for a correct use, are particulary difficult to use.

Finally, the fourth one is represented by the cockpit configuration that often has stairs with elevations very marked, especially with regard the area of lateral walkways and the area for access in the lower deck.

The analysis of these elements has influenced, in the next design phase, also in the design of entire boat layout allowing, moreover, the development of new design solutions that improuve the quality of life on board in navigation.

## 3.2    Identification of "Limit" Users

In the second meta-project phase, in order to define with precision the list o users considered as "at limit" of the autonomous fruition of the boat, a new survey tool was developed and used. The tool called "Ability/Difficulty Table" ("A/D

Table") is strictly referred to the DfA Approach, since allow to define, starting from an hyerarchical analysis of different activities, the one that shows the most problems in terms of required capabilities for the autonomous fulfillment of the task. With the use of this tool it has been possible objectively dicover the "limit" users.

In particular, the activities considered as most critical were six: "getting on board from the quay", "getting on board from another ship", "getting on board from the sea", "moving on board on the cockpit", "moving on board on the lateral walkways" and, finally, "steering the boat".

The use of A/D Table has thus highlighted that the main most disadvantaged psychophysical conditions during the carrying out of activities are mainly due to conditions of physical nature and only in minumous part of cognitive and sensorial conditions; in specific the analysis highlighted that the most critical conditions are: tetraplegia, upper limb impairments, hand prehensility difficulties, and hemiplegia.

## 3.3    Definition of Design Requirements List

A summary scheme was after used as a point of start for the definition of the specific design requirements of the products.

In the last phase, indeed, from the analysis of users considered as potentially most critical, those that had a reasonable probability to use autonomously the boat were selected. The main characteristic of this phase is the action of "choice"; indeed, according to the DfA Approach is the designer that choice, in as much as possible objective and critical way, the user to take into account for make a product the most inclusive as possible for the limit user of the project (Accolla, 2009). Needs and design requirements were organized according to various disabling conditions investigated having, at the end, a detailed list that directly links in one-to-one a need and a design requirement. Such list after has allowed to work on the product starting from the resolution of difficulties expressed by each user.

## 4    DESIGN RESULTS

Design results gained from the research have allowed to design a 10 meter daysailer "for All". The study was articulated in two areas of interest. The first one (4.1) has concerned the definition of a new layout that resolves the requirement of mobility on board through the design of various elements such as: the boat layout, the sailing rig, the area of access on board anf the living area. The second one (4.2) has developed the steering area.

## 4.1    Design Solutions for Improving the Mobility on Board

The improvement of mobility on board was achieved through the combination of different solution that, collaborating synergically, characterize the boat aspect and its organization. In particolar, the solutions focused on three components of the boat. These elements, distribuited in precise parts, provide the level of hospitality on board for intended users (Figure 1).

A first solution worked on redefining the deck layout through the choiche of a "catboat" rig (rotating mast positioned at the prora of the boat), with which it was possible clear the cockpit of any obstructions to allow a general organization for the conditions of users detected in the survey. The cockpit, ideated as an open space, open at the stern, smooth and continuous until the prora was designed to contain (with the help of some kinematic solutions) all the functions required for a minimum habitability for a boat of 10 meters.

Figure 1 General overview of the 10 meters Daysailer "for All".

A second solution has concerned the living area organization and, in particular, the kitchen area (Figure 2). For this solution, located on right side, was ideated an integrate solution with table and worktop that can be opened. The main characteristic of this solution is to be easily manageable and usable also by people with reduced mobility. Ideated for be utilized by four people has a sink, a fridge, a worktop and a compartment for collecting dishes. With the kitchen, four rollway chairs were designed, they removable from the floor but easily extractable with a single gesture, can adapt respect to the diversity of users of the project.

Finally, a third solution concerns the accessibility on board that is assured through the use of an electric footbridge integrated into the floor, that allows to access on board from different altitudes, such as, from an high quay, from another boat or directly from the sea (Figure 3). Such solution, integrating itself and not appearing as a tool for disabled, proves to be a valid tool also for semplify the access on board also to people that, even if don't have walking difficulties, could suffer some discomfort conditions (considering, for example, elderly, obese people, pregnants, but also people that transports packs or nautical equipment, maintenance personnel, etc.)

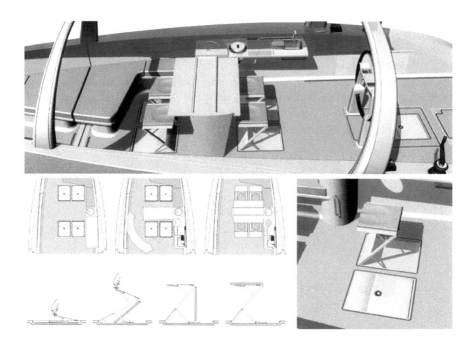

Figure 2  Solutions adopted for the kitchen and the living area. On top a general overview off the entire living system; below on left, exemplifications related to the table and the chairs; below on right, a view of the chair.

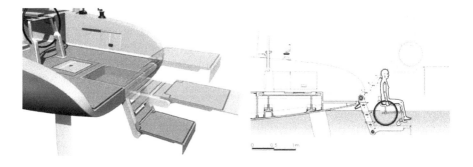

Figure 3  The solution adopted for the electric footbridge at stern. A general overview of the maximum range (250°) (left) and a detailed drawing (right) that shows some hypothetical accesses on board from a quay and from the sea.

To complete the cockpit organization was drawn, continuing the kitchen shape, a sundeck for users' relax. The area is interrupted before the end of deck where is forecast an area for maintenance the board equipments (mast, boom vang, anchor, ropes, etc.).

## 4.2    Design Solution for the Steering Area

The solution proposed for the steering area consists in a "neutral balance" platform that allows the limit users to navigate in total safety also when the boat is heeled. This considering also the different conditions that usually a sailboat deals, such as rolls and different tilt angles related to possible points of sailing (Figure 4). The insert of a swinging platform (maximum angle of 20°), guarantees always a correct stability and the right posture of the helmsman, even during a tack, a jibe, or, with respect to the change of direction.

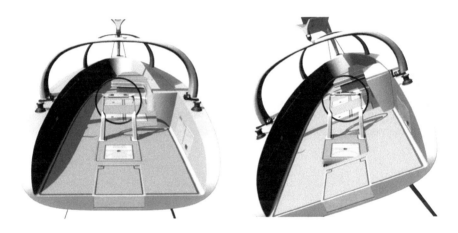

Figure 4  Conceptual functioning of the steering platform system "on rest" (left) and "in navigation" (right).

The platform (Figure 5) integrates a series of solutions that enables an adequate management and utilization of the boat also for people with reduced mobility.

The first solution is the adoption of a steering wheel passing on two side pedestals that contain transmission systems necessary for the conduct. The particolar shape of the system, without the central element, improves the navigation also to people that use a wheelchair (Figure 6).

The second solution concerns the aggregation in the navigation panel of all electronic instruments of propulsion (both those for wind power than those for motor power), of communication, of course, etc.

Finally, the third solution integrates into the platform a rollway chair that makes itself adaptable respect to the possible diversity of users.

508

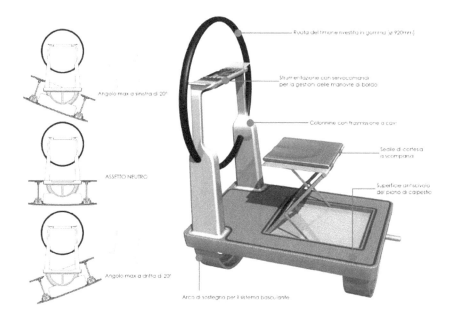

Figure 5 An overview related the design solution for the steering area. From left to right: the conceptual functioning of the platform in navigation when we change point of sail (left) and a general overview of the platform when the integrated chair is opened (right).

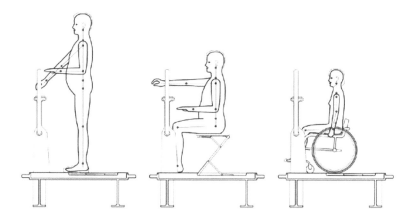

Figure 6 Anthropometric trials that show the flexibility of use demonstrated by the morphological configuration of the steering area during navigation postures. From left to right: man 95th percentile (left and middle) and woman 50th percentile on wheelchair (right).

# 5 CONCLUSIONS

The issue of mobility on board, although characterized by technical and spatial problems, remains one of the most important problems to take into account during the design process of a sailboat. Moreover, the design process seems even more problematic when we consider the diversity and heterogeneity of potential users who might want to use a boat.

In the case study presented here, the use of DfA Approach, an approach based on the valorization of meanings such as human diversity, social inclusion and equality, has allowed to define a 10 meters daysailer that shows original technological and sustainable solutions to improve the life of people on board in a view of "easy sailing". In particolar, the design process here presented has allowed to develop interesting product solutions concerning the themes of mobility and navigation: an original proposal of deck layout, a new and alternative solution for access on board, a specific solution for the living area and, finally, an original design solution for the steering area; the latter is characterizated for the innovative solution that semplify the activity of management and navigation not only to users with physical problems of mobility.

## CREDITS

This paper refers to the results achieved within a M.Sc. Thesis in Architecture entitled *"Daysailer a vela di 10 metri "for All". Soluzioni tecnologiche sostenibili per la conduzione facilitata"* (advisor: Prof. G. Di Bucchianico, Ph.D.; co-advisor: Prof. M. Di Nicolantonio; technical consultant: A. D'Onofro; candidate: E. Rossi) edited in the *"Interior Design of Sustainable Living"* Degree Laboratory, academic year 2008/2009 at the School of Architecture of the University of Chieti-Pescara in Italy.

All images reported here are taken from the above-mentioned M.Sc. Thesis.

The various paragraphs of the present paper can be considered the consequence of a common discussion and a collective review between authors. In particular, the writing of various paragraphs can be attributed to: Emilio Rossi (3 and 4), Giuseppe Di Bucchianico (1 and 5) and Massimo Di Nicolantonio (Abstract and 2).

## REFERENCES

Accolla, A. 2009. *Design for All. Il progetto per l'utenza reale.* Milano: Franco Angeli.
Accolla, A., and L. Bandini Buti. 2005. Design for All ed Ergonomia. *Ergonomia,* 3: 5-13.
EIDD – European Institute for Design and Disability. 2004. *The EIDD Stockholm Declaration,* document of the EIDD Annual General Meeting, Stockholm, Sweden.
Larsson, L., and R. Eliasson. 2000. *Principles of Yacht Design – 2nd edition.* Camden, Me.: International Marine.
Rossi E. 2010. Daysailer a vela di 10 m "for All". Soluzioni tecnologiche sostenibili per la conduzione facilitata. M.Sc. Thesis in Architecture. University of Chieti-Pescara, Italy.

# CHAPTER 53

# Discussion of Issues Relevant to the Ergonomics of a 50 Ft. Sailing Yacht

*Mario Biferali*
*B.Eng. Yacht Designer and Naval Architect*

Mario_biferali@hotmail.it

## 1  INTRODUCTION

A 50 Ft. Fast Cruising Sailing Yacht has been designed as a final year project of the "*B.Eng. (Hons) Yacht ad Powercraft Design*" course at the *Southampton Solent University (UK)*. As an engineering degree, the project covers engineering and naval architect features relevant to a preliminary design of a sailing yacht. The project was awarded for the best project by the RINA (The Royal Institution of Naval Architects) and BAE System prize. It was achieved for the good combination of design, engineering and naval architect features covered in the project.

Even if the project had a more technical background, ergonomic considerations were later considered: they depended on the designer's sensibility in order to produce comfortable and functional spaces for the user of the yacht.

In this paper an ergonomic review and analysis is carried out through the whole project. It is important to underline that there are two kinds of ergonomic intervention: the ones of correction, carried out after the design development, and the others of conception, carried out during and together the design development. (Badini Buti, 2011).

In this paper an ergonomic review and analysis "*a posteriori*" is carried out through the whole project, in order to perform an ergonomic correction in the project. So it means that corrections occur whereas ergonomic considerations are not satisfied.

Suggestions and modifications are thus the results of the research which aims to a better ergonomic product.

Figure 1: The 50 Ft. Fast Cruising Yacht Sailing Upwind

## 2 OBJECTIVES

The aim of the paper is to analyse the non-ergonomic aspects of the project and correct them where necessary. In fact, in the preliminary design, ergonomics was considered by the designer's experience and sensibly while designing the exterior and exterior spaces rather than with a proper ergonomic design approach.

For example, dimensions relative to men and women were provided in order to design reasonable and comfortable spaces for all users. So, anthropometric data referred to man and woman of 95% and 5% percentile respectively were considered. Both 5% and 95% percentile dimensions were used to design interior systems and equipment: the 5% woman percentile were used when designing part accessible to smaller size passengers (galley cupboards, shelves etc.) and the 95% man percentile to provide space even for larger size passengers (beds, doors, ceiling height etc.).

Nevertheless some interior designed parts resulted as not fully ergonomic due to the special attention on engineering features. Thus the goal of the paper: the stairway and the master cabin bed's step/seat are the system discussed and analysed in the ergonomics research.

Even if both systems were designed regarding the percentile dimensions provided, non-ergonomic aspects have been found.

The stairway and the master cabin's bed step/seat dealt with functionality, comfort and with the space available on board. A complex combination of space and system's functionality was effectuated; the main reason of the lack of space on board is due to the low sole level position which led to a reduction of sole surface due to the hull shape. Reduction in surface is greater in the master cabin, because of the shallow hull section forcing to shift the bed up to fit a larger bed size; thus the need of bed steps.

A four steps stairway connects the 1.8 m height gap between the cockpit and the below deck spaces. The choice of a four step system was guided to reduce the encumbrance as low as possible.

Anyway a steep and uncomfortable stairway was thus designed due to the high step height, the short step depth and especially due to the lack of rails, decreasing its functionality and safety.

## 3   METHODS

The ergonomic review of the project was mainly based on the considerations reported in the Chapter *"User-Centred Approach for Sailing Yacht"* (Di Bucchianico and Vallicelli, 2011). The latter considers four fundamental requirements about ergonomics applied to yacht design, and these are respectively the safety, functionality, usability and pleasantness of a product. The chapter and especially the enounced requirements, were the guideline of the whole ergonomic investigation providing suggestions to the designer regarding the non-ergonomic aspects present on the project.

However the four requirements group can briefly introduced:

## Safety

It concerns the interaction between user and product installed on board.

Since the yacht is subjected to continuous longitudinal and transversal movements (pitching and rolling), impact on the utilization of an equipment installed or designed parts of the yacht can occur. It increases when sailing in rough condition, thus attention must be paid while designing interior space according to passengers safety on board.

## Functionality

Due to the less space available on board, any single interior part or equipment installed must be functional and easy to be used. A wrong position demands deeper efforts to the users causing sea sickness especially if activities are ongoing.

## Usability

This requirement deals with two conditions: the first one refers to the minimum space available on board where specific works or living activities are required. The second condition concerns some specialized equipment installed on board, where posture has a fundamental role. The latter is well tied with the task of the equipment and thus a good position is required in order to let the user have a correct posture avoiding any feeling of irritation and pain.

## Pleasentness

The pleasantness concerns the positive sensations and emotions produced in each person by the shape, material, colour, texture, luminosity etc. of a material or object on board. It depends on the user's cultural, sensorial or temporal type.

The designer analysed his project according to the requirements introduced above. The stairway and the master cabin bed's step/seat are thus discussed and modified in order to eliminate the non-ergonomic aspects regarding their safety, functionality, usability and pleasantness.

Furthermore a 3D Mannequin, shown below [Figure 2], was used to simulate the real user's movements while using both analysed systems. It thus allowed the verification and the comparison of the passengers' movements before and after ergonomics correction of both system took place.

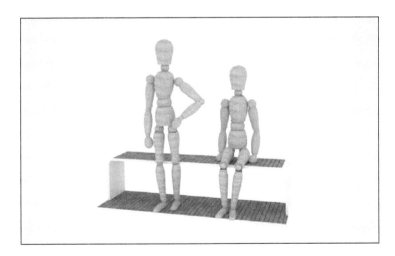

Figure 2 : 3D Mannequin used to simulate the real user's movements in the ergonomic research.

## 4 RESULTS

The results that the ergonomics correction made to the stairway and the master cabin are introduced below respectively.

### 4.1 Stairways:

The preliminary designed stairway was analysed. It had a recessed form and each step was 375 mm height, 290 mm depth and 632 mm wide. On a civic stairway, the height is normally around 200 mm high; on yachts different values are achieved due to the less space available. Indeed the reason of the higher height was guided by the need to design a stairway with the lowest encumbrance as possible. It is known that the higher the step height, the more steep the stairway is, but the less space it occupies. Furthermore the system was not installed with rails thus effecting its functionality, usability and safety.

Therefore, in rough weather condition, it requires deeper effort to the user which thus results in feeling of irritation and discomfort.

The 3D Mannequin simulated the postural position of the passenger on the stairway. A wrong posture was adopted to jump the step and it was due to excessive step height, step depth and the lack of support. Therefore, with the Mannequin simulation in addition to the ergonomic studies, it was possible to carry the ergonomics correction on the latter system.

A new stairway was designed: five steps allow having lower height between each step with depth surface of 295 mm. Of relevance is the effective step's depth of 220 mm which is visible when coming down the stairway.

According to the usability, the user will have a correct posture position in both going up and down the stairs avoiding any effort and feeling of pain. The installation of rails results in an overall increase of ergonomics: with the latter, the passengers will not require extra effort. Furthermore, the step surface has an inclined shape at both edge of 15° improving safety, usability and functionality when sailing heeled.

The ergonomic modifications can briefly summarized as shown on the table below:

| Modification | Reason |
| --- | --- |
| Step height | A lower height between steps ease climbing them reducing the effort of the user and then increasing its functionality, usability and overall pleasantness |
| Surface Section | Inclined surface on both step side ensuring "flat" surface when sailing heeled guarantying safety and functionality in rough sea |
| Reduction Corners | Reduce the risk of shocks with sharp corners increasing the safety on board |
| Rails | Provides additional safety to the user when sailing in all conditions |

Table 1. Modifications made in the Stairway

The pictures below show the ergonomics corrections of the discussed systems.

Figure 3: Stairway Before Ergonomics Correction

Figure 4: Stairway After Ergonomics Correction

## 4.2 Master Cabin Bed step-seat:

The master cabin is located in the fore part of the hull, where the hull section gets shallower. The bed was lifted up in order to achieve a larger hull breadth and fit a reasonable bed size. The height difference of 810 mm between the bed and the sole level was then achieved, thus steps were required to facilitate climbing on the bed.

The preliminary idea was to install on both hull side seats next to the bed which could also act as steps. Those had space issues too regarding the hull section, thus were lifted up having a height of 590 mm.

The ergonomics research led to a clear exposition of the non-ergonomics aspects relative to those designed parts. The main issue regards their height: their 590 mm heights prevent any use neither as a step nor as a seat.

The simulation of the positions by the 3D Mannequin confirmed what was enounced before: issues occur when leaving and jumping on the bed due to the excessive height of it. The modifications that the ergonomic research and analysis led consisted in the installation of side steps on both hull sides, with a height of 250 mm from the sole level. Matched with shelves at same height of the bed base, those have multifunctional roles: can be used as steps, seats and bed side shelves. In this way the passengers will easily reach and leave the bed, as the 3D Mannequins show. Overall reduction of effort is then achieved.

In term of pleasantness, the use of the same material used for the sole level, well mark the role of those designed interior parts emphasizing their functions in relation to the sole level.

The modifications made can be summarized below.

| Modification | Reason |
|---|---|
| Seat height | The seat is rearrange in the cabin in order to have a suitable and comfortable height and depth for letting the user be seated according to the hull section. |
| Additional Step | An additional Step between the seat and the bed height is fitter. It will ease climbing on the bed without requesting any effort to the users |

Table 2. Modifications made in the Master Cabin

The pictures below show the ergonomics corrections of the discussed systems.

518

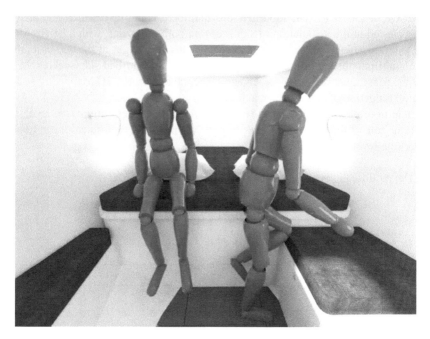

Figure 5: Master Cabin Seats and Steps Before Ergonomics Correction

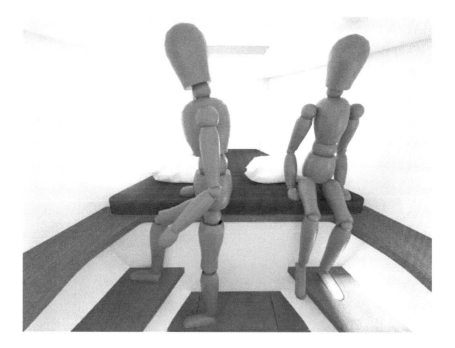

Figure 6: Master Cabin Steps After Ergonomics Correction

# 5 CONCLUSION

An ergonomics correction was so made through the project of a 50 Ft. Fast Cruising Sailing Yacht. The non-ergonomic aspects of the yacht were discussed and modifications were applied in order to improve it ergonomically satisfying the requirements (safety, functionality, usability and pleasantness) which guided the ergonomic research.

Thus dissociate results were found on the stairway and on the master cabin bed's step/seat due to the excessive attention on technical features on the area where these are installed. However the research led on improvement of the preliminary design in term of ergonomics.

Obviously, different results would have been achieved if ergonomics conception was considered from the beginning of the design of the yacht.

However it is interesting to see how ergonomics can guide and wholly effect a design of a yacht, especially in term of its functionality, safety, usability and pleasantness between user-yacht interaction.

# 6 REFERENCES

Badini Buti, L. 2001. Ergonomia e prodotto. Milano: Il Sole 24 Ore.

Di Bucchianico G. And A. Vallicelli. 2011. User-Centered Approach for Sailing Yacht Design. In: Human Factors and Ergonomics in Consumer Product Design: Uses and Application, Chapt. 27: 445-463. Boca Raton, FL: CRC Press-Taylor & Francis Group

# UCD vs ECD: from "User" to "Experiencer" Centered-Approach in Sailing Yacht Design

*Giuseppe Di Bucchianico, Stefania Camplone, Andrea Vallicelli*

## ABSTRACT

Yacht design is a complex design field, in which highly specialized branches of learning converge with as many specific planning contributions that the yacht designer finds himself having to co-ordinate to establish the formal values of the artefact-boat, full of symbolic and functional aspects.

In particular, sailing yacht design deals with greater difficulties, because it must integrate the technological and technical innovation and figurative and cultural codification aspects, with the demands of liveability of the areas in continual movement and of the overall efficiency of the system-boat expressed by the final users. Therefore, the methods, techniques and instruments developed by applied ergonomics offer to yacht design a significant operative support into direct planning action with awareness towards a morphological and spatial research which is both elegant in its language and fluid in its articulation and functionally efficient and pleasant in its use. In fact, the UCD - User-Centered Design's approach, developed by the ergonomics, with its theoretical and applicative scope, is able to place the individual and his requirements at the center of the yacht design, reflecting him on the main levels of the requirements of the nautical product: from those regarding the safety of equipment and areas, to their functionality, ease of use, up to the requirement for pleasantness of the environment and finishing.

This paper proposes an initial reflection on the peculiarity and multidimensional complexity of this design field, above all referred to the sailing yacht design, underlining the aspects on which the UCD ergonomic approach can provide planning indications which are useful to the anthropocentric management of the project. In particular, starting from a known classification of the ergonomic requirements of the products (safety-functionality-usability-pleasantness), it outlines a synthesis from it attributed to the nautical product, with the aim of identifying the various levels of a possible contribution of applied ergonomics to the yacht design. Through some experiences of ergonomic research applied to sailing yacht design, it is demonstrated how ergonomics until now could contribute to identify highly innovative solutions orientated towards user's wellbeing, even in such a sector as pleasure boat sailing, highly connoted by a complex system of semantic and technological ties.

With the recent change of perspective suggested from holistic ergonomics, which under an applicative point of view flows into the Design for All approach, we are however witnessing the transition from the centrality of the individual to the centrality of man in his totality. This issue, addressed In the conclusive part of the paper, raises the question of '"user" that turns into an "experiencer". In the perspective of sustainable development, in fact, issues of usability and pleasantness of use are surpassed by those of a conscious use and of an "experientiality" enhancing in an inclusive manner the diversity among individuals. This certainly has consequences even in a specialized design field like yacht design, starting a project research on issues that only recently started to be explored.

**Keywords**: yacht design, UCD, user, experiencer

## 1 INTRODUCTION: MULTIDIMENSIONALITY OF NAUTICAL PROJECTS AND THE UCD APPROACH

Yacht design is the applicative field of industrial design which deals with the planning of pleasure boats, equipped with an interior space and destined to offshore sailing, sporting or cruise.

This is a design field characterized by a particular complexity, as it must coordinate multi-disciplinary skills, extremely varied and interacting, ranging in scientific fields of engineering, architecture, ergonomics, marketing, ecology, with their disciplinary specialized articulations. The multidimensionality of the design approach increases if one refers to sailing yachts, which are real wind-propelled machines extremely complex, as they move between two fluids (air and water), stressed by static and dynamic loads. Therefore, their study involves several branches of learning among which there are naval architecture (study of resistance to the movement of the hull), aerodynamics (efficiency of the appendages and sails), structural engineering and material technology. If on the contrary we refer specifically to the genesis of the project of a pleasure craft, it can be said that the greater complexity comes from the need to coordinate the internal symbolic and functional values of a "house", a "refuge" (stability, strength, safety, privacy), and the external ones of a "vehicle" (lightness, dynamicity, maneuverability).

The dimensions of the object have an important role in this: in general, it can be retained that so much bigger the boat is, so much narrower does the relationship become with the dimensions of the housing (also as a consequence of its greater effective "stability") and, consequently, its relation to the sea becomes weaker.

Thus, if in big ships the comparison with civil architectural areas is generally immediate and spontaneous, in small crafts, instead, above all sailing ones, one often witnesses compromises between the spatiality of the living area and that of the binnacle. Furthermore, in contrast with other means of "habitable" locomotion, the crafts are forced to compare with a continuous movement: "(…) A boat is represented as an object which moves even when it is still. When a camper van is still, it is static like a house; a craft is always in movement even when it is still" (Spadolini, 1987). A boat, therefore, can be considered as an unstable object that moves inside an unstable element.

Moreover, for the planning of sailing crafts it is necessary to consider the need to integrate the technological and technical innovation aspects, in continual evolution, with a specific figurative and cultural codification, at times millenarian, particularly present in this field of the project. This happens considering the complexity both of "deck surfaces" and the "underdeck" areas, where several elements or distributive typologies are an expression of functional results or simple formal heritage tied in with nautical tradition.

In the evolutionary panorama to which we refer, that is constantly orientated towards research and experimentation, the nautical designer carries out, therefore, a primary and coordinating role between the various areas of competence involved, for a typological and functional redefinition, both of the spatial areas and of positions and single equipment, which take into consideration, on the one hand, the evocative-cultural result of tradition and, on the other, the technological progress obtained in the field of materials, building and production techniques, and electronic miniaturization.

The methods and techniques developed during the last decades by ergonomics applied to design represent a valid operative and instrumental support for Yacht Design, in order to direct the planning action with awareness towards morphological and spatial research which is both elegant in its language and fluid in its articulation, and functionally efficient and pleasant in its use. In particular, we refer to the UCD (User-Centered Design) ergonomic approach, which is based on the idea of planning artifacts that can be utilized by users with maximum efficiency and minimum physical and mental discomfort. It is an approach that allows the acquisition and assessment of users' requirements by means of structured and verifiable methods, to turn them into planning instruments.

Yacht designer, therefore, finds a valid aid in the UCD methods and instruments to assess and plan above all the man-boat interaction with greater awareness: he manages to do this at all levels referable to the users' demands and regarding the different scales of the project, in order to identify the specific requirements of products, from single equipment on board to more complex spatial environments.

Recently, ergonomic literature has placed four different requirement levels referable to the users' demands in a close hierarchical relationship (Jordan, 2000). It is possible to identify an analogous articulation for the nautical product too (Di Bucchianico and Vallicelli, 2011). Thus, the requirements concerning the safety of equipment and environment are identified, those regarding their functionality, ease of their use, including the requirements that concern their pleasantness, the physical and mental "pleasure" that is experienced in interacting with the products and environment. In particular, for what concerns "safety" on board, ergonomics provides the most suitable instruments to analyze the characteristics of individuals and the limits of their psychophysical abilities, even in extreme environmental and postural conditions; with respect to "functionality", ergonomic research can relate directly to the typological evolution of the equipment and the various parts of the craft; with regard to "usability" of the equipment and to postural comfort, ergonomic practice allows the identification of more innovative solutions also by means of observation of organization and structure of tasks on board; finally, on the "pleasantness" plane, ergonomics has defined the most useful instruments and methods to assess the psychosensory interactions of individuals with components, equipment and environment.

## 2    OBJECTIVES: FROM UCD TO ECD

In all four typologies of ergonomic requirements mentioned above with regard to the yacht design, the multidisciplinary approach of ergonomics to the design and the availability of methods, intervention procedures and operative instruments which it offers, allows the study of the requirements of the user's wellbeing to be faced whether in relation to the single product/equipment, or to the task/position or to the environment/context in which he acts. For each one of them, in fact, ergonomics is already able to offer a precious contribution in relation to the nautical project of environment, positions, equipment and finishing, as well as, in a figurative sense, to establishing all those tasks and activities, even apparently secondary, which are carried out on the craft.

As mentioned, however, the tendency is to approach to the different aspects in a "particle" way, that is by parts, with a clear division of investigated tasks, roles and aspects. This trend is also strengthened by the particularly complex and multidisciplinary nature of a nautical project as it has been mentioned above.

This contribution therefore aims to highlight how, with the recent change of perspective suggested by holistic ergonomics, which flows into Design for All, has inevitable repercussions on how the users are considered and on the value of "experience" that comes through the '"use" of products, environments and systems.

In fact, recently a significant contribution towards a critical review of the operational principles of ergonomics has been offered by some reflections on "holistic ergonomics" (Bandini Buti, 2008). This new perspective refers directly to the "holism"[1].

The idea is that, if in the scientific-mechanistic approach all the attention was for the detection of the constitutive "bricks" of the reality, with the holistic approach, also known as "systemic approach", it is recognized the importance to study

---

[1] From greek ὅλος, meaning "the whole", it refers to a philosophical position based on the idea that the properties of a system can not be explained solely through its components.

especially the "fabric that connects". That is, no longer taking care to the "quantity of items", but to the "quality of relations" between them.

The "holistic" approach of the ergonomic discipline determines a significant transition from the centrality of man to that of the individual in its totality. The "individual", in fact, prevails on the "user", as the experiential values of different cultures, different emotions and moods, lead us to consider the "enjoyment" of an environment or of a simple equipment with a broader and involving view respect to their simple "use".

Thus, if the UCD approach has so far helped to identify specific relevant requirements for design validation of products and systems "used" by their "users", it appears particularly promising to start reflecting on an extension of it to less tangible aspects because they are more referred to the complexity of the individual as a whole. Consequently, the traditional concept of UCD - "User" Centered Design, would extend to that of ECD - "Experiencer" Centered Design, a new concept expanded from the simple "user" to the more complex "individual.

## 3    METHOD: TWO PRACTICAL EXPERIENCES

In support of the potentialities expressed by an eventual not simple transition from "user" to "individual" in the conception of the "experiencer" of products, with particular reference to the nautical field, here are briefly presented two different experiences of ergonomic research applied to the sailing yacht design, which show how UCD Approach of ergonomics already contributed to identifying highly innovative solutions orientated towards the wellbeing of the user even in such a sector as pleasure boat sailing, that is highly connoted by a complex system of semantic and technological ties.

In particular, it deals with assessment tests whose experimental results have allowed the definition of certain guidelines useful for orientating project choices on several aspects regarding the shape of pleasure sailing boats, facilitating in actual fact their identification of innovative solutions at a functional and aesthetic level.

The first experience (paragraph 3.1) reports the results of an observation test attributed to tasks and posture of the tailer, who has the decisive role of maneuvering and regulating the sails on crafts), highlighting its principal elements of postural discomfort and defining some guidelines regarding postures, places and equipment.

The second experience (paragraph 3.2) refers to the development and execution of an assessment test of the tactile pleasantness of the finishes of the deck surfaces, in order to provide designers with some planning guidelines useful in avoiding a simply "decorative" approach to design and the placing of the "macro-grips".

## 3.1    Analysis of the tasks and posture of the tailer of a 17-meters sailing yacht[2]

Among the most important roles on board of a sailing yacht, the "tailer" is the determinant subject taking care of furling and setting the sails. In this case, therefore, the target of the research was to point out the most important aspects of a tailer's postural discomfort, in order to define some guidelines referring to design of his different operative conditions (postural, organizational and positional), as well as to the different sets of rigging that he uses.

The research was experimentally organized into three phases: during the first phase data and information were collected concerning both in the large the navigation with a sailing yacht as long as the one studied, and the role, the riggings and the tasks of a tailer; in the second phase theories and methods were deeper examined in order to analyze tasks and postures of subjects working together in organized systems and to apply these methods in an original way to the theoretical evaluation of a tailer's tasks and postures; in the third phase a direct observation was planned in order to verify the rightness of theoretical data of the preview steps and in case to point out any critical points. In this way it was possible to define some guidelines in order to direct and favor the design ideas concerning the tailer's postures, locations and equipment while using winches, allowing an interesting and innovative application within the same design experience from which the research was inspired (Figure 1).

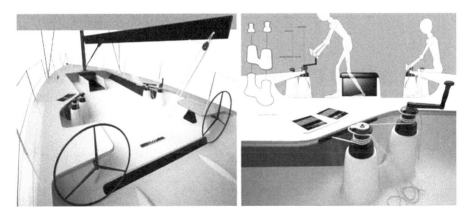

Figure 1 Innovative cockpit solution deriving from the analysis of the tailer's tasks and postures

---

[2] This paragraph refers to an observation test concerning the '"analysis of tasks and postures of the tailer of a sailing vessel of 17 meters", developed as part of a Degree Thesis in Industrial Design held at the Faculty of Architecture 'University of Chieti-Pescara (thesis title: "Naked 55. 17-meters open-sea sailing craft"; supervisor Prof. A. Vallicelli, co-examiner Prof. G. Di Bucchianico, graduand Marco Scuderi). The research has already been widely documented in numerous publications, so we save here an extended discussion.

## 3.2    Assessment of the tactile pleasantness of the antiskid deck surfaces of a sailing yacht[3]

The research objective was to control and verify in the most objective way the aspects determining the tactile pleasure of some antiskid deck surfaces which the users on board are constantly in contact with. Therefore, an evaluation test was planned and carried out. It was referred to the well known Sequam method - Sensorial Quality Assessment method (Bandini Buti, Bonapace, and Tarzia, 1997), though with some methodological adaptations, necessary to the feature of the object to be surveyed and to the limits imposed by the application context. The experience, which was divided into three phases and in a subsequent critical reading of the collected data, helped to identify some guidelines useful to the design of antiskid deck surfaces for sailing boats. In particular, two groups of guidelines were distinguished, particularly referring to morphological features of the grip:

- as trample plane, that is connected to the dynamic activity of walking;
- as supporting plane for the body, that is connected to the sitting, lying or simply standing positions.

Later, the obtained guidelines were used to make hypotheses on some possible macro-grip design solutions of deck surfaces of a sailing boat (Figure 2).

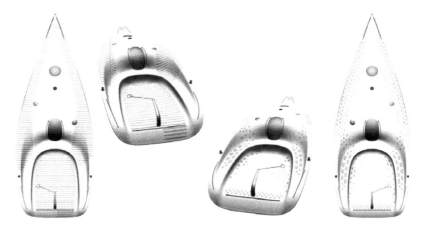

Figure 2  Different macro-grip design solutions of deck surfaces

---

[3] Here is reported a research experience evaluating the tactile pleasantness developed as part of a Degree Thesis in Industrial Design held at the Faculty of Architecture 'University of Chieti-Pescara (thesis title: "10-meters day cruiser sailing craft"; supervisor Prof. A. Vallicelli, co-examiner Prof. G. Di Bucchianico, graduands Francesco Merla and Grazia Patruno). Also in this case, as in the previous one, the research has already been widely published. That's why here are briefly reported only some results obtained with the test.

# 4    CONCLUSIONS

Until now, ergonomics contributed to identify highly innovative solutions orientated towards the wellbeing of the user even in such a sector as pleasure boat sailing, highly connoted by a complex system of semantic and technological ties.

The many empirical experiences of research conducted in different fields of application of "maritime ergonomics", security-related or referred to functionality, usability and physical and mental pleasantness experienced when interacting with equipment and environments, have significantly contributed to improving their overall quality.

Both experiences here presented, therefore, show the effectiveness of the traditional ergonomics approach referred to the scientific-mechanistic analytical breakdown, which also the UCD approach refers directly to.

In a sustainable development perspective, however, all design disciplines, including ergonomics, are required to have an emblematic change of perspective in the theoretical definition and articulation of practical activities related to the ergonomic design, that go far beyond their simple "interdisciplinarity". This in order to face the complexity, the extent and the versatility of the problems that the transition towards sustainability places. In fact, in this perspective, both the issues of usability and pleasantness of use are surpassed by those of a conscious use and a "experienceness" that enhances in an inclusive way the diversity among individuals.

A first promising path therefore appears that promoted from the DfA approach. In fact, considering the diversity among individuals as a resource, an opportunity for product innovation rather than as a constraint, the DfA has already successfully tested the first attempts to to recover the value of a unitary study on individual's needs, starting just from "diversities" among individuals (Figure 3). This, also in a specialized field of the project as yacht design is, where only recently it has been started a research activities on issues that relate to those aspects, apparently more intangible, connecting the design activity with the "extended scene" of a society oriented towards sustainability.

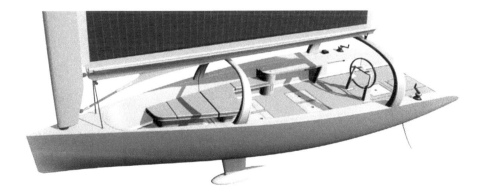

Figure 3  Daysailer "design for all" solution making mobility on board easier and inclusive

## CREDITS

Giuseppe Di Bucchianico is author of the paragraphs 1 ("Introduction: multidimensionality of nautical projects and the UCD approach") and 2 ("Objectives: from UCD to ECD"). Stefania Camplone is author of the paragraph 3 ("Method: two practical experiences"). Andrea Vallicelli wrote the Abstract and paragraph 4 ("Conclusions").

Furthermore, this chapter reports some results of two researches carried out in the hambit of some different degree theses in industrial Design carried out at the Faculty of Architecture in Pescara. In particular, in paragraph 3.1 reference is made to the Thesis entitled "Naked 55. 17 meters open-sea sailing craft" (supervisor Prof. A. Vallicelli, co-examiner Prof. G. Di Bucchianico, graduand Marco Scuderi). In paragraph 3.2 reference is made to the thesis entitle "10-metre day cruiser sailing craft" (supervisor Prof. A. Vallicelli, co-examiner Prof. G. Di Bucchianico, graduands Francesco Merla and Grazia Patruno). Finally, figure 3 refers to the Degree Thesis dissertation in Industrial Design carried out at the Faculty of Architecture of Pescara titled "Daysailer of 10 m for All. Sustainable technological solutions for easier navigation" (supervisor prof. G. Di Bucchianico, co-exhaminer prof. M. Di Nicolantonio, graduand Emilio Rossi).

## REFERENCES

Spadolini, P.L. 1987. *Le analisi delle funzioni d'uso e loro relazioni, gli spazi minimi ed i problemi dimensionali nelle imbarcazioni da diporto a vela e a motore*. In. *Architettura imbarcazioni da diporto*, Vol 1°, ed A. Vallicelli. Firenze: Cesati.

Jordan, P. W. 2000. *Designing Pleasurable Products. An Introduction to the New Human Factors*. London: Taylor & Francis.

Di Bucchianico, G., and A. Vallicelli. 2011. *User-Centered Approach for Sailing Yacht Design*. In. *Human Factors and Ergonomics in Consumer product Design. Uses and Applications*, eds. W. Karwowski, M. M. Soares and N. A. Stanton. Boca Raton (FL): CRC Press, Taylor & Francis Group

Bandini Buti, L., L. Bonapace, and A. Tarzia. 1997. *SEnsorial QUality Assessment: a method to incorporate perceived user sensations in product design. Application in the field of automobiles*. Proceedings of the IEA 1997 Meeting, Helsinki, Finland.

# CHAPTER 55

# Safety and Comfort as Design Criteria for High Speed Passenger Craft

*Ermina Begovic, Carlo Bertorello*

University of Naples Federico II
Napoli, Italy

## ABSTRACT

At the moment, the research and the development of very high comfort standards for fast passenger ferries are most interesting for designers, ship owners and builders. Comfort, in the case of passenger transportation is strictly connected with safety and both represent the synthesis of man-ship relationship. Comfort on board is one of the key factors for successful passenger ship design; it is strictly connected to ship main characteristics and layout, and deeply influences design choices. While ergonomics are generally considered in passenger ship design, cognitive ergonomic, generally, is still ignored or neglected. Cognitive ergonomic is the analysis of environment human perception; this is a dynamic and complex process where the human mind selects and reads sensorial stimuli to get a significant image of the world. Only recently environmental psychology through cognitive ergonomics has been considered in passenger ship design procedure.

Within this frame the optimisation of the man-ship relations according to living environment characteristics, to safety issues by International Codes, and to human biomechanical behaviour aboard has been considered.

This work is one of the steps of a research program concerning ship comfort and safety carried on at the Naval Architecture Department of the University of Naples Federico II.

**Keywords**: High Speed Craft design, comfort, safety

# 1    INTRODUCTION

Comfort on board is a very complex problem including human psychology, physiology and environment perception; it affects the ship's main characteristics, layout and very often can be a limiting factor for ship performances. Living environment on board refers both to ship design and to ergonomics in the widest meaning. Significant efforts have been produced by designers and shipbuilders to achieve the highest possible comfort standards.

In this research program the investigation has been limited to the relation between the passenger and the ship - excluding the crew interaction with the ship - of high-speed ferries. The scenario of passenger ship relation has been considered through the main systems detailed in the following. In Figure 1 the design modulus based on the considered approach is shown.

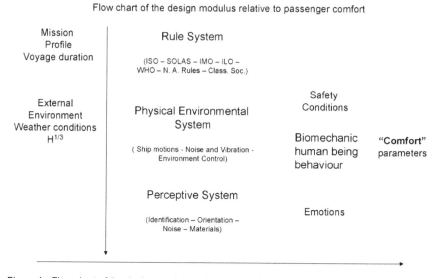

Figure 1 - Flow chart of the design modulus relative to HSC passenger comfort .

Before going into details of each step some general comments are proposed. Input data are divided into two groups. The first is relative to "mission profile";in this case essentially the length of the route can be decided and in some case modified. The second cannot be influenced or modified and is most important for High Speed Craft. The complying to the Rule System is mandatory and stringent; it will result in safety conditions, but it will strongly influence ship layout and consequently both Physical Environmental and Perceptive System. Finally the biomechanical and psychological behaviour of the human being have to be considered to evaluate and to assess the comfort level. It is immediate to observe that this procedure is very far from the standard traditional passenger ship design mostly based on ship performance optimization.

## 2    MISSION PROFILE EXTERNAL ENVIRONMENT

If the length and the site of the route are chosen as "mission profile" we indirectly assume the sea states the ship will encounter on her way. There are two important steps of sea state description: the regular waves and the principle of superposition in linear field, which is used for calculations of ship responses for each harmonic excitation, and the realistic marine environment described by significant wave height ($H^{1/3}$) and period of zero crossing or with the different spectra of energy. Generally the ship responses in regular waves is "multiplied" by sea spectra representative for every part of route and the results will be statistical response of ship. At this step, the appropriate criteria are imposed on ship motions criteria, ship acceleration or Motion Induced Interruptions (MII) or Motion Sickness Index (MSI). These calculations will provide the set of maximum significant wave height approaching the ship from different headings when some of the criteria are exceeded as well as the limiting ship speed.

## 3    PHYSICAL ENVIRONMENTAL SYSTEM

Ship motions are a limiting factor for any passenger ship and specially for HSC and must be considered with the other significant disturb effects that are noise and vibration; in the standard design procedure they are well known and carefully considered. Appropriate regulations issue sustainable and comfort limits.

Environment Control is performed through climatization, illumination, smell control, and onboard space design. These factors have been generally considered according to Rule System and to ship economic performance optimization. Only recently environmental psychology has been implemented among design criteria to take into account environment control.

### 3.1    Environmental psycology

The relation between persons and the surrounding environment has been systematically studied firstly in USA in the late fifties. Environment has been considered according to the different aspects: artificial, natural and social. First studies and experiences have been addressed to identify and describe the relation among architectural design and human behaviour into psychiatric hospitals. Among the first results the proof of the existence of environmental characteristics able to discourage or encourage the social interaction.

Then the connection between psychologists and architects became closer. In the sixties the researches of city planner Kevin Lynch on town image proposed a new design approach starting from the "image" of the built object.

The main arguments considered by environmental psychology are

- Environmental assessment that considers the evaluation of environment emotional characteristics, how much the environment helps in targets achievement, the interaction between human activity in a given environment and the information provided by the environment.
- Cognitive mapping that is the ability to obtain and use space perception and knowledge.
- Environmental stress that is the excess of simultaneous input from the environment.
- The spatial behavior (interaction between behavior and surrounding space), "the defensible space" mostly related to privacy.

In the environmental psychology applied research two possible fields can be considered.

- The study of artificial habitat (pleasure living, crowding).
- The relations among large environments and the human community.

These concepts can be tailored to passenger ships where the environment is perceived by passengers for a limited period (trip or holiday). In this case the environment is moving together with the user (customer) and the living spaces defines microhabitats.

It is very difficult when studying perceptive elements to separate the individual from his environmental context. In this process we can recognize perceptive conditions that have common rules.

- The individual, when perceiving an environment is not outside from it but interacts with it. He uses a space co-ordinate system, where, instantaneously, he is one of the possible centers.
- The environment perception happens through the simultaneous activation of several sensorial paths: sight, smell, and touch. Very important are perceptive signals concerning position, equilibrium, temperature, comfort, discomfort.
- The information, is not static but can be considered as a continuous flow.

The environmental information is a unitary system. When the environment is not readable the difficulty in the cognitive efforts gives discomfort to the individual that is addressed toward a negative environment evaluation.

## 4    PERCEPTIVE SYSTEM

The study of Perceptive System has been soon recognized as most important when dealing with environmental psychology. In the case of small and complex habitat as passenger ship are, perceptive system is considered mainly as Space perception, that is:

- Identification;
- Orientation (sense of direction);
- Noise;
- Materials and their characteristics.

## 4.1 Environmental perceptive models

Two are the most widely used perceptive models. The Brunswik model considers the individual as an active information processor, that constructs the perceptions from interaction of present sensations and last experiences. The other one is the ecological approach to visual perception by Gibson. In such theory the biological aspects of human behavior in a given environment are highlighted; the record of the events got by our senses is correct because our senses have evolved to allow us to survive. The experience does not interact with perception because this working model is innate. In this second model the so-called pattern theory has been introduced by Gibson into environmental psychology. In this theory the perceptive information is selected according to existing patterns so that our attention is addressed to some specific or peculiar aspects.

From several experiences carried on with human beings it has been highlighted that the remind and the record of a certain environment depends on the attention level and consequently on the reasons of the permanence in that environment.

The environment space elements are recorded in the individual memory as a map called "cognitive map".

## 5 RULE SYSTEM

Mandatory Rules and regulations to comply with are:
- ISO, SOLAS, IMO (International Marittime Organization)
- ILO (International Labour Organization)
- WHO (World Health Organization)
- National Authorities Rules
- Rules by chosen Classification Societies:

Rule system covering the maritime safety distinguish clearly two parts for them; one part is the promotion of maritime safety as shore-based service, the other part consists of the working routines on board a vessel. Beyond regulatory needs the requirements for the construction and maintenance of ships and associated equipment are defined in part in international standards but primarily in Class Rules issued by Classification Societies.

As regard safety on working routines the marine industry is highly competitive and costs are tightly controlled employing the minimum number of suitably skilled staff. Maritime Organization are concerning mainly the safety procedures in performing operations on board and set of operational limits according to:
- safety, comfort and workability criteria
- structural loading and responses and
- machinery and propulsion loading and response prescribing the maximum allowable value for each or number of occurrence in one hour.

The critical values of each operational limit have been developed in 70s by Olson (1978), Allen (1978), Comstock et al (1980), Mandel (1979), the values actually used by STANAG or NORDFORSK are still the same values.

# 6 THE BEHAVIOR OF HUMAN BEING ABOARD

## 6.1 Environment and emotions

The emotion is a complex psychological state. To define it [2], generally, four components are considered: sentimental, cognitive, physiological and behavioural.

Emotions and environmental preferences are very close connected. An environment gives pleasant or unpleasant emotions only in relation with past experiences or present aims.

There are several accepted environmental preference models. The widely used are that from Kaplan, in which to get a positive evaluation of certain environment "conditions" relative to "dimensions" must be fulfilled.

The other one is given by Purcell (discrepancy model 1986-87). It is connected to mental scheme theory. In this model the discrepancy between environment and the relative mental scheme is considered according to the individual past experiences, with the aim to identify an optimal state. Whitefield showed that the closer to our mental scheme an external stimulus is, the more positive the evaluation will be. Purcell model is more representative for young people while Whitefield's is meaningful for older people.

Three types of factors influencing environmental preference can be identified. They are related to

Person: age, personality, state of mind, knowledge, expectations;

Environment: natural or built habitat, luminosity, color, noise;

Person Environment relation: environment functionality, presence of other persons or animals.

Other important factors for the identification of influencing elements in the environment perception are the concepts of personal space and privacy.

Personal space can be considered as a regulation factor of interpersonal relations. Standard minimum distances among persons considered in Western Countries are:

- Minimum distance from 0.15 to 0.45 m;
- Personal distance from 0.45 to 1.20 m;
- Social distance from 1.20 to 3.60 m;
- Public distance from 3.00 e 6.00 m.

These distances have been analysed in detail by Hall highlighting the defence (or the invasion) of a private space Modifying elements of the preceding values as age, sex and individual socio-cultural characteristics have to be considered.

Differences in the behaviours relative to personal space are due to cultural diversities of people. Individuals of Mediterranean culture (Arabians, southern Europeans, south Americans) present larger privacy in spatial relations in comparison with northern culture that have larger separation (distance).

These concepts have been developed in several field of human sciences and are already applied to civil architecture and city planning. They can be used for the design of passenger accommodation also, so that cognitive ergonomics appears very useful for the improvement of ship comfort and of passenger wellness.

## 6.2    Biomechanic human being behaviour

The first step of Human postural stability is of great importance in many aspects of everyday life and has application in several scientific and technological fields. These range from medicine, neurophysiology, to sports, biomechanics, robotics, etc. For over fifty years, researches and funds have been devoted to understand postural stability principles and the strategies employed to maintain it, with conceptual and computational model being developed.

Human upright stance is inherently unstable, with most of the body's mass concentrated above the lower extremities, higher up in the trunk; the erect posture is maintained over a relatively small base of support with the pivot point at certain height from the sole. Even in the absence of additional environmental disturbances (movements of the supporting surface, external forces, etc.), a small deviation from upright body orientation is enough to result in an increase of a destabilizing gravitational component, that accelerates the body further away from the upright position. Corrective torque must be generated to counter the destabilizing torque due to gravity; moreover, such a torque generation must be dynamically regulated and changed as the environmental conditions change. The complete nature of the human postural control is unknown yet, since the human body and underlying control system are so complex that a unified and universally accepted theory, explaining how they work, does not exist. Many researchers, in different sectors, have worked towards explaining the mechanism at the basis of upright stance, and a variety of models has been proposed and continues to be developed.

Mathematical models can aid in understanding and explaining complex systems; their purpose is to capture salient features that may differ according to the specific questions they have to address. Their complexity depends on the accuracy required, but, in general, a trade-off between simplicity and accuracy is desirable. While added complexity usually improves the realism of a model, it can make the model more difficult to understand and analyze; in addition, the uncertainty would rise since each single part induces some amount of variance into the whole complex system. The computational cost of adding a huge amount of detail could hinder from using effectively the model, in particular if it were part of a wider simulation/prediction system. Therefore, in order to get a more usable and flexible model it is sometimes appropriate to make some approximations and reduce the model to a suitable form.

Human postural stability is a critical matter as far as the naval and commercial ship sector is concern. The maintenance of an upright stance is a "complex task" per se, and ship motion can make it more difficult and cause even routine duties to be hard, demanding and hazardous, or, in harsh conditions, impossible to perform. Weitheim outlined that motion primarily reduces motivation due to motion sickness, increases fatigue due to increased energy requirements, and creates balance problems. Research into using postural stability models for predicting the effects of ship motions on human task performance is still in its infancy. Currently, the model used to predict the probability that ship motions may cause a person to lose balance or slide on the deck assumes the individual to react as a rigid body having geometrical and inertial proprieties of a human.

## 6.3 Articulated postural stability model

Articulated dynamic modelling of humans has been introduced in the naval sector by Langlois and his group.

Two planar models were proposed for predicting MII frequency as a function of ship operating conditions. The idea comes from models used for human quiet standing, which generally divide human body into two perpendicular planes: the sagittal and the coronal planes.

The median sagittal plane cuts the body into left and right halves passing through the centreline (a bilateral symmetry is assumed); within this plane, the model moves in the front/anterior and back/posterior (A/P) direction.

The mid-coronal or frontal plane bisects the body into front-back or anterior-posterior sections. In this plane, the body moves in the left-right or medio-lateral (M/L) direction.

## 6.3 Comfort indexes

Ship motions and more in particular vertical accelerations are the principal causes of "experiencing" ship as comfortable or not. Two effects are influenced by ship motions and depending on ship type are gaining importance in ship design. The first is a "Motion Sickness Incidence (MSI)" defined by O'Hanlon and McCauley (1974) as percentage of subjects who vomited within two hours journey. The second is "Motion Induced Interruptions (MII)" defined by Baitis, Woolaver and Beck (1983) as an occasion when a crewman would have to stop working at his current task and hold on to some convenient anchorage to prevent loss of balance.

## 7 REPRESENTATIVE PARAMETERS DATABASE

A rank of different existing ship and the development of new design alternatives can be performed according to design features apt to evaluate design performances, called attributes. They do not cover all the ergonomic aspects before mentioned but can be used in a multiattribute procedure for comparing existing designs and for the development of optimal new ones. To this aim a design modulus based on equations able to provide the attribute values for the considered type of ships has to be assessed. In previous works the following characteristics and ratios have been considered as design attributes:

   a. Cruising speed
   b. Passenger Deck area / Passenger number
   c. Passenger Volume /Passenger number
   d. Available window area for passengers / Passenger Deck Area
   e. Available window area for passengers / Passenger number

As regards HSC a separation between mono and multihull has to be done. It is possible to get regression formulas from an adequate number of similar ships. As an example the following equations providing reference values according to ship length for b and e are reported.

b mono(m$^2$) = 0.1039 LPP 0.6238
b multi(m$^2$) = 0.0998 LPP 0.6498
e (m$^2$)= 0.1615 LPP 0.0055

Further attributes related to environmental psychology and cognitive ergonomic can be considered and assessed mostly through passenger interview.

# 9    CONCLUSIONS

The recent marketing oriented design approach should provide full answers to customer (passenger) aspirations and desires as well as references for ship owners commercial choices.

The marketing approach considers customer choices based on cognitive processes; customer final choices are connected to the emotions and perceptions that influence the state and the evolution of the aspirations and expectations.

Cognitive ergonomic can become an useful tool for cruising ship design optimisation through dedicated procedures. To such extent a suitable design model has been identified.

To better consider cognitive ergonomics in the design procedure, data from personal interview of passengers will be used in the next future as well as a wider range of attributes including noise, vibration level, spatial proportion, materials and colours.

## ACKNOWLEDGMENTS

The This work has been financially supported by University of Naples "Federico II" within the frame of 2010-2011 research program.

## REFERENCES

Alvarez-Filip, L., N. K. Dulvy, and J. A. Gill, et al. 2009. Flattening of Caribbean coral reefs: Region-wide declines in architectural complexity. *Proceedings of the Royal Society, B* 276: 3019–3025.
Boyd, J. and S. Banzhaf. 2007. What are ecosystem services? *Ecological Economics* 63: 616–626.
Allan G. 2002. Humane or Human? *Conference on the Human Factors in Ship Design and Operation*, RINA, London.
American Bureau of shipping. 2001.*Guide for passenger comfort on ships*.
Baroni Maria Rosa. 1998. *Psicologia Ambientale*, il Mulino, Bologna.
Berger John. 2000. *Modos de ver*, Editorial Gustavo Gili, Barcelona.
Bertorello C., 2001. *Design Trends of a Small Size Fast Passenger Ferries*, Proc. of Hiper 2001, , pp.79-91
Bonnes M. Secchiaroli G. 1992. *Psicologia Ambientale*, La Nuova Italia Scientifica, Roma.
Edward T. Hall.1996. *La dimensione nascosta*, Bompiani, Milano;
Tosi Francesca. 2001. *Progettazione ergonomica*, Il Sole 24 Ore, Milano.

Hanson M. A. 1998. *Contemporary ergonomics*, Taylor & Francis, Londra-Philadelphia-Philadelphia.

Russo Krauss Giulio. 2001. *Il Progetto della Nave*, appunti del corso, Napoli

Dobbins, T., Rowley, I., Campbell, L., 2008. *High Speed Craft Human Factors*

Alexandrov, A.V. Frolov, A.A., Horak F.B., Carlson-Kuhta, P., Park S. 2005, *Feedback Equilibrium Control During Human Standing*. Biol Cybern, vol. 93(5), pp. 309–322. National Institute Of Health.

Beck, R., Reed, A. 2001. *Modern Computational Methods for Ships in a Seaway*, Transactions, Society of Naval Architects and Marine Engineers, Vol. 109, pp. 1-52.

Borg, F.G. 2005. *An Inverted Pendulum with a Springy Control as a Model of Human Standing*, Website: arXiv:physics/0512122v1.

Cnyrim, C., Mergner, T., Maurer, C. 2009. *Potential roles of force cues in human stance control*. Exp Brain Res. Springer-Verlag. DOI 10.1007/s00221-009-1715-7.

Colwell, J. L. 1989. *Human factors in the Naval environment: A review of motion sickness and biodynamic problems*. DREA Technical Memorandum 89/220.

Colwell, J.L.2005. *Modeling Ship Motion Effects on Human Performance for Real Time Simulation*. Naval Engineers Journal, Winter 2005, pp. 77-90.

Colwell, J. L., 1989. *Human factors in the Naval environment: A review of motion sickness and biodynamic problems*. DREA Technical Memorandum 89/220.

Colwell, J.L.,2005. *Modeling Ship Motion Effects on Human Performance for Real Time Simulation. Naval Engineers Journal*. Winter 2005, pp. 77-90.

Colwell, J.L.,2006. *Human Performance at Sea. Presentation at the ABCD Symposium on the Influence of Ship Motions on Biomechanics and Fatigue*. 25-26 April, 2006. Panama City, Florida.

Dobbins, T., Rowley, I., Campbell, L. 2008. *High Speed Craft Human Factors Engineering Design Guide*. ABCD-TR-08-01 V1.0.

Dobie, T. G. 2003. *Critical Significance of Human Factors in Ship Design*. Proceedings of the 2003 RVOC Meeting, 8 – 10 October, 2003. Large Lakes Observatory, University of Minnesota.

Faltinsen, O.M., Zhao, R.. 1991. *Numerical predictions of ship motions at high forward speed*. Phil. Trans. R. Soc. Lond. A, vol. 334, pp. 241–252.

Faltinsen, O.M., 1990. *Sea Loads on Ships and Offshore Structures*. Cambridge University Press.

Khalid, H., Turan, O., Bos, J.E., Kurt, R.E., Cleland, D. 2010. *A Comparison of The Descriptive and Physiological Motion Sickness Models in their Ability to Predict Seasickness Aboard Contemporary High Speed Craft*. International Conference on Human Performance at Sea HPAS 2010, Glasgow, Scotland, UK, 16-18 June, 2010.

Langlois, R.G., MacKinnon, S.N., Duncan, C.A. 2009. *Modelling Sea Trial Motion Induced Interruption Data Using an Inverted Pendulum Articulated Postural Stability*. The Transactions of The Royal Institution of Naval Architects, International Journal of Maritime Engineering, vol. 151(A1), pp. 1-9. ISSN 1479-8751.

McCauley, M.E., Pierce, E. 2008. *Evaluation of Motion Sickness and Motion Induced Interruptions on FSF-1*. Pacific International Maritime Conference. Sydney, Australia, 29 January - 1 February, 2008.

Stockwel, C.W., Koozekanani, S.H., Barin, K., 1981. *A Physical Model of Human Postural Dynamics*. Annals New York Academy of Sciences.

# Section IX

## Changes at the Organizational Level

# CHAPTER 56

# Ergonomics Aspects of CSR in System Shaping the Quality of Work Environment

*Adam Górny*

Poznan University of Technology, Faculty of Management Engineering
Poznan, Poland
adam.gorny@put.poznan.pl

## ABSTRACT

Companies aiming to become more socially responsible have to manage relations with surrounding environment including clients, shareholders and the society as a whole. Moreover, building a competitive edge over market competitors entails creating propitious internal conditions, including adequate workplace health and safety. Part of those requirements is ergonomic criteria which influence evaluation of whether a company operates in lines with CSR requirements, beneficial to business.

**Keywords**: Ergonomics, CSR, Quality, Work environment, Continuous Improvement

## 1    INTRODUCTION

Companies actively participating in the marketplace, seek competitive advantage to reinforce their balance sheets and consequently improve growth potential. Measures they elect to take, have to comply with transformations taking place worldwide.

Currently, corporate responsibility has become an important aspect of every economic activity. It involves managing relations with surrounding environment including clients, shareholders and the society as a whole. Social responsibility guidelines position businesses within the marketplace and are crucial to their development. Acknowledged as particularly important has been assurance of propitious internal conditions, especially creating a safe working environment, posing no threat to employees' health. It can be assumed, that the work environment is a vital element influencing working conditions, centrepiece to pursuing intended economic benefits (Cempel, 2009; Górny, 2008).

Thus, new opportunities have to be sought for in a bid to develop that business area. At the same time, emphasis has to be put on shaping systemic approach to quality of work environment, corporate social responsibility and the role of development factors traditionally identified with ergonomic requirements.

## 2    QUALITY WORK ENVIRONMENT (SYSTEMIC APPROACH)

Workplace health and safety has been traditionally associated with working conditions, work organisation and employee behaviour, all assuring the required level of health and life protection against potential threats at the workplace. Equally as often, health and hygiene at work are defined as the process of shaping both the working environment and conditions thus assuring required safety. The requirements encompass factors of organisational, technical, physiological, physical, biological and chemical nature.

Characteristic approach to the issues of quality assurance at the workplace leads to conclusion that safety is the fundamental element assuring minimum quality of working conditions. In that context, complying with workplace health, safety and ergonomic guidelines is no longer perceived as a positive gesture of goodwill. It becomes a requirement transforming into technical and economic category in its own right, determining given company's growth potential, influencing output, product quality, profitability and efficiency.

In those circumstances safety becomes a formal obligation. In order to assure high quality working conditions inherent to good quality output, is high level of workplace responsibility and awareness among employees. Measures taken by all businesses stem from (Górny, 2008; Mazur, 2009):

- strong belief that work consistent with safety regulations protecting life and health, benefits the company and its employees,
- optimistic perception shared within the company, that workplace health and safety can be improved and procedures can comply with current standards,
- viewing life and health as precious assets needing protective measures worthy financing,
- collective sense of responsibility shared among employees to assure workplace health, safety and secure working practices,

- obligation to put in place observation and measurement procedures verifying whether workplace practices comply with current safety standards, which incentivise elimination of hazardous practices.
- belief that investigating hazardous non-injury incidents and subsequently designing preventative measures is equally important as investigating accidents,
- belief that only immediate repair or replacement of faulty equipment assures adequate working conditions and workplace safety.

All those measures have to be concentrated on human - the entity ultimately subject to working conditions (Cempel, 2009).

The definition of quality, used by solutions to working conditions issues is given by ISO 9000 i.e. a degree to which a set of inherent characteristics fulfils requirements.

Those requirements in context of working conditions are ergonomic, hygienic and physiological standards specifying compulsory regulations concerning working environment. Interested parties are company employees reaping benefits from proper working conditions and indirectly, immediate and broader business environment. Hence, quality working conditions should be perceived as company's ability to comply with legal and ergonomic regulations concerning work environment (Górny, 2008; Nowak and Pacholski, 2009).

The quality of life within an organisation in the context of quality work environment is a free, conscious and intentional compromise between technical, economic, organisational and sociotechnical actions. They should foster physical and personal development consistent with dynamic model of a human. An employee spends a substantial amount of his time in an organisation. Thus, providing good working conditions is a priority task for the employer (.

Guidelines for systemic development of a quality work environment originate from determinants specifying the above-mentioned guidelines from relevant management standards. Total Quality Management could give guidance, since it is the byword for streamlining human resources management and efficient resource management. TQM is a Continuous Improvement process, taking account of working conditions (Dahlgaard and all, 2000; Lancucki, 2001). Hence the statement that compliance with current regulations, standards and recommendations is the ultimate step in achieving the intended workplace health and safety is not correct. Such process relies on creativity to devise measures furthering the case. Furthermore, employees must be motivated to seek new procedures and practices minimising their exposure to negative impact of the work environment. In practice, achieving satisfactory organisational and technical solutions with respect to working conditions depends on (Górny, 2005):

- tangible working conditions, comprising tangible factors (machines and basic devices, auxiliary equipment, workrooms), physical factors (lighting, temperature, noise, mechanical vibrations, radiation), chemical, organic and non-organic factors (vapour, gas and aerosol),
- work organisation, including workload and working time distribution, managing working practices.

- economical working conditions,
- social working conditions, determined above all by social relations.

In order to guarantee success of taken actions i.e. assure adequate quality of the work environment, it is necessary to enrich each area with ergonomic requirements. All of those areas require Continuous Improvement. Employer and employee awareness has to be raised as well.

Based on systemic approach (consistent with ISO 9001 or OHSAS 18001), any system should put the human at the centre, thus guaranteeing healthy relationship-building. According to the requirements, human resources - characterised by education, skills and experience - ought to be managed with special attention. Those factors often play a crucial role for efficiency of the entire system. All the more, since human factor determines how effectively corporate goals are pursued. It determines the ability to implement changes, and to effectively manage every business area. In context of resources, human factor can be viewed as a constituent of an organisation's economic value. Above all though, human factor is centrepiece to efficiently satisfy client needs and expectations, thus playing an important role in the context of social responsibility determinants. Introducing human factor to strategic planning enables given business to more effectively and efficiently achieve its objectives.

In order to successfully introduce systemic approach (Grudowski, 2003; Hamrol, 2005; Lancucki, 2001):

- at the design stage, elements of the system which are to guarantee the organisation and system to operate efficiently have to determined,
- at the stage of implementing the system, adversaries have to be "converted" to realise the importance of introduced changes, which would require them to change their habits and beliefs,
- at the stage of maintaining and streamlining the system, previously made decisions have to be substantiated, including shifting power and responsibilities between competent parties (employees).

Beyond doubt, it also crucial to bring to employees' attention the relationship between workplace health and safety, and both written and unwritten rules, which are not legal obligations sensu stricto, but oblige all interested parties to comply - otherwise sanctions could be imposed. Employee-centred employee-employer relation should be perceived as a long-term investment benefiting all parties involved, including employees.

# 3    CORPORATE SOCIAL RESPONSIBILITY

## 3.1    General guidelines for developing social responsibility

As intangible assets gain significance, the company-society dialogue also becomes ever-more important. Corporate social responsibility can be perceived as economic, legal, ethic or philanthropic self-imposed obligation towards internal and external social groups. Furthermore, as a rational action on sound footing it has potential to become source of competitive advantage.

When introducing corporate social responsibility, emphasis should be put on how the human factor is introduced to business strategy. It determines the effectiveness and efficiency of achieving business goals, which often involve finding a fine balance between commercial, social and ecological objectives. In doing so, companies create a sense of solidarity and cohesion between interested parties (*ISO and...*, 2010; ISO 26000).

Social responsibility does not only boil down to meeting intended expectations, but also through taking account of people, resources and stakeholder relations advancing beyond them. Bear in mind that a stakeholder could be a single individual or a group of people, which have power to influence given company or the reverse.

Corporate Social Responsibility enables businesses to increase their social capital, which is perceived as intangible source of tangible efficacy allowing achieving set goals through interaction and collaboration. It also has potential to be a contributing factor to assure sustainable development, whilst at the same time reinforcing capacity for innovation and competitiveness (*Implementing...*, 2006; *Working...*, 2005).

Modern businesses ever-often view social responsibility as being strategically significant, due to introduced social actions translating to competitive edge. Along with that, it fuels corporate identity. In this context, corporate social responsibility should be viewed as an action intended to boost business valuation, thus keeping the company afloat.

Approaching employees' health in that manner can yield the following (Górny, 2003):

- comfortable working conditions, guaranteeing to preserve employees' health,
- workplace safety complemented by good employee-employer relations and observance of human rights,
- free access to different forms of employment,
- social responsibility and measures counteracting negative impact of commercial activity.

Key here is strong trust, which lies at the heart of developing a safe workplace and convincing employees to change their hazardous practices.

## 3.2 Social responsibility and quality work environment

Following guidelines of corporate social responsibility (CSR) allows maintaining loads experienced by employees at the workplace optimum. They comprise all areas important from the point of view of sustainable development. They assure economic growth and good life quality to employees, their families, local communities and the entire environment (*Implementing...*, 2006; *ISO and...*, 2010).

In order for the CSR to ultimately bring intended effects, it has to be part of corporate strategy and its elements have to be tied in with core business areas. Only then a company is capable to operate efficiently and effectively. It should be assumed, that some of those requirements are criteria for quality work environment. While seeking to find the sweet spot between stakeholder expectations (employees) and business owners' (employers) room for change, it has to be done in an honest, ethical and dignified manner (i.e. in a socially responsible manner). Consequently good image in eyes of business environment is also gained.

There is a straightforward relationship between the process of improvement and working conditions. Those requirements put great emphasis on workplace environment factors. Impact of workplace environment factors is a vital evaluation criterion for assessing work environment's consistency with optimum requirements for particular professional activities.

In order to include work environment factors in social responsibility guidelines, they have to be combined to address needs and expectations of interested parties. It concerns factors traditionally perceived as:

- onerous, substantially decreasing working comfort,
- harmful, posing threat to employees' health,
- hazardous, posing threat to employee safety.

## 4     ERGONOMICS IN DEVELOPING A QUALITY WORK ENVIRONMENT

### 4.1     Impact of ergonomics on quality of work environment

In order for an organisation to commercially succeed, it has to operate in a transparent manner. The overwhelming part of every success story is the human. Thus good operating conditions have to be provided. Apart from being safe, they also have to fulfil stipulations concerning load optimisation.

Ergonomic criteria play a major role in work environment specification. They have been traditionally identified as ergonomic standards or standards stipulating ergonomic requirements, which could be associated with ergonomic criteria (Górny and Dahlke, 2005).

Ergonomic criteria, above all concentrate on human factor, its role in commercial success and contribution to effectiveness of actions taken. Bearing fruits as a consequence of safe and healthy working conditions whilst having failed at factoring in ergonomic criteria is either extremely difficult or impossible all together. Hence, it is desirable or indeed obligatory, to include ergonomic criteria during development of social responsibility. Any economic venture can operate effectively and succeed, given it provides workplace environment factors optimum for work.

It needs to be become a given, that ergonomics is one of more important elements creating quality, through pro-humanistic, corporate approach towards business-environment, and business-internal client relationships. All those measures are implemented to find best directions for development and to optimise loads an employee is subject to at the workplace.

Scope of requirements needed to be considered during designing working conditions, factoring in ergonomic requirements has been given in tab. 1. Ergonomic standardisation is crucial to all listed measures. It concerns requirements and guidelines for developing working conditions and environment.

Table 1. Characteristics of ergonomic requirements

| Area (category) of requirements | Scope of ergonomic requirements |
|---|---|
| Anthropometric requirements | guidelines for adapting technical equipment to dimensions and weight of a human body, or indeed its parts, both in static and dynamic setting, in order to ensure e.g. a rational working position. |
| Physiological requirements | guidelines for adapting technical equipment to psychological needs, factoring in optimum load on muscles, skeletal, respiratory and cardiovascular systems, joints and limbs. |
| Psychophysical requirements | guidelines for adapting technical equipment to human senses: sight, hearing, smell, touch and taste. |
| Hygienic requirements | guidelines for adapting the workplace to human needs and limitations in order to diminish negative impact of workplace environment factors and to provide comfort in the working environment. |

Assuring ergonomic working conditions and optimum loads on human body entail, above all, factoring in human factor, perceived as the main focus of all actions. Human factor is becoming the ever-important element of corporate development. Failing to do so hinders developing modern, pro-growth concepts. However, in order to succeed, an organisation would have to deploy goal-oriented measures indicative of its commitment and engagement.

## 4.2    Ergonomics in successful, CSR-consistent measures

Ergonomics is inextricably linked with working conditions, including development of intended workplace health and safety.

Ergonomic criteria:
- are source of ergonomic information,
- initiate the solutions-generating mechanism, producing ergonomic solutions,
- identify opportunities to deploy ergonomic solutions.

In order to sufficiently substantiate them and achieve intended results, they have to be included in job description, specifying how given workstation has been adapted to the needs of prospective worker.

It is important to create optimum conditions factoring in current requirements. In order to successfully address CSR requirements, human needs have to be understood. Providing a quality work environment entails creating an environment satisfying needs and expectations of interested parties (Górny, 2008; Mazur, 2009).

Including ergonomic requirements in development of work environment involves taking account of so called ergonomic risk factors, which through using standards concerning ergonomics enable all process participants to access interdisciplinary knowledge, necessary to optimise business processes. Work environment developed in line with ergonomic criteria is harmonised with psychophysical human ability, thus effectively safeguards employees from potential hazards and consequently improves safety indicators, including psychological aspects of accident frequency rate.

Taking on board ergonomic criteria brings about improved safety through improving the working conditions and job performance. Moreover, working conditions become better adapted to somatic and psychological human qualities relative to physical, social and technological environment at the workplace (Górny, 2008).

It has to be carved in stone that people are the greatest asset a company can hold, creating its commercial successes. Hence, physical and psychosocial working conditions not only impact health, but also performance and job satisfaction of workers. Thus, this aspect of developing work environment should be perceived as crucial to economic success of any organisation. An employer should pay great interest in it through e.g. fulfilling obligations dictated by ergonomic requirements.

Identification of ergonomically relevant hazardous areas enables to:

- draw up requirements for technical equipment with respect to their ergonomic quality (including workstations and work systems),
- draw up requirements for work environment with respect to its ergonomic quality,
- implement solutions reducing accidents and occupational conditions, improving working environment (conditions), boosting job performance and reducing health related absence.

In practice the above-mentioned objectives often fail to realise intended occupational ergonomics as they are achieved thorough a.o. legally substantiated:

- identification, evaluation and monitoring of workplace risk factors,
- engineering, organisational and administrative preventative measures against risk factors,
- staff and employer training expanding knowledge on occupational risk factors at the workplace.

Although there is a legal obligation to deploy the above-named measures, their scope in line with ergonomic requirements and CSR determinants should be substantially broadened beyond legally imposed employer duties and thus reach the essence of ergonomic risk factors.

# 5    CONCLUSIONS

Introducing change to corporate operations can create organisational culture, friendly working atmosphere for employees and management, compel employees to identify themselves with the company and put its resources to a better use, consequently having positive impact on financial results. It is synonymous with embracing ergonomic requirements - which play an important part in development of good working conditions - as part of corporate strategy, which enables to reap benefits from quality work environment.

# REFERENCES

Cempel, W.A. 2009. Human aspects in the model of creation and implementation the enterprise strategy. In *Macroergonomics vs. social ergonomics.* ed. L. Pacholski. Publishing House of Poznan University of Technology, Poznan.

Dahlgaard J. J., and K. Kristensen, G. Kanji. 2000. *Podstawy zarządzania jakością.* Wydawnictwo Naukowe PWN, Warszawa.

Górny A. 2008. Ergonomic requirements in system management of industrial safety, *Foundation of Control and Management Science* 11: 127 – 138.

Górny, A. 2003. Human Factors in Occupational Health and Safety Management. In *Mind and Body in a Technological Word.* Proceedings of 35th Annual Conference of Nordic Ergonomics Society. Reykjavik, Island.

Górny A. and D. Dahlke. 2005. The OHS management trough using of the TQM strategy elements. In *Ergonomics and work safety in information community. Education and researches.* eds. L. M. Pacholski, J. S. Marcinkowski, W. Horst W., Institute of Management Engineering, University of Technology, Poznan.

Grudowski P. 2003. *Jakość, środowisko i bhp w systemach zarządzania.* AJG, Bydgoszcz.

Hamrol A. 2005. *Zarządzanie jakością z przykładami.* Wydawnictwo Naukowe PWN, Warszawa.

*ISO and social responsibility.* 2010. International Organization for Standardization, Geneva.

ISO 26000:2010. *Guidance on social responsibility.*

ISO 9000:2005. *Quality management systems – Fundamentals and vocabulary.*

ISO 9001:2008. *Quality management systems – Requirements.*

*Implementing the partnership for growth and jobs: Making Europe a pole of excellence on corporate social responsibility.* Communication from the Commission to the European Parliament, the Council and the European Economic and Social Committee. COM (2006), European Commission, Brussels.

Lancucki J. 2001. *Podstawy kompleksowego zarządzania jakością TQM,* Wydawnictwo Akademii Ekonomicznej, Poznań.

Mazur A. 2009. Shaping quality of work conditions, In *Health protection and ergonomics for human live quality formation.* Eds. G. Dahlke and A. Górny. Publishing House of Poznań University of Technology. Poznań.

Nowak W. and L. Pacholski. 2009. Human factors in improving process of the company organizational culture. In *Macroergonomics vs. social ergonomics.* ed. L. Pacholski. Publishing House of Poznan University of Technology, Poznan.

OHSAS 18001:2007. *Occupational health and safety management systems – Requirements.*

*Working together for growth and job.* Communication to the spring European Council: A new start for the Lisbon Strategy, Commission of the European Communities, COM (2005) 24 final, Brussels.

CHAPTER 57

# A Process-Driven Socio-Technical Approach to Engineering High-Performance Organisations

*David Tuffley*

Griffith University
Brisbane, Australia
D.Tuffley@griffith.edu.au

## ABSTRACT

For organisations, achieving the transition from competence to high-performance can be an elusive goal. The transition requires a systematic, consistent improvement that is difficult to sustain over time. Process models are a recognized way to achieve consistent, repeatable results, yet such models have had limited success when applied to the so-called "soft" areas of organisational behavior. Recent work done in the Software Engineering process improvement discipline has seen a new category of process reference model come into existence, one which is designed to address soft areas of organisational performance. This paper introduces a *Reference Model of Organisational Behavior* for the leadership of complex virtual teams that has proved effective in industry trials. While it deals with leadership, the approach outlined can be generalized to other aspects of organisational behavior such as innovation and competencies / capabilities, creativity and resilience to adversity, to name a few.

**Keywords**: process, process model, Sociotechnical, organisational behavior, software engineering process improvement.

## INTRODUCTION

The *Reference Model of Organisational Behavior* (RMOB) approach outlined in this paper is a hybrid of Sociotechnical and Software Engineering methods. RMOB's seek to optimise the interaction between people and technology within the organisation context.

Reference Models of Organisational Behavior take a previously difficult to describe "soft" concept such as Leadership and deconstructs it so that its underlying components can be described in process-oriented terms and used for capability assessment and process improvement purposes (Tuffley, 2010). It does this by using proven process modelling techniques derived from the Software Engineering domain.

In Sociotechnical terms, this approach applies a process-derived understanding of leadership to the important matter of optimising organisational work design in a way that is sensitive to the complex interactions between people and processes/structures in the organisation. Beyond Leadership, this approach could arguably be applied with good effect to other soft organisational issues like Culture, Innovation and Capabilities. These important concepts must be understood if we are to model and manage all of the important activities that occur within the modern Organization (Tuffley, 2010).

The outcome of the on-going research project that underlies this paper supports this conclusion. In preliminary trials, the *Reference Model of Organisational Behavior* appears to be an effective tool for describing organisational behavior. A model such as this defines organisational behavior that, if performed repeatedly, will result in consistently achieving the prescribed purpose.

There is no reason why a process model that describes organisational behavior should be restricted to the software engineering domain. There is nothing in the model that would restrict its scope in such a way. It is generic and applicable to a broad range of Design disciplines and domains, making it of potential interest and use to the business community at large.

In earlier software engineering-related process models the focus was on conformance to prescribed activities and tasks. The Leadership RMOB re-focuses attention from prescriptive conformance to a focus on the demonstration of desired organizational behavior.

## 1. CAN LEADERSHIP BE DESCRIBED AS A PROCESS?

It is reasonable to ask this question, since leadership is a difficult topic to define, much less describe in process-oriented terms (Takeda eta al, 1980). Many people doubt that leadership is something that can even be learned, subscribing as they do the view that leaders are born not made.

Leadership has been observed and studied for countless generations, yet interestingly little consensus exists as to what constitutes true leadership. It has been the subject of intense and on-going controversy among psychologists, sociologists, historians, political scientists and management researchers (Yukl, 1994). It is interesting to note that no universally accepted definition of leadership has been developed. The operational definition of leadership has much to do with the purpose of the researcher (Yukl, 1994).

Yet observers like Peter Drucker (1996) and Warren Bennis (1994, 1985) are clearly of the opinion that leadership can indeed be learned by those prepared to make sufficient effort.

Can leadership be defined as a process? W. Edwards Deming (2000) famously observed that, *if you cannot describe what you are doing as a process, then you don't know what you are doing.* Assuming that the leadership factors could be identified from a broad literature review, then a RMOB is a logical way for these factors to be formalized and applied in real situations.

## 2. COPING WITH ORGANISATIONAL COMPLEXITY

Managing complex projects across dispersed geographical locations has never been more difficult, given the rising complexity of the global economic environment and the multi-national corporate entities that now inhabit this world. There is a clear need to find improved ways of managing this often difficult process now and into the future (Herbsleb & Moitra, 2001).

RMOB's are an implementation of the Software Engineering discipline known as *Model Based Process Improvement* (MBPI) which provides a suitable framework around which the definition of leadership processes can occur. This approach has not been used (to the knowledge of the author) to address leadership, though there is arguably a sound basis for thinking that it can be used in this way.

MBPI aims generally to improve the performance and maturity of an organisation's processes (Heston & Phifer, 2009). It combines the discipline of process improvement with the several international standards and frameworks now in use (i.e. ISO/IEC 15504, Capability Maturity Model, Integrated). Combining this awareness of process performance with internationally recognized standards is advantageous to organisations. It affords a structured and comprehensive framework as a way forward and prescribes in general terms the scope of activities required to systematically improve their process maturity.

Heston and Phifer (2009) ascribe the following organisational benefits to MBPI:

- *Improving consistency and repeatability*: consistency and repeatability assist with minimising process variation, a major source of product defects. It also allows project staff to move into and out of projects more easily by having clearly defined roles and responsibilities.
- *Improving communication*: achieved through the adoption of a common vocabulary with clearly prescribed meanings that allows project staff, clients and business partners to communicate with less ambiguity.
- *Enabling more improvement*: process improvement programs create an environment which is conducive to further improvement. Beyond consistency and repeatability comes the ability to measure and record process performance. This performance data can then be used to plan further improvements and to benchmark against best practice.
- *Providing motivation*: objective targets, for example being assessed at a certain level of maturity, become a visible motivator for project staff to maintain their efforts to improve process performance.

# 3. RMOB ARCHITECTURE & CONTENT

## 3.1 Architecture

The format and content of Process Reference Models in Software Engineering are prescribed by certain international standards (ISO/IEC 15504-2:2003 and ISO/IEC 24774:2007) and these have been faithfully followed in this project.

A survey of the leadership and process model literature in academic library catalogues showed little research covering the specific topic of this project. The search included Biology, Life Sciences, and Environmental Science; Business, Administration, Finance, and Economics; Chemistry and Materials Science; Engineering, Computer Science, Mathematics; Medicine, Pharmacology, Veterinary Science; Physics, Astronomy, and Planetary Science; Social Sciences, Arts, and Humanities.

Leadership of complex teams in virtual environments therefore represents a largely vacant intersection between the areas of teams, virtual teams, complex teams and effective leadership. The following architecture for the *Reference Model of Organisational Behavior* (RMOB was devised.

1. **Generic Leadership Skills**. There is a generic set of leadership skills/qualities that will apply in both face-to-face and virtual team environments. This generic set is identified and distilled from the wealth of leadership research over time.
2. **Specific examples of practices for integrated (complex) teams**. The integrated teaming goals and practices of CMMI-Integrated Product & Process Development extension (2011) constitute leadership criteria by default in the sense that someone has to give effect to them, and that will be the responsibility of the leader.
3. **Specific Virtual Environment Challenges for Leaders**. The virtual teaming challenges outlined by Bell & Kozlowski (2002) will be successfully met by an effective leader. These factors have been hypothesised by Bell & Kozlowski as being specific factors influencing the success of virtual team leaders.

This RMOB architecture also theoretically allows for application to virtual teams only, and integrated teams only by using the generic leadership layer plus the relevant virtual or integrated factor layer.

This RMOB architecture would also be applicable to the generic leadership capability of a conventional co-located team that is neither virtual nor integrated.

Maximum flexibility is desirable in this project to allow for the widest range of future research possibilities and practitioner applicability.

## 3.2 Content

Using the structure and content headings obtained from the two ISO/IEC standards mentioned above, the following sets of leadership factors, as derived from the literature are incorporated to produce the V0.1 RMOB.

The leadership factors are represented in mind-map format deliberately to imply that there is no particular order that these should be performed, rather that they comprise constellations of factors that collectively represent leadership.

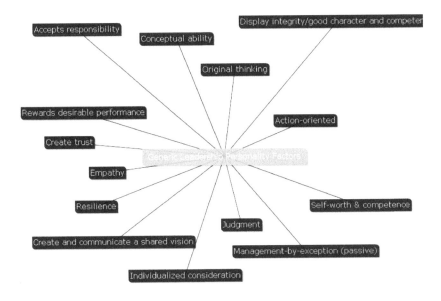

Figure 1: Generic Leadership Factors

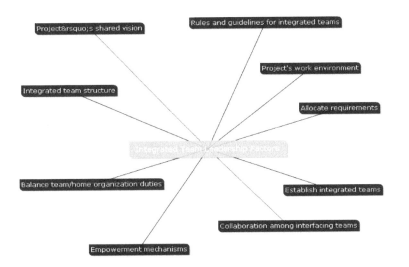

Figure 2: Integrated Leadership Factors

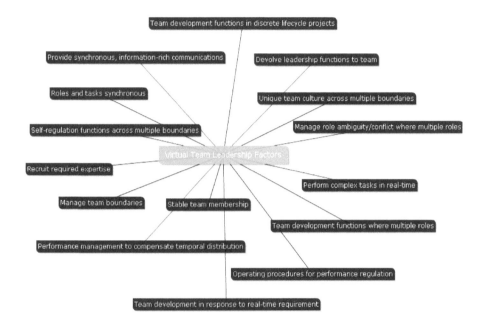

Figure 3: Virtual Leadership Factors

The leadership characteristics, seen as elements in the cluster diagrams above are grouped according to the three categories (generic, integrated, virtual) and rendered into the prescribed Process Reference Model format, a sample of which appears below:

### Create a shared vision

**Purpose**: to perceive a guiding principle/idea that captures the imagination of members to create a shared vision and inspire them to realise that vision. An aspect of charisma.

**Outcomes**: as a result of the successful implementation of creating a shared vision:

1. The leader perceives and formulates a unified vision of what is to be accomplished, ideally seen as an accomplished fact.
2. The leader develops a strong commitment to the achievement of that vision, based on a sense of rightness and timeliness, such that they have sufficient resilience to overcome goal frustrating events.
3. The leader develops a clear and unambiguous set of objectives or goals that are concrete and achievable.

**Elaboration**: the shared vision is a clear and unambiguous expression of an envisioned future. It is the basis for a common understanding among stakeholders of the aspirations and governing ideals of the team in the context of that desired outcome. Conditional on being effectively communicated by the leader to the team, the shared vision grounds the team's governing ideas and principles and allows for appropriate objectives to be derived.

Highly effective groups are often convinced they are engaged in important work, sometimes nothing short of being on a 'mission from God'. The work becomes an abiding obsession, a quest that goes well beyond mere employment. This intensely shared vision and sense of purpose endows cohesion and persistence.

After the six reviews, the *Create a shared vision* process was improved to become:

### Vision
**Purpose**: The purpose of the vision process is to create and communicate a shared vision in ways that inspires people to realise that vision.

**Outcomes**: As a result of the successful implementation of shared vision process:
1. A vision of the goal(s) is created.
2. The vision of the goal(s) is communicated to the team.
3. Commitment by team to the shared vision is gained.

### Informative Notes
Outcome 1 -- the vision of the goal is seen by the leader as achievable. The goals will still be abstract at this point. The goal(s) become concrete when translated into objective(s).

Outcome 2 – the shared vision should be communicated in a way that creates positive expectation among the team.

Outcome 3 – the way in which the shared vision of the abstract goal(s) is communicated should generate strong commitment to the achievement of the goal(s)

### General:
The shared vision is a clear and unambiguous expression of an envisioned future. It is the basis for a common understanding among stakeholders of the aspirations and governing ideals of the team in the context of that desired outcome. Conditional on being effectively communicated by the leader to the team, the shared vision grounds the team's governing ideas and principles and allows for appropriate objectives to be derived.

Highly effective groups are often convinced they are engaged in important work, sometimes nothing short of being on a 'mission from God'. The work becomes an abiding obsession, a quest that goes well beyond mere employment. This intensely shared vision and sense of purpose endows cohesion and persistence.

Creating and communicating a compelling vision of the future is an aspect of charisma; inspirational motivation, optimism, individualized consideration and contingent reward all appear to optimise team performance by creative a positive affective climate.

In summary when promulgating a shared vision, the following factors should be considered:
1. the project's objectives

2. *the conditions and outcomes the project will create*
3. *interfaces the project needs to maintain*
4. *the visions created by interfacing groups*
5. *the constraints imposed by outside authorities (e.g., environmental regulations)*

*project operation while working to achieve its objectives (both principles and behaviors)*

## 4. DATA COLLECTION METHOD

As mentioned, a Reference Model of Organizational Behavior (RMOB) is a formal expression of a set of behaviors that if performed over time will bring about desirable outcomes. A RMOB has a statement of purpose and outcome, with the outcomes collectively achieving the purpose. V0.1 RMOB emerged from the literature review, formatted in the way shown below.

## 4.1 Draft Process Reference Model (V0.1 RMOB)

The draft RMOB that emerged from the literature review is developed in compliance with ISO/IEC 15504-2:2003 and ISO/IEC 24774:2007 (the two standards that collectively prescribe how RMOB's in software engineering should look). An example is seen in the table below. In addition to the process name, purpose and outcome(s), informative material is included to provide context and clarification of the purpose and outcomes to assist the user. The V0.1 RMOB is then ready for review (data collection).

---

*1.1 Create and communicate a shared vision*

---

**Purpose**: to perceive and communicate a guiding principle/idea that captures the imagination of members to create a shared vision and inspire them with the enthusiasm to realise that vision. An aspect of charisma.

**Outcomes**: as a result of the successful implementation of creating a shared vision:

1. The leader perceives and formulates a unified vision of what is to be accomplished, ideally seen as an accomplished fact.
   *Activities and/or artefacts to support:*

2. Leader communicates shared unified vision with team, ideally seen as an accomplished fact.
   *Activities and/or artefacts to support:*

3. Leader develops strong commitment to achieving vision, based on a sense of rightness and timeliness, such that they have sufficient resilience to overcome goal frustrating events
   *Activities and/or artefacts to support:*

---

4. The leader develops a clear and unambiguous set of objectives or goals that are concrete and achievable.
   *Activities and/or artefacts to support:*

5. Leader engenders hope/optimism towards achieving the objectives.
   *Activities and/or artefacts to support:*

*Elaboration: the shared vision is a clear and unambiguous expression of an envisioned future. It is the basis for a common understanding among stakeholders of the aspirations and governing ideals of the team in the context of that desired outcome. Conditional on being effectively communicated by the leader to the team, the shared vision grounds the team's governing ideas and principles and allows for appropriate objectives to be derived.*

Table 1: Extract from draft RMOB to illustrate data collection method

## 6.2 Data collection

Data collection is by a series of walk-through interviews with project manager participants from organisations operating integrated virtual teams. The purpose is to investigate whether there is *objective evidence* of process performance (base practices and work products) to indicate that the outcomes are being achieved.

If an outcome can be substantiated, it remains in the RMOB. If objective evidence in the form of an activity or artefact cannot be found, then according to the normative ISO standards that outcome cannot remain in the RMOB. In this event, the outcome can be reframed or merged with another of a similar nature. Evidence of artefacts and/or activities is recorded on the space provided on the interview form.

A total of six reviews were performed:

**Stage 1 review**: four interviews with project manager participants from organisations operating virtual teams in which the V0.1 RMOB is walked-through looking for objective evidence for the performance of each outcome. These were drawn principally from the defence contracting and general commercial IT industry sectors.

**Stage 2 review**: four interviews with different project manager participants from organisations operating virtual teams in which the V0.2 RMOB is walked-through looking for anything that does not make sense. These were drawn principally from the defence contracting and general commercial IT industry sectors.

**Stage 3 review**: researcher only (performing ISO/IEC 24774 compliance analysis on the V0.3 RMOB).

**Stage 4 review**: researcher performing a Behavior Tree notation analysis on the V0.4 RMOB (Behavior Tree notation is part of a broader formal method known as Behavior Engineering, developed by Dromey (2006) to verify software requirements by imposing a formal syntax on the expression of said requirements. This method has been adapted for use as a verification tool for model developers).

**Stage 5 review**: Expert Panel review on the V0.5 RMOB.

**Stage 6 review**: researcher performing a Composition Tree notation analysis on the V0.6 RMOB producing the V1.0 RMOB at the end of the process. (Composition Tree notation is another part of the formal method Behavior Engineering, developed by Dromey (2006) to verify software requirements by developing a vocabulary of terms used in a specification to identify synonyms, redundancies and other ambiguities).

## 5. PRELIMINARY TRIALS

Subsequent to the development and baseline release (V1.0 RMOB) of the Process Reference Model, a detailed review session with four project managers of virtual teams was conducted. Substantial improvements were identified which collectively contributed to the development of V1.1 RMOB. The consensus view of the four participants was that the RMOB was a useable tool that they could apply in their own practice. Further reviews and research will be conducted to further validate the usefulness of the V1.1 RMOB leading to further improvements and subsequent releases. These preliminary results have been encouraging and point the way to future trials.

## 6. CONCLUSIONS

The process reference model approach outlined in this paper is broadly consistent with the Sociotechnical perspective. It takes a previously difficult to describe "soft" concept such as Leadership and deconstructs it so that its underlying components can be described in process-oriented terms and used for capability assessment and process improvement purposes.

In Sociotechnical terms, this approach applies a process-derived understanding of leadership to the optimising of complex organisational work design in a way that is sensitive to the complex interactions between people and processes/structures in the organisation. Beyond Leadership, this approach could arguably be applied to other soft organisational issues like Culture, Innovation and Capabilities. These important concepts must be understood if we are to model and manage all of the important activities that occur within the modern Organization.

The outcomes of the research project that underlies this paper support the conclusion that a *Reference Model of Organisational Behavior* is an effective tool for describing organisational behavior, though this question needs further exploration. A model such as this defines organisational behavior that, if performed repeatedly, will result in consistently achieving the prescribed purpose.

There is no reason why a process model that describes organisational behavior should be restricted to the software engineering domain. There is nothing in the model that would restrict its scope in such a way. It is generic and applicable to a broad range of industries.

In earlier software engineering-related process models the focus was on conformance to prescribed activities and tasks. The Leadership RMOB re-focuses attention from prescriptive conformance to a focus on the demonstration of desired organizational behavior.

The method outlined in this paper may therefore be of considerable interest to researchers and professionals from many disciplines and domains by providing the means by which any desirable behavior may be defined in a process model and subsequently applied in the organisation. In this project, the difficult to define topic of leadership has been distilled into a generic model for leadership, and subsequently found to be useful by project managers operating virtual teams. The same model could be applied in any project environment to improve the leadership abilities of the project manager.

## REFERENCES

Bennis, W. and Nanus, B., (1985). *Leaders: the strategies for taking charge*. New York, Haper and Row.

Bennis, W. (1994). On Becoming a Leader, *What Leaders Read 1*, Perseus Publishing, p 2.

Bell, B.S., Kozlowski, S.W. (2002). *A Typology of Virtual Teams: Implications for Effective Leadership.* Group and Organisational Management, Vol. 27, No.1 pp. 14-19.

Deming, W.E., (2000). *Out of the Crisis*, MIT Press, Cambridge MA.

Dromey, R.G. (2006). *Climbing Over the 'No Silver Bullet' Brick Wall*, IEEE Software, Vol. 23, No. 2, pp.118-120.

Drucker, P. (1996). *Managing in a Time of Great Change*, Butterworth Heinemann, London.

Heston, K.M, Phifer, W, (2009). *The Multiple Quality Models Paradox: How Much 'Best Practice' is Just Enough?* Software Process Improvement and Practice. Wiley InterScience.

Herbsleb, J.D. Moitra, D., (2001). *Global Software Development*, IEEE Software, Vol 18, Issue 2, (16-20).

ISO/IEC 15504 (2003). *Information Technology: Process Assessment*. Joint Technical Committee IT-015, Software and Systems Engineering. This Standard was published on 2 June 2005.

ISO/IEC TR 24774 (2007). *Software and systems engineering -- Life cycle management -- Guidelines for process description*. This Standard was published in 2007.

SEI (Software Engineering Institute) (2008) *Capability Maturity Model, Integrated* (CMMI). Accessed 24 December 2012. Available at http://www.sei.cmu.edu/cmmi/solutions/dev/

Takeda, H., Veerkamp, P., Tomiyama, T., Yoshikawam, H. (1990). *Modeling Design Processes*. AI Magazine Winter: pp 37-48.

Tuffley, D. (2010), Reference *Models of Organisational Behavior: A new category of Process Reference Model?*, in Proceedings of the 10th International SPICE Conference on Process Assessment & Improvement, Pisa, Italy, 19-20 May 2010.

Yukl, G., (1994). *Leadership in Organisations*. Englewood Cliffs, N.J. Prentice-Hall.

CHAPTER 58

# Work in Organization: Key Factors of the Reistic Approach

*Janusz Rymaniak*

Gdansk School of Banking
Gdańsk, Poland
jrymaniak@wsb.gda.pl

## ABSTRACT

The paper provides a new concept of reistic approach to work, which assumes the equivalence of all elements for the effects of work. The concept has been compared with existing methodological approaches to work and to the social-economic conception, which exclude the exclusiveness of human agent (Rymaniak, 2011). The key elements of the concept are presented. Specifically, the theoretical analysis of a relation between different elements of work, their relative position in organisation and models of economical effectiveness; is presented. The development of the concept provides the opportunity to create models and reistic theory of work.

**Keywords**: work design, reistic approach, resource-based view (RBV)

## 1    THEORETICAL APPROACHES TO WORK

The review of relevant literature indicates that there are two major trends in research on work. One of them, called the anthropocentric (subjective) approach, identifies man with work, i.e. gives work a purely human aspect. The physiological concept concerns the relation between the potential (abilities) and the input (effective use) of man's physical and mental powers to an activity (Karwowski 2010). The psychological concepts deal with the expenditure of mental effort on different activities; the sociological concepts refer to work as a set of necessary interactions between employees and teams, which form the basis for social, professional and functional diversification, and for stratification systems; economic concepts relate to the need of generating income with the aim of satisfying demands. The mechanistic, motivational, cognitive and biological archetypes of work systems take shape (Corder, and Parker 2007). This trend takes into consideration the individual dimension of (man's) work and the conditions for defining the determinants of the employee's desired economic behaviour. The other, sociotechnical (mixed) approach is based on the assumption that there is 'a common sphere' for the social and technical elements, though it stresses the predominant role of human creativity. The main achievements within this trend are: a dynamic development of team structures in work organization, and research on the complexity and the multi-level character of work in an organization (Rymaniak 2011).

## 2. REISTIC APPROACH – NEW CONCEPT OF WORK

### 2.1. Definition and characteristics

The concepts presented incline towards a new paradigm and a new concept of work, designed for present-day needs. The author proposes the reistic approach, by which work is an organizational system of conversion of measurable components (i.e., human capital, machines, buildings, structures, devices, tools, materials, raw materials, information, area). The system operates in a physical environment and within a specified period of time to convert the components into end products (goods or services) on order, in accordance with their technical norms and the specifics of the contract. The characteristics of this approach are outlined in Table 1.

The mechanism and components, i.e.:

a) the organisational dimension
b) conversion
c) the diverse nature of components and conversion
d) the space and time in which conversion takes place

are included in the definition of the concept.

Every sort of work is of an organizational nature, irrespective of the amount and kind of components, or the actual type of work. The organizational nature of work can be understood in two ways. Firstly, it determines the parameters for measures that need to be taken to prepare the methods and resources which are necessary for obtaining a product in accordance with the terms of the contract. Hence, it involves diverse organizational measures of an economic nature or context (e.g. specification, purchase, hire, leasing, contracts, work plan, budget etc.).

Table 1 Characteristics of the organizational work mechanism

| Attribute | Content |
|---|---|
| Equivalence | The necessity of the presence of specified resources which are used in the product realization process |
| Resource Structuralism | The amount, method of use (conversion) of resources and its quantitative, evaluative and qualitative structures |
| Accumulativeness | The accumulative nature of product development with respect to materials and costs |
| Complementarity | The mutual complementation of components within a resource (a module) or between resources (modules) |
| Systemic Conception | Sets of relations between components within and between modules |
| Substitution | Replacing components with other components from the same resource (module) or from other resources (modules) |
| Modularity | Sets of resources creating subsystems which realize subsequent technological phases of product development |
| Measurability | Taking into account all tangible and legal assets (i.e. eligible for registration and accounting) |
| Duality | Basic activities which create costs but also generate income, and others, which only create costs |

Secondly, the organizational nature implies achieving organizational goals (i.e., core business), which include: profit in market systems or optimization of budget use (allocated funds) in non-market activities (Pitelis 2010).

A relevant distingushing feature is the complexity of the process of obtaining end product, which determines the technological – technical division of labour. One can distinguish between phases, or sets of arranged activities of a similar character which form separate stages in the production process (Jasiński 2011). They create modules in the work process and constitute a basis for planning, provision of resources and evaluation of results, and sometimes for creating the elements of the organisational structure (sections, departments etc.). The difference between a haircut (hairdressing services) and a ship (shipbuilding) depicts the possible level of diversity. In craft, one or at most a few workers perform all the necessary activities to obtain the end product and the difference between individual phases (e.g. washing, haircutting, dyeing, backcombing etc.) is small. In the case of shipbuilding, there is a big generic diversity of operational phases. The number of operations and their types rises, which forces division within the company (departmentalization) and also a division of work in the form of internal cooperation of work types and time frames as well as external cooperation (outsourcing). The complexity of actions contributes to an increased contract risk, which is minimised by calculative (time frames, prices), resource and formal coordination, mainly in contracts and methodology (Crooper et al. 2008).

Work organisation is demand-related. Therefore, work which is directly oriented towards the end result (a profit or output maximization within the budget framework) is of fundamental importance, as this work provides for a demand for company products, which has been caused by its offer and which brings the company profits from the fulfilled contracts. All the other sorts of work have an auxiliary character and consist in realization of core busines processes and running the company. As opposed to the core activity, it only involves a lot of expense.

Modularization is thus a relevant element in the process of creating an organizational resource base. It enables, technically and operationally (management), to ensure a successful production in response to a demand (an order) or the standard of the economic system. Because, as regards technology, order realization is multi-stage, the definition of work is based on the phenomenon of conversion.

Conversion is predicated on the assumption that as the technological process advances, the product accumulatively takes shape. The share of individual components in a specified work mainly depends on its type, as well as the specifics of the order and the deployed technical resources, which determine the choice of a production technology. An intellectual activity, like that of a designer's in a construction office, a manager's or a doctor's, involves a lot creative work. The share of other kinds of resources (designer, manager) is relatively small or in general short-term, e.g. medical examination with expensive specialist equipment. In the case of a machine operator, an assembly crew, a plate welding department or a driver there is, however, hardly any share of creative work. The work is predominantly repetitive, routine and defined by procedures, with a considerable role of machines and devices. Order specifics is another determinant of component share. It implies realization of a product with a set of operational parameters, which is defined as typical, i.e. standard. There might also appear the alternative to do some additional tasks which result from the customer's specific needs. It often entails using other machines and devices than those specified in standard parameters, i.e. non-standard. Yet another main determinant is the company's resources, which allow for a more material-intensive or time-consuming production technology as the technical potential determines the choice of technology (production method.

Accumulativeness is also very relevant. It reflects in the order of actions, the pace of product development and the technical form. The latter aspect is of great significance as concerns workflow on the production line, coupled operations or actions which are dependent on the stage results of the preceding modules. Every single break in work automatically brings about an 'empty cost' for the entire module or even the consecutive modules, which are dependent on the output of the preceding ones. In practice, preventive measures are taken against potential hazards. Whereas a machine breakdown is difficult to foresee (spare parts as buffers), additional sources of energy become activated when a supply failure occurs.

## 2.2.  Resources as components of conversion

The conversion of particular resources has a diverse character as well. People use their knowledge and skills to perform particular operations, activities, and tasks in the work process. In literature on the resource approach (RBV) and HRM it is stressed that doing a particular sort of work allows to extend one's intellectual capital (knowledge and skills) or at least to maintain it by establishing a routine through systematic and frequent carrying out of the same tasks and activities. An efficient use of the intellectual capital is dependent on the physical capital, i.e. the psychophysiological potential and abilities. Man constitutes a complex whole and one's behaviour also depends on individual abilities (Lindberg et al. 2005). The authors stress that „the uniqueness of human capital stems from the fact that people cannot be separated from their knowledge, skills, health or values in the way they can be separated from their financial and physical assets" (Becker, 2008; Wright, and McMahan 2011). It is the employee's sole responsibility to retrain, however from the perspective of the presented work concept, shaping up workers in the multi-skill arena is of vital importance – in particuar, as concerns carrying out tasks in process modules (core business). There are some examples of practical use of that factor. For instance, in TESCO supermarkets, accountants and other administrative workers sit at the checkouts during the peak hours to step up customer service. It implies increased flexibility as regards the use of human resources within an organisation.

Machines and devices (including equipment, technological lines of machines, buildings and structures) are another sort of resources. They are resources exploited in work, i.e. ones which also wear out physically. The extent of the wear is, however, hard to estimate. Therefore, the producer determines the standard period of manufacturing capacity, i.e. service life. It is usually measured in the number of service hours of the expected, i.e. 'normal' manufacturing capacity. In economic research, the price of the resource is used to assess the amount of monthly instalments of depreciation write-off. With each month, the write-offs reduce the value of a given device. In this way, they symbolize its wear-out and determine the time limit for its use (and a purchase of a new device). From the viewpoint of the concept that is being presented in this paper, having multifunctional machines at disposal is of key importance, as it boosts the supply potential without changes in costs (investment, hire, leasing, etc.).

Yet another kind of conversion is the use of raw materials, materials and semi-finished products. These are resources which get entirely consummed, i.e. ones which become part of the developing product. They are the material for conversion and include raw materials, materials and semi-finished products as well as electricity, water or gas supplies, on which conversion is technically dependent. Their optimum wear-out is estimated according to the technical norms for individual machines, devices and technologies.

Information is similar with respect to the results of its use. We gain a piece of information, process it or create new information on this basis or irrespective of it. There are thus different ways of creating information. Moreover, information has different functions, from evidence to commercial and business information (patents, know-how, formulas, technologies etc.). Therefore, information may of a different value to a company. The value is assessed internally or determined by the market. In general, all the components of a given piece of information, i.e., the original and the secondary ones, stay in the company. It stems from the specific role of information, which is both the basis for the functioning (procedures, formalization of activities and organization, decision making) and for the registration and settlement of business units. It creates the need to 'materialize' or file information.

Area is the last of the resources mentioned in this paper. It is a component which can be used like machines and can be improved in quality like human capital. Construction of a hall on a purchased estate is an example of such activities. It is a basis for resource allocation and requires physical space (office design, workspace, flowspace). We also need to bear in mind the currently-expanding virtual space. It is information-based so its components are communities, platforms, networks, social media (Deprez, and Tissen 2011). In general, the concept includes different sorts of space as cost drivers, taking into account the need for a minimization of resource allocation and the functions of space (manoeuvrability, lack of resource concentration, space density etc.).

The systematization results in the following outline of the organisational work concept scheme:

a) an organisation has at its disposal diverse resources (including purchased services)
b) part of the resources form modules which achieve successive goals in the production process
c) part of the resources is used for realization of business processes within the company
d) the above-mentioned components $a$, $b$ and $c$ create costs and component $b$ also generates income
e) processes $b$ and $c$ are organised into modules, which facilitates selection optimization and the assessment of the efficiency of resource use
f) the activity is planned, performed and evaluated in the evaluated in the organization time. i.e. the hours of company functioning

Hence, the above-mentioned outline includes the 'black box' mechanism or the inner functioning of the company (the method of conversion of input into output), which so far has been solely left to the managers. It also proposes an introduction of modules for a standardization of the processes, which will streamline management.

## 3. DISCUSSION

Whereas the first feature of the organizational dimension implies a praxeological, three-dimensional understanding of an organization (Kotarbiński 1995), the determinant of an organizational goal can indicate two types of organizational work: core activity, i.e. the realization of an organizational goal and other activities (assistance, services), which are required by the standards of business conduct. The activities comprise minimization of technical and technological hazard (working conditions, machine and device certificates, the quality of materials and raw materials, qualification standards etc.) as well as economic and financial transparency, which enable a business to function in accordance with the rules of business conduct (tax settlement).

It means that as an activity becomes more complex, three types of work take form within an organization: realized work (products delivered to the consumer and settled in the form of payment or receipt in a non-market system, work that is being realized, i.e. work in progress, stored products and purchased work (outsourcing). They are specificated by the stage of product development. Realized work is a cost element which has generated income (budget assessment). Work that is being realised is an element which keeps creating inner expenses. Purchased service is cost imported from the outside.

In this concept, work is a conversion of supply (resources) into a product which satisfies the demand. Thus, at the organizational level work is of a demand-related nature and productivity should be assessed from this viewpoint, as resources are the main cost drivers. It does not stray from the concepts which are present in relevant literature. A differenciation between individual and organizational work is adopted. Individual work (jobs) can be defined as the tasks an individual employee is appointed to by the organization. Organizational work can be defined as a set of activities that is undertaken to develop, produce, and deliver a product or a service (Sinha, and van de Ven 2005). Organizational work has the form of product realization processes. Work division, i.e. dividing the process into comprehensive sub-elements (technological units of processes), is of great significance, both for economic (monitoring the cost of realization of stage results) and for organizational reasons, i.e. coordination, which is related to authority, especially control and responsibility. The accuracy of work division within the company is therefore of fundamental importance.

It implies system thinking, which integrates the diverse components into an entirety. Hence, universal or systemic theorizing seem to assume a standardized or standardizing workplace, work for teams, processes etc., in which technology, science, and managerial discourse aims at creating common methodologies. System thinking is common to all deductive reasoning, which builds on abstract concepts, in which context-dependent rationality is suspended for the purposes of building ideal or pure typologies (Smith 2006).

In the reistic approach, the sub-activities are of a modular nature. The literature relates to manifold aspects of modularity, i.e. product, organization and process modularity as well as dynamic interfaces. In organizational literature there are more units of analysis: production system modularity and organizational design modularity (Campagnolo, and Carnuffo 2010). We are interested in process modularity. Modules can be of diverse nature and organizational context. Identification of modualrity with autonomy is the predominating trend. For both horizontal and vertical links within, and between, organizational units, modularity may be characterized by a relative absence of the need for interaction. At the same time, within the development function, a development team that has a clearly defined set of tasks that do not depend on the completion of tasks by other teams is autonomous and therefore modular (Sako 2003). Modules can also be of an individual nature, but in the vast majority of cases they will comprise complex elements creating module levels (at least a team) or an organization. Therefore, also human factor researchers propose the development of a "meso theory," which would attempt to systematically specify the relative importance of different job characteristics in predicting outcomes in different contexts and circumstances (Fried, Levi, and Laurence 2008). A complex, integrative approach towards resources in the presented concept needs a reconstruction of this notion, which clearly refers to human resources.

RBV (resource-based view) concepts stress the significance of resources for the competitive edge (external influence). But at first the firm is presented as a collection of productive resources. Theoreticians stress that 'The important distinction between resources and services is not their relative durability; rather it lies in the fact that resources consist of a bundle of potential services and can, for the most part, be defined independently of their use, while services cannot be so defined, the very word 'service' implying a function, an activity. Ideally, the size of a firm for our purposes should be measured with respect to the present value of the total of its resources (including its personnel) used for its own productive purposes' (Penrose 1995).

There is a need to rationalize the types and amount of obtained resources because it influences the cost of their obtaining and maintenance. It is, however, productivity, i.e. the capacity for producing output from a resource, that matters the most. Organizational literature names it 'organizational capability'. The authors consistently differentiate between resources and abilities. 'Resource refers to an asset or input to production (tangible or intangible) that an organization owns, controls, or has access to on a semi-permanent basis. An organizational capability refers to the ability of an organization to perform a coordinated set of tasks, utilizing organizational resources, for the purpose of achieving a particular end result' (Helfat, and Peteraf 2003; Maritan, and Peteraf 2011).

The RBV is a theory about the nature of firms, as opposed to theories such as transaction cost economics which seeks to explain why firms exist. In effect, the RBV is a statement about how firms actually operate (Lockett, Thompson, and

Morgenstern 2009). From the perspective of the inside of an organization, i.e. the before-mentioned 'black box', construction of modules means specificating resources with features which are completely different from the classic elements of the VRIN, i.e. **V**alue, **R**areness, **I**nimitability, **N**on-substitutability concept (Barney 1991; Truijens 2003). Modular and intermodular relations within specified types of resources and between resources are of significance. We do not reap profit from particular attributes of resources, but from the accurate selection and allocation and from the organizational method. By extending the previous trend of 'structural flexibility', these elements create a new prospect of an organization's productiveness (Iravani et al. 2005).

## 4.    CONCLUSIONS

The author presents an outline of a new concept, which formulates a new design and architecture of work. The work system is considered from the integrative perspective, as an effect of the conversion of components which optimize their obtaining and use. A process approach is proposed, with the use of normative modules in the planning phase (the standard economic capacity) and preventing failures in the realization phase (cost not bigger than planned allowing for buffers within the plan). The concept proposes a switch from resource supply (HRM, stock management, etc.) into organizational demand, i.e. creating products which the organization has on offer by reasonable management of limited resources. The proposed combination of the achievements of economic and organizational sciences restores the economic sense to work and activities.

## REFERENCES

Barney, J.B.2001, Resource-based theories of competitive advantage: a ten-year retrospective of the resource-based view, *Journal of Management,* 27:6, 643-650.

Becker, G.S. 2008. Human capital. In. *The Concise Encyclopedia of Economics*. Library of Economics and Liberty. Retrieved from the WorldWide Web: http://www.econlib.org/library/Enc/HumanCapital.html, accessed 11 August 2010

Campagnolo, D. and A. Carnuffo, 2010. The Concept of Modularity in Management Studies: A Literature Review. *International Journal of Management Reviews,* 12:3, 259-283.

Crooper, S., M. Ebers, C. Huxham and P.S. Ring. 2008. The field of inter- organizational relations: A jungle or on Italian garden. In. *Oxford Handbook of Inter- Organizational Relations,* eds. S., Crooper, M., Ebers, C., Huxham and P.S., Ring, Oxford University Press: Oxford, 719-738.

Cordery, J. and S.K. Parker. 2007. Work organization. In. *Oxford Handbook of Human Resource Management*, eds. P. Boxall, J. Purcell and P. Wright, Oxford University Press: Oxford, 187-209.

Deprez, F.L.and R. Tissen, 2011. *Developing spatial organizations: a design based- research approach (part I).* Neyenrode Research Paper no. 11-01. Neyenrode Business Universiteit, Breukelen

Fried, Y., A.S. Levi and G. Laurence. 2008. Motivation and Job Design in the New World of Work. In. *The Oxford Handbook of Personnel Psychology*, eds. S. Cartwright, C.L. Cooper, Oxford University Press: Oxford, 586-605.

Helfat, C. E. and M.A. Peteraf. 2003. The dynamic resource-based view: Capability lifecycles. *Strategic Management Journal*, 24: 997-1010.

Iravani, S.M., M.P. van Oyen and K.T. Sims, 2005. Structural Flexibility: A New Perspective on the Design of Manufacturing and Service Operations, *Management Science*, 51:2, 151-166.

Jasinski, Z., 2011, Istota, elementy i zasady organizacji działalności operacyjnej. In. *Podstawy zarządzania operacyjnego*, ed. Z. Jasiński, Oficyna a Wolters Kluwer business, Warszawa, 13-27.

Karwowski, W. 2010, *Zarządzanie wiedzą o czynnikach ludzkich w organizacji*, Szkoła Główna Handlowa w Warszawie, Warszawa

Kotarbiński, T. 1955. *Traktat o dobrej robocie*. Ossolineum:Łódź

Lindberg, P., E. Vinga, M. Josephson. L. and Alfredsson. 2005. Retaining the ability to work—associated factors at work, *European Journal of Public Health*, 16:5: 470–475

Lockett, A., S. Thompson and U. Morgenstern. 2009. The development of the resource-based view of the firm: a critical appraisal, *International Journal of Management Reviews*, 11:1, 9-28.

Maritan, C.A. and M.A. Peteraf. 2011. Building a Bridge Between Resource Acquisition and Resource Accumulation. *Journal of Management* 37:5, 1374-1389

Penrose, E.T. 1995. *The Theory of the Growth of the Firm*. Third edition. Oxford University Press: Oxford

Penrose, E.T. 2003. The Theory of the Growth of the Firm. In. *Resources, firms, and strategies: a reader in the resource-based perspective*, ed. N.J. Foss, Oxford University Press: Oxford, 27-39.

Pitelis, Ch.N. 2010. Economics: Economic Theories of the Firm, Business, and Government. In. *The Oxford Handbook of Business and Government*, eds. D. Coen, W. Grant and G. Wilson, Oxford University Press: Oxford, 35- 62.

Rymaniak, J., 2011. O nowy paradygmat pracy – ponad i poza człowiekiem, *Humanizacja Pracy* 1:23-44.

Sako, M., 2003, Modularity and Outsourcing. The Nature of Co-evolution of Product Architecture and Organization Architecture in the Global Automotive Industry In. *Business of System Integration*, eds. A. Prencipe, A. Davies and M. Hobday, Oxford University Press , 229-253.

Sinha K.K. and A.H. de Ven. 2005. Special Issue: Frontiers of Organization Science, Part 1 of 2: Designing Work Within and Between Organizations, *Organization Science* July/August 16:389-408.

Smith, Ch. 2006. Beyond Convergence and Divergence: Explaining Variations in Organizational Practices and Forms. In. *The Oxford Handbook of Work and Organization*, eds. S. Ackroyd, R. Batt, P. Thompson and P.S. Tolbert, Oxford University Press: Oxford, 602-625.

Truijens, O. 2003. A Critical Review of the Resource-based View of the Firm, University of Amsterdam, Netherlands. *Sprouts: Working Papers on Information Systems*, 3(6). http://sprouts.aisnet.org/3-6, accesed 20 January 2012

*Worlds of Work*. 2008, eds. P. van Baalen, F. Go, E. van Heck, M. van Oosterhout, et al. Erasmus University, Rotterdam

Wright, P.M. and G.C. McMahan. 2011. Exploring human capital: putting human back into strategic human resource management, *Human Resource Management Journal*, 21:2, 93-104

CHAPTER 59

# Prescribing Knowledge for Making Arrangements Regarding Human Factors in Management of Change: Toward Incorporating Human Factors Considerations into System Design Process Model

*Toshiya Akasaka, Yusaku Okada*

Keio University
3-14-1 Hiyoshi Kohoku-ku Yokohama, JAPAN
{to48_a, okada}@ae.keio.ac.jp

## ABSTRACT

When a change is made in an organizational process, factors shaping workers' performance at behavior level should be identified and incorporated into the design along with well-formulated factors such as business process. The challenge here is to identify such performance shaping factors. The goal of our research is to develop an explicit way to identify such factors with the aim of assisting the management of change in organizational processes. In this paper, we present our approach to identify performance shaping factors via guiding the concretization of human behavior models. We also show how our approach can eventually make it possible to integrate factors affecting human performance with the systems development.

574

Finally, we explain how the management of performance shaping factors benefits organizations.

**Keywords**: organizational process, human behavior, human performance

# 1    INTRODUCTION

When people work as an organization, systematic business processes are defined laying down how work is supposed to be done. Although the granularity and the degree of abstraction varies from case to case, such organizational processes are formulated and prescribed to ensure that workers achieve the results expected by those who designed the processes. However, such formulated processes alone do not completely shape how workers behave and work. For example, people following the same procedure can have different ways of getting the work done. Some may prepare necessary tools and equipments well before they start following the procedure so that they will not have to pause while implementing the procedure. On the other hand, others may begin to take actions without any proactive preparation, bringing any necessary tool as needed. They follow the same process,

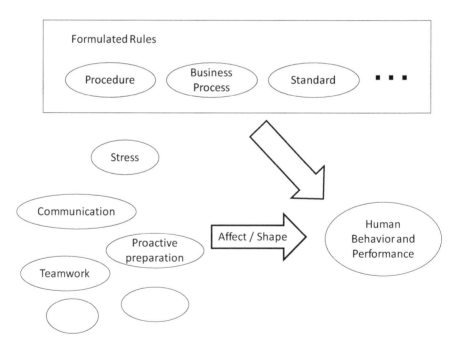

Figure 1. An illustration of how various factors affect human behavior and its performance. One cannot tell how work is actually done at human behavior level only by looking at formulated rules. Informal, local, and temporal factors also play a major role in shaping human performance of their performance.

but their behaviors differ in terms of preparedness, which could affect the quality and preciseness of their work. In other words, it contributes to the shaping If you have a behavioral viewpoint, you can find many factors shaping people's behavior and thus affecting their performance. Such performance shaping factors are what greatly contribute to the formation of the way the work is actually done at the human behavior level (See Figure 1).

Careful considerations should be made for performance shaping factors when changing an organizational process. When an existing process has been already in place, people are likely to have their own way of doing their jobs. It is not easy to ask them to change long-established practices and behaviors. This is especially true if the proposed new process brings bad factors affecting workers' performance adversely. For example, if the new process requires preparing a lot more tools in advance, proactive preparations, which help improve performance, will become difficult, making the workers reluctant to accept the new scheme. Also, such a performance damaging process cannot be ensured to work as expected even after cutting its way through peoples' reluctance. After all, a change in an organizational process cannot be accepted and effective unless it is designed taking into account what factors are shaping workers' performance.

The question is how to identify performance shaping factors in human behavior that are relevant for the scheme being considered. Performance shaping factors can be very local, temporal, and subtle, difficult to capture and articulate. This is all the more true because one's performance of behavior is shaped through interactions between the human and his/her surrounding environments. In the field of systems development, many business process models and engineering models have been proposed and put into practical use, but the factors shaping performance of human behavior have been left out of focus so far. With no or few guiding models available, incorporating performance shaping factors in human behavior into the design and implementation of an organizational process is still in the realm of tacit knowledge.

The goal of our research is to investigate an explicit way to help identify performance shaping factors in human behavior with the aim of assisting the design of organizational processes. In the field of human factors, ergonomics, and applied psychology, there are many insightful models about human behavior. The problem is that they are not necessarily practice-oriented; they do not immediately lead to the identification of performance shaping factors that are concrete enough to suggest us what actions to take. Therefore, our immediate goal is to develop a guiding technique to concretize models of human behavior. This will eventually pave the way for the specification regarding human behavior and performance. With formulated specifications, considerations for human behavior can now be integrated with systems development considerations. Our long-term vision is to incorporate the considerations for performance shaping factors into a systems development process. In this paper, we present our approach of concretizing models of human behavior as well as how that efforts benefit organizations.

## 2    CONCRETIZATION OF HUMAN BEHAVIOR MODEL

To identify performance shaping factors in human behavior, our approach concretizes models about human behaviors. Models and concepts about human behaviors are abundant, from casual sketches to rigorous mechanistic models. Most of them are descriptive; the main aim of those models is to describe the behavior of humans as it is, with little intention to link the description to the consideration on how human behaviors can be assisted. Still, many of those models successfully identify subtle elements which have critical impacts on the formation of human behavior and performance. Our idea is to make those elements the landmarks around which we investigate factors affecting performance. In other words, we pose questions for each element, like "Why does s/he fail in that element?" (See Figure 2 for an illustration)

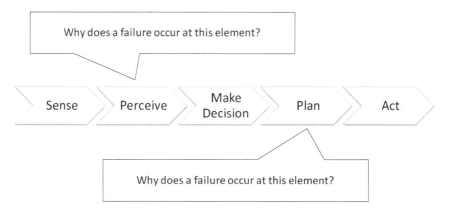

Figure 2  An illustration of our approach. The figure uses an imaginary human behavior model. For each element in the model, we pose a question like the one in the figure and investigate what factors contribute to any possible failure.

However, human behavior is not shaped logically, and looking for causes and reasons for a certain behavior is not straightforward. For example, there is a subtle difference between "Why couldn't you see it?" and "Why didn't you see it?" The latter question implies that the person did not have the intent to see it. The object the person had to see might be just in front of him/her, being occluded by nothing. Obviously, "They did not see it because it was not there" is an insufficient investigation. They need to know that they have to see it. They need to have the intent to see it. If they do not want to see it, they do not see it. If they assume that they do not have to see it, they do not see it. Likewise, there are many reasons and conditions why people behave in a certain way. Let us call them collectively "Human Logic." Human logic is much more subtle and varied than the causal logic. If we are to identify factors shaping human performance, we have to take into account such a wide variety of human logic.

Therefore, the key to identify performance shaping factors is to know as many

**Table 1** Types of human logic with example performance-reason connections for each type.

| Type of Logic | Example performance-reason connections |
|---|---|
| Knowledge | People (do not) do something because they (do not) know that they should/can do it. People do a wrong thing because they have wrong knowledge. |
| Assumption | People (do not) do something because they assume that they should/can (not) do it. People do a wrong thing because they have a misleading assumption. |
| Motive | People (do not) do something because they (do not) want to do it. |
| Accessibility | People (do not) use something because they (do not) have a good access to it. |
| Availability | People (do not) use something because it is (not) there when needed. |
| Attention | People (do not) do something because they (do not) pay attention to it. |
| Detectability | People (do not) see/hear something because it is (not) (easily) visible/hearable. |
| Skill | People (do not) perform something because they can(not) (easily) do it. |

types of human logic as possible. Table 1 summarizes the types of human logic that we have been able to identify so far. When posing questions for each element in a human model, we do so with performance-reason connections like the ones shown in the table in mind. In the following subsection, we pick up one particular human model and discuss how these viewpoints help derive performance shaping factors from elements in the model.

## Case Study on Rasmussen's Decision Ladder Model

Decision Ladder Model (Rasmussen, 1986) proposed by Jens Rasmussen is a model to describe the cognitive process of human operators of major physical systems such as chemical plants. Rasmussen, who had long studied human-machine interface in such major systems, discovered that operators seemed to travel between different modes of information processing. According to his view, operators are in the mode of pattern recognition when they overview a pattern of lighting alarms. Once they know that the pattern is something they have never seen, they now delve themselves into a higher cognitive mode where they imagine the structure of the whole system and diagnose what is happening inside. Representing human's information processing as a travel between different modes of thinking is an insightful idea, which could suggest possible improvements on the methodology of human-machine interface design.

As an illustration of how our approach works, we now would like to show how an element in Decision Ladder Model can be concretized. Let us use the example we mentioned above where operators have to understand a pattern of information that they have never seen. Therefore, the element we will investigate is "the understanding of a situation after recognizing they do not have any repertoire of immediate response to the pattern of information before them." What we are going to do is to identify what factors affect the performance in this element. We do so by finding reasons with human logic in mind. An example result is shown in Figure 3. The figure shows how human logic plays a guiding role in investigating why they fail to show proper performance. After coming up with possible reasons (e.g., after listing up the blocks in the middle column in the figure), we still have to concretize each reason so that we can know what actions to take. However, identifying concrete factors from each reason is much easier than doing the same thing from the scratch.

## Toward Specifiable Factors and Integration into Systems Development Process

Our vision is to develop a well-defined way to identify performance shaping factors that are concrete enough to be used as a specification. As we have mentioned in the first section of this paper, current engineering models for systems development only take into account well-formulated factors such as business processes and written procedures. This is partly because there is no way to specify what factors about human performance should meet what conditions. Our technique to guide the identification process of performance shaping factors can be the first step to a well-defined way to make such factors specifiable. With specifications about performance shaping factors, one can ensure that workable organizational processes be designed and developed.

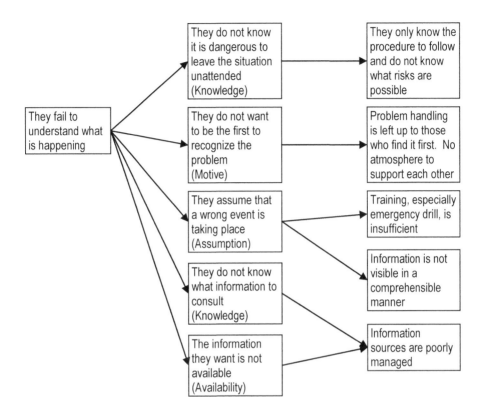

Figure 3  An example result of investigation with human logic in mind.  Human logic plays a guiding role in listing up the reasons for the first failure (the leftmost block).  The reasons are further investigated for possible reasons behind them.  This chain continues until we find factors that are concrete enough to suggest actions to take.  Having a guide in the first round of investigation makes a difference to the following investigations.

# 3    APPLICATION OF HUMAN MODEL CONCRETIZATION

As discussed in the first section of this paper, identifying performance shaping factors is important for designing organizational processes.  This need is especially high in safety-conscious industries such as chemical manufacturing, because any bad performance in human behavior can lead to serious consequences like injury, death, and facility damage.  Therefore, there is a growing demand for an effective way to prevent human-caused troubles.  Many organizations today recognize the importance of collecting reports of potential incidents.  A potential incident is an event that did not cause any substantial damage but might have caused one had some conditions been different.  Because potential incidents often result from common patterns of problems, successful analysis of their reports could help identify performance shaping factors pervading that organization.  This could in turn show how their organizational processes can be improved. However, with no guiding knowledge, most organizations end up in collecting superficial descriptions of potential incidents, failing to identify what common factors are behind each case.

Our technique to identify performance shaping factors, which we discussed in the previous section, can be a guide for that effort. In cooperation with an oil manufacturing company in Japan, we launched a project to develop a reporting scheme that could collect information leading to identification of performance shaping factors. The objective is to link descriptive information of each potential incident to performance shaping factors behind the case. The project is still underway, but we have successfully identified performance shaping factors that can be useful input for analysis.

First, we selected models based on which we would identify performance shaping factors. The selection of models can be based on high-level purposes about what kind of factors to elicit. In this case, we attached importance to eliciting factors related to lack of proactive measures taken by individual workers. Many veteran workers are well aware of what risks lie in a certain situation and able to take proactive measures before performing tasks. Workers in younger generation, on the other hand, tend to lack such awareness of risk, often experiencing troubles due to poor preparation. In order to collect factors behind such troubles, we decided to identify performance shaping factors based on the two models, namely, SRK model (Rasmussen, 1986) proposed by Jens Rasmussen, and Endsley's theoretical model of Situational Awareness (Endsley, 1995). The former model well captures the differences between the experience-driven behavior of veteran workers and inference-intensive behavior of young workers. The latter model describes how people with a good awareness of situation perceive and interpret environments surrounding them. By selecting models based on the purpose, we can ensure that factors derived from the models will meet the ends.

The next step is the concretization of the models to derive factors affecting performance in the behavior that each model shows. Human logic discussed in the previous section played a guiding role in this phase.

The detail of the project will be appearing in another paper by Tanuma and the authors of this paper. The paper will be in this series of proceedings.

## 4    CONCLUSION

In this paper, we described our approach of letting human logic guide the concretization of a human behavior model with the aim of identifying performance shaping factors. Identifying performance shaping factors is beneficial for designing organizational processes, and our approach has successfully helped identify performance shaping factors in the real-world project. However, our approach still cannot guide the whole investigation thoroughly. We need to develop a well-established methodology to make our approach more explicit. Eventually, the establishment of an explicit way to identify performance shaping factors will make it possible to integrate factors affecting human performance into the design of systematic organizational processes.

## REFERENCES

Rasmussen, J. 1986. *Information Processing and Human-Machine Interaction: An Approach to Cognitive Engineering.* New York: Elsevier Science Publishing Co., Inc.

Endsley, M.R. 1995. Toward a theory of situation awareness in dynamic systems. *Human Factors* 37: 32-64.

# CHAPTER 60

# Socio-technical Integrity in Maintenance Activities

*Małgorzata Jasiulewicz-Kaczmarek*

Poznan University of technology
Poznan, Poland
Malgorzata.jasiulewicz-kaczmarek@put.poznan.pl

## ABSTRACT

Sustainable maintenance is a new challenge for enterprises realizing concepts of sustainable development. They can be defined as pro-active maintenance operations striving for providing balance in social dimension (welfare and satisfaction of maintenance operators), environmental and economic (losses, consequences, benefits). It requires introducing broad analysis concerning loss or putting into risk continuity of enterprise performance and making optima decisions from the costs and benefits point of view, hence understanding influence of stakeholders (internal and external) on business decisions. One of the aspects of maintenance is introduction of social aspects (after economic and environmental) to analysis of service activities.

The objective of the paper is to explore the specific of socio-technical maintenance management system and to analyze its key elements in sustainable development perspective.

**Keywords**: maintenance, socio-technical system, sustainable

## 1    INTRODUCTION

The main purpose of maintenance engineering is to reduce the adverse effects of breakdown and to maximize the availability at minimum cost, in order to increase the performance and improve the dependability level. Thus, to maintain a manufacturing system means to implement objectives (cost, delay, quality, etc.) fixed by the direction of the production, taking surrounding events (risks, perturbation, etc.) into account. In every case the maintenance system must:
–  ensure smooth and reliable operation of the production line,

- integrate the human and technical resources, technical knowledge management and capitalization,
- be able to reconfigure manufacturing systems for the production of innovative products.

New concepts and approaches to corporate management, such as Lean Manufacturing, new challenges connected with natural resources management (Green Manufacturing), and thinking in a sustainable way (Sustainable Manufacturing) resulted in developing new approach to maintenance management as well (fig. 1).

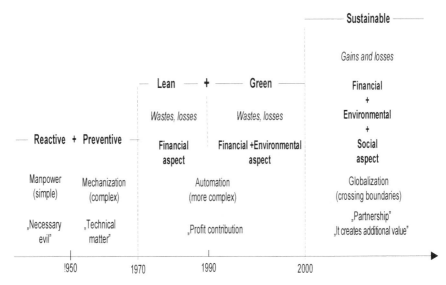

Figure 1 The evolution of maintenance on a time perspective. (From M. Jasiulewicz-Kaczmarek. 2012. Sustainability – orientation in maintenance management, Eco Production and Logistics. Emerging Trends and Business Practices, Series: Eco Production, eds. Paulina Golinska, Springer-Verlag Berlin Heidelberg (in press))

Contemporary maintenance is not only a set of operations focused on dealing with breakdowns and failures and conservation of machines and devices (traditionally the scope of maintenance activities has been limited to the production vs. operation chase). Nowadays, it is more like long term strategic planning which integrates all the phases of a product lifecycle, includes and anticipates changes in economic, environmental and social trends, benefits from innovative technologies (f.ex. e-maintenance, e-diagnostic). Whereas the goal of maintenance is increasing profitability and optimization of total lifecycle cost without disturbing safety and environmental issues (aspects: loss, consequences, benefits).

Thus, it is necessary to include sustainable development category (sustainable development (SD) is about reaching a balance between economic, social, and environmental goals, as well as people's participation in the planning process in order to gain their input and support) into processes and activities realized in the area of enterprise's technical infrastructure maintenance.

## 2    MAINTENANCE MANAGEMENT

### 2.1    Maintenance as a socio-technical system

Maintenance exists, because disregarding the branch of a company and products it provides, it has resources (technical objects) which need to be maintained. Maintenance is thus an important input for the core process of manufacturing companies, the production process. Maintenance system is a set of organizational units and relations between them defined by maintenance processes accordingly to technologies accepted and used (the technical and social factors interact to influence maintenance outcomes). Thus, maintenance system has the characteristics of socio-technical systems. The technical subsystem comprises the devices, tools and techniques needed to transform inputs into outputs in a way which enhances the economic performance of the maintenance. The social system comprises the employees (at all levels) and the knowledge, skills, attitudes, values and needs they bring to the work environment as well as the reward system and authority structures. Striving for general efficiency of maintenance system increasing requires common (shared) building and shaping social and technical subsystems in order to achieve their best fitness with respect to goals and requirements of both subsystems and the system as a whole. Technical solutions can increase efficiency of activities undertaken and improve information flows enabling operators, mechanics, automatics, logistics etc. taking right decisions, whereas building knowledge, skills and transfer of responsibilities will increase commitment to proper use and improvement of available technical solutions

### 2.2    Maintenance for sustainability development

Sustainability is most commonly accepted as a triple bottom line of economic profitability, respect for the environment, and social responsibility. Recently, technology is being viewed as a fourth dimension for sustainability (Meier, Roy and Selige,. 2010).

The first step in transforming a company into a sustainable business is to develop a sustainability vision and sustainability strategies that include sustainability objectives. To provide ability of realization and success of the strategy, it needs to be based on resources available (mostly human), experience and competences of a company and processes creating internal supply chain. One of the key elements creating internal supply chain in an organization is maintenance.

Creating a sustainable production environment requires, among other things, the elimination of breakdowns and other sources of energy waste. The inadequate maintenance can result in higher levels of unplanned equipment failure, which has many inherent costs to the organization including rework, labor, and fines for late order, scrap, and lost order due to unsatisfied customers (More, 2006). This has been one of the decisive drivers for changing the perception that people normally have maintenance from "fail and fix" maintenance practices to a "predict and prevent" mindset.

In contemporary maintenance it is impossible to take only financial aspects into consideration. Balance between environmental (green) and social aspects should be found and kept in all actions taken and systematic approach should be applied actions taken, their consequences and outcomes, as well as benefits company can achieve.

Economic, environmental and social dimensions of maintenance are interrelated and any change in the objectives of a dimension greatly influences the other two dimensions. Taking systematic approach as a key principle for sustainable maintenance enables finding relations between all the dimensions and taking optimal decisions in terms of costs and benefits, hence understanding influence of business decisions on internal and external stakeholders (fig. 2).

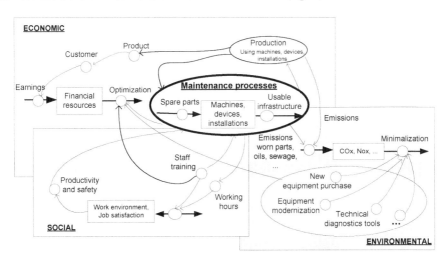

Figure 2  General model of interdependencies in sustainable maintenance. (From M. Jasiulewicz-Kaczmarek. 2012. Sustainability – orientation in maintenance management, Eco Production and Logistics. Emerging Trends and Business Practices, Series: Eco Production, eds. Paulina Golinska, Springer-Verlag Berlin Heidelberg (in press))

Thanks to that, both managers and employees at operational level understand better how decisions made influence each dimension of sustainable development.

Sustainable maintenance is a new challenge for companies. It can be defined as pro-active maintenance operations focused on providing balance in social (welfare and satisfaction of operators and maintenance staff), environmental (6R) and financial (losses, consequences, benefits). It requires analysis of loss or putting under risk continuity of enterprise functioning (in economic, environmental and social aspects), if maintenance policy (corrective, preventive,...) and actions taken do not provide expected technical condition of technical infrastructure (machines, devices, installations).

According to Maggard and Rhyne (1992) the maintenance can represent between 10 and 40 percent of the production cost in a company. Coetzee (2004) means that the numbers should be 15-50 percent and Bevilacqua and Braglia (2000) state that maintenance costs can represent as much as 15-70 percent of the total production cost. The basic models of maintenance costing divide the maintenance costs into direct and indirect costs. Direct costs consist of labor costs, spare parts, and other costs that clearly are directly linked to maintenance activities. Indirect costs include the cost of recovering for lost production (due to equipment failures) cost of insufficient quality etc. Global trade, increased levels of automation and ambitions to apply lean production increases the demand for effective maintenance. Still, the maintenance function is often regarded as having mainly a tactical role for the existing assets. This often leads to the view of maintenance as an expense and therefore a subject for cost reduction programs. Lean production and maintenance are both essential and interconnected concepts. Maintenance must improve its ability to improve the value adding capability by delivering stabilized process and equipment performance to reduce unplanned events and waste and by delivering optimized performance to reduce quality defects, cost and delivery lead times. Lean thinking can help maintenance by the application of its proven tools and techniques to target the reduction of waste and non-value added maintenance activities (Willmott, 2010).

Practical realization of environmental aspect of SD approach to maintenance depends on numerous internal factors (f.ex. maintenance strategy, service planning mode, applied service technology and materials, competences and attitude of employees), as well as external ones (f.ex. maintainability of equipment).

Thus, realization of environmental aspects of maintenance (green maintenance – GM), enterprise should consider environmental issues each time decision concerning new equipment purchase is made. Assessment of project should be referred to total lifecycle cost, thanks to which financial benefits or losses emerging from predefined service practices and maintenance undertaken to achieved required level of reliability, accessibility and maintainability can be identified. At the stage of exploitation of a technical object, GM approach requires, except from running traditional analysis, also collecting data in the context of adequacy of maintenance strategy applied and its impact on natural environment. It makes monitoring of condition of production equipment is a key element of efficient GM. Most machines can be monitored continuously, without stoppages.

Advance of solutions available, with reference to vibrations, stability of work, balancing and thermography helps in achieving and maintaining proper condition of machinery. Using these technologies enables "listening" to what machine says on its state and early detection of failures before they lead to breakdowns. If there is a need for service activities, these tools enable their fast and efficient performance, without any negative influence on production process.

Fitness of accepted frequency of service activities taken, their range, materials, components and parts should be updated with respect to minimization of resources used and utilization of potential waste ("6R").

Period of exploitation of technical objects is accompanied with various modifications and modernizations, which are supposed to adjust functionalities of a given technical object to changing requirements of an organization (concerning quality, production process, safety and others) and on the other hand renovate it improving its reliability and prolonging its lifecycle.

Maintenance system supporting sustainable development idea requires not only including technical aspects, but also social ones. Except from technical objects, there are also employees and work environment which require keeping "in good condition". Social sustainability in maintenance activities (similar to other aspects of enterprise functioning) is realized in concepts such as preventive occupational health and safety, human-centered design of work, empowerment, individual and collective learning, employee participation, and work-life balance. Safety and hygiene of work, including psychical (stress), are reflected in quality of work done and satisfaction from work. Human errors emerging from lack of knowledge, improper work conditions and tools of work in machines operating environment.

Safety and hygiene of work, including social aspects (stress) influence quality of work done and satisfaction from work as well. Human errors emerging from lack of knowledge, improper work conditions and tools, both in work environment (setups, conservation, repairs, etc) and during operations performance results in hazards for both, operators and employees staying in their closest environment .numerous actions striving for safety in a company are taken after incidents happen, which means that actions to prevent problems are taken after damages appear (procedures are developed, trainings are organized, protection means are introduced). However that refers only to situations and events that have happened, which means they represent reactive safety. Sustainable maintenance promotes active safety culture based on safety awareness and responsibility for oneself and others (interdependent organization) (Jasiulewicz-Kaczmarek and Drożyner, 2011).

## 2.3 Integrity and sustainability in maintenance management

Integrity is a concept of consistency of actions, values, methods, measures, principles, expectations, and outcomes. Socio-technical integrity in maintenance activities from sustainable maintenance point of view refers to, among others:
– maintenance strategy and goals (their decomposition at all the levels of management maintenance management activity level – fig. 3),
– processes and relations between them (methods of realization focused on goals and including a high quality of work life for participants), stakeholders (including production) and their requirements (fig. 4)
– people (including knowledge, skills, values and expectations),
– technology (machines and devices necessary to perform service activities and supporting IT and communication), and
– outcomes (referred to financial, environmental and social dimensions) and system of achievements measurements, including measures and indicators which enable concurrent and periodical assessment.

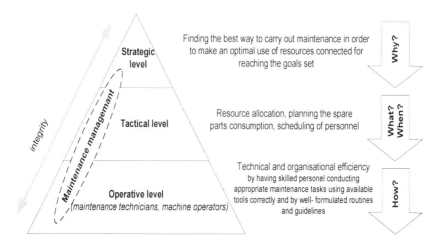

Figure 3. The scope of maintenance management.

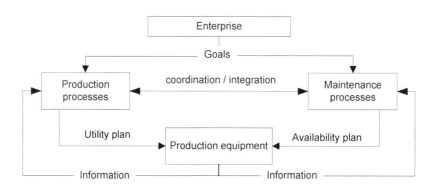

Figure 4. Linkages between production, maintenance and enterprise goals.

The integrator is maintenance strategy and goals emerging from this strategy. The purpose and goals of maintenance have to be defined with respect to the goals of the production and the overall strategic goals of the company. The literature points to strong linkages between business strategy and manufacturing maintenance strategies (Pinjala et al., 2006; Rosqvist et al., 2009). As such, there is a need for a well designed and implemented organizational system to manage maintenance and related performance aspects from a strategic perspective. To take responsibility for ensuring that company meets goals defined by managers, maintenance department needs a tool which enables monitoring of results of service activities and results of maintenance.

What is more, for maintenance managers it is crucial to have the knowledge on relations between inputs of maintenance process and outcomes in terms of their total input in productivity and realization of strategic objectives of the company. The real challenge is how to cascade vertically downwards and aggregate the measures upwards, and how to integrate activities amongst various departments within organizations horizontally so that total maintenance effectiveness and desired business objectives are achieved.

The balanced scorecard developed by Kaplan and Norton (1992, 1996), is the most popular and balanced PM framework, used by the most of the industries all over the world. The BSC is powerful communication tool for providing a sharp focus on factor that are important to maintenance in making contributions to business success of the company. It enables holistic assessment of unit performance and guards against sub-optimization because all the key measures that collectively determine the total performance of maintenance are monitored (Alsyouf, 2006). This framework integrates perspectives of both financial and non-financial types, like; financial, customer, internal business, and innovation and learning. To ensure financial performance, the other perspectives act as drivers and need to be given equal weighting.

Sustainability in maintenance activities can be described through each of the four perspectives of the balanced scorecard (fig. 5):

- from a financial, sustainability means creating an acceptable return for investors.
- from maintenance stakeholder, sustainability means satisfying and providing value for the production people of safety and sustainability-conscious consumers.
- from maintenance process, sustainability means managing materials, energy, and waste in the most eco-efficient way possible.
- from employee education and safety, sustainability means creating a culture that values sustainability, reflected in the choices that employees make every day.

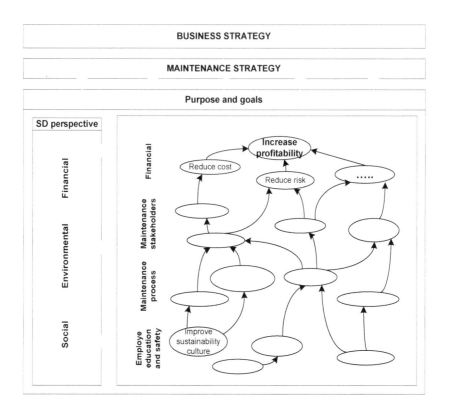

Figure 5. The evolution of maintenance on a time perspective.

This way of presentation can help in communication of strategy to both, managing staff and employees and thanks to creating better links between results of measure and maintenance strategy, enables identification of actions and channeling efforts of employees to strategies realization.

Each of perspectives is reflected in a set of indicators (performance indicator – PI) which refer to dimensions of sustainable development and which in quantitative way demonstrate and measure the effect of maintenance on sustainable manufacturing. Maintenance PIs could be used for financial reports, for monitoring the performance of employees and their satisfaction, the health, safety and environmental rating, and overall equipment effectiveness (OEE), as well as many other applications.

The tool supporting developed system of maintenance achievements measure system is ICT. At present, e-technologies start to play a crucial role to support maintenance decision-making. Thanks to availability of unique e-maintenance solution by ICT use, industry can have an efficient maintenance performance measurement system from the server-based software applications, latest embedded internet interface devices and state-of-the-art data security. The advent of ICT with e-maintenance application can facilitate ease of real time data collection, analysis and monitoring maintenance performance with different level of data at all hierarchical level for effective decision making. It includes different views of maintenance and different expertise (i.e. technical, administrative and managerial), allowing the electronic interaction and collaboration between actions and activities ensured by different actors (human or software). Whereas strategy is material integrator of technical and social maintenance sub-systems, ICT technologies such as e-maintenance are virtual integrator of this areas.

## 3    CONCLUSIONS

Socio – technical integrity in maintenance is the necessary conditions for efficient management of this area of enterprise and realization of challenges which contemporary organizations are facing. Such integrity can be achieved by systematic and process-based shaping of activities and their results.

An understanding of the systemic nature of the maintenance operation is vital to understanding how one's individual actions affect the whole organization, while applying process approach to organize maintenance system actions allows to meet the most important needs of contemporary organization.

## REFERENCES

Alsyouf, I. 2006. "Measuring maintenance performance using a balanced scorecard approach". *Journal of Quality in Maintenance Engineering*, Vol. 12: 133 – 149.

Bevilacqua, M. and M. Braglia, 2000. "The Analytic Hierarchy Process Applied to Maintenance Strategy Selection". *Reliability Engineering and System Safety*, 70(1): pp. 71–83.

Coetzee, J. L. 2004. "A Holistic Approach to the Maintenance Problem". *Journal of Quality in Maintenance Engineering*, 5(3): 276-280.

Jasiulewicz-Kaczmarek, M. 2007. "Process approach in maintenance". [in.] M. Fertsch, K. Grzybowska, A. Stachowiak (Ed.) Logistyka i zarządzanie produkcją – nowe wyzwania, odlegle granice, Publishing House of Poznan University of Technology.

Jasiulewicz-Kaczmarek, M. and P. Drożyner, 2011. Preventive and Pro-active Ergonomics Influence on Maintenance Excellence. *Ergonomics and Health Aspects of Work with Computers,* eds. Michelle M. Robertson, LNCS 6779, Springer-Verlag Berlin Heidelberg: 49-58

Jasiulewicz-Kaczmarek, M. 2012. Sustainability – orientation in maintenance management, *Eco Production and Logistics. Emerging Trends and Business Practices, Series: Eco Production,* eds. Paulina Golinska, Springer-Verlag Berlin Heidelberg (in press)

Maggard, B. and D. Rhyne, 1992. "Total productive maintenance: a timely integration of production and maintenance", *Production and Inventory Management Journal*, Vol. 33: 6-10.

Meier, H., R. Roy, G. Seliger. 2010. Industrial Product-Service Systems—IPS2, CIRP Annals - Manufacturing Technology 59: 607–627.

Moore, W. J. and A. G. Starr, 2006. "An intelligent maintenance system for continuous cost-based prioritization of maintenance activities". *Computers in Industry*, 57(6): 595-606.

Pinjala, S.K., L. Pintelon, and A. Vereecke, 2006. "An empirical investigation on the relationship between business and maintenance strategies". *International Journal of Production Economics*, Vol. 104: 214-29.

Rosqvist, T., K. Laakso, and M. Reunanen, 2009. "Value-driven maintenance planning for a production plant", *Reliability Engineering & System Safety*, Vol. 94: 97-110.

Willmott, P. 2010. Post the Streamlining. Where's Your Maintenance Strategy Now? Maintworld No. 1: 16-22.

# CHAPTER 61

# Socio-technical Systems Engineering

*Gordon Baxter, Ian Sommerville*

School of Computer Science
University of St Andrews
St Andrews, UK
Gordon.Baxter@st-andrews.ac.uk

## ABSTRACT

It is widely accepted that socio-technical approaches to system development lead to systems that are more acceptable to their users and deliver better value to stakeholders. These approaches have had limited impact, however, so we have proposed the creation of the broader field of socio-technical systems engineering (STSE) which builds on the (largely independent) research of groups investigating work design, information systems, computer-supported cooperative work, and cognitive systems engineering. We illustrate the ideas underpinning STSE using examples from large projects we have worked during the past decade. The first shows how we addressed dependability issues in neonatal intensive care using cognitive task analysis, the second describes responsibility modeling, a method we developed for analyzing vulnerabilities in socio-technical systems, and the third describes how we intend to apply our ideas to analyze the use of cloud computing.

**Keywords**: socio-technical systems, systems engineering, software engineering

## 1    INTRODUCTION

The vast majority of computer-based systems can be described as *socio-technical systems* after Emery and Trist (1960). In other words, these systems involve a complex interaction between people, technology and the context in which they operate—organizational, physical and so on. When developing socio-technical systems it is therefore important to give appropriate consideration to people, technology and context and the way in which these different factors interact and are interdependent.

Socio-technical systems design (STSD) methods were developed as a way of ensuring that both technical and organizational aspects of a system were considered together. These methods help to increase the understanding of how human, social and organizational factors affect the ways that work is done and technical systems are used. The methods can therefore be deployed in the design of organizational structures, business processes and technical systems.

The rationale for adopting any STSD approach tends to be expressed in negative terms: a failure to employ STSD may increase the risk that a system will not make the expected contribution to an organization's goals. Many systems that satisfy their technical requirements are considered failures because they do not appropriately support the ways that real work is performed by the organization. The underlying reason for this failure is that techno-centric approaches to systems design do not take appropriate account of the complex relationships between the organization, the people who carry out the business processes, and the computer-based system that supports these processes (Goguen, 1999, Norman, 1993). Even where developers claim to be following a socio-technical approach, they often begin by decomposing the system into a social part and a technical part, thereby ignoring how the two are interrelated.

Despite the widespread acknowledgement of the importance of socio-technical issues, STSD methods have had limited impact. The main reason for this lack of continued use may be that they are not always easy to use: they tend to be less prescriptive than many software engineering methods, for example. In addition, it is not always clear how the methods can be applied to particular issues of technical engineering and individual interaction with technical systems.

Elsewhere we have argued for the need for a pragmatic approach to the design of socio-technical systems based on the gradual introduction of socio-technical considerations into existing software procurement and development processes (Baxter and Sommerville, 2011b). Our long-term goal is to develop the field of *socio-technical systems engineering* (STSE) which encompasses the systematic and constructive use of socio-technical principles and methods across all stages of the development life cycle for complex computer-based systems.

STSE is inherently interdisciplinary, and draws on research into work and workplace design (e.g., Eason, 1988, Mumford, 1983); information systems (e.g., Taylor, 1982); computer supported co-operative work, and particularly ethnographic studies (Ackroyd et al., 1992, Bentley et al., 1992, Clarke et al., 2003, Heath et al., 1994, Heath and Luff, 1992, Rouncefield, 1998); and cognitive systems engineering (Hollnagel and Woods, 2005, Woods and Hollnagel, 2006). To date there has been relatively little interaction between these disciplines. Mumford's (2006) review of STSD, for example, has no references to CSCW or cognitive systems engineering, and only a few to the information systems literature.

We also draw on the field of human-computer interaction (HCI) where some researchers have been influenced by socio-technical ideas. This has often been limited to sensitization to socio-technical issues (e.g., Dix et al., 2004 has a chapter on this topic), however, and there has been little analysis of how socio-technical issues directly affect the user interface design for complex software systems

We believe that for STSE to succeed it has to be pragmatic. There is little point in analyzing a system from a socio-technical perspective and then just presenting the results to the software engineers. STSE has to be able to help develop and evolve systems, whilst fitting in with existing software design methods and tools, because many companies have already invested heavily in them. STSE also has to use a terminology that is familiar to engineers, provide an approach that they can readily use, and generate value to the organization that is proportionate to the amount of time spent learning and using STSE.

For more than a decade now we have been involved in large projects investigating the dependability of computer based systems (www.dirc.org.uk) and the engineering of large scale complex IT systems (http://lscits.cs.bris.ac.uk/). The first of these involved collaboration between computer scientists, psychologists, sociologists, and statisticians; the second involves collaboration between groups that analyze systems at different levels of abstraction from architectural details up to policy making. Below we present three examples from these projects--one completed, one ongoing, and one future—to illustrate how we have been applying socio-technical approaches, which led us to develop and refine the notion of STSE.

In section 2 we describe a case study where we used a combination of methods to inform the design and integration of a new system into the neonatal unit of a major teaching hospital in the UK. We then describe a technique we have developed called responsibility modeling which we have used to identify a category of vulnerabilities that affect socio-technical systems. In section 4 we briefly look at how we can apply socio-technical ideas to the rapidly developing area of cloud computing. Finally we draw some conclusions based on our experiences in using social technical ideas to develop STSE.

## 2 DECISION SUPPORT IN NEONATAL INTENSIVE CARE

The neonatal intensive care unit (NICU) of St James' Hospital in the UK was looking to introduce a decision support system, FLORENCE (Fuzzy Logic REspiratory Neonatal Care Expert). The main aim of FLORENCE was to assist junior doctors, in particular, in managing the mechanical ventilation of babies born prematurely. It was important that the new system was both usable and acceptable to staff, as well as making sure that the dependability of neonatal intensive care was not reduced by the introduction of the new technology.

Many medical systems are never used because they fail to fit in with the way that staff normally work or require changes to procedures that affect other people's responsibilities (e.g., Berg, 2001). Part of our role was therefore to understand the broader work context of the NICU, including identification of the various stakeholders and their roles, responsibilities and relationships. In order to do this, we carried out a cognitive task analysis (CTA) (Chipman et al., 2000) that comprised a lightweight rich pictures analysis of the work context(Monk, 1998), an analysis of the decision making of staff involved in the NICU using the critical decision method (CDM) (Klein et al., 1989), and observation of work in the NICU (see Baxter et al., 2005b for full details and justification of the choice of methods).

## 2.1     Domain and context familiarisation

Before beginning the CTA we spent some time familiarizing ourselves with the domain of neonatal intensive care, and the working environment of the NICU. This comprised several visits to the NICU and meetings with staff. In any working context it is important to be able to communicate with the staff, which means being able to understand the terminology that they routinely use to describe their work.

## 2.2     Lightweight rich pictures analysis

The lightweight rich pictures analysis highlighted the importance of communication in the NICU and the extensive use of written records. It also showed that work is normally structured hierarchically, with junior doctors and nurses doing most of the decision making, but deferring to registrars and then consultants for the more difficult cases. The daily ward round is attended by most staff and provides the formal handover from the night shift to the day shift. In addition to the formal handover, the nurses, registrars and junior doctors also each have their own informal handovers, and there are frequent meetings of the nursing sisters on the unit.

Figure 1 shows an extract from one of the rich pictures highlighting the communication links in the NICU.

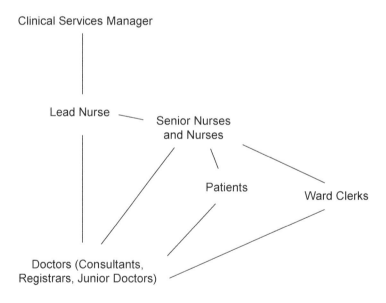

Figure 1. Extract from rich picture showing who communicates with whom for the staff that are normally present in the neonatal unit. The arrowheads indicate the direction of the communication.

The main implication that came out of the rich pictures analysis was the need to make sure that the decision support system did not detract from communication between staff. Other implications include logistical issues, such as the need to find space to accommodate FLORENCE in what was already a quite crowded area, and the need for a power supply for FLORENCE, which would potentially introduce a trailing power cable into the NICU.

## 2.3    Critical decision method

The CDM was used to analyze the way that staff in the NICU make decisions about changing the mechanical ventilator settings. These decisions are very important. If a baby is given too little oxygen, for example, it can suffer brain damage, and if it is given too much it can go blind.

The main implications arising out of the CDM were the need to ensure that FLORENCE fits in with existing working practices. It is important that staff understand and recognize the limitations of FLORENCE which should not be used when dealing with more complex cases. One of the roles of FLORENCE is to help educate junior doctors, too, so it is important that FLORENCE be able to explain why it is making particular suggestions about changes to the ventilator settings. One other important result that came out of the CDM was the way in which the decision making hierarchy flattens when dealing with acute problems: all staff are encouraged to contribute to solving such problems when they arise.

## 2.4    Observation

Observation in the NICU had to be carried out opportunistically. FLORENCE is intended to be used to help in the treatment of premature babies with respiratory distress syndrome (RDS). RDS usually occurs within the first few days after birth, and only lasts about 72 hours. Not all premature babies suffer from RDS.

The implications arising from the observation included the importance of having a distinct audible alarm for FLORENCE, and making sure that staff knew how to respond to this alarm. The NICU is a noisy context in which alarms are constantly sounding, often only intermittently. Staff learn which alarms can be dealt with quickly (sometimes by just glancing at the baby) and which require intervention.

## 2.5    Conclusions

Using a CTA helped us to understand how the NICU at St James' Hospital operated on a day to day basis. By using a combination of methods to analyze the roles played by staff and technology, and the normal structure of work, we were able to make a set of detailed recommendations covering aspects of how the user interface for FLORENCE needed to be designed, through procedures for ensuring the integration of FLORENCE into existing working practices, down to the logistics of deploying FLORENCE in the NICU. The complete list of implications for the design of the system and its impact on work in the NICU are described elsewhere (Baxter et al., 2005b, Baxter et al., 2005a)

# 3  RESPONSIBILITY MODELLING FOR RESILIENCE

Systems often fail for socio-technical reasons, rather than purely technical reasons. Based on this observation, we have been analyzing systems in terms of responsibilities, rather than functions, where we define a *responsibility* as

> *a duty, held by some agent, to achieve, maintain or avoid some given state, subject to conformance with organizational, social and cultural norms.*

Responsibilities are allocated to *agents* (generally people or organizations, but sometimes technology or software applications), and those agents have to use some *resources* (time, people, information or physical resources) to discharge them. Responsibilities are discharged in a context which includes factors that determine how they are discharged.

We have developed a graphical representation technique called Responsibility Modelling (RM) for analyzing systems and organizations. We have used RM to analyze major event contingency plans (Lock et al., 2009), which are drawn up in advance, and are usually long unwieldy documents, written by several authors at different times. They therefore tend to be inconsistent and incomplete. Delivering a resilient response to a major incident, however, is highly dependent on the way responsibilities are allocated and discharged.

Failures in socio-technical systems can manifest themselves in many different ways. Some will occur during operation, for example, when people interact with the technology; when processes do not work as documented; and when people try to provide data (or information) to the users. We have identified one class of failures that are associated with responsibility and agent vulnerabilities:

1. Unassigned responsibilities: a responsibility has not been assigned to any agent.
2. Duplicated responsibilities: different agents believe that they have been assigned a particular responsibility and all try to discharge it.
3. Uncommunicated responsibilities: a responsibility is assigned to a role, but the agent that is assigned that role is not told about the responsibility.
4. Misassigned responsibilities: the agent does not have the capabilities or resources to discharge the assigned responsibility.
5. Responsibility overload: the agent does not have sufficient resources to discharge all its responsibilities in a timely manner.
6. Responsibility fragility: a responsibility is assigned to an agent, but if that agent is not available, there is no back-up agent who can discharge that responsibility.
7. Conflicting responsibilities: where an agent fulfils multiple roles and some of the roles have responsibilities that are antagonistic to responsibilities associated with their other roles.
8. Social diffusion of responsibility: where several agents are assigned a responsibility and each assumes that another agent will discharge it.

Although duplicated responsibilities are listed as a vulnerability, the resilience of a system may be increased by assigning back-up agents for particularly important responsibilities. These secondary agents, however, need to understand how and when they should act. It should also be noted that a misassigned responsibility may not necessarily be discharged ineffectively (if at all). People are inherently flexible and adaptable, so they will often find a way to discharge a responsibility, using workarounds as appropriate.

Figure 2 shows an extract from a responsibility model for flood evacuation plans in Carlisle, UK. The responsibilities (in shaded rectangles) are allocated to agents (shown in angle brackets). There can be several levels of responsibility (in the figure, *Evacuation* has four lower level responsibilities, for example). For large socio-technical systems it is often best to represent the system as a set of responsibility models, reflecting these different levels. These models are often organized hierarchically, but may also form a network-based structure.

There are vulnerabilities associated with resources and how they are provided, accessed and used. These can be particularly important in situations where plans for emergency situations are stored on-line, for example, and may not be accessible in the event of a fire or a flood.

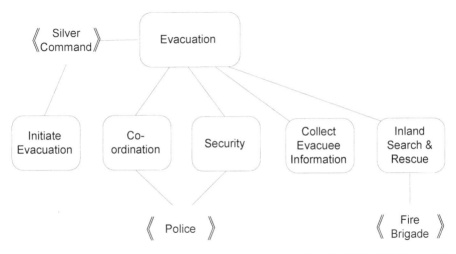

Figure 2. Extract from a responsibility model for flood evacuation plans in Carlisle, UK. The model shows that at the highest level responsibility for *Evacuation* lies with the agent (in this case, an organization) called *Silver Command*. It also shows an example of an unassigned responsibility, *Collect Evacuee Information,* has not been assigned to any agent.

In addition to vulnerabilities that can be identified statically, there are tactical considerations too. Using RM in conjunction with a keyword based approach like that of Hazard and Operability Studies (HAZOPS) (Kletz, 1999) allows you to ask questions such as "What if this responsibility is discharged too late?" and "What if there are not enough resources available to discharge this responsibility?".

Responsibility models can be used to describe an existing system, and by enabling system descriptions and plans to be drawn up during design (before the system exists). In this way vulnerabilities can be identified and appropriately managed.

We have found responsibility models to be particularly useful in situations involving inter-group, and multi-agency working. In these situations, different agencies often having different views about particular responsibilities are, and who is supposed to discharge them. There can also be emergent responsibilities that arise out of the collaboration between agencies. Responsibility models help because they are a concrete representation that allows the various stakeholders to discuss, and resolve the issues associated with assigning and discharging responsibilities.

We have successfully used RM to analyze the emergency flood plan for Cumbria (Lock et al., 2009), to identify information requirements for a system that co-ordinates agencies to handle emergencies (Sommerville et al., 2009) and to investigate problems arising out of the use of e-counting systems in the Scottish elections in 2007 (Lock et al., 2008).

We have recently been applying RM to the analysis of critical infrastructure to help facilitate responses to events (Baxter and Sommerville, 2011a, Boin and McConnell, 2007). RM is intended to be a relatively lightweight method that is easy to learn. We have been developing checklists to help ensure that responsibilities are appropriately assigned, resourced and made visible to all the required agents so that the responsibility can be discharged in a timely and effective manner. It identifies a class of vulnerabilities that are not addressed by other methods for identifying system vulnerabilities.

## 4    FUTURE WORK

We are now looking at using STSE in analyzing the use of cloud computing. We have already identified a role for RM. When a customer moves to using software as a service, for example, issues like the responsibility for security start to become somewhat blurred. The customer will have direct contact with the service provider, so they will probably expect that service provider to be responsible for ensuring security of use of the service. The service provider, however, will probably expect the cloud provider to be responsible for ensuring the security of use of the service whilst the service is running on the cloud providers' hardware.

## 5    CONCLUSIONS

Our experiences have highlighted the interdisciplinary nature of systems development. STSE is intended to be a broad field, which will allow practitioners to use the best method available to carry out the job at hand. The provenance of the method is less important than its appropriateness and its relative ease of use and compatibility with existing methods and tools. Here we have illustrated this point by using a range of methods from different disciplines that have all proved to be useful in analyzing and informing the design of socio-technical systems.

## ACKNOWLEDGMENTS

The DIRC and LSCITS projects were both funded by the UK Engineering and Physical Sciences Research Council (EPSRC).

# REFERENCES

ACKROYD, S., HARPER, R., HUGHES, J. A. & SHAPIRO, D. (1992) *Information technology and practical police work,* Milton Keynes, UK, Open University Press.

BAXTER, G., FILIPE, J. K., MIGUEL, A. & TAN, K. (2005a) The effects of timing and collaboration on dependability in the neonatal intensive care unit. . IN REDMILL, F. & ANDERSON, T. (Eds.) *Constituents of Modern System-safety Thinking: Proceedings of the Thirteenth Safety-critical Systems Symposium.* London, UK, Springer-Verlag.

BAXTER, G., MONK, A. F., TAN, K., DEAR, P. R. F. & NEWELL, S. J. (2005b) Using Cognitive Task Analysis to facilitate the integration of decision support systems into the neonatal intensive care unit. *Artificial Intelligence in Medicine,* 35, 243-257.

BAXTER, G. & SOMMERVILLE, I. (2011a) Responsibility modelling for resilience. IN HOLLNAGEL, E. & RIGAUD, E. (Eds.) *Proceedings of the fourth Resilience Engineering Symposium.* Sophia Antipolis, France, Presses des MINES.

BAXTER, G. & SOMMERVILLE, I. (2011b) Socio-technical systems: From design methods to systems engineering. *Interacting with Computers,* 23, 4-17.

BENTLEY, R., HUGHES, J. A., RANDALL, D., RODDEN, T., SAWYER, P., SHAPIRO, D. & SOMMERVILLE, I. (1992) Ethnographically-informed systems design for air traffic control. *Proceedings of the 1992 ACM conference on Computer-supported cooperative work.* Toronto, Ontario, Canada, ACM.

BERG, M. (2001) Implementing information systems in health care organizations: myths and challenges. *International Journal of Medical Informatics,* 64, 143-156.

BOIN, A. & MCCONNELL, A. (2007) Preparing for critical infrastructure breakdowns: The limits of crisis management and the need for resilience. *Journal of Contingencies and Crisis Management,* 15, 50-59.

CHIPMAN, S. F., SCHRAAGEN, J. M. & SHALIN, V. L. (2000) Introduction to cognitive task analysis. IN SCHRAAGEN, J. M., CHIPMAN, S. F. & SHALIN, V. L. (Eds.) *Cognitive task analysis.* Mahwah, NJ, Lawrence Erlbaum Associates.

CLARKE, K., HUGHES, J. A., MARTIN, D., ROUNCEFIELD, M., SOMMERVILLE, I., GURR, C., HARTSWOOD, M., PROCTER, R., SLACK, R. & VOSS, A. (2003) Dependable red hot action. *Proceedings of the eighth conference on European Conference on Computer Supported Cooperative Work.* Helsinki, Finland, Kluwer Academic Publishers.

DIX, A., FINLAY, J., ABOWD, G. D. & BEALE, R. (2004) *Human Computer Interaction, 3rd ed.,* Harlow, UK, Addison-Wesley.

EASON, K. (1988) *Information technology and organisational change,* London, UK, Taylor & Francis.

602

EMERY, F. E. & TRIST, E. L. (1960) Socio-technical systems. IN CHURCHMAN, C. W., & VERHULST, M. (Ed.) *Management Science Models and Techniques* Oxford, UK, Pergamon.

GOGUEN, J. (1999) Tossing algebraic flowers down the great divide. IN CALUDE, C. S. (Ed.) *People and ideas in theoretical computer science.* Berlin, Germany, Springer.

HEATH, C., JIROTKA, M., LUFF, P. & HINDMARSH, J. (1994) Unpacking collaboration: the interactional organisation of trading in a city dealing room. *Computer Supported Cooperative Work,* 3, 147-165.

HEATH, C. & LUFF, P. (1992) Collaboration and control: Crisis Management and Multimedia Technology in London Underground Line Control Rooms. *Computer Supported Cooperative Work,* 1, 69-94.

HOLLNAGEL, E. & WOODS, D. D. (2005) *Joint cognitive systems: Foundations of cognitive systems engineering,* Boca Raton, FL, CRC Press.

KLEIN, G. A., CALDERWOOD, R. & MACGREGOR, D. (1989) Critical decision method for eliciting knowledge. *IEEE Transactions on Systems, Man, and Cybernetics,* 19, 462-472.

KLETZ, T. (1999) *HAZOP and HAZAN: Identifying and assessing process industry standards* Rugby, UK, Institution of Chemical Engineers.

LOCK, R., STORER, T., HARVEY, N., HUGHES, C. & SOMMERVILLE, I. (2008) Observations of the Scottish elections, 2007. *Transforming Government: People, Process and Policy,* 2, 104-118.

LOCK, R., STORER, T., SOMMERVILLE, I. & BAXTER, G. (2009) Responsibility modelling for risk analysis. *Proceedings of ESREL 2009.*

MONK, A. (1998) Lightweight techniques to encourage innovative user interface design. IN WOOD, L. (Ed.) *User interface design: Bridging the gap from user requirements to design.* Boca Raton, FL, CRC Press.

MUMFORD, E. (1983) Designing human systems for new technology - The ETHICS method.

MUMFORD, E. (2006) The story of socio-technical design: reflections in its successes, failures and potential. *Information Systems Journal,* 16, 317-342.

NORMAN, D. A. (1993) *Things that make us smart: Defending human attributes in the age of the machine,* Boston, MA, Addison-Wesley.

ROUNCEFIELD, M. (1998) An ethnography of 'everyday admissions work'. Lancaster, UK, Lancaster University.

SOMMERVILLE, I., LOCK, R., STORER, T. & DOBSON, J. E. (2009) Deriving information requirements from responsibility models. *Proceedings CAiSE 2009: 21st international conference on advanced information systems engineering.* London, UK, Springer.

TAYLOR, J. C. (1982) Designing an Organization and an Information-System for Central Stores - a Study in Participative Socio-Technical Analysis and Design *Systems Objectives Solutions* 2, 67-76.

WOODS, D. D. & HOLLNAGEL, E. (2006) *Joint cognitive systems: Patterns in cognitive systems engineering,* Boca Raton, FL, CRC Press.

# CHAPTER 62

# The Labour Market and Material Environment Design

*Katarzyna Lis*

Department of Labour and Social Policy
Poznań University of Economics
Al. Niepodległości 10, 61-875 Poznań, Poland
katarzyna.lis@ue.poznan.pl

## ABSTRACT

The purpose of this paper is to analyze changes in the labor market and the importance of material environment. With regard to these issues the paper has been divided into four parts. The introduction is characterized by changes in the labor market and an increasing share of services sector. The second part of the article analyzes data on employment in particular sectors. In the third section describes the changes that occur in the design of material work environment. The conclusions of the abovementioned subject are in last part of the paper.

**Keywords**: labour market, ergonomic

## 1    INTRODUCTION

The concept of labour market is defined as a mechanism within which economic transactions take place. The transactions consist, on the one side, in hiring people to work, and on the other – in selling one's work (or their availability for work) (Mortimer-Szymczak H., 1995). The market is shaped by both demand and supply factors. The factors affecting the demand for labour include: structural trans-formations in the economy, economic growth rate, nature of the economic growth and internal transformations within businesses, whereas the factors influencing labour supply comprise: demographic situation, emigration and immigration, social policy, labour market policy and employee mobility (Makać W., 2010).

Issues related to labour market division are based on an assumption that it is internally heterogeneous. The specialist literature provides several criteria for labour market division, and those most often applied include: economy sector, geographical location, profession, employee qualifications and their "kind", e.g.: age, gender (Unolt J., 1999; Oleksyn T., 2001; Jarmołowicz W. 2008; Lis K., 2010). Due to the production technology advancement and the labour efficiency growth, for many years we have been observing evolutionary changes in the employment sectoral structure, the theory of which was originated by A.B.G. Fisher, C. Clark and J. Fourastie in the 30s of 20th century. In accordance with that approach, the division criteria for the labour market are the three economic sectors: agriculture, industry and services. According to this theory, the labour market is characterised by the following trends (Dach Z., 2008):

- advancement of economic growth is accompanied by employment decrease in the agricultural sector,
- industrialisation growth results in rising (to a certain point) employment in the industrial sector, which is accompanied by decreasing employment in agricultural sector,
- limiting the employment in agriculture followed by stabilised employment in the industry sector results in employment increase in the service sector and servicisation of the economy.

Apart from technology, the factors affecting the evolving labour market also comprise labour supply, including a bigger share of mature age employees (population ageing), employees with moderate disabilities, as well as the increasing percentage of working women. As a result of those changes, flexible forms of employment and work organisation are developed. More and more often, employers tend to hire temporary and part-time workers, also, they increasingly apply new technologies. In the context of those changes it is advisable to take a deliberate approach to workplace design, including the ergonomic aspect of designing the tangible working conditions. Employers must be aware that the less onerous and harmful working conditions their employees have, the longer and the more efficiently they will be working.

## 2. EMPLOYMENT IN THE PARTICULAR SECTORS

As it was already mentioned in the introduction, in accordance with the theory originated by A.B.G. Fisher, C. Clark and J. Fourastie, the employment sectoral structure changes lead to increased numbers of employees in the service sector. Work in the service sector is characterised by:

- Activity or process whose results cannot be stored – it is at most possible to increase the employee work efficiency as a result of experience accumulation.
- Work is not a simple transfer of an ownership title to the product, as it is the case with selling tangible goods - rendering a service requires taking into account the conditions and standard of performance, it also

involves service variability, its scope, quality and duration.

- Simultaneously, participation of both the provider and recipient of the service creates conditions for the relations to be individualised in the area of the service provision (Orczyk J., 2007; Bowen D.E. et al., 1990; Looy B. et al., 2003).

Moreover, services being marketable merchandise may be provided in different forms. In relation to the form, it is possible to distinguish the services that have features of distribution (transport), features of production (financial agency services), or features of private entrepreneurship (trade, hotel and restaurant industries).

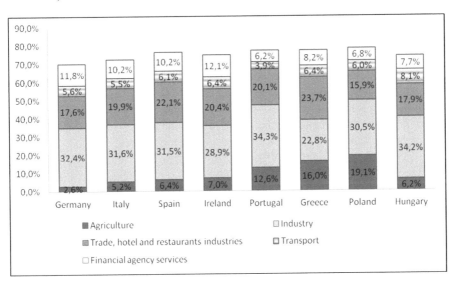

Figure 1  Employment structure by sectors in selected countries (Pomianek T. et al., 2005)

Figure 1 presents the employment structure by labour market sectors in selected countries, while the service sector was divided into three groups with respect to various forms of distribution:

- Trade, hotel and restaurants industries,
- Transport,
- Financial agency services.

As the diagram shows, the biggest number of employees is found in the first group which facilitates the development of the other services. It should be noted that the service sector often requires changing the work principles that functioned in the industrialisation era and were based on explicitly specified standards. Moreover, the possibility to control the work process itself is diminished in this sector due to the place and conditions of the performance (Orczyk J., 2007). Therefore, it is necessary to pose the question who will be the guarantor of the liability for the processes / mechanisms found in the service sector. Sometimes it is professional associations that become such a guarantor, however, in many a case it is important that the employee should realise what the correct job performance is, i.e. the manner of the work performance and organisation, as well as creating an ergonomic workplace. One of the methods to solve this problem is developing some training materials which may be included in the scheduled trainings for employees and employers.

In view of the changing situation on the labour market in the European Union, an employment strategy was developed, which is based on four pillars (Lis K., 2010; Noga M. et al., 2009):

1. Employability – means the efforts made so that the unemployed are able to return to work, and promoting the idea of the labour market open to everyone. The aim of this pillar is counteracting the long-term unemployment through enhancement of continuous learning, support of training and retraining programmes, developing advisory programmes, assisting in education programmes adjustment to the changing requirements of the labour market.

2. Entrepreneurship – means efforts made in order to facilitate business establishing and running by the EU citizens, and consequently creating jobs for others. The issue particularly emphasised here is creating new jobs, including motivation for self-employment.

3. Adaptability of employees and employers – aimed at:
- promoting work modernisation and organisation,
- providing more flexible working hours,
- applying various forms of employment,
- modernising employment contracts,
- maintaining the adaptability of enterprises through employee trainings conducted at the workplace.

4. Equal opportunity – this pillar focuses on finding the balance between work and family life, in an attempt to solve the problem of unequal treatment of men and women in the workforce. Another aspect is facilitating a return to the labour market after a long time of inactivity, and also creating work conditions for the disabled.

The European Council has adopted detailed principles of the employment policy, which include:
- active and preventive measures to help unemployed and inactive persons,
- creating jobs and entrepreneurship,
- promoting the adaptability and mobility on the labour market,
- promoting the development of human capital and continuous learning,
- increase in labour supply and promotion of active ageing,
- policy of equal opportunity for men and women,
- supporting the integration and fighting the discrimination of people in unfavourable positions on the labour market,
- developing financial incentives for getting employed,
- limiting the grey labour market,
- decreasing the regional variations in employment.

The major methods for implementing the employment strategy in the EU include the open coordination being the method for European integration, a modified system of education and trainings adjusted to the needs of the labour market and corresponding to the continuous learning requirements, a modified system of benefits and taxes, using the employment potential of the knowledge-based economy and in the service sector, relying on flexible forms of employment (fixed-term employment contracts, part-time employment contracts, reducing the overtime, etc.), improving the employees qualifications so that each of them has gained knowledge enabling him or her fast and easy access to information (Lis K., 2010; Pomianek T. et al., 2005).

Apart from the presented solutions, changes in employment that occur on the labour market must also take into account the possibility to design safe and ergonomic workplaces, which may become one of the factors that enable maintaining and improving occupational activity.

## 3. CHANGES IN DESIGNING THE TANGIBLE WORK ENVIRONMENT

The process of ergonomic design begins with specifying social needs which become an impulse for the initiation of the designing process (Górska E., 2002). In the case of ergonomic design of workplaces for the disabled, the social needs will often pertain to a specific workstand which must be adjusted to the concrete person, taking into account the concrete aspects of his or her disability. As it was already mentioned in the preceding sections, due to the changes taking place on the labour supply market, designing any workplaces should first and foremost account for the structure of those changes. Therefore, the first thing to be taken into account is the changes in demographic structure regarding labour supply, such as a greater number of mature-aged employees, employees with moderate disabilities and the increasing number of women in the workforce. Moreover, the growing role of the service sector should be taken into consideration.

On the one hand, considering the growing average life expectancy, the population's health may seem good (Bugajska J. et al., 2008). However, as shown in the report of the Organization for Economic Cooperation and Development (OECD), the health improvement does not affect the average retirement age. What is more, an unfavourable trend has been observed – the average retirement age has actually lowered. This process has resulted in a decrease in labour market resources and problems with the pension systems in many countries. Consequently, it was necessary to increase the employment ratio as early as in mid-90s of 20th century. The way to address the situation was raising the retirement age. Lengthening the occupational activity period by several years pertains mainly to women, as in many countries decisions were taken to level the retirement age for men and women. Designing workplaces for mature age people, including women, must take into account the variability of physical and mental capabilities conditioned by the rate of biological ageing process, diseases, environmental conditions and lifestyles in the population (Bugajska J. et al., 2008).

Table 1 Impact of physiological changes on the human body function

| Physiological changes | Human body function |
|---|---|
| Decrease in muscle mass (sarkopenia) | Decrease in muscle strength and endurance |
| Impairment of bone tissue structure | Decrease in bone tissue endurance |
| Slower heart rate | Weakness, tendency to fainting |
| Decreased blood vessel flexibility | Atherosclerotic changes in arteries |
| Increase in volume of the respiratory tract, decrease in gas exchange surface area | Decrease in oxygen uptake |
| Decrease in thermoregulation processes | Heat tolerance impairment |
| Decrease in angiospastic reaction | Increased sensitivity to cold |
| Impaired eye accommodation | Blurred vision of near objects |
| Impaired lens translucency | Impaired ability of the eye to adapt to darkness |
| Hearing impairment | Difficulty with speech understanding and hearing sounds |
| Changes in the nervous system | Impaired ability to remember and process information and impaired concentration |

As shown in Table 1, advancing age may bring – at different rates – physiological changes in the body, which lead to a decrease in the ability to make physical or mental efforts, which in turn means there is an impairment of the body physical or mental function. It must be emphasised here that the term "inability to work" is not tantamount only to a bad health condition, i.e. physical or mental impairment. Currently, a bad health condition itself does not constitute an inability to work, as the definition of inability to work features an important element connected with the lack of ability to work. This approach to people with a body function impairment enables their occupational insertion.

The Finnish Institute of Occupational Health (FIOH) in Helsinki developed the Work Ability Index which makes it possible to specify how the workers are able to perform their work in terms of their physical and mental capabilities. The index is an element of the work ability model, in which four levels are distinguished:

- The first level pertains to health defined as the functional capacity comprising the physical and mental capacities as well as social functionality.
- The second level comprises the competence specified as knowledge and skills.
- The third level presents the values affecting the motivation and attitude.
- The fourth level is the work characteristics: its kind, organisation and the tangible work environment.

In accordance with the work ability model, a human should keep a proper balance between the changing capabilities which impair with age and the work that accounts for those changes. One of the possibilities to keep this balance is ergonomic design of the tangible work environment. Analysing the Work Ability Index, in accordance with the studies presented by J. Laurier ability to work decreases over time (Laurier J., 2011).

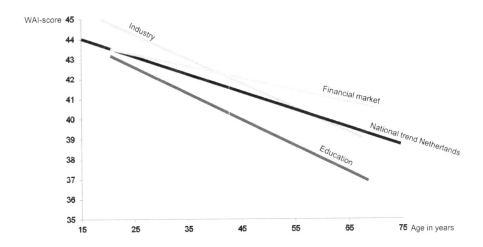

Figure 2  WAI score by sector (Laurier J., 2011)

As can be seen in Figure 2, the work ability index in various labour market sectors shows a decreasing trend. Interestingly, far lower work ability is found in the education sector compared to the industry. This may result from many factors having an impact on the contents and process of work. However, the specialist literature does not provide a more detailed sectoral analysis of the work ability. The available data pertain to the sectoral analyses of occupational accidents, occupational diseases and the number of people employed in hazardous conditions.

The literature on ergonomics provides a set of principles regarding workplaces for mature age persons (Jasiak A. et al., 2005), indicating which work is inappropriate for such people and what principles must be obligatorily observed to enable work performance. The first group of jobs that excludes mature age employees comprises: physically demanding work, work hazardous to health, piece-work, forced-pace work, work at height, work requiring an unchanged body posture with limited movements, work requiring quick and precise action in combination with good visual and motor coordination, as well as work demanding considerable psychic and mental focus. The principles necessary for work performance include in particular:

- visual comfort – appropriate light intensity,
- comfortable microclimatic conditions,
- providing extra sensory information,
- sequence of providing the information,
- enhancing and lengthening the signal intensity,
- organising the work to enable continuous monitoring.

In reply to the social diagnosis, the designer of the tangible work environment will have to take into consideration the fact that the service sector will be employing more and more mature age people with impaired body functions and decreasing work ability. Therefore, it will be necessary to find solutions that can be introduced to a wide range of workplaces. The workplaces will have to be easily adaptable to the most common limitations connected with the human body impaired function resulting from physiological changes caused by the ageing process.

## 4    CONCLUSIONS

Ergonomic design makes it possible to reduce or even eliminate the factors that are onerous and harmful at the workplace. It is possible due to a homocentric criterion in the approach to the designed object. A designed object in ergonomics is defined as the following system: a human – a technical object – the environment. This approach accounts for the work of humans with machines and technical equipment, and it contributes to a considerable decrease in the number of people employed in hazardous conditions. Consequently, it enables creating safe working conditions, thus decreasing the number of occupational accidents and diseases. Therefore, on the one hand it may be assumed that ergonomic design of the tangible work environment has an impact on the decrease in the number of people unable to work, since work inability is directly connected with the occupation. Secondly, ergonomic design contributes to occupational insertion of people unable to work also for reasons other than occupational, accounting for a tiny percentage of the society. The provided data show that in the case of the majority of the people unable to work their inability is only moderate, so from the perspective of economy and social welfare it is important that this group should be inserted to work. The current research results (Brzezińska A. et al., 2008) on occupational activity of people with impaired abilities indicate there is unequal access to the labour market.

Changes in the labour market result in a decrease in the number of people employed in agriculture and industry, accompanied by employment increase in the service sector. Moreover, due to the population ageing, on the labour market there will be more and more people with moderate inability to work. A great challenge will be to create a competitive labour market for those people. Due to the physiological ageing process and the decreasing body functionality it will be essential to adjust the workplaces to the wide range of employees. The service sector growth may become the one to employ the most of mature age people. A considerable impact may also be made here by wider and wider use of new technologies, in particular IT.

## REFERENCES

Bowen D.E., Cummigs Th.G. 1990. *Suppose We Took Service Seriously?* Service Management Effectiveness.

Brzezińska A., Kaczan R., Piotrowski K., Rycielski R. 2008. *Aktywność osób z ograniczeniami sprawności na rynku pracy: czynniki wspomagające i czynniki ryzyka.* Academica Wydawnictwo SWPS. Warszawa.

Bugajska J., Makowiec-Dąbrowska T., Konarska M. 2008. *Zapobieganie wczesnej niezdolności do pracy – założenia merytoryczne.* Wydawnictwo CIOPiPIB. Warszawa.

Dach Z. 2008. *Rynek pracy w Polsce. Aspekty ekonomiczno-społeczne.* Wydawnictwo Uniwersytetu Ekonomicznego w Krakowie. Kraków.

Górska E. 2002. Projektowanie stanowisk pracy dla osób niepełnosprawnych. Oficyna Wydawnicza Politechniki Warszawskiej. Warszawa.

Jarmołowicz W. 2008. *Przemiany na współczesnym rynku pracy.* Wydawnictwo Forum Naukowe. Poznań.

Jasiak A., Swereda D. 2005. *Ergonomia osób niepełnosprawnych.* Wydawnictwo Politechniki Poznańskiej. Poznań.

Laurier J. 2011. *The workability of the Dutch labour force uncovered.* Older Workers & Work Ability Conference, Australia.

Lis K. 2010. *Zmiany rynku pracy a wypadkowość.* [W :] Wybrane kierunki badań ergonomicznych w 2010 roku (pod red. J. Charytonowicz). - Wrocław: Polskie Towarzystwo Ergonomiczne. Wrocław. p. 153-160.

Lis K. 2010. *The social role of ergonomics and material environment design* / red. P. Vink, J. Kantola. [W :] Advances in Occupational, Social, and Organizational Ergonomics. - Boca Raton, London, New York : CRC Press Taylor & Francis Group. p. 786-791.

Makać W. 2010. *Statystyka ekonomiczna. Wybrane mierniki makroekonomiczne.* Wydawnictwo Uniwersytetu Gdańskiego. Gdańsk.

Mortimer-Szymczak H. 1995. *Rynek pracy i bezrobocie. Acta Universitatis Lodziensis*, Folia Oeconomica 135.

Noga M., Stawicka M. K. 2009. *Rynek pracy w Polsce w dobie integracji europejskiej i globalizacji.* CeDeWu. Warszawa.

Oleksyn T. 2001. Praca i płaca. Międzynarodowa Szkoła Menedżerów. Warszawa.

612

Orczyk J. 2007. *Rozwój usług a zmiany stosunków pracy*. Polityka Społeczna. nr 2. pp.1-6.

Pomianek T., Rozmus A., Witkowski K., Przywara B., Bienia M. 2005. *Rynek pracy w Polsce i Unii Europejskiej*. Wyższa Szkoła Informatyki I Zarządzania z siedzibą w Rzeszowie. Rzeszów.

Unolt J. 1999. *Ekonomiczne problemy rynku pracy*. BPS - Śląsk. Katowice.

Van Looy B. Gemmel P., Van Dierdonck R. 2003. *Services Management. An Integrated Approach*. Prentice Hall.

# CHAPTER 63

# The Emerging Human Performance Model for Homeland Security

*Bonnie B Novak[1], Thomas B. Malone, Ph.D[2], Janae Lockett-Reynolds, Ph.D.[1], Darren P. Wilson[1], and Katherine Muse Duma[3]*

[1]Department of Homeland Security
Science & Technology Directorate
Human Factors/Behavioral Sciences Division
Washington, DC USA

[2]Carlow International
Sterling, VA USA

[3]Booz Allen Hamilton
McLean, VA USA

## ABSTRACT

The Human Systems Research and Engineering (HSRE) program in the U. S. Department of Homeland Security (DHS) Science and Technology Directorate (S&T), Human Factors/Behavioral Sciences Division is developing a Human Performance Model (HPM) with broad applicability to the range of technology and systems employed by DHS Component. The objectives of the HPM are to (a) support the identification of human performance requirements for DHS threats, systems, and operations specific to a particular domain area or mission space (e.g., aviation security, border security, cyber security, first response, etc.); and (b) serve to focus the HSRE research agenda on capability gaps associated with human performance requirements within these domain areas. The core element of the HPM is a set of top level Mission Essential Tasks (METs) that are performed by end users, and derived from threats, systems, and operations encountered by human operators in each domain. In the HPM, the METs are represented at two levels of specificity (task and task element). Human performance requirements for each

MET include MET performance standards, MET performance conditions, MET user interface requirements, MET manpower, personnel and training requirements, and applicable MET performance data. The HPM research requirements include, for each MET, the capability gap, criticality of the gap, research issues and variables, and user performance impacts. The HSRE program has completed the initial version of the HPM for the aviation security domain area and is currently addressing the models for the remaining operating environments.

**Keywords**: U.S. Department of Homeland Security, human factors, human performance model

# 1 STATEMENT OF THE PROBLEM

The U. S. Department of Homeland Security (DHS) has the mission of securing the Nation against the many threats it faces. This requires the dedication of many DHS employees, ranging from Transportation Security Officers (TSOs), Customs and Border Patrol (CBP) agents, and cybersecurity analysts in the aviation, border, and cybersecurity domains, respectively, to first responders tasked with critical, time sensitive emergency response activities. The duties are wide-ranging, but the goal is clear – to keep America safe.

DHS activities cover a wide range of responsibilities, including: counterterrorism for a variety of threats; maritime and border security; preparedness, response and recovery from man-made and natural catastrophes; immigration enforcement; citizenship and immigration services; transportation security; asset security; and cyber-security. Each of these responsibilities has associated with it a myriad of systems, facilities and technologies that enable attainment of mission objectives. There are also a wide variety of human users, including operators, maintainers, controllers, supervisors, first responders, first receivers, passengers, residents, and the public that use a broad range of DHS systems, facilities and technology.

Wilson et al (2009) stated that, although similar to the Department of Defense (DoD) in the mission to enhance the safety and security of the Nation (Department of Defense, 2008), the mission space of DHS differs greatly in that the systems developed and deployed by DHS are used within the borders of the United States, and they affect all citizens. The users of DHS technology and systems represent a far more diverse population in terms of skills, anthropometry, age, training quantity and quality, intelligence, and readiness, than those in the military user community. Not only are the users and affected communities as diverse as the Nation, but the performance requirements and operating environments associated with homeland security systems and technology end-users are also more diverse and more constrained. Where issues of technology accessibility and privacy are not a major concern for the military, they can be a design constraint, and a driver of requirements in the development of homeland security systems and technology.

The Human Systems Research and Engineering (HSRE) program in the DHS S&T Human Factors/Behavioral Sciences Division is working to address the issue of designing technology and systems to maximize human performance. These efforts have been directed at institutionalizing the application of Human Systems Integration (HSI) in DHS to maximize the performance of the wide variety of human users. These users are performing required tasks with a broad range of technology and systems to address a multitude of missions and operational environments. The HSRE program is developing processes and best practices to support the application of HSI to any DHS systems and technologies that depend on human performance for their success. All DHS systems are designed, built, operated or maintained by humans. To fail to address those risks and issues associated with the human in the loop would be a failure to address a mission critical component of the system.

The HSRE program plans to focus the HSI analytical effort in support of technology or system design to ensure that human performance requirements are met in terms of ensuring product usability, supportability, reliability (reduction of human error), affordability, safety, acceptability, and accessibility. To further support these design efforts, HSRE has identified the need to develop a research roadmap to generate the performance data to successfully integrate the human element with other components of the technology or system. The problem is how to address these needs with the diverse end-users, systems, missions, and operating conditions that are part of the DHS sphere of responsibility. To meet these interrelated objectives, the HSRE program has initiated the effort to develop the DHS Human Performance Model.

## 2    OVERVIEW OF THE HUMAN PERFORMANCE MODEL

At its core, the Human Performance Model (HPM) is a task database with a set of guidelines to support the definition of technology or system end-users categorized by DHS domain areas, and to identify mission essential tasks, task requirements, user interface, and research requirements. It will describe requirements for performing mission essential tasks (METs) in terms of the appropriate DHS domains, the threat, the counter threat system, and the end-user. The METs are generic top-level tasks that describe activities for end-user/system combinations, are associated with specific end-users, and that define required capabilities for end-users. The MET performance requirements in the model will include the requirements that serve as drivers for task performance (information, decision, execution, skills, and time), the standards of successful performance of the task, and the conditions (environmental, personal and tactical) under which the task must be performed for the system or technology. The DHS HPM will be populated by results of ongoing HSRE research projects, interviews with subject matter experts in the Component, and with inputs from existing system and technology descriptions. The model will provide outputs to HSI standards, human performance requirements analysis (HPRA) in system acquisition and technology development projects, human performance research requirements, and the HSRE research agenda. The DHS HSI process as it applies to analysis of human performance requirements and human-centered design of user interfaces has been described by Novak et al (2010), Malone et al (2010), Lockett-Reynolds et al (2009), Wilson et al (2009a), Wilson et al (2009b), Malone et al (2008), and Malone et al (2004).

For the purposes of the Human Performance Model, the scope of DHS projects has been categorized into a number of domains characterized by a functional grouping of missions, systems, environments and end-users. The HPM is structured around the following domains which are aligned with various DHS mission areas:

| | | | |
|---|---|---|---|
| 1. | Aviation Security | 6. | Cyber-Security |
| 2. | Mass Transit Security | 7. | Public Security |
| 3. | Border Security | 8. | Physical Security |
| 4. | Maritime Security | 9. | Emergency Response |
| 5. | Infrastructure Security | 10. | Immigration Enforcement |

The relationships of the HPM with other activities of the HSRE program are depicted in Figure 1. The top portion of this Figure is concerned with the human performance research activities of HSRE. The lower portion is directed at applying HSI in the acquisition of systems and development of technology. Central to these is the HPM.

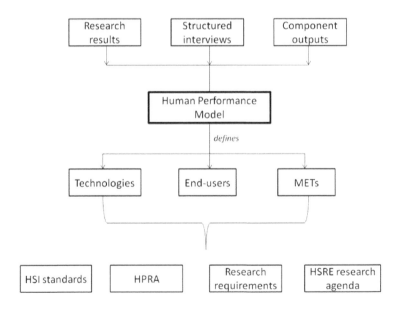

Figure 1    Relationship of the Human Performance Model (HPM) with research and acquisition activities within the Human Systems Research and Engineering (HSRE) program in the DHS S&T Human Factors/Behavioral Sciences Division.

# 3    HUMAN PERFORMANCE MODEL OBJECTIVES

The objectives of the HPM are as follows:
*   Provide a database of (METs) for each DHS domain;
*   Provide guidance for identification of the end-users within specific DHS domains;
*   Provide templates for defining the human-allocated functionality of technology and systems in terms of METs which will be included in HSI inputs to the Concept of Operations (CONOPs) for the emerging system or technology;
*   Provide support in the definition of performance requirements, standards of performance and conditions for individual METs;
*   Provide guidance on user interface requirements associated with METs; and
*   Provide guidance on identification of research gaps and requirements for each MET.

Essentially, the model will allow for the characterization of human performance for all DHS end-users when performing mission critical tasks across all systems, technologies, and environments.

# 4    OPERATION OF THE MODEL

The HPM will be used within HSRE to initially locate a system of interest within the structure of DHS domains, and to describe the expected end-users of the system or technology. The next step will be to list the METs associated with end-user performance with the systems/technologies. The expected performance requirements, performance standards and conditions under which METs are performed will then be identified as will user interface requirements for performance of specific METs for the system or technology. Finally, relevant research human performance requirements will be developed for specific METs.

The information developed by the HPM will (a) provide inputs to the human performance requirements analysis implemented in the application of HSI to system acquisition and technology development; (b) provide inputs to the CONOPs for the emerging system or technology; (c) drive the roadmap supporting the HSRE research agenda; (d) support user centered design of user interfaces; (e) serve as a source of human performance metrics for conducting HSI test and evaluation; (f) serve as the basis for end-user training requirements identification; and (g) support the organization of future DHS HSI standards.

# 5    DEVELOPING HUMAN PERFORMANCE REQUIREMENTS IN THE HPM

The human performance requirements begin with the identification of a given threat for a particular mission area within the domain, and the class of systems or technologies existing in the mission area or being proposed for the mission area. The end-user is identified as the human operator, controller, maintainer, manager, first responder, first receiver, and appropriate members of the public (passengers, travelers, bystanders). The METs are identified for the end-user and system/technology combination as the top level human activities judged to be critical for a specified mission in terms of required performance on the part of the end user(s) for the system/technology. METs contain three critical components:

1) functional activity, in the context of the mission scenario;
2) conditions under which the functional activity can be expected to be required (the range of scenarios that apply);
3) criteria for successful performance of the functional activity in the selected mission scenario, and the measures to be used to assess performance of the functional activity.

The METs are described at two levels of abstraction, the MET and the MET sub-task. The performance requirements for MET sub-tasks are identified which include (a) information required from the technology and the human to enable performance of the MET; (b) capabilities required by the system to execute the MET; (c) decision requirements for MET performance; (d) time required to enable the system to perform the MET; and (e) skills, knowledge, and tools required of the end-user. MET performance standards are identified to include the criteria for successful performance of the functional activity (MET or MET sub-task) in the selected conditions (scenarios), and the measures to be used to assess performance of the functional activity. MET Performance Conditions are identified, which include environmental, operational, personal, and tactical conditions under which MET sub-tasks will be performed, and which form the basis for selected scenarios. Requirements for the user interfaces (including human computer interfaces, controls and displays, alarms and alerts, procedures, and workspace) are identified for conducting METs and MET sub-tasks. Finally, manpower, personnel and training requirements associated with performance of METs and MET sub-tasks are identified. As MET and MET sub-tasks are identified, it may be the case that one MET/MET sub-task may apply to more than one domain area. This will ultimately allow for the characterization of domain specific tasks versus tasks that cut across various domain areas.

Once MET and MET sub-tasks are identified, guidance will be developed and contained within the HPM for selection of METs and MET sub-tasks. This guidance consists of standard activities that apply across domains, and are specific for DHS systems and technology. It is expected that the HSI analyst will use the MET guidance to identify generic activities that apply to the system or technology under development. The HPM guidance will include generic human performance requirements and research requirements for each of the standard METs and MET sub-tasks. A selection of top-level METs and MET sub-tasks are as follows in Table 1.

Table 1  A selection of METs and MET Sub-tasks

| MET | MET Sub-task |
|---|---|
| Receive and Use Information | Obtain image |
| | Interpret Image |
| | Detect visual target |
| | Detect auditory target |
| | Detect/recognize anomalies |
| | Recognize patterns in images |
| | Maintain intelligence on adversary behavioral patterns |
| | Obtain and use intelligence |
| | Fuse and integrate information from many sources |
| | React to augmented alarms |
| | Use adaptive interfaces |
| | Use brain-based sensors |
| | Use wearable electronics/displays |
| Provide Command and Control | Control robotic and tele-operated systems |
| | Provide process control |
| | Use multi-modal workstations |
| | Use advanced displays and graphic user interfaces |
| | Use a common operating picture |
| | Conduct command and control |
| Conduct Surveillance | Maintain vigilance |
| | Examine for presence of makers |
| | Conduct search and inspection |
| | Monitor status of sensors |
| | Control sensors |
| Interact with Automation | Supervise automation |
| | Monitor automation |
| | Intervene into automated processes |
| | Use intelligent tutors |
| | Use intelligent aids |
| | Control robot/telerobots |

# 6 IDENTIFYING RESEARCH REQUIREMENTS IN THE HPM

The identification of research requirements within the HPM begins with a determination of the capability gaps associated with specific METs. A capability gap is defined as a deficiency in a required capability for which research is required to characterize the gap and/or fill the gap. The criticality of the gap is then established as the importance of the research for DHS mission success. Research issues and variables include the issues or problems associated with the gap, for the specific MET, and research variables needed to conduct the research. User impacts include the expected impact of the research on HSI domains, specifically manpower, personnel, training, human factors engineering, habitability, health and safety, and survivability. The required research is then determined to be generic or unique, in terms of the extent to which the gap, and research to fill the gap, is unique to a domain or generic across several domains. Finally the research will be classified according to the extent to which it is directed at: 1) Enhancement (improving end-user performance); 2) Sustainment (sustaining end-user performance through knowledge and skill retention); and 3) Acquisition (knowledge and skill acquisition).

# 7 STATUS OF THE HPM DEVELOPMENT

The full HPM for the Aviation Security domain has been developed, with the assistance of the Transportation Security Administration (TSA). . Efforts are underway within the HSRE program to develop the human performance models for the remaining domains. For these domains, the HSRE program will develop a database of METs and MET sub-tasks, and a set of generic human performance requirements associated with METs and MET sub-tasks as they apply to each class of end users. These requirements will support the conduct of the human performance requirements analysis (HPRA) process for emerging system acquisitions and technology developments. These Aviation Security METs will be validated, and the requirements associated with METs for individual classes of end users will be developed on the basis of:
- interviews with personnel of each end user class within each domain;
- reviews of requirements documents for already developed and developing systems within the domain; and
- results of HPRAs already conducted for systems within this domain.

# 8 CONCLUSIONS

Characterizing the nature of performance of the DHS end-user presents a unique challenge. Unlike that of many other federal agencies, the types of end-users within DHS are especially diverse. As mentioned previously, the DHS end-user represents a wide range of responsibilities with regard to the various domain areas in support of DHS missions (e.g., TSOs, CBP Agents, Intelligence Analysts, First

Responders, Coast Guardsmen, etc.). The basic utility of the HPM is that it will provide a systematic means of classifying the various types of end-users and their respective performance requirements across domains. Specifically, this will facilitate a more streamlined process by providing a common set of tasks, scenarios, conditions, performance standards, and requirements. Ultimately, this effort will enable a more effective and efficient means for identifying both human performance and research requirements to be leveraged across the Department.

## REFERENCES

Department of Defense, (December 2008). Operation of the Defense Acquisition System. Department of Defense Instruction Number 5000.02.

Lockett-Reynolds, J., Malone, T.B., Wilson, D.P., and Muse-Duma, K. (2009) "TDRA and Modeling and Simulation", Proceedings of the ASNE 4th Human Systems Integration Symposium, Annapolis, MD.

Malone, T.B, Lockett-Reynolds, J., Wilson, D.P., Novak, B.B., and Avery, L.W., (2010) "Human Performance in Maritime Security", Proceedings of the International Conference on Human Performance at Sea, Glasgow, Scotland.

Malone, T.B., Pharmer, J., Lockett-Reynolds, J., and Muse-Duma, K., (2008), "Integrated HSI", Proceedings of the 52nd Annual Meeting of the Human Factors and Ergonomics Society.

Malone, T.B., (2004) Template for applying the HSI Top Down Requirements Analysis (TDRA) in a ship or ship system acquisition program, prepared for NAVSEA 03.

Novak, B.B., Malone, T.B., Lockett-Reynolds, J., Wilson, D. P. , (2010) "Human Systems Integration in the US Coast Guard's Offshore Patrol Cutter (OPC) Acquisition Program" Proceedings of the International Conference on Human Performance at Sea, Glasgow, Scotland.

Wilson, D.P., Malone, T.B., Lockett-Reynolds, J., and Wilson, E.L., (2009a) "A Vision for Human Systems Integration in the U.S. Department of Homeland Security", Proceedings of the 53rd Annual Meeting of the Human Factors and Ergonomics Society.

Wilson, D.P., Lockett-Reynolds, J., Malone, T.B., and Wilson, E.L., (2009b) "Small Unit Performance in DHS", Paper presented and published at the Interservice/Industry Training, Simulation and Education Conference (I/ITSEC).Burke, L. and J. Maidens 2004. *Reefs at risk in the Caribbean.* Washington D.C.: World Resources Institute.

# CHAPTER 64

# Determinants of Integration Strategies in HRM after Mergers & Acquisitions in Pharmaceutical Sector

*Anna Stankiewicz-Mróz, Jarosław P. Lendzion*

Technical University of
Łódź, Poland
anna.stankiewicz-mroz@p.lodz.pl
jledzion@p.lodz.pl

## ABSTRACT

The paper presents the research results on 5 M&A transactions in the companies of a pharmaceutical sector in the years 2007-2009. In the research there was a conviction that the evaluation of the integration effects should be done at least 2-3 years after the transaction. The diagnosis of the range and of the pace of integration in the personnel field was the research aim. There was a hypothesis that M&A processes influenced the "modernization" of a personnel function which is proven by the introduction of modern methods, techniques and personnel tools.

**Keywords**: Post merger integration, Human Resources Management, pharmaceutical sector

## 1   POST MERGER INTEGRATION

A post merger integration is "a continuous, interactive process, in which people from a transferee and a transferor company learn how to work together and how to cooperate in the transfer of strategic capacities (Haspeslagh, P.C., Jemison, 1991). It is treated as a lever important to generate values in mergers and acquisitions processes and as the most important determinant in the implementation of acquisition synergy (Habeck, M.M. and others 2000). Simultaneously, the integration is one of the most hazardous stages of the whole transaction. A. Pablo notices that it is connected with the implementation of changes in all the functional areas, in the organizational structures and in the cultures of the merging companies in order to facilitate their consolidation into one functioning totality (A. Pablo, 1994). The adopted integration strategy defines the guidelines for a company's activities in a new post merger structure. A transaction aim, a kind of capital engagement and a source of the value in a transferee company, social and legal commitments, culture differences and sizes of merging companies are the factors which determine the choice of the integration strategy.

The adopted strategy defines the priorities for the integration process. The definition of the range and depth of the integration process (the depth of changes) maintaining the proper proportions between interdependence and independence remains a key issue. The integration range and depth depend first of all on the aims defined while taking a decision on a merger. It is important to plan this process carefully and to define a procedure to control its implementation (O'Rourke, 1989). S. Sudarsanam, though, pays attention to the fact that the integration process, in order to keep up with the changes, should be characterized by flexibility, visible also in the plans (Sudarsanam, 1998). Three leading integration strategy chosen by the majority of the organizations can be pointed out (M. Stuss, 2008):

1. *Stand alone or preservation approach.* This strategy is applied in conglomerate mergers, where there is a need for the autonomy of single units or a small need of interdependence.
2. *Strategic or partial integration.* This strategy is suitable for vertical mergers where there is a need for the interdependence and keeping limits between the units which merge.
3. *Complete integration or absorbing approach.* This strategy is applied in horizontal mergers when there is a need for the units' consolidations in all the areas of activities.

Among the factors which exert influence on the choice of the strategy, the attention is paid to a company's market situation (whether a company's position on the market is endangered or not) and to the differences between the partners which can refer e.g. to the sizes of the merging companies. This factor can imply serious difficulties for the integration process which result from the fact that the transactions partners have the different organizational forms, structures and procedures. Chakrabarti, Hauschildt and Sueverkruep underline that a merger of small companies with the big ones causes many serious organizational and technical problems (Chakrabarti and others 1994). Gerpott underlines on the basis of the conducted research that companies with similar sizes were more likely to succeed (Gerpott, 1995). The integration of the different human resources management systems which derive from the different national and organizational conditionings and which create uncertainty is also problematic (Aguilera, Denckner, 2007). Time factor is particularly important from the point of view of a M&A success (Stankiewicz-Mróz, 2010)In the literature related to this issue, there is no simple opinion on an optimal pace of integration. On one hand it is sometimes underlined that shortly after a merger the atmosphere to implement changes is the most friendly and it is advisable to use this moment (Von Schewe, Gerds, 2001). On the other hand it must be remembered that a quick integration is accompanied by a threat that the necessary integration activities will not be implemented either totally or partially as the result of the wrong decisions or a loss of the integration control.

## 2   INTEGRATION IN THE FIELD OF PERSONNEL FUNCTION

Mergers  and acquisitions definitely imply a necessity to implement  changes in the field of human resources management. Their range is dependent on the adoption of one of  four approaches to the integration process: *portfolio, blending, new creation* and *assimilation* (Schuler and others, 2004). The portfolio concept takes for granted a big range of managers' freedom in two transaction partners. In the new post merger structure, the separate organizational cultures are maintained. In the blending approach a new organization chooses the best elements from the cultures of the transaction partners and the members of both organizations learn the procedures and culture of the partner. This concept is used by the organization which operate in the same field. The new creation concept is implemented when partners create a new company, with a new organizational culture and with a new name. The concept of assimilation takes for granted that one of the partners is dominant and this partner imposes their organizational culture. This process can take place in two directions (when a  transferee company adjusts to a transferor company it is *"absorption"* whereas when a transferor company adjusts to a transferee company it is defined as  *a reverse merger*).

## 3   PHARMACEUTICAL INDUSTRY IN POLAND

Poland is the sixth largest pharmaceutical market in Europe in terms of the value of the sold drugs which achieved the threshold of 20.1 billion zlotys in producer net prices (an average yearly change of the value of the pharmaceutical industry between 2009 and 2010 was +6.4%) (IMS Health, 2011). The Polish pharmaceutical market is characterized by a considerable participation of generic drugs (in this respect Poland stands out from other European countries). In the qualitative approach , the generic drugs constitute 65% of the market and in the quantitative approach even up to 86.5%. For comparison, the generic drugs in Germany constitute 22.7% of the market as far as the value is concerned and 41% as far as the quantity is concerned, in the Czech Republic  respectively 32% and 55%. In Poland there are more than 450 pharmaceutical companies out of which 62 are the innovative ones. The pharmaceutical companies employed in 2010 about 31.1 thousand people which is about 0.23% of the total number of the employed in Poland. The innovative pharmaceutical companies provide all together about 11 thousand work places which is about 35% of the total number of the employees in the pharmaceutical industry  in Poland. The ten biggest companies stand for 50% of the value of pharmaceutical industry and the first thirty companies for almost 83% of the market. The remaining 17% of the market is divided between 415 companies (www.egospodarka.pl 2011). In the structure presented above, a potential for consolidation treated as a tool for the development and growth can be noticed.

## 4 THE PROCESS OF THE PERSONNEL FUNCTION INTEGRATION IN THE RESEARCHED COMPANIES (THE PERSPECTIVE OF A TRANSFEROR COMPANY)

The research aim was to define the pace and the range of the integration in the field of human resources management (HRM). It was conducted in five deliberately chosen companies of the pharmaceutical sector which were taken over in the years 2008-2009. The research was conducted among the drug producers in which the transactions had been done 2-3 years prior to the research[1]. It is assumed that this time perspective allows to evaluate the effects of the transaction which was carried out (Frąckowiak W., (edited) 2009). The structure of the sample was as follows:

1. A big international company took over a private national company (of a middle size) buying 100% of its shares. The year of the transaction: 2008. Organizational form after a merger: a concern.
2. A small private company (Polish-Slovenian) takes over a middle-sized employee-owned company, buying 100% of its shares. The transaction was carried out in 2008. The structure after a merger: a company.
3. A big international organization takes over a big company with Slovenian capital. The transaction was done in 2009. The organizational form after a merger: a concern.
4. A big private company takes over a joint-stock company wholly owned by the Treasury buying 85% of their shares (15% of the shares remains in the hands of the employees). The year of the transaction: 2009. The structure after a merger: a concern.
5. A big international company takes over in 2008 a big international company. Organizational form after a merger: a concern.

In the research there was a conviction that the level of compliance (similarity) is a factor which influences the range and the depth of post transaction integration. Five areas were analyzed: strategy- capital resources- technology, organizational culture and a management style corresponding to it.

---

[1] The research was conducted using a few research techniques: free-form interview with the consultants of the consulting companies which deal with M&A processes, in-depth interviews with presidents/ directors of the transferor companies, surveys with the managers of personnel departments, panel interviews with the participation of presidents/ directors of companies, directors/ financial managers, organizational managers, managers (specialists) of personnel units.

**Table 1** The level of compliance of the transaction partners and the transaction motives versus the range of integration

| Transaction motive | Level of compliance at the beginning of the transaction | Integration plan | Range of integration |
|---|---|---|---|
| **Transaction 1** | | | |
| Business expansion<br><br>Increase of the market position | Low level in the area: strategy- capital resources- technology. Medium compliance in the field of organizational cultures and management styles. | Very general, prepared after a transaction | High only in the field of marketing, sales and budgeting. |
| **Transaction 2** | | | |
| Increase of the value | Low level of compliance in all the areas | Very general, prepared after a transaction | Integration of organizational structures and brands. High level of integration in the field of marketing and sales, distribution and finances |
| **Transaction 3** | | | |
| Business expansion<br><br>Increase of the market position | Medium level of compliance in all the areas | Very general, prepared after a transaction | Integration of structure, technology, brands, IT systems, financial and accounting procedures, supply, R&D |
| **Transaction 4** | | | |
| Diversification of export activities<br><br>Development of R&D unit<br><br>Usage of the complementary effect of the drugs portfolio | Low level of compliance in all the areas | General, prepared before a transaction.<br><br>Made detailed after a transaction. | Integration of IT systems, financial and accounting procedures. High level of integration in the field of marketing and sales. |
| **Transaction 5** | | | |
| Increase of market shares<br><br>Increase of the value | Medium | Quite general, prepared after a transaction | Integration in marketing and sales, IT systems, financial and accounting procedures, integration of technology, brands, supply, R&D |

It results from the data in the chart that in the companies in which the research was conducted the level of the compliance in the moment of entering the transaction (analyzed in the areas: strategy- capital resources- technology- organizational culture- management style) was in two cases low and in the other cases was "medium". The low level of partners' compliance was declared in case of the transaction done by the company with national capital which took over a big, joint-stock company wholly owned by the Treasury and in case of the company with Polish-Slovenian capital which took over an employee-owned company. In one of the researched companies a moderate level of compliance was underlined in the area of culture- management style and low level of compliance in the other areas. The integration priorities are defined first of all by the transaction motives. Since in each case the motive of the transaction was to "increase market shares" and "increase values", marketing and sales were treated as the key areas of integration and in these areas the integration was the quickest and its range was the biggest. In each of the analyzed cases the sales departments were moved to Warsaw and functioned in the headquarters. Generally, in each case the emphasis was put on the integration of tough factors. The integration in the field of personnel function was relatively late.

**Table 2 Changes in the field of personnel function and HR areas and the pace of post transaction integration**

| Changes in the field of personnel function after M&A | HR areas which were integrated and the pace of integration |
|---|---|
| **Transaction 1** | |
| 1. Changes in the regulations of remuneration. <br> 2. Changes of the names of posts, adequately to the tasks performed. Definition of the range of duties. <br> 3. Changes in the system of awarding bonuses. Introduction of new criteria for awarding bonuses. <br> 4. Modification of training policy. <br> 5. Introduction of Ethic Code. <br> 6. Implementation of flexible working time for some groups of positions. <br> 7. Changes in the organization of HR activities. | 1. Staff evaluation (after 12 months of functioning in a post merger structure) <br> 2. Training policy of an organization (after 12 months) |
| **Transaction 2** | |
| 1. Changes in work regulations. <br> 2. Changes in remuneration regulations <br> 3. Changes in the description of positions, creation of profiles connected with a company's key values. <br> 4. Implementation of periodic performance assessment system <br> 5. Changes in the practices of choosing the personnel (external choice, complex procedures for sales departments with full AC sessions) | 1. The choice of personnel for sales positions (after 6 months of functioning in post merger structures). <br> 2. Periodic performance assessment system for all employees (after 18 months). |

| Transaction 3 | |
|---|---|
| 1. Creation of competencies matrix<br>2. Reconstruction of the remuneration system (corporation benchmarking rule).<br>3. Changes in the system of awarding bonuses<br>4. Periodic performance assessment for all the employees<br>5. Creation of career paths for all the employees<br>6. Research on employees' satisfaction<br>7. Implementation of a program for *High Potential* group (a global program for top managers, separate program for managers of lower level and specialists).<br>8. Modifications in the field of training policy<br>9. Introduction of "employment diversification" for production workers.<br>10. Introduction of the measurers of HRM effectiveness. | 1. Remuneration system (after 24 months of functioning in post-merger structure)<br>2. Research on employees' satisfaction.<br>3. Cascading competencies (after 24 months).<br>4. Periodic performance assessment for all the employees (after 24 months)<br>5. *High potential* program for top managers (a global program implemented in the company under research after 12 months) for managers of lower level and for specialists after 24 months<br>6. Changes in the training policy (modifications for the purpose of implementing goals).<br>7. ROI, analysis of absenteeism rate, level of implementation of corrective actions, share of external recruitment, time of attracting employees (after 28 months) |
| **Transaction 4** | |
| 1. Introduction of position descriptions and competencies profiles | Lack |
| 2. Job evaluation | Lack |
| 3. Changes in qualification scales and pay scales | Lack |
| 4. Changes in remuneration system | Lack |
| 5. Changes in employment practices (no prolongation of temporary contracts with employees). | Lack |
| 6. Implementation of personnel changes using the services of external companies | Transferee company used outsourcing of personnel activities (after 6 months) |
| **Transaction 5** | |
| 1. Preparation of competencies profiles<br>2. Choice for the position of medical sales representatives is outsourced<br>3. Modifications of the training policy<br>4. Cascading competencies<br>5. Usage of Assessment Center and Development Center as evaluation tools and personnel development<br>6. The usage in a company of | 1. Training policy (after 12 months of functioning in a post merger structure).<br>2. Procedures for the choice of personnel for sales and managerial positions (after 18 months)<br>3. Research on employees' engagement (after 24 months)<br>4. Development center as a method |

| | | |
|---|---|---|
| "Management by objectives" technique results in the intensification of evaluation functions<br>7. Introduction of the measurers of HRM effectiveness<br>8. Change of the role of personnel department (is a partner for business) | | of evaluation and personnel development (after 18 months)<br>5  Introduction of the measurers of HRM effectiveness |

The summary presented in table 2 shows that M&A process exerted influence on the changes in human resources management. However, just in one case these changes were implemented relatively quickly (the first one already three months after the functioning in a new post merger structure). In other companies the integration activities in the field of personnel function were spread in time (in two cases with one year transition period). **In all the researched companies it was underlined that the pharmaceutical industry is characterized by a long-term perspective and that is why M&A processes are not treated as tools for the short-term implementation of aims.**

## 5  CONCLUSIONS

An integration phase is the most risky stage in M&A processes. The integration will go more smoothly if the partners of transaction have common features. In the companies which participated in the research, although they belonged to the same branch, the level of compliance in the area: strategy- capital resources- technology- organizational culture- management style was in two cases low and in the other moderate. **The research showed that in the pharmaceutical companies, although the motives to carry out the transactions are similar, there is no one leading integration strategy. Generally the accent is put on internal integration in hardware areas. The integration of the activities in the field of  human resources management was generally the last stage.** The range of changes implemented in this area is big and it provides new quality for personnel activities but only some of them are common for the merged organizations.

## REFERENCES

Aguilera R.V., Dencker J.C 2007., The role of Human Resource  Management in cross-border mergers and acquisitions, [in:]M.E.  Mendenhall, G. R.oddou, G.H. Stahl, Readings and Cases in International Human Resource Management, Reutledge, London  and New York.

Branża farmaceutyczna  a  polska  gospodarka,  2011.Accessed  September  13 www.egospodarka.pl.

Chakrabarti, Hauschildt , Sueverkruep,1994. Does it pay to acquire technological firm [in]R&D Management, nr 24.

Frąckowiak W. (edited)2009. Fuzje i przejęcia, PWE, Warszawa: 465.

Gerpott, Successful integration of R&D functions after acquisition: an exploratory empirical study,1995. [in] R&D Management nr 25.

Habeck, M.M. ,Kröger, F. , Träum, M.R. ,2000. After the Merger – Seven Rules for Successful Post-Merger Integration. Pearson Education Limited: Edinburgh Gate.

Haspeslagh, P.C., Jemison, D. 1991.Managing acquisitions: Creating value through corporate renewal, New York; The Free Press: 106.

Health, Dane Narodowe, 2001 Mai.

O'Rourke, Post merger integration [in]; The Arthur Young Management Guide to Mergers and Acquisitions, 1989.(edited) .R. Bibler, New York, John Wiley and Sons.

Pablo A., Determinants of Acquisitions Level: a Decision making Perspective, 1994."Academy of Management Journal" nr 4:.803- 836.

Sudarsanam S.,Fuzje i przejęcia 1998, WIG Press.

Schuler R.S., Jackson S.E., Lou Y.2004. Managing human resources in cross- border alliances, Routledge, London : 90.

Stankiewicz-Mróz A., Post merger situation. Analysis of the process of concentration in a knowledge –based organization,[w]:D.Lewicka (edited),2010.Organisation Management.Competitiveness, Social Responsibility, Human Capital, AGH University of Science and Technology Press.

Stuss M., Zarządzanie nową organizacją po fuzji lub przejęciu, [in]:Fuzje, przejęcia..Wybrane aspekty integracji, A.Herdan (edited), 2008, Wydawnictwo Uniwersytetu Jagiellońskiego, Kraków :52.

Von Schewe G., Gerds J.,2001, Erfolgsfaktoren von Post Merger Integration; Ergebnisse einer pfandanalytischen Untersuchung, [in] Zeitschrift für Betriebswirtschaft", nr 3.

# CHAPTER 65

# Exploring Resilient Team Processes in Control Room Teams of a Nuclear Power Plant

*Cornelia Kleindienst, Jonas Brüngger and Frank Ritz*

University of Applied Sciences and Arts Northwestern Switzerland
School of Applied Psychology
Olten, Switzerland
Cornelia.kleindienst@fhnw.ch

## ABSTRACT

The notion (or concept) of "resilient team processes" is yet not well established. An important aim with this explorative research is therefore the discovery and description of relevant characteristics of team processes in the light of resilience. One of the strengths of naturalistic observations is that they support such a discovery process. Therefore the aim of this research project in the control room of a nuclear power plant is the detection of specific team processes within the operating team that serve for safety oriented coping in abnormal situations by using observational methodology. The analysis is based on videorecordings of a simulator training based on scenarios of critical and abnormal situations. The exploration of these resilient team processes will serve as a basis for the demand-oriented construction of simulator training.

As a starting point for this project served different models of team performance in high-risk domains (air traffic control: Malakis, Kontogiannis and Kirwan (2010); medical domain: e. g. Manser et al., 2009; nuclear power plants: O'Connor, 2008). These models gave input for developing the first structure ("sketch") of a taxonomy for naturalistic observations of resilient team processes in the control room team. Furthermore we relied on existing methods for behavioral observation of interaction processes (Kolbe et al., 2011).

The results of this research and development project will contribute to an observational method for the analysis and training of resilient team processes.

**Keywords**: resilience, team processes, adaptive coordination, critical situation, nuclear power plant, control room team, observation method

# 1    RESILIENT TEAM PROCESSES

## 1.1 Resilience

In high-risk organizations like nuclear power plants, safety is a basic organizational goal that have to be achieved by the design of the organizational structure and by the working processes. For achieving safety it is not sufficient to combine reliable components and processes (Fahlbruch et al., 2008), but to continually manage the process of constructing safety. This is also the perspective of Weick & Sutcliffe (2001) when they speak of safety as a dynamic non-event.

The concept of resilience is stated to conceptualize the properties and processes of dynamic "safety producing" systems (Hollnagel et al., 2006). Resilience is defined by Hollnagel (2011) as follows:"The intrinsic ability of a system to adjust its functioning prior to, during, or following changes and disturbances, so that it can sustain required operations under both expected and unexpected conditions." Hollnagel (2006) highlights especially the aspect of system adaptation to situational variation: "By resilience we understand capacity of organizational systems to function adequately under environment variations." (see Vidal et al., 2009). Resilience ist the ability to adapt or absorb disturbance, disruption and change especially to disruptions that fall outside of the set of disturbances the system is designed to handle (Woods et al., 2007). Emergencies and abnormal situations represent critical situations close to the margins of safe operation that challenge the operators practices. In the middle of resilient team processes there should be the focus how teams recognize and adapt to variations, changes and surprises by adequate problem-solving and decision-making.

For looking at resilient team-processes the following questions seem important: How can the properties of a resilient system be operationalized or translated in observable behavior of teams and team members? What properties of team processes are critical for the resilience of a working-system? Which team processes are functional for adaptation? Of what kind have team processes to be that we can call them resilient team-processes?

## 1.2  Models of teamperformance in high-risk industries

The starting points of this explorative project looking for resilient team processes are different models of team performance developed in high-risk industries like air traffic control and the medical domain. These models are stating relevant team processes for team performance in high-risk industries and it also have been developed taxonomies for observational methods stemming from these models.

From the air traffic domain there is the model and taxonomy of Malakis, Kontogiannis and Kirwan (2010) which was developed for investigating team performance in abnormal situations. The taxonomy for describing team processes in aircontroller teams was developed on the basis of three earlier frameworks of non specific skills for teamwork (Flin et al. 2003; Fletcher et al., 2004; Salas et al., 2005). The following relevant concepts for team-processes are included: shared mental models, communication, cooperation, coordination.

From the medical domain we refer especially to research done by the work group of Gudela Grote at the ETH Zürich. Their work on team performance is related to different aspects of team processes, like the influence of standardization as a task characteristic (Grote et al., 2003; Zala-Mezö et al., 2009), the importance of adaptive coordination in managing abnormal situations (Burtscher et al., 2010) and the development of observational methods (Manser et al., 2008; Kolbe et al., 2009). Referring on these methods and findings is highly valuable for exploring resilient team processes in the operating team of a nuclear power plant. In the domain of the control room of a nuclear power plant O´Connor et al. (2008) developed taxonomy for describing team processes in the operating team.

However, the composition and culture of air traffic teams or medical teams in the operation theatre are different from those of control room teams in nuclear power plants, especially regarding the composition of the team, the number of team members and their roles within the team, and furthermore in the coordinating function of the shift supervisor.

In the following section we concentrate on the relevant team processes in adapting and managing critical and abnormal situations.

## 1.3 Managing abnormal situations in the control room

One fundamental factor for successful team work in high-risk environments like nuclear power plants is the right balance between standardization on the one hand and flexibility and openness to changes in unexpected situations on the other (Grote et al., 2010). For exploring resilient team-processes the adaptation of the team and of team processes to abnormal not standardized situations or emergencies seems crucial. The breaking point between the highly regulated work-process structured by standard operating procedures and the necessity of problem solving in abnormal situations seem especially interesting for assessing adaptive capability in teams.

In these kinds of situations there is thought that coordination activities have to be changed to face the challenges of new and unstructured situational demands. Burtscher et al. (2010) defined adaptation or adaptive coordination as "a team´s ability to change its coordination activities in response to changing situational demands, such as the occurrence of unexpected events and varying task charakteristics (e. g. level of task load, degree of standardization, time pressure). Burke et al. (2006) described as one mode of adaptation the adjusting process, i.e. changes in coordination mechanisms, decision making, and communication patterns in response to unexpected events. For example Zala-Mezö et al. (2009) have shown that teams in healthcare adapt to changing situational demands. The relationship between adaptation and team-peformance has to be investigated more deeply.

For the exploration of team processes in nuclear power plants the context of control room teams has to be taken in account. For example Carvalho et al. (2005) described the normative process of problem-solving and decision-making in the control room team as follows: information gathering from all available sources, appropriate resource utilization, valid interpretation of information, and valid selection based on this information, review of potential consequences and probabilies, development of a plan of action, quick interventions.

The composition of the control room team is dominated by the shift supervisor, who has a leading function. Vidal et al. (2009) described that during problem-solving situations, the strictly centralized control mode is modified and all team members submit hypothesis and participate in a collective decision-making process supplying information. However, all the information ist submitted to the shift supervisor who has the ultimate authority to make decisions. Sharing information is the main way team members use to cooperate in control room environments. The coordination of this information sharing process by the shift supervisor seems to be essential for the adaptation of the team.

Vidal et al. (2009) highlighted the special situation of the complex expert organization "nuclear power plant". They look at this organization as a distributed cognition environment, where nobody has complete information about everything that is happening at a particular moment, so the operators need to share their cognition to be able to operate the plant in a more resilient way. They make the distinction between informational redundancies and human redundancies. These information sharing situations are materialized in actual work interactions, both written and spoken. Especially in this environment the explicit coordination of information sharing seems essential for good team-performance in problem-solving and decision-making.

With regard to resilient team-processes we found the following questions of interest (see also Kolbe et al., 2011):

- How do group members communicate in order to coordinate information exchange and decision-making in order to adapt to critical or abnormal situations?
- How can human groups be effectively coordinated during their decision process in order to minimise process losses and to optimise decision quality simultaneously?

## 2 OBSERVING COLLECTIVE WORK IN THE CONTROL ROOM

### 2.1 Description of the context

The aim of this explorative research project in the control room of a nuclear power plant is the detection of specific team processes within the operating team that serve for safety oriented coping in abnormal situations ("safety interaction patterns", Ritz & Rack, 2009). The exploration of these team processes (esp.

interaction processes for problem-solving and decision-making) in control room teams will contribute to the context-oriented description of resilient strategies in practice, and as a basis of the demand-oriented construction of concepts for simulator training.

The explorative study is conducted together with a Swiss nuclear power plant. The control room team consists normally of two operators (reactor operator and secondary circuit operator) and the responsible shift supervisor who is supported by a second shift supervisor ("foreman"). Additionally there is a shift engineer who supported the control room team in special situations or emergencies. There exists standard operation procedures (SOP) as well as emergent operation procedures (EOP). The work process normally is managed and standardized by these procedures. The functional role of shift supervisors comprises different roles: to follow procedures by checking the actions performed by the reactor Operator and secondary circuit operator; to think about the process in conjunction with the foreman and the shift engineer; to authorize the use of different auxiliary procedures, especially in case of failure of some equipment; to communicate with people around the plant to authorize/suspend interventions (Carvalho et al., 2005).

Simulator training of critical situations and emergencies consists of three scenarios, that varied in their potential of crisis. There was one very critical scenario with high time-pressure, the other two scenarios also included non-routine and abnormal situations, but were not as critical in consequences and timepressure as the first one. The simulator shows exactly the situation in the realistic control room.

The use of simulators is supported by the ecological approach (Proctor and Vu, 2010) and it is believed that a person's contribution/behavior can be understood best in the context of the emergent properties that arise from the interaction with an environment. It is believed that the environmental features strongly influence a person and his or her actions. Therefore, the more realistic a simulated environment is, the more natural the behaviour of persons acting in that environment will be. The research of teams in high hazard industries is enabled by highly sophisticated simulators, which are used to simulate critical situations in which team members can be trained to safely interact whith each other in a realistic environment. Without simulators the investigation of behaviour in critical situations in high risk industries would be impossible because critical situations happen too rarely and an intentional provocation would, for obvious reasons, not be feasible. Modern simulators let people act as if they were in a real world situation not just in an artificial simulation. One reason for using simulator based research methods in high hazard industries is obvious. Through simulation it becomes possible to confront operators with situations that in non-simulation circumstances would be impossible to create without danger to human integrity. In addition to this obvious advantage there are also others. Simulators often offer observers possibilities like recording or different angles of view to observe what is happening in a critical situation that are not possible in a non-simulated environment.

We recorded the simulator trainings and used this material for iterative development of a first version of an observation method for analysing resilient team-processes in the control room team.

## 2.2 Observation Method

One of the primary strengths of naturalistic observations is that they support a discovery process. They serve to draw attention to significant phenomena and suggest new ideas (Roth & Patterson, 2005). Field observations allow a realistic view of the full complexity of the work environment. Observation methods enable researchers to reveal and document cognitive and collaborative demands imposed by a domain and the strategies that practitioners have developed in response to those demands.

Scientific observation methods (Greve & Wentura, 1997) are characterized by (i) intention including a plan to achieve certain objectives, (ii) selection of certain aspects while neglecting others, (iii) results orientated and systematic analysis. In complex systems, state of the art research in team interaction in high hazard industries often focuses on observation methods (e.g. Fletcher et al. 2003, Künzle et al., 2010, Stout et al., 1997). Advantages of observation methods are primarily the possibilities of a direct measure of actual behavior, no subjective intentions and elimination of reporting bias. Additionally, certain data can only be collected through observation (personal interaction, gestures, facial expression). Disadvantages of observation as technique for data collection are in general its time and cost intensity. In addition, the main disadvantage of observation is the selective perception of the researchers (Greve & Wentura, 1997) and the inability in capturing cognition and perception of the observed subjects. Even though observation as a research method has many advantages, its disadvantages cannot be easily dismissed. Scientists often lack the knowledge of subject matter experts (SME) – experts of a certain domain - to completely understand what they observe, as their expertise is in interpreting patterns and behavioral processes, but not in their knowledge about the content of the observed tasks. However, such knowledge might be crucial to the correct interpretation of certain behavior.

For these reasons we conducted group-discussions with the control room teams for verifying our understanding of the team and work processes during the scenarios. These group-discussions were structured by looking for relevant situations within the video recordings and asking questions for a better understanding.

## 2.2 First steps: Exploring resilient team-processes in the control room team

The main goal of this research project is to detect resilient team pocesses ("safety interaction patterns" - SIPs, see Ritz & Rack, 2009) within the operating team of a control room in a nuclear power plant. Resilient team processes serve as successful, safety oriented coping mechanisms in critical situations for which no pre-defined action planning exists. Of special interest is a process starting with the detection of a deviation, passing over in the development of a collective strategy up to the regulation of collective actions for coping with the critical situation. Especially the process of sharing information will be focussed.

To identify these team processes, an explorative field study is conducted. Teams in a Swiss nuclear power plant (N=36 in 5 teams) worked on three realistic emergency scenarios in a control room simulator. To analyze the team processes, a combination of quantitative measures (team work rating scales and simulator instructors' ratings of the quality of results) and field observations is used. This combination enables a realistic view of the full complexity of the work environment and a documentation of practitioners' strategies to respond to specific emergency situation demands.

The exploration process will start with looking at the quality and structure of team-processes of two extreme groups (best vs. least rating of simulator instructors). The description of the team-processes will be done by using the taxonomy for explicit coordination activities by Kolbe et al. (2011). The focus will lie especially on the explicit coordination activity of the shift supervisor for structuring the process of sharing information. A second focus will be the sharedness of information and decision-making within the team.

Preliminary results will point out relevant abilities of teams in building up and keeping on high levels of situation awareness in dynamic critical situations.

As an implication, the description of resilient team processes will provide ideas for further investigations of safety-relevant team strategies. Additionally, they serve as an idea pool for simulator training concepts to teach control room teams relevant non-technical skills.

## 3  EXPECTED RESULTS AND OUTLOOK

One prospective result will be a prototype of an observation method focussing on explicit coordination in the control room team. Furthermore, there will be descriptions of resilient team interaction patterns in the control room for problem-solving and decision-making in coping with abnormal situations. These descriptions will provide ideas for further investigating safety-relevant team strategies and as a pool of ideas for the development of training concepts to teach relevant non-technical skills in simulator trainings for control room teams.

For exploring more deeply the preconditions and processes of the information sharing acitivities within the control room team we will extend the perspective by including the concept of team situational awareness (TSA). One methodological approach of getting information about expert´s mental models, esp. in complex environments, high-risk environments and the resulting way of decision-making of these experts, is Team Situation Awareness (TSA) in the area of Natural Decision Making. Situation awareness involves interpreting situational cues to recognize that a problem exists which may require a decision or action. Crews must go beyond merely noticing the presence of cues; they must appreciate their significance. Recognizing and defining the nature of a problem encountered in a dynamic environment is the first and perhaps most critical step in making an effective and safe decision. Situation awareness allows teams to plan ahead (Schöbel and Kleindienst, 2001). How team situation awareness is built by coordinated information sharing within the control room team should be of special importance for successful adaptations to environmental variations.

## ACKNOWLEDGMENTS

The authors would like to acknowledge our colleagues Dr. Oliver Rack, Dipl. Psych. Anna Tschaut and Stefanie Gobelie for their contribution and especially Dr. Michaela Kolbe from ETH Zürich for sharing her experience with regard to observation methods.

## REFERENCES

Burke, C.S., Stagl, K.C., Salas, E., Pierce, L. and Kendall, D. 2006. Understanding Team Adaptation: A Conceptual Analysis and Model. *Journal of Applied Psychology* 91: 1189 – 1207.

Burtscher, M.J., Wacker, J., Grote, G. and Manser, T. 2010. Managing Nonroutine Events in Anesthesia: The Role of Adaptive Coordination. *Human Factors* 52: 282-294.

Carvalho, P.V.R., dosSantos, I.L. and Vidal, M.C.R. 2005. Nuclear power plant shift supervisor´s decision making during microincidents. *International Journal of Industrial Ergonomics* 35: 619 – 644.

Fahlbruch, B., Schöbel, M., Domeinski, J. 2008. Sicherheit. In *Human Factors. Psychologie sicheren Handelns in Risikobranchen*, eds. P. Badke-Schaub, G. Hofinger and K. Lauche. Heidelberg: Springer.

Fletcher, G., Flin, R., McGeorge, P., Glavin, L., Maran, N. and Patey, R. 2003. Anaesthetists' Non Technical Skills (ANTS): evaluation of a behavioural marker system. *British Journal of Anaesthetics* 90: 580-588.

Fletcher, G., Flin, R, McGeorge, P., Glavin, L., Maran, N. and Patey, R. 2004. Rating non-technical skills: developing a behavioural marker system for use in anaesthesia. *Cognition Technology and Work* 6: 165 – 171.

Flin, R., Martin, L., Goeters, K., Hoermann, J., Amalberti, R., Valot, C. and Nijhuis, H. 2003. Development of NOTECHS (Non-technical skills) system for assessing pilot´s CRM skills. *Human Factors and Aerospace Safety* 3: 95 – 117.

Greve, W. and Wentura, D. 1997. Wissenschaftliche Beobachtung. Eine Einführung. Weinheim: Psychologie Verlags Union.

Grote, G., Zala-Mezö, E. and Grommes, P. 2003. Effects of standardization of coordination and communication in high work-load situations. *Linguistische Berichte* 12: 127-154.

Grote, G., Kolbe, M., Zala-Mezö, E., Bienefeld-Seall, N. and Künzle, B. 2010. Adaptive coordination and heedfulness make better cockpit crews. *Ergonomics* 53: 211 – 228.

Hollnagel, E. 2006. Resilience – the challenge of unstable. In. *Resilience engineering. Concpts and precepts*, eds. E. Hollnagel, D.D. Woods and N. Leveson. Aldershot, UK: Ashgate.

Hollnagel, E., Paries, J., Woods, D. D. and Wreathall, J. (2011). *Resilience Engineering in Practice. A Guidebook*. Ashgate e-Book.

Kolbe, M., Künzle, B., Manser, T., Zala-Mezö, E., Wacker, J. and Grote, G. 2009. Measuring coordination behaviour in anaesthesia teams during induction of general anaesthetics. In. *Safer Surgery: Analysing Behaviour in the Operating Theatre (pp. 202-220)*, eds. Flin, R. and Mitchell, L. Aldershot: Ashgate.

Kolbe, M., Strack, M., Stein, A. and Boos, M. 2011. Effective Coordination in Human Group Decision Making: MICRO-CO: A Micro-analytical Taxonomy for Analysing Explicit Coordination Mechanisms in Decision-Making Groups. In. *Coordination in Human and Primate Groups,* ed. M. Boos. Berlin: Springer.

Künzle, B., Zala-Mezö, E., Kolbe, M., Wacker, J. and Grote, G. 2010. Substitutes for leadership in anaesthesia teams and their impact on leadership effectiveness. *European Journal of Work and Organizational Psychology* 19: 505-531.

Malakis, S., Kontogiannis, T. and Kirwan, B. 2010. Managing emergencies and abnormal situations in air traffic control (part II): Teamwork strategies. *Applied Ergonomics* 41: 628-635.

Manser, T., Harrison, T. K., Gaba, D. M. and Howard, S. K. 2009. Coordination patterns related to high clinical performance in a simulated anesthetic crisis. *Anesthia and Analgesia* 108: 1606 – 1615.

O´Connor, P., O´Dea, A., Flin, R. and Belton, S. 2008. Identifying the team skills required by nuclear power plant operations personnel. *International Journal of Industrial Ergonomics* 38: 1028-1037.

Proctor, R. W. and Vu, K.-P. L. 2010. Cumulative Knowledge and Progress in Human Factors. *Annual Review of Psychology* 61: 623-651.

Ritz, F. and Rack, O. 2009. Steigerung der Sicherheitsleistung von Teams durch systematische Optimierung kooperativer Arbeitsprozesse. Beitrag zur 51. Fachausschusssitzung Anthropotechnik: Kooperative Arbeitsprozesse vom 27. und 28.10.2009 in Braunschweig. Bonn: Deutsche Gesellschaft für Luft- und Raumfahrt, pp. 295-310.

Roth, E.M. and Patterson, E.S. 2005. Using observational study as a tool for discovery: Uncovering cognitive and collaborative demands and adaptive strategies (pp. 379-493). In. *How Professionals Make Decisions*, eds. Montgomery, H., Lipshitz, R. and Brehmer, B. Mahwah, N.J.: Lawrence Erlbaum.

Salas, E., Sims, D.E. and Burke, C.S. 2005. Is there a "Big Five" in Teamwork? *Small Group Research* 36: 555-599.

Schöbel, M. and Kleindienst, C. 2001. The Psychology of Team Interaction. In. *Risk Control and Quality Management in Neurosurgery,* eds. Steiger, H.-J. and Uhl, E. Wien, New York: Springer.

Stout, R. J., Salas, E. and Fowlkes, J. E. 1997. Enhancing teamwork in complex environments through team training. *Journal of Group Psychotherapy, Psychodrama & Sociometry* 49: 163-186.

Vidal, M.C.R., Carvalho, P.V.R., Santos, M.S. and dos Santos, I.J.L. 2009. Collective work and resilience of complex systems. *Journal of Loss Prevention in the Process Industries* 22: 516-527.

Weick, K. and Sutcliffe, R. 2001. *Managing the unexpected. Assuring high performance in an age of complexity.* San Franciso, CA: Jossey-Bass.

Woods, D. D., Patterson, E. S. and Cook, R. I (2007). Behind human error: taming complexity to improve patient safety. In. *Handbook of Human Factors and Ergonomics in Health Care and Patient Safety,* ed. P. Carayon. Hillsdale, NJ: Lawrence Erlbaum.

Zala-Mezö, E., Wacker, J., Künzle, B., Bruesch, M. and Grote, G. 2009. The influence of standardisation and task load on team coordination patterns during anaesthesia inductions. *Quality & safety in health care* 18: 127-130.

# CHAPTER 66

# Human Cooperation Assessment Methodology

*Venétia Santos, PhD.\*, Maria Cristina Zamberlan, PhD\*\**

\* Pontifícia Universidade Católica do Rio de Janeiro - Engenharia Industrial, \*\*Human Reliability and Ergonomics Research Group for the Oil, Gas and Energy Sector, CNPq Instituto Nacional de Tecnologia / MCT (National Institute of Technology)

## ABSTRACT

The evolution of ergonomics methodology has become necessary due to the dynamics imposed by the work environment, by the increase of the need of human cooperation and by the high interaction between various sectors within a company. In the last 25 years, as of studies made in the high risk process control, we have developed a methodology to evaluate these situations that focus on the assessment of activities and human cooperation, the assessment of context, the assessment of the impact of work of other sectors in the final activity of the operator, as well as the modeling of existing risks.

**Keywords:** Cooperation, Complex Environment and Work Analysis

## 1    INTRODUCTION

The sociotechnical systems in the continuous process industries have increased their complexity. The events are not predictable in number, in relation to its order of occurrence or content, imposing an important variability to the human work which is nowadays also done in a context of strong temporal imposition. Operators are responsible for the adjustments of all the processes, searching to guarantee a good performance of the system and establish modalities of action in order to face the imposed variability (SANTOS; ZAMBERLAN; PAVARD, 2009).

Control rooms operators can recover fails and adjustments in most cases (SANTOS; ZAMBERLAN, 1995,VICENTE, 1999, HOLLNAGEL, 2004) and their success depends on many factors: the degree of work intensification; the operator´s competence, the existing cooperation conditions amongst operators, as well as their physical, mental and psychic conditions.

Considering the human performance in the control of continuous processes, our studies have a focus on the study of the cooperation between operators, on the informal mechanisms of regulation, on the operational and organizational strategies and on the spaces of autonomy of the operators for the accomplishment of necessary adjustments.

## 2    METHODOLOGY

With the developed methodology (SANTOS; ZAMBERLAN; PAVARD, 2009) it is possible to identify the existing risks, fails and obstructions in the cooperation, and we can also state that the activities in the process control are the result of this interaction amongst elements of the system. This methodology is based on the results obtained from the assessment and from the ergonomic design of 40 control centers in different sectors (nuclear power, oil, energy production and distribution, and oil transportation). The systematic observation of work activities considers normal, degraded, and emergency situations. The risks discussed are represented into a modeling of the human interaction with the process.

The methodology developed to assess the process control activities focuses on the understanding of cooperation and mapping of the context and it is based on the following phases:

## 2.1    The work analysis done in many different geographical points and in work or cooperation plans (SANTOS, 2002)

Considering the problem it is necessary to analyze the geographical points and establish priority observation plans in which certain work activities are developed in accordance with the hypothesis formulated.

We named it work plan or cooperation plan, the collective space of execution of activities that can be defined by the work of different enterprises sections, or by the teams that perform many activities simultaneously, aiming to the same result.

## 2.1.1    The work systematic analysis in which it is detailed the chronological analysis of activities performed in certain periods throughout work day and a shift change

In order to detail the operators activities the video and audio registers are recorded by the ergonomics team. It is recorded the information exchange (in person, verbal, by radio or telephone); the visual exploration and detection of information and screen navigation, the operator´s displacement, the alarm recognition, as well as the verbalizations made during the work performance. Yet, besides mapping the activities performed throughout this period, it is possible to quantify the interference between them in the chronological assessment (SANTOS; ZAMBERLAN, 2005).

## 2.1.2  The analysis of interference in the main activity

Due to the dynamic of processes, Control Centers are exposed to the execution of multiple tasks and, consequently, to the interferences between the activities to be executed. **This** interference rate is important in dynamic work situations subjected to great variability (SANTOS, 2002) because it determines the intensification of activity and the risks of losses or misunderstanding information, disclosing the origin of fails in the human cooperation.

In many Control Rooms, we noticed that there is an important frequency in the use of radio (to exchange information with the field), in which the operator looks into the supervising system screen and also recognizes the alarms. It should be mentioned that 20 years ago we could not imagine that we would expose the operator to an excessive number of alarms to be recognized, demanding a continuous attention and a performance of a double task. We still verified in these situations, when the operator performs double activities, that he can be interrupted by phone calls, and by people entering the Control Room to discuss a different issue. Facing this situation, the operators establish strategies to assure the execution of the activity.

With this approach it is possible to identify the strategies and measure the impact of interferences in the human cooperation and in the risks to the process control.

## 2.1.3  The analysis of the communication content

Besides mapping the interferences, an analysis of content of communication is made. The information exchange amongst operators is recorded and the operator´s conversations are transcribed to fill in the data collected in the chronological analysis of activities.

The identification of problems and flaws during cooperation is made possible by the transcriptions of the communication content.

By observing the communication throughout the degradation of the situation we conclude that not only the interferences must be considered to show the existing cooperation. Based on the studies, we have seen that there is an evolution in the information exchange throughout the degradation of a work situation .The priority axis of decision-making and exchange of information throughout the situation is altered and redirected.

The figure 2 shows the differences in receiving information in three different phases detected throughout the degradation of the situation. In phase A (maintenance of normal condition), the operator receives information in the field and the panel. In phase B (signs of losses in the normal condition) the cooperation process of the operators of field and the operators of the Control Center starts. In phase C (loss in the normal condition) there is a greater search for information in the field and a greater cooperation process between the Control Center operators and the field operators.

Thus, it is verified that operators establish strategies to solve problems (SANTOS; SANTOS, 2003). These strategies are based on the search of information (questioning dialogues / QD), and on the exchange of ideas and discussing about the information itself, what is called explanation dialogues (ED). The latter (ED) is very important because it allows operators to identify, treat and elaborate information to understand certain situations. Therefore, considering the evolution of the situation there is an increase on the information dialogues (in order to check information with the plant or area) and an increase in the explanation dialogues (ED) that shows the level of established cooperation.

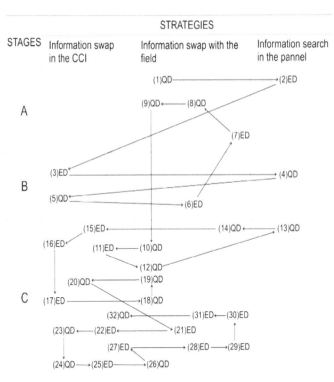

Figure 1   The questioning dialogues (QD) and explanation dialogues (ED) during the 3 stages observed throughout the degradation of the situation: A - Maintenance of normal condition; B - Signs of losses in the normal condition; C - Loss in the normal condition

Throughout the oscillation in the amount of work, our methodology forecasts measuring the elasticity of the exploration process of information in the work team, i.e., the variation on searching, sharing, and treating the information individually or collectively. Depending on the context, the operators can access, restrict, share and distribute much information.

As far as it shows, communication system forecast today in the Control Centers, are not compatible with this dynamics as we will see ahead in the results obtained.

## 2.2    The reconstruction of cooperation scenarios

Further to the systematic analysis made throughout the performance of operator´s activities in the Control room, we chose the mapping of cooperation considering other places of work where there is an interface with the activities studied.

We verified the need for comprehension and visibility of other activities that were being done simultaneously, in other places. These activities done in certain places were hidden to the systematic analysis of work .To recover all these data and the activities it would be necessary many research teams in many places. As of the necessity to assess cooperation in the workplace we decided to use what we named as: Retrospective Assessment of Scenario (to be applied after systematic registering in certain places).

The information about the interaction and cooperation of the operators during this period is recovered by the events detected from defined scenarios. This data is important as far as it shows the dynamics necessary for many teams to perform the activities in field and in the Control Room. The data recovery is obtained by the confrontation of the analysis made, by interviews with many teams that participated in the performance of the event in different sites within the plant.

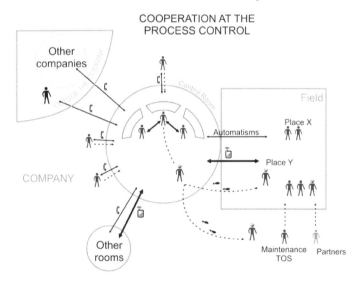

Figure 2   Modeling the Human cooperation in petrochemical plants. Collective Work representation in the Control Room and in the Field.

According to the data we obtained the modeling of the cooperation during the shift, we registered the main interactions that nowadays are performed through the exchange of information using the radio with the field, using the telephone with other units and sectors, and even in person amongst many operators (see figure 2).

## 2.3 Mapping the impact amongst many levels of companies (see figure 3) on the final activities of operators

Besides the assessment mentioned above, it is necessary to detail the context and its impact on the activities performed.

In order to recover some elements of the system and work environment that cause an impact on the final activity in the Control Rooms, we considered an analysis of the three defining levels in the company work process. (Adapted from MARMARAS; PAVARD, 1999, p. x).

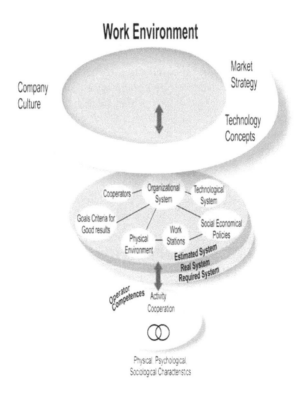

Figure 3 Representation of the defining levels in the company work process. (Adapted from MARMARAS; PAVARD, 1999, p. x).

We considered:

- in the first level. The work environment in which we studied: goals, definition of human work, strategies, definition of technology, and the company culture;

- in the second level. The organizational system in which we studied: quality management, human resources, physical work environment, equipment, and its maintenance;

- in the third level, the activities performed in which we studied the cooperation amongst people inside a work system, where competences are defining factors for the work.

To collect the data in relation to the three levels, we chose to make interviews with many sectors of the company. First, we presented the results of the analysis of the final activities (e.g. Control Room), then we outlined the representation that each sector has of its work and, also, the possible interferences of their actions upon the work of other sections.

This assessment in different levels is important to understand:

- The construction of the final activity of operators, because in many occasions there is an influence of others factors that arise from the context that might introduce an intensification, a stress, a performance of a double task, and errors into the final activity of operators.

- in order to build an approach to human error now seen as a result of the context influence, considering the process as a whole

The results obtained from the collection of data and interviews with different sections complement the data collected in the systematic observation of work. With this assessment, we identify the factors that have negatively influenced the final activity of operators.

## 2.4    Operator discourse

Finally, it is necessary to approach operator discourse, difficulties, demands, and experiences with high-risk situations, critics and proposals for improvement of work situation.

## 3    MODELING THE IMPACT UPON THE FINAL ACTIVITY OF OPERATORS

The figure 5 shows that many factors: work organization, equipment condition, maintenance, automation, excellence management and training can cause an impact upon the final activity of operators which can be or become stressful, intensified with the presence of a double task and errors.

Our studies show that the activity of operators can be different and it can bring major benefits to the system performance as well as to their health. For example, if there is a significant amount of false alarms (maintenance problem) to manage in the Control Room, instead of having the operators foreseeing the incidents, we will have the operators that persecute the alarms during the shift. The operators will have greater mental demands and if they are obliged to work extra hours, beyond their work journey, they will have their level of attention reduced. In this case, based on the context, the operator uses strategies to minimize his action, i.e. the accumulation of alarms to be recognized in a group. In this case, a true alarm status can arise and mix with others (false alarm status), and the first can be hidden and may not be detect by the operator, as it already happens.

Some studies made in Ergonomics, based on the ergonomic work analysis, restrict their observations to the real activity of work, the final actions of the operators, without considering the construction of this activity which could be different. Our aim is to analyze the elements that collaborate to the final construction of the activity, in other sections of the company, and in recent past.

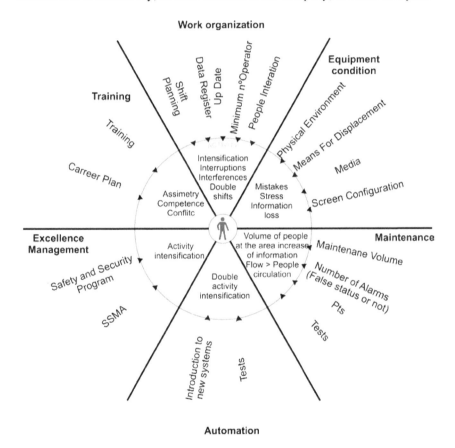

Figure 4   E.g. Impact of work organization, equipment condition, maintenance, automation, excellence management and training can cause an impact upon the final activity of operators.

# 4    RESULTS

Applying this methodology in the study of Control Centers, we have verified that there are recurring problems in innumerable situations.

Actually, we believe that this is the result of the misconception of the development of automation that did not consider human intervention as a core element in system adjustments. It was believed that systems could work alone without a great human participation, and the cooperation between operators was not necessary, what was not verified in practice. Our results indicate that not only operators, but also other actors responsible for the performance of the processes, make continuously adjustments of the systems.

In the centers studied, we have observed that nowadays there are great issues that affect the performance of human work:

**a) Flaws in the human cooperation**. Automation leads to an increase in maintenance, information flow, and these carry out an interface amongst the team to diagnose and solve problems. However, the conception of communication systems in this process is still centered in radio systems that are inefficient and not adapted to the existing needs. With the radio, the exchange of oral information is made from the speech of one operator followed by an interruption, so as to make possible for the other operator speak. Thus bringing restrictions in the exchange of explanation dialogues (ED) amongst the teams. In many occasions, there might be flaws in receiving the information, noises, misunderstanding and the need to repeat the information.

**b) Interaction problems amongst the teams.** The operational fields do not have adequate coverage of TV circuit systems, and in many occasions, there is a lack of visibility on important areas on the operational aspect and no information about the physical position of the field teams that are interacting with the team in the control room.

**c) Intensification of activities due to the limited number of operators facing the system complexity**. Multiple tasks are performed with the presence of activities that interfere one another. Many people interact simultaneously, the exchange of verbal information is superimposed and much information is lost. The performance of multiple activities and the superposition of those take an important time of work for the operators, in which a double task is performed. A chronological analysis of the activity shows the human interaction in systems and the performance of multiple activities, as: exchange of verbal information, acting in the supervision system, alarms recognition, etc.

**d) The amount of extra activities added to the process control** (updating procedures, quality programs, etc) contributes to the densification of the activity of operators who waste time interacting with computers, and, simultaneously controlling the production data, performing actions related to the company quality management, along with other activities.

**e) Flaws on the exchange of information from one shift to another.** Due to the complexity of systems and unpredictable events, the operators must face a great amount of information. This information must be registered and passed into the next team of operators in certain periods of time, and it counts on the memory capacity of these operators. The great amount is significant, and we should introduce information technology to help the operators in the storage of this information.

**f) Renewing the teams, the distribution between field and Control room of** the experienced operators and the beginner ones produces **asymmetry**, in the group work, that is not considered in the organization process of work. Many times, the teams using their own initiative develop strategies for rotating and allocating operators in the shift, considering their level of experience that favors the training of the beginner operators and improves the balance of competences.

## 5     CONCLUSION

Therefore, our challenge now is to promote the human interaction with the systems, to develop cooperation systems, and other facilities, considering that: man is part of technology, and also that his actions and strategies have an impact on companies' reliability.

## REFERENCES

Hollnagel, E. Barriers & Accident Prevention. Aldershot: Ashgate, 2004, 242 p.

Marmaras N.; Pavard B. Problem-Driven Approach to the Design of Information Technology Systems Supporting Complex Cognitive Tasks. Cognition, Technology & Work. Londres: Springer-Verlag London Limited; 1999; 222-236.

Santos, V.; Zamberlan, M.C.P.L, PAVARD, B. Confiabilidade Humana e Projeto de Centros de Controle de Alto Risco. Rio de Janeiro: Synergia / IBP / CNPq / INT, 2009, 296 p.

Santos V.; Zamberlan M. C. P. L. Projeto Ergonômico de Salas de Controle. São Paulo: Fundación Mapfre Sucursal Brasil, 1995, 136 p.

Santos V. A Ergonomia e a Intensificação do Trabalho nas Centrais de Atendimento: A Gestão Temporal de Múltiplas Tarefas e de Atividades Interferentes. Rio de Janeiro: UFRJ, 2002. 299 f. Tese (Doutorado em Engenharia de Produção) - Coppe, Universidade Federal do Rio de Janeiro, Rio de Janeiro, 2002.

Santos, P.; Santos, V. Estudo Ergonômico de uma Refinaria de Petróleo. 2003. Documento Interno da Ergon Projetos, Rio de Janeiro.

# CHAPTER 67

# Organizational Trust and Trust in Automated Systems as Predictors for Safety Related Team Performance - Results from a Cross-cultural Study

*Frank Ritz*

University of Applied Sciences and Arts Northwestern Switzerland
School of Applied Psychology
Olten, Switzerland
frank.ritz@fhnw.ch

## ABSTRACT

Technological advance is making fundamental changes in the design of complex socio-technical systems. Hazardous side effects are the rising of complexity and dynamics. In critical situations, safety can be stabilized by team interactions that cope with unknown challenges. This contribution focuses on underlying predictors of collective interactions toward safety. The construct of team performance resembles conceptually the cultural dimension collectivism. Assuming that teams in collectivistic cultures show a higher level of team performance, their behavior can be used as reference system of team performance. Team performance in the operating theatre is the subject of this study. Team members (n=279) of Chinese and German operating theatres participated in an investigation by questionnaire. T-Tests and structural equation modeling reveal that (i) the team performance of Chinese team members is significantly higher compared to German team members and (ii) a transcultural model of the safety related team performance could be created, which includes organizational trust and trust in automated systems as predictors.

**Keywords**: safety, team performance, organizational trust, trust in automated systems, cross-cultural study, transcultural model, Germany, China

# 1    BACKGROUND: CLOSING THE GAP FROM RELIABILITY TO SAFETY

Controlling high hazard systems is a central problem of the design and prioritization of human control over technical subsystem (Grote, 2005). Many authors emphasize that technology is changing faster than the engineering techniques to deal with the new technology (e.g. Rasmussen, 1997). The given concept of reliability is only suitable to handle work processes as planned or to support the overcome of known unexpected situations. New technology introduces unknown variables into systems and brings more complex relationships between humans and automation. Digital technology has changed the nature of accidents (Leveson, 2004) by accelerating system processes and increasing interaction complexity. That leads to the assumption that we are designing systems with potential interactions among system components (human and technology) that cannot be thoroughly planned and their control still remains to be a challenge. Rapid changes are leading to new types of hazard (e.g. mode confusion) among which the behavioral context is centered. High levels of complexity and system dynamics provoke unexpected situations in which hazardous phenomena like "tight coupling" (Perrow, 1999) appear and lead to events, which are potentially safety relevant.

Nevertheless a general strategy that can be observed to solve the problem is trying to achieve system safety by increasing the reliability of individual system components. Therefore the design of the human subsystem is dominated by the trend of standardization of processes and the use of more technology to control the system behavior. Human advantages like adaptability and inductive reasoning (Fitts, 1951) are very often ignored. Considering the "ironies of automation" (Bainbridge, 1983), this can be characterized in terms of "replacing human manual control, planning and problem solving by automatic devices and computers". In other words: the aim is to reduce complexity by designing more complex systems.

We seem to be caught in a paradigm of "human error" (Reason, 1990), which describes humans as latent source of unsafe acts. In this context human performance is associated with unreliability that has to be eliminated by standardization of working processes and technical components. But there is a problem; obviously future is not fully predictable. All standardized procedures naturally base on experience. Due to the high degree of complexity, it is not possible to plan all prospective conditions of a system, much less its outcome. A specific situation determines its conditions during its occurrence. So, preparation in terms of reliability seems not to be enough to prevent safety critical events. By analyzing safety related events (e.g. Three Mile Island 1979, Alaska Airlines 2000, Forsmark 2006) safety research points out that a system can be reliable and unsafe or safe and unreliable, which makes clear that safety and reliability can be described as different system components: "...making the system safer may decrease reliability and enhancing reliability may decrease safety" (Leveson, 2011, p. 55).

On the one hand we have this dangerous "basic assumption" (Schein, 1990) that more reliability leads to more safety and on the other hand it seems quite clear that such complex systems and the resulting unexpected situations can be managed successfully by teams of human actors only, using technology. According to Fahlbruch and Wilpert (1999, p. 11) safety is defined as "... quality of a system that allows the system to function without major breakdowns under predetermined conditions with an acceptable minimum of accidental loss and an unintended harm to the organization and its environment".

In this context safety related teamwork can close the gap between reliability and safety. Teams are able to fulfill a collective switch during a work process. They can adapt and further develop reliable standard procedures in unexpected situations. Reliable standards are a very useful orientation system. According to Leveson (2004) this reliable standards base upon the concept of "constraints". Inadequate control can occur either because control actions do not adequately identify hazards or because the control actions do not adequately enforce the constraints. Both result from "...inconsistent or incorrect process models...or by inadequate coordination among multiple controllers and decision makers (Leveson, 2004, p. 261). Therefore teams are challenged by (i) identifying unexpected situations that affect the system control, (ii) anticipating hazardous situations and related constraints, (iii) creating operating solutions concerning the situational context adequately and (iv) reporting the situation including contributing factors to an organizational learning system. The question is: how success critical criteria can be found to cope with unexpected situations appropriately.

## 1.1   Team performance, a collective process toward safety

The advantage of teamwork is to provide a wide base of knowledge and strategies for coping with unexpected situations. A team, as organizational unit, has a rich competency for the challenging mission of system control that can be described by its unique feature: being flexible in order to adapt cooperation processes to unexpected contextual needs and situational dynamics by using shared mental models (SMM, Orasanu and Salas, 2001). SMM can be defined as "...knowledge structure held by members of a team that enable them to form accurate explanations and expectations for the task, and, in turn to coordinate their actions and adapt their behavior to demands of the task and other team members" (Cannon-Bowers, Salas and Converse, 2001, p. 228). In unexpected situations the coordination of the work process becomes more important and the quality of team performance can be described as "regulation of collective action" (Ritz, 2008). Therefore, the adequate use of communication and other non-technical skills (Flin et al., 2003) are important resources. The construct of team performance describes the process of collective influence on the execution of a joint activity by using SMM that focuses two objective targets: (i) to ensure action and (ii) to avoid mistakes with hazard potential. The quality of the regulatory process that leads to a high level of team performance depends on its goal orientation. Concerning safety as a moving target, the development of an adequate collective process bases on decision making and communication. The aim is to avoid blindspots in terms of dealing with "expected but unpredictable contingencies in settings that were organized, demanding, hazardous, interdependent, unpredictable, bound by norms, and threatened by events that have to be precluded" (Weick, 2011, p. 21). However, the question occurs, which underlying predictors can be used to evaluate team performance. This question will be answered in close association to the research field of this study.

## 1.2    Team Performance in the Operating Theatre

The operating theatre (OT) is a prime example of a complex socio-technical system with high dynamics and mixed structured workflows. Many studies underline the high hazard potential of the OT (e.g. Kohn, Corrigan and Donaldson, 1999; Vincent, Neale and Woloshynowych, 2001). In this context communication appears to be the most important factor of team performance (Ritz and Friesdorf, 2003). According to Dickinson and McIntyre (1997) a team consists of several individuals working together towards a shared goal. This shared goal is "the well-being of the patient" (Schaefer, Helmreich and Scheidegger, 1995). To achieve this goal, individual team members have to coordinate their work with other team members of different disciplines in a way that relevant information is shared. Communication leads to SMM, which enable controlling the goal oriented behavior concerning its situational needs (Ritz and Schoebel, 2003). Concerning the results of the Human Factors Gruppe (1997) for the observation of the team process in the OT three dimensions of communicational behavior seem to be needed: (i) team concerns, (ii) decision making and (iii) management of the work situation.

This leads to the following questions: first, which predictors are influencing communication effectively and second, which reference system is adequate to verify the success of collective behavior. Some studies point out the function of trust in teams and automated systems being the relevant factors. The conceptualization of team performance can be compared with collectivism as cultural dimension.

## 2    TRUST

Trust is relevant in all kinds of social processes, therefore also in teamwork. Concerning sociological terms, trust is a mechanism to reduce social complexity (Luhmann, 1968). According to Reason (1997) and Weick and Sutcliffe (2001) communication based on trust is one of the most important conditions that influences safety culture (e.g. Schein, 1990) positively. Trust can be defined as expectation of a person that an interaction partner is able and willing to bring her or himself forward (Koller, 1988). Hence, trust seems to be a relevant variable that influences team performance. Trust concepts with potential influence on team performance in socio-technical systems are "organizational trust" (Cummings and Bromiley, 1996) and "trust in automated systems" (Jian, Bisantz and Drury, 2000).

Organizational trust in relation to the organization form team is based on three dimensions: (i) comply with agreements, (ii) honest negotiations and (iii) avoid excessive benefits at the expense of other team members. Trust in automated systems is a scale that describes a person's expectation in the effectiveness of a technical system concerning its reliability. This reliability bases on the experience, a person has made with a specific machine during work processes.

# 3    CULTURE, A PRECONDITION OF COLLECTIVE BEHAVIOR?

The construct of team performance, which by communication can be observed on the behavioral level, is conceptually quite similar to the cultural dimension of "collectivism". (e.g. Hofstede, 1980; Triandis, 1995). Cultural studies point out that the historical cultural divergence between ancient Chinese and Greeks has led to different cognitive processes in individualistic cultures like Germany and collectivistic cultures like China within the past 2500 years (e.g. Nisbett et al., 2001). Basically and essentially two fundamental differences can be described. In collectivistic cultures a more interdependent self-construction and a more holistic reasoning can be found. In contrast to that independent self-construction and analytic reasoning is more common in individualistic cultures. Concerning to this, many evidences could be found to lead to the assumption that teams in collectivist cultures could be an appropriate reference model for team performance, e.g.: people socialized in collectivistic cultures compared to people in individualistic cultures, are more focused on group goals (Kitayama et al., 1998), show a more context-dependent and goal-orientated communication (Haberstroh et al., 2002), are more likely to resist the illusion of control (Langer, 1977) and make less fundamental attribution errors (Nisbett, 2003).

According to this following questions arise: (i) Do teams of collectivistic cultures really show better team performance, which is suitable to lead to a basic behavioral pattern aiming safety and (ii) how can the cause-effect relationship between trust and team performance be systemized.

# 4    METHOD

The present study uses as a team performance questionnaire a further development of the observation form KOMSTAT (communication status, Human Factors Gruppe, 1997), assuring compliance with its three dimensions (see above). It is combined with an inventory of organizational trust (OTI-SF, Cummings and Bromiley, 1996) that is focused on the team as organizational form, with regard to its three dimensions (see above) and a scale of trust in automated systems (Jian, Bisantz and Drury, 2000) that focuses on the automatic blood pressure measurement as a specific technical component.

The questionnaire consists of statements. The statements have to be rated by using seven-point Likert scales from "I strongly agree" to "I strongly disagree". The survey was answered subsequent to an operation in an OT.

For the use in this cross-cultural study the questionnaire was translated from German into Chinese language and retranslated twice. The two iteration loops of translation allow a higher comparability of the results of both cultural groups.

In this cross-cultural study the teams of Chinese (n=120) and German OT-teams (n=159) participated in the questionnaire investigation. Chinese OT-team members of four Chinese hospitals constitute a sample of collectivistic socialized people. German OT-team members from four German hospitals act as an individualistic sample. The cross-cultural comparison aims at (i) exploring differences of team performance between both cultures and (ii) building a transcultural model of team performance including its predictors, as assumed organization trust in the team and trust in automated system of blood pressure measurement. This can be realized by the method of structural equation modeling (SME, Joereskog and Soerbom, 1999). For this purpose the scale trust in automated systems is parallelized into two subscales (see illustration 1).

# 5    RESULTS

In a first step t-tests are used to compare the mean scores of the two independent cultural samples on the given variables. According to the expectation, the team performance of the members of Chinese teams is significantly higher on all three dimensions, compared to the members of German teams (see table 1).

**Tabel 1  Results of t-tests between Chinese and German team members**

| Dimension | Germany | | | China | | | pair-wise comparison | | | |
|---|---|---|---|---|---|---|---|---|---|---|
| | n | M | s | n | M | s | t | df | d | p* |
| team concerns | 159 | 4.15 | 1.20 | 119 | 5,58 | 1.06 | 10.315 | 276 | 1.25 | <0.001 |
| decision making | 158 | 4.33 | 1.10 | 119 | 5,75 | 0.84 | 11.763 | 275 | 1.43 | <0.001 |
| management of the working situation | 158 | 5.38 | 1.14 | 119 | 6,41 | 0.80 | 8.380 | 275 | 1.02 | <0.001 |

*=double-sided

With regard to the construct of organizational trust in the team, Chinese team members rate significantly higher on the dimensions (i) comply with agreements (t=4.084, p<0.001) and avoid excessive benefits at the expense of other team members (m=5.047, p<0.001). Regarding the dimension honest negotiations (t=-0.753, p<0.452) there is no significant difference between both cultural groups.

On the scale trust in automated systems the team members of the partial sample of Germany show a significantly higher level of trust in the automated system of blood pressure measurement (t=-3.876, p<0.001).

Using the method LISREL (Joereskog and Soerbom, 1999) a transcultural model of perceived team performance could be realized. As predictors of team performance (i) organizational trust in the team and (ii) trust in automated systems were confirmed (Figure 1).

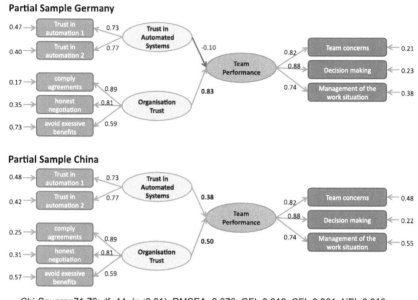

Chi-Square=71.76; df=44; (p<0.01); RMSEA=0.076; GFI=0.918; CFI=0.961; NFI=0.910

Figure 1 Illustration of the causal connections of the final model 3 as completely standardized solution, using culture as constraint for a multiple-group comparison and differentiated by partial samples China and Germany

The insignificant influence of trust in automated systems on team performance in the partial German sample is the most remarkable difference between the cultures. (see Figure 1, insignificant path of -0.10 from trust in automated systems to team performance). The quality of the resulting model can be discriped as good, as tabel 2 points out.

**Tabel 2  Godness-of-fit Indicies by models and in comparison with null model**

| Model | Goodness-of-fit indices | | | | | | Comparison with null model | |
|---|---|---|---|---|---|---|---|---|
| | $\chi^2$ | df | GFI | CFI | RMSEA | NFI | AIC model | AIC null model |
| 1 | 99,030 | 54 | 0,896 | 0,931 | 0,087 | 0,870 | 135,030 | 860,678 |
| 2 | 86,565 | 52 | 0,906 | 0,950 | 0,078 | 0,891 | 126,565 | 860,678 |
| 3 | 71,756 | 44 | 0,918 | 0,961 | 0,076 | 0,910 | 127,756 | 860,678 |
| 4 | 62,228 | 36 | 0,933 | 0,963 | 0,081 | 0,922 | 134,228 | 860,678 |

($\chi^2$ = "Normal Theory Weighted Least Squares CHI-Square")

# 6    CONCLUSIONS

In conclusion, the results allow the assumption that teams of collectivistic cultures show better team performance, in terms of communication relating to safety. It is important to control the teamwork process by building up adequate shared mental models, which allow an adaption on unexpected situations. Organizational trust in the team could be identified as transcultural predictor of team performance. Trust in automated systems is only in collectivistic teams suitable to predict the team performance. Due to the fact that trust in automated systems is objectified by the automated system of blood pressure measurement only, a generalization of this result should be object of further research.

It was shown that cross-cultural research, using the dimension collectivism – individualism, is a good approach to further exploration of the team performance. In the context of globalization, it seems relevant to derivate knowledge about the potential opportunities and threats of the cooperation in multicultural teams, in further studies. This knowledge could be very useful to the future trend of socio-technical systems, especially in managing safety.

A model of team performance including its predictors organizational trust in the team and trust in automated systems could be generated successfully. According to this, it seems indicated that further efforts in exploring team performance as a source of safety is necessary. Therefore, the development of valid success critical criteria is important. Concerning safety and reliability as different system properties, it seems important to support the active perception of turning points in system dynamics and to organize the allocation of attention on different system components. The realized facts could be a basis for the collective decision making. Identifying and using of relevant situational information could lead to an adequate problem solution. Misleading predefined processes, which provoke unsafe acts, could be avoided. In order to manage such unpredictable challenges in a situational context, high hazard systems have to rely on human skills like adaptability and inductive reasoning. Hence, the investigation of these skills, especially in the team

658

as organizational form should be subject of further research. Regarding this, in the further development of socio-technical systems human performance could be considered as a potential source of safety. This new paradigm seems to be needed, particularly in times in which increasing technical control steadily leads to higher system complexity.

## ACKNOWLEDGMENTS

The author is grateful to Bernhard Wilpert for the helpful discussions, Wu Jing for the support in China and all participants of the sample in China and Germany.

## REFERENCES

Bainbridge, L. 1983. Ironies of Automation. *Automatica 19*: 775-779.
Flin, R., L. Martin, and K. Goeters, et al. 2003. Development of the NOTECHS (non-technical skills) system for assessing pilots' CRM skills. *Human Factors Aerospace Saf* 3: 95-117.
Cannon-Bowers, J.A., E. Salas and S. Converse. 1993. Shared mental models in expert team decision making. In *Individual and group desicion making* (pp. 221-246), ed. N.J. Castellan. Hillsdale (NJ): Lawrence Erlbaum.
Cummings, L.L. and P. Bromiley. 1996. The Organizational Trust Inventory (OTI). Development and Validation. In *Trust in Organizations: Frontiers of theory and research* (pp. 303-327), eds. R.M. Kramer and T.R. Tyler. London: Sage.
Dickinson, T.L. and R.M. McIntyre. 1997. A conseptual framework for the development and Validation. In *Team Performance assessement and measurement* (pp. 19-34), eds. M.T. Brannick, E. Salas and C. Prince. Mahwah: Lawrence Erlbaum Associates.
Fahlbruch, B. and B. Wilpert. 1999. System Safety – An Emerging Field for I/O Psychology. *International Review of Industrial and Organisational Psychology* 14: 55-93.
Fitts, P.M. 1951. *Human engineering for an effective air navigation and traffic control system*. National Research Council: Washington, DC.
Grote, G. 2005. Menschliche Kontrolle über technische Systeme – Ein irreführendes Postulat. In *Beiträge der Forschung zur Mensch- Maschine-Systemtechnik aus Forschung und Praxis* (p. 65-78), eds. K. Karrer, B. Gauss and C. Steffens. Düsseldorf: Symposion.
Haberstroh, S., D. Oysermann, and N. Schwarz, et al. 2002. Is the interdependent self a better communicator than the independent self? Self-construal and the observation of conversational norms. *Journal of Experimental Social Psychology* 38: 323-329.
Hofstede, G. 1990. Empirical models of cultural differences. In *Contemporary issues in cross-cultural psychology*, eds. N. Bleichrodt and P.J.D. Drenth. Amsterdam: Swets & Zeitlinger.
Human Factors Gruppe. 1997. *Kommunikations-Status (KOMSTAT)*. Department Anästhesie Kantonspital Basel. Basel: Universität.
Jian, J.J., A.M. Bisantz, and C.G. Drury. 2000. Foundations for an empirical determined scale of trust in automated systems. *International Journal of Cognitive Ergonomics* 4 (1): 53-71.
Joereskog, K.G. and D. Soerbom. 1999. *Lisrel 8.30 and Prelis 2.30*. Chicago: Scientific Software International.

Kitayama, S., S. Duffy, and T. Kawamura, et al. 2003. Perceiving an object and its context in different cultures: A cultural look at the New Look. *Psychological Science* 14: 201-206.

Kohn, L.T., J.M. Corrigan, and M.S. Donaldson. 1999. *To err is human: building a safer health system*. Washington DC: National Academy Press.

Koller, M. 1988. Risk as a determinant of trust. *Basic and Applied Social Psychology* 9: 265-276.

Langer, E. 1975. The illusion of control. *Journal of Personality and Social Psycholoy* 32: 311-328.

Leveson, N.G. 2004. A new accident model for engineering safer systems. *Safety Science* 42 (4): 237-270.

Leveson, N.G. 2011. Applying systems thinking to analyse and learn from events. *Safety Science* 49: 55-64.

Luhmann, N. 1968. *Vertrauen*. Stuttgart: Lucius & Lucius.

Nisbett, R.E. 2003. *The geography of thougt – How Asians and Westerns Think Differently...and Why*. New York: The Free Press.

Nisbett, R.E., K. Peng, and I. Choi, et al. 2001. Culture and systems of thought. *Psychological Review* 108: 291-310.

Orasanu, J.M. and E. Salas. 1993. Team decision making in complex environments, In *Decision making in action: Models and methods* (pp. 327-345), eds. G.A. Klein, J. Orasanu, R. Calderwood, & C.E. Zsambok. Norwood, NJ: Ablex Publishers

Perrow, C. 1999. *Normal Accidents: Living with High-Risk Technologies*. Princeton: University Press.

Reason, J.T. 1990. *Human Error*. Cambridge: University Press.

Reason, J.T. 1997. *Managing the Risk of Organizational Accidents*. Aldershot: Ashgate

Rasmussen, J. 1997. Risk management in a Dynamic Society: A Modeling Problem. *Safety Science* 27: 183-213.

Ritz, F. 2008. *Einflussfaktoren auf die Teamleistung im interkulturellen Vergleich - Untersuchung in deutschen und chinesischen Operationssälen*. Saarbruecken: VDM.

Ritz, F. and W. Friesdorf. 2003. Team performance in high risk environments - Communication in the operating room. In *Quality of Work and Products of the Future*, eds. H. Strasser, K. Kluth, H. Rausch & H. Bubb. Stuttgart: Ergonomia.

Ritz, F. and M. Schoebel. 2003. Teamleistung im OP. In *Entscheidungsunterstützung in der Fahrzeug- und Prozessführung*, ed. M. Grandt. Bonn: Deutsche Gesellschaft für Luft- und Raumfahrt e.V.

Schaefer, H.G., R.L. Helmreich, and D. Scheidegger. 1995. Safety in the operating theatre – Part I: Interpersonal relationships and team performance. *Current Anesthesia and Critical Care* 6 (1): 48-53.

Schein, E. 1990. Organizational Culture. *American Psychologist* 45 (2): 109-119.

Triandis, H.C. 1995. *Individualism-Collectivism*. Boulder: Westview Press.

Vincent, C., G. Neale, and M. Woloshynowych. 2001. Adverse events in British hospitals: preliminary retrospective recors review. British Medical Journal 320: 745-749.

Weick, K.E. 2011. Organizing for Transient Reliability: The Production of Dynamic Non-Events. *Journal of Contingencies and Crisis Management (*19): 21-27.

Weick, K.E. and K.M. Sutcliffe. 2001. *Managing the Unexpected: Assuring High Performance in an Age of Complexity*. San Francisco: Jossey-Bass.

# Section X

New Ways of Work

# CHAPTER 68

# The Influence of Expectations and Pre-experiences on Comfort at Work

*C. Bazley, P. Vink , A. De Jong*

TU Delft
Delft, The Netherlands
cbazley@jimconna.com
pvink@tudelft.nl
adejong@tudelft.nl

*A. Hedge*
Cornell University
Ithaca, USA
ah29@cornell.edu

## ABSTRACT

Comfort research often includes a variety of factors e.g., environmental conditions, physiological, psychological, emotional levels, pre-experiences, expectations, and sustainability. This paper examines the effects pre- experiences and expectations may have on comfort levels over-time. Sixteen participants from a professional office completed a daily survey on their comfort levels throughout the workday for a four-day workweek. The survey measured comfort levels upon waking and the effects pre- experiences and expectations have on, before arriving to work, and throughout the workday. The results confirm the time of day relationships between pre-office experiences and intellectual comfort levels. Participants showed high expectations levels for work performance from co-workers, managers, and themselves. Additionally, the majority of participants said that "people" made them feel comfortable and uncomfortable.

**Keywords:** comfort, workplace, expectations, pre-experiences

# 1    INTRODUCTION

A "holistic" approach in workspace and workplace design includes health, wellness and comfort. Optimal internal and external environmental conditions include good quality air, water, and climate, physical comfort and health, adequate space, smooth flow of work, general safety and safe working conditions. Other factors to consider beyond the physical environment are the mental and emotional health. These factors help promote growth, flexibility, creativity, and well-being in the workplace (INQA-Buero, 2005).

Comfort is a term that is often associated with a physical state or an environmental factor. Richards (1980), stresses that comfort involves the sense of subjective well-being. That is the reaction a person has to an environment or situation. According to Looze, Kuijt-Evers &  Dieën (2003) some issues are generally accepted in comfort literature are; (1) comfort is a construct of a subjectively defined personal nature; (2) comfort is affected by factors of various nature (physical, physiological, psychological); and (3) comfort is a reaction to the environment. Comfort has other meanings as well. For example, "This room is comfortable" or "Being with you makes me comfortable". Comfort relates to ease, which relates to satisfaction, contentment and finally to pleasure, for example, "The pleasure of your company", has a warm feeling to it, although it rarely has to do with the physical temperature. Frequent interaction leads to the feeling of familiarity, which breeds fondness (Bornstein, 1989; Bornstein, 1999). By combining comfort with pleasure, we encounter balance and sustainability. The aspects of pleasure and comfort are in product and game design (Green & Jordan, 2002).

A cognitive process is required in answer to "Are you comfortable?" It is necessary to think about who, what or why something is comfortable or not. Emotions are quick to overwhelm cognition particularly in the case of pleasure. Sensibility and a healthy balance of all comfort levels is the intended optimum result (Green and Jordan, 2002). Slater (1985) defines comfort as a pleasant state of physiological, psychological and physical harmony between a human being and his environment.

On the other hand, Helander & Zhang, (1997) note that discomfort due to sitting is related to more physical aspects like pressure points and stiffness, while comfort relates more to luxury and refreshment. Jordan (1997) discusses how environmental sustainability interrelates between comfort, pleasure and the usability of products.

Today, there is a significant amount of research projects focusing on human health and the connection between the environments and human psychological responses. The fundamental hypothesis of this body of research is,"since the earliest evolutionary phases of human life, we have had a visceral, survivalist need to be sensitive and responsive to our surroundings" (Bilchik, 2002, p.10).

The surveys used in this paper investigate pre-comfort experiences along with incremental comfort level assessments for a group of professionals throughout the workday and a workweek. The survey for this study was modeled after Konieczny (2001), who considered three main elements to access pre-experiences and attitudes for flight travel; **Hardware** (airport signs, walking distance, toilets…), **Software** (waiting and boarding times) and **Life ware** (staff competencies and personal support). A modified version for pre-experience and comfort expectations is possible by exchanging hardware for physical comfort, software for intellectual comfort and life ware for emotional comfort. Adding pre-comfort experiences to the study of comfort deepens the understanding and formation of personal comfort realities.

Taking into account the many definitions for comfort, this paper combines environmental elements and psychological and emotional factors, measured over time to establish changes in comfort levels.

The three questions asked:

Do pre-experiences have an effect on comfort levels throughout the day?

Do comfort levels decline overall throughout the day?

Are physical conditions the most important factor influencing comfort?

## 2    METHOD

In response to the three hypotheses, a Four-Day survey on comfort was administered to a group of 16 people. The survey was given to 16 professionals working four, ten-hour days a week. The workweek begins on a Monday and ends on a Thursday. Eleven males and five females were surveyed, twelve Engineers, two Designers and one Manager, and one Administrative Assistant. The age of the participants ranged was between age 22-60 and 13 of the participants were between the ages of 40-60.

The participants working at a professional office received a packet of four surveys and began the survey the following day. The workweek for this office is four, ten-hour days.

The first set of survey questions asked for a rating 1-5, (1, being excellent – 5, being very bad), on each waking comfort level (physical, intellectual and emotional). In addition, participants were asked to circle a word best describing the level of comfort upon waking. They were also asked to circle a color that best described their present comfort level. A "yes" or "no" question was included in this set of questions asking if the participant was looking forward to going to work for the day.

The next set of questions again asked for comfort levels (physical, intellectual and emotional, each rated 1-5), a word and a color to best describe comfort levels. Additionally, they were asked to circle the word that best described the most significant thing they saw on the way to work and another question asked for the most significant thing they heard before arriving to work.

The last part of the survey was completed at the end of workday and participants were again asked to rate their comfort level (physical, intellectual and emotional, each rated 1-5) and circle a word that best described their comfort level at that time and another question to circle a color to describe the comfort level. In addition, they were to circle a word that best described what made them most comfortable in the office setting and an additional question asked them to circle a word that best described what was most uncomfortable in the office setting.

The survey began on a Wednesday and ended the following Tuesday. A two-factor repeated measures ANOVA with DAYS and TOD as factors was performed. Results were marginally significant for the intellectual comfort level.

## 3    RESULTS

The results were marginally significant for the intellectual comfort level. On average for the four days, the most significant visual that participants saw was, "nature", followed by "roadway". This is consistent with the route taken to the worksite.

The most significant thing that was heard was "singing", followed by "talking". Most participants listened to music, the news, or had conversations with carpool members on the way to work.

The "temperature" of the office was circled as most uncomfortable in the workplace was either too hot or too cold, followed by "people" at the office that made them uncomfortable. On the other hand, the "people" in the office also made them the most comfortable, followed by the "tasks" or nature of the job.

The physical, intellectual and emotional comfort levels declined comfort throughout the day, with the exception of Thursday, when all comfort levels improved (Figure 1).

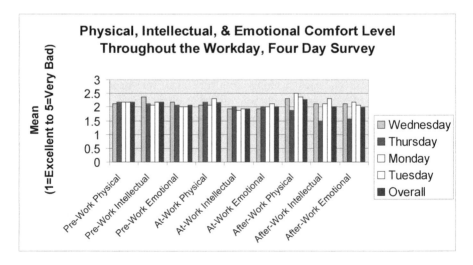

Figure 1 Four Day Survey physical, intellectual and emotional comfort levels throughout the workweek

There were no significant effects for physical or emotional levels for a two-way repeated measures ANOVA test with DAYS and TOD as factors, only a marginally significant F(p 0.051) DAYXTOD interaction was observed for the intellectual comfort level suggesting that levels slightly improved.

## 4 DISCUSSION

### 4.1 Pre-experiences and the effect on comfort levels.

Konieczny (2001) concluded that the changing environment of comfort at work and in service is dependent on the pre-experience and attitude and considered three main elements to access pre-experiences and attitudes for flight travel; Hardware (airport signs, walking distance, toilets…), Software (waiting and boarding times) and Life ware (staff competencies and personal support). He concluded that pre-experiences and attitude toward flight correlated with the flight outcome. A modified version for pre-experience or pre-comfort experiences for comfort studies is possible (Table 1) by exchanging hardware for physical comfort, software for intellectual comfort and life ware for emotional comfort.

Table 1. A modified version of Konieczny's (2001) table of pre-experiences for flight shows an example of elements for pre-comfort experiences. Pre-experience for work elements may occupy more than one comfort type

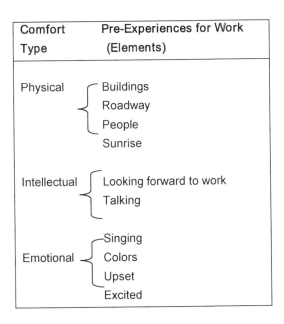

## 4.2    Comfort levels decline during the day

The participants physical comfort level was consistent with studies and declined throughout the working day (Vink, 2005). Declining physical comfort levels are observed throughout the day in offices where people have sedentary jobs, often sitting for up to 10 hours a day (Figure 1). Physical discomfort is often reported from sitting in an improperly fitted chair, as well feeling too hot or too cold from poor regulation of room temperature. Fatigue from eyestrain can occur from inadequate lighting, screen glare or and lighting. Temperature was the most uncomfortable element in the office for participants in both studies. Studies show that fluctuations in temperature, too hot, too cold have a negative effect on health, well-being and productivity (Haynes, 2008, ASHRAE, 2004).

The intellectual comfort level improved from the beginning of the day to the middle of the day, and then declined at the end of the day to about the same level as the beginning of the day. Anticipation about cognitive work completed can cause a decline in comfort due to anxiety and the nature of cerebral work (Murata, Uetake & Takasawa, 2005). This may attribute to the marginally significance ANOVA result in intellectual comfort level. Emotional comfort levels were consistent with studies by Vink et al., (2009) throughout the day and work week. Although, Thursday showed a marked improvement in all comfort levels by the end of the day and the last workday. This is consistent with studies that participants are ready for the weekend and exhibit excitement and positive expectations for the upcoming days off (Integrated Media Measurement, 2009, Sonnentag et al., 2008).

On average throughout the week, intellectual comfort levels improved from beginning to mid-day and declined again at the end of day. Interestingly 48% said the tasks of the office made them most comfortable. The physical comfort levels declined throughout the day and the emotional comfort levels remained consistent or improved by the end of the day.

The words "relaxed" and "calm" were circled to represent the waking comfort level, and "tired" was circled describing the end of the day comfort level, though technically the end of day comfort levels were similar to those upon waking.

## 4.3    Physical conditions are the most important factor in influencing comfort

Interestingly, the number one factor that made the participants most comfortable was the "people" in the office. People have an influence on the office environment and can change a seemly physical environment comfortable or uncomfortable. Comfort and discomfort includes the interaction with co-workers or clients as well as factors and experiences outside of work, such as family, friends, health and personal business matters.

The survey found that people made them most comfortable and the office temperature the most uncomfortable. This was dependent on the mood and responsibilities for each day. A workweek has variations in personal duties, interactions, and biorhythms. According to Trump, (2009) working with people you like makes a difference in comfort at work. Oftentimes we try to improve the physical conditions and then measure comfort (Vink, 2005). Introducing people into an environment has influences on all three-comfort levels.

The natural environment affects not only the physical health but also the psychological health of human beings (Sundstrom, Bell, Busby & Asmus, 1996). There are many studies in environmental psychology research related to the effect of natural elements. For example, moreover, in offices, the window size and sunlight penetration influences the worker's mood (Sundstrom, Bell, Busby & Asmus, 1996).

In this study, the physical "temperature" was the most uncomfortable factor (ASHRAE, 2004). However, "people" were found to be the second most uncomfortable factor.

Physical conditions are an important comfort factor according to studies by Vink (2005). The discomfort pyramid (Bubb, 2008) demonstrates that if the environment is unpleasant the physical conditions will ultimately overrule all other factors if not addressed (Figure 2).

Figure 2 The discomfort pyramid of Bubb (2008)

However, having said that, the emotional connections may be more important for sustainability of comfort.

## 5    CONCLUSION

The results confirm the time of day relationships between pre-office experiences and comfort levels and 12 out of 16 participants said they were looking forward to work for the day. A positive attitude towards work is beneficial for the health of the company as well as the worker (Trump, 2009). The word circled by most participants to describe physical waking comfort was "relaxed", the word for mid-day comfort was "calm", and the word for end of day comfort was "tired". Feeling tired and having lower comfort levels at the end of a working day is consistent with findings by Vink et al. (2009).

The intellectual comfort level improved from the beginning of the day to the middle of the day, and then declined at the end of the day to about the same level as the beginning of the day. Anticipation about cognitive work completed can cause a decline in comfort due to anxiety and the nature of cerebral work (Murata, Uetake & Takasawa, 2005). This may attribute to the marginally significance ANOVA result in intellectual comfort level.

Emotional comfort levels were consistent with studies by Vink et al., (2009) throughout the day and work week. Although, Thursday showed a marked improvement in all comfort levels by the end of the day and the last workday. This is consistent with studies by Sonnentag, Mojza, Binnewies, & Scholl (2008) that participants are ready for the weekend and exhibit excitement and positive expectations for the upcoming days off

By providing optimal internal and external conditions, the concept of "comfort in an office" equates to health and wellness in and out of the workspace and workplace to improve overall productivity and well-being. Often we try to improve the physical conditions and then measure comfort (Vink, 2005). However, this study shows the importance of the pre office experience and expectations and the fluctuations in the day of the week.

## ACKNOWLEDGMENTS

This study would not have been possible without the survey participants. Thank you for your time and support.

## REFERENCES

ASHRAE. (2004). ANSI/ASHRAE Standard-55-20004: Thermal environmental conditions for human occupancy. American Society of Heating, refrigerating and Air-Conditioning Engineers, Atlanta.

Bilchik, G. S. (2002). "A Better Place to Heal" *Health Forum Journal*, 45(4), July-August, Chicago: Health Forum Inc.

Bornstein, R. F. (1989). Exposure and affect: Overview and meta-analysis of research, 1968-1987. *Psychological Bulletin*, 106, 256-289.

Bornstein, R. F. (1999). Source amnesia, misattribution, and power of unconscious perceptions and memories. *Psychoanalytic Psychology*, 16, 155-178.

Bubb, R. (2008). Sitting comfort, presentation at IQPC aircraft interior innovation.

Green, W. and Jordan, P. (2002). Pleasure with Products: Beyond Usability, New York, NY: Taylor & Francis.

Haynes, B. (2008). The impact of office comfort on productivity. *Journal of Facilities Management*, 6, 37-51.

Helander, M.G. and Zhang, L. (1997). Field studies of comfort and discomfort in sitting, *Ergonomics*, 40: 895-915.

INQA-Buero (2005). Well-Being in the Office. Federal Institute for occupational Safety and Health (ENWHP) Dortmund-Dortsfeld, ISBN: 88261-469-9, pp.2-36.

Integrated Media Measurement, Inc. Dvr Primetime Multi-tasking Skyrockets as the Workweek. Retrieved June 4, 2009, from http://www.immi.com/marketTests.html.

Konieczny G. (2001). The measurement an increase of the quality of services in the Flugzeugcabine a contribution for customer-oriented airplane development, Dissertation, TU Berlin.

Looze M.P. de, Kuijt-Evers L.F.M., Dieën J.H. van.(2003). Sitting comfort and discomfort and the relationships with objective measures. *Ergonomics*, 46: 985–997.

Murata, A., Uetake, A. & Takasawa (2005). Y. Evaluation of mental fatigue using feature parameter extracted from event-related potential. *International journal of industrial ergonomics* 35(8) pp. 761-770.

Richards, L.G. (1980). "On the psychology of passenger comfort." In D.J. Oborne and J.A. Levis, eds., Human Factors in Transport Research, vol. 2. London: Academic Press, 15–23.

Sonnentag, S., Mojza, E., Binnewies, C., Scholl, A. (2008). Being engaged at work and detached at home: A week-level study on work engagement, psychological detachment, and affect, *Work & Stress*, 22:3, pp. 257-276.

Sundstrom, E., Bell, P. A., Busby, P. L. & Asmus, C. (1996). *Environmental psychology Annu. Rev. Psychol*, Camino Way Palo Alto CA: Annual Reviews Inc.

Slater,K. (1985). Human comfort (Springfield, IL: Thomas).

Statistical Package for Social Sciences® (2003). [SPSS for Windows, Rel. 12.0.0] Chicago: SPSS, Inc.

Trump, D. It's Good Business to Work with People You Like. Retrieved July 2, 2009, from http://www.trumpuniversity.com/mynetwork/inside-trump-to-tower/issue09.cfm.

Vink, P. (2005). Comfort and design: principles and good practice, Boca Raton: CRC Press.

Vink, P., Konijn I., Jongejan B., and Berger M. (2009). Varying the Office Work Posture between Standing, Half-Standing and Sitting Results in Less Discomfort. *Ergonomics and Health Aspects*, HCII 115-120.

# CHAPTER 69

# Analysis of Changes in Work Processes

*Aleksandra Kawecka-Endler, Beata Mrugalska*

Poznań University of Technology
Poznań, Poland
aleksandra.kawecka-endler@put.poznan.pl, beata.mrugalska@put.poznan.pl

## ABSTRACT

Over the last two decades much attention has been directed to the notion of work process and its changes. New forms of employment and work, which have aroused those days, result from the introduction of changes in the method of people employment. The traditional form of employment based on full-time work is replaced with part-time jobs. Performance of such work determines a fast development of self-employment. This paper draws on data and case studies collected mainly among Polish enterprises. The paper examines the role of small enterprises and its role in economy. Furthermore, it presents different aspects of work safety. Finally, the contribution tries to answer the question on how new work practices influence enterprises' performance.

**Keywords**: work process, human factors, work conditions

## 1    INTRODUCTION

Profound political, socio-economic and demographic transformations and progressing globalization, registered in the surrounding environment of both Polish and European enterprises, are the reason of changes occurring in their organization and methods of activities.

In Poland the effects of these transformations are fundamental changes in a structure and organization of enterprises and applied forms of employment. In order to fulfil varied and often changing requirements companies have to do current tasks effectively and efficiently. Such way of functioning makes the enterprise perform constant analysis of threatens existing in the surrounding environment and requires a possibly fast response. On the other hand, a systematic and rational introduction of changes is a basic condition of the enterprise success which should result in:

- flexible organization which enables fast adjustment to changeable market requirements,
- safety work conditions,
- good quality and modernity of products and services,
- improvement of efficiency and productivity,
- increase of client satisfaction.

The characteristic feature of economic structure of member states of the European Union, among which Polish economy is numbered, is a huge participation of small enterprises. In this group there are also microenterprises which employ up to 10 people according to Commission Regulation of 25 February 2004 (EC No 364/2004). Their annual turnover or annual balance sheet total does not exceed 2 million euro. However, independently from the fact that they are business entities, which have a small manufacturing potential (work sources, profit stream and capital sources), they are highly significant for economy as:

1. the sector of these business entities belongs to the biggest group of employers in Poland,
2. the functioning of microenterprises, next to small and medium-sized enterprises (SMEs), is conducive to healthy competiveness and is an expression of entrepreneurship in society.

In the European Union aaccording to data released in 2007among ca. 20.5 million of enterprises, microenterprises constitute 18.8 million and their contribution is 91.9%.

In Poland enterprises, which belong to the sector of SMEs, represent 99.86% of all enterprises (3784 thousand). The detail contribution of enterprises according to employment size is the following:

- microenterprises 94.86 %,
- small enterprises 4.2 %,
- medium enterprises 0.8% (Lubińska-Kasprzak, 2009).

According to data (2009) the number of establishing new microenterprises is 300 000 in one year, whereas 250 000 are liquidated in one year. The index of survival calculated for these companies is 66%.

## 2 SMALL ENTERPRISES AND THEIR SIGNIFICANCE FOR ECONOMY

One of widely used forms of employment is a so called self-employment which is associated with the establishment of a one-person company creating a one-person business entity. This form concerns a situation where a natural person conducts a business entity on his own account and risk.

It can be assumed that such a decision is taken by three groups:

1. people who want to build their own company from the ground up, develop it and achieve specific earnings and professional satisfaction from this activity,

2. people who want to run a freelancer business and work for many customers having a free hand in organizing their own work,

3. people who are often induced to run such a business by employer because of expected economic benefits and then they take up or continue their professional career formally within the confines of their own business entity (Kawecka-Endler, 2004; Skulska, and Kawecka-Endler, 2011).

Running such business entity is not an easy task. It requires current knowledge about agreements, declarations, terms of payment for insurance and all types of taxes from the future entrepreneur. The fulfilment of these obligations is difficult as law regulations are changed very often in Poland (Kawecka-Endler, 2005).

In order to get more precise information about this form of employment the PIT.pl portal group did a research in 2010. The results of it show that the greatest fear of self employed is connected to a small size of a business and the risk resulting from this fact. However, it is not a reason for striving to turn a small company into a bigger one. Thus, it can be assumed that self-employment is rather a matter of choice and it results from positive assessment of own possibilities and potential. The characteristics of problems connected to the practice of running such a business are illustrated on Figure 1 (www.samozatrudnienie.pit.pl).

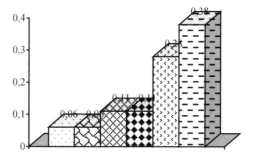

□ I am happy that I do have work and the rest is not important for me

☒ Self-employment facilitates me to find work and even in a few companies at the same time

☒ Employers force self-employment for their own benefits

☐ It makes me savings but it is not ethical

☒ Self-empoyment is burdened with a great economic risk

□ Contribution rates for the Social Insurance Institution are firmly too high

Figure 1 Results of statements of self-employed about the problems of this form of employment (Adapted from (www.samozatrudnienie.pit.pl))

# 3    NEW FORMS OF EMPLOYMENT AND WORK

Contemporary enterprises function in conditions of continuous and often turbulent transformations. In the result requirements and principles of competitions for companies, managers and employees have changed.

New forms of employment and work result first and foremost from the changes in the way of employing people. In this case the traditional form of employment based on full-time work is replaced with part-time jobs. It often relies on transfer of work out of the company and the employee is employed on a contract basis. Such way of performing work determines a fast development of self-employment.

The progress of this type of work is conditioned by the following activities:

- an aspiration of enterprises to reduce costs and improve competiveness,
- an attempt to increase chances of employing certain professional groups,
- an alleviation of situation on labour market which relies on joining partial employment with partial unemployment benefit.

The mentioned activities result from first of all the need of adjusturization of employers to the needs of restructurization and continuous and difficult to forecast changes in the enterprise environment. Moreover, taking up such activities brings a lot of benefits. First and foremost, it leads to more flexible and efficient management of possessed resources in the answer to unpredictable economic conditions. The example of changes on labour market is the growing number of work done within temporary contracts, part time contracts or contracts with an indefinite period (Krzyśków, 2009).

According to the statistical data in the end of 2011 year in Poland 27% of workers were employed temporarily.

In the opinion of the Polish Employee's Organization in present conditions statutory regulations of working time are anachronism and new forms of employment should be legalized as they are used for a long time in the West European countries and they turned out to be useful in practice. However, at present flexible systems of working time cannot be used in Poland because of complex and incomprehensible legal regulations which are obsolete and do not fulfill present needs of enterprise. Thus, the organization postulates to give up statutory regulation of working time and introduce longer accounting period and possibility of flexible planning of rest. Theoretically such solutions applied into practice constitute a chance and should lead to the increase of efficiency and competiveness of enterprises (Piechowiak, 2011).

The problem of efficiency and competiveness concerns both production and service companies. With reference to product production in the most of enterprises a lot of activities were taken up, different concepts and methods were implemented (e.g. lean production, six sigma lean management, reengineering, benchmarking, controlling, HR etc.), which introduced a significant increase in efficiency and improvement of competiveness.

In the forecast elaborated in 1996 *Handy* wrote that: *„now time of service is approaching where productivity is invariable for a long time ago"* (Handy, 1996).

This statement is still very actual as the years pass/go. The participation of service activities in the market increases systematically, however, it does not reflect in clearly perceptible improvement of productivity and increase of competiveness of service companies.

Nowadays in all well developed countries productivity is one of the basic criteria of assessment of functioning of enterprises. The effects of small productivity are always high prices and increase in demand for production factors (materials, energy, production area etc.). Low productivity influences existing economic and social problems fundamentally (it causes increase of inflation and unemployment, deterioration of quality of life among other things). The increase of productivity is due to fall of cost of business activities and increase of profits what next allow to increase individual salaries, influence development and motivation of employees positively (satisfaction from work and its effects), improves social satisfaction generally (Durlik, 1993).

Nowadays the following forms of work can be differentiated:

1. Employment contract with definite period of time – known form but now it develops very fast i.e. in the EU countries 9.3% of people are employed for a definite period of time. This form is not favourable for employees as it does not offer most of privileges concerned in traditional employment and the contract comes to the end by virtue of law upon lapse of time which it was concluded.

2. Part-time employment – connected with partial activity of employees, its character is voluntary and in the EU countries it concerns ca. 15% of employees whereas it taken up by 26% and 19% of people in Japan and the USA, respectively. There can be distinguished two forms of work:
   - *job-sharing* – arrangement where typically two (or more) people perform a job at one workplace and as a team they share compensation voluntarily and other privileges of this workplace,
   - *work-sharing* – method, which relies on reducing number of working hours, concerns a group of workers and is used in order to avoid laying off employees.

3. Temporary employment – in Poland there are no regulations, in the EU it is used really. In practice it may be connected with:
   - transfer of management staff – organized by temporary work agency. Workers do various works, companies reduce their costs, and the situation of the employee is worse,
   - employee transfer – one or a few employees are eligible for transfer or lend another company.

4. Work on-call up, demand or telephone – the obligation of the employee is to wait for an employer's demand and then come to work.

5. Outwork or cottage industry – the worker is employed full-time and the work is done at home.

6. Telework – long-distance work facilitated by computers and telecommunication. It is a good chance for disabled and women who bring up children. The main disadvantage of this type of work is lack of contact with people. In Poland this form of work concerns only 2% of all employed,

however, its trend is growing. The main reason of small interest in telework is rather mentality than law. The workers surveyed by the Polish Agency of Enterprise Development are afraid of lack of interpersonal contacts, responsibility for work effects and lack of possibility of on-going consultation of problems with their boss. They are also wrongly anxious about lost of working privilege resulting from work at home because teleworkers have the same rights and duties as the workers employed in stationary mode (Wójcik-Adamska, 2012).

7. Employment without employer – often used in private companies. The reason of such an arrangement is the desire to avoid working costs and contribution rates paid to the Social Insurance Institution. Both sides decide that the employers will do their tasks on the basis of a different legal arrangement. The employer establishes a one-person business which provides service for this company (one of the forms of self-employment).

All these arrangements were previously regarded as atypical but now they are increasingly becoming the norm (Baruch, and Smith, 2002). New ways of working are perceived as continual opportunities to be flexible and to routinely adapt to fit current needs. Moreover, employees are directly influenced by changes towards more flexible and less hierarchical working methods such as flexibility in working time and location, multiple skills, flexibility in job content and greater use of part-time and fixed term contract staff (Shapiro, 2001). These forms are often popular with employees as they allow them to have more control over their work and personal lives. In the consequence it can decrease conflict between the choice of these two and increase job satisfaction (Thomas, and Ganster, 1995; De Route, 1995; Creagh, and Brewster, 1998). Furthermore, in the literature it is shown that flexibility in work arrangements leads to getting the work done rather than how, where and/or when it is done (Lobel, and Kossek, 1996; Shapiro, 2001). It can contribute to such benefits for employers as lower absenteeism and turnover and increases in productivity (Dalton, and Mesch, 1990).

According to the opinion of the Employee's Organization the following flexible forms of employment should be taken into account in the Polish Labour Code:
1. Flextime as an arrangement where an employee is at employer's disposal in a specific time (e.g. from 11a.m. to 14 p.m.) and the rest of working time can be made up for any time during the day e.g. after 18 p.m.
2. Flexplace as a type of flextime, but it does not concern working time but working place e.g. an employee work 5 hours per day off-site location and three hours in the office. Both forms are very often combined.
3. Job-sharing as an employment arrangement where one job is divided between two workers who share their compensation, duties and work in different time e.g. every two week.
4. Compressed working week as an arrangement which allows to divide number of week working hours into smaller amount of days e.g. 48 hours can be worked for in 4 days instead of 6 days (Piechowiak, 2011).

# 4    WORK SAFETY

The growing number of accidents at work in Poland is undoubtedly the effect of lifestyle change, technical progress, transformation of economic systems, pursuit of work, fast living and lack of time for rest and bad working conditions. The accident risk increases each year in all human business units (Kawecka-Endler, 2007; Pawłowska, 2009).

The negative phenomena, particularly intensively growing in last years (connected to the world economic crisis), are earning irregularities. Labour inspectors affirmed that in 2010 as much as 36% of controlled employers (in 2009 – 32%) did not pay all due workers' compensations, and every fourth of them did not pay it on time. National Labour Inspectoriate conducted more than 95thousand inspections at about 70 thousand employers in 2010 (Wójcik-Adamska, 2012). They investigated that in 337 thousand cases there were violated safety and work health regulations. Over 9.5 thousand from among them included orders to stop work tasks immediately due to imminent hazards to health and safety, and 8 thousand decisions ordered workers to stop operation of machines and equipment. Furthermore, in 2010 the district labour inspectors issued 33 decisions which ordered to stop business activities what is an extreme in practice (Zając, 2011).

In practice, it is shown how it is difficult to reconcile economic development with assurance of safety and hygiene work conditions in accordance with regulations. In spite of introduction of many modern protections of workplaces, systematic controls and preventive activities the number of accidents is still increasing. The situations of threatens are still often belittled or even there is lack of awareness of potential hazard of accident to both employees and employers. Thus, the existing low condition of safety and work hygiene culture should be systematically increased in Polish enterprises. It should be also remembered that in case of unawareness of possibility of existence of accident hazards or in the situation of risk underestimation of accident occurrence, it is indisputable to generate any motivation to detect hazards or take up preventive actions (Skulska, and Kawecka-Endler, 2011).

In that case it is very crucial and vital to aware both employees and employers of negative consequences of disobeying safety standards.

In the last years the reasons and circumstances of accidents at work are the subject of interest of numerous research and analysis. They are basically carried out in order to set causes of accidents and then introduce preventive actions. There is no doubt that the accidents at work have a compound character and are the results of combination of technical, environmental, human and organizational event occurring.

The analysis of statistics of accidents at work allowed to indicate the most common causes of their occurrence. The research showed that the causes of accidents are known for years and unfortunately they recur (e.g. not using personal protective equipment, bad psycho-physical condition of employee). One of the factors, which influence accident rates significantly, is work organization at workplace. Among organizational causes such as improper work organization, lack of appropriate supervision, disregard for OSH rules, improper training etc. were indicated (Zając, 2011).

Working conditions about which safety decides a considerable number of varied factors, can also influence accident threat condition (in a different degree and scope). In practice the probability of occurrence of accident threats can be often influenced by one specific particularly important factor.

New forms of work and employment implicate a number of challenges to safety and work conditions.

Well and safely done work (at each workplace) is an integral part of each enterprise and it directly influences different aspects of its business such as its quality, competitiveness and productivity. The awareness of existence of such a relation should be a basis to take a decision connected with:

- design, shaping and assessment of work processes,
- application of certain technical and organizational solutions,
- technical condition of machinery, technical devices, tools and instrumentations,
- assurances of safety and work hygiene at particular workplaces and in the whole enterprise,
- application of defined labour law.

The decisions connected with the mentioned tasks directly influence work safety in the enterprise and should be in accordance with provisions in force (for employee and employer). In order to realize them in practice a specific knowledge about ergonomics, work and production organization, human resource management etc. is required.

New forms of work and employment have also an impact on change of real needs and solutions used in practice (for entrepreneurs and employees) with reference to knowledge about work safety.

Nowadays a systematic development of interest and participation of employees in safety and work hygiene management called direct participation is noticed (Lubińska-Kasprzak, 2009). In Poland a concept of corporate social responsibility (CSR), which is a form of corporate self-regulation of social and environment interests, begins to develop (Kaźmierczak, 2009).

## 5    CONCLUSIONS

It is not difficult to notice that the changes concerning new forms of work and employment, which were registered in the last years, were predicated much earlier. The example of it can a forecast made by Charles Handy in the ninetieth for a new century which aim was to describe a direction of changes of employees, organizations and the work itself. The author wrote then that: "there must be a significant change in a way of our thinking about organizations. They are organizers, employers and are minimalistic" (Handy, 1996).

Nowadays the unusual accuracy of this statement can be assessed. Handy also predicted that the number of workers employed in large enterprises would decrease whereas the number of workers employed in smaller companies and self-employed would increase. In the result of changes much more employees will not have a job at all because of lack of specialist knowledge needed both inside and outside companies (Handy, 1996).

All these changes also concern with the nature of work itself, its division, scope and way of realization. Change and changeability concern each activity of human being regardless of the role which he fulfils in work process. It involves all objects of business activities such as production and service. The interest in work refers to work effects by the assumption that the performed work is interesting and allows to achieve satisfying remuneration. The measure is a specific work result called productivity or work efficiency (Kawecka-Endler, and Mrugalska, 2010).

The increase of work efficiency is possible by the assumption that in productive and effective processes:

- employees find safety and free from threats working conditions,
- work allows employees to educate, improve abilities and develop (Bullinger, Warnecke, and Westkämper, 2003).

It is also necessary to eliminate perceived irregularities (which are unfortunately quite popular among employers) such as: not paying salary on time, irregularities in working time evidence and calculation of overtime, lack in transfer of premium for social employee insurance (Piechowiak, 2011).

In practice the most crucial characteristic of new forms of activities of business entities is their openness to introduction of changes and flexibility which allow to consider the above mentioned assumptions.

## REFERENCES

Baruch, Y., and I. Smith. 2002. The legal aspects of teleworking. *Human Resource Management Journal* 12(3): 61–75.

Bullinger, H.J., H.J. Warnecke, and E. Westkämper. 2003. *Neue Organisationsformen im Unternehmen* (2 ed.). Berlin-Heidelberg: Springer-Verlag.

Creagh, M., and C. Brewster. 1998. Identifying good practice in flexible working. *Employee Relations* 20(5): 90–503.

Dalton, D., and D. Mesch. 1990. The impact of flexible scheduling on employee attendance and turnover. *Administrative Science Quarterly* 35: 370–387.

De Route, F. 1995. Reconciliation of family and work. In. *Workshop Proceedings on the Reconciliation of Family and Work*, Athens, Greece.

Durlik, I. 1993. *Inżynieria zarządzania*. Katowice: Wyd. Naukowe amp., Andrzej Matczyński Publisher.

Handy, C. 1996. *Wiek paradoksu. W poszukiwaniu sensu przyszłości*. Warszawa: Dom Wydawniczy ABC: 161–162.

Lobel, S.H., and E. Kossek. 1996. Human resource strategies to support diversity in work and personal lifestyles: Beyond the 'family-friendly' organization. In. *Managing diversity: Human resource strategies for transforming the workplace,* eds. E. Kossek, and L.S. Blackwell. Oxford, UK: 221–244.

Kawecka-Endler, A. 2004. Innowacyjność w sektorze małych i średnich przedsiębiorstw. In. *Controlling w małych i średnich przedsiębiorstwach,* eds. P.D. Kluge, and P. Kużdowicz. Zielona Góra: Wyd. Drukarnia „Druk-Ar": 37–46.

Kawecka-Endler, A. 2005. Analiza wpływu wybranych czynników innowacyjności na rozwój małych przedsiębiorstw. In. *Uwarunkowania rynkowe rozwoju mikro i małych przedsiębiorstw,* ed. A. Bielawska. Szczecin: Wydawnictwo Naukowe Uniwersytetu Szczecińskiego. *Rozprawy i Studia* 571: 61–70.

Kawecka-Endler, A. 2007. Safety and hygiene of work as a basis of the enterprises strategy. In. *Ergonomics in Contemporary Enterprise,* eds. L.M. Pacholski, and S. Trzcieliński. Proceedings of the Eleventh International Conference on Human Aspects of Advanced Manufacturing: Agility and Hybrid Automation (HAAMAHA). Poznań, Poland.

Kawecka-Endler, A., and B. Mrugalska 2010. Contemporary aspects in design of work. In. *Advances in Human Factors, Ergonomics and Safety in Manufacturing and Service Industries,* eds. W. Karwowski, and G. Salvendy. CRC Press Taylor & Francis Group. Boca Raton, USA: 401–411.

Kaźmierczak, M. 2009. Bezpieczeństwo i higiena pracy, a rozwój koncepcji społecznej odpowiedzialności biznesu. *Bezpieczeństwo Pracy* 5: 10–13.

Krzyśków, B. 2009. Ochrona pracowników zatrudnionych w nietypowych stosunkach pracy. *Bezpieczeństwo Pracy* 5: 17–19.

Lubińska-Kasprzak, B. 2009. *Raport o stanie sektora MŚP w Polsce.* Warszawa: PARP.

Pawłowska, Z. 2009. System zarządzania bezpieczeństwem i higieną pracy, a rozwój koncepcji społecznej odpowiedzialności biznesu. *Bezpieczeństwo Pracy* 1: 13–15.

Piechowiak, Ł. 2011. „Nowe formy zatrudnienia – szansa czy zagrożenie?" Accessed February, 2012, http://www.bankier.pl/wiadomosc/Nowe-formy-zatrudnienia-szansa-czy-zagrozenie-2321601.html

Skulska, M., and A. Kawecka-Endler. 2011. Wpływ projektowania ergonomicznego stanowisk pracy na zmniejszenie liczby wypadków przy pracy. In. *Praktyczne aspekty projektowania ergonomicznego w budowie maszyn,* eds. A. Kawecka-Endler, and B. Mrugalska. Poznań: Wydawnictwo Politechniki Poznańskiej.

Shapiro, G. 2001. Exploring the impact of changes in work organisation on employees. *AI& Society* 15: 4–21.

Thomas, L., and D. Ganster. 1995. Impact of family-supportive work variables on work-family conflict and strain: A control perspective. *Journal of Applied Psychology* 80: 6–15.

Wójcik-Adamska, K. 2012. Praca w domu nadal mało popularna. Accessed February 27, 2012, http://www.rp.pl/artykul/807236.html?print=tak&p=0

Zając, T. 2011. Główny Inspektor Pracy podsumował miniony rok. *Bezpieczeństwo Pracy* 5: 3–5.

CHAPTER 70

# Towards the Ideal Loungeworkseat

*Annechien Verkuyl*

Intespring B.V., The Netherlands
annechien@intespring.nl

*Peter Vink*

Delft University of Technology, The Netherlands
p.vink@tudelft.nl

## ABSTRACT

In many offices and terminals lounge areas are introduced. However, there are no lounge seats available which support a person with a hand held device to work in an optimal way. This paper is about an experiment in which requirements for a loungeworkseat are established. 15 subjects worked with a laptop in different positions. The posture was recorded and questionnaires on comfort and discomfort were completed. The self-chosen position resulted in highest comfort ratings. It was interesting to see that the ideal comfortable angles of the seat pan and back rest were connected like in a synchromechanism of an office chair. The laptop was preferred on the lap. A foot rest is preferred as well also to compensate for body length. It is recommended to build a seat based on these requirements and test this seat again with different hand held devices.

**Keywords**: lounge, chair, comfort, office, productivity, notebook

## 1    INTRODUCTION

### 1.1    More lounge areas in the office

"New ways of work" is getting much attention in the Netherlands. Often the following elements of this new way of work are observed: flexible working times, working from home or while travelling, Information Communication Technology

(ICT) which facilitates distant work, a new way of leader-ship, more empowerment and an office with shared desks and an interior design that stimulates informal meetings (Blok et al., 2009). Elements of this new way of work differ depending on the definition and how the organization designed and implemented the program. Mostly a small office is used with less work stations than the number of employees working at the office. This is possible due to the shared desk principle. An effect of introducing the "new ways of work" is that employees are less connected to the company (Hengst et al., 2008). Hengst et al. (2008) evaluated six cases in which the "new ways of work" principle was introduced. Three were governmental organizations and three were commercial organizations. One of the main concerns in these six cases was the fact that employees feel less connected than in the traditional office. As a consequence employees could be less loyal and accept easier a job in another organization. One of the measures companies take to enlarge the employee connection is to make the office interior more attractive by including lounge areas. Another consequence of the new ways of work is that work is also done while travelling. For the waiting time while travelling lounge areas in airports, train stations and in hotels are introduced, where work can be done as well.

## 1.2    More laptops and tablets

Another aspect of "new ways of work" is that new information and communication technology is introduced. At the HFES 2011 congress Walline (2011) presented that in 2008 the number of sold laptops was for the first time more than the number of sold desk tops (see fig. 1). The number of sold tablets also increased significantly since the iPAD 2 was announced in February 2011. This means that laptops and tablets will be used more often also in the office. However, most office interiors are now mainly occupied for desk top work. Also, the ergonomic guidelines focus mainly on desk tops.

Figure 1. Sales of desk tops, laptops and tablets presented by Walline (2011).

## 1.3    The need for a lounge work chair

The fact that more lounge areas are introduced and the need for environments where employees can work with laptops and tablets, made us decide to develop a loungeworkchair. There are many seats and chairs on the market for lounge areas (Vink, 2009). Rosmalen et al. (2009) developed a lounge chair for watching a screen, which facilitates to sit in various positions (see fig.2). Subjects participating in the tests of this new seat liked the comfort and the possibility of varying their posture. However, they missed the possibility to work with a laptop. Other studies show that there is a difference in position reading and typing (Groenesteijn et al., 2009). So for an ideal loungeworkchair various positions supporting different tasks should be possible. Another requirement is that the change in position should be realized with intuitive controls. In a study of Vink et al (2007) it is shown that 63% of all office workers never adjust their chair. One of the reasons was that users did not understand the adjustment controls. This indicates the importance of intuitive adjustment controls.

Figure 2. Variations in postures found while watching a screen in lounge chair developed by Rosmalen et al. (2009).

## 1.4    Requirements for a lounge work chair

The above mentioned literature findings were input for the following requirements for the design of a lounge work chair:
- it should look attractive as it should fit within lounge areas (at offices) and should attract employees
- it should facilitate working with a hand held device and feel comfortable
- it should be adjustable to increase performance during reading and typing
- adjustability should be intuitive

The intention was that based on these requirements a concept could be designed. However, in designing the concept the lack of guidelines for reading and working in a lounge position was a problem. Therefore, an additional study was developed. A research chair was developed, which could be adapted to lounge work positions and an attempt was made to find ideal positions for subjects for reading and working with a laptop. Based on a model two ideal positions were established by the designer and the subjects were asked to define an ideal position for themselves (the 3th position).

## 1.5    The model for defining two postures

A biomechanical model was made to limit the number of possible positions. The model consists of several segments, as shown in figure 3 (left). The seven segments indicate the respective rear of the body. The dimensions of the segments are determined on the basis of data from the online website DINED (www.dined.com, selected for: 2004 Dutch adults ages 31-60, P50 men). The ranges of the several joints in the model that are used are defined by Lange et al. (1988). In the model the angles of the elements could be adjusted manually (see figure 3 right).

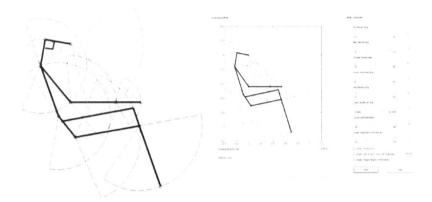

Figure 3. Segments in the biomechanical model to limit the choice of positions (left) and the model in matlab used to define ideal sitting postures while using a laptop (right).

For the angles between seat and backrest 110° and 120° were chosen. The 120° is based on the results of Zenk (2008) who showed that sitting with a reclined back with a good arm support resulted in the lowest discomfort. This corresponds with Wilke (1999), who showed that the pressure in the intervertebral disk is also low in a reclined backrest position. The 110 ° was chosen as Knijnenburg (2007) recorded 109 ° as the optimal back rest position for watching a screen.

The model is made in matlab and the user can change all body segments angles, within range of motion ( see fig 4). The model will determine arm segments angles (by using inversed kinematics) by setting each angle on the same percentage of the maximal range of motion per joint. By using several scenarios of the eye viewing angle the ranges of different positions were collected. These different positions where checked whether shear forces stay within the range defined by the  model of Goossens et al. (1995). There are different positions possible to get an angle of 110° or 120°  between the seat and back. The figure 4 model showed two optimal postures for each condition, given that the shear forces are kept as low as possible. Therefore, in the test set up and the research chair these positioned were chosen with the angles:
- Position 1: Seat angle 20° back angle 40°
- Position 2: Seat angle 22° back angle 52°

In both positions the laptop is positioned just above the legs, resulting in an eye viewing angle of 36°.position).

## 1.6   Research question

The research question for this study is: which of these two positions is most comfortable and seen as most productive according to end users and what position do end users choose if they are free to adapt the chair after experiencing these positions?

## 2   METHODS

## 2.1   Procedure

15 students of the Delft University of Technology aged between 22 and 31 years old (6 female, 9 male) participated in the research.

Table 1 The three conditions of this study

| Condition | Backrest angle | Seat angle | Laptop | Head rest |
|---|---|---|---|---|
| 1 | 130° | 20° | on tray table | 90° |
| 2 | 142° | 20° | on tray table | 90° |
| 3 | free | free | free | free |

Each experiment started with  an explanation of the research chair and the goal of the research were given and a measurement of the body measures. Then in the experiment three different conditions were studied (see table 1), random selected. The subjects took each position for 30 minutes and performed three tasks, each 10 minutes: reading, writing and gaming. The comfort and discomfort experiences were measured three times just after each condition, every ten minutes, by completing the questionnaire. After experiencing all three conditions participants were asked to complete a final questionnaire with questions on preferred positions and angles.

During the experiment, the subjects could change their position without changing the supporter angles. The subject was able to adjust the supporter (all lengths, angles and the kind of laptop support) to their preferred position at the beginning of the third condition.

The 30 minutes per condition were recorded on four video camera's : one recording laterally the side of the subject, one the front of the subject , one camera focused on the arms of the subject and the last camera focused on the pressure distribution on screen (see fig 4).

All subjects had seven markers on their body. These markers were positioned by palpating the body (acromion, lateral epicondyle, ulnar styloid, trochantor major, lateral condyle, lateral malleolus, and two on glasses to determine the head position (see fig 5).

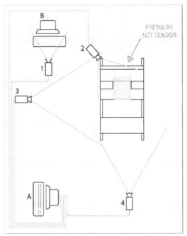
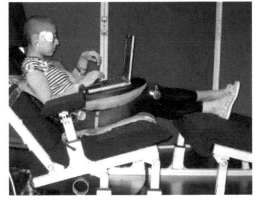

Figure 4. The experimental setup    Figure 5. The subject with markers on the joints.

## 2.2    Data analysis

The average scores of the local postural discomfort (=LPD) of all participants per condition and time of recording were calculated. For the LPD a summation of all the regions was used. Additionally, the first LPD score was subtracted from all the others to indicate the change of the LPD while sitting in the supporter. Finally, the body location showing the highest LPD was established. By using the Friedman Nonparamatic test significances between conditions were tested ($p < .05$), since the data were not normally distributed. In the photographs the markers were connected in the program solidworks (see figure 7). In the third condition the dates were analyzed using frequencies in spearman's rho correlation (bivariate, $p < 0.05$).

Table 1  Characteristics of the subjects

|  | Age | Length (cm) | Weight (kg) | Upper leg (cm) | Body length while seated (cm) |
|---|---|---|---|---|---|
| Mean | 26.5 | 174.9 | 70.7 | 48.7 | 88.2 |
| SD | 2.8 | 11.9 | 16.2 | 5.4 | 5.1 |
| P50 male | X | 182 | x | 50 | 94.6 |

## 3    RESULTS

Table 2 shows an overview of the characteristics of the 14 subjects. Condition 1 shows an increase in discomfort over time, while in the other conditions the discomfort is unchanged in time. The third position had the lowest discomfort ratings. This third condition was also considered as the most productive position according to the test subjects, but showed much variation (see fig 6). In the neck, most people experienced the highest discomfort. This discomfort in the neck is in the third position the lowest (p=0.001). The subjects closest to the P50 showed a lower discomfort in every condition.

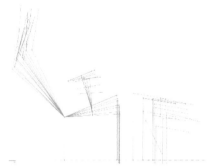

Figure 6. A plot of all connected lines between the markers.

In the third condition the head support was positioned in such an angle that it always followed the tilt of the seat and back (p=0.008,r=0.721). If the backrest is tilted backwards the angle between headrest and backrest is smaller.

The larger the subjects the larger the length of the backrest is chosen (r=0.602). However, 12 people did change the head support angle instead of the backrest length.

An interesting finding was that there was a relationship between preferred back angle and seat angle: A rotation of the backrest of 5° backwards corresponds to a seat angle 3° upwards of the front of the seat (p=0.027,R=0.667). The mean angle of the relative angle between seat and back is 118° (SD 7.5). Results also indicate a relationship between seat height and leg support height, however the correlation is not high (r=0.459). This indicates large variations on preferences within the group.

Regarding the armrest the preference was to connect it to the supporter and have it fixed in the position it was just above the lap. Only four subjects changed the armrest height. The horizontal position was not changed.

The position of the laptop support was varied a lot during the research. The support connected on top of the right arm rest was sometimes chosen, but often a position on the lap (in 11 out of 14 situations) was preferred as well. It was clear that dependent on the task a position was chosen. For reading a higher viewing point was chosen than for typing. It should be noted that in a laptop the position of the hands and viewing point are coupled. In third condition changes in the position of the foot rest is used to give enough support for the lower and upper legs.

The open questions supported the outcomes. Not only LPD values were lowest in the third self-chosen situation, it was also scored as the most comfortable position. Additionally, most subjects mentioned that they preferred the laptop on their lap, because of stability.

# 4    DISCUSSION

This study did shine some light on the research question "which of these two positions is most comfortable and seen as most productive according to end users and what position do end users choose if they are free to adapt the chair after experiencing these positions". The third condition was most often preferred by the test subject, but also showed much variation. Additionally, for the subjects close to the P50 condition 1 and 2 was rated comfortable. For the whole group the self-chosen condition was most comfortable. This position was also considered as the most productive position according to the end users.

The results show that there are two areas where discomfort is experienced by the users. These are the neck and the lower back areas. This is found in other VDU studies as well (Young et al., 2012; Groenesteijn et al., 2009). Neck discomfort can be caused by neck bending, which is needed because the viewing point is low (Vink, 2005). A solution for this problem could be to heighten the object and support the neck. By this neck support the user needs less muscle force to balance the head above the spine. This is in agreement with the research of Franz et al. (2012).

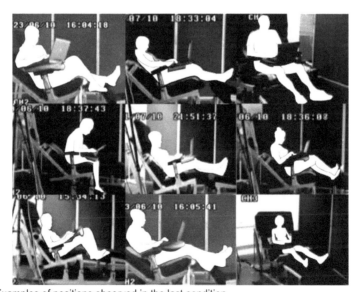

Figure 7. Examples of positions observed in the last condition

The third position was for every person different (see fig. 7). This diversity in position was in agreement with the research of Rosmalen (2009) and Leuder (2004). However, we found some interesting characteristics.

The comfort is better when the head rest follows the tilt of the seat and back. Moreover, by changing the headrest instead of the backrest length the chair became more comfortable. Franz (2012) did find the same result. Adding a head rest which has contact with the head reduced the neck discomfort.

The found relationship between the back and seat makes it possible to design a chair that is easier to adjust. No separate controls for seat pan and back rest are needed. This reduced degree of freedom reduces the separate adjustment actions the user needs to take to adjust the support, which is preferable as the adjustment can be kept intuitive. Moreover, the seat depth could be fixed at one length since most of the people will use the foot rest to increase their comfort, thereby avoiding pressure on the thighs and inside of the knees. In the study of Rosmalen the subjects also used foot rests in many different positions (Rosmalen et al., 2009), showing the importance of a foot rest.

The results on the relationship between seat height and height of the footrest was not significant. However, the analysis of the videos showed that people have a different posture of their feet when the feet rest is at the same height as the seat. This is more difficult in other situations. To stimulate 'motility' (small differences in posture) the same height is still preferred.

The support of the notebook is an important requirement of the chair. The user wants a solid base for the laptop. However, during the test the people were limited in changing their laptop position. Future research should focus more on the laptop position, which is a conclusion of the study of Gold et al., (2011) as well.

The hypothesis was that the back angle would be larger compared with working at a desk. However, the mean of the chosen back angles was more upright than expected (still larger than on normal office seats), being 133° (SD 5.5). The mean of the seat angle is 14.7° (SD 6.1). A reason for this upright back angle could be the position of the laptop and the tasks on the laptop. All subjects fixed the laptop on their knees using a cushion. The low position of the laptop results in a neck flexion. A large back angle would result in even a larger neck flexion which is also concluded by Harrison et al (2000). He showed that sitting with a large back angle will bent the neck too much, when looking at a viewing point located below eye level. In conclusion, the position to work with "handheld devices" is driven by the viewing position and hand position.

During the research the subjects placed the laptop further away when they had to read a large text. This result is in agreement of Ellegast et al. (2012) who concluded that the effects on posture are more dependent on the task than on the configuration for comfort of a supporter.

The aim of the project was to design a chair in which a variety of users could work and launch in a comfortable and productive way. Based on the results a new concept could be proposed. The requirements are still not specific enough to make a complete design. Therefore, if a new seat is designed, it is probably wise to test this new version first.

A shortcoming of the study was the fact that the chair did not follow the movements of the person in the chair (active and passive working position). This could have affected the comfort experience. Moreover, other hand held devices should be involved in the study to find out if there is one preferred position for all the handheld devices (including the tablet, laptop, smart phone, mobile phone etc). Therefore, a new research needs to take this points in to consideration. And with this inputs the concept could be improved and build into a real working device.

Conclusion: This study shows that it is possible to gather additional requirements to design a lounge work seat.

## REFERENCES

Blok, A.M., E. de Korte., L. Groenesteijn, M. Formanoy, P. Vink. 2009. The effects of a task facilitating working environment on office space use, communication, concentration, collaboration, privacy and distraction. *Proceedings of the 17th World Congress on Ergonomics*, Bejing, cd-rom.

Ellegast, R. P., K. Kraft, L. Groenesteijn, F. Krause, H. Berger, P. Vink 2012. Comparison of four specific dynamic office chairs with a conventional office chair: Impact upon muscle activation, physical activity and posture. *Applied Ergonomics* 43: 296-307.

Franz, M., A. Durt, R. Zenk, P.M.A. Desmet, 2012. Comfort effects of a new car headrest with neck support. *Applied Ergonomics* 43: 336-343.

Goossens, R.H., C.J. Snijders. 1995. Design criteria for the reduction of shear forces in beds and seats. *J. Biomech*, 28(2): 225-30.

Groenesteijn, L., P. Vink, M.P. de Looze, F. Krause. Effects of differences in office chair controls, seat and backrest angle design in relation to tasks. *Applied Ergonomics* 40: 362-370.

Hamberg-van Reenen, H.H., Physical Capacity and Work related Musculoskeletal Symptoms (diss. Amsterdam VU), Amsterdam 2008.

Harrison, D.D., et al.. 2000. Sitting biomechanics, Part II: Optimal car diver's seat and optimal driver's spinal model. *J. Manipulative Physiol Ther* 23(1):p.37-47.

Hengst, M. den, J. de Leede, M.P. de Looze, F. Krause and K. Kraan. Working on distance, virtual teams and mobile work (in Dutch) TNO, 2008 Hoofddorp.

Walline, E. 2011. Presentation at the Human Factors and Ergonomics Society 55th Annual Meeting - 2011 HFES 2011 congress.

Knijnenburg, R. 2007. Future passengers side truck seat for DAF, confidential, 2007 TU-Delft.

Lange, W. 1988. Kleine ergonomische datensammlung. 5 verand. Auf tuv Rheinland Koln 1988.

Lueder, R. 2004. Ergonomics of seated movement. A review of the scientific literature. Humanics ergosystems, Encino.

Rosmalen, D.M.K. van, L. Groenesteijn, S. Boess, P. Vink. 2009. Using both qualitative and quantitative types of research to design a comfortable television chair, *J. of Design Research* 8: 87 - 100.

Vink, P. 2009. Aangetoonde effecten van het kantoorinterieur. (proven effects of the office interior (in Dutch): Alpen aan den rijn: Kluwer.

Vink, P., 2005. Comfort and design: principles and good practice, CRC Press, Boca Raton.

Vink, P., R. Porcar-Seder, A. Page de Poso, F. Krause. Office chairs are often not adjusted by end-users. In: Proceedings of the Human Factors and Ergonomics Society (HFES) 51st Annual Meeting, Baltimore, October 1-5, 2007. CD-ROM.

Wilke, H.J., et al. New in vivo measurements of pressures in the intervertebral disc in daily live. Spine (phila pa 1976). 1999, 24(8): P.755-62.

Young, J.G., M. Trudeau, D. Odell, K. Marinelli, J.T. Dennerlein, J.T., 2012. Touch-screen tablet user configurations and case-supported tilt affect head and neck flexion. Work 41: 81–91.

Zenk, R., 2008. Objectivierung des Sitzkomforts, und seine automatische Anpassung (Dissertation TU München).

# CHAPTER 71

# New Ways of Working in a Notebook Manufacturing

*Symone A. MIGUEZ [1], Carlos R.PIRES [2], Jefferson L.R. DOMINGUES[3]*

[1]Delft University of Technology,
Faculty of Industrial Design Engineering
The Netherlands
Ph.D Student
symone@ergosys.com.br

[2,3]Maintenance Technicians
Mechatronics Students
Esatec - Brasil

## ABSTRACT

In order for manufacturing companies to increase the productivity and the quality of products, they need to incorporate the quality of life of workers in their production process. However, this cannot be done without difficulty, especially when there is no budget for ergonomic improvements or people specialized in ergonomics. Thus, this study was developed to show new ways of working; it was carried out in a multinational company in Brazil, it lasted two months and had four steps. The first step of this study involved a 16-hour training for maintenance technicians of the company. The second step consisted in inviting two maintenance technicians to develop ergonomic devices for the notebook sector. The third step was to allow these technicians to identify opportunities for ergonomic improvements in the area. The fourth step was to develop, partnered with the technicians, ergonomic devices at no cost, using existing material in the company, such as MDF boards and pieces of tube pipe. The first ergonomic device was a garbage collection tipping cart. This device facilitates the collection of garbage, preventing movement of the shoulder above 90 degrees. The second ergonomic device was a cart to separate and distribute notebook labels. This device has decreased the time of separation of the labels by 30% and removed the manual loading of boxes. In order to check posture improvement, RULA instrument was used before and after the implementation of the ergonomic devices. The conclusion of this study showed positive outcomes, increasing the quality of life and productivity of workers. We can say that the training was the basis for the success of this study, but what about the new way of working? For us the new way of working in manufacturing provides for exchange of knowledge that can be translated into three words: Ergonomic Team.

**Keywords**: notebook; manufacturing; design; ergonomic device

# 1    INTRODUCTION

The impacts of globalization and the easy access to technology have changed both work relations and environments. Flexible working hours and home office are ever so common in our day-to-day. This novel panorama may be described as "New Ways of Working (NWW)" and it consists of changes within four aspects of work: 1) physical space, 2) technology, 3) organization and management, and 4) culture at work (Block et al., 2012). The aim of this paper is not to discuss the aspects of these transformations but to demonstrate that in order for one to have the technology needed for the NWW one cannot refrain himself from investing in the quality of life of workers in the electronics industry, who need to produce more and more computers in order to meet the increasing demand for these devices. The sales of notebooks in Brazil has been growing every year. The Brazilian Association of the Electronics Industry (ABINEE) estimates that the trading of computers will reach 16.7 million units in Brazil in 2012, which represents a growth of 9% when compared to last year, when 15.3 million units were sold. From this amount, 6.2 million units correspond to sales of desktop computers whereas the sales of notebooks and netbooks reached 9.1 million (ABINEE, 2012).

The preference for notebooks and netbooks is ascribed to the accessible prices and mobility of these devices.

The present 4-step study was initiated due to this universe that comprises the high productivity of notebooks, lack of funding granted to ergonomic investments and the willingness of maintenance technicians to acquire knowledge about ergonomics. The first step included the training of two professionals from the maintenance area who joined the task force called "Ergo Development" within the Ergonomic Committee of the company. The remaining steps were able to provide the new members of the committee with opportunities to apply their recently acquired knowledge in ergonomics, which resulted in two ergonomic devices. The first device was designed to collect recyclable material and the second one was made in order to facilitate the distribution of notebook labels in the assembly line. Both devices prevent work-related musculoskeletal disorders.

## 2    METHOD

The place where this study was conducted was a 3000-employee company in the electronics industry in Brazil. Workers volunteered to participated in this research, which lasted two months and was divided into 4 steps for didactic purposes.

**Steps of Method**

**Step 1: Training**
The 16-hour ergonomic training for the members of the ergonomic committee at the company takes place once a year. It is divided into four 4-hour sessions once a week and it is carried out by an outsourced ergonomics consultant, who possesses ergonomics certification. In addition to the basic concepts of ergonomics, the concepts of anthropometry are widely disseminated in order to meet the practices of the maintenance area. According to Dall'Oca & Sampaio (2001), educational actions possess a potential that is accommodative at times and transforming at others. It is the perspective of transformation that stimulates discussions as well as pro-active attitudes within the health of the workers. In face of these statements, we have decided to stimulate changes in people's behavior by suggesting a new way of working.

**Step 2: Invitation to the maintenance technicians**
Silverstein & Carcamo (2006) report that the modifications in productive environments are usually made by engineers and maintenance technicians. Therefore, inviting these technicians to join the Ergonomic Development group in the Ergonomic Committee after training was a fundamental milestone for the development of this study. Staff members were motivated by the training and willing to put their knowledge into practice. The technicians were personally invited by the ergonomist to participate in the development process; she informed them that all the ergonomics activities would take place during regular work time, that no overtime would be necessary and that it was approved by both the area management and coordination.

**Step 3: Identification of improvements in the notebook sector**
After participating in the ergonomic training, both maintenance technicians chose the place where the ergonomic improvement should occur. There was no specific demand such as musculoskeletal complaints or requests from workers in order to select the activity in which the ergonomic intervention would be carried out. However, the choice could not have been more appropriate because we usually focus on ergonomic improvements within the production cells or assembly lines and tend to neglect those activities that give support to the production process itself. If we paid more attention to the support areas, we would notice how these are essential and may end up being the solution for drawbacks in the production process.

**Step 4: Development of ergonomic devices**
Two ergonomic devices were developed (cases 1 and 2) in order to bring about a practical ergonomic intervention. Leftover materials such as MDF and pipes were employed, aiming at minimizing costs and recycling materials, thus contributing to sustainable actions.

**Subjects**: **Case 1**: A participant of this study was a general services assistant in the notebook production, male, in his fifties, 1.67m tall, with primary education. The subject works in fixed shifts of 8 hours a day, from Monday through Friday.
**Subjects**: **Case 2**: Another participant of this study was a notebook assembly line operator, male, in his mid twenties, 1.75m tall, with secondary education. The subject works in fixed shifts of 8 hours a day, from Monday through Friday.

**Instruments case 1 and 2:** The ergonomic analysis of the task was done through direct observation of postures, unstructured interviews with the workers, and pictures. Comparisons of data before and after the ergonomic intervention were made through RULA (Rapid Upper Limb Assessment), which is "a screening tool that assesses biomechanical and postural loading on the whole body with particular attention to the neck, trunk and upper limbs." (McAtamney & Corlett ,1993).

## 2.1    CASES

### 2.1.1    CASE 1 -  GARBAGE COLLECTION TIPPING CART

**Description of the <u>activity</u> <u>before</u> intervention:** The activity of collection of recyclable waste material throughout the six notebook assembly lines is performed several times a day by a sole general services assistant. The tasks involved in this activity are: 1) place a large paperboard box onto a cart, see figure 1; 2) walk along the entire notebook area and collect the recyclable waste; 3) deposit all collected material into an exterior dumpster for recycling and reinitiate the activity. There are no determined cycles for these tasks because the worker is responsible for establishing his own pace of work.

Figure 1 Cart for the transportation of recyclable material

**Description of the <u>problem</u> <u>before</u> intervention:** The large amount of material deposited in the cart rendered its removal to the external dumpster a hard task. In addition, the worker needed to move his shoulders above 90 degrees, figure 2.

Figures 2 and 3 Large amount of material deposited in the cart and dumped into external area

## 2.1.2 CASE 2 - CART TO SEPARATE AND DISTRIBUTE NOTEBOOK LABELS

**Description of the <u>activity before</u> intervention:** This activity consists of distributing the labels throughout the six notebook assembly lines and is performed by a sole operator several times a day. The tasks involved in this activity are: 1) collect the empty paperboard boxes from a separate area close to the assembly line and manually bring them to the area where the sealed boxes of labels are; 2) place the labels in a box according to the requests from the production; 3) distribute the labels in the assembly line boxes and transport them manually or using a simple cart, see figure 3; 4) wait in front of each assembly line approximately six minutes until the labels are checked and, if necessary, receive the leftover labels and relocate them into the cabinet. The cycles of this task were 1 hour long so all six assembly lines could be served, which reached an average of 10 minutes per line.

Figure 4 Lifting box to be placed onto cart and pushing the cart.

**Description of the <u>problem</u> <u>before</u> intervention:** During an informal 10-minute interview with the worker the following difficulties were reported: 1) difficulty to grab the boxes since they do not possess any kind of hand support , figure 4; 2) mixing and loss of material because they were not fixed in place in the box; 3) constant complaints by supervision and quality control staff due to the wrong placement of labels in the assembly line process.

Figure 5 Boxes in which the labels were transported.

# 3    APPROACH AND RESULTS

## 3.1 Case 1 – Garbagge collection tipping cart

A cart (height=1m; width=0.60m, length=1.26m) with a tilt   (height=0.78m; width=0.76m, length=1.11m) for collecting recyclable material was designed. The worker no longer had to position the box on his shoulders but to tilt the cart in order for the material to slide down, this action is aided by a small rake for removing the material left on the bottom of the box, figures 6 and 7.

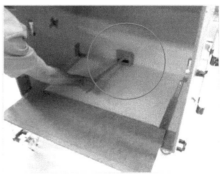

Figure 6 New tipping cart                    Figure 7 Auxiliary small rake

Each cart cost U$160.00. The acceptance by the worker was satisfactory. RULA instrument returned a significant improvement outcome, see table 1.

Table 1 - RULA Scores CASE 1 -  For the posture adopted when collecting waste material from the box and dumping it outside

| BEFORE INTERVENTION | **Grand Score (right/left): 7/7** A score of seven or more indicates investigation and changes are required immediately. |
|---|---|
| AFTER INTERVENTION | **Grand Score (right/left): 3/3** A score of three or four indicates further investigation is needed and changes may be required. |

## 3.2    Case  2  -  CART  TO  SEPARATE  AND  DISTRIBUTE NOTEBOOK LABELS

The cart (height=1.36m; width=0.60m, length=1.34m) for the separation of labels was designed in such a way to facilitate the task for the worker as well as the storage of the labels, figure 8. The new cart allowed for: 1) more space in the storage area; 2) improvement of quality of manufactured notebooks through the elimination of the issue of mixing the materials; 3) storage to be done on the cart itself, eliminating the need for the worker to wait 6 minutes at the assembly line; 4) less physical and mental effort; 5) the decrease of the time for separating labels by 30%. The making of the cart cost U$ 285 per unit. RULA instrument returned a significant improvement outcome, see table 2.

Figura 8 Label cart.

Table 2 - RULA Scores Case 2 - For postures adopted when lifting the boxes and placing them onto the cart

| BEFORE INTERVENTION | Grand Score (right/left): 5/5 A score of five or six indicates investigation and changes are required soon. |
|---|---|
| AFTER INTERVENTION | Grand Score (right/left): 3/3 A score of three or four indicates further investigation is needed and changes may be required. |

## 4    CONCLUSIONS

In order to remain competitive in the market, companies in the most diverse segments are always looking for cost reduction without affecting product quality or client's satisfaction. However, this can only be done effectively with the aid of a multidisciplinary team (Berk, 2010); without it, the endeavor may damage both employees' health and the company's financial health (labor problems, absenteeism, among others) in the short and medium terms. We have initiated slight changes in the areas of support to the production process and we have observed a direct impact therein: the waste material was collected more adequately and there a 30% gain in the separation of labels, which benefited the notebook assembly lines. We knew that good results would not come about solely through the knowledge of ergonomic principles. Literature is clear when it states that trainings should focus mainly on people's behavior (Lavender, 2006). We believe the learning of ergonomic concepts, along with the opportunity to put into practice that knowledge acquired through the development of ergonomic devices are the aspects that contributed most significantly to the positive outcomes. We must emphasize that the basis for our results stems from participatory ergonomics. Therefore, the interaction between the ergonomist, members of the ergonomics committee, those responsible for the evaluated areas and the workers expresses a new way of working within the manufacturing areas. Not the utopian way, but the healthy one that involves constructing together, gaining together; the way that can be translated into two words: Ergonomic Team.

## ACKNOWLEDGMENT

The authors want to thank the company, the workers, Mr. Carlos Pierre and his team, for supporting this study.

# REFERENCES

"ABINEE – Associação Brasileira da Indústria Elétrica e Eletrônica", accessed February 01, 2012, http://www.abinee.org.br.

Blok,M., Groenesteijna, L., Schelvisa, R. and Vink, P. 2012. New Ways of Working: does flexibility in time and location of work change work behavior and affect business outcomes? *IOS Press Work*, 41: 2605-2610.

Berk, J. 2010. Organizing a Cost-Reduction Program, in *Cost Reduction and Optimization for Manufacturing and Industrial Companies*, eds. John Wiley & Sons, Inc., Hoboken, NJ, USA.

Dall'Oca,A.V. and Sampaio.M.R. 2001. As Intervenções Educativas da Fundacentro - ERMS no Campo da Segurança e Saúde do Trabalhador. In *Trabalho-Educação-Saúde: um mosaico em múltiplos tons,* eds. Fundacentro, pp.33-53.

Lavender,S.A.2006.Training Lifting Techniques. In *Interventions, Controls, and applications in occupational ergonomics,* eds. W. S. Marras and W. Karwowski. Second Edition, pp. 23-1-24-15, CRC Press.

McAtamney, L. and Corlette.E.N. 1993. RULA: A survey method for the investigation of work-related upper limb disorders. *Applied Ergonomics*, 24 (2): 91-99.

Silverstein, B., Spieholz,P. and Carcamo,E. 2006. Practical Interventions in Industry Using Participatory Approaches. In *Interventions, Controls, and applications in occupational ergonomics,* eds. W. S. Marras and W. Karwowski. Second Edition, pp 3-1-3-27, CRC Press.

# CHAPTER 72

# New Ways of Work: Task Specific Train Seat Design

*Liesbeth Groenesteijn, Merle Blok,*
*Suzanne Hiemstra-van Mastrigt and Peter Vink*

TNO Healthy Living, P.O. Box 718, 2130 AS Hoofddorp, The Netherlands
& Delft University of Technology, The Netherlands
liesbeth.groenesteijn@tno.nl

*Cedric Gallais*

SNCF Research and Design, Paris, France

## ABSTRACT

Working in the train is a part of New Ways of Working and is also a way to attract more passengers from the car into the train. However, it is unknown which is the ideal working position. Moreover, the ideal position for reading and relaxing is also unknown. Therefore, in this study the observations, and later on performed experiments, will lead to requirements for future train seats to enable working, relaxing and reading in comfortable postures.

**Keywords**: train passenger activities, postures, comfort, seat design

Figure 1 Example of working in a train

# 1 NEW SEAT DESIGN FOR MOBILE WORKERS IN TRAINS

The way we work is changing (Manoochehri & Pinkerton, 2003). Nowadays information technology enables new ways of working. For example in the UK the number of teleworkers is estimated to more than double in popularity in eight years to 2.4 million workers (Office of National Statistics, 2006), indicating that telecommuting continues to become an omnipresent work arrangement (Golden, 2006). Telecommuting is working outside the company office building and can be done at home, at an external location but also while travelling. In the US more specified numbers are presented in the WorldatWork 2006 Telework Trendlines (2007) for working while travelling; of 149.3 million workers in the US labour force, 10.6 million (7%) had worked on an airplane, train or underground railway. For both employer and employee it is efficient that travel time can be used to perform work tasks. For employees this means that they can balance their work and private lives better (Beauregard and Henry, 2009). An advantage of the train with respect to the car is that working is easier with PDA, smartphone or laptop especially now internet access is possible in trains.

A potential disadvantage of working while travelling is that this mobile workplace does not provide an optimal working posture and that this is less comfortable and less productive for the worker compared to the office workplace. Moving places are still designed for transporting employees, not for performing their work (Vartiainen and Hyrkkänen, 2010)

The question for train seat design then is what the ideal seat is for a train to optimally support productive work and also relaxation. Studies regarding postures and activities performed in the train were conducted in the past (e.g. Branton & Grayson, 1967; Bronkhorst et al, 2005), but the way of working and telecommuting possibilities has been changed.

To design seats for the future new knowledge on postures and activities are needed. Kamp , Kilincsoy and Vink (2011) recently published the activities performed and the associated postures adopted, while in semi-public/leisure situations and during transportation, as inputs for seat design in cars. People have diverse behavior and perform different activities. For creating a comfort experience it is important to take the behavior and also the diversity in sizes of users into account. SNCF (French railway operator) has defined the research project DYnamic Activity POSture (DYAPOS) to solve the gap between dynamic and postural comfort. They approached TNO for their expertise in studying comfort in seating. The objective of the total study with TNO is to define scientifically based train seat requirements to provide comfortable seats for current and future travelling by train. The objectives for the study with TNO were:

1. to define what train's passengers mainly performed activities were and which corresponding postures were adopted, and to evaluate the effects of different morphology groups;
2. to evaluate the required seat adjustments to provide a comfortable posture adapted to the activity, postures and the various morphologies.

The first phase, presented here, was to determine the main activities performed by the passengers, their mainly adopted postures.

## 2    METHODS FOR TRAIN PASSENGERS OBSERVATION

## 2.1    Observation types

Two types of observation were performed; short term to collect observations of a large population and long term to observe durations of performed activities and number of  different performed activities in a journey. The observations were all made in 4 different high speed train types with 5 different seat types in both first and second class cars. These seats were selected after having measured and compared various seat parameters of all SNCF seats mounted in commercial trains. Observations were done  mainly in France, but also in Belgium, the Netherlands and United Kingdom. Comfort questionnaires data were collected of most of the observed subjects.

*Type 1 Short observation*
Short observation of momentary activities and corresponding postures to measure many passengers (aimed at 500-1,000 passengers). Every person was observed only once, in order to get as many different persons' postures and activities. The short term observations were performed in the period of November and December 2010.

*Type 2 Long observation*
The long observation consists of an observation of approximately 1-2 hours in order to measure the number of activities/postures used by passengers during a journey (aimed at 50 passengers). The passengers' activity and postures were determined at T=0, and after that real-time activity changes, posture changes and micro movements (short movements without an actual posture change) were measured. The long term observations were performed in January 2011.

## 2.2    Observation measurement system and configuration

*Observation measurement system*
The observations were performed with handheld PDAs and a fully configured observation protocol. The observers were guided through the observations by this configuration and protocol. Every activity was indicated as a new data row in the database.

The following variables were observed:
- Main characteristics: train, car and type of seat;
- Main characteristics observed person: seat position, sex, age category, morphology category (according to SNCF earlier analysis on distinguishing morphology categories);
- Equipment: book, laptop and position on table, lap or bag;
- Main activities: 10 activities;

- Corresponding seat contact of body parts;
- Corresponding postures of body parts.

In the long observation also macro- and micro movements were observed.

## 2.3    Protocol

Questionnaire and PDA data were connected by a code, compromised of seat number and configuration. Questionnaires were handed out after the observations per car were finished. A pre-set of observation schemes for random selection, of which seats to observe, was made to avoid selection by preference of observer.

*Protocol short observations*

The observation started with entering the car specifications. There was no specific selection of observed persons, other than age (children and adolescents were excluded from observation).

The corresponding posture and corresponding seat contact were selected and registered in the PDA .

*Protocol long observations*

With the long observations, the observers followed activities and postures of 2 to 3 persons simultaneously. Micro movements, activity changes, posture changes and partial changes in posture were entered at the occurring moment during the observation period.

## 2.4    Data analysis

*Short observation analysis*

The data of the observation of seat contact, seat posture and activity will be combined per person, resulting in a specific code. The aim was to select the activities with the highest percentage of observation in each morphological group, and leave activities and postures with a low percentage of total observation out of the analysis.

The following analysis steps were made:

1.    Removal of incomplete/faulty data files;
2.    A general overview of all activities and total morphology was made;
3.    Selection of the 4 main activities was done by selection of the highest frequencies of observed activities;
4.    Selections of the main postures corresponding to the 4 main activities were made by selection of the highest frequencies. A reduction of observed categories was done by selection of both posture - contact codes that represent the most important body parts and contact areas (head position, backrest contact, back posture, buttock contact) in relation to seat design.
5.    The posture-contact codes that cover 60% for each of the 4 main activities lead to a top 8 of postures. For these postures was checked whether this was representative for the morphology distribution of all observation data.

*Long observation analysis*

The aim of the long observations was to observe the number and duration of different activities that passengers perform during 1-2 hours in the train seat.

The following analysis steps were made:

1. Removal incomplete/faulty data files;
2. The number of observed changes in activities and variation in activities per subject are counted;
3. The frequency and duration of activities over all subjects are counted;
4. The number of posture changes, partial posture changes and micro movements per subject are counted over the total observation duration per subject;
5. The variety in body part postures is shown over the 4 main activities.

# 3 RESULTS OF SHORT AND LONG OBSERVATION

## 3.1 Main characteristics of short observations

After removal of incomplete/faulty data files, 786 observations were used for further analysis distributed in:

- 287 female and 499 male;
- 702 adults (18-60) and 84 elderly (>60);
- 293 first class and 494 second class passengers.

Figure 2 shows the observation distribution in morphological groups. The largest group by far is the "middle male or female" category.

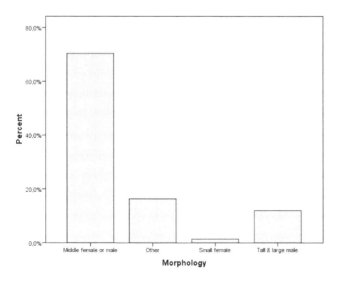

Figure 2    Morphology distribution of the observed population (n=786)

## 3.2 Main activities of short observations

A first top 4 main activities selection was made based on the observations and in discussion with SNCF for Reading, Staring or sleeping, Talking, and Working on laptop. This selection of activities covers 70% of all observations. Distribution of the observed activities is shown in figure 3.

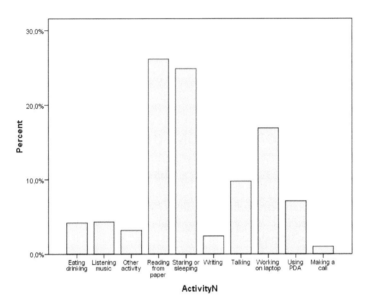

Figure 3     Activities distribution based on frequencies of short observations

## 3.3 Corresponding postures to four main activities

The posture-contact codes that cover 60% for each of the 4 main activities lead to a top 8 of postures. It was verified that for this selection of 8 postures, the morphological group had a distribution similar to the overall tested population. Figure 4 shows the observed posture-activity combinations and the corresponding comfort scores (Questionnaire item: How do you evaluate your comfort on your seat to practice this activity? Scale 1-10: not comfortable at all - very comfortable) . A variation is seen in comfort rating of postures with performing different activities.

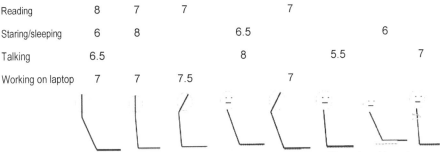

| Reading | 8 | 7 | 7 | | 7 | | |
| Staring/sleeping | 6 | 8 | | 6.5 | | 6 | |
| Talking | 6.5 | | | 8 | 5.5 | | 7 |
| Working on laptop | 7 | 7 | 7.5 | | 7 | | |

Figure 4 Main activities, corresponding postures and comfort scores (Question: How do you evaluate your comfort on your seat to practice this activity? Scale 1-10: not comfortable at all - very comfortable)

The top 8 of most observed postures is used as input for the experimental variations in the following research parts.

## 3.4 Main characteristics of long observations

Out of 48 data files 30 observations contained useful data for analysis. The excluded data files did not contain considerable data because the measurements were stopped early. This was mainly caused by a disturbance in observation (for example ticket control).

The distribution of observed persons and classes was:
- 9 female and 21 male;
- 25 adults (18-60) and 5 elderly (>60);
- 21 middle/female, 4 tall large male and 5 others;
- 8 first class and 22 second class passengers.

The observation time varied from 16 minutes to 2 hours 5 minutes and the average was 1 hour 11 minutes.

## 3.5 Main activities of long observations

During the time observed, subjects changed in activities within a range of 2-26 times and with a variation 2-6 different activities. There is a lot of variation between subjects. The total time per activity and the total number that activities are scored, are counted (see Figure 5).

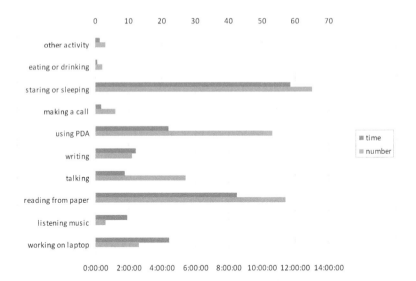

Figure 5    Total number (frequency) and total time duration (hours) of activities of 30 subjects

The top 4 main activities according to the total duration of 30 subjects is:
- Staring/sleeping ( average 29 minutes, range 1 minute-1 hour 29 minutes);
- Reading (average 28 minutes, range 1 minute-1 hour 8 minutes);
- Using PDA (average 15 minutes, range less than 1 minute-56 minutes);
- Working on laptop (average 53 minutes, range 14 minutes-1 hour 52 minutes).

The top 4 main activities according to the total number (frequency) is:
- Staring/sleeping;
- Reading;
- Using PDA;
- Talking (average duration 17 minutes, range 1 minute-36 minutes).

The top 3 according to the total number is similar to the top 3 in total duration. In duration, the activity 'working on laptop' is more prominent and in number, the activity 'talking' is more prominent.

Although not exactly comparable, because we did not analyze as far as in the short term observations to top 8 postures, the highest frequency scores for the body parts seem similar to the short term observations.

## 4    DISCUSSION AND CONCLUSIONS

Based on the observations and questionnaires we have selected the most observed activities and corresponding postures.

The short term observation results based on frequency of occurrence led to 4 main activities presenting 70% of all activities performed: Reading, Staring/sleeping, Talking, and Working on laptop. Associated with these 4 activities, 8 different postures were selected based on the variations in head position, back posture and buttock contact. This selection is still representative for the morphology of the total observed population.

In the long term observations there is a broad variation between passengers in number of activities performed, type of activities and duration of activities. The 4 main activities for total frequency of occurrence are Staring/sleeping, Reading, Using PDA, and Talking. The 4 main activities expressed in total duration are Staring/sleeping, Reading, Using PDA, and Working on laptop. The top 4 activities is different on one activity between short term and long term observation. Using PDA is not in the top 4 in short term observation, but is on a fifth place. Although the analyses was not identical, first results showed that long term and short term observation tends to present similar postures for the 4 main activities.

Concerning the large population of the short term observations compared to the small population of the long term observations, the frequency results of the short term observations will be leading in conducting experimental conditions. The frequency results of the long term observations do not differ so much that we need to make rigorous adjustments to main activities and selection of postures. The activity Using PDA might be interesting to take into account as the usage of this is growing.

The selection procedure postures in the present study excluded arm and legs postures, because variability was really small for these variables. As input for the onward performed experiments the corresponding arm and legs postures were added based on percentages of frequencies of the total observed population. Also the comfort scores and the morphology distribution is used as experimental input.

There were no statistical analyses performed between observed postures as the variety in group sizes based on frequencies was too diverse and capriciously divided.

Kamp , Kilincsoy and Vink (2011) observed passengers in German trains which resulted in a different order, but a similar top 4 of activities with Talking/Discussing, Relaxing, Reading and Sleeping. The study only took into account the number (frequency) of the activities and not the duration. Kamp et al. observed an equal number as 5th activity in using smaller and larger electronic devices with can be assumed to be PDA's and laptops. For most observed postures a full comparison cannot be made as the observation categories and analyses are different. The first two mainly observed postures seem comparable to the postures found in this study though.

The goal of this observation study and the later on performed experiments is to give directions for the design of train seats. As several researchers have shown (Corbridge and Griffin 1991, Khan and Sundström 2004, 2007, Krishna Kant 2007, Bhiwapurkar et al. 2010), a dynamic situation often influences the chosen activities. Vibrations and unexpected movements of the train have an influence on the comfort experience of passengers and should therefore be studied as well in onward experiments. For the development of comfortable train seats that allow New Way of Work it is important to take into account the different activities passengers want to perform and the difference in anthropometry that should be addressed.

710

## ACKNOWLEDGMENTS

The authors would like to acknowledge Valérian Courtaux and Hugo Marcus for their support with collection observational and questionnaire data.

## REFERENCES

Beauregard, T.A., Henry, L.C., 2009. Making the link between work-life balance practices and organizational performance. Human Resource Management Review. 19, 9 -22

Bhiwapurkar, M.K., Saran, V.H., and Harsha, S.P., 2010. Effect of multi-axis whole body vibration exposures and subject postures on typing performance. International Journal of Engineering Science and Technology, 2 (8), 3614–3620.

Branton, P. and Grayson, G., 1967. 'An Evaluation of Train Seats by Observation of Sitting Behaviour'. Ergonomics, 10: 1, 35 - 51

Bronkhorst, R.E. and Krause, F., 2005. Designing comfortable passenger seats. In: P. Vink, ed. Comfort and design, principles and good practice. Boca Raton: CRC Press, 155–167

Corbridge, C. and Griffin, M.J., 1991. Effects of vertical vibration on passenger activities: writing and drinking. Ergonomics, 34 (10), 1313–1332.

Golden, T.D., 2006. Avoiding depletion in virtual work: Telework and the intervening impact of work exhaustion on commitment and turnover intentions. Journal of Vocational Behavior 69, 176-187

Kamp, I. , Kilincsoy, U. and Vink, P., 2011. Chosen postures during specific sitting activities, Ergonomics, 54:11, 1029-1042

Khan, S. and Sundstro¨ m, J., 2004. Vibration comfort in Swedish inter-city trains – a survey on passenger posture and activities. In: Proceedings of the 17th international conference in acoustics (ICA), Kyoto, Japan, 3733–3736

Khan, S.M. and Sundstro¨ m, J., 2007. Effects on vibration on sedentary activities in passenger trains. Journal of Low Frequency Noise, Vibration and Active Control, 26 (1), 43–55

Krishna Kant, P.V., 2007. Evaluation of ride and activity comfort for the passengers while travelling by rail vehicles. Thesis (Master of Technology inMechanical Engineering). India: Indian Institute of Technology Roorkee.

Manoochehri, G., Pinkerton, T., 2003. Managing Telecommuters: Opportunities and Challenges. American Business Review. 21(1), 9 -16

Office of National Statistics. Labor market trends (October 2005), http://www.statistics.gov.uk/StatBase/Product.asp?vlnk=550, 27 September 2006.

Vartiainen, M. and Hyrkkänen, U., 2010. Changing requirements and mental workload factors in mobile multi-locational work. New Technology, Work and Employment 25:2, 117-135, 2010

'WorldatWork 2006 Telework Trendlines', 2007, Commissioned from the Dieringer Research Group. http://www.workingfromanywhere.org/ (accessed 19 July 2008)

CHAPTER 73

# A New Way to Establish Ergonomics Expertise in Manufacturing Locations Without Ergonomists

*N. Larson and H. Wick*

3M, USA
Email: nllarson2@mmm.com

## ABSTRACT

The primary objective of an ergonomics program in many companies is to reduce exposure to work-related musculoskeletal disorder (WMSD) risk factors, thereby resulting in reduced WMSD illness severity and incident rates. Establishing an effective ergonomics program in a global manufacturing company with over 190 manufacturing locations in over 30 countries requires new thinking and strategies about both human and technical resources. This case study describes the process used at 3M Company to develop knowledgeable and skilled ergonomics resources at each manufacturing location. These individuals identify unacceptable ergonomics risk exposures associated with manufacturing jobs in a facility and then implement solutions that are effective at reducing or eliminating the risk exposures. For these individuals to be successful, additional education about ergonomics and expertise in ergonomics job assessment is needed. To obtain this expertise, each individual completes ergonomics training and demonstrates knowledge attained and skills developed through participation in an internal ergonomics job assessment (EJA) Certification process.

**Keywords**: ergonomics training, ergonomics program, MSD, job analysis, WMSD

# 1 INTRODUCTION/BACKGROUND

The intent of the Ergonomics Program at 3M is to use ergonomics expertise consistently throughout the company to improve employee health and safety through development and utilization of standardized program elements, job analysis tools, and training programs. 3M has over 190 manufacturing locations in over 30 countries, with plant sizes ranging from less than 50 employees to over 1,500 employees. Since the vast majority of manufacturing locations have fewer than 400 employees, hiring professional ergonomists at each location is not always feasible. However, each location does have a professional safety, industrial hygiene, and/or occupational health resource. The corporate ergonomics program relies on these location resources to lead and implement the ergonomics program at each location (Larson and Wick 2012).

In 2003, as part of a company-wide ergonomics program revision, the Ergo Job Analyzer (EJA) (Auburn Engineers 2003) was adopted as the job assessment tool used to identify and assess risk exposure for all manufacturing locations. At this time a training and certification process was created so that location-based health and safety staff, as part of their formal job responsibilities, could learn to accurately identify ergonomics issues and implement effective solutions. The formal training and EJA Certification process validates that core ergonomic knowledge has been attained and verifies ergonomics job assessment skills are developed.

# 2 CERTIFICATION PROCESS

In support of the new ergonomics process, each location designates a health or safety person to become the EJA Resource and participate in a corporate-established EJA training and certification process. The EJA Certification process is the quality assurance that the EJA tool is used accurately to identify unacceptable ergonomics risk exposure. The EJA Certification process has three phases.

First, the EJA Resources, since few have formal ergonomics training, need to learn the basics of biomechanics, physiology, anthropometry, and workstation design. Each person pursuing certification completes several online training modules. The first module describes how the human body works and how musculoskeletal disorders develop. Another module introduces general principles of anthropometry and how work layout impacts employee postures and workstation design. Each module has a test and a homework assignment that must be completed prior to the next phase of the Certification - the EJA Workshop.

The second phase of EJA Certification is participation in the three-day hands-on EJA Workshop. At the Workshop, participants learn about and apply the EJA Tool to analyze manufacturing jobs for unacceptable WMSD risk exposures. They also practice using measurement tools such as force meters, goniometers, and pinch gauges, and learn how to take video to further analyze jobs. By conducting the Workshop at manufacturing locations, participants practice using the EJA Tool in a real-life setting. They also get to see other manufacturing processes in operation and experience valuable networking with others with similar job responsibilities.

The three-day EJA Workshop agenda includes the following topics:

- Company Ergonomics Program Overview
    - Ergonomics Risk Reduction Process Strategy and the EJA
- Review of Pre-Workshop Training Modules Homework
- EJA Tool
    - EJA Risk Factors
    - How to Use Job Assessment Equipment
- EJA Job Assessments - Jobs from Manufacturing Areas
    - Repetitive Motion Job
    - Manual Material Handling Job
- Brainstorming Solutions, Options and Choosing the Best Solutions
- Engineering, LEAN, and Ergonomics
- Cost Justification Techniques
- Participant Case Study Discussions

In the third phase of Certification, each EJA Resource submits evidence of accurately-completed EJAs. Three completed job improvement projects are submitted to the corporate ergonomics staff for review. Each job must include a completed baseline EJA, risk exposure conclusions, solutions implemented, and a follow-up EJA to verify that implemented changes were effective. To assist the corporate staff in the review, video of the job being performed is also provided. EJA Certification is completed with the presentation of a Capstone Project. This presentation summarizes one job improvement project and includes a problem statement, the baseline risk assessment findings, solutions considered and chosen, justification, and follow-up risk assessment conclusions. The Capstone Projects are presented at corporate-led ergonomics web conferences, supporting sharing of best practices throughout the company (Larson and Wick 2012).

## 3    RESULTS

Most 3M manufacturing locations have a designated EJA Resource with the knowledge and skills to lead ergonomic improvements. Forty EJA Workshops have been conducted in over 12 countries. Almost 100 employees have completed the EJA Certification process as part of their professional development and over 80 continue in the Certification process.

Over 2,500 jobs have been analyzed and improvements put in place to reduce 73% of the identified unacceptable MSD risk exposure, resulting in a 60% reduction in recordable MSDs in the US locations alone.

# 4     CONCLUSION

The EJA Certification process supports every 3M manufacturing location having the capability to identify and resolve their location's specific ergonomic concerns. One standard ergonomics assessment and improvement process: supports consistent and effective communication with management and employees; provides operational efficiency; facilitates the identification and management of WMSDs (Larson and Wick 2012); and finally, protects the assets of the corporation, including employee safety and health, product–quality and productivity, and the company reputation.

# REFERENCES

Auburn Engineers, Ergo Job Analyzer User Guide, Auburn Engineers, 2003.

Larson, N., Wick, H., 30 Years of Ergonomics at 3M: A Case Study, IEA 2012: 18th World Congress on Ergonomics, Work: A Journal of Prevention, Assessment and Rehabilitation, 41, Supplement 1/ 2012, 5091 – 5098.

CHAPTER 74

# Stimulation and Assessment of Physical Activity at Office Workplaces

*Rolf Ellegast[1], Britta Weber[1], Rena Mahlberg[1] and Volker Harth[2]*

[1]Institute for Occupational Safety and Health of the German Social Accident Insurance (IFA),
Alte Heerstraße 111, 53757 Sankt Augustin, Germany
[2]Institute for prevention and occupational medicine of the German accident Insurance (IPA),
44789 Bochum, Germany

## ABSTRACT

Prolonged physical inactivity and static work tasks may lead to serious health risks. Permanent low levels of physical loads are associated with musculoskeletal, cardiovascular, metabolic and mental disorders. A lack of physical activity (PA) is often a problem at office workplaces. However, PA interventions at office workplaces that aim at a quantifiable increase of PA have so far been performed rarely and the intervention effects on the PA behavior are often studied only through the methods of subjective self-assessment. For this reason a method inventory was developed consisting of objective PA assessment methods. The developed method inventory has been tested in a pilot intervention study at office workplaces. The current paper presents and discusses parts of the applied inventory. Several positive intervention effects could be observed: the intervention subjects spent more time standing and less time sitting during the work day, they showed a slight increase in physical activity intensity and felt better. No significant improvements concerning energy expenditure could be measured. The intervention subjects primarily used the breaks to perform additional PA. Since break times are limited there is still a lack of opportunities to perform PA directly at VDU workplaces. Therefore new innovative concepts are needed that directly increase PA and energy expenditure during office work.

**Keywords:** VDU workplaces, intervention study, measurement, physical activity, energy expenditure

# 1    INTRODUCTION

Today more than 40 % of all employees in the EU are working at visual display units (VDUs) (Parent-Thirion et al., 2007). Many office workers are sustained sitting without effective breaks and without body movements of sufficient intensity and duration. Permanent low levels of physical loads are associated with musculoskeletal, cardiovascular, metabolic and mental disorders (Pate et al., 1995; Aarås et al., 2000; Straker and Mathiassen, 2009). These health risks can be reduced by preventive strategies which promote physical activity (PA) at sedentary workplaces. Accurate and reliable measurement of PA behavior and its attributed health outcomes is considered an essential component of health promotion research and evaluation practice (Baumann et al. 2006). Past effect analysis often rely exclusively on self-reports. To determine effects on the movement behavior, hardly any objective measurement methods were used (Dishman et al., 1998; Dugdill et al., 2008; Proper et al., 2003). Furthermore, intervention effects were rarely quantified on the basis of physiological parameters. The aim of this study was to develop and test a comprehensive assessment inventory for PA within a pilot intervention study at office VDU workplaces. In this article results of the evaluation of parts of the assessment inventory for PA will be presented.

# 2    METHODS

## 2.1    Subjects and experimental design

25 experienced office workers (6 women, 19 men) volunteered as subjects in a randomized controlled trial at VDU workplaces. The intervention group (IG) (n=13) was introduced to a wide-ranging package of PA promoting measures, whereas the control group (CG) (n=12) continued their usual office work. The package consisted of measures aiming at the workplace design (e. g. sit-stand tables) and the behavior (e. g. pedometers as activity feedback, face-to-face motivation for lunch walks etc., an incentive system for bicycle commuting or sports activities). The intervention lasted 12 weeks.

## 2.2    Assessment of physical activity

During the intervention phase, daily occupational PA was assessed by activity logs and a simple activity measurement system (AiperMotion 320). The activity logs provided the daily time spent sitting, standing and walking whereas the AiperMotion gave information on the amount of time spent actively and number of steps.

Precise assessments before and during the intervention were conducted with an expert measurement system (CUELA Activity System) (Weber et al. 2009). This person-centered measuring system consists of seven inertial motion sensors (3D accelerometers and gyroscopes) as well as a miniature data storage unit with a flash memory card, which can be attached to the subject. The sensors were positioned at the thoracic spine (Th3), lumbar spine (L5/S1), the upper arm of the dominant arm, the thighs and lower legs. From the measured signals (sampling rate: 50 Hz) body and joint angles and physical activity intensities (PAI) are calculated. PAI values are determined by calculating a sliding root mean square of the high-pass filtered vector magnitude of the 3D acceleration signals. The CUELA software automatically identifies various activities and body postures and determines energy expenditure. The measurement data can be depicted with the software together with the digitalized video recording of the workplace situation and a 3D animated figure.

## 2.3    Assessment of well-being

Before and after the intervention a multidimensional mood state questionnaire (MDBF, "Mehrdimensionaler Befindlichkeitsfragebogen") was applied to assess the emotional well-being (Steyer et al., 1997). The MDBF consists of 24 items (each with five-step rating scale) for measuring three bipolar dimensions (scales) of the current mental state: good mood-bad mood, awake-tired, calm-nervous. For analysis, mean values and standard deviations of the scales are calculated and compared with the values of the norm sample. For this purpose, the scores are converted in percentile ranks developed on the basis of the norm sample. A percentile rank value of 60 means for example that 60% of the norm sample has worse values for the corresponding scale.

## 2.4    Data processing and statistical analysis

Two-way ANOVAs were performed for the group comparisons of the pre-post differences. Day-to-day data (activity logs and AiperMotion) were analyzed by Wilcoxon signed-rank tests.

## 3    RESULTS

## 3.1    Assessment of physical activity

The PA assessment revealed several significant positive intervention effects: The activity logs showed that the intervention subjects spent more time standing and less time sitting during the whole period ($p \leq 0.001$) compared to the control subjects (see Table 1).

The intervention group reported on average additional standing times of $64.4 \pm 18.7$ min in comparison to the control subjects. The daily time spent in sitting postures was correspondingly reduced for the intervention subjects (mean difference $58.3 \pm 19.3$ min) in comparison to the control group.

The AiperMotion activity measurement system found higher step numbers for the intervention group ($p \leq 0.001$). The intervention group reached a target of 5000 steps per workday in most cases, whereas the control group missed it most times.

The expert CUELA measurements revealed significant differences in PA behavior change (see Table 2): A reduction of sitting ($p \leq 0.001$) and an increase of standing ($p \leq 0.01$) and walking ($p \leq 0.05$) as well as increased PAI levels of the upper and lower extremities and the trunk (each $p \leq 0.01$) were measured for the intervention group, whereas no significant effects were found for the control group. For the energy expenditure, no significant group differences were found ($p = 0.742$).

Table 1: Daily standing, sitting time and steps per hour measured by AiperMotion (group means per week ± SD) for the intervention group (IG) and control group (CG) during the intervention weeks 1 to 12

| Interven-tion week | Standing time (min/d ± SD) | | Sitting time (min/d ± SD) | | Steps (steps/h ± SD) | |
|---|---|---|---|---|---|---|
| | IG | CG | IG | CG | IG | CG |
| 1 | 100.4±20.3 | 68.7±37.2 | 345.5±42.1 | 355.6±47.6 | 637±44 | 512±26 |
| 2 | 115.6±34.8 | 77.7±22.6 | 305.6±90.8 | 362.4±45.4 | 524±58 | 618±72 |
| 3 | 119.8±45.3 | 61.9±18.8 | 342.2±30.0 | 384.4±12.3 | 540±58 | 482±63 |
| 4 | 117.9±17.5 | 41.5±17.0 | 331.9±37.0 | 392.7±27.1 | 589±46 | 497±46 |
| 5 | 104.3±20.6 | 38.0±15.3 | 359.9±19.9 | 403.9±21.5 | 613±42 | 509±34 |
| 6 | 111.7±9.4 | 31.8±12.4 | 349.4±33.5 | 419.1±38.0 | 618±63 | 570±79 |
| 7 | 90.5±20.6 | 23.6±9.6 | 351.6±43.9 | 414.2±9.0 | 602±72 | 587±90 |
| 8 | 130.7±26.1 | 29.7±4.3 | 326.2±13.4 | 409.3±20.8 | 642±54 | 514±57 |
| 9 | 94.8±18.4 | 39.7±8.3 | 360.0±25.5 | 420.5±17.8 | 579±67 | 498±66 |
| 10 | 104.2±13.5 | 43.7±15.1 | 353.7±31.0 | 411.7±34.8 | 708±62 | 491±52 |
| 11 | 114.8±28.2 | 36.5±11.8 | 348.3±51.6 | 425.8±30.7 | 649±51 | 461±78 |
| 12 | 110.0±33.5 | 49.4±13.3 | 331.8±27.7 | 402.3±12.9 | 623±72 | 476±69 |

Table 2: Pre and post task percentages. levels of PAI of different body regions and energy expenditure (mean ± SD per group, IG= intervention group; CG= control group) measured by CUELA and related p-values

| | IG | | CG | | P-value |
|---|---|---|---|---|---|
| | Pre | Post | Pre | Post | |
| Sitting [% of shift ± SD] | 78.9±9.8 | 52.5±16.3 | 77.1±9.7 | 75.8±11.4 | 0.001 |
| Standing [% of shift ± SD] | 12.9±8.1 | 35.0±15.1 | 14.5±7.9 | 15.4±8.4 | 0.002 |
| Walking [% of shift ± SD] | 8.2±3.4 | 12.4±3.5 | 8.4±2.9 | 8.8±4.5 | 0.020 |
| PAI Legs [g ± SD] | 0.11±0.02 | 0.14±0.03 | 0.12±0.02 | 0.12±0.03 | 0.013 |
| PAI Trunk [g ± SD] | 0.06±0.01 | 0.08±0.02 | 0.07±0.02 | 0.07±0.02 | 0.003 |
| PAI Whole Body [g ± SD] | 0.08±0.02 | 0.11±0.02 | 0.09±0.02 | 0.09±0.02 | 0.004 |
| Energie Expenditure [Mets/$\frac{kcal}{kg\,h}$ ± SD] | 1.70±0.30 | 2.02±0.38 | 1.80±0.35 | 2.07±0.37 | 0,742 |

## 3.2    Assessment of well-being

The applied methods showed some significant positive intervention effects regarding well-being. Generally the MDBF assessment led to better rank values (> 50 percentile rank) for the intervention and control group compared to the values of the MDBF norm sample. For the intervention group increases in all dimensions were documented. In the pre-post comparison the subjects showed improved mood ($67.1 \pm 35$ to $69.5 \pm 34.3$ percentile rank); they also felt more awake ($65.1 \pm 31.5$ to $70.2 \pm 27.5$ percentile rank) and calm ($86.5 \pm 14.2$ to $89.1 \pm 11.2$ percentile rank). For the control group decreases in all subjective perceived well-being dimensions were found in the pre-post comparison.

## 4    DISCUSSION AND CONCLUSIONS

The intervention group showed improvements of the perceived well-being. The subjects stated better mood and felt more awake and calm. The MDBF questionnaire turned out to be an applicable instrument in an assessment inventory for PA.

According to the analyses of the activity logs and CUELA measurements, the intervention group stood significantly longer and sat less during VDU work. This different behavior is referable to the use of the sit-stand tables in the intervention group, which was not available for the control group. Regarding the body dynamics, the intervention group showed also significant increases in physical activity compared to the control group. These were reflected by higher step counts measured with AiperMotion and elevated PAI levels of all body areas in the pre-post comparison of CUELA measurements. In addition, the CUELA pre-post comparison revealed a slight increase of the percentage of walking in the intervention group. This increase in body dynamics is probably caused by the preventive measures aimed at the behavior. However, the changes in body dynamics were not strong enough to produce a measurable change in energy expenditure. This was due to the fact that the intervention subjects primarily used the breaks between the computer works to perform additional PA. Since break times are limited the absence of significant group difference in energy expenditure can thus be explained by the lack of opportunities to perform PA at VDU work-places. It is therefore important, to develop and implement further PA enhancing measures, which lead to a substantial increase in energy expenditure and physical activity at VDU workplaces. The use of dynamic office chairs seems not to be appropriate to solve the problem. Previous studies revealed no significant differences in muscular activation and physical activity intensities between the use of specific dynamic chairs and a conventional chair (Ellegast et al., 2012). By contrast, the performance of different office tasks strongly affects muscle activation, postures and physical activity intensities (Groenesteijn et al., 2012). Therefore new innovative concepts should aim at combining computer tasks with physical activity. Examples of new dynamic (or active) computer workstations that are developed to stimulate PA during office work are walking workstations, ergometer facilities or portable pedal exercise machines. Research is needed to quantify the effects in increasing PA of office workers while using these dynamic workstations. In this context important research questions are: What intensity of physical activity is acceptable for office workers and how does the physical exercise influence the mental performance of office workers while performing computer tasks?

The preliminary results of this pilot study suggest that the tested methods were suitable to quantify PA intervention effects at VDU workplaces and can so be applied to evaluate dynamic workstations.

## ACKNOWLEDGEMENTS

This project was initiated and funded by the German Social Accident Insurance (DGUV).

## REFERENCES

Aarås, A.; Horgen, G.; Ro, O.: Work with visual display unit: health consequences. International J. Human-Computer Interaction 12 (2000), 107-134.

Baumann, A.; Phongsavan, P.; Schöppe, S.; Owen, N.: Physical activity measurement – a primer of health promotion. Promot. Educ. 2006; 13:92-103.

Dishman, R. K.; Oldenburg, B.; O'Neal, H.; Shepard, R. J.: Workplace physical activity interventions. Am. J. Prev. Med. 15 (1998), 344-361.

Dugdill, L.; Brettle, A.; Hulme, C.; McCluskey, S.; Long, A. F.: Workplace physical activity interventions: a systematic review. Int. J. Workplace Health Management. 1 (2008), 20-40.

Ellegast, R. P.; Kraft, K.; Groenesteijn, L.; Krause, F.; Berger, H.; Vink, P.: Comparison of four specific dynamic office chairs with a conventional office chair: impact upon muscle activation, physical activity and posture. Appl. Ergonomics 43 (2012), 296-307.

Groenesteijn, L.; Ellegast, R. P.; Keller, K.; Krause, F.; Berger, H.; de Looze, M. P.: Office task effects on comfort and body dynamics in five dynamic office chairs. Appl. Ergonomics 43 (2012), 320-328.

Parent-Thirion, A.; Fernández, M.E.; Hurley, J.; Vermeylen, G.: Fourth European Working Conditions Survey. European Foundation for the Improvement of Living and Working Conditions, Dublin 2007

Pate, R.R.; Pratt, M.; Blair, S.N.; Haskell, W.L.; Macera, C.A.; Bouchard, C. et al.: Physical activity and health. A recommendation from the Centers for Disease Control and Prevention and the American College for Sports Medicine. JAMA (1995), 273, 402-7.

Proper, K. I.; Koning, M.; van der Beek, A. J.; Hildebrandt, V. H.; Bosscher, R. J.; van Mechelen, W.: The Effectiveness of Worksite Physical Activity Programs on Physical Activity, Physical Fitness and Health. Clin. J. Sport Med. 13 (2003), 106-117.

Straker, L. and Mathiassen, S. E.: Increased physical work loads in modern work – a necessity for better health and performance. Ergonomics 52 (2009), 1215-1225.

Steyer, R.; Schwenkmezger, P.; Notz, P. and Eid, M.: Der Mehrdimensionale Befindlichkeitsfragebogen (MDBF). Hogrefe, Göttingen 1997

Weber, B.; Hermanns, I.; Ellegast, R.; Kleinert, J.: A person-centered measurement system for quantification of physical activity and energy expenditure at workplaces. In: Karsh B-T, editor. Ergonomics and Health Aspects, HCII 2009. Berlin: Springer; LNCS 5624, 121-130.

CHAPTER 75

# Towards an Adaptive Office Environment: Effects of Sound and Color of Light on Performance and Well-being

*Elsbeth M. de Korte [a,b], Lottie F.M. Kuijt-Evers [a]*

[a]TNO,
Hoofddorp, The Netherlands;
[b]Delft University of Technology, Faculty of Industrial Design Engineering,
Delft, The Netherlands.

## ABSTRACT

To study the effects of color of light and sound in the office environment on performance and well-being, an experiment was set up. Fourteen subjects participated in the study. Six different environments were presented while they performed a standardized dual task: three colors of light (i.e. blue, red and a dynamic pattern of lights), two sound conditions (classic piano music and office noise) and a standard condition with no color of light and no sounds. Results show no effects of color of light and sounds on performance. However clear effects on well-being were shown: the objective measures of facial emotions showed significant less negative emotions in the color light conditions compared to the other conditions. The subjective measures showed that the blue and red color light conditions were rated in general more positive. The results of the study are promising and support the necessity to build individually adaptive smart workstations.

**Keywords**: office, light, sound, performance, well-being

## 1    INTRODUCTION

Many organizations are seeking ways to effectively design office spaces suitable to meet the current needs of office work. The goals that these organizations are striving for with office innovations are: improved image, improved performance, group collaboration, better communication, increased workers satisfaction, increased comfort experience, efficient use of office space and reduced psychological and physical stress (Van der Voordt, 2003; Brauer et al., 2003; Robertson and Huang et al., 2006; Huang et al., 2004). Examples of these changes are working with innovative office concepts like shared workspaces, open plan offices and teleworking (De Croon ea, 2005; Brauer ea, 2003; Lee & Brand, 2005).

Studies show that office environments may impact on office worker health, performance and well-being, either in a negative or a positive way (De Croon ea, 2005; Lee & Brand, 2005). For instance, open plan offices may improve communication (Lee & Brand, 2005), but on the other hand, a disadvantage may be reduction in concentration due to the increased distractions, which may have a negative effect on the productivity (Lee & Brand, 2005; Banbury & Berry, 2005; Voordt, 2003). The ambient features of the physical environment including noise, light, temperature, existence of windows, and others, influence employee attitudes, behaviors, well-being and performance (Lee & Brand, 2005). This increases the need for adapting the office environment to personal needs. Office workers would like to personalize their own workstation (ambient experience), supported by technology.

The aim of the present study is to investigate how we are able to influence well-being and performance by changing ambient features of the office environment. In order to achieve this goal, an experiment was set up in which we varied color of light (blue, red and alternating pattern of colors) and noise (classic piano music and office noise) and a standardized condition with no color of light and sound.

Color of light was selected to include in the experiment, because recently light color has received attention in relation to perception, well-being and arousal. So far, most studies focus on light or on color of the interior space (Küller et al., 2006).

Research into the color of light is relatively new, especially color of light in offices. So far, studies show that subjects feel more alert in blue light than in green light and that also higher cognitive functions in the brain like arousal are modulated by light color (Cajochen, 2007). Knez (2001) investigated the influence of color of light ('warm', 'cool' and 'artificial 'daylight' white light) on subjects self-reported mood, cognitive performance and room light estimation. Although no effects on mood were found, the results showed that subjects performed better in the 'warm' than in the 'cool' and artificial 'daylight' white light. Knez (1995) did find an effect of color temperature of light ('warm white'; 3000K versus 'cool white'; 4000K) on mood in earlier work. Even though we have not found any information in the literature on this type of color of light we decided to include an alternating pattern of light colors in our experiment.

Sound was selected to include in the experiment, because literature shows clear effects of noise (Lee & Brand, 2005; Leather et al., 2003; Schlittmeier et al., 2008).

In literature, auditory distraction or noise in the office environment is often mentioned as the most problematic distracting factor in relation to concentration (Lee & Brand, 2005). Also, distraction is related to a negative perception of the work environment, irritation, job satisfaction, well-being and absenteeism (Lee & Brand, 2005; Banburry & Berry, 2005; Leather et al., 2003; Schlittmeier et al., 2008). According to Hongisto (2005), auditory distraction is particularly a problem in complex tasks. Noise with relevant information has more influence on performance than irrelevant information. Especially background speech affects concentration, to a lesser degree distinctive or salient sounds and not so much the level of noise (Hongisto, 2005; Banburry & Berry, 2005; Schlittmeier et al., 2008).

To cope with office noise in open plan offices, employees often turn to use their Mp3 players, to listen to music, thereby shutting out auditory distractions coming from the office environment. There has been considerable interest into whether productivity can be increased in the presence of background music at work (Furnham & Strbac, 2002; Cassidy and MacDonald, 2007). Experimental literature on the potential benefits and drawbacks of background music at work has noted the interaction between type of task, type of music and individual differences in understanding the distractibility of music at work. It is hypothesized that music may have a positive effect on routine tasks as it serves to reduce tension and boredom, but for complex tasks, music may act as a distracter (Furnham & Strback, 2002; Cassidy and MacDonald, 2007).

## 2    METHODS

In this experiment, the effect of the office workstation environment (color of light and sound) was investigated on performance and well-being of the office worker. Six different environments were presented while the subjects performed a standardized dual task: three colors of light (i.e. blue, red and dynamic), two sound conditions (classic piano music and office noise), and a standard condition with no colored light and no sounds (Figure 1).

## 2.1    Participants

Fourteen participants volunteered in this study (9 female, 5 male; mean age 34.9 years ± 11.7). They all give their written informed consent. None of them did report color blindness nor mental illness.

## 2.2    Experimental settings

The Ahrend 750 was used as work station in a grey/white color combination (design Bas Pruijser). The computer screen and the computer keyboard were located on the table. The participants could vary the distance towards both devices in order to create a comfortable position.

Behind the table and the computer screen, a reflective white projector screen was placed. Four LED par cans were mounted on a stand system above the head of the participant. Blue (RGB 0:0:255), red (RGB 255:0:0) and dynamic light was emitted on the projector screen, providing a colored area of 80 cm x 120 cm (Figure 1). The illuminance in gaze direction was about 90 lux at eye distance from the projector screen for all light color conditions. The light color was controlled using Laptop with FreeStyler software linked with DMX-controller. All windows were completely covered to shut out natural daylight. We chose to use LED lights instead of traditional theater light in order to avoid temperature increase, which could affect the participant's performance and comfort. (Hedge et al., 2005).

Two sound conditions were presented in this experiment: music and office noise. In the sound conditions, a sound sample of classical piano music (110 bpm) was played (65dB). Office noise was simulated by taped office noise containing speech also at 65dB. Several studies have shown that speech impairs the performance of reading and short-term memory. However, it is not the sound level of speech that determines its distracting power but its intelligibility (Hongisto, 2005).

Figure 1 Experimental set-up of the workstation, with red light condition, blue light condition, dynamic light condition. The last picture represents the neutral condition, as well as the office noise condition and the music condition.

The task which was performed by the participants was a dual task to achieve a certain level of mental effort. A visual memory task as described by Capa et al. (2008) was presented to the participants. During 3 seconds two letters were shown on the computer screen. The participants had to compare this set of letters with four letters in the recognition set, which was on the screen for 3 seconds. The task was to indicate if any letter in the memory set was in the recognition set by pressing the "yes" key or -if none of the letters were in the recognition set- by pushing the "no" key with the right and left index finger respectively. The "yes" and "no" key were the ctrl keys at the right and the left of the space bar of a computer keyboard. The keys were indicated with a green (along with the written text "yes") and red label (along with the written text "no") respectively. The participants had 3 seconds to answer.

The recognition set changed every time and the probability that a letter appeared in the recognition set or not was equal. At the end of each trial, participants received feedback on their reaction time, which concerned the response speed and the type of error. To make the task more difficult, a counting task was added. The recognition set was presented in either red or green color. Participants were asked to count the number of red recognition sets while carrying out the visual memory search task. One condition consisted of 64 trials. The number of red recognition sets was form 29 to 34. Each condition had a fixed number of red recognition sets. Participants were instructed to react as quickly as possible without making errors. Simultaneously, they had to count the number of red recognition sets. At the end of each block of 64 trials, the participants had to indicate the number of red recognition sets. After that, they received feedback on their performance for both the visual memory search task (average reaction speed and number of errors) as well as the counting task (correct number of red recognition sets).

## 2.3 Measurements

Performance was measured with a dual, visual memory search task: average reaction speed and number of correct answers in the recognition part of the task and deviation of the correct number of the memory part of the task.

Well-being was indicated by the FaceReader 2.0 (VicarVision/Noldus) which indicates six basis emotions and a neutral, and subjectively by a questionnaire asking for emotion and perception of the workstation environment (Uyl & Kuilenburg, 2005). The FaceReader classifies the face expression into 6 basis emotions: happy, sad, angry, surprised, disgusted and scared and a "neutral" state. Furthermore, perception of the workstation and well-being of the participant was indicated by a questionnaire. For both purposes a VAS scale was used with verbal anchors on both sides. The participant's emotion was rated on seven items ("I feel ….. ": unpleasant-pleasant, happy-unhappy, relaxed-tensed; sad-happy, fatigue-fit, bad-good, bored-challenged). The perception of the workstation was indicated by rating five items ("I rate the office workstation environment as….": pleasant-unpleasant; cheerful-cheerless; distracting-elevating concentration; tiring-stimulating; ugly-beautiful).

## 2.4 Procedures

The participants were first informed about the experiment. Then, the participant performed the dual task for 6 minutes in order to get familiar with it. After a 5 minute rest, the first condition started. The participant performed the dual task. During the task, the FaceReader registered the emotional expressions on the participant's face. After 64 trials of letter recognition, the task was finished. The participant told the experimental leader the number of red sets of letters he counted. Next, the participant rated well-being and perception of the workstation on the VAS scales. After the participant filled out the questionnaires, he had a 5 minutes rest break. This procedure was repeated for all conditions. The order of conditions was systematically varied among the participants in order to avoid fatigue effects.

## 2.5    Data analysis

Performance was calculated  by average value of the reaction time, percentage of correct answers on recognition and deviation of the correct number of red recognition sets was.

As a measure of well-being, average emotions were calculated for each condition for the entire task duration. Also, ratings on the VAS-scales for well-being and perception of the workstation were obtained by measuring the distance from the left of the VAS scale (negative anchor) to the mark which indicates the rating of the participant on a scale of 10, representing a score from 0 to 100.

Differences in performance and well-being between the six conditions were tested using a SPSS General Linear Model (GLM) analysis ($p < 0.05$).

## 3    RESULTS

## 3.1    Performance

There were no significant differences between measurement conditions for all task performance parameters. On average, participants correctly answered 94% of the trials, their reaction time was 1.1 seconds and they had an average deviation of 2.2 from the correct number of red recognition sets.

## 3.2    Well-being

Well-being was objectively measured with facial emotions. During the dual task participants mainly showed angry and sad emotions, respectively 34% an 24%. The facial emotions neutral and disgusted differed between measurement conditions (figure 2 and 3). During red and dynamic there was significantly more neutral emotion then during standard, music and office noise. For the disgusted emotion, the color conditions (blue, red and dynamic) were significantly lower than the other conditions (standard, music and office noise). All other emotions did not significantly differ between the measurement conditions.

Well-being was measured subjectively by seven items on participant's emotion and by five items on perception of the workstation in a questionnaire. GLM repeated measures did not show any differences between de conditions for emotion. However, differences were found for perception of the workstation on the items pleasantness, distraction, stimulus and beauty (Figure 4).

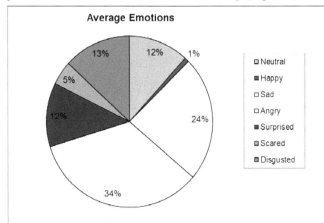

Figure 2. Average Emotions of participants for all conditions

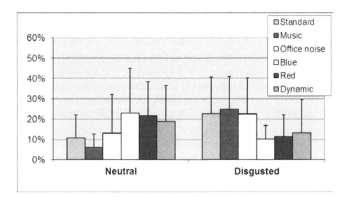

Figure 3. Results from the neutral and disgusted emotion for the measurement conditions.

For pleasantness, the standard condition was rated significantly more pleasant compared to office noise. Dynamic light was rated significantly more unpleasant compared to standard, blue and red. For distraction, the dynamic condition was rated significantly most distracting compared to all other conditions. Standard and blue were rated significantly less distracting compared to dynamic light. For stimulation, the condition with dynamic light was significantly less stimulating compared to blue light. For beauty, music was rated most beautiful compared to standard and office noise. The standard condition was rated significantly less beautiful than music, blue and red. The red condition was rated significantly more beautiful compared to standard and dynamic.

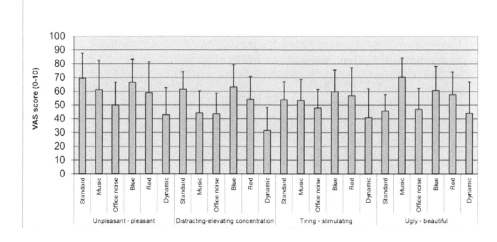

Figure 4. VAS scores (0-100) for perception of the workstation environment.

## 4    DISCUSSION

For well-being, we did find differences between the conditions. The Face Reader data showed that during the blue, red and dynamic condition, the contribution of "disgusted" in the facial expression was significantly less than in the standard condition. Based on these findings, it seems that the participants showed less negative emotions when working in a colored environment. For the subjective measurements, differences were found for the workstation perception on four items (pleasant-unpleasant, distracting-elevating concentration, tiring-stimulating and ugly-beautiful). It is a remarkable result that the dynamic condition for all these items, was rated significantly lower than one or more of the other conditions, which means that the perception of the dynamic workstation was worse, while the facial emotion measured by the FaceReader showed "less disgusting" compared to the standard, office noise and music condition.

In our study, we did not find significant differences between the conditions on performance parameters. Ainsworth et al (1993) did also not find significant differences between three colors of walls in an office interior on performance. Elliot et al. (2007) describe several studies on the effects of color environment on performance. They do not mention color of light, but for instance, painted walls. They conclude that in the present literature no clear evidence is found for a color effect on performance attainment. Kwallek et al (1996) found differences of three interior color schemes on worker performance, but only taking into account individual differences in environmental sensitivity: high screeners versus low screeners. This refers to individual differences in arousability; individuals who are most adept at screening less relevant stimuli of their environments are referred to as high screeners, while individuals who typically cannot screen incoming sensory information as well are referred to as low screeners. An important implication of the study of Kwallek et al (1996) is that altering a specific interior environment to enhance productivity may be quite difficult, because the impact on that interior may vary from person to person depending on each person's characteristics.

As described before, Varkevisser et al (2011) found physiological differences between different light conditions. However, they state that an illuminance of 200 lux is necessary before physiological response happens. In our study, we exposed the participants to an illuminance of 90 lux, which is far below the illuminance level proposed by Varkevisser. It is possible that we could have found effects by using a higher illuminance level.

In our study, relatively short task durations were used. As we know that in experience research it is important to give users the opportunity to test a product for a longer period of time (Looze et al., 2003; Vink, 2005), it might be advisable to study long term effects.

What is clear from our results is that large inter-subjects differences exist. This was already indicated by Capa et al (2007), who split their subjects into two separate groups: approach driven and avoidance driven persons. The change in task difficulty affected the mental workload (measured by HRV) in different ways. Furthermore, Knez (1995) found gender to be an important factor for emotional and cognitive responses to indoor light. This is a really important result, as it supports the necessity to build an individually adaptive smart workstation as people do show different physiological values when they get stressed.

## 5    CONCLUSION

No effects of ambient features were found on performance. However, well-being effects were found in emotion (facial expression) and perception of the environment. The results of our study are promising and support the necessity to build individually adaptive smart workstations.

## ACKNOWLEDGEMENTS

The authors acknowledge Koninklijke Ahrend NV, in particular Nico Prins and Bas Pruijser, for supporting this study.

## REFERENCES

Ainsworth, R.A., Simpson, L., Cassell, D., 1993. Effects of three colors in an office interior on mood and performance. Perceptual and Motor Skills, 76:235-241.

Banbury, S.P., Berry, D.C.: Office noise and employee concentration: identifying causes of disruption and potential improvements. Ergonomics 48(1) (2005) 25-37.

Brauer, W., Lozano-Ehlers, I., Greisle, A., Hube, G., Keiter J. & Rieck, A. (2003). Office 21 -push for the future- better performance in innovative working environments. Cologne/Stuttgart: Fraunhofer Baldonado

Cajochen, C., 2007. Alerting effects of light. Sleep Medicine Reviews 11: 543-464.

Capa, R.L., Audiffren, M., Ragot, S., 2008. The interactive effect of achievement motivation and task difficulty on mental effort. Int. J. Psychophysiol. 70 (2), 144-50.

730

Cassidy, G., MacDonald, R.A.R., 2007. The effect of background music and background noise on the task performance of introverts and extraverts. Psychology of Music 35(3): 517-537.

Croon EM de, Sluiter JK, Kuijer PPFM, Frings-Dresen MHW (2005). The effect of office concepts on worker health and performance: a systematic review of the literature. Ergonomics 48(2), 119 – 134

Elliot, A.J., Maier, M.A., Moller, A.C., Friedman, R., Meinhardt, J., 2007. Color and psychological functioning: the effect of red on performance attainment. Journal of Experimental Psychology. 136(1): 154-168.

Furnham, A., Strbac, L., 2002. Music is as distracting as noise: the differential distraction of background music and noise on the cognitive test performance of introverts and extraverts. Ergonomics 45(3):203-217.

Hedge, A., Sakr, W., Agarwal, A., 2005. Thermal effects on office productivity. HFES 49th Annual Meeting Proceedings, Orlando HFES 2005: 823-827.

Hongisto, V.: A model predicting the effect of speech of varying intelligibility on work performance. Indoor Air. 15 (2005) 458-468.

Huang, Y-S., Robertson, M., Chang, K., 2004. The role of environmental control on environmental satisfaction, communication, and psychological stress. Environment and behavior 36 (5) : 617-637

Knez, I., 1995. Effects of indoor lighting on mood and cognition. Journal of Environmental Psychology. 15: 39-51

Knez, I., 2001. Effects of colour of light on nonvisual psychological processes, 2001. Journal of Environmental Psychology, 21: 201-208.

Küller, R., Ballal, S., Laike, T., Mikellides, B., Tonello, G., 2006. The impact of light and colour on psychological mood: a cross-cultural study of indoor work environments. Ergonomics 49(14):1496-1507.

Leather, P., Beale, D., Sullivan, L.: Noise, psychosocial stress and their interaction in the workplace. Journal of Environmental Psychology 23 (2003) 213-222

Lee, S.Y., Brand, J.L.: Effects of control over office workspace on perceptions of the work environment and work outcomes. J. Environmental Psychology 25 (2005) 323-333.

De Looze, M.P., Kuijt-Evers, L.F.M., Van Dieën, J.H., 2003. Sitting comfort and discomfort and the relationships with objective measures. Ergonomics. 46, 985 997.

Robertson, M., Huang, Y-S., 2006. Effect of a workplace design and trianing intervention on individual performance, group effectiveness and collaboration: the role of environmental control. Work. 27: 3-12

Uyl, M.J.d., Kuilenburg, H.v.: FaceReader: an online facial expression recognition system. In: Proceedings of the 5th international conference on methods and techniques in behavorial reseearch, pp. 589–590 (2005).

Varkevisser, M., Raymann, R.J.E.M., Keyson, D.V., Nonvisual effects of LED coloured ambient lighting on wellbeing and cardiac reactivity: preliminary findings. HCI International, Orlando (Fl.), 13 July 2011.

Vink P., 2005. Comfort and Design: Principles and Good Practice, CRC Press, Boca Raton.

Voordt, DJM van der (2003). Costs and benefits of innovatieve workplace design. Center for People and Buildings, Delft. ISBN 90-807720-3-8

CHAPTER 76

# Requirements for the Back Seat of a Car for Working While Travelling

*Sigrid van Veen, Peter Vink*
Delft University of Technology, The Netherlands
p.vink@tudelft.nl

*Suzanne Hiemstra van Mastrigt*
TNO Hoofddorp, The Netherlands

*Irene Kamp, Matthias Franz*
BMW, Munich, Germany

## ABSTRACT

Vehicle interiors for working while travelling have certainly room for improvement. In this paper, travelers in the rear seat of a car using a laptop, a tablet and a book were observed and questioned. Additionally, EMG recordings were made in these three conditions. It appeared that the seat dictates the posture; the discomfort is low after 30 minutes driving. However, EMG shoulder muscle activity was high in tablet work compared with the other conditions and neck bending was often seen. It is advised to study the posture further and develop armrests that support the tasks.

**Keywords**: tablet, notebook, discomfort, posture, EMG, working while traveling, rear seat.

## INTRODUCTION

New ways of work also extends to working while traveling. However, many interiors are not designed for work. The opinion of three tablet users in an airplane (see fig.1) on their comfort illustrates this clearly. The tablet user   pictured on the right in the back complaints about neck discomfort, while watching a video, "but at least I have my hands free", he said. The person in the front placed the tablet on the table. After holding it in the hands 30 minutes the arms and hands were getting tired. The person on the left side in the figure misses an arm support on the right side as the neighbor is using that arm rest and the hand is getting tired after an hour.

Figure 1. Three passengers using a tablet, each in a different way.

This is just anecdotic evidence and not a scientific approach. However, in the literature also indications are found on posture effects of tablet use. Young et al. (2012) found that head and neck flexion angles during tablet use were greater, in general, than angles previously reported for desktop and notebook computing. A higher display location often leads to reduced neck flexion that approaches more neutral postures, while lower viewpoints often increase the flexed posture which is associated with an increase in neck extensor activity and discomfort (Ariens, 2001; Straker et al, 2008a). Hamberg (2008) showed that this neck discomfort increases the chance of getting neck pain by more than two times. A typical difference between working with a desk top and tablets is that the hands (often thumbs) are positioned at the location where you have the viewpoint on the tablet, while in the desk top it is possible to type blind and have the keyboard in the position creating optimal hand/arm postures and the screen facilitating optimal neck positions. The laptop is in between as screen and keyboard are connected, but the position of the keys is not so close to the viewing point.

For the neck, a high position of the screen is preferable. However, this means that the hands and arm should be elevated. Hedge et al. (2011) showed that an arm support reduced the shoulder complaints among 1504 office workers from 24% to 16% and Zhu and Shin (2011) showed that an arm support reduced the muscle activity and discomfort while working with a touchscreen. So, there are indications that an elevation without arm support should be avoided or find an optimum between neck bending and arm raising.

This knowledge will become of importance for aircraft, train and car interiors as the tablet sales is increasing. Notebook sales were for the first time higher than the desk top sales in 2008 and tablet sales is growing rapidly. Also, for the car driver seat this could be of importance. A trend in the automotive industry is developing features that support the driver and partly take over control like the automatic distance control and lane detection. These developments are the first steps towards autonomous driving. Therefore the future driver is potentially able to do the same activities as the rear seat passengers do now.

The question is how an interior for working with a laptop should be designed and what the requirements are. However, there are not much data available on how passengers in the rear seat behave working with a laptop or tablet. Therefore, in this study more information is gathered on this issue. Kamp (2012) showed that pictures and observations of passengers in rear seats do not give enough information as often the vision is blocked by glare. Therefore, in this study we asked passengers to perform the task in the back while driving.

The research question is:

*What are posture characteristics of passengers using a laptop and a tablet in the rear seat and what is their opinion on the comfort?*

## METHOD: THE DRIVING TEST

To observe the postures and gather information on what passengers prefer in working with a laptop and a notebook, subjects were asked to participate in a driving test. In the driving test, 14 men and 12 women of different nationalities (European, American and Asian) participated. Their average age was 29.4 years (20-67 years). Their average weight was 71.2 kg (50-105 kg) and their average height was 175.6 cm (163.0-193.0 cm). Two participants were invited in Delft at the same time and received an introduction on the study. When sitting in the BMW 7series car the control group received information on a game in the left rear seat. The subject in the right rear seat was instructed to do one of the following tasks: reading a book, working on a laptop or playing a game on a tablet pc. When someone indicated (severe) motion sickness, the task least likely to cause sickness was chosen by the subject him-/herself. In total nine subjects played a game on a tablet pc, nine subjects read a book and eight subjects worked on a laptop during the ride. The subject in the left rear seat always used the game seat. After the instructions the two subjects in the rear seats were driven for approximately 30 minutes by one researcher and observed by another researcher sitting in the front row. At the beginning (T=0), twice during (T=10, T=20) and at the end (T=30) of the drive subjects indicated on a body map their perceived discomfort by means of the local perceived discomfort (LPD) method as described by Grinten (1991). The researcher monitored the subjects, indicated when they had to start the game (left seat) and when to fill out the LPD body map. The game was played for five minutes and alternated with five minutes rest. The other activity was done constantly.

During the travel several pictures were made and after 30 minutes the participants changed places and their subjective experience was asked in a questionnaire with questions on whether they felt relaxed, refreshed, tired etc. Next, the participants switched places and tasks (gaming vs. working on a laptop/using a tablet/reading a book). Then the same procedure as at the start of the test was followed and after approximately 30 minutes driving the final part of the questionnaire was completed again. The driving track was the same for all travels and consisted mainly of a highway as this is probably where most laptop and notebook work will be done. The questionnaire consisted of questions related to comfort (Likert-scales) for both the game and the three tasks (reading, working on laptop and gaming on tablet).

## METHOD: EMG

To get an indication on which muscles are active during the different activities four subjects aged between 20-21 years old (three male, one female) participated in an EMG measurement. EMG while driving is difficult as the vibrations of the car could disturb the signal. Therefore, measurements were taken while the car was not driving. The average weight of the subjects was 73.5 kg and their average height 183.5 cm.. Surface electrodes were placed with an interelectrode distance of 20 mm and located according to Hermens et al. (2000). Before the electrodes were applied, the skin was shaved, scrubbed and cleaned with alcohol. Skin resistance was not measured. The muscle activity of the following muscles was measured:
-        Lower back: m. erector spinae (m. longissimus)
-        Upper back: m. erector spinae
-        Shoulder: m. trapezius pars transversalis
-        Neck/shoulder: m. trapezius pars descendens
A reference electrode was placed on the C7 spinous process.

When all electrodes were placed, subjects were asked to sit in the right rear seat of the test car and to perform four different tasks: reading a book, working on a laptop and playing a game on a tablet pc. During every activity the EMG signal was recorded twice for approximately 10 seconds. The raw EMG signals were sampled during the test contraction with a sample frequency of 2000 Hz and band-pass filtered (10-400 Hz). The raw signals were rectified and averaged over the 10 seconds recording time.

# RESULTS: THE DRIVING TEST

In figure 2 examples of the recorded postures were visualized. For working with the laptop only one dominant posture was found. For both other tasks (reading and tablet) two dominant positions were observed (see table 1). When working with the laptop, the subjects sit upright in the seat with both legs together supporting the laptop and both hands active in typing. In gaming on the tablet the head was bent forward, one arm was always supported. The body was upright or leaning to the left in the seat, the legs were apart from each other. The tablet was on the lap or on one leg. This supporting leg was always the left leg. Subjects were either using one hand to hold the tablet and the other to touch the screen or the tablet was held in the two hands using the thumbs to control the functions. While reading a book the arms were alongside the body, the trunk upright or leaning to the left and the book was held by two hands or only the right one. The book was on the lap or only supported by the hand(s). The legs were apart.

The LPD scores were very low. Far below the level Hamberg et al (2008) mentioned.

Table 1  Typifying of the observed postures

| | book | | laptop | tablet | |
|---|---|---|---|---|---|
| legs | legs apart | | legs together | legs apart | |
| trunk | leaning to left | upright | upright | upright | leaning to left |
| arm | alongside body | | alongside body | one arm supported | alongside body |
| hands | one on book | both on book | on keyboard | one hand holding tablet, one touching | both thumbs controlling the pad |
| device | on lap | supported by 2 hands | on lap | on left leg | on lap |
| head | bent forward | | bent forward | bent forward | |

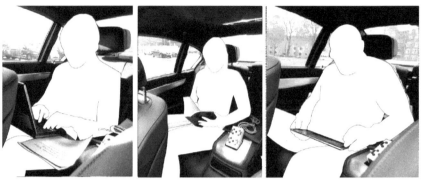

Figure 2. Examples of postures in the three tasks.

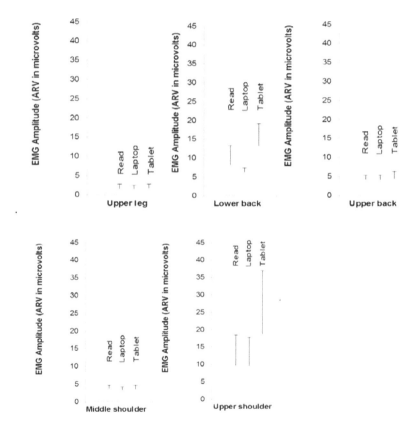

Figure 3. The mean amplitude of the muscle activity expressed as the average rectified value (ARV) for all muscles averaged over all subjects in the various conditions (reading, working on laptop and gaming on tablet).

# RESULTS: EMG

In figure 3 the mean amplitude of the muscle activity rectified value (ARV) for all muscles averaged over all subjects in the various conditions (reading, working on laptop and gaming on tablet) is presented. The only muscle where the activities (especially gaming on tablet) show higher muscle activity and variability is the upper shoulder (neck) muscle (see fig. 3).

# DISCUSSION

Regarding the research question on the posture characteristics of passengers using a laptop and a tablet in the rear seat it is clear that most of the posture is dictated by the seat itself. Compared to other studies on postures and activities, this study shows less diversity in postures. Gold (2012a) found for instance 22 self selected different postures when using a laptop at home. During train journeys and leisure situations at least eight different postures when using mall and larger electronic devices were observed by Kamp et al (2011). On the other hand this study confirms some characteristics of using a touchscreen shown by Zhu & Shin 2011) like the fact that the arms demanded support or the neck bent forward.

The discomfort during the driving test was very low for all activities, even after 30 minutes driving. This can be explained by the fact that the discomfort is already low in the BMW7-series seat during driving. Another explanation could be that the subjects did not all had equal postures, which results in a variation and low LPD scores.

The higher muscle activity and variability in the upper shoulder (neck) muscle is probably due to the head of the subject which is bent forward in the activities (reading, working on laptop, gaming on tablet), therefore muscle activity in the upper shoulder (neck) muscle is needed to counteract the gravity force working on the head. For the tablet condition, the extra increase in muscle activity and variability can be explained by the need to use the arm while playing the game on the tablet. Gold (2012b) also observed in 91% (n=782) of all subjects a flexed neck when using a mobile device. Straker et al (2008b) emphasize the importance of display height on neck and upper limb muscle activity for different tasks (working on a computer and book use) in a desk setting.

Based on this study several recommendations for car interior design supporting desired tasks can be formulated. There is a need for a solid laptop support which will enable posture variation for the user, prevent shifting and minimize the difference in vibrations between user and laptop. An arm support is needed for various tasks: for the laptop arm support is needed while typing, for using books, e-readers, tablets and smart phones arm support is needed to operate the devices and to keep them up in order to prevent a flexed neck. People also looked for possibilities to turn/lean to one side while seeking support, to enable variation in posture, create more leg room, have conversations with fellow passengers or look out the window; the future seat should offer more flexibility to do this. Finally there

738

is a need for varying the back rest angle in relation to the level of activity(low level, relaxed activities vs. high level, intense activities).

## REFERENCES

Ariens, G.A.M., 2001. *Work-related risk factors for neck pain*. PhD thesis Vrije Universiteit, Amsterdam, 2001lvarez-Filip, L., N. K. Dulvy, and J. A. Gill, et al. 2009. Flattening of Caribbean coral reefs: Region-wide declines in architectural complexity. *Proceedings of the Royal Society, B* 276: 3019–3025.

Gold, J.E., J.B. Driban, V.R. Yingling, E.Komaroff, 2012a. Characterization of posture and comfort in laptop users in non-desk settings. *Applied Ergonomics*, Vol 43, 392-399.

Gold, J.E., J.B. Driban, N. Thomas, T. Chakravarty, V. Channell, E. Komaroff, 2012b. Postures, typing strategies, and gender differences in mobile device usage: an observational study. *Applied Ergonomics,* Vol. 43, 408-412

Grinten van der, M.P., 1991. Test-retest reliability of a practical method for measuring body part discomfort. In Y. Quéinnec and F. Daniellou (Eds.), *Designing for Everyone* (1) (pp. 54-57). London: Taylor & Francis

Hamberg-van Reenen, H.H., A.J. van der Beek, B.M. Blatter M.P.van der Grinten, W. van Mechelen, P.M. Bongers, 2008. Does muculoskeletal discomfort at work predict future musculoskeletal pain? *Ergonomics* 51: 637-648

Hedge, A., J. Puleio, V. Wang, 2011. Evaluating the Impact of an Office Ergonomics Program. In: *Proceedings of the Human Factors and Ergonomics Society* 55th Annual Meeting - 2011. Santa Monica: HFES, 2011:cd-rom.

Hermens, H.J., Freriks, B., Disselhorst-Klug, C., Rau, G., 2000. Development of recommendations for EMG sensors and sensor placement procedures. *Journal of Electromyography and Kinesiology*, Vol. 10 (5), 361-374.

Kamp, I., 2012. *Comfortable car interiors: experiments as a basis for car interior design contributing to the pleasure of the driver and passengers*. Doctoral thesis; in progress.

Kamp, I., Ü. Kilincsoy, P. Vink, 2011. Chosen postures during specific sitting activities. *Ergonomics,* Vol. 54 (11), 1029-1042.

Straker, L., R. Burgess-Limerick, C. Pollock, K. Murray, K. Netto, J. Coleman, J., 2008a. The impact of computer display height and desk design on 3D posture during information technology work by young adults. *J Electromyogr Kinesiol.* 18(2): 336-349

Straker, L., C. Pollock, R. Burgess-Limerick, R. Skoss, J. Coleman, 2008b. The impact of computer display height and desk design on muscle activity during information technology work by young adults. *Journal of Electromyography and Kinesiology*, Vol. 18, 606-617.

Young, J.G., M. Trudeau, D. Odell, K. Marinelli, J.T. Dennerlein, J.T., 2012. Touch-screen tablet user configurations and case-supported tilt affect head and neck flexion. *Work* 41: 81–91.

Zhu, X., G. Shin, 2011. Effects of Armrest Height on the Neck and Shoulder Muscle Activity in Keyboard Typing Program. In: *Proceedings of the Human Factors and Ergonomics Society* 55th Annual Meeting - 2011. Santa Monica: HFES, 2011:cd-rom.

CHAPTER 77

# Household Work: An Ergonomic Perspective

*Hema Bhatt, M. K. Sidhu, R. Bakhshi and P. Sandhu*

Department of Family Resource Management,
Punjab Agricultural University, Ludhiana, India
hemabhatt2000@gmail.com

## ABSTRACT

For women the household work is very time consuming and drudgery prone activity. They seldom realize the cost of energy and other physiological costs incurred due to wrong posture. Poor posture increases the physiological cost of work and energy expenditure. The static muscular efforts and incorrect posture if sustained for a long period of time can give rise to various types of health and musculoskeletal problems. Due to the dual responsibility at home as well as at work places outside the home, there is a greater pressure for productivity enhancement, quality work and profitability. Ergonomically sound kitchen workstation layout saves cooking time, distance travelled and postural change, therefore reduces the women drudgery. This paper presents the results of the study undertaken to do ergonomic assessment of kitchen workstation for females engaged in cooking activities, with objectives; to study the problems faced by women in existing workstation and to evaluate the design of selected work station in-terms of ergonomic standards. For this purpose a survey of eighty homemakers from Punjab state was conducted. Results revealed that in urban areas all respondents had closed, standing kitchen with either 'L' or 'U' shaped counters. Kitchen area ranged between 6.7 to 8.4 sq. m. Storage facility was found neglected in many of kitchens as it was beyond comfortable reach of homemakers. Most of the activities were performed in standing and sitting posture and very few activities like sieving where squatting posture (38.75 per cent) was adopted by the respondents. While performing the various kitchen activities mostly fatigue was felt in forearm and wrist.

**Keywords**: Ergonomics, women, kitchen, drudgery

# 1 INTRODUCTION

In recent years, there has been a trend of more and more women getting employed outside the home, in addition to their traditional domestic work, to share the financial burden of the family, and also to gainfully utilize their professional expertise (Varghese, 1995). Experience had shown that most of the people keep a kitchen for approximately two decades or even more and do not get it repaired or renovated with change in demand or change in the fashion. That's why it pays to carefully consider the needs and wants of the homemaker while designing the kitchen. Poorly designed kitchen work surfaces, storage spaces, material and dimensions cause permanent body damage besides increasing the work cost. This justifies that dimensions of kitchen work surfaces and storage spaces should be given careful attention thereby, minimizing stress on cardio-vascular, muscular and respiratory system. Therefore "work space must get considerable attention in the designing of the layout". Considering the importance of designing it was felt that an evaluation is needed of the existing kitchen situation, so that recommendations/ guidelines for improvement can be made to suit the Indian style of kitchen.

# 2 MATERIALS AND METHODS

Study included the field survey of eighty respondents, forty each from east and west zone of Ludhiana district in Punjab was done. A self structured interview schedule was used for collection of data. The interview schedule consisted of three parts. First part dealt with socio-economic status of the family which gathered information related to occupation, education, income, family type and family size. The second part of the interview schedule specific information like kitchen type, kitchen size, dimension of work counter etc. were all analyzed. Further information regarding the problems faced by the homemaker at the kitchen workstation was collected. Data for the study were collected through personal interview method. Equipments like accutape, measuring tape, anthropometer were used to record data. The data collected were tabulated and suitable statistical tool such as frequency, averages, percentages, correlation coefficient and standard deviation were used for analysis of data.

# 3 RESULTS AND DISCUSSION

## 3.1 Socio-economic profile of the respondents and their family

Data revealed that the average age of the selected respondents was found out to be 38 years, the average height 156 cm, and the average weight 58 kg. Majority of respondents were graduate i.e. 70 per cent, followed by respondents who had qualification up to matriculation (16.25 per cent) and the least number (2.50 per cent) of respondents were having professional degree of doctorate. Further it was observed that most of the respondents were housewives (56.25 per cent), 35 % were self employed and about 8.75 % were in government or private jobs.

**Table 1 Personal profile of the respondents**

| Respondents' profile | Number (n=80) | Percentages |
|---|---|---|
| **Age (yrs)** | | |
| 26 – 35 | 27 | 33.75 |
| 35 – 45 | 36 | 45.00 |
| Above 45 | 17 | 21.25 |
| Average age: 38 yrs | | |
| **Height (cm)** | | |
| 145 – 155 | 37 | 46.25 |
| 155 – 165 | 28 | 35.00 |
| Above 165 | 15 | 18.75 |
| Average height: 156 cm | | |
| **Weight (Kg)** | | |
| Below 55 | 10 | 12.50 |
| 55 – 65 | 46 | 57.50 |
| 65 – 75 | 24 | 30.00 |
| Average weight: 58 Kg | | |
| **Qualification** | | |
| High school | 13 | 16.25 |
| Graduate | 56 | 70.00 |
| Post Graduate | 9 | 11.25 |
| Doctorate degree | 02 | 02.50 |
| **Occupation** | | |
| Housewife | 45 | 56.25 |
| Self employed | 28 | 35.00 |
| Pvt. Job | 03 | 03.75 |
| Govt. job | 04 | 05.00 |

## 3.2 Kitchen size

It was observed that majority of the kitchens of the selected sample had an area from 6.7 to 8.4 sq. m (56.25 per cent) followed by kitchens with area in the range of 8.4 sq m. to 10.1 sq. m (23.75 per cent) and minimum number of kitchens were with area 5 to 6.7 sq. m (20 per cent). Though the area of the majority of kitchens was found within the limits but in a considerable number of kitchens i.e. 20.00 per cent, it was less than the minimum recommended kitchen area. These findings are similar to the findings of the Verma (2001) who also reported that area of kitchen in urban areas was less than the recommended area.

## 3.3 Kitchen type

It was observed that all of the respondents (100 per cent) had closed and standing type of kitchens. It may be due to the obvious advantages of the standing kitchen type as compared with the older way of cooking while sitting. The style of kitchen i.e. standing or sitting style has an effect on the home maker's performance level, time spent and her physical fitness (NBC of India, 2001). In standing type of kitchens as the worker maintains the standing posture while working, we can move quickly from washing to cutting to cooking areas, thereby saving our time and energy.

**Table 2 Features of the selected kitchens**

| Kitchen features | Number (n=80) | Percentages |
|---|---|---|
| **Kitchen Size (sq. meter)** | | |
| 5.0 - 6.7 | 16 | 20.00 |
| 6.7 - 8.4 | 45 | 56.25 |
| 8.4 - 10.1 | 19 | 23.75 |
| Minimum and maximum recommended kitchen area size :8 to 15 sq. m. Grandjean,1973) | | |

## 3.4 Shape of kitchen counter

Not any one type of kitchen is more ideal than another. The room and interior character should itself dictate the design of an efficient kitchen layout (Conran, 1986). From the Table 3, it is evident that only 'L' and 'U' shaped kitchen counters were common in the selected kitchens, with majority of respondents having 'L' shaped kitchen counter (61.25 per cent) and 'U' shaped (38.75 per cent). Similar were the observations of Mittal (1971), Grandjean (1988) and Sumangala (1995) who all reported that 'L'-shaped kitchen arrangement is the best as it is found to be the most efficient for performing kitchen work.

**Table 3 Shape and construction material of the work stations in the selected kitchens**

| Shape of kitchen counter | Number (n=80) | Percentages |
|---|---|---|
| 'L' shaped | 49 | 61.25 |
| 'U' shaped | 31 | 38.75 |
| **Material of kitchen work station** | | |
| Granite | 36 | 45.00 |
| Marble | 26 | 32.50 |
| Cement | 18 | 22.50 |

## 3.5 Material of kitchen counter

Regarding the material used for the kitchen counters, it can be seen from Table 3 that most of the kitchens had counters of granite (45 per cent) followed by kitchen counters of marble (32.5 per cent) and only (22.5 per cent) had cemented counters in their kitchens.

## 3.6 Posture adopted while performing different activities

Table 4 revealed that most of the kitchen activities were performed in standing and sitting posture by the selected homemakers. Very few activities like sieving were performed in squatting posture (38.75 per cent) by selected the respondents. Verghese *et al* (1989) also observed that while studying the posture adoption during kitchen work all the homemakers preferred to do cooking activities in standing posture as compared to sitting or squatting.

**Table 4 Posture adopted while performing different activities**

| Activity | Sitting | | Standing | |
|---|---|---|---|---|
| | Frequency | Percentage | Frequency | Percentage |
| Boiling | 00.00 | 00.00 | 80.00 | 100.00 |
| Sieving | 49.00 | 61.25 | 15.00 | 18.75 |
| Kneading | 69 .00 | 86.25 | 11.00 | 13.75 |
| Peeling | 65 .00 | 81.25 | 15.00 | 18.75 |
| Cutting | 65.00 | 81.25 | 15.00 | 18.75 |
| Washing vegetables | 00.00 | 00.00 | 80.00 | 100.00 |
| Rolling | 00.00 | 00.00 | 80.00 | 100.00 |
| Puffing | 00.00 | 00.00 | 80.00 | 100.00 |
| Stirring | 00.00 | 00.00 | 80.00 | 100.00 |
| Grinding | 04.00 | 5.00 | 76.00 | 95.00 |
| Grating | 13.00 | 16.25 | 67.00 | 83.75 |
| Dish wash | 00.00 | 00.00 | 80.00 | 100.00 |

## 3.6 Various activities in which fatigue is felt by the homemaker

Table 5 gives the information regarding the feeling of fatigue while performing different kitchen activities by the homemakers. It was observed that 40.00 per cent homemakers felt fatigue while grating followed by kneading (38.75 %), dishwashing (37.50 per cent) and minimum number of respondents (6.25 %) complained fatigue while performing activities like puffing chapattis (8.75 per cent), stirring (12.50 per cent) and sieving (21.25 per cent). It was observed that while performing different kitchen activities fatigue was felt by the homemakers in their forearm and wrist. The maximum respondents felt fatigue, while performing activities like grating (40.00 per cent), kneading (38.75 per cent), dish-washing (37.50 per cent) and rolling chapattis (33.75 per cent). During these activities trouble spots disclosed were mainly the wrist, fore arm, followed by discomfort in the shoulders, upper arm, lower back and neck.

**Table 5 Feeling of fatigue while performing various kitchen activities**

| Activity | Yes | | No | |
|---|---|---|---|---|
| | Frequency | Percentage | Frequency | Percentage |
| Grating | 32 | 40.00 | 48 | 60.00 |
| Kneading | 31 | 38.75 | 49 | 61.25 |
| Dishwashing | 30 | 37.50 | 50 | 62.50 |
| Rolling | 27 | 33.75 | 53 | 66.25 |
| Cutting | 21 | 26.25 | 59 | 73.75 |
| Sieving | 17 | 21.25 | 63 | 78.75 |
| Stirring | 10 | 12.50 | 70 | 87.50 |
| Puffing | 07 | 08.75 | 73 | 91.25 |
| Grinding | 05 | 06.25 | 75 | 93.75 |

## 4   CONCLUSIONS

Today, the urban Indian homes are modern in architecture, stocked with latest appliances and well attuned to change. The opportunity we have in our home that we do not always have in our work area is the ability to create a user-friendly environment. So the physical amenities and designing of the house especially the kitchen should be planned as per the needs, comfort and dimensions of the homemaker.

## ACKNOWLEDGMENTS

The authors would like to acknowledge from her heart to benevolent advisor Dr. M. K. Sidhu for valuable, cooperative and constant suggestions and benevolent criticism throughout the course of investigation.

# REFERENCES

Conran T, 1986. *The kitchen Book.* Mitchell Beazley Publishers Limited, London.

Grandjean E, 1988. *Fitting the task to the man.* 4th ed. Pp 35 – 36. Taylor and Francis, London.

Mittal M, 1971. *Development of guides for setting up storage cabinets at the serving and cleaning.* M.Sc. Dissertation. Haryana Agricultural University, Hisar, India.

Sumangala P R, 1995. *Developments of standards based on anthropometric measurements and their implications in the designing of kitchen and storage.* Ph.D. Dissertation, M S University, Baroda.

Verma S, 2001. *Ergonomic study of existing kitchen designs of rural and urban homes of Ludhiana District.* Unpublished Ph. D. Dissertation, Punjab Agricultural University, Ludhiana, India.

Verghese M A, Chattarjee L, Aterya N and Bhatnagar A, 1989. Ergonomic Evaluation of Household Activities UGC Report . Deptt. of Family Resource Management, SNDT, Bombay, India.

Verghese M A, Chatterjee L, Aterya N and Bhatnagar A. 1995. Rapid appraisal of occupational workload from a modified scale of perceived exertion. *Ergonomics* **37**: 485 – 491.

# Section XI

---

## User Experience, Comfort and Emotion

# CHAPTER 78

# Objects from Prehistory, a Study from the Point of View of Ergonomics – a Reference in Brazil

*Mariana Menin 1, José Carlos Plácido da Silva 2,*

*Luis Carlos Paschoarelli 3*

UNESP – Univ Estadual Paulista
Bauru, Brazil
*mariana_menin@yahoo.com.br*

## ABSTRACT

The Ergonomics has the exact date of its birth but the events preceding the formal launching of today are little explored. Specifically this issue in Brazil is still little explored in books or conference proceedings that area. The article aims to reflect and discuss the development of ergonomics in the prehistory, specifically in Brazil through sample of lithic artifacts from museums of the cities in the State of Sao Paulo. Thus, archaeological remains of instrument from chipped stone are important to research the history that precedes the formal launching of Ergonomics in Brazil and the world and to understand the criteria that the man was sought, unconsciously, comfort, safety and ease in building their tools.

**Keywords**: Ergonomics, Prehistory, tools

## 1    INTRODUCTION

Ergonomics is one of the sciences, if not the only one, with an exact date for its birth – July 12, 1949 – however, the events that precede the official appearance of this science are, until today, little explored. In Brazil, specifically, this theme is still little explored in books or annals of congresses of the referred area.

Prehistory of Ergonomics is directly related to the development of tools by man, that is, it is related to the prehistoric tools. Thus, archaeological remains of flint tools are important tools for investigating the history that precedes the official appearance of Ergonomics in Brazil and the world – and for the comprehension of the criteria man would already seek, unconsciously: comfort, safety and ease in the construction of their tools.

This article aims to reflect and discuss the birth of Ergonomics in the prehistory of Brazil through samples of lithic artifacts from Brazilian museums, specifically those found in the museums of the cities of Presidente Prudente and Jaú, in the State of São Paulo.

## 2    PREHISTORY AND ERGONOMICS

Coincidentally, the prehistory of Ergonomics is linked to human prehistory. According to Sanders and McCormick (1993), Ergonomics goes back to prehistory and is related to the creation of the first tools by man, that is, when primitive man feels the need to adapt tools to facilitate their daily tasks. Authors such as Vidal (2000) and Iida (2005) also share this idea. Moraes and Mont'Alvão (2000) point out that since ancient civilizations man sought to improve the tools, instruments and utensils for use in their daily activities to provide a more comfortable use. Figure 01 illustrates the statement of the authors. It is important to point out that, one of the most notable characteristics of man – and that distinguishes him from animals – is their ability to make and utilize tools.

FIGURE 01 – an example of grip on a brazilian lithic instrument (from approximately 5.000 years ago).

Iida (2005 p.03) supposes that the beginning of ergonomics happened probably "[...] with the first prehistoric man who chose a rock with a format that best fit the shape and movements of their hand, to use it as a weapon".

Meirelles (1991 as cited in Therrien and Loiola, 2001) explains that there is evidence of concerns about ergonomic aspects since the Upper Paleolithic and that the artifacts utilized for the work went on gradually being specialized for the use and handling and also miniaturized, thus demonstrating the need for adaptation of objects to productive use by means of a specialized shape and a size befitting the use and a more comfortable and easier handling.

Menin, Silva and Paschoarelli (2010a) state that the trajectory of products development begins along with the story of man, when people sought to combine comfort, safety, maneuverability and ease in the construction of their tools to perform daily activities such as hunting, cutting, scraping, grinding, sewing, breaking seeds and fruits and also for his defense. On the other hand and before, the weakness of man in relation to nature and their need for survival are the main reasons for the human being to have started utilizing and manufacturing tools.

Thus it is evident the importance of object manipulation in human evolution, being the hand and brain the main determiners of the development of man, their culture and civilization. Childe (1973) reaffirms that man could survive and multiply mainly by the perfecting of instruments handled.

The appearance of man on earth is indicated by such tools (Childe, 1965) and the first tools were made from pieces of wood, bone and stone, slightly sharpened or accommodated to the hand by breaking or chipping. However, the tools that are commonly found today through archaeological excavations are made of stone and bone, because according to Prous (2006) these materials are more easily preserved. As for the lithic tools (lithos - Greek word that means stone) man begins to make them 2500 years ago (Childe, 1973; Leakey and Lewin 1981; Leakey, 1994; Cook, 2005). However, these early tools were very simple and rudimentary, their shape being determined by the need of the moment and mainly by the raw materials, therefore without an established project as we know for current objects.

Primitive man had to learn which of the stones were more suited to the manufacture of tools and how to chip them correctly. Thus, a scientific tradition was built on which of the stones were the best, where they could be found and how they should be treated (Childe 1965). These techniques were culturally transmitted, which for Cook (2005) proves that man is not genetically programmed to make stone tools, they had to learn with other humans. Childe (1973) highlights that the creative act in was individual its first moment, but was passed on to the generations after, which characterizes the transmission of culture.

Thus, Ergonomics does not come with the first tools made by man, but when man develops mental ability to establish a design for the tools targeted to their needs, i.e. the time at which the form of these tools become more complex and requires skills for its preparation. This happens around 1.4 million years ago (Menin, Silva, and Paschoarelli, 2010b). Observe in Figure 02.

752

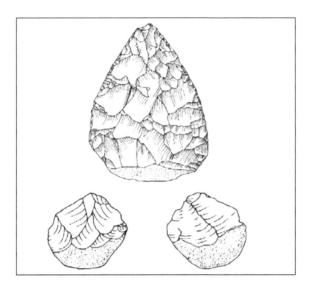

Figure 02 - Instruments Stone. The instrument line above is of tool made 2.5 million years ago, the two tools on the line below date from 1.4 million years.

Leakey (1994) states that the first manufacturers had no specific forms of individual artifacts in mind, i.e., when they were making them, and that forms were most likely determined by the original form of raw material. However, about 1.4 million years ago in Africa, it appeared a new kind of collection of lithic artifacts that show evidence that there was a mental model of what they wanted to produce, and that they intentionally imposed a form the raw material they utilized.

For Toth (1985 as cited in Leakey, 1994) the manufacture of this type of artifact requires the coordination of cognitive and motor skills. Leakey and Lewin (1981) state that the manufacturers of these (more sophisticated) tools had a clear conception of the implements they were making, as it is possible to observe a consistency of models within a set of artifacts.

Man arrives in Brazil a long time after that date, around 12,000 years ago (Prous, 2006). In Brazil it is possible to find a great amount of lithic instruments distributed by archaeological sites nationwide. Around between 12 thousand and 5 thousand years ago there were groups of people in almost the whole Brazilian territory. The southeast region of Brazil was dominated by hunter-gatherers (lifestyle of hunting and gathering, hunting was the occupation of men and the collection of vegetables, seeds, roots, were of women (Leakey, 1994)), with the earliest occupation of the State of São Paulo dating from 14.200 years BP (Funari and Noelli, 2006).

In São Paulo, the State covered in this article, the oldest resident discovered so far, is the skeleton of an adult man with approximately 1.60 m in height, called Luzio and dated 10180 – 9710 years BP (Eggers, Parks, Grupe, and Reinhard, 2011).

## 2.2 Lithic tools in Brazil

In Brazil it is possible to find a great amount of lithic tools throughout archaeological sites in all parts of the country.

During the period between 12 and 10 thousand years ago, human population in Brazil showed great diversity of lithic industries (Funari and Noelli, 2006), being important to point out that even in following periods (of Polished Stone and ceramics), chipped stone tools were utilized and produced.

The stone was utilized for making artifacts such as tools, weapons and ornaments. And its uses were the most varied, as cutting, scraping, drilling, grinding, planing, sawing, fishing, and hunting, among others. Chipped stone tools produced in Brazil were of various types, such as projectile points, scrapers, slugs, punches, chisel, chipping.

In this study, pieces of Brazilian museums are presented, specifically those found in the museums of the cities of Presidente Prudente and Jaú, in the State of São Paulo.

The pieces from the Municipal Museum of Jahu can be seen in Figure 03, these are around 5000 years. Tool A is a scraper, tools from figures B and C are called "slugs" and are used with plane movements and pieces D and E are Scrapers /punches.

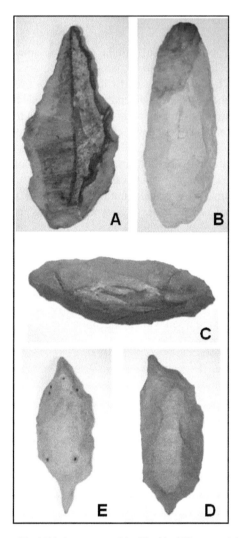

Figure 03 – Lithic instuments of the Municipal Museum of Jahu

The pieces from the Centre of Museology, Anthropology and Archaeology (CEMAARQ), located in the city of Presidente Prudente, are (Figure 04): A is a concave scraper, B is a Chopping Tool, C is a Zinken (an artifact in the shape of a claw which was commonly utilized to cut or scrape), D is a scraper, E is a convex scraper and the artifact of letter F is a slug.

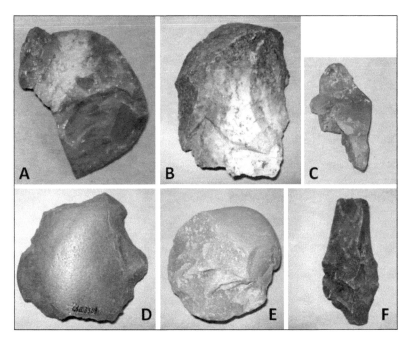

Figure 04 – Lithic instruments from CEMAARQ – Presidente Prudente

# 3    DISCUSSION

The history that precedes the official birth of Ergonomics is until now little explored. We know that man begins to make their lithic tools 2,500 years ago. But it was possible to understand that these early tools were very simple, being its shape determined by the need of the moment and especially by raw materials, i.e., without a previous project.

When we look at lithic artifacts made millions of years ago we conclude that these humans began a segment of inventions and technology until today uninterrupted.

Analyzing the Brazilian artifacts mentioned above, we can state that prehistoric man planned grip areas for their instruments, and that these handles made various fittings for the hands possible – and consequently, different types of grip required for each type of specific activity.

We can come to this conclusion by observing that in the same instrument it is possible there to be different types of grip. As an example, we cite the case of the tool D from the Municipal Museum of Jau where to perform the movement of drilling it is possible to utilize hand grip and a pinch grip, depending on the material one wants to pierce and pinch grip to perform the movement of scraping, being important to point out that all types of grips for different instruments were shown to be comfortable and with a perfect fitting for the hands, which made them very safe for the performance of movements.

This statement corroborates the statements of Moraes and Mont'Alvão (2000), that since ancient civilizations man is interested in adapting the form of the handles of the instruments to the form of the human hand.

Therefore, unconsciously, Brazilian prehistoric man utilized concepts of Ergonomics when crafting the first tools and utensils, seeking for a more comfortable, safe and effective way to perform their daily activities.

Thus, this study is a contribution to the writing of the history of Brazilian Ergonomics along with prehistoric man.

## ACKNOWLEDGMENTS

This study was supported by CAPES

## REFERENCES

Childe, G. 1965. Man makes himself. Londres: C. A. Watts & Co.

Childe, G. 1973.What happened in history. Inglaterra: Penguin Books.

Cook, M. A. 2005. A Brief History of the Human Race. Rio de Janeiro: Jorge Zahar, [in Portuguese].

Eggers S, Parks M, Grupe G, Reinhard KJ (2011) Paleoamerican Diet, Migration and Morphology in Brazil: Archaeological Complexity of the Earliest Americans. PLoS ONE 6(9): e23962. doi:10.1371/journal.pone.0023962.

Funari, P. P. and F. S Noelli. 2006. Prehistory of Brazil. São Paulo: Contexto, [in portuguese]

Iida, I. 2005. Ergonomics – Design and Production. São Paulo: Edgard Blücher, [in portuguese].

Leakey, R. 1994 The Origin Of Humankind. New York: Perseus Book.

Leakey, R. and R. Lewin, 1981. Origins: what new discoveries reveal about the emergence of our species and its possible future. Inglaterra: Penguin Books.

Menin, M., J. C. Silva, and L. C. Paschoarelli,. 2010a. The Prehistory of Design and Ergonomics. In: Silva, J. C.P.; J. C. Paschoarelli; Design: Issues research (pp. 9-16). Rio de Janeiro: Riobooks. [in portuguese]

Menin, M.; J. C. P. Silva and L. C. Paschoarelli. 2010b. Ergonomics and lithic artefacts: reflecting on the emergence of ergonomics in prehistory. Proceedings of the XVI Brazilian Congress on Ergonomics, Rio de Janeiro. [in portuguese]

Moraes, A. and C. Mont'alvão. 2000. Ergonomics: Concepts and Applications. Rio de Janeiro: 2AB, [in portuguese]

Prous, A. (2006). The Brazil Before The Forum: The Prehistory of Our Country. Rio de Janeiro: Jorge Zahar. [in Portuguese]

Sanders, M. S., and E. McCormick. 1993. Human Factors in Engineering and Design. New York: McGraw-Hill.

Vidal, M. C. 2000. Introduction to the Ergonomics. PEP/COPPE. UFRJ. Rio de Janeiro [in Portuguese].

# CHAPTER 79

# Representing Traditional Culture - Poetry Applying Elements on Product Design

*Chi-Hsien Hsu* [1], *Mo-Li Yeh* [2], *Po-Hsien Lin* [3], *Rungtai Lin* [4]

National Taiwan University of Arts
Ban Ciao District, New Taipei City 22058, Taiwan
assah16@gmail.com

## ABSTRACT

Chinese culture is rich and diversified. Chinese styles and culture are becoming focal points in the world in recent years, so poetry cultural elements will certainly become an accelerating force for the pursuit of cultural and creative industries. The essence of cultural product design lies in refining culture elements and transferring them symbolically through metaphors so as to attach new aesthetic significance to the product design thereby providing a moving experience for consumers. Therefore, the main purpose of this paper is to build a transformation model of cultural product design. The approach includes two phases. In phase one, the transformation model for design development is constructed by design principles, design process and literature review. In phase two, we undertake design practices based on the transformation model in order to prove its effect.

**Keywords**: Poetry, cultural product design, Chinese character

## 1    INTRODUCTION

As the demands of life increase, the consumer market advances in an era of experience-focus and aesthetic economics. The distinctness of local culture and the structure of innovation-knowledge become the national core competency. It's seen within the trend of the promotion of cultural and creative industries as an important strategy for economic development in each country. For example, the transformation of the pewter-manufacturing industry in the UK (Yair, Press, &

Tomes, 2001) and Oktoberfest in Munich of Germany aim to integrate their the cultural assets in order to enhance the national image and improve competency of their industries. By the transformation and migration abroad of industrial structure, we have to endeavor to increase the value and added-value of the product through the design activities of cultural innovation (Lin, & Lin, 2009).

Recently, Chinese culture has caught the attention of the world not only in the field of design application but also the style of movies. The multi-cultures and friendly people in Taiwan have been deemed as our characteristic. Products with distinctness are popular in the consumer market but that distinctness is gradually disappearing because there are too many products with the same function and style. As a result, consumers start to ask for a product with cultural distinctness and recognition. The characteristics of traditional culture in Taiwan bring potential application-value in the design field. However, most cultural design products simply copy a form or are decorated with cultural totems. Products without the spirit of the culture will not help to upgrade the life culture (Lin, 2005; Li, 2007).

Therefore, it is critical to design a product with cultural value and localized characteristics. Therefore this exploration studies the feasibility of poetic rhyme transformed into cultural product design. In order to construct the model for transforming cultural elements into product design, we've investigated the literature and surveyed current applications of cultural innovations of Chinese characters. Following that model, we invest the traditional culture into modern product design. Therefore, this exploration will serve as reference for future investigation of Chinese characters and the design of cultural products.

## 2    CULTURAL ELEMENTS AND PRODUCT DESIGN

The design and development of cultural products help to improve life quality and the social culture level. Along with technology progress, designing "feeling" into products to present the emotional communication of user experiences became a design trend in the 21st century. Design should not only focus on function and elegant appearance but also on the heritage and connection of the culture concerning problems in our society in order to redefine people's life style.

## 2.1   Three Cultural Levels

Culture generally refers to styles of human activity and symbolic structures. Moreover, culture has been described as the evolutionary process that involves language, customs, religion, arts, thought and behavior. From the design point of view, Leong and Clark (2003) developed a framework for studying cultural objects distinguished by three special levels: the outer 'tangible' level, the mid 'behavioral' level, and the inner 'intangible' level. Furthermore, the UNESCO (2005) thinks that the cultural product possesses an economic and cultural nature and becomes a carrier of culture characteristics, values and meaning through the experience of use and the preservation of cultural heritage.

Based on previous studies (Leong & Clark, 2003; Hsu, 2004; Lin, 2007), a framework for studying cultural objects is summarized in Figure 1. As shown in Figure 1, culture can be classified into three layers: (1) Physical or material culture, including food, garments, and transportation related objects, (2) Social or behavioral culture, including human relationships and social organization, and (3) Spiritual or ideal culture, including art and religion. These three culture layers can be fitted into Leong's three culture levels given above. Since cultural objects can be incorporated into cultural design, three design features can be identified as follows: (1) the inner level containing special content such as stories, emotion, and cultural features, (2) the mid level containing function, operational concerns, usability, and safety, and (3) the outer level dealing with colors, texture, form, decoration, surface pattern, line quality, and details. In each layer of exploration, this figure can help to focus on the key point when we design a cultural product.

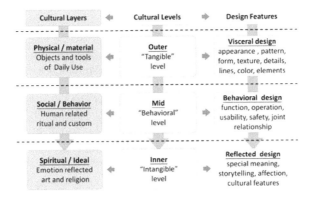

Figure 1. Three layers and levels of cultural objects and design features (Hsu, 2004; Lin, 2007)

## 2.2 The Distinctness of the Culture of Chinese Character

The character is the most important tool to record the history and culture. Among well-known cultures, the Chinese character is the only one to be used till now, and which is deemed as a most antique character like Sumerian Cuneiform and the Egyptian Hieroglyph. The Chinese character presents not only the heritage and development of the culture but also the symbol of the traditional esthetics. The Chinese Character originated from illustration by transforming the tangible materials into images on the plane surface. After a long development, structure and proportion have become more esthetic. As the Chinese character contents the cultural value and the calligraphy shows the esthetic of art, people always like it as a beautiful painting even they don't understand it.

For instance, the logo of the 2008 Olympics held in Beijing made people see the various applications and transformations of the charming Chinese stamping culture. This logo was designed in Chinese stamping style with red color, the basic color in

Chinese tradition. The Chinese character 「京」 was writing by seal character and looked very dynamic. It helped the people around the world to re-know the application and variation of Chinese characters. Since 1980s, modern calligraphy became popular in Taiwan. In 2001, Taiwanese calligrapher Hsu Yung-Chin, was retained by our Tourism Bureau. He designed the symbol for the Taiwanese tourism industry. He boldly wrote down "Taiwan" using calligraphy. The Taiwanese cultural custom shown up in this word has become the vivid symbol of our culture. And the artist, Tsai Yu, surprised the public with her design of "Chinese Character Jewelry" by the combination of the humanity of Chinese characters and the fashion of jewelry. In the international auction in Beijing in 2010, the art work "Dreams Come True" was designed according to the Chinese character 「夢」 which it was made of jadeites and diamonds (Tsai, 2009). The Taiwanese cultural innovation of Chinese character was successfully to be seen in the global art market.

## 2.3 Design Theories Relevant to Cultural Products

Thanks to the improvement of technology, it is easier to implement innovation. In order to satisfy consumers' demands, the design should be accomplished by the enhancement of investigation of the product and human-product interaction. In the future Industrial Design should concern aesthetics of humanity. Norman (2004) also suggested that a successful design should consider the suitability, practicability and aesthetics of the product in which emotion is the most important factor. The design concept for emotion can be accomplished by the aesthetic value of the product. As a result, the intention of the design gradually focuses on humanity with the consideration of the consumers' feelings when they use the product.

Cultural features are considered to be unique characteristics that can be embedded into a product for both the enhancement of its identity in the global market and individual consumer experience (Handa, 1999; Yair, Tomes, & Press, 1999). They could trigger a cultural reflection by consumers through design. In general, the common discussion of cultural applications to the product is the theory of product semantics. For example, Lin and Huang (2002) classify the logic of figurative designs whose forms are based on some reasonable visual connections. It defines visual connections such as metaphor, simile, allegory, metonymy, and analogy borrowed from linguistics, and then systematically analyzes these elaborate relationships between products and the signs. In addition, Butter (1989) suggested that the design process can be seen as somewhat linear with clearly distinguishable phases and suggested eight steps for the systematic generation of semantically relevant design concepts. Based on this literature, an approach was undertaken to integrate the design theories and provide assistance for cultural product design.

A good understanding of the cultural attributes will benefit articulating the context between the culture and product design and therefore accelerate concept development. Based on the cultural product design framework and process, the cultural product is designed using scenario and semantics approaches. And according to the literature review and expert opinions, design guidelines are developed based on the research of consumers' needs, cultural content and design theories. Hsu, Lin and Lin (2011) provided a practical design process in four phases and ten steps which are used to design a cultural product, namely: identification (telling a situation), investigation (setting an objective), interaction (writing an analysis), and implementation (designing a product).

# 3    RESEARCH METHOD

We can see the traditional localized culture in our life; the distinctive totem or image intangibly shows up in our life materials. Globalization brings people convenience in life but their localized culture drives people to look for a prosperous quality in the life and mind-set. In order to avoiding the application arbitrarily of product design from the distinctive localized characteristics. As a result, it is important to understand well the cultural elements and product languages in the field of cultural product design.

When we design, we should consider the characteristics in various aspects like tangible materials, traditional custom and spirit. The understanding and analysis will help us to consider the design transformation in order to add the cultural elements into the cultural product. In general, there's a procedure and model in the development and manufacturing process in each industry. We are wondering if there's a procedure to apply which can help in the deliberation and transformation during the design process. Lin (2007) taught students to explore the three levels and model of design using the ceramic pot of the Rukai Tribe in Taiwan. In Figure 2, A is the hand bag transformed by the appearance and decoration of the clayware pot which is deemed as the transformation of appearance. B is fruit bowl made of stainless steel transformed by the function and use behavior as the transformation of behavior. C is a hollow candleholder. When we light up the candle, there will be human shadows swaying by the wind like the campfire surrounded by aborigines.

This exploration is to build up the model of the transformation of cultural product design integrating consumers' expectations and design behavior though Hsu, Lin and Lin's (2011) step six to ten of the design procedure of cultural products and Lin's (2007) exploration in regards to cultural products derived from the clayware pot. Having analyzed and synthesized a literature review, we construct a model for transferring cultural elements into product design. The five implementation steps are: (A) Analyzing and selecting the cultural elements; (B) Connecting the reasonable relationships between cultural elements and design concepts; (C) Selecting the rational concept; (D) Developing the concept; (E) Completing the design. Steps A to C refer to the process which connects reciprocally the cultural elements and the product. Step D and E address the rationality, and embodiment of the product and maturity of the transformation of the cultural elements.

In order to design a modern product with traditional  poetry and cultural elements, we adopt "The Pipa Tune" of Bai Ju-yi, a poet of the Tang Dynasty, as the source of inspiration for considering different cultural levels and the method of transfer, transit and transform of the cultural elements of this poem (Figure 2). This model is a reference for designer(s) which can lead logically to an adequate design transformation and expose the cultural elements properly no matter whether the design is transformed from a tangible or intangible culture.

When we design a cultural product, we must analyze its property first in order to discover the reasonable connection between cultural elements and the product in each culture. During the design procedure much uncertainty exists which is hard to control by designers. The "design features" and "transformation of design consideration" in Figure 2 helps us not only to understand and explore the cultural elements but also to consider the application scope of a cultural product. We will be able to design a product in which there's cultural transformation of application and product design with cultural elements instead of merely copying the appearance of relics or totem.

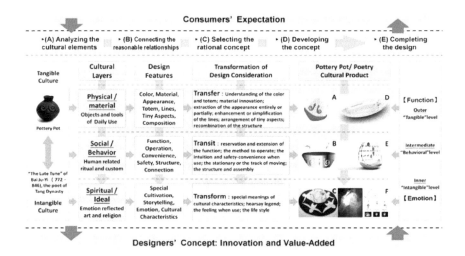

Figure 2. The Transformation Model of Cultural Product Design

## 4    THE DESIGN PRACTICES BY POETRY

In Taiwan, the promotion of cultural and creative industries has been the focus of development. Localized cultural elements, products and tours have gradually become an opportunity for animating the regional economy. Various types of souvenirs and the use of cultural product spring up all around. However, most of them lack distinctive and localized cultural elements or are limited to craft works. These cause less recognition from the consumers and become less popular in life.

In order to make cultural elements to enhance life quality, it is necessary to develop modern products with cultural intention. In Taiwan, we've initially seen the improvement of the combination of traditional craft artworks and cultural innovation. However, the majority of cultural innovation products are still limited to the application of material culture. It is worthwhile to explore the intangible cultural assets like Chinese characters. The current design application or research of Chinese characters is focusing on the transformation of the font (Lin, Lin, & Hsieh, 2005). The Chinese character has fostered much precious poetry and many tunes. The Chinese character culture is unique and deemed to be an innovation cause in the fields of art, fashion and the design. It also leads the fashion trend of eastern style.

Literary works as rhymes derived from Chinese character, such as classical poetry, tune, are very touching and worthwhile exploring. Sometimes the wording of Chinese poetry describes concrete articles but other times the feeling hiding behind the articles or circumstances. As a result, the reference information for design will be rich with good understanding of the poetic rhymes. We even can manage the design by the transformation of different cultural levels (Figure 2). We give the following as an example in order to explain the design of a cultural product from Chinese poetry.

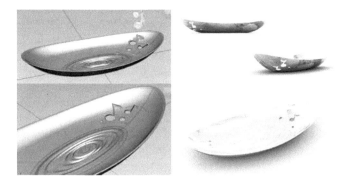

Figure 3. "Jade Plate" utensil

## 4.1   Material Application of Cultural Product

The design of material application is the transfer of cultural elements. We consider the design by the enhancement of color and understanding of totem and application, innovation of the material, extraction of the appearance entirely or partially, enhancement or simplification of the lines, arrangement of tiny aspects, recombination of the structure, etc (see D in Figure 2).

This design was inspired by the poem "The Pipa tune" of Bai Ju-yi (772-846), a poet of the Chinese Tan Dynasty. We selected several phrases in this poem which describe the sound of Pipa: *"The thick strings loudly thrummed like the pattering rain, and the fine strings softly tinkled in murmuring strain. When mingling loud and soft notes were together played, it was like large and small pearls dropping on a jade plate."* The mingling loud and soft sound of the Pipa is like the sound of pearls dropping on the plate. The concept of this product design is to combine the article in the jade plate with the wording describing the intangible music played by Pipa. We can imagine the product surrounded by the aerial music. The stainless steel shows the aesthetics of an ancient jade plate and modern innovation. Therefore, we conclude this design as the transfer of the appearance (Figure 3).

## 4.2 The Use Behavior Application of Cultural Product

The design application of use behavior is the transit of cultural elements. We can explore it by the reservation and extension of the function, the method of operation, the intuition and safety-convenience when in use, the stationary or the track of movement, the structure and assembly, etc (see E in Figure 2).

The design concept in Figure 4 was also originated from phrases of "The Pipa tune" of Bai Ju-yi. In the past, people used to put precious article on a jade plate which itself is of high value. We transit it into a life product as a piggy bank. When we insert different size of coins, the tinkling sound of rolling coins makes us recognize the beautiful wording and aesthetic Pipa sound in this poem. "Pearls dropping on the jade plate" describes the regular tinkling music played by Pipa and the classical melodious sound makes us imagine and think about the ancient times. Consumers not only can feel the fun of the sound "Pearls dropping on the jade plate" but also recall the memory of the pinball machine in childhood.

The visual fun makes users enjoy the use of it because they can see the saving money and the moving of coins through the transparent material. This product was designed based on the material and experience of use. Consumers will easily feel the aesthetics of Chinese poems in their life and the renewed design innovation for the traditional culture. When we use it, on the other hand, we are also inspired with new recognition of the product. This is the design transit of use behavior.

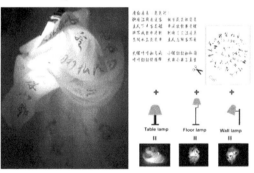

Figure 4. "Pearls dropping on the jade plate" Piggy Bank

Figure 5. "The Pipa Tune" lampshade

## 4.3 The Ideology Application of Cultural Product

The design transformation of ideology is the transform of cultural elements. We can explore it through cultural characteristics containing special meaning, hearsay legend, the feeling when we use it, the life style, etc (see F in Figure 2).

We can sense the distinct scenario in Chinese calligraphy and poetry. The concept of the product in Figure 5 is also derived from the "The Pipa tune" of Bai Ju-yi: *"One night by the riverside I bade a friend good-bye, In maple leaves and rushes autumn seemed to sigh. ... Without flute songs we drank our cups with heavy heart, the moonbeams blended with water when we were to part. Suddenly over the stream we heard a pipa sound, I forgot to go home and the guest stood spellbound. ... Listening to her sad music, I sighed with pain, hearing her story, I sighed again and again. Both of us in misfortune go from shore to shore, meeting now, need we have known each other before?"*

In "The Pipa tune" of Bai Ju-yi, the cantatrice from Chang-An talked about the joy and happiness in youth with sadness. She is now not only old but also has a vagrant life. The author suddenly revealed the sadness of his demotion from official life. We design the product as a plain lampshade using foggy material. It will be solid after swirling cutting. Then the user can give it different shapes of his choice. When there's the light, the twisted characters and the shadow are like large and small pearls dropping on a jade plate. The room will look Chinese Style and be full of nostalgia. It will enrich the consumers' feeling when they use the product in this way through the transformation of ideology.

## 5    CONCLUSION

Recently, Taiwan dedicated itself to the promotion of cultural and creative industries. However, the design application for now is to transform mostly the concrete products of traditional life into creative applications in industries. We can see prosperous characteristics of Chinese cultural materials in which Chinese characters are the most distinctive around the world. Chinese characters not only has distinctive features but are also loved by people worldwide. If we make good use of this precious Chinese character, the ideas of innovation and practical applications will be more prosperous and distinctive.

Regarding the application of cultural innovation, the design is not only to fit consumers' needs but also to imbed the cultural assets and to show it in an aesthetic way through the product. Chinese poetry describing not only concrete articles but also emotional expression inspired by the scenery, inspires people to continuously aftertaste, investigate and study. The design of cultural innovation derived from Chinese poetry helps consumers to experience both the history and the modern creation when they use this modern life product. The idea to transfer, transit, and transform intangible cultural into cultural and creative industries will enhance the distinctness of industries or even create the cultural and creative industries of Taiwanese style.

## REFERENCES

Butter, R. (1989). Putting theory into practice: An application of product semantics to transportation Design. *Design Issues,* 5(2), 51-67.

Handa, R. (1999). Against arbitrariness: Architectural signification in the age of globalization. *Design Studies,* 20(4), 363-380.

Hsu, C. H. (2004). *An application and case studies of Taiwan aboriginal material civilization conferred to cultural product design.* Unpublished master's thesis, Chang Gung University, Taoyuan. (in Chinese, semantic translation)

Hsu, C. H., Lin, C. L., & Lin, R. T. (2011). A study of framework and process development for cultural product design. In P.L.P. Rau (Ed.), *Internationalization, Design, HCII 2011,* LNCS 6775, pp. 55–64. Heidelberg: Springer-Verlag.

Hsu, Y. C. (2001). *Taiwan Image–Touch Your Heart.* Tourism Bureau, M.O.T.C. Taiwan, from http://admin.taiwan.net.tw/. (in Chinese, semantic translation)

Leong, D., & Clark, H. (2003). Culture-based knowledge towards new design thinking and practice- A dialogue. *Design Issues,* 19(3), 48-58.

Li, J. C. (2007). Communicative objects and communications among objects: The development strategy of cultural commodity from the museum's View. *Technology Museum Review, 11*(4), 53-69. (in Chinese, semantic translation)

Lin, H. Y., Lin, R., & Hsieh, H. Y. (2005). Exploring the possibilities of transforming Chinese characters into product design. *Journal of Design, 10*(2), 77-87. (in Chinese, semantic translation)

Lin, M. H., & Huang, C. C. (2002). The logic of the figurative expressions and cognition in design practices. *Journal of Design, 7* (2), 1-21. (in Chinese, semantic translation)

Lin, R. (2007). Transforming Taiwan aboriginal cultural features into modern product design: A case study of cross-cultural product design model. *International Journal of Design,* 1(2), 45-53.

Lin, R. (2005). Combination of technology and human: Cultural creativity. *Science Development,* 396, 68-75. (in Chinese, semantic translation)

Lin, R., & Lin, P. H. (2009). A study of integrating culture and aesthetics to promote cultural and creative industries. *Journal of National Taiwan College of Arts, 5*(2), 81-106. (in Chinese, semantic translation)

No author (2008). *Logo of 2008 Olympic.* Beijing Olympic City Development Association, from http://www.beijing2008.cn/

Norman, D. A. (2004). *Emotional design: Why we love (or hate) everyday things.* New York: Basic.

Tsai, Y. (2009). *Chinese Character Jewelry.* Tainan: Phoenix Cultural Artifacts Center. (in Chinese, semantic translation)

UNESCO (2005). *Convention on the Protection and Promotion of the Diversity of Cultural Expressions.* Retrieved October 20, 2005, from United Nations Educational, Scientific and Cultural Organization Web site: http://unesdoc.unesco.org/images/0014/001429/142919e.pdf.

Yair, K., Press, M., & Tomes, A. (2001). Crafting competitive advantage: Crafts knowledge as a strategic resource. *Design Studies,* 22 (4), 377-394.

Yair, K., Tomes, A., & Press, M. (1999). Design through making: Crafts knowledge as facilitator to collaborative new product development. *Design Studies, 20*(6), 495-515.

# CHAPTER 80

# A Study of Applying Saisait Tribe's Tabaa Sang (Buttocks Bell) into Cultural Creative Industry from a Cross-Disciplinary Perspective

*Po-Hsien Lin 1, Jao-Hsun Tseng 2, Chih-Yun Tsou 3*

National Taiwan University of Arts
Ban Ciao District, New Taipei City 22058, Taiwan
t0131@ntua.edu.tw

## ABSTRACT

The Tabaa Sang (Buttocks Bell) of Saisait Tribe is not only a kind of dance property but also a special musical instrument used in the Pas Taai Festival (Short Spirit Festival). Using Tabaa Sangs in the dancing ceremony has its ritual meaning of calling back the soul of the dead and is said to be able to ward off evil spirits. This study intends to explore the feasibility of applying Saisait Tribe's Tabaa Sang to cultural creative industry with an emphasis on cross-disciplinary integration. An ergonomic approach of anthropometry and biomechanics will be employed to examine the appropriate weight and operational position of the Tabaa Sang. A redesign of the Tabaa Sang is suggested for the purpose of performance. The ultimate purpose of the study is to demonstrate a cross-disciplinary model of transforming Saisait Tribe's Tabaa Sang into cultural creative products integrating design and performing art.

**Keywords**: Cultural creative industry, Saisait Tribe, Tabaa Sang (Buttocks Bell)

# 1    INTRODUCTION

Taiwan is entering a new era of aesthetics economy driven by cultural creative. The new design concept defines itself as follows: In addition to convenience for production and functionality for usage, design serves as a creation of lifestyles, the experience of various tastes, and the realization of life values, with an essential core of culture. Taiwan is a multicultural society. The rich vitality, roughness, and mystery from aboriginal arts had been popular themes borrowed by modern art designers, providing the designers prodigious stimulation and inspiration.

Among the heritage of Taiwanese aboriginal crafts, Saisait Tribe's Tabaa Sang (Buttocks Bell) is a classic. The Tabaa Sang is not only used as a prop for dancing, it is, more importantly, the only heard musical instrument in the traditional festival of pas'taai (worship festival for dwarf spirits). The bells produce crispy and loud sounds, thus become agitating and highly touching to its audience. Swinging with the rhythmical dancing movements, the bell fulfils worshiping function and expresses artistic aesthetics. It is, therefore, an outstanding achievement in Taiwanese aboriginal crafts, exhibiting extraordinary cultural features.

# 2    PURPOSE OF THE STUDY

The purpose of this study is to explore the feasibility of applying Saisait Tribe's Tabaa Sang to cultural creative industry with an emphasis on cross-disciplinary integration. An ergonomic approach of anthropometry and biomechanics is employed to examine the appropriate weight and operational position of the Tabaa Sang to comply with the performance. Moreover, a series of derivative products are created and examined, in support of the promotion for Tabaa Sang performance. The objectives of this study are multi-folded:

(1) with literature review and interviews, investigate Saisait Tribe's traditional culture, study the inner meaning of their rituals and ceremonial vessels, and explore the feasibility of applying Tabaa Sang to the cultural creative industry.

(2) integrate theories and practice of cultural creative design, and construct a design model for aboriginal cultural products.

(3) explore appropriate ergonomic design for operation of Tabaa Sang so as to establish solid principles for manufacturing the performance art props.

(4) apply the design model for aboriginal cultural products to implementing Tabaa Sang derivative products so as to provide a concrete illustration for promoting aboriginal cultures.

# 3    CULTURAL HERITAGE OF THE SAISAIT TRIBE

With a population of around 7000, the Saisait tribe mostly inhibit in the shallow mountain areas, around 500 to 1500 meters above sea level, in northern and central Taiwan. The Saisaits is a humble people showing extreme respect to the Heaven. The tribe believes in and fear the spirits in their traditional culture. From their daily life to ceremonial festivals, there are a lot of taboos. This is especially evident in the worship festival of pas'taai (Chu, 1995; Tian, 2011).

## 3.1 Legends and Totems

The traditional attire for the tribe is kayba'en. Men and women wear sleeveless gown with the length reaching their calves, and an outer vest. Two major colors employed are read and white. In the fiber, the tribe employ vast amount of 卍-shaped patterns and horizontal dot patterns (Figure 1). The special patterns signal two major characters in the festival song. The horizontal dots represents the ta'ay (dwarf or pigmy) spirit while the 卍-shaped pattern the thunder lady (Li, 1998).

The worship festival of pas'taai roots in a tribal legend. The ta'ay used to live neighboring the Saisait tribe and taught the tribe how to grow their crops for good harvest. However, after an incident in which a Saisait woman was assaulted by the ta'ay, the Saisait tribe was infuriated and acted for revenge. The bridge leading to the ta'ay's residence was cut, leading to the drowning of all but two of the ta'ay. The two survivors composed a festival song and taught it to the Saisait, asking them to remember the painful lesson and regularly perform memorial rituals. After the departure of the two survivors, the Saisait changed the original post-harvest festival into the pas'taai, a memorial festival for the ta'ay, with the purpose of commemorating the ta'ay and praying for blessing (Chu, 2005; Lin, 2000).

The 卍-shaped pattern on the Saisait attire represents the Thunder Lady. In the legend, the Thunder God sent his daughter Wa-an to earth for teaching the tribe how to grow crops. Wan-an fell in love with the Saisait warrior Dar-in, but they were childless after years of marriage. Dar-in's father blamed Wa-an for spending all her time farming in the mountains but never cooked as a responsible daughter-in-law. One day, with the insistence of her father-in-law, Wa-an went to the kitchen reluctantly, but leaving after her a loud crack in the kitchen. Dar-in and his father rushed there, only to observe a lightening scraping the sky. Wan-an disappeared, leaving a banana tree on the kitchen floor. As a result, the tribal women put the 卍-shaped pattern into their woven cloths so as to convey their appreciation and gratitude to the Thunder Lady (Chu, 2005; Tian, 2011).

Figure 1  The traditional Saisait attire      Figure 2  The dancers bearing the Tabaa Sang

(From Indigenous Peoples Commission of Taipei City Government, 2009)

## 3.2    Features of the Pas'taai and the Tabaa Sang

To the Saisaits, pas'taai is a specific manifestation of true Saisait spirit. The festival is held every other year, displaying the three major stages of entertaining the spirits, welcoming the spirits, and leave-taking of the spirits (Figure 2). The stage of entertaining the spirits allows outside spectators, while the later two stages are exclusively for the tribal members. It is believed that the festival has been held for hundreds of years, evidenced by relevant documentary films completed in the Japanese-occupancy era. (Tian, 2011).

The Tabaa Sang looks like a cloth pack attached to a suspender. A traditional Tabaa Sang is composed of mainly thin bamboo vessels and decorative beads. The beads are fruitage of wild barley, patched with rattan stalks and peels. In the past, the main body used woven flex nets, adding beautifying decorations such as cloth covers, decorative mirrors, and beads. Along the advance and change of time, the Tabaa Sang now are produced massively using mainly modernized materials—first cloth pads are stitched in triangle or Trapezoid shape, then in rows, from top to bottom, mini-mirrors, bead-strings, bamboo or mental vessels, sometimes additional four or five rows of bells, are sewn to the pad (Figure 3) .

Figure 3  Conventional (left) and modern (right) Tabaa Sang

## 4    AN ERGONOMIC ANALYSIS OF THE OPERATION OF THE TABAA SANG

With an emphasis on cross-disciplinary integration of craft and performance, this study intends to examine the appropriate weight and operational position of the Tabaa Sang during performance so as to apply and promote the bell for performing art.

## 4.1    Experiment Design

The approach of quasi-experimental research was used in this study to gain the ergonomic data. The prototype used in this study is a Tabaa Sang, which was provided by Aboriginal Cooperative Association of New Taipei City, in the shape of a backpack hanging a couple of rows of cooper pipes. The first independent variable (W) is the weight of the Tabaa Sang including three levels of heavy (w1=1560g), middle (w2=1240g), and light (w3=900g), by adding or reducing cooper pipes. The second independent variable (H) is the operational position of the Tabaa Sang including three levels of height established by adjusting the position of the Tabaa Sang in accordance with each performer's body (h1=the position of the tailbone, h2=10cm lower than h1, h3=20cm lower than h1). The dependent variable (X) was participants' responses to the effect of manipulating the Tabaa Sang using a five-point Likert scale. The framework of ergonomic test is shown in Figure 4.

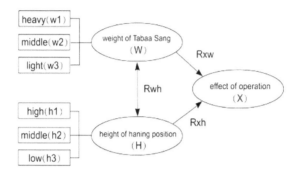

Figure 4  Three positions of ergonomic test

## 4.2    Experiment Outcomes and Analyses

The participants of the study were 38 students of Dance Department. The outcome of a Two-way Repeated Measures ANOVA (Table 1) shows that there is no significant interactive effect between two independent variables—weight and operational position—of the Tabaa Sang (F=.696). However, the main effect is significantly occurs alone in the second independent variable of operational position suggesting that it is the position (F=10.045, p<.01) but not the weight (F=.008) that affect the manipulation of a Tabaa Sang.

**Table 1  Two-way ANOVA for the Effect of Manipulating the Tabaa Sang**

| Source of Variance | SS | DF | MS | F |
|---|---|---|---|---|
| Weight ( W ) | .017 | 2 | .009 | .008 |
| Height ( H ) | 33.453 | 2 | 16.726 | 10.045*** |
| Interaction ( W×H ) | 2.735 | 4 | .684 | .696 |
| Error | 149.265 | 152 | .982 | |

***p < .001        ( Post-hoc: h1 > h3,  h2 > h3 )

A further examination of One-way ANOVA was employed aiming at the variable of operational position. The outcome shows that, to each weight level, the effect of manipulating the Tabaa Sang is significantly different in each position. A post-hoc analysis suggests that the best manipulation position is h2 when cooper pipes were located 10 cm under the tailbone of the performer. Figure 5 is the means plot of the effect of manipulating the Tabaa Sang. The figure shows that the mean score of h1 is 3.50, h2 is 3.54, and h3 is 2.86, suggesting that to locate cooper pipes right on the height of the tailbone (h1) is acceptable. The effect of manipulating the Tabaa Sang locating cooper pipes 20 cm under the tailbone (h3) is significantly worse than the other two positions.

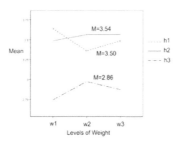

Figure 5  Mean Comparisons of the Effect of Manipulating the Tabaa Sang

## 5    THE TABAA SANG AS AN ILLUSTRATION FOR CULTURAL CREATIVE PRODUCT DESIGN

This study aims at designing and promoting cultural products. With initial literature review and interviews, we conducted a thorough exploration of the Saisait culture and traditions and obtained a profound understanding of the inner meaning of their rituals and festival artifacts. Then, through creative expression in combination with modern elements, transferring the rich, colorful and unique culture into commercially valuable cultural products could be made feasible.

## 5.1    Design Concepts for the Cultural Creative Industry

Taiwan is encountering a new era of aesthetics economy driven by cultural creative. The advancement of technology, automation, and digitalization has made production in small scale but with great variety possible and brought innovative concepts to modern design. Design nowadays takes not only convenience for production and the functionality for usage into consideration, it is more significantly a creation of lifestyles, the experience of various tastes, and the realization of life values. As social science scholars tell us, life itself is culture.

The essence of cultural creative design lies in extracting cultural elements and transforming cultural symbols. As German philosopher Cassirer (1962) indicates, culture is the externalization and materialization of humans, is the realization and concretization of symbolic activities, the core element being symbols. He asserts that symbolic form is the form for all human cultures. Through extracting cultural elements and transferring them into symbols, design endows the products with new aesthetic significance and brings the consumers an empathic sensation.

Langer (1953) suggests that art is the creation of human emotional symbols. A good product is a craft which exercises discourse with people through its sensational image and brings inspiration to them. Empathy, what the aestheticians stress, is the essential strength in bringing a heart-moving experience. The foundation of empathy lies usually on the reflections of individual daily life experiences.

Cultural creative industry should stress on heart-felt affection. As Lin (2009) indicates, culture is a lifestyle, design is a life taste, creativity is recognition of a touching experience while industry is a medium, means or method in realizing cultural creative. When automation and digitalization make possible production in small scale but with great variety, the craftsmanship of quality industrial products cannot be overstressed. As American design expert Norman (2002) states, the emotional side of design may be more critical to a product's success than its practical elements. Norman's concepts, representing a new direction for modern design philosophy, also emphasize the great value of craftsmanship in cultural creative industries.

## 5.2    Design Models for Cultural Creative Products

The changes in the 21$^{st}$-century design philosophy have offered a rich theoretical framework for cultural creative design. Norman (2002) presents in his book "Emotional Design" three levels of emotional design-visceral, behavioral and reflective. The visceral level involves direct feelings when in touch with a product, including shape, style, tactile impression, material and weight. The behavioral level is non-conscious, including the pleasure after exercise, or the delight after a shower. The reflective level presents conscious behaviors such as the pop culture or style and tastes.

Leong and Clark (2003) suggest a brief framework for investigating cultural product design, dividing it into three space structures as the external concrete tactile level, middle behavioral level and the inner invisible spiritual level. Hsu, Lin and Chu (2004) expand the three levels, offer more detailed explication, and provide a cultural creative design model which could further facilitate comparison, application and thinking for design (Figure 7). This theoretical framework could serve as the focal points when investigating individual stages for cultural product design.

For practice in design, Hsu (2004) studied products with aboriginal cultural-elements. He conducted factor analysis on cultural information, product content, design check-up and design evaluation, integrated two additional recognition facets on consumer expectation and design behavior, and then derived a ten-step model for cultural product design concept. This initial model now has been modified by Hsu himself, inducting the ten steps into four stages of describing current status, goal-setting, compiling analysis and product design (Hsu, Lin, & Lin, 2011).

As mentioned above, the essence of cultural creative design is to extract cultural elements and transform cultural symbols. In the ten steps presented in Figure 8, the first two identify the fundamental starting points for product design, steps three to five emphasize setting design principles from the consumer viewpoints, steps six to eight concretely and thoroughly depict the process in capturing cultural elements, and the final two steps stress the transformation of cultural symbols. As a guideline, the first five steps could be applied to general product design, while the latter five outlines the true essence for cultural creative product design.

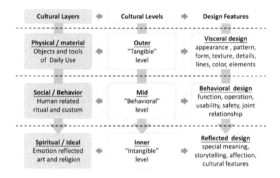

Figure 7  Three layers and levels of cultural objects and design features (Hsu, Lin, & Chiu, 2004)

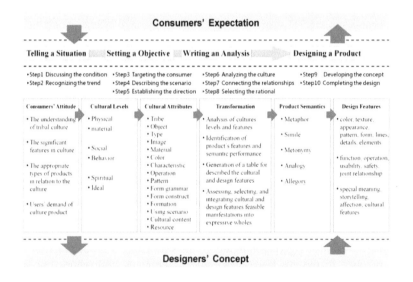

Figure 8  Design framework and process for cultural products (Hsu, Lin, & Lin, 2011)

## 5.3    Examples of Derivative Tabaa Sang products

Following the above-mentioned design model, this study employs Tabaa Sang as our design focus. With concept transformation from (1) the functions, the style and decorative elements of Tabaa Sang, (2) the Saisait legend and its totem elements, and (3)the colors of the Saisait attire, the products enhance the cultural content, bring the users emotional involvement through empathy, and make themselves a closer emotional extension and experience for the users. To clearly illustrate the style, a 3-D simulation of the form was created for each product. The design process for the Tabaa Sang derivative products is presented in Figure 9.

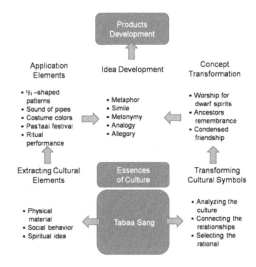

Figure 9  The design process for the Tabaa Sang derivative products

### 5.3.1    Design case A: The Tabaa Sang key pouch

(1) Design concept: The key pouch is made in the shape of the Tabaa Sang, with the sounds produced by bamboo or brass vessels transferred to the striking sounds among keys. A traditional Tabaa Sang is possessed by individual family-name units, non-members of the family-name units are forbidden to touch the sacred instrument. Therefore, the association is transformed to the shape of a key pouch so that personal keys could be well kept there. The Thunder Lady pattern is used for decoration, while the three most dominant colors of red, black, and white are employed on woven cloth for this product.

(2) Materials: woven cloth, mental rings (Figure 10).

### 5.3.2    Design case B: The Tabaa Sang bar stool

(1) Design concept: Representing the look of a person carrying a Tabaa Sang, the stool chair paired with the person sitting on it resembles the Tabaa-Sang-bearing dancer in the festival. Bamboo vessels are added to the back of the stool, forming a resemblance of Tabaa Sang bearing when a person sits on it. In harmony with the person's movement on the stool, the bamboo vessels would strike against each other and produce pleasant sounds.

(2) Material: wood, stainless steel (Figure 11).

### 5.3.3 Design case C: The Tabaa Sang rocking chair

(1) Design concept: Bamboo vessels used for traditional Tabaa Sang could produce crispy pleasant sounds. Thus, a harmonious movement of humans and nature could be composed with the dancer's movement, their melodious songs, and the Tabaa Sang. With the rocking swing of the rocking chair, the wooden vessels create soft sounds, offering natural music together with peaceful relaxation for enjoyment. The rocking chair is made of wood, to be used and imprinted by time gradually. It will tell the story of the Tabaa Sang and denote the memory and inheritance of the Saisait culture.

(2) Materials: wood, strings of mental beads (Figure 12).

Figure 10  Key pouch          Figure 11  Bar stool          Figure 12  Rocking chair

### 5.3.4 Design case D: Tabaa Sang stem glass cups

(1) Design concept: This is a transformation of the sound produced by Tabaa Sang when the bamboo and brass vessels strike against each other. The dangling decoration would swing during usage, while the rustling produced when the glass is lightly tapped is similar to that of a Tabaa Sang when the glass is lightly tapped. In an occasion when the stem glasses are raised and tapped against each other, the crispy striking sound would evidence the joyful moments shared by each other, serving an identical function the Tabaa Sang fulfils in the festival for calling the spirits to the festival. The three most commonly used colors of red, black and white are employed. The reflective mental material represents one of the Tabaa Sang elements, the mini mirrors. The black glass and stainless steel symbolize the mystery of the ta'ay spirits and highlight the three colors of red, black and white.

(2) Materials: black glass, stainless steel (Figure 13).

### 5.3.5 Design case E: The Tabaa Sang matching cups

(1) Design concept: The design is a transformation of a legend about the Thunder Lady. The Thunder-Lady pattern is a much worshipped totem to the tribe, commemorating the goddess who came to earth to teach the tribe farming for bountiful harvest. In remembering the Thunder Lady, we presented on the woven cloth the conception of lightening. The two-hue matching cups is an extension of the love story of the Thunder Lady and the Saisait warrior. The love between a goddess and a human, though without a perfect happy ending, could go down the history endlessly along time.

(2) Materials: glass, stainless steel, mental alloy (Figure 14).

### 5.3.6    Design case F: The aboriginal-dance fruit plate

(1) Design concept: Entertaining the spirits exhibits the most joyful moments in the festival, with the Saisaits, the Tabaa-Sang–bearing dancers, and the participating guests dancing and singing on the prairie. This plate, with forks for fruit-sharing, reproduces the merrymaking atmosphere. The plate represents the field for the festival, with the forks signifying the Saisaits bearing the Tabaa Sang. The use of the forks reminds the users of the prairie where the festival was held. The engraved Thunder Lady pattern expresses the appreciation for the goddess as well as the gratitude towards the culture and legends.

(2) Material: stainless steel (Figure 15).

Figure 13  Stem glass cups     Figure 14  Matching cups        Figure 15  Fruit plate

## 6     CONCLUSION

The rich aboriginal traditional culture is a precious inspiration source for promoting cultural creative industry in Taiwan. To the aboriginal peoples, however, the cultural resources are their inheritance from ancestors and the reality in indigenous daily life. The Tabaa Sang was originally a musical instrument used exclusively in a worship festival and therefore taboos for its usage have been observed in traditional. In recent years, some Saisait artists challenge the taboos about not making publicly accessible the festival song. Bringing the songs and dances public, they remove the mysterious mask of the ritual song and dance, and provide their peer tribal members a stage for self-expression of the tradition.

Two issues that concern us are, for one, the foreign majority culture has deprived the indigenous cultures their living space and, for the other, the overconsumption of the traditional cultural resources by modern business and industries has neither pay back the possessors of the indigenous cultures nor facilitate the conservation or renewal of the existing culture. We look forward to seeing from aboriginal designers their indigenous design with global view, and that concepts and methods of cultural creative design could be handed to their hands and enable them, employing their own cultural resources, to raise the value of their products and set up business. In addition, with the products circulating in the market, the indigenous cultural content would be promoted to the general public. In essence, design is the source for shared prosperity of economy and culture.

This study intends to explore the feasibility of applying Saisait Tribe's Tabaa Sang to cultural creative industry with an emphasis on cross-disciplinary integration. An ergonomic approach of anthropometry and biomechanics was employed to examine the appropriate weight and operational position of the Tabaa Sang. A series of derivative products are created to comply with the performance. We look forward to seeing not only fruitful results of added value from design of cultural creative, but also the facilitation of the development of aboriginal cultural creative industry through the devotion and integration of design education.

## ACKNOWLEDGMENTS

This study was partly sponsored with a grant, NSC99-2410-H-144-010, from the National Science Council, Taiwan.

## REFERENCES

Cassirer, E. 1962. *An essay on man: An introduction to a philosophy of human culture.* London: Yale University Press.

Chu, F. S. 1995. *Saisait people.* Hsinchu: Hsin Chu County Wu Feng Town/ Saisait Festival Management Committee.

Chu, F. S. 2005. *The legend and custom of the Saisait tribe.* Hsin Chu: Hsin Chu County Government.

Hsu, C. H. 2004. *An application and case studies of Taiwanese aboriginal material civilization confer to cultural product design* (Master Thesis). Chang Gung University, Tao Yuan, Taiwan.

Hsu, C. H, Lin, R.T., & Chiu W. K. 2004. Taiwanese aboriginal product design. *International Innovation Design Symposium Thesis,* 157-164.

Hsu, C. H., Lin, R. T., & Lin, C. L. 2011. A study of framework and process development for cultural product design. HCI International 2011, July 9-14, 2011, Orlando, Florida, USA.

Indigenous Peoples Commission of Taipei City Government 2009. *Taiwanese aboriginal tribe's cultural knowledge network.* Retrieved from http://www.sight-native.taipei.gov.tw/public/MMO/TIPC/portal.html

Langer, S. K. 1953. *Feeling and form: A theory of art developed from philosophy in a new key.* NY: Charles Scribner's Sons.

Leong, D., & Clark, H, 2003. Culture-Based Knowledge Towards New Design Thinking and Practice—A Dialogue. *Design Issues, 19,* 48-58.

Li, S. L. 1998. *The culture of costumes and adornments of Taiwanese aborigines: Tradition, meaning, and illustration.* Taipei: Nan Tian Bookstore.

Lin, R. T. 2009. Skill could not be abandon, art could not be discarded—The possibility of aesthetic expression of modern arts. *Taiwan Crafts, 33,* 6-12.

Lin, S. C. 2000. *The history of Taiwanese aborigines: The Saisait Tribe history.* Nan Tou: Taiwan Historica.

Norman, D. A. 2002. *Emotional design.* NY: Basic Books.

Tian, Z. I. 2011. *Taiwanese aborigines: Saisait Tribe.* Taipei: Tai Yuan Publisher.

# CHAPTER 81

# Proposal for a *Kansei* Index Related to the Uniqueness of a Product

*Yusuke Ohta, Keiko Kasamatsu*

Department of Industrial Art, Graduate School of System Design
Tokyo Metropolitan University
6-6 Asahigaoka, Hino, Tokyo, 191-0065 Japan
ota-yusuke@sd.tmu.ac.jp

## ABSTRACT

Consumption behavior is considered to be one of the human forms of self-expression. For example, as seen with the iPhone, it is thought that the consumption behavior of purchasing new products and nigh-novelty products is one of the forms of self-expression that satisfies the desire to express one's self better. Kano (1984) proposed the concept of the "must-be quality" and the "attractive quality." In order to increase the attractive quality of a product, it is necessary to include a Kansei value in a product. One of the key on *Kansei* values is product design—the appearance of the product—as design has a major impact on consumers. Besemer & O'Quin (1986) investigated product creativity and developed the Creative Product Semantic Scale (CPSS), based on the Creative Product Analysis Matrix (CPAM theory) (Besemer & Treffinger, 1981). CPSS can be used to evaluate the creativity of a creative product using three factors: Novelty, Resolution, and Elaboration and Synthesis. While the sense of creativity is one of the points that contributes to the attractiveness of a product, "Uniqueness" can also be considered as one of the attractiveness points of a product. The purpose of this study is to clarify the psychological of what consumers want in order to develop a product mechanism to aid in the design of attractive products. Therefore, this report focused on "Uniqueness," which was considered to be one of the points of the attractiveness on the product. "uniqueness components" were identified to enable the proposal of a Uniqueness Scale for products. The experimental method was as follows: the images of products were presented on a 24-inch display screen. The eight products for evaluation were selected from products nominated by the Japan Good Design

Award over the past 10 years. Forty four subjects evaluated each image for evaluation items on the 7-point scale method. 20 evaluation items were 31, broken down as: 14 items extracted by brainstorming, and 13 items selected according to the CPSS. 5 items ware regarded consumer appetite and 1 item of comprehensive evaluation "UNIQUE". The experimenter interviewed the subjects regarding the taste of the design of the products and the frequency of purchasing and other factors. As the result of this research, the elements contained in the complicated concept of Kansei "Uniqueness" were clarified, and a Uniqueness Scale for products was proposed. This scale is expected to prove to be an effective indicator for product design, that will be applicable to other products.

**Keywords**: *Kansei*, uniqueness, product design,

# 1    BACKGROUND

Consumption behavior is considered to be one of the human forms of self-expression. For example, as seen with the iPhone, it is thought that the consumption behavior of purchasing new products and high-novelty products is one of the forms of self-expression that satisfies the desire to express one's self better. Kano (1984) proposed the concept of the "Must-Be Quality" and the "Attractive Quality." "Must-Be Quality" is a quality element that it will be thought that it is natural if it is satisfied, but it causes dissatisfaction if not sufficient. "Attractive Quality" is a quality element that it will give satisfaction if it is satisfied, and even if not sufficient, it is received that there is no help. Figure 1 shows the relation of both quality elements. In order to increase the attractive quality of a product, it is necessary to include a Kansei value in a product. One of the key on Kansei values is product design—the appearance of the product—as design has a major impact on consumers. Besemer & O'Quin (1986) investigated product creativity and developed the Creative Product Semantic Scale (CPSS), based on the Creative Product Analysis Matrix (CPAM theory) (Besemer & Treffinger, 1981). CPSS can be used to evaluate the creativity of a creative product using three factors: Novelty, Resolution, and Elaboration and Synthesis. While the sense of creativity is one of the points that contribute to the attractiveness of a product, "Uniqueness" can also be considered as one of the attractiveness points of a product. As much as product is Unique, people are easy to stay eyes. However, it may not be connected directly with a consumption action.

Figure 1 Must-Be Quality and Attractive Quality. (From Noriaki Kano, Nobuhiko Seraku, Fumio Takahashi, Shin-ichi Tsuji : Affective Quality and Must-Be Quality, The Japanese Society for Quality Control, 1984, 14(2), pp.39-48. (In Japanese))

## 2    PURPOSE

The purpose of this study is to clarify the psychological of what consumers want in order to develop a product mechanism to aid in the design of attractive products. Therefore, this report focuses on "Uniqueness," which is considered to be one of the points of the attractiveness of a product. "uniqueness components" were identified to enable the proposal of a Uniqueness Scale for products.

## 3    METHOD

The products for evaluation were selected from products nominated by the Japan Good Design Award over the past 30 years. Selected products are 8 home electronics (Figure 2).

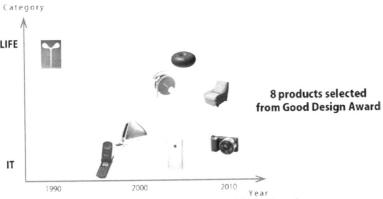

Figure 2 Selected products and that categories.

The experimental method was as follows: Images of products were presented on a 24-inch display screen. Figure 3 shows sample of PC view. In the middle of the screen will present the image of the product and also shown by the image color variations, the insignia, and product name.

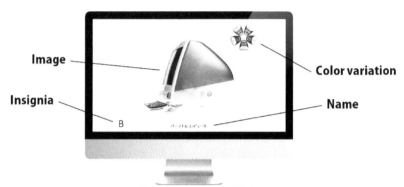

Figure 3  Sample of PC view.

Forty four subjects evaluated each image for evaluation items on the 7-point scale method. There were 31 valuations items, broken down as: 14 items extracted by brain storming, 13 items selected according to the CPSS of "NOVELTY" dimension, 5 items regarding consumer appetite and 1 item of comprehensive evaluation "UNIQUE" (Table 1).

The experimenter interviewed the subjects regarding the taste of the design of the products, lifestyle pattern, important matter and tendency at the shopping, the frequency of purchasing and other factors. The subjects was made to answer about reason for "UNIQUE" evaluation, and whether the product known.

Table 1  31 items of questionnaire.

| | | | | | |
|---|---|---|---|---|---|
| Want to show | — | Want to conceal | Zippy | — | Bland |
| Fine | — | Coarse | Fresh | — | Overused |
| Suggestive | — | Non-Suggestive | Novel | — | Predictable |
| Individual | — | Public | Unusual | — | Usual |
| Good | — | Bad | Original | — | Commonplace |
| Delicate | — | Rugged | Startling | — | Stale |
| Interesting | — | Boring | Astonishing | — | Commonplace |
| Attractive | — | Unattractive | Astounding | — | Common |
| Well-crafted | — | Crude | Unexpected | — | Expected |
| Eccentric | — | Conventional | Trendsetting | — | Warmed over |
| New | — | Old | Revolutionary | — | Average |
| Distinctive | — | Ordinary | Radical | — | Old hat |
| Surprising | — | Customary | Want to buy | — | Not want to buy |
| Shocking | — | Ordinary | Want to use(imagine) | — | Not want to use(imagine) |
| Exciting | — | Dull | Easy to use | — | Hard to use |
| | | | UNIQUE | — | Non-UNIQUE |

# 4 RESULTS

## 4.1 EVALUATION ITEMS

The subjects were forty four, broken down as: 18 men and 25 women (Average: 23.5 age, standard deviation: 4.00 age). A lot of items get evaluation more than 4 point (not which) generally. Particularly, "Interesting", "Well-crafted", "Want to show", "Individual", "Distinctive", "Good", "Attractive", "UNIQUE" item are get high evaluation. On the other hand, 3 items: "Zippy", "Want to buy", "Easy to use(imagine)" get low evaluation.

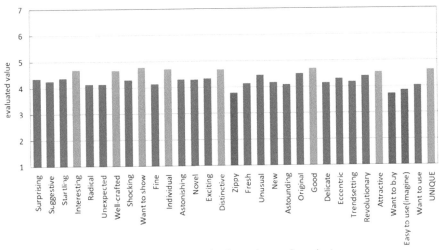

Figure 4  Average evaluation value on 8 products.

## 4.2 FACTOR ANALYSIS

Factor analysis by maximum likelihood method (promax method) was used to identify a sensitivity factor about "UNIQUE". Object variable were selected 28 items that were excepted "Want to buy", "Easy to use (imagine)", "Want to use" from the inside of 31 items. As the result, 5 factors are obtained, cumulative contribution ratio was about 69.9%. Table 2 shows factor analysis including factor loadings.

Factor 1 includes 13 items of "Distinctive", "Novel", "Astonishing", "Original", "Individual", "Eccentric" and so on. Factor 2 includes 6 items of "Attractive", "Good", "Want to show", "Trendsetting" and so on. Factor 3 includes 3 items of "Radical", "New" and "Fresh". Factor 4 includes 3 items of "Fine", "Delicate" and "Zippy". Factor 5 includes 2 items of "Interesting" and "Suggestive". From the above, 5 Factor were given meaning names, Factor 1: *"Feeling of non-daily life"*, Factor 2: *""*, Factor 3: *"Cutting-edge"*, Factor 4: *"Legerity"*, Factor 5: *"Interesting & Suggestive"*.

784

Table 2  Result of factor analysis

|  |  |  | Factor Analysis | | | | |
|---|---|---|---|---|---|---|---|
| Factor 1 | Feeling of non-daily life | Distinctive | .837 | .422 | .500 | -.001 | .372 |
|  |  | Novel | .826 | .526 | .664 | .039 | .484 |
|  |  | Astonishing | .804 | .334 | .481 | .019 | .470 |
|  |  | Original | .797 | .477 | .521 | -.043 | .329 |
|  |  | Individual | .791 | .325 | .405 | -.045 | .397 |
|  |  | Eccentric | .787 | .366 | .467 | -.085 | .391 |
|  |  | Shocking | .769 | .511 | .580 | .067 | .479 |
|  |  | Unusual | .757 | .231 | .379 | -.097 | .248 |
|  |  | Astounding | .710 | .430 | .577 | -.014 | .277 |
|  |  | Unexpected | .706 | .293 | .507 | -.056 | .548 |
|  |  | Revolutionary | .704 | .628 | .557 | .086 | .286 |
|  |  | Startling | .686 | .407 | .582 | -.011 | .624 |
|  |  | Surprising | .650 | .297 | .498 | -.072 | .572 |
| Factor 2 | Captivated | Attractive | .413 | .881 | .589 | .162 | .298 |
|  |  | Good | .366 | .851 | .555 | .264 | .336 |
|  |  | Want to show | .471 | .788 | .555 | .196 | .468 |
|  |  | Trendsetting | .575 | .762 | .616 | .247 | .300 |
|  |  | Well-crafted | .321 | .735 | .511 | .508 | .373 |
|  |  | Exciting | .694 | .696 | .619 | .095 | .441 |
| Factor 3 | Cutting-edge | Radical | .567 | .648 | .880 | .159 | .468 |
|  |  | New | .500 | .619 | .877 | .156 | .280 |
|  |  | Fresh | .609 | .629 | .870 | .130 | .373 |
| Factor 4 | Legerity | Fine | .172 | .534 | .337 | .776 | .261 |
|  |  | Delicate | .147 | .657 | .379 | .658 | .189 |
|  |  | Zippy | .145 | .043 | .080 | .181 | .082 |
| Factor 5 | Interesting & Suggestive | Interesting | .718 | .603 | .613 | .016 | .722 |
|  |  | Suggestive | .119 | .170 | .099 | .125 | .345 |
| Cumulative contribution ratio (%) | | | 45.416 | 57.473 | 62.199 | 66.230 | 69.948 |

## 4.3   MULTIPLE REGRESSION ANALYSIS

The multi collinearity by VIF was not confirmed. Multiple regression analysis using the step wise method was conducted. The dependent variable was "UNIQUE", and independent variable was 5 factor *"Feeling of non-daily life"*, *"Captivated"*, *"Cutting-edge"*, *"Legerity"* and *"Interesting & Suggestive"* from result of factor analysis.

As a result, it was provided below a multiple regression equation (1).

$$\text{"UNIQUE"} = 4.656 + 0.935 \times \textit{"Feeling of non-daily life"}$$
$$+ 0.410 \times \textit{"Captivated"} \dots (1)$$
$$(R^2 = 0.520)$$

## 4.4 CORRELATION COEFFICIENT OF EACH PRODUCT

Then, the correlation coefficient between "UNIQUE" and other items was calculated. Table 3 shows correlation coefficient between "UNIQUE" and other items for each 8 products.

**Table 3  Correlation coefficient between "UNIQUE" and other items for each 8 products**

| Product | Surprising | Suggestive | Startling | Interesting | Radical | Unexpected | Well crafted | Shocking | Want to show | Fine | Individual | Astonishing | Novel | Exciting | Distinctive | Zippy | Fresh | Unusual | New | Astounding | Demand | Good | Delicate | Eccentric | Trendsetting | Revolutionary | Attractive |
|---|---|---|---|---|---|---|---|---|---|---|---|---|---|---|---|---|---|---|---|---|---|---|---|---|---|---|---|
| A | | 129 | | 409 | 417 | 267 | 388 | 662 | 538 | | 606 | 600 | 423 | 507 | 608 | 258 | 419 | 461 | | 409 | 630 | 258 | 078 | 403 | 569 | 475 | 397 |
| B | | | | | 142 | | | | | 161 | 114 | | 464 | | 110 | | | | | 334 | 406 | 611 | 482 | | | | 527 |
| C | 616 | 409 | 624 | | 590 | 473 | | 430 | 633 | | 406 | 600 | 550 | 556 | 590 | | 616 | 638 | 537 | | 587 | | | 416 | | 570 | 538 |
| D | 547 | | 437 | 429 | 450 | 514 | | 139 | | | 516 | 443 | 338 | | 134 | | 506 | | 406 | 498 | | | 654 | | 423 | |
| B | 500 | 130 | 651 | 410 | 475 | | 044 | | 291 | | 438 | 273 | 500 | | 250 | | 479 | 251 | 404 | 453 | 416 | 191 | 013 | 554 | 422 | 423 | |
| b | | | 504 | 517 | 544 | 519 | 528 | | | | | | 542 | | | 621 | 575 | 545 | 650 | 641 | 479 | | 610 | 017 | 340 | 476 | |
| c | 009 | 131 | | 017 | 439 | | 400 | 455 | 530 | | | 480 | 222 | 396 | | 439 | | 436 | 416 | | 062 | 306 | 432 | | 324 | 527 | |
| d | 615 | | | 662 | 376 | 409 | 666 | 511 | | 614 | 603 | 664 | 524 | 646 | | 453 | 644 | | 426 | 497 | 609 | | 592 | 624 | 573 | 429 | |

correlation coefficient : ☐ 0.0-0.2  ☐ 0.2-0.4  ☐ 0.4-0.7  ■ 0.7-1.0  ■ : p < 0.01

Colored points are the item that were obtained significant correlation coefficient. The correlation coefficient on almost items appeared significantly. Particularly, "Interesting", "Individual", "Astonishing" , "Novel", "Distinctive", "Trendsetting" item are get high correlation. However, "Suggestive", "Fine" and "Zippy" items got few products that has significance of correlation coefficient for all products..

## 5    DISCUSSION

From evaluation items, there were many products which chosen was felt relatively "UNIQUE". The product tends to receive a good evaluation generally from evaluation items counting and there is "Attractive", "Interesting" and "Good". However the tendency that was hard to get impression "Want to buy". It was suggested that it might not want to buy it though it was an attractive, good product. It is necessary to investigate the characteristic of a product that for letting consumer to buy.

To clarify relations between 5 factors that were obtained by factor analysis and "UNIQUE", multiple regression analysis was used. From equation (1), "Uniqueness" of home electronics was supposed that it is strongly received an impression by *"Feeling of non-daily life"* that that product had. Then, it was supposed that it was received an impression by *"Captivated"* that this feeling has about half impressions of *"Feeling of non-daily life"*. *"Cutting-edge"* , *"Legerity"* and *"Interesting & Suggestive"* were not adopted in a multiple regression equation, so it was thought that affect for "UNIQUE" of product was extremely small. Because a signification was confirmed, equation (1) presents as Uniqueness Scale of home electronics preliminary version.

Correlation between "UNIQUE" and other items on each 8 products ware checked. Then there was many item of high correlation broken down as:"Individual", "Astonishing", "Novel", "Distinctive" was included in the factor 1," *Feeling of non-daily life"*. And "Distinctive" was included in the factor 1," *Feeling of non-daily life"*. And, a product existed that was indicating high correlation that "Trendsetting" which was the factor 2, an item included in *"Captivated"*. However, a product existed that was indicating high correlation that "Interesting" which was the factor5, an item included in *"Interesting & Suggestive"*.

From these result, the possibility was thought that according to the genre of the product, an impression of evaluation standard for "UNIQUE" was varied. In addition, as the product which release time of the product is new, the high correlation items tend much. From this result, the possibility has was thought that relation with the product influenced of the impression evaluation for the product.

From result of the facesheet, possibility is thought that the recognition of the product affects to impression of the product also. Particularly, clamshell mobile phone, subjects was divided into two ways of tendencies that "I did not feel "UNIQUE" so this is old simply" and "This is very old clearly but watched again this year, it feels fresh. So it's "UNIQUE"". The average age is 23.5 years old this time, and it follows that it relatively targeted the young group. There was one subject that was a little old, but did not receive impression that is different largely. But consideration is necessary for use this scale under the present conditions. Because one possibility could not deny that the age group of experimenter or generation affect about how to feel about recognition of the product and "Uniqueness".

In addition, the tendency was seen that there were few products felt "UNIQUE" outstanding from the answer of reason for "UNIQUE" evaluation in each product in free description. This reason is guessed that it let a product produce constant deflection that all eight products which chosen from winning higher prize of good design award. For constitution of the Uniqueness Scale of product, it is necessary to collect products feeling more "UNIQUE" in a shape and function. In addition, an impression was received from this results that have a difference of reason about "UNIQUE" evaluation between categories in home electronics form the answer of facesheet also. These will be the problems that should be inspected in future.

# 6    FUTURE WORKS

As the result of this research, the elements contained in the complicated concept of Kansei "Uniqueness" were clarified, and a Uniqueness Scale for home electronics was proposed.

In the future, should make a model of home electronics, and do evaluation experiment for inspect this Uniqueness Scale. This time, make a real thing mock-up model and AR model of CG, and discuss for difference of both models in the evaluation experiment. In addition, push forward consideration about relations between this scale and consumer appetite is expected.

This Uniqueness Scale is expected to prove to be an effective indicator for product design, that will be applicable to other products.

## REFERENCES

Besemer, S. P. & O'Quin, K :  Analyzing creative products: Refinement and test of a judging instrument,  Journal of Creative Behavior, 20(2), 1986, pp.115-125.
Besemer, S. P. & Treffinger, D. J. Analysis of creative products: Review and Synthesis. Journal of Creative Behavior, 1981, 15(3), pp.158-178.
Hiroyuki Umehiro (2009) : Affective Quality, Japan Standards Association. (In Japanese)
Noriaki Kano, Nobuhiko Seraku, Fumio Takahashi, Shin-ichi Tsuji : Affective Quality and Must-Be Quality, The Japanese Society for Quality Control, 1984, 14(2), pp.39-48. (In Japanese)

CHAPTER 82

# Assessment of Product Developed with Emphasis on "Emotional Design"

*Rachel de Oliveira Queiroz Silva[1], Caio Márcio Almeida e Silva[2],*
*Wellington Gomes de Medeiros[1], Maria Lúcia Okimoto[3]*

[1]Universidade Federal de Campina Grande
Campina Grande, Brasil
kelqueiroz@gmail.com
wellington@ddi.ufcg.edu.br

[2] Universidade Federal do Paraná | Lactec
Curitiba, Brasil
caiomarcio1001@yahoo.com.br

[3] Universidade Federal do Paraná
Curitiba, Brasil
lucia.demec@ufpr.br

## ABSTRACT

This paper presents a study aiming to establish a comparison between two irons: one that is available in shops, and another one as a concept which design was developed based on the ideas of "emotional design". The second one was presented as a prototype. The experiment was conducted with thirty participants, 18 women and 12 men. The differential semantic technique was used, and the results were tabulated starting from the sum of the punctuations of the positive segment of the scale, considering their weights. The findings show that the prototype has raised more positive associations about its semantic qualities then the existing one.

**Keywords**: emotional design, semantics, product design

# 1    INTRODUCTION

The interaction of users with products involves a series of aspects, such as practical and emotional. Design and emotion has being widely investigated, and tools to uncover the particularities of this aspect of design and interaction are now available (Nagamachi, 1995; Jordan, 2000; Bonapace, 2002; Jordan 2002; Norman, 2004; Desmet, 2004; Hancock et al, 2005; Medeiros, 2007; Hassenzahl, 2007; Damazio, 2008; and Neves, 2011). The existing studies indicate that emotional issues could alter the way we interact with products, reverberating in design processes and assessment of products.

One of the approaches that consider the evaluation of emotions was developed by Desmet (2004). The method provides a tool to assess users' emotional reactions during the interaction with a product. The responses can be classified in fourteen different types: seven are positive, such as desire, pleasant surprise, inspiration, diversion, satisfaction, admiration and fascination; and seven negatives, such as indignation, unpleasant surprise, despise displeasure, dissatisfaction, boredom and frustration.

In opposition, Jordan (2000) investigates pleasure in product interaction which could be clustered as emotional, hedonic or practical. Jordan's approach is primarily theoretical.

Regarding emotion, ergonomic issues are explored from the perspective of Hedonomia.. Hancock et al (2005) consider Hedonomia as an area of knowledge that places pleasure between the individual and the technology. From this point of view, Being like this, what is verified or measured are not emotions, but pleasure triggered during interaction.

Neves (2011) presents a chronology of the experiences starting from different types of interaction with products. The author has investigated four experiences: the mental model of a product, the visualization of an image of that product, the first contact with the physical product, and interaction with product during a period of time. The evaluation uses scales of differential semantic technique to measure. the values of each adjective in each one of the four experiences. Afterwards, a comparison was accomplished among the averages of each one of the adjectives.

In this paper, we use the method developed by Neves (2011), for the comparative evaluation involving an existing product and another one which is in the level of prototype.

# 2    METHODOLOGY

This paper aims to establish a comparison between two irons: one existent in the market and the other one designed by Queiroz and Medeiros (2009). The design process has considered equally elements referring to pragmatic and emotional dimensions. The tests with the two products were performed at the laboratory for design studies Laboratório de Análise e Produção Sensível (LAPS) at the Federal University at Campina Grande (UFCG), state of Paraíba, Brazil. Two questionnaires and the semantic differential scale were applied in a specially prepared room. The tests were conducted with the presence of a mediator.

Thirty people participated, among them, professors, students and employees of UFCG. The experiment followed the ethical principles determined by "Norm ERG-BR 1002 – National Ergonomics Code of Ethics Certificate (ABERGO, 2003). The participants signed a Research Statement of Consent. The experiment began with the signing of the Consent Statement Afterwards, the participants were instructed by the moderator about all the procedures of the experiment.

There were a table and a chair in the test scenario for use of the participants. The two irons were on the table covered with two black cloths. The mediator would reveal the first product (the iron existent in the market). The participant was oriented to observe the product for one minute. Then he should fill out a semantic differential scale proposed by Neves (2011), and composed of thirty pairs of adjectives (ten related to usability, ten related to functionality and ten related to pleasure). After the participant filled out the protocol, it was collected and the product covered. Afterwards, the other product (developed by Queiroz and Medeiros) was uncovered, and the procedures of observation and evaluation were repeated. After that, the test was concluded.

The experiment was accomplished on a white table, on which were the irons hidden by boxes of paper kraft, both in the same size (Fig. 1). The participant sat down in a chair that was positioned among the two boxes, and answered the questionnaire using a ballpoint pen. The experiment was made with 18 women and 12 men.

Figure 1: Place of Experiment.
Source: The authors (2012)

Initially, a pilot test was conducted. It was detected the need of some modifications. The first one was the need to display each iron separately, without the view of the other one. As one iron is regularly available in markets, and the other was a prototype, it was thought that the first could interfere in the assessment of the second one. The other problem detected refers to some bipolar adjectives such as dirty/clean and silent/noisy. It was noted that they would trigger meanings more to do with usability than with the shape of the products, which was the main feature of the products we were interested to explore in the research.

 The experiment began with the participant reading and signing the term of consent. Afterwards, the researcher gave directions about the experiment and the main characteristics of the two irons. The first direction explained about the experiment as a whole:

*Please, look at the form of the two irons and put into words which sensations these products convey. At first, the two products are hidden. After that, the first iron will be uncovered and you will observe it for one minute.*

*Soon afterwards, a semantic differential scale, composed by twenty-eight bipolar adjectives, should be filled out.*

*That completion should happen in the following way: mark a "X" below the the number that corresponds to your choice, that are 0, 1, 2 or 3, regarding neutral, I agree little, I agree, and I agree a lot, respectively, relating the numbering to the adjectives and the observed product. Repeat that procedure in each line of the table. See example below.*

| | I agree a lot | I agree | I agree a little | neutral | I agree a little | I agree | CONCORDO MUITO I agree | |
|---|---|---|---|---|---|---|---|---|
| | 3 | 2 | 1 | 0 | 1 | 2 | 3 | |
| TRADICIONAL | | | | | | X | | INNOVATIVE |
| TECHNOLOGICAL | X | | | | | | | RUDIMENTARY |

*After the completion, the questionnaire will be collected and the product will be covered. Soon afterwards, the other product will be uncovered and the observation procedures and evaluation will be repeated to finish the test.*

*Thank you for participating.*

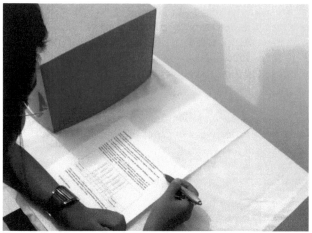

Figure 2: Participant reading the protocols.
Source: The authors (2012)

After that, the participant has received other two pages with the main characteristics of the irons, with illustrative images of each product separately.

**Product A**

*MAIN CHARACTERISTICS:*
*- Wireless iron;*
*- Button selector in the superior part of the product;*
*- Instructions of the fabric type in the superior part of the product, through acronyms of the fabrics, displayed in different colors (in a crescent chromatic scale from a cold color to a hot color). Each color is illuminated when the iron reaches the necessary temperature for the chosen fabric;*
*- Storage of the iron in the vertical, through a base that is fit in in the support of the iron and can be fasten on the wall or on some surface;*
*- Protection cover, that opens when in use and is fixed with a button in the support;*
*- A support that works as an extension to transmit temperature for the iron;*
*- The support of the iron presents a thread that is stored in its interior when the product is not in use.*

**Product B**

*MAIN CHARACTERISTICS:*

*- Iron with wire in the lateral;*
*- Temperature selector on the top of the product;*
*- Instructions of the fabric type on the top through acronyms;*
*- Storage of the iron horizontally or vertically on some surface;*

Figure 3: The two products.
Source: The authors (2012)

The following adjectives were on the semantic differential scale: traditional and innovative; technological and rudimentary; easy to clean and difficult to clean; manual and automatic; capricious and negligent; complicated and simple; fragile and resistant; easy to use and difficult to use; immodest and honest; difficult and practical; modern and antiquated; uncontrollable and controllable; anti-ergonomic and ergonomic; surprising and banal; anti-ecological and ecological; stupid and intelligent; comfortable and uncomfortable; durable and fleeting; wasteful and economical; I hold and insecure; pleasant and unpleasant; boring and stimulant; disorganized and organized; weak and strong; multifunctional and basic; inefficient and efficient; stopped and dynamic; ugly and beautiful.

## 3  RESULTS

The results were measured with emphasis in the positive adjectives. It was made considering the sum of the number of subjects who *agreed* or *agreed a lot*, according to their respective weights (2 and 3). Starting from that tabulation, the following table was built:

Tabela 1: Tabulation of the punctuation of the two adjectives.

| Positive adjective | Product A | Product B |
| --- | --- | --- |
| Innovative | 73 points | 0 points |
| Technological | 64 points | 7 points |
| Easy to clean | 59 points | 20 points |
| Automatic | 40 points | 5 points |
| Capricious | 72 points | 10 points |
| Simple | 64 points | 51 points |
| Resistant | 35 points | 65 points |
| Easy to use | 66 points | 33 points |
| Honest | 57 points | 40 points |
| Practical | 52 points | 21 points |
| Modern | 79 points | 2 points |

| Controllable | 56 points | 31 points |
|---|---|---|
| Ergonomic | 63 points | 15 points |
| Surprising | 55 points | 0 points |
| Ecological | 38 points | 6 points |
| Intelligent | 60 points | 11 points |
| Comfortable | 63 points | 10 points |
| Durable | 42 points | 60 points |
| Economical | 49 points | 4 points |
| Safe | 56 points | 26 points |
| Pleasant | 76 points | 4 points |
| Stimulating | 60 points | 3 points |
| Organized | 71 points | 14 points |
| Strong | 46 points | 59 points |
| Multifuncional | 27 points | 2 points |
| Efficient | 56 points | 49 points |
| Dynamic | 46 points | 2 points |
| Beautiful | 83 points | 2 points |

That table was transformed in comparative graph, presented to proceed:

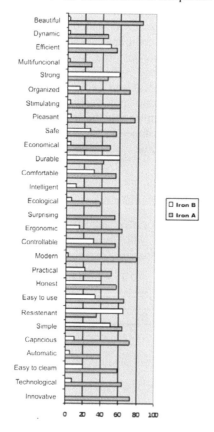

Figure 4: Comparative graph among the products.
Source: The authors (2012)

The table shows that the product A transmits much more beauty, dynamism, multifuntionality, organization, pleasure, safety, economy, comfort, intelligence, ecology, ergonomics, modernity, practicality, easiness in the use, care, automation, easiness of cleaning, technology and innovation that the product B.

Conversely, adjectives such as efficiency, durability, and simplicity were evaluated similarly for products A and B. Concerning resistance, product B was evaluated better than product A.

In that case, as we compared the products without the actual use, the design of the product and their semantic characteristics were shown important for the evaluations. In that sense, the participants of the experiment identified that characteristics of usability were appropriately explored in product A, design with emphasis in the "emotional design."

## 4    CONSIDERATIONS

This paper described an experiment which main objective was to establishing a comparison among two irons: one availed in market, and another designed based on "emotional design" considerations. The experiment was accomplished with thirty participants, 18 women and 12 men. The semantic differential technique was used, and the results were tabulated regarding the sum of the punctuations of the positive segment of the scale, considering their weights.

The results point to the conclusion that the objective of the article was accomplished. The participants of the experiment were asked to choose the adjectives that both products trigged to the participants. With this experiment it can be said that the shape of product A, designed according to the emotional design approach, was evaluated more positively than product B.

Another aspect to be considered was the use of the replication of the method designed by Neves (2011). The author has applied his method to analyze a wash machine regarding diverse types of experiences. In the study presented in this paper, we used the semantic differential to compare two different products. The results indicate the value of the method.

For futures works we recommend that these products should be assessed considering ergonomic aspects, such as usability, and consumption issues.

## REFERENCES

Bonapace, L. Linking product properties to pleasure: the sensorial quality assessment method - Sequam. In: Green, W. S; jordan, p. W. (ed.) Pleasure with Products: Beyond Usability. London: Taylor and Francis, 2002. p.189-217.

Buccini, m.; Padovani, S.. Uma introdução ao design experiencial. Estudos em Design. v. 13, n. 2, p. 9-29, abr. 2006.

Desmet, P. M. A. et al. Product personality in physical interaction. Design Studies, v.29, n.5, p. 458-477, sep. 2008.

Desmet, P.. A multilayered model of product emotions. The Design Journal, p.1-13, 2003.

Desmet, P.; Hekkert, P.. Framework of product experience. International Journal of Design, v.1, n.1, p.57-66, 2007.

Green, W. S.; Jordan, P. W. (ed.), Pleasure with products: beyond usability. London: Taylor and Francis, 2002.

Hancock, Peter A. et al. Hedonomics: the power of positive and pleasurable ergonomics. Ergonomics in Design, Santa Monica, California, v.13, n.1, p.8-14, 2005.

Hassenzahl, M. The hedonic/pragmatic model of user experience. In: Law, E. et al (ed.). Towards a UX manifesto. Lancaster: Mause, 2007. Disponível em: http://141.115.28.2/cost294/upload/506.pdf. Acessado em 16/09/2011.

Iida, I.; Barros, T.; Sarmet, M.. A conexão emocional no design. In: Moraes, D.; Krucken, L. (orgs.) Cadernos de Estudos Avançados em Design. Belo Horizonte: Santa Clara, 2008

Jordan, P.W. Designing pleasure products. London: Taylor and Francis, 2000.

Löbach, B.. Design industrial: bases para a configuração dos produtos industriais. São Paulo: Edgard Blucher, 2000.

Medeiros, Wellington Gomes de. Meaningful Interaction: a Proposition for the Identification of Semantic, Pragmatic and Emotional Dimensions of Interaction with Products. Tese defendida em 09/2007. 325 folhas, volume em capa dura, Universidade de Staffordshire, Inglaterra.

Medeiros, Wellington Gomes de. Interação Significante (IS): Dimensão Semântica da Interação de Usuários com Produtos,URL: <http://www.design.ufpr.br/ped2006/home.htm> .

Meyer, G. C.; Damázio, V.. Elementos para um método de análise da relação emocional entre indivíduos e objetos. In: Congresso Internacional de Pesquisa em Design, 4, 2007, Rio de Janeiro. Rio de Janeiro: ANPEDesign, 2007.

Meyer, G. C.. Diferentes perspectivas sobre os estudos das emoções: comentários da Ergonomia e da Psicologia. In: Congresso Brasileiro de Pesquisa e Desenvolvimento em Design, 5, 2008. São Paulo: AEND/Brasil, 2008. p. 3981-3985.

Mont'alvão, C.; Damazio, V. Design Ergonomia Emoção. Rio de Janeiro:Mauad,2008.

Nagamachi, M.. Kansei Engineering: A new ergonomic consumer-oriented technology for product development. International Journal of Industrial Ergonomics, v.15, n.1, p.3-11, jan. 1995.

Nagamachi, M.. Kansei engineering as a powerful consumer-oriented technology for product development. Applied Ergonomics. v.33, n.3, p.289-294, mai. 2002.

Nielsen, J. Usability engineering. Boston: Academic Press, 1993.

Norman, D. A.. Emotional design. Perchè amiamo (o odiamo) gli oggetti di tutti i giorni. Milano: Apogeo Editore, 2004.

Pereira, C. A. A.. O Diferencial semântico: uma técnica de medida nas ciências humanas e sociais. São Paulo: Ed. Ática, 1986.

Queiroz, S. G.; Cardoso, C. L.; Gontijo, L. A. A linguagem do produto na relação emocional entre usuários e objetos. In: Congresso Brasileiro de Pesquisa e Desenvolvimento em Design, 8, 2008. São Paulo: AEND/Brasil, 2008. p. 3610-3615.

Schlemper, P. F.; Gontijo, L. A. Design experiencial – uma oportunidade para aumentar o valor de marca através do design. In: Congresso Internacional de Pesquisa em Design, 4, 2007, Rio de Janeiro. Rio de Janeiro: ANPEDesign, 2007.

# CHAPTER 83

# Emotions Ergonomics in the Network Society

*Wojciech Bonenberg*

Poznan University of Technology
Poznan, Poland
Wojciech@Bonenberg.pl

## ABSTRACT

The purpose of this article is a scientific reflection on the contemporary transformations of the product creation environment. It is author's interpretation of the requirements which should be met by a product with regard to networking the links between designers, producers and clients.

This article's subject relates to industrial design, design and some aspects of architectural work. The article has a review character - there has been assessed the influence of emotions on product's ergonomics in the network society.

**Keywords**: emotions ergonomics, network society, design

## 1    THE NETWORK SOCIETY

The network society is the term used to describe social relations in the era of global information exchange. The links which are durable features of each society, in our times gained their special features, qualitatively different from the situation in the past (Castells, 1996). The reason for such changes is the growing importance of information in economic, politics, culture, habits and in private life. Information has become a strategic commodity of similar significance as heavy industry products several dozen years ago. World Wide Web is a significant element of the infrastructure and plays a similar role as was played for centuries by road connections between cities. It has similar features as other types of infrastructure, and as such, it does not bring any significant economic benefits, but makes it possible for specialized production branches determining economic growth to function (Buhr, 2003).

The distinctive features of the information and media infrastructure include but are not limited to: flexibility, easy access offering possibility to connect at any time and place, no distance obstacle, large development ability, relatively low construction costs as compared with a traditional infrastructure. Development of WWW affects spatial development. Some of the urban functions become dematerialized. Service functions are replaced with e-services, trade with e-trade, work with e-work, leisure with e-entertainment. These spheres of activity do not longer need material objects such as school buildings, office buildings, concert halls and entertainment facilities and the network of roads and streets connecting them. All these functions are now available from one spot, most often from home. One does not have to waste time to travel in crowded streets as everything is available in the Internet at any time. It is followed by marginalisation of traditional, one-sided mass-media which are gradually replaced with interactive media (virtual fora for opinions exchange and discussion). Earlier standardization of inhabitants' behaviours and attitudes is now replaced with emotional approach related to individualization of attitudes and quest for originality. So, there is the question, how these changes affect clients' preferences as regards products ergonomics, and then, which new challenges designers and producers have to face?

## 2 EUROPEAN TRADITION

In the European tradition the art of designing combined the form of a product and its functionality. This thinking pattern, developed in the methodology of designing, may be found in the papers by Bąbński (1972), Dietrych (1974), Miller (1990). The sources of this approach should be searched for in the Kant's work: *The Critique of Pure Reason* (1998). The traditional position, dated back to the times of Aristotle, assumed the definition of a product as something existing objectively in connection with the theoretical thinking context. In the Greek tradition, extending its influence on the European theory, a product, although technically developed should follow the model of nature. The objects surrounding a man – furniture, equipment, buildings did, to some extent, follow the model of nature and the rights existing in nature. To achieve perfection, one should submit to these rights. An object was deemed functional when it was "made according to the rules". The measure of this value was not originality and beauty but perfection of execution. As soon as it has been achieved, it should be repeated without changes, according to the rules. Creation of a product was a routine activity. It consisted of three key elements: material (provided by nature), work (craftsmanship) and form (established, the only one and timeless). Designing input did not provide for „manufacturing originality", even in such fields as sculpture, architecture or something which is nowadays called functional art. Many of us may be surprised that Greeks were not aware of the original aesthetical values they had created and which we admire today.

In the Christian philosophy, since the times of Augustine, this designing approach has not been questioned, provided, however, that nature, as the prototype, was replaced with the likeness and image of God. Creators did not look for functional and compositional perfection – this perfection was to be found in the God's nature. Their creative approach was related to the craft and technical knowledge required to produce an object. An object did not only imitate nature, but it also created reality consistent with the ideological message.

It is how the repeated, common motif, *topos*, was born, being the material reflection of cultural integrity. It is seen in the Gothic, Renaissance and Baroque architecture and design and some part of this approach has survived till our times. The forms of furniture, clothes, daily necessities, leave no doubts as regards the period in which they were created. This process of shaping the designing canon was noticeable until the end of the Age of Enlightenment.

The notion of a canon is one of these features of a design work which were most vehemently opposed to in the designing philosophy at the end of 18th century - the period of development of new manufacturing and construction methods, in particular the period of industrialization of production. The problem of functional and aesthetical values of the daily necessities produced in large numbers was addressed by the initiators of the movement *Arts and Crafts*: William Morris and Philip Webb, inspired by the views of John Ruskin. It was a kind of a prologue to modernism, where rationally shaped form, deprived of any historical influences, was applied in designing and in architecture. The search for original stylistic image at the end of 19th century and in the beginning of 20th century may be found in the achievements of Werkbund, an organization founded on the initiative of Hermann Muthesius in 1907. While the breakthrough in the views on the modern method of interpretation of a product has taken place thanks to Bauhaus. The school founded in Weimar in 1919 has left its significant mark on architecture, design and interiors design. Bauhaus propagated the idea of design based on the integrity of the fine arts, science and technology. A distinctive mark of the style was integration of architecture, design, furniture and aiming to technical, functional and aesthetical perfection in any field. The reference of this approach in designing was a man, with the wide range of psychophysical, emotional and pragmatic conditions.

Space, arranged by functional divisions, colour, texture, furniture, according to the Bauhaus' approach was to shape the behaviour and the method of people's thinking. Following this definition, a designer was a director of peoples' life space, and not only the creator of objects and structures. In the creative process one had to take into consideration not only the looks of the object, but also its ergonomics, sociology and biology combined with execution technology and production costs. According to the old Vitruvius rule, a product was deemed perfect when it combined technical, functional and aesthetical perfection. Therefore, designing was treated as combination of science and the fine arts. Product designing has become a team work which requires a group of interdisciplinary specialists. Beside technical abilities, a designer had to have the ability to coordinate the team work.

This approach had significant influence on the method of space development and formation of objects completing it for further dozens of years of 20th century. The designing reference was no longer a set of abstractive ideas included in a canon, but a man, with a wide range of pragmatic functional requirements. The canon was replaced with scientific theories, which defined the shape and intended use of a product in the form of a system of hierarchical criteria. These included such criteria, well known to every engineer, as safety, reliability, durability, economics and quality.

World Wide Web defines the new requirements which significantly diverge from the Bauhaus way of product qualification. The motto saying that a product is functional when it meets the criteria defined by engineers and artists - has become outdated.

Cultural, social and economic changes so characteristic for the network society have redefined the expectations related to products' features and requirements imposed on their creators. It is related to three characteristic features of WWW:
- defragmentation of design requirements,
- the influence of virtual reality on product's features,
- transformation of the designing process.

# 3    DEFRAGMENTATION OF DESIGN REQUIREMENTS

For several centuries the form of a product was shaped by a uniform idea combining functional and aesthetical values. Fragmentary criteria of selection of designing solutions were coherent and resulted from this idea. In the network society, this relation has been broken. The idea saying that a product should combine functional, compositional and constructional perfection was deconstructed. It is confirmed by the examples of network designing, showing exceptions to this rule, and still being subject to admiration, interest and common acceptance. The network society is ready to accept many functional mistakes, lack of constructive logic and unnecessary costs, just to make sure that a product triggers off emotions, is noticeable and widely commented in Internet fora.

One may say that the new rule is the lack of rules which could define a coherent idea of the product. Spontaneously created designs are changeable, ambiguous and unspecified, completely individualized. Discussion groups, blogs, social networks' fora determine changing vogue and short-lasting trends.

The standards are no longer defined by scientists, engineers or designers. This is directly made by ordinary network users clustered around WWW topic hubs. There are no limitations in the network, in virtual reality anything is possible. Previous requirements related to reference to coherent rules are onerous and unnecessary. Deeper reflections require effort, time and specialist knowledge - most frequently missing in the collective users of the network society. A noticeable effect of this phenomenon is virtual visualization of the product, defining new designing standards.

Functional designs do not have to be durable and reliable. They should, first of all, trigger off emotions, and durability and reliability are their vices because they do not cause quick consumption or wear and tear being grounds to their replacement with new items. The demand is defined by the network's opinion. Traditional designing strategies are no longer effective.

# 4 THE INFLUENCE OF VIRTUAL REALITY ON PRODUCT'S FEATURES

Alternative reality created in the virtual network has influence on the social attitudes in the real world. Characteristic features include emotions which should interest the participants of the network society. Membership in discussion groups and Internet fora is connected with living specific experiences.

These experiences are then transferred to the real world. Reality shaped under the influence of the network is also a place of experiences (stimulating specific emotions). Among these experiences, the prevailing number constitute the most shallow values, such as the will to satisfy own pleasure, trivial entertainment, fun. It has a specific border of superficial aesthetics which, to constantly growing extent, defines the form of the mass culture - creates *styling* of objects and of the environment. Shops' and restaurants' interiors are designed to induce the feeling of pleasure, elegance and good taste. We are expected to be affected by our presence in a car showroom and during our visit at the hair dresser's (nowadays referred to as hair studio). We succumb to the directed experiences during our stay in a shop, in the office or at home. People are looking for experiences during holidays, visiting popular health resorts, properly styled.

Public spaces in cities are subject to superficial aesthetics in order to trigger off a whole range of different emotions – it is the prerequisite to attract clients, tourists and investors. A good illustration to this is the architecture of Las Vegas Strip with hotels and casinos in the shape of Egyptian pyramids and fairy-tale castles duplicating the forms from Walt Disney animated films.

Interesting looks attracting attention, elegant surroundings, well-dressed, smiling people create visual climate of our environment – a hyper-aesthetical scenery being a kind of the reality "face lifting". In a product there is valued style, class, vogue and the fact whether and how it is commented on Twitter.

In this world full of relativity, the traditional rules of ergonomics are pushed to the sidelines. Liability, safety, knowledge and abilities are valued less, much more valuable is vogue, impression, showing off and appearance. *Styling* supposed to trigger off specific emotions becomes the main criterion of products quality.

# 5 TRANSFORMATIONS OF THE DESIGNING PROCESS

Designing process is also subject to characteristic changes - computer simulation becomes more significant. It is no longer used for optimisation of technical and functional features, but becomes an independent process of product creation, first of all focused on stimulating emotions. Constructional and functional features of a product are only derivatives of the styling. To find the solutions one does not longer need advanced engineer expertise, it is sufficient to be fluent in operating a software and to have the ability to use modules libraries available online. Construction starts to resemble building with Lego blocks. These "Lego-constructions", after superficial aesthetisation, create a design of a new product in a cheap and quick manner. Digital media become indispensable marketing tools.

Certainly, this kind of designing is directly related to economy. A consumer buying a product buys, first of all, the emotional surrounding, and the product itself is a side element. And here we are arriving at an important conclusion: an object with its functional features is less important than the emotional "packaging", appearance is valued more than reality. Practically speaking, a product is an additive, whereas the emotional surrounding is the main object of trade.

It is the virtual reality created by mass-media, what shapes the value of goods to the extent going far beyond the actual material advantages. Even poor quality product may be sold in attractive stylistics after successful promotional and advertisement action.

The thesis on the advantage of emotions triggered off by the product over the product itself is marketing confirmed.

# 6    CONCLUSION

The network society has thoroughly changed the relations between "hard" rules of products construction and their "soft" equivalents in the virtual network. Product's functionality has become less important than the emotions having influence on the recipient.

Aesthetics focused on stimulation of emotions have become elements of virtual environment modelled according to specific rules which, following market demands, are then transferred to the real world. It is no longer the rational combination of functionality, durability and safety with the traditional composition canons what defines designing standards, but a superficial styling aimed at stimulating emotions. The emotions which are supposed to be triggered off by the product concern not only the construction but the entire designing and marketing strategy.

In the network society the looks of a product stimulating emotions determine the main designing strategies and are the stimuli of technological transformations. The network society is craving for experiences and sensations, and this is what it expects from the designers. It redefines the traditional role of ergonomics. We are witnesses to the development of a new ergonomic specialty – ergonomics of emotions – focused on understanding and examination of emotional relations between a man and products, environment and world web sites organisation systems.

## REFERENCES

Bąbński, C. 1972. *Elementy nauki o projektowaniu.* Warszawa: Wydawnictwo Naukowo-Techniczne.

Buhr, W. 2003. *What is Infrastructure?* Discussion Paper No. 107-03. Department of Economics, School of Economic Disciplines. University of Siegen.

Castells, M. 1996. *Rise of The Network Society (The Information Age: Economy, Society and Culture, Volume 1)*. Oxford: Blackwell Publisher, Ltd.

Dietrych, J. 1974. *Projektowanie i konstruowanie*. Warszawa: Wydawnictwo Naukowo-Techniczne.

Kant, I. 1998. *The Critique of Pure Reason*. Cambridge: Cambridge University Press.

Miller, D. 1990. *Wpływ wiedzy projektanta na formułowanie problemu projektowego*. Polska Akademia Nauk, Instytut Filozofii i Socjologii. Wrocław: Ossolineum.

# CHAPTER 84

# Emotional Video Scene Detection from Lifelog Videos Using Facial Feature Selection

*Hiroki Nomiya, Atsushi Morikuni and Teruhisa Hochin*

Kyoto Institute of Technology
Goshokaido-cho, Matsugasaki, Sakyo-ku, Kyoto, Japan
{nomiya, hochin}@kit.ac.jp

**ABSTRACT**

For the purpose of the retrieval of impressive video scenes from lifelog videos, an efficient scene detection method is proposed in this paper. A video scene could be assumed to be impressive when it includes a person with some kinds of emotions. Therefore, the impressive scenes are detected based on facial expression recognition. The proposed recognition method is considerably efficient by introducing several types of facial features easily obtained from the positional relationships of facial feature points such as the end points of eyebrows, eyes and a mouth. By using them, multiple facial expression recognition models are constructed and integrated based on an ensemble learning approach. Additionally, an efficient emotional video scene detection method is introduced. It detects emotional scenes by finding a beginning and an ending frames of emotion expression from the recognition results of all frame images. Several emotional scenes are integrated by hierarchical clustering so that the video scenes retrieved have appropriate lengths. The detection performance of the proposed method is evaluated through an experiment using lifelog videos.

**Keywords**: lifelog, facial expression recognition, machine learning

# 1   INTRODUCTION

As a means of recording the activity of a person's life, lifelog has been gaining attention (Gemmell et al., 2002)(Kono and Misaki, 2004). Lifelog could be recorded as various types of data such as video, audio and physiological data. Specifically, it becomes very easy in recent years to record the activity of one's life in the form of video data because of rapid improvement of multimedia recording devices. As lifelog video data increase, however, a critical problem that they are rarely utilized arises from the fact that retrieving useful data from large-scale databases is quite cumbersome. We therefore intend to retrieve impressive lifelog video scenes accurately and efficiently. Since most of lifelog videos containing persons and their emotions will change in impressive scenes, we propose an emotional scene detection method for lifelog videos based on facial expression recognition.

Facial expression recognition has been widely studied because it is useful but difficult. For instance, a recognition method using 3D models of a face has been proposed (Dornaika and Davoine, 2005). The models could capture slight changes on a face and this leads to accurate recognition, but relatively high computational cost is required. An efficient facial expression recognition method by introducing Gabor features has been proposed (Littlewort et al., 2006). The computational cost of obtaining the features is relatively low and this will contribute the recognition efficiency. The learning method using AdaBoost is, however, rather complex and the recognition efficiency could be improved by simplifying the learning process.

We have proposed an emotional video scene detection method based on facial expression recognition using facial feature points (Nomiya and Hochin, 2012). Several types of facial features easily obtained from the points are introduced for efficiency, but some of the facial features are still time-consuming. Our previous method has another problem that the lengths of emotional scenes are too short to display as a scene retrieval result. Furthermore, it was evaluated only on artificial video data consisting of deliberate facial expressions while lifelog videos are mainly composed of emotional scenes with spontaneous facial expressions.

We propose in this paper a more efficient emotional scene detection method by introducing facial features based on only positional relationships of facial feature points. The scene detection method is also improved to produce appropriately long scenes by introducing a clustering algorithm to integrate several emotional scenes into a single scene if they are close to each other. The proposed method is evaluated on lifelog videos including spontaneous facial expressions.

This paper is organized as follows: Section 2 describes the facial features. Section 3 presents the facial expression recognition method. Section 4 refers to the emotional scene detection method of our previous work. Section 5 elaborates the clustering method to produce appropriate emotional scenes. Section 6 shows an emotional scene detection experiment. Finally, Section 7 concludes this paper.

## 2 FACIAL FEATURES

### 2.1 Facial Feature Points

Facial features used to recognize facial expressions are defined based on the positional relationships of facial feature points. The facial feature points are located on a face and associated with expressing various types of facial expressions. A total

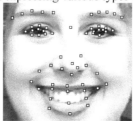

Figure 1 Facial feature points indicated by squares (This facial image is from Cohn-Kanade Facial Expression Database (Kanade, 2000)).

of 59 facial feature points are used as shown in Figure 1. The facial features are located around eyebrows (10 points), eyes (22 points), a nose (9 points), nasolabial folds (4 points), and a mouth (14 points). They are extracted from a facial image by using the face recognition software Luxand FaceSDK 4.0 (Luxand Inc., 2012).

### 2.2 Facial Features

By combining two or three facial feature points, we define four types of facial features. In our previous work (Nomiya and Hochin, 2012), facial features are similarly defined based on the facial feature points. Some of them, however, have a crucial problem of high computational cost caused by the calculation of intensity values of a large number of pixels. To solve this problem, the facial features in the proposed method are calculated using positional relationships of facial feature points. This could drastically reduce the computational cost because the facial features are obtained only by geometric calculations. The facial features used in the proposed method are as follows:

- Distance based feature ($F_1$)
  Assuming that $p_i$ is a facial feature point extracted from a neutral face and $p_i'$ is the corresponding one extracted from an emotional face, $F_1 = l_1'/l_1$ where $l_1$ ($l_1'$, respectively) is the Euclidean distance between $p_1$ and $p_2$ ($p_1'$ and $p_2'$, respectively). An example is shown in Figure 2 (a).
- Angle based feature ($F_2$)
  $F_2 = |\theta_2' - \theta_2|$ where $\theta_2$ ($\theta_2'$, respectively) is the angle between two vectors $\overrightarrow{p_2 - p_1}$ ($\overrightarrow{p_2' - p_1'}$, respectively) and $\vec{v}$. The vector $\vec{v}$ is represented as $\overrightarrow{p_L - p_R}$ generated from the center points of left and right eyes, denoted by $p_L$ and $p_R$ respectively. An example is shown in Figure 2 (b).

- Area based feature ($F_3$)

  $F_3 = S_3'/S_3$ where $S_3$ ($S_3'$, respectively) is the area of a triangle generated by connecting three facial feature points $p_1, p_2$ and $p_3$ ( $p_1', p_2'$ and $p_3'$, respectively). An example is shown in Figure 2 (c).

- Interior angle based feature ($F_4$)

  $F_4 = |\theta_{41}' - \theta_{41}| + |\theta_{42}' - \theta_{42}| + |\theta_{43}' - \theta_{43}|$ where $\theta_{41}, \theta_{42}$ and $\theta_{43}$ ($\theta_{41}', \theta_{42}'$ and $\theta_{43}'$, respectively) are the interior angles of a triangle generated by $p_1, p_2$ and $p_3$ ($p_1', p_2'$ and $p_3'$, respectively). An example is shown in Figure 2 (d).

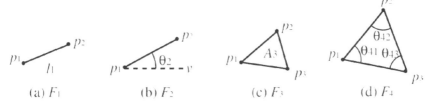

(a) $F_1$     (b) $F_2$     (c) $F_3$     (d) $F_4$

Figure 2 Four types of facial features.

The above facial features are defined for all possible combinations of facial feature points. Thus, the total number of facial features is 68440 since the number of $F_1$ and $F_2$ is $_{59}C_2 = 1711$ and that of $F_3$ and $F_4$ is $_{59}C_3 = 32509$.

## 3     FACIAL EXPRESSION RECOGNITION METHOD

### 3.1     Learning Dataset

A facial expression recognition model is constructed via machine learning in accordance with our previous work (Nomiya and Hochin, 2012). Assuming that the number of training examples in a training set is $n$, the training set is expressed as $\{(x_1, x_1', y_1), \dots, (x_n, x_n', y_n)\}$, where $x_i$ is a neutral face image and $x_i'$ is an emotional one. Here, $y_i$ is a class label, that is, the type of facial expression expressed in $x_i'$. When the number of types of facial expressions is $C$, $y_i \in \{1, \dots, C\}$.

### 3.2     Low-level Recognition Model

Since the distributions of facial feature values could be approximated by normal distributions, we define the facial expression recognition models $L_{ci}$ as equation (1) for each facial expression $c$ ($c = 1, \dots, C$) and facial feature $f_i (i = 1, \cdots, 68440)$.

$$L_{ci}(x, x') = \frac{1}{\sqrt{2\pi}\sigma_{ci}} \exp\left\{-\frac{(f_i(x, x') - \mu_{ci})^2}{2\sigma_{ci}^2}\right\} \tag{1}$$

where $f_i(x, x')$ is the feature value of $f_i$ obtained from the images $x$ and $x'$. The values $\mu_{ci}$ and $\sigma_{ci}^2$ are the mean value and unbiased variance of $f_i$ of the training examples whose class labels are $c$, respectively. We call these models *low-level recognition models* because of their simplicity.

## 3.3 Feature Selection

Many of facial features will not contribute the recognition performance and will increase computational complexity. We thus introduce a feature selection method based on the usefulness of facial features to select a small number of useful ones.

The usefulness is defined as the ratio of between-class variance to within-class variance (Yildiz and Alpaydin, 2000). It is calculated from the facial feature values of training examples. Since higher between-class variance and lower within-class variance lead to more accurate recognition, the usefulness $U_i$ of a facial feature $f_i$ is defined as equation (2).

$$U_i = \frac{B_i}{W_i} \tag{2}$$

where $B_i$ and $W_i$ are the between-class and within-class variances of $f_i$ respectively. The between-class variance $B_i$ is given by equation (3).

$$B_i = \frac{\sum_{c=1}^{C} n_c (v_i - \mu_{ci})^2}{C - 1} \tag{3}$$

where $n_c$ is the number of training examples whose class labels are $c$, and $v_i$ is the mean value of mean values of $f_i$ of training examples whose class labels are $1, \ldots, C$. That is, $v_i = \frac{1}{C} \sum_{c=1}^{C} \mu_{ci}$. The within-class variance $W_i$ is calculated by equation (4).

$$W_i = \frac{\sum_{c=1}^{C} \sum_{k=1}^{n} \{I(y_k = c) \cdot (f_i(x_k, x_k') - \mu_{ci})^2\}}{n - C} \tag{4}$$

In the above equation, function $I(\alpha)$ returns 1 if $\alpha$ is true, 0 otherwise. For all facial features, the usefulness values are calculated and then $m$ most useful features are selected. In this paper, we experimentally set the value of $m$ to 1500.

## 3.4 High-level Recognition Model

By integrating the low-level recognition models of selected features, a few *high-level recognition models* are constructed. The number of high-level recognition models is equal to the number of types of facial expressions (i.e., $C$) since the difficulty of recognition problem will be proportional to $C$. Finally, the recognition result is determined by combining the outputs of all high-level recognition models.

The feature vectors $v_c$ ($c = 1, \ldots, C$) used to train a high-level recognition model are composed of the output values of selected low-level recognition models. This feature vectors are generated for each training example. The feature vector $v_c(x_k, x_k')$ of the $k$-th training example is given by equation (5).

$$v_c(x_k, x_k') = [L_{cu_1}(x_k, x_k'), \ldots, L_{cu_m}(x_k, x_k')] \tag{5}$$

where $u_i$ ($i = 1, \ldots, m$) represents the $i$-th most useful facial feature.

By using $C$ feature vector sets $\{v_1(x_1, x_1'), \ldots, v_1(x_n, x_n')\}, \ldots, \{v_C(x_1, x_1'), \ldots, v_C(x_n, x_n')\}$ as training examples, a total of $C$ high-level recognition models $\{H_1(x, x'), \ldots, H_C(x, x')\}$ are trained by Support Vector Machine (SVM) (Cortes and Vapnik, 1995). The SVM-based learning system is implemented by LIBSVM

(Chang and Lin, 2011) using quadratic polynomial kernel.

The facial expression $y$ of an example $(x, x')$ is estimated based on voting algorithm by all high-level recognition models according to equation (6).

$$y = \underset{c}{\mathrm{argmax}} \sum_{i=1}^{C} I(H_i(x, x') = c) \tag{6}$$

That is, $y$ is the facial expression that has the largest number of votes.

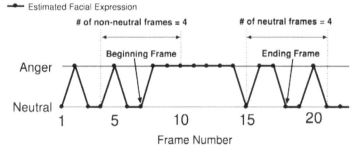

Figure 3 An example of a beginning and an ending frames.

## 4 FINDING EMOTIONAL FRAGMENTS

To find emotional scenes, facial expression recognition is first performed for each image obtained from each frame. Then, a pair of a beginning frame of emotion expression and an ending one is detected (for simplicity, we refer to them as *a beginning frame* and *an ending frame* respectively) according to our previous work (Nomiya and Hochin, 2012). We call the scene including the frames between beginning and ending frames *an emotional fragment*. An emotional fragment is a very short scene (generally several seconds) of emotion expression. Since emotion expression will frequently be observed in a video sequence, emotional fragments are exhaustively searched by finding all the pairs of beginning and ending frames.

A beginning frame is defined as a frame that meets the following conditions. Note that a beginning frame is referred to as the $P$-th frame.

- For a certain natural number $N$, there are at least $(N + 1)$ emotional (i.e., non-neutral) frames between the $(P - N)$-th frame and the $(P + N)$-th one.
- There are no emotional frames between the $P'$-th frame and the $(P - 1)$-th one where $P'$ is the first frame or the ending frame which is nearest to the $P$-th frame and $P' < P$.

Figure 3 shows an example of a beginning frame when $N = 3$. The horizontal axis represents the frame number and the vertical axis means the estimated facial expression of each frame. In this example, the above conditions are met when $P = 7$. The beginning frame is thus the 7th frame.

An ending frame is similarly defined as a frame that meets the following conditions. Note that an ending frame is referred to as the $Q$-th frame.

- There are at least $(N + 1)$ neutral frames between the $(Q - N)$-th frame and the $(Q + N)$-th one.
- There are no neutral frames between the $Q'$-th frame and the $(Q - 1)$-th one where $Q'$ is the beginning frame nearest to the $Q$-th frame and $Q' < Q$.

Figure 3 represents an example of an ending frame when $N = 3$. In this case, the ending frame is the 18th frame because the above conditions are met when $Q = 18$.

The facial expression $E$ in an emotional fragment is determined via a voting by all frames in the emotional fragment according to equation (7).

$$E = \operatorname*{argmax}_{c} \sum_{i=P}^{Q} I(E_i = c) \tag{7}$$

where $E_i$ is the facial expression in the $i$-th frame estimated by equation (6).


## 5    EMOTIONAL SCENE DETECTION

In our previous work (Nomiya and Hochin, 2012), each emotional fragment is output as a result of emotional scene retrieval. The length of an emotional fragment is, however, generally so short that it is inappropriate to display a single emotional fragment as a retrieval result. To solve this issue, we introduce a clustering method to integrate several emotional fragments into an emotional scene. A set of emotional fragments is regarded as a cluster and adjacent clusters could be integrated if they are close to each other. Since it is difficult to determine the appropriate number of clusters because of its dependency on a given video, we adopt a hierarchical clustering method to adaptively determine the number of clusters.

In the clustering process, a cluster set $S = \{s_1, \dots, s_M\}$ is first generated. Here, $s_i$ corresponds to a single emotional fragment and $M$ is the number of emotional fragments detected. Then, the distances of all possible pairs of clusters are calculated and the pair of nearest clusters is integrated into a single cluster. The distance $d(s_i, s_j)$ between two clusters $s_i$ and $s_j$ is defined as equation (8).

$$d(s_i, s_j) = B(s_i) - E(s_i) - 1 \tag{8}$$

where $B(s)$ and $E(s)$ are frame numbers of beginning and ending frames respectively, and $B(s_i) < E(s_i) < B(s_j) < E(s_j)$. That is, the distance is identical to the number of neutral frames between $s_i$ and $s_j$.

When integrating two clusters $s_i$ and $s_j$ into a cluster $s_k$, $B(s_k)$ and $E(s_k)$ are represented by equations (9) and (10).

$$B(s_k) = B(s_i) \tag{9}$$
$$E(s_k) = E(s_j) \tag{10}$$

A pair of nearest clusters is repeatedly integrated while a cluster set contains more than one cluster. From the cluster sets derived during the clustering process, the cluster set $S^*$ given by equation (11) is selected as the best cluster set.

$$S^* = \underset{S}{\mathrm{argmax}} \left( \frac{w(S) - w(S_{min})}{w(S_{max}) - w(S_{min})} \cdot \frac{u(S_{max}) - u(S)}{u(S_{max})} \right) \qquad (11)$$

where $S_{max}$ is the set of clusters including the largest number of elements (i.e., the initial cluster set) and $S_{min}$ is the one including only one element (i.e., the final cluster set). The functions $w$ and $u$ are given by equations (12) and (13) respectively.

$$w(S) = \sum_{i=1}^{|S|} (E(s_i) - B(s_i) + 1) \qquad (12)$$

Table 1 Result of emotional scene detection.

| Scene no. | Beginning – ending frame numbers | Length (min:sec) | #Detected | #Correct |
|-----------|----------------------------------|------------------|-----------|----------|
| 1 | 8830 – 8910 | 0:03 | 1 | 1 |
| 2 | 10820 – 10870 | 0:02 | 1 | 1 |
| 3 | 12200 – 14560 | 1:34 | 4 | 4 |
| 4 | 20520 – 27220 | 4:28 | 8 | 8 |
| 5 | 30100 – 31260 | 0:46 | 2 | 2 |
| 6 | 34910 – 38130 | 2:09 | 4 | 4 |
| 7 | 40670 – 52680 | 8:04 | 13 | 12 |
| Total | 25580 frames | 17:03 | 33 | 32 |

$$u(S) = \begin{cases} \sum_{i=1}^{|S|-1} d(s_i, s_{i+1}) & \text{if } |S| > 1 \\ 0 & \text{otherwise} \end{cases} \qquad (13)$$

Finally, the emotional scenes are determined as the clusters included in the best cluster set $S^*$.

# 6    EXPERIMENTS

## 6.1    Dataset

We prepared some lifelog video clips including the scenes that a subject (a male university student) was playing cards with his friend. They consist of the three types of video clips recorded by a fixed camera to record playing scene and two wearable cameras to record the subject's face and field of view. We used two video clips of the subject's face (we call them "video A" and "video B") in this experiment. The numbers of frames in videos A and B are 57670 and 53600 respectively. The frame rate is 25 frames/sec and the resolution is 480×640 pixels for both video clips. We picked out frame images from videos A and B every 10 frames, and used first 500 frame images in video A as training examples and 5360 frame images in video B as test examples. The facial expression to be detected is only "smile" because the video clips do not contain emotional scenes other than scenes of smile.

## 6.2 Accuracy of Emotional Scene Detection

Table 1 shows the result of emotional (smile) scene detection. In this table, the column "Scene no." represents the scene number. The column "Beginning – ending frame numbers" means the detected frame numbers of beginning and ending frames. The column "Length" is the length of the emotional scene described by the format "min:sec." The column "#Detected" represents the number of emotional fragments detected. The column "#Correct" is the number of detected emotional fragments which actually contain smiling faces. One of the authors determined if each emotional fragment was correctly detected or not.

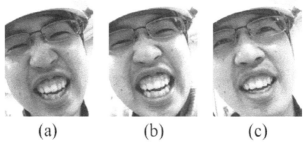

(a) (b) (c)

Figure 4 Examples of frame images in the scenes 3 (a), 4 (b) and 7 (c).

Table 2 Processing time for emotional scene detection.

| Processing | Processing time (in sec.) |
| --- | --- |
| Facial expression recognition | 6.6 |
| Scene detection | 1.2 |
| Total | 7.8 |

From the result that most of detected emotional fragments are correct, the precision of emotional scene detection seems to be adequate. In addition, the lengths of most of emotional scenes are basically appropriate although the scenes 1 and 2 are too short. Here, we show three examples of the frame images in the scenes 3, 4, and 7 in Figure 4 (a), (b) and (c) respectively. Note that (a) and (b) are included in correctly detected emotional fragments and (c) is included in an incorrectly detected one.

While the precision of emotional scene detection was fairly high, the recall of that was insufficient since there were 20 emotional fragments that were not included in emotional scenes because of detection failure. Thus the detection performance of the proposed method should be improved so that most of emotional fragments could be found.

## 6.3 Efficiency of Emotional Scene Detection

In order to evaluate the efficiency of the proposed method, we show the detection time in Table 2. The row "Facial expression recognition" corresponds to the processing time for the facial expression recognition for all frames. The row "Scene detection" represents the processing time for the detection of emotional scenes. The computer used in this experiment has Xeon W3580 CPU (3.33GHz) and 8GB memory. Note that multi-thread processing was not used and the processing time to extract facial feature points is not included.

Considering that the length of the video clip is over 30 minutes, the proposed method achieves fully high efficiency. Additionally, its efficiency is considerably better than that of our previous method (Nomiya and Hochin, 2012) because its processing time for facial expression recognition is about 50 seconds while its detection accuracy is almost equal to that of the proposed method.

# 7 CONCLUSIONS AND FUTURE WORKS

We proposed an emotional video scene detection method for lifelog videos. By introducing concise facial features, the proposed method could perform emotional scene detection taking the tradeoff between accuracy and efficiency into consideration. From the result of the emotional scene detection experiment, it is confirmed that the proposed method is considerably efficient compared with our previous method.

Since the recall of the proposed method is still insufficient, we intend to improve the detection performance by introducing more effective facial features and/or improving the learning algorithm to build the facial expression recognition models. In addition, performing further experiments using various lifelog video data is also included in the feature work.

## ACKNOWLEDGMENTS

This research is supported by a *kakenhi* Grant-in-Aid for Young Scientists (B) (22700098) from Japan Society for the Promotion of Science.

## REFERENCES

Chang, C. and Lin, C. 2011. LIBSVM: A Library for Support Vector Machines, *ACM Trans. Intelligent Systems and Technology* 2(3): 1–27.

Cortes, C. and Vapnik, V. 1995. Support-vector Network, *Machine Learning* 20(3): 273–297.

Dornaika, F. and Davoine, F. 2005. Simultaneous Facial Action Tracking and Expression Recognition using a Particle Filter, *Proceedings of IEEE International Conference on Computer Vision*, 1733–1738.

Gemmell, J., Bell, G., Luederand, S., Drucker, S. and Wong, C. 2002. MyLifeBits: Fullfilling the Memex Vision, *Proceedings of the 10th ACM International Conference on Multimedia*, 235–238.

Kanade, T., Cohn, J. F. and Tian, Y. 2000. Comprehensive Database for Facial Expression Analysis, *Proceedings of the 4th IEEE International Conference on Automatic Face and Gesture Recognition*, 46–53.

Kono, Y. and Misaki, K. 2004. Remembrance Home: Storage for Re-discovering One's Life, *Proceedings of Pervasive 2004 Workshop on Memory and Sharing of Experiences*, 25–30.

Littlewort, G., Bartlett, M. S., Fasel, I., Susskind, J. and Movellan, J. 2006. Dynamics of Facial Expression Extracted Automatically from Video. *Journal of Image and Vision Computing* 24(6): 615–625.

Luxand Inc. 2012. "Luxand FaceSDK 4.0," Accessed February 21, 2012, http://www.luxand.com/facesdk/

Nomiya, H. and Hochin, T. 2012. Efficient Emotional Video Scene Detection Based on Ensemble Learning, *IEICE Transactions on Information and Systems* J95-D(2): 193–205 (in Japanese).

Yildiz, O. T. and Alpaydin, E. 2000. Linear Discriminant Trees, *Proceedings of the 17th International Conference on Machine Learning*: 1175–1182.

# CHAPTER 85

# Research of the Application According to User's Experience in 2011IDA CONGRESS Taipei - Light of Fashion Design

*Mei-Ting Lin1, Rungtai Lin2*

1 Graduate School of Creative Industry Design, College of Design, National Taiwan University of Arts, Taiwan
gua_gua@mail2000.com.tw
2 Dean/College of Design, National Taiwan University of Arts, Taiwan
rtlin@ntua.edu.tw

## ABSTRACT

"Made in Taiwan" is the proud code for many Taiwanese. It means the guarantee of the products quality. Now Taiwan is entering the period of the esthetic economy obeying the fashion design after the promotion of government and enterprises. Many products are made from OEM to self-creative brand. While in illuminate industry, the expression of the creative illuminators is weaker than other fields. In the end of 2011, "2011 IDA congress Taipei" was held. In this design show, it brought us the world wild consciousness of the design industry. Make us understand that design needs both to be progressed by challenging ourselves and the interaction of cross-fields.

In this research, we searched as many as the creative lighting products at 2011 IDA Congress Taipei and tested the degree of users' satisfaction for these products. By questionnaire, interview and analysis, we had some conclusions.
1.  Multi-functional products are not the prior expect point.

2.    Special materials make the extremely feelings, much favorite and hate.

3.    Large volume products don't fit most consumers.

4.    The products with the lighting source which could not be switched might be the skippable producs.

5. Interactive function products would not be the ones for home setting.

Interactive function products would not be the ones for home setting. After analyzing the data, we understood how the users accept the requirement and cultural value from them. At last, we transfer the user experience to design concepts for illuminative designers. This could combine both the user's requests and the cultural esthetic to make progress of Taiwan illuminate industry.

**Keywords:** User Experience, Light of fashion design, fashion design.

# 1    PREFACE

Technical products made by Taiwan amazed the whole world so it is also named high-tech island. According to the data analyzed by ITRI, so far the market share of the LED products made by Taiwan takes the second place all over the world.

The global main lighting market still falls on traditional fluorescent light. LED is the beginning industry on lighting market because of the unfriendly price. On the lighting market in Taiwan, Philips is the most famous brand. Compared with it, other manufacturers are unknown with their brand name. It is a pity that we have developed fine lighting source but not the lighting equipments.

Design idea is from living culture. The most purpose is to increase life quality and cultural level for human being.Thus, this research would collect the relative data of these lighting products to provide the lighting industry the reference of design behavior.

# 2    CULTURAL CREATIVE PRODUCT FEATURES

In Taiwan, Government leads the cultural creativity industry with three steps politics: "local cultural industry" which was promoted during 1990-2000, "cultural industry"which was promoted during 1995-2003 and "cultural creativity industry"which is promoted until now. (Lee-Chun Lan,2009) The features of the products are from local featured to national featured and the last the personal creativity featured. They expressed the highly merged combination with life and art. Cultural creativity industry must include four features: creative idea, intellectual property rights, increasing employment and rate and quality, creating new life styles. (Wu,Yu-Hsing,2011) There are three factors to reach concrete expression of products which influence the creative and artistic value: originality, esthetic value and completeness. (Sobel & Rothenberg,1980)

The productions of cultural creativity industry are neither the usual national crafts nor the simplified arts. They are between these two objects. This industry emphasizes the esthetic creation and technology emphasizes the real science development. Consumers purchase the products indeed according four features and three factors. They are affected by the emotional identification not the rational one. Many factors do affect the purchasing willing, like brands, popularity and cultural identification. That is the reason now many corporations and academy use the international award as the marketing promotional tool to expand the brands visibility by mass media..

## 3 USER EXPERIENCE

UX or UE(user experience) is applied within the interaction with the objects and humans. By the model set according to UX, we got response of the difficulty issue for using the products from most people who had used them. Through this model we could understand the practicability and the service of a product. For the project developer, this model undoubtedly could help the designers to consider more aspects. "project definition" →" concept design" →" concrete design" →" detail design" is the process for the projects(Pahl and Beitz,1993). designers could understand the difficulty issue for the users in developing products.

Designers and engineers take the biggest value from consumers' feedbacks as the most purpose. The created products might be accorded with the formation, function or emotional environment. All the designers could have the personal demand, bias or prejudice. Using the UI research data,it could provides the designers detached point of view. It could help young designers to know the inner problems as the abundant experienced ones. They could even create relative and continued products that are created by UX analysis(Ding-Bang Luh,2007). In future, the designers must take it as a trend to create the idea to improve people' s life.

By three factors of value and four features of cultural creativity industry, we could use the figure of user experience relations to understand their relations clearly. Artistic production is the creation of artistic emotional expression. However, cultural creativity product is the creation for end user. We could use this relationship to create the products close to consumer's need.( Figure 1)

818

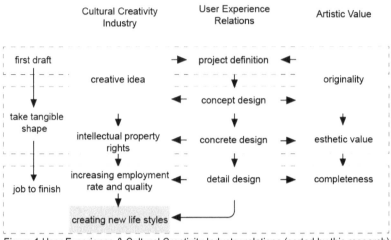

Figure 1 User Experience & Cultural Creativity Industry relations (sorted by this research)

## 4    2011IDA CONGRESS TAIPEI - LIGHT OF FASHION DESIGN

Cultural creativity industry in Taiwan is lifted to international style after the 2011 IDA exhibition. Our commercial industry, creative thought, design ability and professional education are brought to a new position.

In this research we reviewed the products with lighting source. We chose the lighting products with eastern characters from mainland China and Taiwan in 2011 IDA exhibition. Between them, the expression of the lighting source creation could be sorted into three classifications: the direct lighting formation, the materials surrounding the lighting and the hollow style of the surrounding. The direct lighting formation is created by bulbs and tubes shape. The materials surrounding the light are the out shelf made by different materials. It is affected by the penetrant light which goes through the materials. The hollow style is created by the cutting skills to express the special lighting source and shadow on the environment. It is between the first and second classifications.

Here are six products chosen according by these three classifications and each one contains two products. The products of the first classification are the "Jungle birds lighting" made by Zhi-Li Liu (Figure 2)and the "Flying blue" made by Jian-Rong Lin(Figure 3). "Jungle birds lighting" use the bulb row effect to imitate the birds row on the electric wire. It uses the birds as the symbol to connect the emotional feeling. "Flying blue" uses the concrete shape of cloud and human. These two products took lighting source as the leading role to show the connective formations as the symbol decoded emotionally (Tung-Long Lin,2005). It would help to get the recognition from audiences.

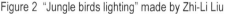

Figure 2 "Jungle birds lighting" made by Zhi-Li Liu        Figure 3 "Flying blue" made by
Jian-Rong Lin

The products of the second classification are "Cherishing lamp" made by Qi-Mei Zhang(Figure 4) and the "Coral lamp" made by Qisda(Figure 5)

Corporation. The "Coral lamp" use special lighting guide technology to hide the LED light inside the center of the formatted acrylic bends. In this classification of lighting, the lighting source don't take the leader role but interaction with materials is the one.

The products of the third classification are the "Stainless steel lamp"made by Fade-design original co. ltd(Figure 6). And the "Accumulating lamp" made by Fang-Yu Zhuang(Figure 7).

The "Stainless steel lamp" material makes people feel the cold and hard formation and it creates the curve outline in these Kind of the hard materials. The light expresses through the hollow sharps. The "Accumulating lamp" combined the recycled plastic bags and to use the colorful specification to create the mood atmosphere.

Here are the all six products with eastern characters chosen in 2011 IDA exhibition. The expression of the lighting source creation is sorted into three classifications. The lighting design mainly provides life requirement and style.

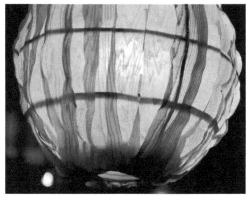

Figure 4 Cherishing lamp" made by Qi-Mei Zhang

Figure 5  Coral lamp" made by Qisda Corporation

Figure 6 Stainless steel lamp"made by Fade-design original co.ltd

Figure 7  Accumulating lamp made by Fang-Yu Zhuang

## 5    DATA OF ANALIZED USER QUESTIONAIRS

In this research, we collected thirty questionnaires and get the feedback of the factors which influence purchasing will. The first four factors are: price, shape, lighting affect and brightness. Three characters stand side by side in fifth factor: function, electricity saving and decoration.

The progress of science and technology helps designers to get more method to design a new product, multi-function, special materials, special shapes and interaction. After analyzing the data with these six products, we have some conclusions.

1.  Multi-functional products are not the prior expect point.
2.  Special materials make the extremely feelings, much favorite and hate.
3.  Large volume products don't fit most consumers.
4.  The products with the lighting source which could not be switched might be the skippable producs.
5.  Interactive function products would not be the ones for home setting.

# 6    CONCLUSIONS

With more than 10 years promotion by Government, the world wild thoughts are brought into Taiwan. After the 2011 IDA exhibition, many fields and designs are reviewed. Inside the industry in Taiwan, it usually focused on lighting source technology and the production with OEM and ODM but not on the lamp design. The lighting source proceeded in recent years, especially after the LED lighting discovered. Lighting express have huge variations to provide both the illumination and the situation. The abundant specification increase the night life activities so it is very important to understand the feedbacks from UX.

In this UX analysis research, we got five conclusions. They showed the user want have simple warm lighting source and warn materials, like paper and fabric. The stainless steel made them feel cold. By the model we could understand the consumers' will and make the design which would more fit their needs to promote the lighting industry in Taiwan.

## REFERENCES

Lee-Chun Lan,Chorng-Ming Chiou and Chun-Chieh Wang.2009.The Development of Cultural Industries within Cultural Policy in Taiwan. CHIA-NAN ANNUAL BULLETIN.Vol. 35 , pp.437-451.

Wu,Yu-Hsing,2011,The Research of Cultural and Creative Industry to Future of Taiwan,Journal of Chinese Trend and Forward Volume7 Number1,May 2011,pp113-119.

Pahl,G.and Beitz,w.1993,edit by Kan Wallance,Engineering Design—A Syatematic Approach,The Design Council

Ding-Bang Luh and Chia-Ling Chang.2007.User Successive Design:Concept and Design Process. Journal of Design,Vol.12,No.2,pp.1-13

Tung-Long Lin and Chia-Fang Yu.2005.Research on the Effects of Symbolic Images on Product Form-The Case of Italian Design Style.Journal of Humanities and Social Sciences,Vol.1,No.1,pp.19-27

Dayton-Johnson J (2000). What's Different about Cultural Products? An Economic Framework.Dalhousie University, Department of Economics. Halifax, Canada.

Towse R (2002). Copyright in the Cultural Industries. Cheltenham (UK) and Northampton, MA(USA), Edward Elgar.

# Author Index

T - #0288 - 071024 - C844 - 234/156/37 - PB - 9780367381073 - Gloss Lamination